D1116909

MECHANICS OF POLYMER PROCESSING

J. R. A. PEARSON
Emeritus Professor of Chemical Engineering,
Imperial College of Science and Technology, London, UK

ELSEVIER APPLIED SCIENCE PUBLISHERS
LONDON and NEW YORK

ELSEVIER APPLIED SCIENCE PUBLISHERS LTD
Crown House, Linton Road, Barking, Essex IG11 8JU, England

Sole Distributor in the USA and Canada
ELSEVIER SCIENCE PUBLISHING CO., INC.
52 Vanderbilt Avenue, New York, NY 10017, USA

British Library Cataloguing in Publication Data

Pearson, J. R. A.
Mechanics of polymer processing.
1. Polymers and polymerization—Rheology
I. Title
668.9 TP1092

ISBN 0-85334-308-X

WITH 187 ILLUSTRATIONS AND 5 TABLES

Printed in Northern Ireland by The Universities Press (Belfast) Ltd.

Preface

This text has grown out of an earlier one, *Mechanical Principles of Polymer Melt Processing*. The main objective remains the same: to describe ways of applying continuum mechanics to polymer processing, which in large part means polymer melt processing. Eighteen years ago, it was said that 'theoretical analysis had had little impact on practical methods of processing, or on the design of processing equipment'. This is no longer true: continuum mechanical analysis is now widely recognized in the more successful parts of the industry as an important aid to understanding and designing new processes.

Other important advances have been made in the last decade. First, useful connections have been made between:

(a) the molecular structure of polymers;
(b) the microscopic morphology of supramolecular units formed from homopolymers, co-polymers, blends or composites; and
(c) continuum rheological behaviour.

It is no longer true that rheological models used for polymer melt mechanics bear little relation to actual behaviour of industrial plastics. Careful fitting of mathematically and mechanically acceptable rheological equations of state to extensive experimental measurements made on selected polymers over a wide range of flow conditions has shown that such models can be suitably chosen. Second, clear connections have been established between morphological and physical characteristics of solid polymers and the deformational and thermal histories of these materials prior to solidification; this is particularly important in the case of crystalline polymers and bears on the subject of orientation in fabricated articles. Lastly, the predictions of relatively elaborate fluid-mechanical analyses for various complex flow processes (often dominated by heat generation and heat transfer effects) have been verified by experiment.

It is thus clear that the approach recommended tentatively in my first monograph can now be regarded as established and successful. Tensors were then introduced almost apologetically; they are now relatively familiar to many graduate students in engineering. No attempt will be made here to develop any of the basic theory of continuum mechanics; many excellent and varied texts are available and, for convenience, this text will be based on the notation and approach clearly and briefly given in *Principles of Non-newtonian Fluid Mechanics* by G. Astarita and G. Marrucci (1974; known henceforth as A & M). A careful study of this latter text would form an excellent introduction for any serious student of the mechanics of polymer processing. It is based on the more encyclopaedic work, *The Non-linear Field Theories of Mechanics*, by C. Truesdell and W. Noll (1965). Some may prefer the approach of A. S. Lodge in the pair of texts *Elastic Liquids* (1964) and *Body Tensor Fields in Continuum Mechanics* (1974) or that given in *Dynamics of Polymeric Liquids, Vol. I, Fluid Mechanics* by R. B. Bird, R. C. Armstrong, and O. Hassager (1977; known henceforth as BA & H). For some parts of the work, the book by K. Walters entitled *Rheometry* (1975) contains a range of useful results and references. The review article by Goddard (1979) can also be recommended. Mention is made throughout of various articles in the literature, but no attempt is made to be exhaustive.

This text is divided into four parts. Much of the first part may be skipped by many, particularly those familiar with the foundations of continuum mechanics. A formal statement of conservation laws is included for completeness, while a brief review is given of standard constitutive relations. Examples and exercises are included to help overcome notational difficulties which arise because of different usages in the best-known literature. A distinction is drawn between constitutive relations that arise from general continuum mechanical considerations, in particular those that have been tested against polymer solutions, and those that relate to observed properties of polymer melts.

The second part is largely concerned with techniques for applying conservation and constitutive relations to actual flows of elastico-viscous, temperature-dependent fluids. The role of dimensionless groups is developed, and the significance of instabilities examined. This part is useful to all who wish to study applications of continuum mechanics to polymer processing.

The third and fourth parts relate to continuous and cyclic polymer-processing flows, respectively. They form the major part of the text and can to a large extent be studied independently, though back-reference is made to earlier results. As far as possible, simple orthogonal coordinate systems are used and only physical components of velocity, rates of deformation and stress appear in the equations derived.

The aim of the text is to present and justify methods of procedure, rather than to give an exhaustive uncritical survey of published work. A full range of examples (sometimes embedded in the text) and exercises is given, to introduce further results and to enable the reader to make progress independently. In practice, computational methods become essential if specific numerical results in complex situations are required, mainly because of the non-linear nature of most of the relevant sets of equations. Those interested in pursuing such computational methods are advised to refer to a recently published text *Computational Analysis of Polymer Processing*, edited by J. R. A. Pearson and S. M. Richardson (1983).

Thanks go to many colleagues, in particular to Christopher Petrie and Stephen Richardson, for the encouragement and assistance they have given in many ways. Finally, my most grateful thanks go to the California Institute of Technology, whose offer of a Sherman Fairchild Distinguished Visiting Scholarship for a full year in 1978–1979 made it possible for me to undertake the writing of this book in the first place and to the Imperial College of Science and Technology whose generous attitude towards its staff enabled me to complete it.

J. R. A. PEARSON
Schlumberger Cambridge Research Ltd,
PO Box 153,
Cambridge CB2 3BE, UK

Contents

PART II—COMPLEX FLOW FIELDS

PART III—CONTINUOUS PROCESSES

A–EXTRUSION

PART IV—CYCLIC PROCESSES

Part I

POLYMERS, MECHANICS AND RHEOLOGY

Chapter 1

Introduction

Synthetic polymers form the most recent addition to the list of materials used industrially in large quantities. Per capita consumption of plastics and rubbers (as they are most frequently called) is now one of the yardsticks of national economic well-being. Tonnage output quadrupled between 1958 and 1971, trebled again by 1977 and despite the present recession is still rising world-wide at significantly more than the relevant GDP's. Details may be conveniently obtained from the January numbers of *Modern Plastics International.*

Natural polymers have been known and used for much longer: natural rubber, textile fibres like cotton and wool, leather, wood and hardened oil paints are familiar to all. Much of the present production of polymers (particularly in terms of value) provides substitutes for such natural products; because of long past experience, these natural products have been so well developed that high quality and performance have been achieved; thus any direct substitutes for them tend to be relatively expensive and specialized.

However, the spectacular growth of polymer production has been caused largely by their use in non-specific and undemanding applications, where their major advantages have been low cost and ease of fabrication. They have competed with metals, glass and ceramics, as well as with paper and wood. Success has come by regarding plastic materials and articles as alternatives and not as direct substitutes. In a very broad sense, the main battle has so far been fought in the field of containers, as opposed to that of machines, where metals remain pre-eminent. The relevant factors have therefore been resistance to damage, from both chemical and mechanical assaults, ability to produce complicated shapes, and lack of damage to surrounding materials.

When viewed in this way, the advantages of modern synthetic polymers are immediately apparent: chemically they are singularly inert, except at temperatures well above normal ambient; they are readily deformed into intricate shapes, either before polymerization or when in a molten or plastic state; many are basically tough and relatively flexible; many are transparent. Their main limitations lie in their deterioration or loss of properties at only moderately high temperatures and their relatively low strength; although some of them are hard, they all tend to fail under conditions of high stress, whether static, periodic or transient. For this reason, they have made relatively little impact as engineering materials, except in special circumstances, though this is now the most vigorous and profitable line of development.

It is now clear that further substantial growth in the use of polymers will depend upon an ability to improve, control and use to best advantage their mechanical properties. There are three basic ways of doing this:

(i) to produce new polymers of intrinsically greater strength and heat resistance;
(ii) to process existing polymers so as to enhance their normal bulk properties; this involves control of crystallization and of molecular or crystal orientation, a technique of great importance in producing textile fibres or thin films;
(iii) to reinforce polymers with other materials such as powders or fibres.

The first of these is a purely chemical problem and will not be discussed in this book. The second is a question of mechanics in the general sense; it is with the problems that arise in this context that this work is ultimately concerned. Very often the grade of polymer chosen for a particular application is selected as much on the basis of its processability (i.e. the ease with which it can be processed) as on its ultimate performance. The last approach is essentially outside the scope of the treatment to be attempted here, although some processing methods will be common to homogeneous bulk polymers and to reinforced compounds. For the latter, interfacial effects will often be dominant whereas we shall be concerned here largely with bulk properties and behaviour.

1.1 INDUSTRIALLY IMPORTANT POLYMERS

The basic polymeric constituents of commercially important plastics and rubber are all long-chain molecules, sometimes branched and sometimes cross-linked. The simplest of these to visualize is linear (or high density) polyethylene,† which consists of a simple sequence of —CH_2— units as shown in Fig. 1.1. The strength of the molecule itself resides in the covalent carbon–carbon bond (which by engineering standards is very strong), but the strength of the material in bulk depends primarily upon the weaker forces that act between chains, namely the Van der Waals forces and those caused by hydrogen bonding. The resistance of polyethylene (and similarly of other organic polymers) to deformation and rupture is therefore dependent on the relative configuration of the individual molecules, and on how labile they are.

At high enough temperatures (assuming thermal degradation in a chemical sense does not ensue), the molecules are sufficiently thermally agitated for the material to be molten and fluid-like (it is thus a thermoplastic). At lower temperatures, a solid, partially crystalline, form occurs in which the conformation of the chains possesses a high degree of order, and they are closer packed than in the more random

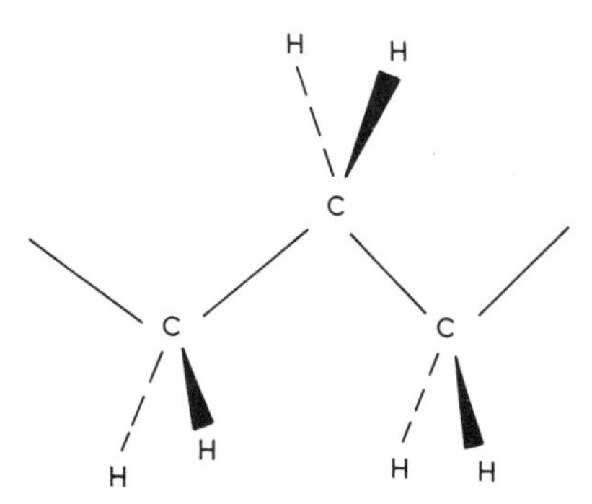

Fig. 1.1. Diagrammatic interpretation of bond structure in planar (average) form for main-chain carbon atoms.

† Often referred to more briefly as HDPE. Similar acronyms for other polymer types are given later in parentheses.

molten form. The single crystals are relatively hard and strong; however, bulk polyethylene, like many metals, is polycrystalline and so its bulk behaviour is determined partly, or even largely, by the properties of the 'amorphous' material and the interfaces between individual crystals. The actual structure involved is complex, and for a full description the reader must be referred to other texts (McKelvey 1962, Chapter 6; Jenkins 1972, Chapters 4 and 11). It is worth noting, even at this stage, that the fine structure and hence the bulk properties of a crystalline polymer such as HDPE are significantly dependent on the way it is formed and cooled from the molten state (Ziabicki 1976, Chapters 2.IV, 3.VII, 6.IV).

Chemically-specific bulk polymer possesses one important variable characteristic, the distribution of chain lengths, i.e. the molecular-weight distribution. In general terms, one can say that the strength and stiffness of material increases as the mean molecular weight increases, for the rather obvious reason that each molecule will on average be 'connected' to more other molecules as its length increases. This general comment applies in both solid and fluid states. It is less easy to describe the effect of increasing the nominal breadth of the molecular-weight distribution, but there is no doubt about its importance. It is customary to characterize a sample of polymeric material by its number average molecular weight ($\bar{M}_n$) and its weight average molecular weight ($\bar{M}_w$).

In the case of HDPE, the submolecular weight of the repeated CH_2 unit is 16 and so we have

$$\bar{M}_n = 16 \times \bar{n} = 16 \int_0^\infty n f(n)\,\mathrm{d}n$$

where $\bar{n}$ is the mean number of repeating units in the sample, $f(n)$ being the fraction of molecules having n units in the chain, and

$$\bar{M}_w = 16 \int_0^\infty \frac{n^2}{\bar{n}} f(n)\,\mathrm{d}n$$

Polyethylene (PE) can exist in a branched form, of lower density, referred to as LDPE. The branching causes the physical properties of the material to be quite different. It has a lower, less well-defined melting point, is less readily crystallizable and appears much more rubbery in its solid form at room temperature. The differences can be attributed to the constraints on molecular ordering that are imposed by the chain branching. Recent developments provide PE with a continuous range of properties.

Polyethylene is the simplest hydrocarbon polymer. Polypropylene (PP) is formed from linear chains of the repeating unit $\lbrack C(CH_3)H—CH_2 \rbrack$ and is in many ways similar to HDPE. If the chains of polyethylene or polypropylene are terminated by hydrogen molecules, they are strictly very-high-molecular-weight paraffins. Other polyolefins with larger repeating units can be made but are not so important commercially.

The structure of polypropylene, as shown in Fig. 1.2, illustrates a further degree of variety possible in long chain polymers: their degree of stereo-regularity. On the one hand, a head/tail property arises in each monomeric unit of a chain because the (CH_3) group could be attached either to the first or the second main-chain carbon atom of that unit. On the other hand, in any given conformation of the main chain, the CH_3 group could be attached to one or other of the two remaining valence bonds of the relevant carbon atom; in simple language it can be on one side or the other of the main chain. In the case of polypropylene as made commercially only the head-to-tail form arises, while successive CH_3 groups are on opposite sides of the main chain, which is the syndiotactic case.

A second important class of long-chain hydrocarbons is obtained when double bonds appear in the main chain of the repeating unit, e.g. as in polybutadiene (PB) $\lbrack CH_2CH{=}CHCH_2 \rbrack$. This class forms the basis of most elastomers, or rubbers. Other examples are the stereoregular cis-polyisoprene (PI) and trans-polyisoprene, both polymers of $CH_2{=}C(CH_3)CH{=}CH_2$, yielding the repeating unit $\lbrack CH_2C(CH_3){=}CHCH_2 \rbrack$. They differ in properties from the stereoregular polyolefins HDPE and PP by being amorphous, i.e. non-crystalline, in the unstrained state, even down to temperatures at which they do not appear to be fluid. They can display large 'elastic', i.e. recoverable, deformations at relatively low stress levels: these large deformations are achieved by the substantial extension that is possible in individual molecular chains, which adopt almost a random conformation in the unstressed state; there are relatively few intermolecular constraints (bonds or entanglements) in uncured rubbers.

(In practice, the situation is confused by the curing of thermoplastic rubber, a chemical process that introduces permanent crosslinks between the polymer chains and so creates a three-dimensional matrix that is characteristic of most manufactured rubber articles. However,

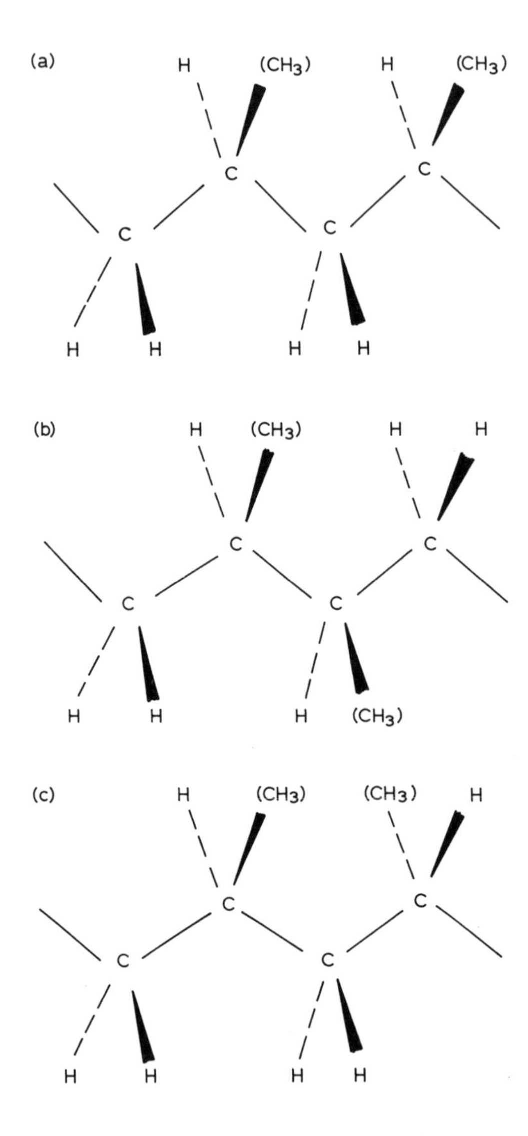

Fig. 1.2. Polypropylene structure. (a) Head-to-tail repetition—isotactic. (b) Head-to-head repetition—isotactic. (c) Head-to-tail repetition—syndiotactic.

the natural or synthetic polymers from which cured rubbers are formed are thermoplastic, exhibiting fluid-like flow. Motor tyres exhibit a further complication in that they are made with a high proportion of finely divided carbon black added to the rubber before curing.)

The earliest and most widely used thermoplastic was polyvinyl chloride (PVC) whose repeating unit is $\left[CH_2—CHCl\right]$. It is made in the head-to-tail isotactic form. The bulky chlorine atom reduces its tendency to crystallize, though the homopolymer forms a hard tough solid (uPVC) at room temperature. It degrades relatively rapidly at or above its melt point, and so much of it is used in a plasticized form (pPVC); addition of relatively-low-molecular-weight plasticizers that are compatible with it (i.e. can be intimately mixed with it almost as in a solution) leads to a greater degree of flexibility and effectively reduces the melt point so that the material can be more conveniently processed. Alternatively, it is mixed with lubricants and anti-oxidants for processing purposes. Atactic polystyrene (PS) based on the monomeric unit $\left[CH(\bigcirc)—CH_2\right]$ is the best example of an amorphous glassy thermoplastic. Although not restrained by crystalline bonds between chains, the bulkiness of the ⬡ side group means that, at room temperatures, thermal motions within molecules are too weak to facilitate substantial chain rearrangement or uncoiling under stress. Polystyrene articles are stiff and brittle.

Improvements in the physical properties of PS, particularly with respect to impact resistance, are achieved by blending with thermoplastic rubbers (giving toughened PS or HIPS) or by copolymerization using other monomers. Thus, we have poly(styrene-butadiene) (SB) which was the basis of the synthetic rubber industry (SBR) and poly(acrylonitrile–butadiene–styrene) (ABS) a relatively expensive but highly-developed thermoplastic with engineering† applications.

Copolymers can exist in various forms. These are shown crudely in Fig. 1.3 as random, alternating, block or graft copolymers. These represent chemical modification of the base polymers, SB being a random copolymer and ABS a block copolymer of the subchains $(S)_m(B)_n$ and $(CH_2CHCN)_p$, though it is sometimes difficult to decide which of the base units imparts the governing physical properties. The range of possibilities is large. The purely physical process of blending

† By 'engineering' is conventionally meant 'capable of sustaining significant intermittent or continuous stress without cracking, breaking, or deforming beyond prescribed limits!'.

(a) ..AABABAAABBABBB..

(b) ..ABABABAB....

(c) ...BBAA...AABB...BBAA..

(i) $(A)_m(B)_n$ (ii) $(B)_{n_1}(A)_m(B)_{n_2}$

(d) ...AAAAA'AAAAAA'AA...

```
                 B          B
                 B          B
                 :          :
                 B          B
                 B          B
```

Fig. 1.3. Copolymerization in a polymer. (a) Random linear chain of A and B monomers (e.g. SBR). (b) Alternating linear chain. (c) Linear block copolymers with subchains of A alternating with subchains of B. The simplest are (i) diblocks and (ii) triblocks. More complicated combinations of (a), (b) and (c) can exist as with ABS. (d) Graft copolymer showing subchains of B grafted (chemically) on to a linear chain of A.

leads to a mixture of different molecules; in the case of polystyrene and butyl rubber the mixing only occurs at a macroscopic level, and so small but distinct lumps of one can be seen to be embedded in the other. This is in contrast to the case of plasticizing of PVC where an intimate molecular mixture (solution) is attained.

Another example of amorphous (glassy and transparent) polymers is provided by the polyacrylates of which polymethylmethacrylate (PMMA) is the best known.

All the polymers we have discussed so far have a 'back-bone' formed of carbon atoms. We now come to those thermoplastic polymers that have other elements, or ring compounds, in the main chain. These tend, in general, to confer more rigidity to the chain molecules themselves, and in some cases lead to intrinsically stronger materials.

The best known is nylon which exists in various related forms, all of which are known as polyamides (PA). A typical example is Nylon 66 with the repeating unit $\text{-[}NH(CH_2)_6NHCO(CH_2)_4CO\text{]-}$. This is a thermoplastic, but with a sharper melting point and less viscoelastic in nature than those mentioned earlier.

Another class is provided by the polyesters of which polyethylene terephthalate (PET) is the most important. It too is a thermoplastic with a well-defined melt point at which a relatively hard solid turns into a basically viscous liquid.

Two relatively expensive thermoplastics are polyformaldehyde, an acetal polymer, and polycarbonate (PC) with repeating structures given by ⁅HCH—O—⁆ and ⁅⬡—CH_3CCH_3—⬡—OCOO⁆ respectively. They are both hard 'engineering' plastics.

Polytetrafluoroethylene (PTFE), a speciality polymer with good high-temperature properties, has the same chain structure as PE but with the hydrogen atoms replaced by fluorine. Its melting point, like PVC, is too high for processing in the melt form and so it is usually formed using lubricants and sintered.

The last important class of thermoplastics are the sometimes misnamed polyurethanes (PU), whose variety and complexity of molecular structure is too great to consider in detail here. They are usually formed from prepolymers which are often themselves polyesters or polyethers, and are joined together by short connecting units. The typical linear structure that can ensue is shown in Fig. 1.4(a). However, in the case of the polyurethanes highly branched polymers can be formed which ultimately crosslink to form thermosetting polymers, which are traditionally rigid. This is shown in Fig. 1.4(c).

Thermosetting polymers form an important part of the polymer industry. They are of less importance in the context of this book because they are not in general subject to melt processing. However, there are certain high-speed processes in which mixing and polymerization are initiated just before or during flow into final solid form; this is particularly true of PUs; their behaviour during that period may be likened to that of molten thermoplastics. More often, they are shaped in powder form, and so the mechanics of the flow of granular media is

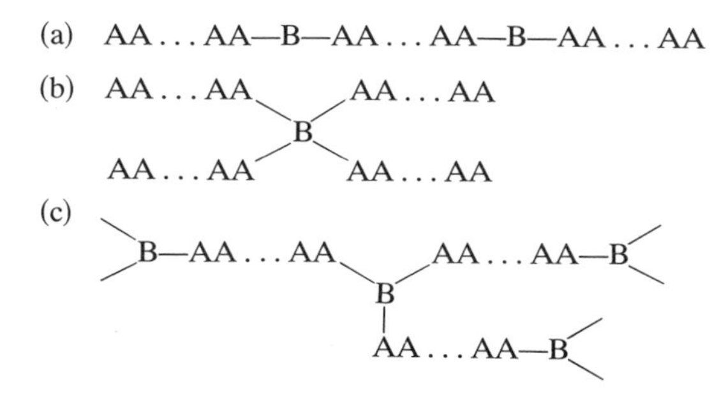

Fig. 1.4. General structure of polyurethanes. (a) Linear chain. (b) Star cluster based on tetrafunctional connecting block B. (c) Cross-linked system based on trifunctional block B.

more relevant to the problems that arise; powder or granule flow is also relevant to the processing of thermoplastics because the latter are first prepared and transported in powder or granule form. The main types of thermosets are urea formaldehyde (UF), phenol formaldehyde (PF) and melamine formaldehyde (MF) resins and certain polyester resins.

Table 1.1 lists various of the most heavily used polymers that have been mentioned above, giving their density and either their melting point or their softening (glass transition) temperature. An indication is given of their use in the developed world. The really bulk polymers PE, PP, PS and PVC cost between 25 and 50 US cents/lb in 1982. The simple thermoplastics and thermosets are thus cheap and used in large quantities. The more expensive polymers (US $ 1–3/lb typically) are made in smaller quantities and are used only where their special properties and strength are particularly important. (It is interesting and perhaps significant that the simple polymers earliest in the field, PVC, PS and LDPE among the thermoplastics and UF and PF among the thermosets have so far maintained their position in terms of volume.

Table 1.1

Basic Data about the Most Heavily Used Polymers

Polymer	*Specific gravity*	*Melting point °C*	*Glass transition temperature °C*	*Approximate†‡ price ¢/lb*	*Approximate‡ consumption × 10^3 tonnes*		
					US	*Western Europe*	*Japan*
ABS	1·01–1·08	—	88–125		400	300	300
PMMA	1·17–1·20	—	90–105		250	120	100
LDPE	0·91–0·92	110–130	—	30	3 400	3 400	1 000
HDPE	0·94–0·96	120–140	—	40	2 200	1 400	600
PP	0·90–0·91	176	—	40	1 800	1 300	1 000
PS	1·04–1·10	—	100	45	1 700	1 100	500
PA	1·03–1·15	194–265	—		150	180	
PU	1·10–1·50	—	—		800	1 000	300
PVC	1·30–1·60	—	75–105	26	2 500	3 200	100
PET	1·34–1·39	258	—		500		
PF	1·25–1·30	—	—		650		
UF-MF	1·5–1·7	—	—		650		

† Price given only where a large market exists for basic homopolymer.
‡ These figures are subject to fluctuations.

The very great efforts that were made after the Second World War to discover and develop completely new polymers have now given way to greater emphasis on blending and modifying the existing polymers, with more careful control of the polymerization process and greater attention to processing.)

1.2 END USES OF PLASTICS AND RUBBERS

The use of cross-linked rubbers is dominated by the motor tyre. This remarkable product exhibits an extraordinary resistance to repeated flexure, impact and abrasion. Its construction is complex, with walls and tread being made of different compounds and essential use being made of cord and steel reinforcement. The cross-linked rubber compounds, heavily loaded with carbon black, of which it is made, are truly engineering materials, for which there are no substitutes capable of equally large elastic deformations at comparable loads. Other uses employ the same properties as in hosing, belting, and flooring.

The thermoplastics industry has grown to its present size largely on the basis of supplying containers and coverings (in the wide sense). One clearly defined area is in thin-film packaging, where HDPE, LDPE, PP, uPVC, and PET all have their place, having largely eroded the market for the artificial transparent cellulosic films and competing strongly with paper and woven sacking. All but HDPE can be conveniently made transparent.

The question of transparency is a complicated one depending on the degree of crystallinity and morphology of the material; this can be altered by changing the production process and, in particular, by altering the cooling and stretching history of the film as it freezes. This is most easily demonstrated in the case of LDPE.

Plastic films can be self-coloured and printed. They can be coated with other polymers and heat-sealed. The choice of film material for a particular application often depends on its permeability to liquids, vapours and gases, in particular water, oils and oxygen. Suitable combinations of plastics are often used to achieve the desired degree of impermeability, and these can be bonded to paper, foil or fabric.

Another obvious use is for slightly more rigid containers or covers for foods, drinks, or small items of hardware. They are cheap, attractive in aspect and need not be reused. The necessary strength and rigidity is often achieved by careful design and the ability to produce intricate shapes. The need for an insulating cup for hot drinks encouraged the use of foamed (or expanded) PS, a development which has spread to other polymers and other uses. PS, uPVC, HDPE, and LDPE for squeeze bottles, are the major thermoplastics used in these applications, though PET is being used more widely despite its higher cost.

More durable items such as kitchenware are made with rather thicker sections. The toughness of LDPE makes it a popular material still, but the greater dimensional stability of the harder thermoplastics and the thermosets makes them more suitable for use in mechanical applications such as mixers, washing machines, vacuum cleaners, or situations involving contact with boiling water. The low density of most polymers makes them particularly attractive in household applications. It is in the more expensive items that the specialty polymers like PAs and PC find important uses.

Thermoplastics have a dominant share now in the field of coatings and coverings. The durability of PVC makes it ideal for flat washable floor coverings, having largely replaced linoleum, and in furnishing, where as pPVC it forms the outer surface of artificial leather cloths. It also dominates the market for moulded shoe soles and cheap moulded shoes. Polyurethanes and certain copolymers of ethylene and vinylacetate are used in similar fashion.

Woven fabrics are now made largely of synthetic, or artificial, fibres. Rayon (a regenerated cellulose), polyamides, polyesters and polyacrylonitrile have been most successful at the top end of the market, partly replacing cotton, linen, wool and silk, while polypropylene has application at the cheaper end of the market competing with jute and sisal. Recent developments anticipate a large growth in non-woven fabrics, such as felts, using heat-bonding techniques to give tough materials; this is a case where virtue is made out of the low softening point of thermoplastics. Tufted and woven carpeting forms an important outlet for synthetic fibres, particularly nylon and PP; indeed carpeting is rapidly taking over from 'vinyl' flooring. A large potential market for paper based on plastic rather than wood fibres has recently been exploited, particularly for art papers.

Special advantage has long been taken in the electrical industry of the very low electrical conductivity of plastics. Thermosets, typically

UF and PF, are used for switches and sockets, while LDPE, pPVC and rubbers are used for cable covering (the dielectric properties of LDPE were the spur to its original development in connection with HF transmission).

Massive increases in the use of plastics are likely to come as more applications are found in the building, furniture and transport industries and as more use is made of existing products. Major markets already exist for LDPE, uPVC, and PP pipes, gutters and drains, where they provide replacements for iron, copper, and ceramics. The space filling and insulating properties of expanded plastics, particularly PS and PU have already been mentioned. The inertness of plastics makes them potentially very suitable for cladding structures; lack of resistance to UV attack has been one of the main difficulties in the past. Plastics are slowly providing a complete range of internal finishes, often imitating the grained effect of wood, the textured surface of fabrics or the ease of application of wallpaper; modern paints are largely based on emulsions of thermoplastics. Flexible foams based on rubber or PU have largely displaced metal springing and felt or lint padding in furniture, particularly because they can be incorporated as backing to tough outer coverings. Mouldings of ABS and PMMA have provided cheap and varied fittings of all types, particularly for lighting. All of these developments have their parallel in the automobile industry.

The rising cost of wood has tempted furniture manufacturers into using plastics as load-bearing elements, as in the seats and backs of chairs (usually using PP or ABS among thermoplastics, or various filled or reinforced thermosetting resins), or in the drawers of desks, chests and cupboards. Most of us are familiar with the decorative sheeting finishes based on epoxy resins that are also the basis of many adhesives.

A very large market already exists and is expected to grow for reinforced plastics. This seems inevitable if the use of polymers in load-bearing applications is to develop and thus begin to erode the dominance of metals, concrete and wood. Glass-fibre and carbon-fibre reinforced resins have made spectacular progress in specific markets, such as boat hulls, even up to the size of fishing trawlers, specialty car bodies and building panels. The technical problems that have to be overcome are complex and are not soluble solely in terms of chemical modification of the base polymer. Deliberate use of the anisotropic properties of reinforced plastics—when the reinforcing is fibrous and can be oriented—seems a necessary factor in successful applications. This places great importance on laying-up and processing techniques, particularly since the polymeric matrix can also be oriented by strain if it is thermoplastic. Problems of adhesion between different elements of composite materials arise, delamination being a particularly undesirable occurrence.

One serious drawback to the widespread use of plastics in buildings is the fact that most of them burn readily once ignited. The production of flame-resistant grades of plastics is seen as increasingly necessary.

1.3 INDUSTRIALLY IMPORTANT PROCESSES

Because growth in the use of plastics has depended overwhelmingly on their low price in finished form, processing of them has had to employ mass-production methods. Common thermoplastics have an immense advantage for processing in that they soften at easily attainable temperatures and can be continuously deformed in a molten or plastic state.

The physical situation is not entirely clear-cut for every material processed in a plastic state. Fibre-forming materials like nylon (PA) and PET exhibit a clear phase change above which they flow readily almost as Newtonian liquids; PE, PS, PP and certain other thermoplastics melt or soften gradually over a range of temperature, forming homogeneous viscoelastic liquids. Both of these groups can be conveniently and simply processed as pure homopolymers. PVC, however, degrades in the molten state and so has either to be plasticized with relatively large quantities of liquids like dioctylphthalate or be lubricated in powder form with small quantities of surface-active materials; PTFE is lubricated, again in powder form, with relatively large amounts of low-molecular-weight hydrocarbons or oils that have to be removed after forming. Thermosetting or cross-linking materials change physically and chemically, while foaming materials develop gas bubbles within the polymeric matrix, during processing. The rheological properties of the inhomogeneous mixtures of chemically changing materials are often extremely difficult to describe or measure.

Economics encourages the use of steady, i.e. continuous, flow processes wherever possible; these usually produce products with cylindrical symmetry (whose cross-section may however be complex). Relatively simple cutting or chopping devices acting repetitively provide items of finite length. If the product is thin and flexible, it can be wound onto cylinders. If more complicated shapes are required, then either the cylindrical elements can be used as preforms for a subsequent forming operation or a stream of molten material can be periodically diverted and pumped into a mould. It is with these continuous and intermittent flow processes that we shall be mainly concerned in this study.

1.4 MATHEMATICAL MODELLING OF PROCESSES

The main objective in this text is to develop useful mathematical models for all the principal polymer-forming processes, which to a large extent means melt processing. The models will be concerned with the mechanics (the kinematics, dynamics and heat transfer) of the processes. Where phase changes or mixing processes are involved, these will have to be modelled also.

The twin starting points of a full analysis will be:

(1) The geometry of the processing equipment, more particularly the internal geometry of the flow channels.

(2) The basic engineering and physics of the process. By this is meant the sequence and intention of the physical processes taking place, e.g. melting, mixing, pumping, forming and solidifying.

Included in the above will be a specification of the boundary conditions on the flowing material, e.g. motions and temperatures of solid boundaries; pressure or flow rates at inlet and exit points.

The first task will usually be to study, often by observation,

(3) The kinematics of the flow. This relates solely to the (vector) velocity field† as a function of time and position. The most important aspects to be considered at this stage are the symmetries of the flow field, and whether it can be viewed as steady. It is usually supposed that the symmetries of the flow boundaries (and their motions) will result in similar symmetries of the flow field; equally that steady boundary conditions will lead to steady flow fields. Such assumptions will lead to very considerable, and usually essential, simplification of any subsequent analysis.

† It should be noted that a full knowledge of the velocity field implies significant constraints on the (scalar) density field and vice versa.

These suppositions are always made in elementary texts and are usually inherent in the initial design of the processing equipment; when the non-linear effects inseparable from polymer-processing flows lead to a breakdown in the basic symmetry or steadiness of the flow, the process is often unworkable or the resulting product unacceptable. Part of our task will be to investigate these undesirable limitations on processes. The next task is to study

(4) The dynamics and thermodynamics of the flow. This introduces the (tensor) stress field and the (scalar) temperature field. Many texts on fluid mechanics separate the dynamics (a study of the force balances) from the thermodynamics (a study of the generation and flow of heat and consequential material changes). It is not realistic to do this when considering many polymer processes, though the procedure will often be adopted in subsequent chapters for illustrative purposes.

It is at this stage that a determinate set of mathematical relations can be written down for the tensor, vector and scalar fields involved. These are based on

(5) The basic conservation laws of physics, i.e. those of mass, momentum and energy, which have the same form whatever the material involved, and

(6) The constitutive relations for the material, which will be specific to any particular material, though similar relations can be expected to hold for all molten polymers. In their most general form these will relate the instantaneous stress-tensor and heat-flux-vector fields to the velocity and temperature fields (or their histories). Much of the difficulty associated with the mechanical analysis of polymer processing lies in choosing constitutive relations that are at once flexible enough to model polymer melt behaviour realistically and yet simple enough to lead to tractable mathematical equations, i.e. capable of solution for the quantities of interest.

It is usual to work with four constitutive functions or equations:

(a) an equilibrium equation of state, relating the density to the pressure and temperature;
(b) a relation for the internal energy as a function of temperature, including any latent heat for phase change;
(c) an equation for the heat flux in a static situation, relating the heat flux vector to the local pressure, temperature and temperature gradient;
(d) a rheological equation of state, relating the local stress tensor to

the local deformation tensor and temperature (or to their histories at the material particle in question).

These are by no means the most general that could and perhaps even should be used: for example, the local density is known to exhibit a dependence on the history of stress and temperature at the particle in question, while the heat flux is not independent of the deformation tensor. In the above description of the rheological equation of state, the use of deformation as a significant kinematic quantity suggests a reference state: for fluids, it can be argued that none such exists and so the rate of deformation (tensor) takes its place.

The rest of Part I will be concerned with providing the basic mathematical and physical background for this formal stage of the investigation. This is done very briefly—because so many other texts are available giving full details—for the underlying notions of frames of reference, coordinate systems, conservation equations and general theories of rheological behaviour. The specific characteristics of polymer melts in general are discussed rather more carefully, with particular reference to modern theoretical ideas. Central to this approach will be a careful selection of special (kinematical) flow fields, so as to simplify the forms taken by the rheological equation of state and to allow experimental measurement of the relevant constitutive functions that are sufficient to describe material behaviour in these chosen situations. These simple, uniform flow fields will be such that the conservation equations are satisfied trivially, it being assumed in most cases that the density and temperature can be treated as constant.

The next stage is to try to analyse flow fields in terms of the basic conservation and constitutive equations. This set of mathematical equations, together with the initial and boundary conditions specific to a flow field of particular geometry represent our starting mathematical model for the given flow. The aim is to obtain as much kinematical, dynamical and thermodynamical information as possible by solution of the set of equations and boundary conditions; more specifically we seek the velocity, stress and temperature fields as a function of position and time. Exact analytical solutions are not in general obtainable; nor is full numerical solution of the general case feasible. The approach taken here is therefore to seek useful

(7) Approximate solutions. Much of the text discusses the approximations that can be made and their significance. These are of various types.

(i) Geometrical approximations, which replace the flow channels in real processing equipment by simpler flow geometries, usually to introduce greater symmetry.

(ii) Physical approximations, which neglect what are relatively unimportant physical effects in particular situations (for example, gravity, inertia and surface-tension forces are often neglected).

(iii) Constitutive approximations, which simplify (and often greatly simplify) the behaviour of the material being processed.

All of these introduce approximations in the mathematical model for the flow field concerned. In many cases the precise nature of the terms being neglected and the quantitative significance of the approximations involved can be investigated by writing the equations in dimensionless form and evaluating the magnitude of the several dimensionless groups that thereby arise. These dimensionless groups form connecting links between the various processing flows to be examined and their physical significance will be emphasized.

Finally, there remain purely

(iv) Mathematical approximations, which are concerned with obtaining analytical or numerical approximations to the already approximate model equations and boundary conditions. These introduce the classic procedures of applied mathematics, particularly those relating to the solution of non-linear systems of equations.

Part II attempts to systematize these approximation procedures and illustrates them for a few common, but non-trivial, flow fields. These are then applied in Parts III and IV to the full range of polymer processes, where the purely mathematical aspects are related to the engineering problems of polymer processing. Relatively little attention will be paid to specific numerical solutions, though references will be given where appropriate. The intention in this text is to present a coherent and systematic approach to mechanical analysis of polymer processing. (The literature is full of *ad hoc* models chosen to suit particular systems, but the connections between them are not always drawn.)

An important and often neglected characteristic of the approximate solutions obtained is their stability (to small disturbances). Careful mathematical analysis—usually by linear perturbation methods—of the stability of approximate solutions has often led to a surprisingly accurate prediction of observable breakdown of expected flow patterns, and has provided a quantitative description of the physical processes leading to breakdown. Several examples are provided in what follows.

Figure 1.5 illustrates schematically the connections between the various factors discussed above.

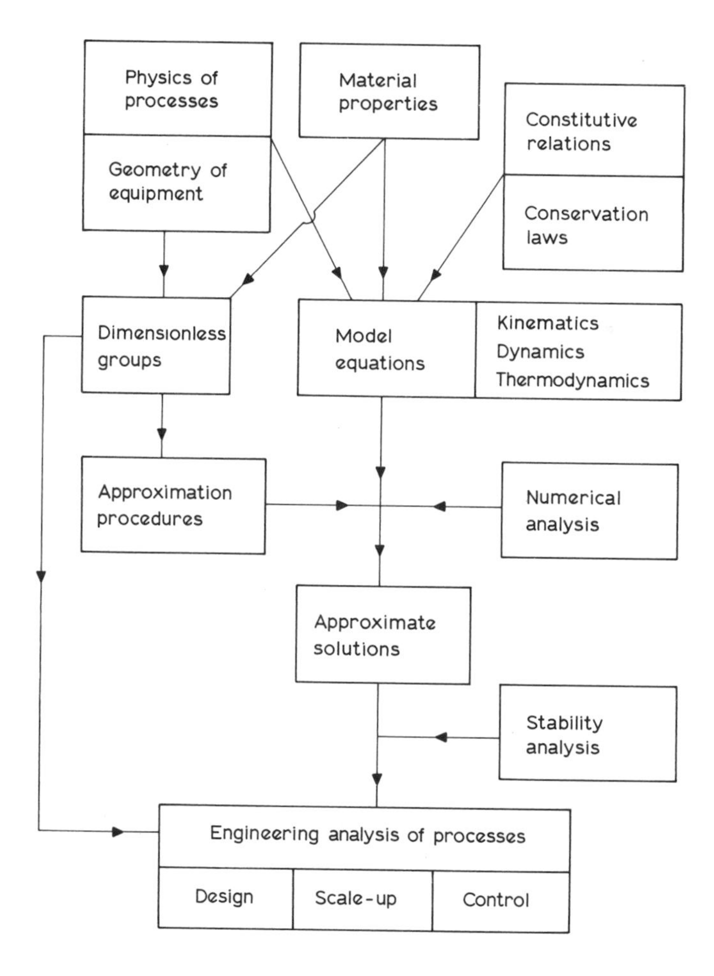

Fig. 1.5. Relationships between concepts to be used in this text.

Chapter 2

Conservation Equations

2.1 FRAMES OF REFERENCE

It is essential, when studying the mechanics of any deformable medium, to be quite clear what frame of reference is being used. We are primarily interested in modelling industrial processes, and so the first obvious and convenient frame of reference (R_0) will be one fixed with respect to the motionless parts of the equipment involved. This is the equivalent of a laboratory frame of reference and will be different from a frame (R_M) fixed in a moving part of the equipment or a frame fixed in a particular portion of the material (R_B) being processed. We shall be using all three types. R_0 is clearly a global reference frame, common to every aspect of the process. An R_M would be chosen to suit a particular portion of the process—for example, the flow in the helical channel of a single screw extruder—but could nevertheless cover an extended region of flow. An R_B would be an essentially local reference frame, usually chosen as suitable for defining the rheological behaviour of a small element of deforming material.

Within each frame of reference, we are free to choose a coordinate system—or several coordinate systems. This choice is usually made by reference to the natural symmetries of the flow system in the case of R_0 and R_M or to those of the body element in the case of R_B.

Unless there is good reason to do otherwise, we shall use rectangular Cartesian coordinate systems for R_0 and R_M: in certain planar or axisymmetric flows, cylindrical polar coordinates fit the boundary conditions most conveniently, while spherical polar coordinates are most suited to other spherically- or axially-symmetric flows; helical (non-orthogonal) coordinates have been used for the analysis of flow fields in screw extruders. In all such cases, the given values of the coordinates (in R_0 and R_M) are regarded as being associated with specific points in space; in R_0, we may regard a fixed point in space as corresponding to fixed values of the three coordinates; in R_M, however, fixed values of the three coordinates correspond to a point moving in space, i.e. a succession of points of space as viewed from R_0. This leads to fairly elementary transformations between R_0 and R_M frames and these will be familiar to most readers. Such relationships describe the geometry of space and the ways chosen to represent it.

A quite different approach is taken in R_B frames by many authors. Here the coordinates refer to particles in the body (particles of the material being processed) rather than to positions in space. The reason for doing this is that mechanical laws, whether of a general type like Newton's laws of motion, or of a specific (constitutive) type relating to particular material behaviour, refer both to material elements and to the 'history' of their positions in space (to their changes of conformation). A basic choice has therefore to be made in the nature of the frame of reference and of its associated coordinate system. Insofar as is possible, space coordinate systems will be used here, even though the frame of reference will be assumed to move and usually to rotate with material particles. (It is wise to defer consideration of any notion of stretching of the frame of reference until such complication becomes essential.)

Much of the difficulty experienced in fluid mechanical analysis of polymer melt flows arises because polymer melts are fluids with memory: the response of any element of material at any given instant of observation depends not only upon its current (i.e. instantaneous) temperature and the current stress acting upon it, but also upon the temperature and stress (or deformation) it experienced in the past. The material is said to be viscoelastic (or elastico-viscous if it is truly fluid-like). Our objective will be to describe its behaviour as accurately as possible. Whether we use body or space reference frames and coordinate systems to do this will be determined by convenience. Others (BA&H) have shown that transformation between one system and another is possible, even if complicated. There is no incompatibility between the approaches. Indeed there is surprisingly little difference in the initial postulates employed or in the interpretation placed upon the results obtained.

Whatever frame of reference is used, certain physical quantities have to be defined and related. These can be:

scalars: e.g. temperature, density, or free energy;
vectors:† e.g. velocity, force, or heat flux;
tensors:† e.g. stress, velocity gradient or rate of strain.

There is seldom any difficulty in dealing with the scalar quantities, except insofar as some have to be defined per unit mass or volume. In the case of vector or tensor quantities, definitions can depend upon both the frame of reference and the metric of the coordinate system used within that frame. Wherever possible, a single symbol will be used to describe the quantity considered, independent of any coordinate system. For the actual solution of processing problems, however, the components of any such quantity in a relevant coordinate system will be the values of interest. The complexities that then arise when writing down algebraic, differential, integral or integro-differential equations for these component quantities are largely the result of the complicated transformation rules that have to be applied to the vectors or tensors relevant for elastico-viscous liquids.

2.2 EQUATIONS OF MOTION AND ENERGY

These are the familiar equations of fluid mechanics. We express them in the fixed frame of reference R_0. $\mathbf{X}$ is a point in space,‡ $\mathbf{v}(\mathbf{X}, t)$ is the velocity and $\rho(\mathbf{X}, t)$ the density of the mass point at $\mathbf{X}$ at time t. $\mathbf{T}(\mathbf{X}, t)$ is the total stress tensor at $\mathbf{X}$ and t. This can be decomposed into an isotropic scalar pressure p (supposedly externally applied) and an extra-stress $\mathbf{T}^E$ due to deformation, according to

$$\mathbf{T} = -p\mathbf{1} + \mathbf{T}^E \qquad (2.2.1)$$

The relationship between p in a deforming system and the usual

† Vectors will as far as possible be written as lower-case bold-face letters; tensors as upper-case bold-face letters; components as lower-case italic letters with one and two subscripts respectively. Thus $(\mathbf{v})_i \equiv v_i$; $(\mathbf{T})_{ij} \equiv t_{ij}$.

‡ This use of an upper-case letter for a vector is exceptional and is not carried right through the text.

thermodynamic pressure is not absolutely clear. From the point of view of polymer melts or solutions, which are viscous or rubbery fluids, the contribution of $\mathbf{T}^{\mathrm{E}}$ to $\mathrm{tr}(\mathbf{T})$ does not seriously affect the use of the traditional equilibrium equation of state for a static fluid

$$\rho = \rho_{\mathrm{e}}(p_{\mathrm{e}}, T) \tag{2.2.2}$$

where T is the temperature, and p_{e} is the thermodynamic pressure or quantity defined unambiguously only for an equilibrium state, under isotropic conditions of stress. There is thus some advantage in definiteness in replacing $\mathbf{T}^{\mathrm{E}}$ (however that happens to be defined) by†

$$\mathbf{T}' = \mathbf{T}^{\mathrm{E}} - \tfrac{1}{3}\,\mathrm{tr}(\mathbf{T}^{\mathrm{E}})\mathbf{1} \tag{2.2.3}$$

and assuming that the p then defined as $-\frac{1}{3}\,\mathrm{tr}(\mathbf{T})$ is equivalent to p_{e} as far as relating ρ and T is concerned. However, careful measurements at relatively high values of p and low values of T show that changes in p at fixed T can lead to slow changes in ρ, i.e. that non-equilibrium situations arise and that the notion of memory applies to dilatation as well as to shear deformation (see Ferry 1980, Chapter 11, C). From henceforth nevertheless we shall use (2.2.1) treating p as an effective thermodynamic pressure and $\mathbf{T}^{\mathrm{E}}$ the quantity that will be determined from a constitutive equation for stress.

$U(\mathbf{X}, t)$ is the internal energy per unit mass; $\mathbf{g}$ we take to be the body force per unit mass, usually equal to the gravitational force; $\mathbf{q}(\mathbf{X}, t)$ is the heat flux vector.

The meaning of U for a deforming elastic fluid is again not entirely clear. However, for many of our purposes we shall suppose its main component to be given by an equilibrium thermodynamical relation

$$U = U_{\mathrm{e}}(\rho, T) \tag{2.2.4}$$

The matter is taken up again later and discussed briefly in Astarita & Marrucci (A&M 1974, Section 1.10), where it is made clear that elastic components of U are necessarily present in an elastic liquid.

The equation of continuity, expressing the principle of conservation of mass can be written

$$\mathrm{D}\rho/\mathrm{D}t = -\rho\,\boldsymbol{\nabla}\cdot\mathbf{v} \tag{2.2.5}$$

† We define $\mathrm{tr}(\mathbf{A})$ as a_{ii} where a_{ij} are Cartesian components of $\mathbf{A}$.

where† $\mathrm{D}(\ )/\mathrm{D}t = \partial(\ )/\partial t + \boldsymbol{\nabla}(\ ) \cdot \mathbf{v}$, and $\mathbf{v}$ is the velocity, is a material derivative, or in the equivalent form

$$\boldsymbol{\nabla} \cdot (\rho \mathbf{v}) = -\partial \rho / \partial t \tag{2.2.6}$$

For many purposes we can treat polymer melts as incompressible, in which case we can use the very simple relation

$$\boldsymbol{\nabla} \cdot \mathbf{v} = 0 \tag{2.2.7}$$

commonly employed in hydrodynamics. However, there are cases where the small compressibility of rubbery materials can have significant fluid mechanical effects, as for example in the unsteady flow from a large pressurized reservoir through a capillary die or nozzle—this will be seen to be important in injection moulding.

The equation of motion, expressing the principle of conservation of momentum, can be written

$$\rho\, \mathrm{D}\mathbf{v}/\mathrm{D}t = -\boldsymbol{\nabla} p + \boldsymbol{\nabla} \cdot \mathbf{T}^{\mathrm{E}} + \rho \mathbf{g} \tag{2.2.8}$$

The principle of conservation of moment of momentum implies that $\mathbf{T}$ is symmetric, i.e. $\mathbf{T} = \mathbf{T}^T$ $((\mathbf{T}^T)_{ji} \equiv (\mathbf{T})_{ij})$ for the non-polar fluids we are considering.

Lastly, the principle of conservation of energy, i.e. the first law of thermodynamics, leads to the thermal energy equation

$$\rho\, \mathrm{D}U/\mathrm{D}t = -\boldsymbol{\nabla} \cdot \mathbf{q} + \mathbf{T} : \boldsymbol{\nabla} \mathbf{v} \qquad (\mathbf{A} : \mathbf{B} \equiv (\mathbf{A})_{ij} (\mathbf{B})_{ji} = a_{ij} b_{ji}) \tag{2.2.9}$$

Derivations of the conservation equations (2.2.5), (2.2.8) and (2.2.9)

† The term $\boldsymbol{\nabla}\phi \cdot \mathbf{v}$ where ϕ is a scalar, and hence $\boldsymbol{\nabla}\phi$ is a vector could equally well be written $(\mathbf{v} \cdot \boldsymbol{\nabla})\phi = \mathbf{v} \cdot \boldsymbol{\nabla}\phi$ (which is the conventional order used in vector fluid mechanics) because taking the inner product of vectors is a commutative operation. However differences arise when $\boldsymbol{\nabla}$ operates on a vector or tensor. For any given definition $\mathbf{a} \cdot \boldsymbol{\nabla}\mathbf{b} \neq \boldsymbol{\nabla}\mathbf{b} \cdot \mathbf{a}$ if $\boldsymbol{\nabla}\mathbf{b}$ is not symmetric. A&M, eqn (1.6.7), use the definition

$$(\boldsymbol{\nabla}\mathbf{v})_{ij} = \frac{\partial v_i}{\partial x_j}$$

where the subscripts i and j refer to any Cartesian coordinate system. However BA&H (Table A.7.1) use the transposed definition

$$(\boldsymbol{\nabla}\mathbf{v})_{ij} = \frac{\partial v_j}{\partial x_i}$$

Thus $\boldsymbol{\nabla}\mathbf{v} \cdot \mathbf{v}$ for one becomes $\mathbf{v} \cdot \boldsymbol{\nabla}\mathbf{v}$ for the other. We have used A&M's form in defining $\mathrm{D}/\mathrm{D}t$ just below eqn (2.2.5).

and alternative forms can be found in A&M (1974, Sections 1.6–1.10) or in BA&H (1977, Section 1.1).

Clearly our three conservation equations which are true for all fluids at all times involve more than three variable quantities, namely ρ, $\mathbf{v}$, p, $\mathbf{T}^E$, U and $\mathbf{q}$, and so they must be supplemented by various equations of state (or constitutive equations), specific to particular fluids in given circumstances. Equations (2.2.2) and (2.2.4) provide two, but at the expense of introducing the temperature T. It is usual to suppose that $\mathbf{q}$ is linearly related to ∇T (as in Fourier's Law) by

$$\mathbf{q} = -\alpha \nabla T \tag{2.2.10}$$

where $\alpha(\rho, T)$ is the scalar thermal conductivity.

General theories of thermodynamics allow $\mathbf{q}$ to depend in a more complicated way on ∇T and the history of deformation. There is evidence that oriented (i.e. anisotropic) polymeric solids can exhibit a non-isotropic, i.e. second-rank tensor, form for the constitutive function α, and so in highly-oriented polymer melts such a departure from the static relation (2.2.10) might be expected. This could be important in such strongly sheared flows as arise in calendering or injection moulding. The matter will be raised again later.

Furthermore, in some situations there may be a contribution from radiation. This can either be included in the relation (2.2.10) for $\mathbf{q}$ or as an additional term in (2.2.9), say $\mathbf{q}_{rad}$.

We therefore still need a further constitutive equation, relating $\mathbf{T}^E$ to ρ, $\mathbf{v}$, p and T, so as to provide a determinate set of equations for analysis of flow problems. Providing such equations is the province of polymer fluid rheology, a subject of much current interest, and one which will be discussed at length in the next section.

2.3 BOUNDARY CONDITIONS

To make our set of equations fully determinate, we must provide a suitable set of boundary conditions. While these are best described when modelling each process, certain common (idealized) forms are worth describing now.

2.3.1 Rigid Boundaries

These will usually be associated with fixed or moving metal parts that contain the polymer flows. At such boundaries the velocity of the wall

$\mathbf{v}_W(\mathbf{X}_W, t)$ can usually be prescribed as a function of position $\mathbf{X}_W$ on the wall and time t. The no-slip condition, which can be assumed to hold in the absence of evidence to the contrary, requires that

$$\mathbf{v}(\mathbf{X}_W, t) = \mathbf{v}_W \tag{2.3.1}$$

at the wall, $\mathbf{v}$ being the fluid velocity. Some materials have been shown to slip. Under these circumstances, although the normal velocity is continuous, i.e.

$$\mathbf{v} \cdot \hat{\mathbf{n}}_W = \mathbf{v}_W \cdot \hat{\mathbf{n}}_W \tag{2.3.2}$$

$\hat{\mathbf{n}}_W$ being the unit outward normal to the wall at $(\mathbf{X}_W, t)$, the tangential component will not be continuous, i.e.

$$\mathbf{v} - (\mathbf{v} \cdot \hat{\mathbf{n}}_W)\hat{\mathbf{n}}_W \neq \mathbf{v}_W - (\mathbf{v}_W \cdot \hat{\mathbf{n}}_W)\hat{\mathbf{n}}_W \tag{2.3.3}$$

It is usual to suppose that

$$\mathbf{v} - \mathbf{v}_W = \mathrm{fn}(\mathbf{T} \cdot \hat{\mathbf{n}}_W)(\mathbf{T} \cdot \hat{\mathbf{n}}_W - \mathbf{T} : \hat{\mathbf{n}}_W \hat{\mathbf{n}}_W \hat{\mathbf{n}}_W) \tag{2.3.4}$$

i.e. that the slip velocity will be parallel to the wall shear stress, but the scalar 'factor of proportionality' $\mathrm{fn}(\mathbf{T} \cdot \hat{\mathbf{n}}_W)$ may depend in a complicated way on the stress at the wall.

It is usual to assume that temperature will be continuous at the wall/fluid interface, though, in circumstances where slip takes place, this may not be a reasonable assumption. If the rigid wall is a good conductor—and it usually will be relative to the polymer fluid—it is often reasonable to suppose that the wall temperature $T_W(\mathbf{X}_W, t)$ can be prescribed. The temperature boundary condition on the fluid then becomes

$$T(\mathbf{X}_W, t) = T_W \tag{2.3.5}$$

If a heat flux $\mathbf{q}_W(\mathbf{X}_W, t) = q_W(X_W, t)\hat{\mathbf{n}}_W$ is prescribed, then we have the boundary condition

$$\alpha(\boldsymbol{\nabla} T \cdot \hat{\mathbf{n}}_W) = q_W \tag{2.3.6}$$

Relations (2.3.5) and (2.3.6) are not strictly speaking alternatives, because they both hold at all times. The element of choice arises as to which of T_W or q_W, if either, can be prescribed in any given problem. The full thermal problem would involve a thermal equation (the heat-conduction equation) for the bulk solid material bounding the polymer fluid, which would be coupled through (2.3.5) and (2.3.6) to the energy equation (2.2.9) for the fluid. In cases where neither (2.3.5) nor (2.3.6) can be prescribed as a function of $\mathbf{X}_W$ and t, the true solution to the solid heat-conduction problem is sometimes subsumed in an apparent wall-boundary-condition

$$\alpha(\nabla T \cdot \hat{\mathbf{n}}_W) = h_W(T - T_W) \tag{2.3.7}$$

h_W being a heat transfer coefficient, and T being the temperature of the polymer at the wall.

2.3.2 Free Boundaries

These will usually be associated with air or water interfaces, across which relatively small stresses can be exerted. The polymer fluid is therefore largely unconstrained at these boundaries and the relevant boundary condition will be on the stress. The main contribution in absolute terms will usually be the ambient external isotropic pressure p_a but, for highly curved surfaces, the surface tension, and, for rapidly moving polymer or external flows, air or water drag may be significant. The relevant boundary condition therefore can be written in general

$$\mathbf{T}(\mathbf{X}_{FB}, t) \cdot \hat{\mathbf{n}}_{FB} = (2\sigma H - p_a)\hat{\mathbf{n}}_{FB} - \nabla_{FB}\sigma - \mathbf{f}_{drag} \tag{2.3.8}$$

where σ is the interfacial surface tension, H is the first principal curvature† of the surface, ∇_{FB} represents a gradient in the surface and $\mathbf{f}_{drag}$ is the force/unit area exerted by the exterior fluid across the interface.

The relation (2.3.8) does not allow for an interface with structure of the type considered in Aris (1962, Chapter 10) involving its own constitutive relations, though the contribution of surface tension could be regarded as a degenerate form of the latter. The implications of boundary condition (2.3.8) are very considerable. When solving a flow problem involving free boundaries, we cannot prescribe $\mathbf{X}_{FB}(t)$. Hence (2.3.8) is implicitly involved in the global solution. Furthermore, unless the symmetry of the solution is very obvious, translation of the seemingly innocuous terms H and ∇_{FB} into useful operational form is a complicated matter. The reader is referred to Aris (1962) and to Joseph (1974) for further discussion of this matter.

The temperature boundary conditions at the free boundary will be similar to those for a rigid wall. However, (2.3.5) and (2.3.6) will not be very helpful, because T_{FB} and q_{FB} will not usually be prescribed.

† This can be written $\nabla_{FB} \cdot \hat{\mathbf{n}}_{FB}$.

Relation (2.3.7) is much more likely to be the most useful general form with h_{FB} requiring careful estimation.

EXERCISE 2.2.1. Show that $\mathbf{T}:\nabla\mathbf{v}=\mathbf{T}:(\nabla\mathbf{v})^{\mathrm{T}}$ if T is symmetric.

EXERCISE 2.3.1. Verify that h_W will be positive for the definition given for $\hat{\mathbf{n}}_W$ following (2.3.2).

EXERCISE 2.3.2. Verify (2.3.8) using a local Cartesian coordinate system in which $\mathbf{n}_{FB}=(1, 0, 0)$, showing that:

(a) $\mathbf{T}\cdot\mathbf{n}_{FB}$ represents the force/unit area acting from the fluid onto the interface;
(b) $-2\sigma H\hat{\mathbf{n}}_{FB}$ is the normal force/unit area caused by surface tension acting on the fluid;
(c) $p_a\hat{\mathbf{n}}_{FB}$ is the direct component of the outside pressure acting on the fluid;
(d) $\nabla_{FB}\sigma$ is the tangential force/unit area caused by variations in surface tension acting on the fluid.

Chapter 3

Constitutive Equations

3.1 OBSERVATIONS AND REPRESENTATIONS

The specification of suitable non-linear elastico-viscous constitutive equations is what distinguishes the mechanics of polymeric fluids from traditional Newtonian fluid mechanics. Qualitatively different flow phenomena arise in the two cases. BA & H (1977, Chapter 3) mention, as characteristics of polymeric fluids:

Shear thinning in tube or channel flow, interpretable as variable viscosity.
Rod climbing as an example of the effect of normal-stress differences.
Extrudate swell and elastic recoil on sudden cessation of tube flows as consequences of stored elastic strain energy.
Secondary flows, in various rotating systems, opposite in direction to the inertially-driven secondary flows characteristic of Newtonian fluids.
The tubeless syphon demonstrating tension in streamlines.
Large pressure losses and recirculating vortices in flow through sudden contractions and the instability of such flows above a certain flow rate.
Drag reduction in turbulent flow of dilute polymer solutions.

Similar descriptions are given in Lodge (1964, Chapter 10). Readers for whom all these ideas are new are recommended to read at least one source describing these qualitative effects.

Most of these effects and several others that will be mentioned later arise during the processing of plastics and rubbers, and mean that an engineering analysis of the flow process cannot be undertaken without

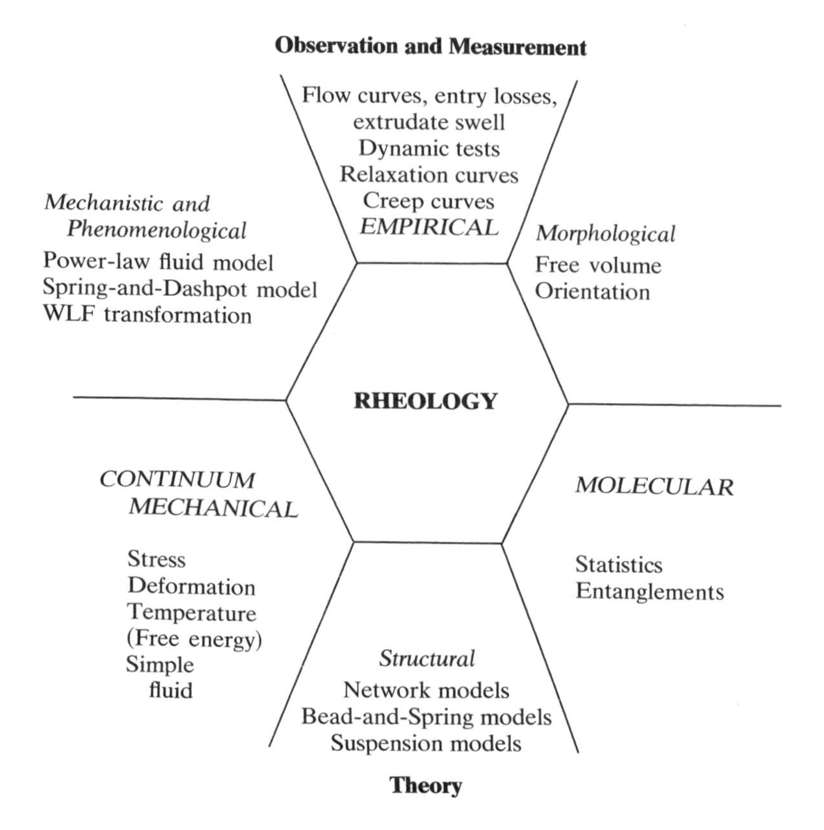

Fig. 3.1. Relationships between various approaches to rheology.

a theoretical framework for describing and predicting these effects. Different workers have attempted to achieve this from different starting points, and this has led, in some cases, to bitter disagreement as to the merit, even acceptability, of the various approaches.

Figure 3.1 presents diagrammatically certain relationships within a categorization of these many approaches. The three main starting points, all clearly distinct, for developing rheological equations of state—relating histories of stress, deformation and temperature—are

1. Empirical observation.
2. Formal continuum mechanics.
3. Molecular theory.

Significant progress can be made from each of these starting points without making much apparent use of ideas derived from the other.

Thus empirical observation can yield:

A. Flow curves relating reservoir pressure, reservoir temperature and output rate in capillary extrusion—the basic flow field in the most simple, reliable and traditional melt rheometer. From these, using capillaries of different length and diameter, can be obtained:
   - end (mainly entrance) pressure losses;
   - wall shear stresses;

   and with a minimum of continuum interpretation:
   - wall shear rates;
   - wall-slip coefficients;
   - shear (and temperature)-dependent viscosities.

   Observation of the extrudate can then yield the
   - Extrudate-swell ratio as a function of wall shear rate and capillary length.

B. Dynamic viscosity† as a function of temperature and frequency using simple oscillatory shearing rheometers.

C. Creep curves using simple extensometers (either the force or the extension can be programmed as a function of time). Stress relaxation curves can be plotted following cessation of creep or simple shear.

All of these experimental observations lead to scalar relations between (easily described) scalar variables. Practitioners are able to express and interpret such results (at least for their own purposes) without recourse to tensor or functional analysis. Such information can be obtained rapidly and serves to distinguish in a fairly simple fashion between different materials, and thus, when required, to maintain the quality of a product. Useful predictions can be made about flowrates, pressure drops, and, as we shall see later, temperature rises, on the basis of these disconnected empirical observations, for processing flows of engineering importance. It would be unreasonable not to recognize the merits of these simple approaches, and indeed much of the analysis given later will use such simple ideas.‡ But there are many examples of

† A complex 'impedance' not to be confused with the more usually defined viscosity or kinematic viscosity.

‡ However, it is important that the limitations of a simple approach be well understood, and that simple notions that are adequate in a simple situation are not carried over, by analogy with Newtonian fluid mechanics for example, to more complicated situations for which they are not adequate.

polymer flows that cannot be predicted or modelled in terms of simple rheological parameters: indeed many technological improvements have been made—or, to put the matter more realistically, technological difficulties have been overcome—only by expensive trial-and-error methods which lead to industrial art, experience or know-how. There is thus good reason for seeking a theoretical background that will enable us to explain apparently unconnected observations on material behaviour as related parts of a connected whole; we seek approximate constitutive equations involving a minimal number of functions or parameters that will model adequately as wide a range of flow behaviour as possible.

Perhaps the most inexplicable phenomena, at least in terms of elementary ideas, are those involving flow instability. Elementary ideas usually imply highly symmetrical and steady flows, thus reducing the dimensionality of the geometrical representation and limiting the number of independent and dependent mechanical variables. Instability, on the other hand, usually implies that the symmetry possessed by the boundaries is not reflected in the flow pattern, which will fluctuate and exhibit unexpected secondary flows. In extreme cases, the flows become chaotic and even discontinuous. This will be taken up later.

At this stage, continuum mechanics, with all its mathematical trappings, becomes a necessary part of our approach. The formal theories of continuum mechanics deal with viscoelastic materials in general and seek to underline similarities in behaviour between materials rather than differences. To this extent the continuum-mechanical approach is complementary to both empirical and molecular approaches. Much of the effort in modern continuum-mechanical theories goes into careful description of deformation histories and into the consequences of the postulates of frame indifference, local action and isotropy; by selecting a variety of specific flow fields—a procedure this approach shares with both the empirical and molecular approaches—very great simplifications are achieved in the generality of possible material response.

Insofar as material behaviour in one flow situation is treated as wholly independent of behaviour in another (for example, complete information on uniform steady simple-shear flow, Section 4.2, need tell us nothing about behaviour in pure extensional flow, Section 4.3), no real advance of engineering significance has been made over the empirical approach. There is no great merit, from our point of view, in having a very elaborate theory leading to perfect agreement with flow curve data if the same theory is wildly inaccurate in its predictions for entry flows—assuming that it makes any predictions at all about the latter flows.

The basic importance of the continuum approach lies in its ability to provide constitutive equations capable of describing behaviour in all flow situations, and hence, when combined with the equations of motion and energy, at least in principle, of predicting flow fields for complicated situations. For our engineering purposes, therefore, we shall always be looking for the most generally applicable approximate constitutive equation, in the hope that flow predictions for polymer melt mechanics may be reduced to the problem of obtaining approximate solutions to a mathematically well-posed set of equations and boundary conditions. The geometry of the flow field and the boundary and initial conditions will specify the process. The parameters of the constitutive equation will specify the fluid.

The distinction between kinematically-general and kinematically-special constitutive equations is often made less clear when physical insight into complicated flow fields is sought, or when solution of the coupled set of flow equations is attempted by analytical approximation methods: in such cases, a complex flow field is often split up into a set of essentially distinct subfields or regions, each of which is chosen to exhibit no more than perturbations about a simple flow pattern. In each of these regions the general constitutive equation chosen for the fluid can be approximated by a limiting asymptotic form which will naturally be equivalent to one of the specialized constitutive equations mentioned earlier. However, any success in matching the flows in the various regions will depend upon deriving the various asymptotic limits as leading terms, in different asymptotic expansions for the same general constitutive equation. There will initially be an interplay between the asymptotic expansions chosen for the constitutive equation and those chosen for the velocity, temperature and stress fields. This point was missed by many earlier authors not concerned with engineering applications, but is now regarded as the key to progress. A simple example is given in Example 3.1.1 relating to Newtonian fluid mechanics.

How then do we decide upon a useful constitutive equation relating the measurable continuum variables

STRESS; DEFORMATION; TEMPERATURE

The first two of these are mechanical variables and the last a thermodynamic one. The classical theories of mechanics deal solely with the first two, and the greater part of work done towards deriving and testing constitutive equations has concentrated on isothermal situations. This emphasis is to be noted in all the texts mentioned earlier. Thermodynamics (and hence temperature) is introduced later almost by stealth.

At the most elementary level, the various material parameters arising in constitutive equations are treated as functions of temperature on an empirical basis; in the case of low-amplitude oscillatory (dynamic) experiments, a time–temperature superposition principle has been proposed and has been remarkably successful, Section 5.1.1. This can be explained physically as meaning that the instantaneous natural timescales of viscoelastic material behaviour are all monotonically related to the local instantaneous temperature in the same way. This powerful and all-embracing assumption keeps the difficulties of thermodynamics from intruding upon the absolute certainties of Eulerian mechanics: time is a primitive variable for mechanics; if temperature can be treated as another primitive variable having local effect only on the timescale, then no damage is done to the postulates of mechanical theory, although there may be some ambiguities in interpretation of the various time differences that arise in particular constitutive equations, as we shall see later.

At the next level, account is taken of the energy equation (2.2.9) and of the stress–power term $\mathbf{T}:\nabla\mathbf{v}$, which couples the constitutive equation to the energy equation. Determination of the global temperature field is of great practical importance in processing flow analysis—indeed it is probably true to say that control of the temperature field is the major objective and problem in polymer processing—and so a satisfactory understanding of the non-equilibrium thermodynamics of elastic liquids is required. The relevant fundamental thermodynamic constitutive equations are best provided by a free-energy functional with temperature and deformation as arguments rather than internal energy with entropy and deformation as arguments. Interpretation of the terms in the energy equation is no longer elementary and so, in practice, some decoupling of the mechanical viscoelastic nature of the fluid and of the thermal equations is adopted, with T as a primitive variable, and U expressed in terms of specific heat, whose value can be obtained from equilibrium (static) measurements. In the absence of a large amount of careful experimental evidence—as has been obtained for elastomeric solids, which can be treated in terms of equilibrium thermodynamics—such apparently crude approximations are as good as any other.

A third level employs the second law of thermodynamics to place constraints on constitutive relations. For example, the viscosity of a Newtonian liquid must be positive. We shall not pursue this matter here.

EXAMPLE 3.1.1. Boundary-layer theory and flow around bluff bodies as an example of asymptotic flow decomposition.

A long-standing problem in Newtonian fluid mechanics is provided by dynamical analysis of flow around a sphere held stationary in an otherwise steady uniform flow. If the Reynolds number of the flow given by

$$\mathrm{Re} = \rho U D/\mu$$

(where ρ and μ are the fluid density and viscosity respectively, U is the flow speed far from the sphere, of diameter D) is large then a thin ($\ll D$ in thickness) boundary layer forms over the upstream part of the sphere; within this layer the vorticity of the flow increases rapidly from a maximum near the surface of the sphere to zero in the outer flow. Near the diametral plane of the sphere that is normal to the flow velocity at infinity, this boundary layer (laminar if the Reynolds number is not too large) separates and moves downstream as a thickening axisymmetric shear layer, which acts as an envelope to a downstream wake (again laminar if the Reynolds number is not too large). Outside the laminar boundary layer and wake, the flow can be modelled by an irrotational flow (with zero vorticity) with streamlines and pressure distribution given by an inviscid ($\mu = 0$) fluid approximation. Within the wake, it has been suggested by Batchelor that a largely inviscid ($\mu \simeq 0$) finite-vorticity approximation might be relevant. Within the boundary layer a largely-plane shear-flow approximation is relevant in which inertial, pressure and viscous forces are all of the same order.

Calculation of the dominant terms in the flow field is performed by solving for each flow field separately, the boundary conditions at 'interfaces' between the outer, large-scale inviscid flows and the boundary or shear layer having to be matched. The problem remains a difficult one to describe analytically, but without such preliminary decomposition no progress whatever could be made.

3.2 THE SIMPLE FLUID WITH FADING MEMORY

The most useful starting point is that of a simple fluid with fading memory. Put briefly in words: this is a fluid, possessing no natural shape but having a natural density whose stress is determined by the history of deformation of an arbitrarily small neighbourhood of the material point in question, the influence of such deformation falling off as time passes. A very general constitutive equation may be written (cf. eqns 4.4.33 and 4.4.42 in A & M)

$$\mathbf{T}(t) = \mathop{\mathscr{H}}_{s=0}^{s=\infty} [T^t(s), \mathbf{F}_R^t(s); T(t), \mathbf{F}_R(t), \rho_R] \tag{3.2.1}$$

Here $\mathbf{T}$ is the total stress, $T(t)$ the temperature and $\mathbf{F}_R$ the deformation gradient tensor with respect to some fixed configuration R for which the density is ρ_R. Here

$$\mathrm{d}\mathbf{X}(t) = \mathbf{F}_R(t) \cdot \mathrm{d}\mathbf{X}_R \tag{3.2.2}$$

where $\mathrm{d}\mathbf{X}(t)$ and $\mathrm{d}\mathbf{X}_R$ are associated instantaneous and reference line elements, in the neighbourhood of a given fluid element, referred to the same reference frame. (The d**X**'s are to be thought of as attached to the fluid.) The notation $T^t(s)$ stands for $T(t-s)$ with positive values of s and it follows that

$$\mathbf{F}_R^t(s) = \mathbf{F}_R(t-s) = \mathbf{F}_t^t(s) \cdot \mathbf{F}_R(t) = \mathbf{F}_t(t-s) \cdot \mathbf{F}_R(t) \tag{3.2.3}$$

The need for the several forms of the basic deformation gradient tensor $\mathbf{F}$, as displayed in (3.2.3), for example, arises from the various possible reference configurations that can be used for the material. In the non-linear mechanics of elastic solids, the best reference configuration is obviously that of the natural (unstrained and unstressed) state, which provides the necessary labelling $\mathbf{X}_0$ of material particles. In the case of fluids, which have no natural rest state, the most convenient reference configuration is the instantaneous state, so $\mathbf{X}_t$ is regarded as a convenient label for particles, that are at position $\mathbf{X}_t$ at time t. Since we know that the stress will depend upon the past history of deformation, we can look backwards from present time, which leads to the deformation gradient $\mathbf{F}_t^t(s)$, defined by (3.2.3) and which is such that

$$\mathbf{F}_t^t(0) = \mathbf{1} \tag{3.2.4}$$

It could have been written alternatively as

$$\mathbf{F}_t(t') = \mathbf{F}_t^t(t-t') \tag{3.2.5}$$

i.e. the deformative gradient at time t' relative to time t of the particle at position $\mathbf{X}_t$ at time t. The definition (3.2.2) makes it clear that

$$\mathbf{F}_{t'}(t) = \mathbf{F}_t^{-1}(t') \tag{3.2.6}$$

i.e. that the deformation gradient at time t relative to time t' is the inverse of the deformation gradient at time t' relative to time t.

We note that all of these deformation tensors refer to various configurations of the fluid material (at various times) viewed in the same reference frame. That reference frame could be an R_0 or an R_M or an R_B as defined in Section 2.1. Within any one of those frames of reference we could choose a variety of coordinate systems. (Some authors, particularly Lodge, point out that the use of $\mathbf{X}_t$ to represent an instantaneous label for both a particle and a position in space† can lead to confusion, but so many others appear to surmount this conceptual difficulty without subsequent loss of precision or flexibility that we shall not introduce body tensors here.) Much of tensor analysis as usually taught refers to changes in coordinate systems within a given frame of reference; in particular, to how the component values for any given vector or tensor fields are related when changing coordinate system. Our notation, so far, is intended to be independent of coordinate system. The important transformations carried out at length in continuum-mechanical derivations refer to changes in frame of reference (see Section 3.3 of A & M). Formally, the processes look alike but conceptually they are different.

The problem is further compounded when it becomes clear that rheological equations of state are most easily described in frames of reference that deform with the body. In these, deformation gradients are replaced by a time-dependent metric tensor for the body coordinate system. The problem then becomes one of transferring from a body frame of reference to a space frame; the rules for doing this were set out by Oldroyd (1950) and can be found in later texts (BA & H, 1977, Section 9.3; Lodge, 1964, 1974; Walters, 1975, Chapter 2).

The various **F**'s are also distinguished by their arguments. Although

† However, we have retained the use of upper case **X** to define the position of a particle in this chapter, instead of lower case **x** which is used elsewhere for the (Eulerian) space coordinate.

all the relations given above are intended to refer to the same particle, the label it bears in some fixed R_0 frame will differ, for example, as we go from $\mathbf{F}_{t'}$ to $\mathbf{F}_t$ because $\mathbf{X}_{t'}$ is in general different from $\mathbf{X}_t$ (this is one issue raised by Lodge, but is not a new one and is dealt with at length in most books on fluid mechanics in connection with the inertial term $D\mathbf{v}/Dt$ introduced in Section 2.2). Furthermore, the time derivative $\partial/\partial t'$ will be different from $\partial/\partial t$ and from $\partial/\partial s$.

What is important is that many of the specific constitutive equations to be developed here, and certainly the general one (3.2.1), involve the addition of tensors defined at different places and different times. If this is to be a meaningful mathematical procedure, then the nature of the tensors to be added is very considerably constrained. This is the basis for Lodge's claims (1974) regarding the primacy of body tensor fields, and for recognizing that the tensors we use here must be, in his definition, Cartesian tensors. Operationally, this means that we have to deal with physical components and gives rise to the considerable complexity that is evident in BA & H.

$\mathcal{H}$ stands for a symmetric tensor-valued functional of the history of temperature $T^t(s)$ and deformation gradient $\mathbf{F}_R^t(s)$, with possible explicit dependence on the current values T and $\mathbf{F}_R$; $s = t - t'$ measures the difference in time between some past time t' and current time t. Explicit dependence on $\mathbf{F}_R(t)$ allows of sudden deformations at time t leading to instantaneous elastic response in terms of an instantaneous change in both T and $\mathbf{F}_R$ between $s = 0-$ and $s = 0$. The reference state has been deliberately left arbitrary in (2.3.1) though it could be chosen so that $\mathbf{F}_R(t-)$ or $\mathbf{F}_R(t)$ were 1 or such that ρ_R is the unstressed density. If the history of deformation is smooth and differentiable, then the need for explicit dependence on $\mathbf{F}_R(t)$ is removed. If further, $T(t)$ is taken to be some fixed T_0, then T_0 becomes a simple parameter. We cannot remove dependence on ρ_R if eqn (3.2.1) is to reduce, in the case of compressible liquids, to the traditional equation of state (2.2.2), i.e. when

$$\mathbf{T} = -p\mathbf{1} \tag{3.2.7}$$

p being the pressure. Knowledge of ρ_R leads to a knowledge of ρ at all other times, i.e. in all other configurations, since

$$\rho(t) = \rho_R/\det \mathbf{F}_R(t) \tag{3.2.8}$$

The principle of material frame indifference leads to the replacement of the relative tensor $\mathbf{F}_t(t')$ by the Cauchy tensor

$$\mathbf{C}_t(t') = \mathbf{F}_t^T \cdot \mathbf{F}_t \tag{3.2.9}$$

or equivalently $\mathbf{F}_t^t(s)$ by $\mathbf{C}_t^t(s) = \mathbf{F}_t^{tT} \cdot \mathbf{F}_t^t$, whence

$$\mathbf{T} = \overline{\mathscr{H}}_{s=0}^{s=\infty} [\mathbf{C}_t^t(s); \rho] \tag{3.2.10}$$

subject to

$$\mathbf{Q}(0) \cdot \overline{\mathscr{H}}_{s=0}^{s=\infty} [\mathbf{C}_t^t(s); \rho] \cdot \mathbf{Q}^T(0) = \overline{\mathscr{H}}_{s=0}^{s=\infty} [\mathbf{Q}(s) \cdot \mathbf{C}_t^t(s) \cdot \mathbf{Q}^T(s); \rho] \tag{3.2.11}$$

where $\mathbf{Q}(s)$ is an arbitrary orthogonal tensor representing rotation of the frame of reference in which $\mathbf{C}_t^t$ is described (see A & M Sections 3.3, 4.3).

Note that expression (3.2.10), particularly if it is written to include $T(s)$ as an argument of the functional $\overline{\mathscr{H}}$, provides an expression for the total stress tensor and not just the extra stress tensor or the deviatoric stress tensor. It is therefore intended to cover purely dilatational deformations as well as shear deformation and allows, for example, for pressure relaxation to the equilibrium thermodynamic value after a rapid change in ρ. These considerations are by no means irrelevant in high frequency propagation of P (pressure) waves, but are usually avoided by most workers in polymer rheology.

On pragmatic grounds, dilations (changes in density) caused by temperature changes are usually regarded as dominating those caused by pressure variations, and a modified incompressibility condition

$$\rho = \rho(T) \tag{3.2.12}$$

allows the effect of temperature changes to be factored out.

In the isothermal case, then, eqn (2.2.1) is assumed to hold, p is regarded as a dynamical variable and (3.2.10) reduces to

$$\mathbf{T}^{\mathrm{E}} = \tilde{\mathscr{H}}_{s=0}^{s=\infty} [\mathbf{C}_t^t(s)] \tag{3.2.13}$$

still subject to (3.2.11) (note the use of the extra-stress tensor). Further progress depends upon making various modelling assumptions.

The formal mathematical development of the simple-fluid model is usually attributed to Noll (1962), though later authors (Lodge, 1974,

Section 11.4) have pointed out that it is implicit in Oldroyd's 1950 paper. Indeed Oldroyd's argument could be interpreted as covering a slightly wider class given by the functional equation

$$0 = \mathop{\mathscr{F}}_{s=0}^{s=\infty} [T^t(s), \mathbf{C}_R^t(s), \mathbf{T}_R^t(s); T(t), \rho(t), \mathbf{C}_R(t), \mathbf{T}_R(t)] \quad (3.2.14)$$

in which $\mathbf{C}$ and $\mathbf{T}$ appear implicitly within the tensor functional $\mathscr{F}$. The use of $\mathbf{T}_R$ takes account of the fact that the stress tensor may be referred either to the instantaneous configuration ($\mathbf{T}$) or to a reference configuration (Truesdell & Noll, 1965, p. 124).

EXERCISE 3.2.1. Prove relation (3.2.8) using definition (3.2.2), with $d\mathbf{x}_R = \boldsymbol{\delta}_i \, dx$, $i = 1, 2, 3$, where the $\boldsymbol{\delta}_i$ are 3 mutually orthogonal unit vectors, and dx is a small element of length. Note that the triple-scalar product $(d\mathbf{x}_1, d\mathbf{x}_2, d\mathbf{x}_3)$ is the volume of a parallelepiped with adjacent sides $d\mathbf{x}_1$, $d\mathbf{x}_2$ and $d\mathbf{x}_3$.

3.3 PHENOMENOLOGICAL OR MECHANISTIC MODELS

We now move away from the purely formal continuum theories back towards empirical observation, but do not enquire at all into the structure of the polymeric materials we seek to model.

If we consider first materials with very short memories, we are led to a class of purely viscous fluids. The most celebrated is the incompressible Newtonian fluid.

3.3.1 Incompressible Newtonian Fluid

For this

$$\mathbf{T}^E = 2\eta_0 \mathbf{D} = \eta_0 \mathbf{A}_1 \quad (3.3.1)$$

where

$$\mathbf{D} = \tfrac{1}{2}\{\nabla \mathbf{v} + (\nabla \mathbf{v})^T\} \quad (3.3.2)$$

is the rate of deformation tensor, $\mathbf{A}_1$ is known as the first Rivlin–Ericksen tensor and η_0 is the Newtonian viscosity.

Following our earlier discussion, we can suppose η_0 to be a parametric function of T and ρ. There will be many analyses reported later based on a Newtonian model. It has importance for three reasons:

(1) Certain polymer melts, such as those of PET and PA, are indeed

Newtonian in behaviour at rates of deformation and temperatures characteristic of many industrial processes involving them, e.g. injection moulding. The same is true for low-molecular-weight waxes.

(2) All viscoelastic materials are Newtonian in behaviour for flows that are sufficiently slow and smooth, and so relation (3.3.1) is a relevant asymptotic limit in retarded-motion expansions (see Section 4.1 in A & M or Section 8.4 in BA & H).

(3) Solutions to the dynamical equations of motion are simpler to obtain for a Newtonian fluid than for any other. A great range of analytical and numerical techniques are available for solving the Navier–Stokes equation that results in the isothermal (or thermally decoupled) case. In practice such Newtonian solutions can often conveniently be employed as starting solutions for more realistic solutions based on non-linear constitutive equations. Alternatively, the nature of various physical phenomena relevant in complex processes can be investigated qualitatively using a Newtonian model before detailed quantitative calculations are carried out for more complex constitutive models.

Shear dependence of viscosity can be accounted for in a generalized Newtonian fluid.

3.3.2 Generalized Newtonian Fluid

For this

$$\mathbf{T}^{\mathrm{E}} = 2\eta(D)\mathbf{D} \tag{3.3.3}$$

with

$$D^2 = 2\mathbf{D} : \mathbf{D}, \qquad D > 0 \tag{3.3.4}$$

$\eta(D)$ being a generalized viscosity. This can be parametrized into a *power-law fluid*† form with

$$\eta(D) = \mathrm{K}D^{\nu-1} \tag{3.3.5}$$

This is often called the Ostwald–de Waele fluid; the parameters K and ν can be obtained rapidly from flow curves plotted using logarithmic axes and form a very convenient way of referring, albeit crudely, and

† It is worth noting at this point that K can be a function of temperature. The approximate form

$$\mathrm{K} = \mathrm{K}_0 \exp\{\zeta_0(T_0 - T)\}$$

will be much used later, ζ_0 and T_0 being related thermal parameters.

over a limited range in D to the viscous response of a polymeric fluid. Values of ν vary from 1 for a Newtonian fluid like hot PET to as low as 0·1 for some natural rubbers. It is between 0·5 and 0·3 for most rubbery polymers and processes. The meaning of ν may best be understood from the relation

$$P \propto \dot{V}^{\nu} \tag{3.3.6}$$

with P the capillary pressure drop and $\dot{V}$ the flowrate in capillary rheometry. (The equivalence of (3.3.5) and (3.3.6) is proved in Example 7.1.1.)

Other forms for $\eta(D)$ are mentioned in BA & H (Section 5.1). The Carreau model

$$\frac{\eta - \eta_\infty}{\eta_0 - \eta_\infty} = [1 + \lambda^2 D^2]^{(\nu-1)/2} \tag{3.3.7}$$

has the advantage that it asymptotes to η_0, a Newtonian viscosity, as $D \to 0$ and to η_∞ as $D \to \infty$, thus avoiding the singularities of (3.3.5). Having four parameters, η_0, η_∞, ν and λ (a characteristic time), the model can be usefully fitted to much melt data over a rather wider range than (3.3.5), and is suitable for numerical work. However, it is still very much an empirical approximation.

The Newtonian approximation is paradoxically the one most removed from the general form (3.2.13) for a simple fluid. To see this we note that

$$\mathbf{D} = -\tfrac{1}{2}(\mathrm{d}\mathbf{C}_t^t/\mathrm{d}s)_{s=0} = \tfrac{1}{2}(\mathrm{d}\mathbf{C}_t(t')/\mathrm{d}t')_{t'=t} \tag{3.3.8}$$

and so (3.3.1) is a very special limit of (3.2.13). The Newtonian fluid does not admit of a sudden strain because that would lead to a singularity in stress. The objective of the general definition (3.2.1) and more especially of (3.2.14) was to allow just such a sudden strain. The mathematical relationship between the two forms has been demonstrated by Coleman and Noll (1961) using a retarded-motion expansion, which leads to the sequence of Nth-order fluids.

3.3.3 *N*th-order Fluids

$N = 1$

$$\mathbf{T}^{\mathrm{E}} = \eta_0 \mathbf{A}_1 \tag{3.3.9}$$

$N = 2$

$$\mathbf{T}^{\mathrm{E}} = \eta_0 \mathbf{A}_1 + \beta_0 \mathbf{A}_1^2 + \gamma_0 \mathbf{A}_2 \tag{3.3.10}$$

$N = 3$

$$\begin{aligned}\mathbf{T}^{\mathrm{E}} = {} & (\eta_0 + \delta_1 \operatorname{tr} \mathbf{A}_1^2)\mathbf{A}_1 + \beta_0 \mathbf{A}_1^2 + \gamma_0 \mathbf{A}_2 \\ & + \beta_1(\mathbf{A}_1 \cdot \mathbf{A}_2 + \mathbf{A}_2 \cdot \mathbf{A}_1) + \gamma_1 \mathbf{A}_3\end{aligned} \tag{3.3.11}$$

etc. where $\eta_0, \beta_0, \beta_1, \gamma_0, \gamma_1$ and δ_1 are constants and the $\mathbf{A}_N$ are the Rivlin–Ericksen tensors defined by

$$\mathbf{A}_N = \overset{N}{\mathbf{C}}_t \bigg|_{t'=t} \equiv \frac{\mathrm{d}^N \mathbf{C}_t(t')}{\mathrm{d}t'^N}\bigg|_{t'=t} \tag{3.3.12}$$

which leads to the more common recurrence relation

$$\mathbf{A}_{N+1} = \frac{\mathrm{D}}{\mathrm{D}t}\mathbf{A}_N + (\boldsymbol{\nabla}\mathbf{v})^T \boldsymbol{.} \mathbf{A}_N + \mathbf{A}_N \boldsymbol{.} \boldsymbol{\nabla}\mathbf{v} \tag{3.3.13}$$

$\mathrm{D}/\mathrm{D}t$ being defined in the previous chapter, after eqn (2.2.5). Example 3.3.1 gives (3.3.13) in various forms.

These Coleman–Noll constitutive equations are characterized both by their relative simplicity—in that no more than six constants are involved for the third-order fluid, for example—and by their inevitable appearance in any slow-motion expansion for a smooth viscoelastic flow field. From our point of view, their importance lies in how well they can model real polymeric fluids over a wide range of behaviour and in how easily they can be substituted in the momentum and energy conservation equations.

The first-order fluid is the Newtonian fluid which has been discussed earlier. The second-order fluid exhibits some viscoelastic properties (through $\mathbf{A}_2$) and a difference of normal stresses (through $\mathbf{A}_1$ and $\mathbf{A}_2$). However, its main disadvantage is that it predicts a constant viscosity in steady simple-shear flow, which is at variance with one of the most important characteristics of real polymer flows in processing situations. Thus, the large number of flow situations that have been analysed in the literature using (3.3.10) as a constitutive equation are of marginal utility to us. The third-order fluid, with suitable choice of coefficients, exhibits (see Table 4.1 with $\delta_1 + \beta_1 < 0$) a viscosity decreasing with shear rate, though not as realistically as that given either by (3.3.5) or (3.3.7); indeed the viscosity goes to zero at a finite shear rate, which clearly limits the utility of the model, as might be expected from the nature of the expansion from which it came; it also exhibits difference of normal stresses and viscoelasticity.

Various authors (Joseph & Fosdick, 1973; Langlois & Rivlin, 1959; Datta & Strauss, 1976) have used third- and even higher-order fluids to predict substantial departures from Newtonian behaviour in relatively complex flow situations, which can be tested experimentally. The predictions are in some cases (Joseph & Beavers, 1977) qualitatively very good and so we can regard (3.3.11) as a potentially useful model constitutive equation.

An equivalent series of approximations can be obtained by using

$$\mathbf{B}_N = -\overset{N}{\mathbf{C}}{}_t^{-1}\bigg|_{\tau=t} \equiv -\frac{\mathrm{d}^N \mathbf{C}_t^{-1}(\tau)}{\mathrm{d}\tau^N}\bigg|_{\tau=t} \tag{3.3.14}$$

instead of $\mathbf{A}_N$. The resulting equations are sometimes known as the White–Metzner expansion.

An alternative approach due to Rivlin and Ericksen leads to a class of model fluids that is obtained when (3.2.13) is replaced by a tensor function of the (instantaneous values of the) first N Rivlin–Ericksen tensors.

3.3.4 Rivlin–Ericksen Fluids

$$\mathbf{T}^{\mathrm{E}} = \mathbf{M}_N(\mathbf{A}_1, \mathbf{A}_2, \ldots, \mathbf{A}_N) \tag{3.3.15}$$

where $\mathbf{M}_N$ is a general isotropic tensor-valued function of $\mathbf{A}_K$, $K = 1, 2, \ldots, N$, and of the principal and joint invariants of the $\mathbf{A}_K$. The first in this sequence is the Reiner–Rivlin fluid.

Reiner–Rivlin Fluid

$$\mathbf{T}^{\mathrm{E}} = \phi_1(\overline{\mathrm{II}}_{\mathbf{A}_1}, \overline{\mathrm{III}}_{\mathbf{A}_1})\mathbf{A}_1 + \phi_2(\overline{\mathrm{II}}_{\mathbf{A}_1}, \overline{\mathrm{III}}_{\mathbf{A}_1})\mathbf{A}_1^2 \tag{3.3.16}$$

where ϕ_1 and ϕ_2 are functions of the two invariants†

$$\overline{\mathrm{II}}_{\mathbf{A}_1} = \mathrm{tr}(\mathbf{A}_1^2) = 2D^2 \tag{3.3.17}$$

$$\overline{\mathrm{III}}_{\mathbf{A}_1} = \mathrm{tr}(\mathbf{A}_1^3) \tag{3.3.18}$$

It has been assumed that the first invariant

$$\mathrm{I}_{\mathbf{A}_1} = \mathrm{tr}(\mathbf{A}_1) \tag{3.3.19}$$

is zero, i.e. that the fluid is incompressible.

Although this fluid model is positively unhelpful in situations where simple shear is the dominant kinematic mode, it is potentially useful in

† BA & H, eqns (A.3.20)–(A.3.24), use the opposite notation to that of A & M given here. Further definitions are given in (3.4.10) and (3.4.11) below.

situations where the vorticity tensor†

$$\mathbf{W} = \tfrac{1}{2}(\nabla\mathbf{v} - \nabla\mathbf{v}^T) \qquad (3.3.20)$$

is zero (see BA & H Section 8.5b).

The next in the sequence (see A & M Section 6.2) contains seven arbitrary functions of the invariants and joint invariants of $\mathbf{A}_1$ and $\mathbf{A}_2$ and is not generally useful. However, it degenerates in the case of steady simple shear flow to a much simpler form, which is usually known as the Criminale–Ericksen–Filbey fluid.

Criminale–Ericksen–Filbey Fluid

This can be written as (see BA & H Section 8.5a)

$$\mathbf{T}^{\mathrm{E}} = \eta(D)\mathbf{A}_1 + \{\Psi_1(D) + \Psi_2(D)\}\mathbf{A}_1^2 - \tfrac{1}{2}\Psi_1(D)\mathbf{A}_2 \qquad (3.3.21)$$

where η is essentially the same as in (3.3.3) and $\Psi_1(D)$ and $\Psi_2(D)$ are the difference-of-normal-stress functions that can be measured in a steady viscometric flow.

The advantage of the constitutive relation (3.3.21) is that the three functions η, Ψ_1, Ψ_2, of the single argument D can be relatively readily obtained by comparison with experiment. In some cases it has proved useful to use power-law representation for all three. The C–E–F fluid model is seen to be quite closely related to the R–R fluid (3.3.16) and to the second-order fluid of Coleman and Noll (3.3.10), but improves significantly in terms of realism on either of them. It has not been widely used in solving flow problems.

3.3.5 Implicit Rate Models

All of the special constitutive relations mentioned so far have expressed the extra-stress tensor $\mathbf{T}^{\mathrm{E}}$ directly in terms of the rate-of-strain tensor $\mathbf{D}$ and various of its convected derivatives. (See A & M Section 3.3 for a concise description of the corotational, upper and lower convected derivatives and the associated relative tensors.) An approach based on (3.2.14) due to Oldroyd leads to similar derivatives of the extra-stress tensor appearing in the constitutive equation. He developed, as a suitably general fluid model, linear in the stress, bilinear in stress and rate-of-strain and quadratic in rate of strain, the Oldroyd 8-constant fluid.

† BA & H, eqn (7.1.3), define $\boldsymbol{\omega} = -2\mathbf{W}$ as the vorticity tensor. Petrie (1979), eqn (2.14), uses $-\mathbf{W}$. All are related to the vorticity vector $\nabla \wedge \mathbf{v}$.

Oldroyd Eight-constant Fluid

$$\mathbf{T}^{\mathrm{E}}+\lambda_1\overset{\circ}{\mathbf{T}}{}^{\mathrm{E}}+\mu_0\operatorname{tr}(\mathbf{T}^{\mathrm{E}})\mathbf{D}-\mu_1(\mathbf{T}^{\mathrm{E}}\,.\,\mathbf{D}+\mathbf{D}\,.\,\mathbf{T}^{\mathrm{E}})+\nu_1(\mathbf{T}^{\mathrm{E}}:\mathbf{D})\mathbf{1}$$
$$=2\eta_0\{\mathbf{D}+\lambda_2\overset{\circ}{\mathbf{D}}-2\mu_2\mathbf{D}^2+\nu_2(\mathbf{D}:\mathbf{D})\mathbf{1}\} \quad (3.3.22)$$

where λ_1, λ_2, μ_0, μ_1, μ_2, ν_1, ν_2 and η_0 are constitutive constants and

$$\overset{\circ}{\mathbf{J}}=\frac{\partial\mathbf{J}}{\partial t}+(\boldsymbol{\nabla}\mathbf{J})\,.\,\mathbf{v}-\mathbf{W}\,.\,\mathbf{J}+\mathbf{J}\,.\,\mathbf{W} \quad (3.3.23)$$

is the Jaumann or co-rotational derivative.

There is an element of arbitrariness as to whether the co-rotational (°) the upper convected ($^\nabla$) or the lower convected ($^\triangle$) derivative be used in eqn (3.3.22), because any one representation can be converted into any other by a suitable adjustment in the values of μ_1 and μ_2. Indeed the form of (3.3.22) suggests that a general derivative

$$\overset{\square}{\mathbf{J}}=F_{abc}\mathbf{J}=\overset{\circ}{\mathbf{J}}+a(\mathbf{J}\,.\,\mathbf{D}+\mathbf{D}\,.\,\mathbf{J})+b(\mathbf{J}:\mathbf{D})\mathbf{1}+c\mathbf{D}\operatorname{tr}\mathbf{J} \quad (3.3.24)$$

could be used to simplify the left-hand side of (3.3.22) to a canonical form

$$\mathbf{T}^{\mathrm{E}}+\lambda_1\overset{\square}{\mathbf{T}}{}^{\mathrm{E}}$$

where a, b and c are suitably chosen constants. However, various physical theories (based on structural models) regard the co-deformational upper-convected derivative (see Exercise 3.3.3) as being of primary importance in affinely deforming systems.

It will be seen that suitable choice of the constants will yield a second-order fluid in either the Coleman–Noll or White–Metzner sequences. It will also yield a range of *Jeffreys fluid* models defined by

$$\mathbf{T}^{\mathrm{E}}+\lambda_1\overset{\square}{\mathbf{T}}{}^{\mathrm{E}}=2\eta_0(\mathbf{D}+\lambda_2\overset{\square}{\mathbf{D}}) \quad (3.3.25)$$

where the derivative ($^\square$) is defined in (3.3.24) above. These exhibit a relaxation time λ_1 and retardation time λ_2.

A significant difference from all the essentially viscous fluids considered so far is obtained if λ_2 and μ_2 are put equal to zero, for then we obtain the class of *Maxwell fluid* models given by

$$\mathbf{T}^{\mathrm{E}}+\lambda_1\overset{\square}{\mathbf{T}}{}^{\mathrm{E}}=2\eta_0\mathbf{D} \quad (3.3.26)$$

These fluids are characterized by being able to sustain sudden (instantaneous) strains.

This can be seen if we regard **D** as a derivative of the strain as is implied by (3.3.12), giving

$$\mathbf{D} = \tfrac{1}{2}\dot{\mathbf{C}}_t \tag{3.3.27}$$

and recognize that $\overset{\square}{\mathbf{T}}^{\mathrm{E}}$ contains a term $\dot{\mathbf{T}}^{\mathrm{E}}$. Thus, a singularity in $\dot{\mathbf{C}}_t$ can be balanced by a singularity in $\dot{\mathbf{T}}^{\mathrm{E}}$, or

$$[\mathbf{T}^{\mathrm{E}}] = G[\mathbf{C}] \tag{3.3.28}$$

where

$$G = \eta_0/\lambda_1 \tag{3.3.29}$$

is an elastic modulus and [ ] indicates a jump discontinuity, $\mathbf{C}$ being suitably defined.

We have chosen to consider the various special fluid models derived from the simple fluid as phenomenological or mechanistic. Although some would argue that many of them follow from general continuum mechanical principles alone, it is more helpful in our context to think of them as chosen to suit the various empirical observations and associated experimental measurements that have been made on polymeric fluids. We have already mentioned the power-law fluid in this connection. Equally important has been the influence of low-amplitude oscillatory measurements which are still usually interpreted at an elementary level in terms of scalar equations derived from spring and dashpot models. In these infinitesimal deformations, all the measures of strain and the associated time-like derivatives developed to deal with large-amplitude motions coincide; simple Fourier (or spectral) decomposition of the dynamic viscosity leading to relaxation (or retardation) time spectra is then a purely formal mathematical procedure, which need bear no relation to any internal structural features of the fluid in question.

Then, we can build on eqn (3.3.26) by writing:

$$\mathbf{T}^{\mathrm{E}} = \sum_{N=1}^{M} \mathbf{T}^{\mathrm{E}(N)} \tag{3.3.30}$$

where

$$\mathbf{T}^{\mathrm{E}(N)} + \lambda_N \mathrm{F}_{abc} \mathbf{T}^{\mathrm{E}(N)} = 2\eta_N \mathbf{D} \tag{3.3.31}$$

which can be shown to be equivalent to

$$\mathbf{T}^{\mathrm{E}} + \sum_{N=1}^{M} \alpha_N F_{abc}^{N} \mathbf{T}^{\mathrm{E}} = 2\eta_0 \left[\mathbf{D} + \sum_{N=1}^{M-1} \beta_N F_{abc}^{N} \mathbf{D}\right] \tag{3.3.32}$$

where the λ_N, η_N, α_N, β_N and η_0 are all suitable constants (see A & M Section 6.4, or Petrie, 1979, p. 28). This is a generalized Maxwell fluid. If however, $\alpha_M = 0$, then we get a generalized Jeffreys fluid.

If in eqn (3.3.30) we had used (3.3.31) for $N > 1$ and put

$$\mathbf{T}^{E(1)} = \mathbf{G}_t \tag{3.3.33}$$

where

$$\mathbf{G}_t = \mathbf{C}_t - \mathbf{1} \tag{3.3.34}$$

is what A & M call the Cauchy strain tensor, we would have defined a viscoelastic solid model. Indeed, it can be seen that $\mathbf{G}_t$ can take the place of $\mathbf{C}_t$ in the definition of a simple fluid.

Various authors report some success in modelling polymer melt behaviour with Oldroyd, Maxwell or generalized Maxwell models. However, comparison with experimental results has shown that a significant improvement on the upper convected Maxwell model can be obtained by letting the constants be functions of the 'shear rate' D. Thus we have the *White–Metzner fluid* given by

$$\mathbf{T}^E + \lambda(D)\overset{\square}{\mathbf{T}}{}^E = 2\lambda(D)G\mathbf{D} \tag{3.3.35}$$

Substitution of any of the (incompressible) model constitutive equations described so far into the equations of motion leads to a set of coupled differential equations (in tensor notation). By choice of a suitable coordinate system and use of the associated components of the relevant vectors and tensors, a coupled set of scalar non-linear partial differential equations results, with fixed space coordinates and time as the independent variables.

In the traditional Navier–Stokes form of eqn (2.2.8) the time derivative and any non-linearity come from the acceleration term $\mathrm{D}\mathbf{v}/\mathrm{D}t$; however, for differential-type viscoelastic fluid models, time derivatives and non-linearities now come from the $\nabla \cdot \mathbf{T}^E$ term as well. The problem of obtaining solutions, whether analytically or numerically (e.g. by finite difference techniques) is enormously complicated.

A further difficulty arises in the boundary conditions at surfaces where fluid is assumed to enter the system. In Newtonian fluid mechanics, only the velocity or the pressure need be specified. However, for our model fluids that involve convected derivatives and which seek to represent fluids with memory, further entry boundary conditions—defining the past history of the fluid—have to be specified. On physical grounds, we are tempted to argue that past history can only affect the present through some current state: the deduction would then be that the relevant boundary condition would be a statement about the entering state. This approach to describing elastic fluids will be taken up later. It does not in any case simplify the basic problem of defining boundary conditions, which is not easy to resolve. We shall defer the matter for special consideration on each occasion that the problem arises.

EXAMPLE 3.3.1. Express (3.3.13) in Cartesian tensor form, and compare it with representations used by other authors.

From the definition given by A & M (Section 4.16) we write, using Cartesian tensors,

$$(\nabla \mathbf{v})_{ij} = \partial v_i/\partial x_j$$

and so (3.3.13) becomes

$$A_{ij}^{(N+1)} = \frac{\partial A_{ij}^{(N)}}{\partial t} + v_k \frac{\partial A_{ij}^{(N)}}{\partial x_k} + \frac{\partial v_k}{\partial x_i} A_{kj}^{(N)} + A_{ik}^{(N)} \frac{\partial v_k}{\partial x_j}$$

The definition (3.3.12) is that given in BA & H, eqns (9.2.28) and (9.2.32), for $\boldsymbol{\gamma}^{(n)}$, and the definition (3.3.13) is that given in BA & H (9.3.20). The difference between (3.3.13) above, which is (3.2.29) of A & M, and (9.3.22) of BA & H, arises because of the different definition of $\nabla \mathbf{v}$ mentioned earlier.

Equation (3.3.13) uses what is usually called the lower convected derivative, defined in eqn (2.17) of Petrie (1979) and implied by eqn (3.3.39) of A & M as

$$\overset{\triangle}{J} = \overset{\circ}{\mathbf{J}} + (\nabla \mathbf{v})^T . \mathbf{J} + \mathbf{J} . (\nabla \mathbf{v})$$

BA & H refer to it as a co-deformational derivative. Lodge (1964 eqn (12.59), 1974, Section 5.4) uses simple time derivatives $\partial^N \gamma_{ij}/\partial t^N$ of the covariant form of the metric tensor to describe the strain-rate tensors $\mathbf{A}^{(N)}$.

EXERCISE 3.3.1. Show that, for $\mathbf{v}$ given by $(v_1(x_2), 0, 0)$ in a Cartesian coordinate system,

$$D = \mathrm{d}v_1/\mathrm{d}x_2$$

using definition (3.3.4) for the shear rate D. Note that this is consistent with the traditional definition $\dot{\gamma}$ (BA & H eqn 5.1.8) and is equivalent to S in A & M eqn (2.4.1).

EXERCISE 3.3.2. Show that (3.3.8) holds by noting that

$$\mathbf{F}_t(t) = \mathbf{F}_t^T(t) = \mathbf{1}$$

and that

$$\mathrm{d}\mathbf{v}(t) = \left(\frac{\mathrm{d}\mathbf{F}_t(t')}{\mathrm{d}t'}\right)_{t'=t} \cdot \mathrm{d}\mathbf{X}_t$$

whence

$$\boldsymbol{\nabla}\mathbf{v}(t) = \left(\frac{\mathrm{d}\mathbf{F}_t(t')}{\mathrm{d}t'}\right)_{t'=t}$$

using A & M's definition (see paragraph after eqn (2.2.5)).

Show also that definition (3.3.2) holds whether the A & M or the BA & H definition of $\boldsymbol{\nabla}v$ is used.

EXERCISE 3.3.3. Verify that the tensors $\mathbf{B}_N$ defined by (3.3.14) are the $\boldsymbol{\gamma}_{(n)}$ of BA & H, eqns (9.2.29) and (9.2.33), and involve the upper convected derivative

$$\overset{\nabla}{\mathbf{J}} = \mathbf{J} - (\boldsymbol{\nabla}\mathbf{v}) \,.\, \mathbf{J} - \mathbf{J} \,.\, (\boldsymbol{\nabla}\mathbf{v})^T$$

as defined by Petrie (1979, eqn (2.16)), which written in Cartesian form yields

$$B_{ij}^{(N+1)} = \frac{\partial B_{ij}^{(N)}}{\partial t} + v_k \frac{\partial B_{ij}^{(N)}}{\partial x_k} - \frac{\partial v_i}{\partial x_k} B_{kj}^{(N)} - B_{ik}^{(N)} \frac{\partial v_j}{\partial x_k}$$

Show that $\mathbf{A}_N$ and $\mathbf{B}_N$ are symmetric.

EXERCISE 3.3.4. Show that the co-rotational derivative $\mathrm{D}\dot{\boldsymbol{\gamma}}/\mathrm{D}t$ used by BA & H in their definition of the Criminale–Ericksen–Filbey eqn (8.5.2), is the same as the Jaumann derivative defined in (3.3.23), and hence the equivalence of BA & H (8.5.2) and (3.3.21).

EXERCISE 3.3.5. Show that, in (3.3.24),

$$F_{100}J \equiv \overset{\triangle}{J}, \qquad F_{-100} \equiv \overset{\nabla}{J}$$

where the upper and lower convected derivatives are defined in Exercise 3.3.3 and Example 3.3.1 above. (Note that (3.3.34), which is eqn (6.4.3) of A & M, with $b = c = 0$, differs from Petrie's (1979) eqn (2.21) by a minus sign.)

EXERCISE 3.3.6. Let $M=2$ in (3.3.30), $a=1$, $b=c=0$ in (3.3.31). Show that (3.3.32) becomes

$$\mathbf{T}^{\mathrm{E}}+(\lambda_1+\lambda_2)\overset{\triangledown}{\mathbf{T}}{}^{\mathrm{E}}+\lambda_1\lambda_2\overset{\triangledown\triangledown}{\mathbf{T}}{}^{\mathrm{E}}=2(\eta_1\lambda_2+\eta_2\lambda_1)\overset{\triangledown}{\mathbf{D}}+2(\eta_1+\eta_2)\mathbf{D}$$

Carry out the same analysis when $\mathbf{T}^{\mathrm{E}(1)}$ obeys (3.3.25) and $\mathbf{T}^{\mathrm{E}(2)}$ (3.3.31).

EXERCISE 3.3.7. Show that $\mathbf{G}_t$ of (3.3.33) is the $\boldsymbol{\gamma}^{[0]}$ of BA & H defined by their eqn (9.2.11) and -2 times the Almansi strain defined by Petrie (1979) in his eqn (2.19b).

3.4 INTEGRAL MODELS

A high degree of elasticity or rubber-like behaviour is invariably associated with the presence of very long-chain molecules. It has therefore always seemed obvious, to chemical physicists at least, that information about the constitutive behaviour (stress/strain laws) of rubbers should be obtainable directly from a consideration of their structure. Indeed a simple and elegant network theory of rubber elasticity has proved remarkably successful in predicting the behaviour of permanently cross-linked rubbers. This theory can be based on the statistics of chain dynamics of long flexible chains and can be expressed in equilibrium thermodynamic form in terms of a strain-energy function.

The theory has been adapted to considering dilute polymer solutions and polymer melts. In the first class of theories, the polymer molecules are treated as linear and as not interacting with one another although they must and do interact with the suspending fluid. The bead-rod-spring theories are well developed (see particularly, Bird *et al.*, 1977) but they are essentially model theories and not truly molecular. Indeed one of the preferred models proves to be a suspension of rigid dumb-bells! Although the constitutive equations derived for these models prove to be useful starting points for polymer melts, we are not primarily interested in dilute solutions here, and so any useful contribution to our understanding of melt rheology through structure must come from a consideration of strongly interacting long-chain molecules.

One relevant class of models was proposed by Lodge and

Yamamoto.† These are network models characterized by having temporary junctions between chain segments. The impetus for developing these models clearly came from the success of the theory of rubber-like (solid) elasticity, which is simply adapted to show stress relaxation after suddenly imposed fixed and finite strain. The body coordinate system of Lodge and Oldroyd proves to be an ideal system in which to describe the elementary physical ideas underlying temporary network models. Again it must be emphasized that these are not truly molecular models except insofar that the mechanical response of the chain segments joining network junctions is derived from the statistics of chain dynamics.

A recent class of more nearly molecular models has been developed by de Gennes, Edwards and various coworkers, of which the most interesting and important has been described by Doi & Edwards (1978), Doi (1980). In these, the interactions between entangled molecules along their entire length are explicitly considered. Any given linear molecule reptates (moves back and forth) along a tube it occupies in a statistical sense during the constrained thermal motion of its convoluted chain segments. The bulk melt can be likened to a dense assembly of active worms, isotropic when at rest. The statistical mechanics of this realistic model are reduced to continuum representation in the form of an explicit constitutive equation for stress in terms of deformation history.

In all these theories, passage from the statistical to the continuum representation is facilitated by assuming Gaussian chain statistics and affine (i.e. strictly homogeneous) local deformation, while stress is assumed to be carried wholly along chains and not between them (except at junctions). (Though see recent papers of Curtiss & Bird (1981) pointing to the difficulties of defining the stress tensor in a kinetic theory.) The resulting constitutive equations are usually simple integrals over past time of some relevant measure of strain. The same type of constitutive equation can be derived from a consideration of the original functional equation (3.2.13) for a simple fluid in the limit of linear viscoelasticity, i.e. small deformations about a rest state, and so it is difficult to say whether the models are structural (molecular theoretic) or phenomenological (continuum theoretic and empirical) in origin. For our purposes here, it is convenient to treat them as the latter.

† See review by Lodge *et al.* (1981). See also Bird (1982).

Thus, we are led to consider the following.

3.4.1 General Linear Viscoelastic Fluid

This is given by

$$\mathbf{T}^{\mathrm{E}} = \int_0^\infty m(s)\mathbf{G}_t^t(s)\,\mathrm{d}s \tag{3.4.1}$$

where

$$\mathbf{G}_t^t = \mathbf{C}_t^t - 1 \tag{3.4.2}$$

and $m(s)$ is the memory function. This can be written in the alternative form

$$\mathbf{T}^{\mathrm{E}} = \int_0^\infty G(s)\dot{\mathbf{G}}_t^t(s)\,\mathrm{d}s \tag{3.4.3}$$

where

$$\frac{\mathrm{d}G}{\mathrm{d}s} = m(s) \tag{3.4.4}$$

and

$$\dot{\mathbf{G}}_t^t = -\frac{\mathrm{d}}{\mathrm{d}s}\mathbf{G}_t^t \tag{3.4.5}$$

$G(s)$ is the relaxation modulus.

One higher approximation is given by

$$\mathbf{T}^{\mathrm{E}} = \int_0^\infty m(s)\mathbf{G}_t^t\,\mathrm{d}s + \int_0^\infty\int_0^\infty \{\alpha(s_1, s_2)\mathbf{G}_t^t(s_1)\cdot\mathbf{G}_t^t(s_2) + \beta(s_1, s_2)\mathrm{tr}[\mathbf{G}_t^t(s_1)]\mathbf{G}_t^t(s_2)\}\,\mathrm{d}s_1\,\mathrm{d}s_2 \tag{3.4.6}$$

where $\alpha(\ )$ and $\beta(\ )$ are material functions. (See A & M eqn 4.3.25.)

However, there is nothing unique about the choice of expansion (3.4.6) for the general simple fluid given by (3.2.13). In the limit $\|\mathbf{G}_t^t(s)\|\to 0$, (3.4.1) is indeed unique in that all choices of strain measure should tend (or be proportional) to the same infinitesimal strain. In that limit, the difference between (3.4.1) and (3.4.6) becomes arbitrarily small. But when the deformation becomes finite, alternative measures of strain could have been used, as for example

$$\mathbf{H}_t^t(s) = \mathbf{C}_t^{t^{-1}} - \mathbf{1}\dagger \tag{3.4.7}$$

† $\frac{1}{2}\mathbf{H}_t^t(s)$ is called the Signorini strain measure by Petrie (1979), eqn (2.18a).

instead of $\mathbf{G}_t^t$ in (3.4.1) (see A & M eqn (6.3.3), or Petrie (1979) eqn (2.26a); this form is used as the starting point for most network theories). Accordingly, the next higher approximation, analogous to (3.4.6), would have contained $\mathbf{H}_t^t$ instead of $\mathbf{G}_t^t$. This arbitrariness, reminescent of the analogies between the Coleman–Noll and White–Metzner expansions (see (3.3.9)–(3.3.14) above), is unsatisfactory. Furthermore, any attempt to use double integrals involving kernel functions $\alpha(\ )$ and $\beta(\ )$ of two variables, as in (3.4.6), in the solution of fluid mechanical problems leads to daunting complexity, and so many authors have attempted to select a 'best' strain measure that will allow the most important finite-strain effects to be included in the single integral.

One non-linear extension of the general linear viscoelastic fluid that is said to correspond well to polymeric fluid behaviour and to have significant advantages is given by BA & H (Section 8.3) as

$$\begin{aligned}\mathbf{T}^{\mathrm{E}} = &-\int_{-\infty}^{t} G_{\mathrm{I}}(t-t')\dot{\mathbf{\Gamma}}(t')\,\mathrm{d}t' \\ &-\frac{1}{2}\int_{-\infty}^{t}\int_{-\infty}^{t} G_{\mathrm{II}}(t-t', t-t'')\{\dot{\mathbf{\Gamma}}(t')\cdot\dot{\mathbf{\Gamma}}(t'')+\dot{\mathbf{\Gamma}}(t'')\cdot\dot{\mathbf{\Gamma}}(t')\}\,\mathrm{d}t'\,\mathrm{d}t'' \\ &+\text{terms involving higher derivatives}\end{aligned} \tag{3.4.8}$$

where $\dot{\mathbf{\Gamma}}$ is the co-rotational form of the strain-rate tensor ($\dot{\mathbf{\Gamma}} = 2\mathbf{D}$ in the notation of A & M, eqn (3.3.20), or equivalently $\mathring{\mathbf{1}}$). This expansion, the co-rotational memory-integral expansion, is attributed to Goddard. The single integral involving $G_{\mathrm{I}}\{\equiv G(s)$ in (3.4.3)$\}$ which can be obtained from small-amplitude oscillatory measurements alone, provides of itself a useful rheological equation of state.

3.4.2 Kaye–BKZ Fluid

An alternative approach is provided by using the most general single-integral model in the form

$$\mathbf{T}^{\mathrm{E}} = \int_0^{\infty} \{\Phi_1(s, \mathrm{I}_{\mathbf{C}_t^t}, \mathrm{II}_{\mathbf{C}_t^t})\mathbf{C}_t^t + \Phi_{-1}(s, \mathrm{I}_{\mathbf{C}_t^t}, \mathrm{II}_{\mathbf{C}_t^t})\mathbf{C}_t^{t^{-1}}\}\,\mathrm{d}s \tag{3.4.9}$$

where Φ_1 and Φ_{-1} are material functions, and

$$\mathrm{I}_{\mathbf{C}} = \mathrm{tr}(\mathbf{C}) = \mathrm{II}_{\mathbf{C}^{-1}} \tag{3.4.10}$$

$$\mathrm{II}_{\mathbf{C}} = \tfrac{1}{2}(\mathrm{I}_{\mathbf{C}}^2 - \mathrm{tr}(\mathbf{C}^2)) = \mathrm{I}_{\mathbf{C}^{-1}} \tag{3.4.11}$$

(It is assumed that $\mathrm{III}_{\mathbf{C}}=0$, i.e. that the material is incompressible.) One special form is particularly well known, the BKZ fluid given by

$$\Phi_1 = 2\frac{\partial \dot{U}}{\partial \mathrm{I}_{\mathbf{C}_t^t}}, \qquad \Phi_{-1} = -2\frac{\partial \dot{U}}{\partial \mathrm{II}_{\mathbf{C}_t^t}} \tag{3.4.12}$$

where $\dot{U}(s, \mathrm{I}_{\mathbf{C}_t^t}, \mathrm{II}_{\mathbf{C}_t^t})$ can be regarded as the time-derivative of a strain-energy function.

There remains the question of prescribing $m(s)$ in (3.4.1), G_{I} and G_{II} in (3.4.8) or Φ_1 and Φ_{-1} in (3.4.9). The phenomenological approach is to compare the predictions of any one of the models with experimental observations and to devise best fits. If we do this, then our integral models will have the same status as the differential models discussed in the previous subsection. Indeed, it can be shown that, in certain circumstances, they are equivalent. Thus, if

$$m(s) = \frac{\eta_0}{\lambda_1^2}\,\mathrm{e}^{-s/\lambda_1} \tag{3.4.13}$$

and $\mathbf{G}_t^t(s)$ is replaced by $(\mathbf{C}_t^t)^{-1}$ in (3.4.1), the resulting single integral model is merely the general solution of the differential Maxwell model (3.3.26) with $(\stackrel{\square}{}) = (\stackrel{\triangledown}{})$, the upper convected derivative (see Exercise 3.3.3 or A & M Section 6.4).

3.4.3 Doi–Edwards Fluid

The alternative, synthetic, approach is to construct a molecular model. That leading to the Doi–Edwards fluid gives a form of the Kaye–BKZ fluid, which is most conveniently written in the form

$$\mathbf{T}^{\mathrm{E}} = G_0 \int_0^{\infty} \mu(s) \int_{\substack{\text{unit}\\ \text{sphere}}} \frac{(\mathbf{F}^{-1}\,.\,\mathbf{u})(\mathbf{F}^{-1}\,.\,\mathbf{u})}{\mathrm{tr}[(\mathbf{F}^{-1}\,.\,\mathbf{u})^2]}\frac{\mathrm{d}\mathbf{u}^2}{4\pi}\,\mathrm{d}s \tag{3.4.14}$$

where $\mathbf{u}$ is a unit vector in orientation space, $\mathbf{F} = \mathbf{F}_t^t(s)$, and

$$\mu(s) = \sum_{k\ \mathrm{odd}} \frac{8}{\pi\lambda_{\mathrm{d}}} \exp(-sk^2/\lambda_{\mathrm{d}}) \tag{3.4.15}$$

Both the forms (3.4.13) and (3.4.15), when interpreted in molecular terms, imply a monodisperse polymer, with a single characteristic relaxation time (λ_1 or λ_{d}). If a polydisperse system is considered, then the stress tensor can be represented by a sum or integral of such contributions, leading to constitutive equations analogous to the empirical form (3.3.30). Having at one's disposal a spectrum of relaxation times obviously makes fitting of experimental curves easier.

All of the mechanistic and molecular theories of polymeric fluid behaviour are based on the phenomenon of strain-induced orientation of the polymeric structure, leading to an instantaneous anisotropic state, which relaxes with time once no further straining occurs. In the simplest interpretation, the extra stress $\mathbf{T}^{\mathrm{E}}$ is itself a measure of the orientation; this arises when all of the stress is carried along polymer chains through a network structure. Formal models relating **F** and **T** directly, as in (3.2.14), are in principle adequate. However, in more complex interpretations, as, for example, in the theory of suspensions of deformable particles subject to Brownian motion (Hinch & Leal, 1975) it proves more convenient to introduce various structural parameters subject to evolutionary equations. The constitutive equation for stress becomes an algebraic function with deformation rate and these structural variables as arguments, while the evolution equations involve first-order derivatives, of the type given in eqn (3.3.24), for symmetric second-rank tensors and with similar but simpler forms for vectors and scalars. We shall not use such forms in this treatment. A related approach based on entanglements is discussed in Liu *et al.* (1981).

Various special forms of these constitutive equations and their connections with molecularly motivated theories of polymer melt behaviour will be discussed later in connection with the observed rheological behaviour of polymer melts. At that stage, we take up again the question of temperature effects.

Recent work has shown that apparently very different models of both rate and integral type, obtained from both structural and empirical considerations, relating to both networks and suspensions, prove to be interchangeable, simply by redefinition of the strain or rate-of-strain measure (see for example Lan & Schowalter, 1981). Hence much of the formal work done on apparently unphysical models as exercises in continuum mechanics is proving to be of real value in modelling and understanding real flows.

EXERCISE 3.4.1. Show that

$$\frac{\mathrm{d}}{\mathrm{d}s}\mathbf{G}_t^t(s) = -\frac{\mathrm{d}}{\mathrm{d}t'}\mathbf{G}_t(t')$$

and that

$$\left|\frac{\mathrm{d}}{\mathrm{d}s}\mathbf{G}_t^t(s)\right|_{s=0} = -\left|\frac{\mathrm{d}}{\mathrm{d}s}\mathbf{H}_t^t(s)\right|_{s=0}$$

Hence or otherwise show that

$$\mathbf{G}_t^t(s) \to -\mathbf{H}_t^t(s) \quad \text{as} \quad s \to 0$$

and hence that the relation (3.4.1) leads to the alternative form

$$\mathbf{T}^{\mathrm{E}}(t) = -\int_0 m(s)\mathbf{H}_t^t(s)\,\mathrm{d}s$$

if the same linear viscoelastic function $m(s)$ is used.

Chapter 4

Basic Flow Fields

Both the rheological equations of state (constitutive equations) of the last chapter and the equations of motion of Chapter 2 imply continuum flow fields, defined kinematically by $\mathbf{v}(\mathbf{x}, t)$, dynamically by $\mathbf{T}(\mathbf{x}, t)$ and thermally by $T(\mathbf{x}, t)$. Before considering fully general and extensive flow fields, we seek those specially simple situations that will be helpful in the description of polymer processes.

It is worth remarking that many texts on the mechanics of non-Newtonian fluids develop flows of progressively greater complexity while neglecting both inertia and body forces and any interactions with the energy equation. We shall adopt the same approach, particularly with regard to neglect of inertia and body forces (but with reservations about the neglect of thermal effects), in order to demonstrate the consequences of rheological complexity and in order to explain in the next section the theory behind the rheometers (devices for measuring rheological functions experimentally) currently used. For most purposes therefore, the conservation equation (2.2.8) can be written

$$\boldsymbol{\nabla} \cdot \mathbf{T} = \mathbf{0} \tag{4.1}$$

Steady Flows

If $\mathbf{v}$, $\mathbf{T}$ and T are independent of t, i.e. if

$$\frac{\partial \mathbf{v}}{\partial t} = 0, \qquad \text{so} \quad \frac{\mathrm{D}\mathbf{v}}{\mathrm{D}t} = \boldsymbol{\nabla}\mathbf{v} \cdot \mathbf{v} \quad \text{etc.} \tag{4.2}$$

then we say that the flow is steady in an Eulerian sense.

This is not the same as being steady in a material (or Lagrangian) sense, which requires that $\dfrac{\mathrm{D}}{\mathrm{D}t} = 0$, or in a rheological sense, which requires that the corotational derivative of $(\boldsymbol{\nabla}\mathbf{v})$ or $\mathbf{T}$ be zero.

Uniform Flows

If $\mathbf{T}$ and T are independent of $\mathbf{x}$, then we say that the flow is uniform.

Clearly such a flow is isothermal. The case when $\mathbf{v}$ becomes independent of $\mathbf{x}$ is a trivial one, being equivalent to a rigid body motion (unless, of course, the independence of $\mathbf{x}$ only holds instantaneously). Except in the case of materials exhibiting a non-zero yield stress, the stress $\mathbf{T}$ then becomes isotropic, i.e. $-p\mathbf{1}$, where p is a scalar constant pressure.

It is clear that a steady and uniform flow is achieved kinematically when the velocity field can be written

$$\mathbf{v} = \mathbf{L} \cdot \mathbf{x} \tag{4.3}$$

in a suitable inertial reference frame with $\mathbf{L}$ a constant.† We see at once that

$$\mathbf{L} = \nabla \mathbf{v} \tag{4.4}$$

and that

$$\mathbf{D} = \tfrac{1}{2}(\mathbf{L} + \mathbf{L}^T) \tag{4.5}$$

$$\mathbf{W} = \tfrac{1}{2}(\mathbf{L} - \mathbf{L}^T) \tag{4.6}$$

are constants in such a uniform field.

We shall not prove here that, for a simple fluid, $\mathbf{L}$, T constant implies $\mathbf{T}$ constant nor shall we consider the possibility that T, $\mathbf{T}$ constant could lead to non-constant $\mathbf{L}$. We need merely point out that the first result can be demonstrated very easily for each of the constitutive equations mentioned in the previous chapter.

If we require our fluid to be incompressible, then

$$\mathrm{tr}(\mathbf{L}) = 0 \tag{4.7}$$

Having particularized our consideration to steady, uniform, incompressible flows, we note that a wide range of flow fields are still permissible. Equation (4.7) has only removed one of the nine degrees of freedom apparently open to us, while arbitrary rotation of the coordinate system can remove at most three more. We are thus left with a basic flow field that is dependent upon five scalar parameters. We may naturally enquire whether any further useful categorization of the flow field is possible, and in particular whether the stress field $\mathbf{T}$

† We use $\mathbf{L}$ for the case $\nabla \mathbf{v}$ constant and not, as do Petrie (1979) and A&M, for the general case.

that follows from **L** can be usefully described in terms of a smaller number of parameters than five. This is very much an open question, but one that has been carefully considered by previous authors, particularly those concerned with specific materials and model equations of state. Much of the work deals with rheologically steady flows—those that are steady and uniform from the point of view of a moving and deforming material particle—and thus covers a wider range of global flow fields than (4.5), namely motion with constant stretch history.

4.1 MOTIONS WITH CONSTANT STRETCH HISTORY (MWCSH)

The basic ideas have been separately enunciated by Oldroyd (1950) and Noll (1962); they are discussed briefly in A&M (Section 3.5) and more thoroughly in Huilgol (1975), Tanner & Huilgol (1975). For our purposes it is sufficient to say that there is a 1:1 correspondence between MWCSH's and those given by (4.3), i.e. that the condition

$$\frac{\mathrm{D}\mathbf{L}}{\mathrm{D}t}=0 \tag{4.1.1}$$

with **L** regarded as $\nabla\mathbf{v}(\mathbf{x}, t)$, which is evidently satisfied by (4.3), implies MWCSH, and that all MWCSH correspond uniquely to a flow of type (4.3).

The primary categorization has been into those flows for which either

(i) $$\mathbf{L}^2=0 \tag{4.1.2}$$

or (ii) $$\mathbf{L}^2\neq 0, \qquad \mathbf{L}^3=0 \tag{4.1.3}$$

or (iii) $$\mathbf{L}^3\neq 0 \tag{4.1.4}$$

We can show that the eigenvalues of **L** and its invariants $\mathrm{I}_{\mathbf{L}}$, $\mathrm{II}_{\mathbf{L}}$ and $\mathrm{III}_{\mathbf{L}}$ are all zero for flows of types (i) and (ii). In the classical canonical (Jordan) form for the matrices [**L**] (obtained when **L** is expressed in terms of rectangular Cartesian coordinates, say), we find (Tanner & Huilgol, 1975) that

(i) $$[\mathbf{M}_1\cdot\mathbf{L}\cdot\mathbf{M}_1^{-1}]=\begin{vmatrix}0 & 1 & 0\\ 0 & 0 & 0\\ 0 & 0 & 0\end{vmatrix} \tag{4.1.5}$$

(ii)
$$[\mathbf{M}_2 \,.\, \mathbf{L} \,.\, \mathbf{M}_2^{-1}] = \begin{vmatrix} 0 & 1 & 0 \\ 0 & 0 & 1 \\ 0 & 0 & 0 \end{vmatrix} \tag{4.1.6}$$

for some non-singular transformations $\mathbf{M}_1$ and $\mathbf{M}_2$. For the case $\mathbf{L}^3 \neq 0$, we can distinguish between

(iii)(a) 3 real and distinct eigenvalues, μ_1, μ_2, $-(\mu_1+\mu_2)$, for which

$$[\mathbf{M}_3 \,.\, \mathbf{L} \,.\, \mathbf{M}_3^{-1}] = \begin{vmatrix} \mu_1 & 0 & 0 \\ 0 & \mu_2 & 0 \\ 0 & 0 & -(\mu_1+\mu_2) \end{vmatrix} \tag{4.1.7}$$

(iii)(b) 1 real and 2 complex eigenvalues -2α, $\alpha+\mathrm{i}\beta$, $\alpha-\mathrm{i}\beta$, for which

$$[\mathbf{M}_4 \,.\, \mathbf{L} \,.\, \mathbf{M}_4^{-1}] = \begin{vmatrix} \alpha & \beta & 0 \\ -\beta & \alpha & 0 \\ 0 & 0 & -2\alpha \end{vmatrix} \tag{4.1.8}$$

(iii)(c) 3 real, non-zero, but not distinct, eigenvalues α, α, -2α, for which either (4.1.7) holds with $\mu_1 = \mu_2$, or

$$[\mathbf{M}_5 \,.\, \mathbf{L} \,.\, \mathbf{M}_5^{-1}] = \begin{vmatrix} \alpha & 1 & 0 \\ 0 & \alpha & 0 \\ 0 & 0 & -2\alpha \end{vmatrix} \tag{4.1.9}$$

where $\mathbf{M}_3$, $\mathbf{M}_4$, $\mathbf{M}_5$ are also non-singular transformations.

Case (iii)(c) gives rise to two possibilities, i.e. two distinct Jordan forms—in the same way that the case $\mu_1 = \mu_2 (= \mu_3) = 0$ gives rise to three distinct Jordan forms, viz. (i), (ii) and the totally degenerate $\mathbf{L} = \mathbf{0}$.

The transformation matrices $\mathbf{M}_i$ $(i = 1, \ldots, 5)$ will not in general be orthogonal matrices—ones for which

$$\mathbf{P}^{-1} = \mathbf{P}^T \tag{4.1.10}$$

—and hence do not correspond to mere rotation of the coordinate system. However, given $\mathbf{L}$, it is possible to find the eigenvalues μ_K $(K = 1, 2, 3)$ and the relevant $\mathbf{M}_i$ by standard algebraic procedures.

From the forms (4.1.5)–(4.1.9) we can derive the relevant forms for:

$$\mathbf{C}_t^t(s) = \mathrm{e}^{-s\mathbf{L}^T} \,.\, \mathrm{e}^{-s\mathbf{L}} \tag{4.1.11}$$

which latter result follows directly from the easily verified relation

$$\dot{\mathbf{F}}_t = \mathbf{L} \cdot \mathbf{F}_t \tag{4.1.12}$$

(see Exercise 4.1.1).

Similarly, we can calculate the Rivlin–Ericksen tensors

$$\mathbf{A}_1 = \mathbf{L}^T + \mathbf{L} \tag{4.1.13}$$

$$\mathbf{A}_2 = \mathbf{L}^T \cdot \mathbf{A}_1 + \mathbf{A}_1 \cdot \mathbf{L} \tag{4.1.14}$$

$$\mathbf{A}_N = \mathbf{L}^T \cdot \mathbf{A}_{N-1} + \mathbf{A}_{N-1} \cdot \mathbf{L}, \quad \text{etc.} \tag{4.1.15}$$

or the White–Metzner tensors, $\mathbf{B}_1, \mathbf{B}_2, \ldots, \mathbf{B}_N$. Furthermore, we see that the general tensor derivatives F_{abc}, defined in (3.3.24) can be put into the form

$$\begin{aligned} 2F_{abc}[\ ] = [\ ] \cdot (\mathbf{L} - \mathbf{L}^T) - (\mathbf{L} - \mathbf{L}^T) \cdot [\ ] \\ + a\{(\mathbf{L} + \mathbf{L}^T) \cdot [\ ] + [\ ] \cdot (\mathbf{L} + \mathbf{L}^T)\} \\ + b[\ ] : (\mathbf{L} + \mathbf{L}^T)\mathbf{1} + c(\mathbf{L} + \mathbf{L}^T)\mathrm{tr}[\ ] \end{aligned} \tag{4.1.16}$$

Thus $\mathbf{T}^{\mathrm{E}}$ can be directly evaluated for all of the specific constitutive equations defined in the previous section, and will be independent of both $\mathbf{x}$ and t. Thus, trivially $\boldsymbol{\nabla} \cdot \mathbf{T}^{\mathrm{E}} = 0$ and the conservation equation (2.2.8) will be satisfied by $\mathbf{L}$ constant in (4.3).

For the case of a simple fluid given by (3.2.13), it can be shown that the functional $\mathcal{H}$ degenerates into a simple function, and the Rivlin–Ericksen representation given in (3.3.15) will be relevant. What we need to discover is how many of the $\mathbf{A}_N$ are independent when $\mathbf{L}$ is constant. A theorem of Wang (1965) shows that at most $\mathbf{A}_1$, $\mathbf{A}_2$ and $\mathbf{A}_3$ are required to prescribe $\mathbf{L}$, and hence

$$\mathbf{T}^{\mathrm{E}} = \mathbf{M}_3(\mathbf{A}_1, \mathbf{A}_2, \mathbf{A}_3) \tag{4.1.17}$$

is a completely general representation. An alternative form using $\mathbf{D}$ and $\mathbf{W}$, the symmetric and anti-symmetric parts of $\mathbf{L}$, has been given by Giesekus (1961) and is quoted as eqn (2.14) in Pearson (1966).

4.1.1 Weak and Strong Flows

A different categorization has been proposed by Tanner (1976) and others. It is based on the eigenvalue of

$$\mathbf{L}_\Lambda = \mathbf{L} - \tfrac{1}{2}\Lambda^{-1}\mathbf{1} \tag{4.1.18}$$

that has the largest real part; Λ is a characteristic relaxation time of the fluid.

If the real part is zero, then the flow is said to be weak; if it is positive, then the flow is said to be strong.

By letting $\Lambda \to \infty$ we can replace $\mathbf{L}_\Lambda$ by $\mathbf{L}$, which has attractions because it decouples our kinematic categorization from the fluid in question. The meaning of the criterion is most easily understood by noting that MWCSH cases (i), (ii) and (iii)(b) (with $\alpha = 0$) above are all weak flows, whereas all other cases (iii) are strong flows. Strong flows all have at least one exponentially increasing component of $\mathbf{C}_t^t(s)$, and of $(\mathbf{C}_t^t)^{-1}$, whereas weak flows have only algebraically increasing (or bounded) components (see Exercise 4.1.1). The argument leading Tanner to his choice of $\mathbf{L}_\Lambda$ as the relevant tensor concerns the behaviour of a classical elastic dumb-bell having a precisely-defined relaxation time Λ immersed in a flow field $\mathbf{L}$. Strong flows extend a dumb-bell without limit; weak flows do not.

A more elaborate analysis is given in Olbricht *et al.* (1982).

EXERCISE 4.1.1. Show that $\mathbf{C}_t^t(s)$ can be written in the following ways for the various canonical forms described by (4.1.5)–(4.1.9).

(i) $\mathbf{1} - s(\mathbf{L}+\mathbf{L}^T) + s^2\mathbf{L}^T . \mathbf{L}$ using (4.1.2)

(ii) $\mathbf{1} - s(\mathbf{L}+\mathbf{L}^T) + \frac{1}{2}s^2(\mathbf{L}^T+\mathbf{L})^2 - \frac{1}{2}s^3\mathbf{L}^T . (\mathbf{L}^T+\mathbf{L}) . \mathbf{L} + \frac{1}{4}s^4\mathbf{L}^{T2} . \mathbf{L}^2$ using (4.1.3)

(iii) (a) $\exp\{s\mathbf{A}\}$ using (4.1.7)

(b) $\exp\{is\mathbf{B}\}$ using (4.1.8) when $\alpha = 0$

(c) $\exp\{s\mathbf{C}\}$ using (4.1.9)

where $\mathbf{A}$ is real and symmetric and $\mathbf{B}$ and $\mathbf{C}$ are real.

Hence deduce that flows (i) and (ii) only lead to algebraically-growing strain, while (iii)(b) is oscillatory when $\alpha = 0$; (a) and (c) will in general yield exponentially-growing strains.

4.2 VISCOMETRIC FLOWS

These are the class for which $\mathbf{L}^2 = 0$ and can all be shown (Lodge, 1964; Coleman *et al.*, 1966) to behave like the steady unidirectional shear flow

$$v_1 = Dx_2, \qquad v_2 = v_3 = 0 \tag{4.2.1}$$

A thorough discussion is given in BA & H I Section 4.1, where it is shown that the most general stress tensor associated with flow (4.2.1) is given by

$$t_{12} = t_{21} = \eta(D)D \tag{4.2.2}$$

$$t_{11} - t_{22} = \Psi_1(D)D^2 \tag{4.2.3}$$

$$t_{22} - t_{33} = \Psi_2(D)D^2 \tag{4.2.4}$$

$$t_{13} = t_{23} = t_{31} = t_{32} = 0 \tag{4.2.5}$$

The three material functions η, Ψ_1 and Ψ_2, are all even functions of D; η is known as the viscosity, $t_{11} - t_{22}$ as the first normal-stress difference and $t_{22} - t_{33}$ as the second normal-stress difference. As $D \to 0$, Ψ_1 and Ψ_2 tend to constants known as the first and second normal-stress coefficients respectively.

We have here our first case of kinematical restriction giving rise to rheological simplification, in the sense that the extra-stress tensor takes a restricted form. Forms for η, Ψ_1, Ψ_2 can be obtained directly for each of the constitutive relations given in Chapter 3. One of these, the Criminale–Ericksen–Filbey fluid, (3.3.21), is just sufficiently flexible to accommodate the generality implied by (4.2.3)–(4.2.5).

EXERCISE 4.2.1. Show, on grounds of symmetry, that the stress components t_{31} and t_{23} will be zero for the viscometric flow (4.2.1).

EXERCISE 4.2.2. Show that, for viscometric flow, the η, Ψ_1 and Ψ_2 of (3.3.21) become precisely the functions defined by (4.2.2)–(4.2.4).

EXERCISE 4.2.3. Verify the results given in Table 4.1 for η, Ψ_1 and Ψ_2 for a third-order and a Reiner–Rivlin fluid model.

EXERCISE 4.2.4. Show that, for viscometric flow of the Jeffreys fluid model (3.3.25), the viscosity function is given by

$$\eta(D) = \frac{1 + \lambda_1\lambda_2\{1 - (a+b)(a+c)\}D^2}{1 + \lambda_1^2\{1 - (a+b)(a+c)\}D^2}$$

Consider the case $b = c = 0$, $a < 1$, and show that $\lambda_2 < \lambda_1$ if $\eta(D)$ decreases with D.

4.3 EXTENSIONAL FLOWS†

These are the class for which $\mathbf{L}=\mathbf{L}^T$, i.e. $\mathbf{W}=0$ from (3.3.20), and are essentially equivalent to the flow

$$v_1=\dot{e}_1x_1, \qquad v_2=\dot{e}_2x_2, \qquad v_3=\dot{e}_3x_3 \tag{4.3.1}$$

where

$$\dot{e}_1+\dot{e}_2+\dot{e}_3=0 \tag{4.3.2}$$

for an incompressible fluid. They are typical of MWCSH case (iii)(a). The most general stress tensor associated with (4.3.1), (4.3.2) is

$$t_{11}-t_{33}=N(\dot{e}_1, \dot{e}_2) \tag{4.3.3}$$

$$t_{22}-t_{33}=N(\dot{e}_2, \dot{e}_1) \tag{4.3.4}$$

$$t_{13}=t_{23}=t_{13}=0 \tag{4.3.5}$$

As with the case of viscometric flow, forms for N can be obtained directly for given constitutive equations. The full generality of (4.3.3)–(4.3.5) can most easily be accommodated by the Reiner–Rivlin fluid (3.3.16).

Two special cases of (4.3.1) have been studied:

4.3.1 Elongational Flow

$$\dot{e}_1=\dot{e}, \qquad \dot{e}_2=\dot{e}_3=-\tfrac{1}{2}\dot{e} \tag{4.3.6}$$

where $\dot{e}$ is known as the elongation rate, and

$$t_{11}-t_{22}=t_{11}-t_{33}=\eta_e(\dot{e})\dot{e} \tag{4.3.7}$$

defines the elongational (or Trouton) viscosity $\eta_e(\dot{e})$, which is generally not quadratic in $\dot{e}$. If $\dot{e}>0$ we have uniaxial extension; if $\dot{e}<0$ we have biaxial extension.

4.3.2 Pure Shear Flow

$$\dot{e}_1=\dot{e}, \qquad \dot{e}_2=-\dot{e}, \qquad \dot{e}_3=0 \tag{4.3.8}$$

which leads to the viscosity function $\eta_{PS}(\dot{e})$ defined by

$$t_{11}-t_{22}=\eta_{PS}(\dot{e})\dot{e} \tag{4.3.9}$$

† These are also called shearfree flows by BA & H.

Once again, kinematical restrictions as in Subsections 4.3.1 and 4.3.2 above have led to very considerable rheological simplification.

In the case of a Newtonian fluid

$$4\eta_e = 3\eta_{PS} = 12\eta = 12\eta_0 \quad \text{(constant)} \tag{4.3.10}$$

where η_e, η_{PS}, and η are defined by (4.3.7), (4.3.9) and (4.2.2) respectively, or more generally,

$$N(x, y) = 2\eta_0(2x + y) \tag{4.3.11}$$

$N(x, y)$ is clearly lacking in symmetry as defined in (4.3.3) and (4.3.4) and must obey a variety of ancillary conditions of the type

$$N(x, y) - N(y, x) = N(x, -x - y) \tag{4.3.12}$$

Furthermore, it will be constrained by the thermodynamic requirement that dissipation must be positive; this, for example, ensures that η_e, η_{PS} and η are all positive.

EXERCISE 4.3.1. Show that for extensional flow (4.3.1), (4.3.2),

$$\overline{\text{II}}_{\mathbf{A}_1} = 8(\dot{e}_1^2 + \dot{e}_1\dot{e}_2 + \dot{e}_2^2) = 4(\dot{e}_1^2 + \dot{e}_2^2 + \dot{e}_3^2)$$
$$\overline{\text{III}}_{\mathbf{A}_1} = -24\dot{e}_1\dot{e}_2(\dot{e}_1 + \dot{e}_2) = 24\dot{e}_1\dot{e}_2\dot{e}_3$$

and that for the Reiner–Rivlin model (3.3.16), it follows that

$$N(\dot{e}_1, \dot{e}_2) = 2(2\dot{e}_1 + \dot{e}_2)(\phi_1 - 2\dot{e}_2\phi_2)$$

where ϕ_1 and ϕ_2 are functions of $\overline{\text{II}}_{\mathbf{A}_1}$ and $\overline{\text{III}}_{\mathbf{A}_1}$, each symmetric in $\dot{e}_1$, $\dot{e}_2$, and N is defined by (4.3.3) and (4.3.4). Hence express ϕ_1 and ϕ_2 in terms of N.

EXERCISE 4.3.2. Show that for the Reiner–Rivlin model

$$\eta_e(\dot{e}) = 3[\phi_1(6\dot{e}^2, 6\dot{e}^3) + \phi_2(6\dot{e}^2, 6\dot{e}^3)\dot{e}]$$
$$\eta_{PS}(\dot{e}) = 4\phi_1(8\dot{e}^2, 0)$$

EXERCISE 4.3.3. Verify the results given in Table 4.1 for $N(\dot{e}_1, \dot{e}_2)$, $\eta_e(\dot{e})$ and $\eta_{PS}(\dot{e})$ for a third-order and a CEF fluid.

EXERCISE 4.3.4. Show that for a Newtonian fluid in pure shear, defined by (4.3.8), the second normal-stress difference

$$t_{11} - t_{33} = 2\eta_{PS}\dot{e}$$

Show also that, in general, there will be a second normal-stress function of $\dot{e}$, characteristic of uniform steady pure shear which yields $(t_{11}-t_{33})/\dot{e}$.

4.4 FOURTH-ORDER FLOWS

These are the weak flows of case (ii), eqn (4.1.3). Specific examples have been discussed by Noll (1962), Oldroyd (1965) and Huilgol (1975) but have not yet been shown to be of particular significance.

4.5 ELLIPTICAL FLOWS

A simple method of visualizing the range of weak and strong flows was given by Astarita (1967) in terms of the 'elliptic' flows

$$v_1 = Wx_2, \qquad v_2 = aWx_1, \qquad v_3 = 0 \tag{4.5.1}$$

where a and W are constants and

$$W > 0, \qquad a^2 \leqslant 1 \tag{4.5.2}$$

without loss of generality. They are all MWCSH. The three eigenvalues of $\mathbf{L}$ are given by

$$\lambda^3 - aW^2\lambda = 0 \tag{4.5.3}$$

whose roots are 0, $\pm a^{\frac{1}{2}}W$.

By our earlier definitions, the case $a > 0$ leads to strong flows, while $a \leqslant 0$ leads to weak flows. If $a = -1$, we have pure rotation; if $a = 0$, we have viscometric flow, and if $a = 1$, we have pure shear, with extension and contraction at angles $\pi/4$ to the coordinate axes 1 and 2. The stress tensor will be given by

$$t_{12} = \eta_A(a, W)W \tag{4.5.4}$$

$$t_{11} - t_{22} = \Psi_{1A}(a, W)W^2 \tag{4.5.5}$$

$$t_{22} - t_{33} = \Psi_{2A}(a, W)W^2 \tag{4.5.6}$$

where η_A, Ψ_{1A} and Ψ_{2A} are material functions of a and W, and are so defined that they become the viscometric functions (4.2.2)–(4.2.4) when $a = 0$. Fuller and Leal (1981) have shown how these flows may be generated using a four-roll mill.

EXAMPLE 4.5.1. The special cases of elliptical flow.

$$a=-1: \qquad \mathbf{L}=W\begin{vmatrix} 0 & 1 & 0 \\ -1 & 0 & 0 \\ 0 & 0 & 0 \end{vmatrix}=\mathbf{W}, \qquad \mathbf{D}=\mathbf{0}$$

It follows, from $\mathbf{D}=0$, that there is no straining and the constant value for $\mathbf{W}$ implies pure rotation. It may readily be verified that the velocity field $v_1=Wx_2$, $v_2=-Wx_1$ represents a rigid-body rotation about the x_3 axis with angular speed W.

$a=0$: This is recognized directly as the simple shear flow (4.2.1) with $D=W$.

$a=1$: If we change to coordinates

$$x_1^*=(x_1+x_2)/\sqrt{2}, \qquad x_2^*=(x_2-x_1)/\sqrt{2}, \qquad x_3^*=0$$

then

$$v_1^*=(v_1+v_2)/\sqrt{2}=W(x_1+x_2)/\sqrt{2}=Wx_1^*$$
$$v_2^*=(v_1-v_2)/\sqrt{2}=W(x_1-x_2)/\sqrt{2}=-Wx_2^*$$

which is immediately recognizable as pure shear.

4.6 UNSTEADY UNIFORM FLOWS

We may relax the condition that $\mathbf{D}\mathbf{L}/\mathrm{D}t=0$ without altering the uniformity condition that $\nabla\mathbf{L}\equiv\mathbf{0}$, and in particular we consider the case

$$\mathbf{L}(t)=l(t)\mathbf{L}_0 \tag{4.6.1}$$

No dynamical constraints will arise if we consider the inertial contributions to stress (in the equation of motion) to be negligible compared to those of the extra stress. This gives us obvious generalizations of viscometric, extensional and elliptical flows. Equivalent generalizations of the relevant material functions into functionals will arise, i.e.

$$f(x, y) \to \underset{t'=-\infty}{\overset{t}{\mathscr{F}}}(x_0 l(t'), y_0 l(t')) \tag{4.6.2}$$

for differentiable histories x, y where x, y might stand for $\dot{e}_1$ and $\dot{e}_2$ in (4.3.3) and (4.3.4) or for a and W in (4.5.4)–(4.5.6). The cases that have been most investigated are oscillatory flows and ramp (suddenly started steady) flows.

4.6.1 Small-amplitude Oscillatory Flows

For these we take

$$l(t) = \mathcal{R}e\{e^{iwt}\} \tag{4.6.3}$$

with

$$D_0/w \ll 1 \tag{4.6.4}$$

where w is an angular frequency. If we take $\mathbf{L}_0$ to be viscometric flow, then we recover the most usual type, namely small-amplitude oscillatory shear flow, with

$$v_1 = \mathcal{R}e\{D_0 x_2 e^{iwt}\}, \qquad v_2 = v_3 = 0 \tag{4.6.5}$$

i.e. $l_{12} = \mathcal{R}e\{D_0 e^{iwt}\}$, $l_{ij} = 0$ all other ij. This gives rise to an oscillatory shear stress

$$t_{12} = \mathcal{R}e\{\eta^*(w) D_0 e^{iwt}\} \tag{4.6.6}$$

of the same frequency, and normal stresses that are displaced and oscillate with frequency $2w$, i.e.

$$t_{11} - t_{22} = \Psi_1^d(w) D_0^2 + \mathcal{R}e\{\Psi_1^*(w) e^{2iwt} D_0^2\} \tag{4.6.7}$$

$$t_{22} - t_{33} = \Psi_2^d(w) D_0^2 + \mathcal{R}e\{\Psi_2^*(w) e^{2iwt} D_0^2\} \tag{4.6.8}$$

$$\eta^*(w) = \eta'(w) - i\eta''(w) \tag{4.6.9}$$

is known as the complex viscosity, with $\eta'(w)$, the dynamic viscosity, being interpreted as a viscous (in-phase) contribution and $G'(w) = w\eta''(w)$, the dynamic rigidity, as an elastic (out-of-phase) contribution. In the limit $w \to 0$, we expect to recover the (real) Newtonian viscosity, i.e.

$$\underset{w \to 0}{\text{Lt}}\ \eta^*(w) = \eta_0 \tag{4.6.10}$$

This matter is exhaustively dealt with in BA & H Section 4.4.

The essentially equivalent theory of small-amplitude (solid) viscoelasticity is well described in Ferry (1980), where emphasis is placed on the complex elastic modulus

$$G^*(w) = G'(w) + iG''(w) \tag{4.6.11}$$

for which $G''(w) = w\eta'(w)$ represents viscous loss and

$$\underset{w \to 0}{\text{Lt}}\ G'(w) = G \tag{4.6.12}$$

the shear modulus of the material.

4.6.2 Ramp Flows

For these we take

$$l(t)=H(t)=\begin{cases}0 & t<0\\ 1 & t\geqslant 0\end{cases} \tag{4.6.13}$$

which represents sudden start of flow at time $t=0$. For suddenly-started simple-shear flow, we have the phenomenon of stress growth,† with

$$t_{12}=\eta^{+}(t;D)D \tag{4.6.14}$$

$$t_{11}-t_{22}=\Psi_1^{+}(t;D)D^2 \tag{4.6.15}$$

$$t_{22}-t_{33}=\Psi_2^{+}(t;D)D^2 \tag{4.6.16}$$

(cf. (4.2.2)–(4.2.4) above). The same sudden start can be made to elongational flow, giving rise to the time-dependent extensional viscosity $\eta_e^{+}(t;\dot{e})$ (cf. (4.3.7) above).

Further complications can be introduced by having

$$l(t)=\sum_{n=1}^{N} H(t-t_n) \tag{4.6.17}$$

but these need not be considered here. We note that the extra stress $\mathbf{T}^{\mathrm{E}}$ is zero until $t=0$ in ramp flows and that

$$\operatorname*{Lt}_{t\to\infty} f^{+}(t;x)=f(x) \tag{4.6.18}$$

where f is a material function.

4.6.3 Stress-relaxation Flows

If we use the inverse Heaviside function, $H(-t)$, then the flow field $H(-t)\mathbf{L}_0$ represents sudden cessation of the steady uniform flow $\mathbf{L}_0$ at $t=0$. In general, the stress field will not decay instantaneously, and so a range of material stress-relaxation functions will be defined by the various special cases of $\mathbf{L}_0$ that we have considered above. Using once again the notation of BA & H Section 4.4, for $t>0$,

$$t_{12}=\eta^{-}(t;D)D \tag{4.6.19}$$

$$t_{11}-t_{22}=\Psi_1^{-}(t;D)D^2 \tag{4.6.20}$$

$$t_{22}-t_{33}=\Psi_2^{-}(t;D)D^2 \tag{4.6.21}$$

† The nomenclature is that of BA & H, eqns (4.4.18)–(4.4.20).

for the case of cessation of uniform shear flow, and

$$t_{11}-t_{22}=\eta_e^-(t;\dot{e})\dot{e} \tag{4.6.22}$$

gives the relaxation function for cessation of elongational flow. Note that

$$f^-(0;x)=f(x) \tag{4.6.23}$$

and that

$$\operatorname*{Lt}_{t\to\infty} f^-(t;x)=0 \tag{4.6.24}$$

EXERCISE 4.6.1. Verify the results given in Table 4.1 for $\eta'(w)$ and $G'(w)$ showing that only the $\mathbf{A}_1$ term and $\partial/\partial t$ terms in $\mathbf{A}_2$, $\mathbf{A}_3$ are relevant.

4.7 ECCENTRIC DISC (ORTHOGONAL RHEOMETER) FLOW

This is a MWCSH of case (iii)(b) with $\alpha=0$. In terms of our usual Cartesian coordinates

$$v_1=Wx_2+AWx_3, \qquad v_2=-Wx_1, \qquad v_3=0 \tag{4.7.1}$$

It is clearly a weak flow. It is one in which all particles trace out circles in planes $x_3=$ constant at the same rate, so that they return to their origin with frequency W. It may be shown (Walters, 1975, Subsection 6.4.2) that, for $A\ll 1$, the Eulerian uniform steady flow field (4.7.1) is essentially equivalent to the oscillatory shear field given by (4.6.5) with $W=w$ and

$$t_{13}=\eta'(W)AW \tag{4.7.2}$$

$$t_{23}=\eta''(W)AW \tag{4.7.3}$$

EXERCISE 4.7.1. Show that, for the eccentric disc flow defined by (4.7.1), the velocity-gradient tensor is given by

$$\mathbf{L}=\begin{bmatrix} 0 & W & AW \\ -W & 0 & 0 \\ 0 & 0 & 0 \end{bmatrix}$$

the vorticity tensor by

$$\mathbf{W}=\begin{bmatrix} 0 & W & \frac{1}{2}AW \\ -W & 0 & 0 \\ -\frac{1}{2}AW & 0 & 0 \end{bmatrix}$$

and the rate-of-deformation tensor by

$$\mathbf{D}=\begin{bmatrix} 0 & 0 & \frac{1}{2}AW \\ 0 & 0 & 0 \\ \frac{1}{2}AW & 0 & 0 \end{bmatrix}$$

Hence that

$$\overline{\mathrm{II}}_{\mathbf{W}}=-\tfrac{1}{2}W^2(4+A^2)$$
$$\overline{\mathrm{II}}_{\mathbf{D}}=\tfrac{1}{2}W^2A^2$$

Consider the case $A \gg 1$.

4.8 SLIGHTLY PERTURBED FLOWS

Many real flows approximate to one or other of the flows described in Sections 4.2–4.7 above. In certain circumstances, it is the dynamical consequences of small departures from the steady uniform approximating flow field that is most interesting. Two basic flow fields have been studied in this regard.

Pipkin & Owen (1967) have given a full treatment of almost-viscometric flows, which has led to various special cases being used for rheometric purposes, while more recently Poutney & Walters (1978, 1979) have studied almost-extensional flow (see also Huilgol, 1981, and Powell & Schwartz, 1981).

4.9 RHEOMETRY

Table 4.1 shows how most of the constitutive relations mentioned in Chapter 3 give rise to a corresponding set of predictions for the material functions described above. Typically, each constitutive equation involves a finite number of material constants and functions of the principal invariants of the rate of strain, which appear in various combinations in the corresponding forms for the material functions.

For example, the third-order fluid (3.3.11) is defined by six constants, η_0, δ_1, β_0, γ_0, β_1 and γ_1. These appear in various combinations in the forms taken for $\eta(D)$, $\Psi_1(D)$, $\Psi_2(D)$, $\eta_e(\dot{e})$, $\eta_{PS}(\dot{e})$, $\eta'(w)$ and $G'(w)$. Clearly, in this case, a knowledge of the material functions mentioned more than suffices to prescribe η_0 to γ_1: it provides a test for the adequacy of the third-order fluid model. Alternatively, we may say that there exist several relationships between the material functions for a third-order fluid, e.g.

$$w^2\Psi_1 = 2G'(w) \tag{4.9.1}$$

or

$$\left(\frac{\mathrm{d}\eta_e}{\mathrm{d}\dot{e}}\right)_{e=0} = 3(\Psi_2 + \tfrac{1}{2}\Psi_1) \tag{4.9.2}$$

It is therefore natural to seek to measure various material functions for real fluids as a means of selecting, where possible, suitable constitutive equations to represent them. This is the province of rheometry, which has so far received more attention than that of processing flows themselves. The basic objective is clear: to set up one of the uniform flows described above and to measure the relevant stresses that ensue.

Except for the general extensional flows (4.3.1), the elliptical flows (4.5.1) and the eccentric disc flow (4.7.1), all the flow fields described are functions of a single variable describing the magnitude of the velocity-gradient tensor. Thus, the material functions of interest will be readily describable for steady or purely periodic flows, and comparison with model predictions should not be difficult.

The simplest way to obtain controllable flow fields is to contain the test fluid within rigid boundaries and to move the boundaries in such a way that the desired flow field will ensue. Various viscometric (simple shear) flows can be approximated in this way. However, it is obvious that no flow field can be infinite in extent, and so there will be difficulties associated with the finite nature of experimentally obtainable flow fields.

For example, if we attempt to obtain steady simple-shear flow by moving a large plate P_1 (at $x_2 = H$) at constant velocity $\mathbf{v}_1$ in its own plane relative to the fixed plate P_2 (at $x_2 = 0$), with the intervening gap filled with test fluid, then we run into difficulties at the edges of the plates. Indeed for any finite-sized plates, the motion can only take place for a finite time before the plates cease to lie above one another at all (at which stage the fluid ceases to be contained). The method will

Table 4.1
Rheological Functions for Some Model Fluids

General function for simple fluid	*Eqn (3.3.1) Newtonian fluid*	*Eqn (3.3.3) Generalized Newtonian fluid*	*Eqn (3.3.11) Third-order fluid*
$\eta(D)$	η_0	$\eta(D)$	$\eta_0+2(\delta_1+\beta_1)D^2$
$\Psi_1(D)$	0	0	$-2\gamma_0$
$\Psi_2(D)$	0	0	$(\beta_0+2\gamma_0)$
$N(\dot{e}_1, \dot{e}_2)$	$\eta_0(2\dot{e}_1+\dot{e}_2)$	$\eta(\dot{e})(2\dot{e}_1+\dot{e}_2)$ $\dot{e}=2(\dot{e}_1^2+\dot{e}_2^2+\dot{e}_1\dot{e}_2)^{\frac{1}{2}}$	$\{2\eta_0-4(\beta_0+\gamma_0)\dot{e}_2+8(2\delta_1+2\beta_1+\gamma_1)$ $\times(\dot{e}_1^2+\dot{e}_1\dot{e}_2+\dot{e}_2^2)\}(2\dot{e}_1+\dot{e}_2)$
$\eta_e(\dot{e})$	$3\eta_0$	$3\eta(\sqrt{3}\,\dot{e})$	$3\{\eta_0+(\beta_0+\gamma_0)\dot{e}+3(2\delta_1+2\beta_1+\gamma_1)\dot{e}^2\}$
$\eta_{PS}(\dot{e})$	$4\eta_0$	$4\eta(2\dot{e})$	$4\{\eta_0+4(2\delta_1+2\beta_1+\gamma_1)\dot{e}^2\}$
$\eta'(W)$	η_0	$\to\eta(0)$	$\eta_0-w^2\gamma_1$
$G'(W)$	0	0	$-w^2\gamma_0$
$\eta^+(t; D)$	η_0	$\eta(D)$	not applicable
$\eta_e^+(t; \dot{e})$	$3\eta_0$	$3\eta(\sqrt{3}\,\dot{e})$	not applicable
$\eta^-(t; D)$	0	0	not applicable
$\eta_e^-(t; \dot{e})$	0	0	not applicable

clearly be more adequate for small amplitude oscillatory shear flow, with $\mathbf{v}_1=\mathbf{v}_{10}\sin wt$ particularly if H is very much less than the lateral dimensions of either plate, because the fluid will remain in contact with both plates and the influence of edge-flow distortion will be small over most of the flow field.

Extensional flow fields can be more easily approximated by sheets of material held along their edges which are displaced according to the

Table 4.1—*contd.*

Eqn (3.3.16) Reiner–Rivlin fluid	*Eqn (3.3.21) Criminale–Ericksen–Filbey fluid*	*Eqn (3.3.22) Oldroyd 8-constant fluid*
$\phi_1(2D^2, 0)$	$\eta(D)$	$\eta_0 \dfrac{\{1-(\lambda_1\lambda_2+\mu_0\mu_2+\frac{3}{2}\mu_0\nu_2 - \mu_1\mu_2-\mu_1\nu_2)D^2\}}{\{1+(\lambda_1^2+\mu_0\mu_1-\frac{3}{2}\mu_0\nu_1-\mu_1^2+\mu_1\nu_1)D^2\}}$
0	$\Psi_1(D)$	$2\{\lambda_2\eta_0-\lambda_1\eta(D)\}$
$\phi_2(2D^2, 0)$	$\Psi_2(D)$	$(\lambda_2+\mu_2)\eta_0+(\mu_1-\lambda_1)\eta(D)$
$2(2\dot{e}_1+\dot{e}_2)(\phi_1-2\dot{e}_2\phi_2)$ see Exercise 4.3.1 for arguments of ϕ_1 and ϕ_2	$2[\eta(\dot{e})-\{\Psi_1(\dot{e})\dot{e}_2 + 2\Psi_2(\dot{e})\dot{e}_2\}](2\dot{e}_1+\dot{e}_2)$ $\dot{e}^2=4(\dot{e}_1^2+\dot{e}_1\dot{e}_2+\dot{e}_2^2)$	very complex
$3\{\phi_1(6\dot{e}^2, 6\dot{e}^3) + \phi_2(6\dot{e}^2, 6\dot{e}^3)\dot{e}\}$	$3\{\eta(\sqrt{3}\,\dot{e})+\frac{1}{2}\Psi_1(\sqrt{3}\,\dot{e})\dot{e} + \Psi_2(\sqrt{3}\,\dot{e})\dot{e}\}$	$3\eta_0 \dfrac{\{1-\mu_2\dot{e}-(3\mu_0-2\mu_1)(\mu_2-\frac{3}{4}\nu_2)\dot{e}^2\}}{\{1-\mu_1\dot{e}-3\nu_1(\frac{3}{2}\mu_0-\mu_1)\dot{e}^2\}}$
$4\phi_1(8\dot{e}^2, 0)$	$4\eta(2\dot{e})$	$4\eta_0 \dfrac{[1+2\{\mu_0(2\mu_2-3\nu_2)-\mu_1(\mu_2-\nu_2)\}\dot{e}^2]}{[1-(6\mu_0\nu_1-4\mu_1\nu_1)\dot{e}^2]}$
$\to\phi_1(0, 0)$	$\to\eta(0)$	$\eta_0\left(\dfrac{1+\lambda_1\lambda_2 W^2}{1+\lambda_1^2 W^2}\right)$
0	$\to\frac{1}{2}w^2\Psi_1(0)$	$\eta_0 \dfrac{(\lambda_1-\lambda_2)W^2}{(1+\lambda_1^2 W^2)}$
not applicable	not applicable	not applicable
not applicable	not applicable	not applicable
not applicable	not applicable	not applicable
not applicable	not applicable	not applicable

desired flow field or, in the case of elongational flow, by columns of fluid whose ends are moved axially at the required relative speed. Here most of the bounding surface is essentially unrestrained and distortions to the flow field will arise where the material is gripped. It follows that such techniques will be most likely to work with very stiff materials.

The principles of rheometry are discussed in many texts and review articles (Walters 1975, Chapters 4–8; Ferry 1980, Chapters 5–8; Eirich

1958, Vol II, Chapter 11, 1960, Vol III, Chapter 2; BA & H Section 4.5; Lodge 1964, Chapter 9, 1974, Chapters 3, 9; Tanner 1976; Petrie 1979) but detailed discussions of particular pieces of equipment and practical aspects of rheometry are not so easy to reference. Viscometry was relatively well covered into the late 1950s by Van Wazer *et al.* (1963), while Ferry (1980) covers small-amplitude oscillatory measurements; for other elastico-viscous measuring devices, individual articles and manufacturers' manuals are the best sources.

The subject of rheometry is a broad one, in that it has to deal with all the departures from reality that have been assumed above. The list of these is daunting; it includes effects due to:

gravity
inertia
surface tension
geometrical inaccuracies in the flow-field boundaries, including play in bearings
heat generation
material and thermal inhomogeneity
unsteadiness in drive mechanisms

There will be further errors arising from the techniques of measurement, particularly of stress.

Because of all these practical difficulties, much of the actual rheological characterization of polymeric materials in industry has been on the basis of empirical tests for fluidity, which usually provide an ordering of materials similar to that provided by the shear viscosity, but are in no sense equivalent to even a crude constitutive equation. Examples of such tests are the Melt Flow Index—BS(UK) ASTM(USA) DIN(FRG)—for thermoplastics and the Mooney plasticity (USA and UK) for curable elastomers.

Almost all of the research into rheological properties of polymeric fluids has concentrated on simple-shear flows, either steady or small-amplitude oscillatory. By using

(a) rotating-cylinder viscometers,
(b) parallel-plate viscometers, and
(c) cone-and-plate viscometers,

reliable measurements of $\eta(D)$, $\eta'(w)$ and $G'(w)$ have been made over a relatively wide range of D or w for dilute solutions, concentrated solutions, melts and for suspensions of particles in the former. A very considerable amount of effort has been devoted to obtaining reliable values for the first normal-stress function $\Psi_1(D)$, but routine determination of $\Psi_2(D)$, the second normal-stress function, is still not practicable (but see recent work of Okobu & Hori, 1980). Improved instrumentation on existing devices makes $\eta^+(t; D)$ and $\eta^-(t; D)$ the next most readily accessible functions.

One alternative to sudden imposition of a constant rate of strain, as a means of measuring transient behaviour of materials, is sudden imposition of a constant stress, with, similarly, sudden removal of the stress instead of sudden cessation of the rate of strain. The consequential deformations, as a function of time, are generally known as creep curves and creep recovery curves, respectively. These are discussed in BA & H Section 4.4. A simpler and briefer account of various uniform-simple-shear experiments is given in Jenkins (1972, Chapter 6).

Exhaustive shear flow information about well-characterized samples of polymer melts can provide sensitive tests for proposed constitutive equations, though there is no agreement yet over whether it suffices to define the constitutive equation, for any particular fluid, completely and satisfactorily. Some workers argue that it is better to make elongational flow measurements as well because these provide more sensitive tests for the predictions of any given model equation in extensional flow fields. Much current work is devoted to this end.

One of the main difficulties in elongational-flow rheometry is to get steady extension over a sufficiently long period to derive a Trouton viscosity, and this has been regarded as a drawback to such rheometry. However, values for $\eta_e^+(t; \dot{e})$ and creep curves are easy to obtain, and recently these have been used as important means for testing constitutive models. An ambitious device for arbitrary biaxial testing has been described by Denson & Hylton (1980), and another by Meissner *et al.* (1981). Relatively high extension rates have been obtained by Hull (1981*b*).

Chapter 5

Rheological Behaviour of Polymer Melts

5.1 EXPERIMENTAL

There is relatively little reliable experimental information in the open literature on polymer melt rheology. Many results have been obtained within industrial laboratories and have either not been published at all or have been released in limited, simplified form.

There is one overwhelmingly good reason for this: polymers are not, in the chemical sense, pure compounds and so it is rare that the products of two companies, or even of two reactors within the same company, ostensibly producing equivalent polymers, are indistinguishable rheologically.

For example, if we take a simple homopolymer, like polyethylene, PE, it can vary independently in terms of

(1) molecular weight (number average),
(2) molecular weight (weight average), and
(3) degree of chain branching,

and still not be uniquely specified because other statistical parameters of the molecular weight distribution are relevant. Rheological properties depend significantly on all these parameters, and so it is impracticable to speak about the rheology of polyethylene as such.

Furthermore, it is now common to market modified polymers, in the sense that they may be random, or block, copolymers of two or more monomers. Thus, for example, properties of what is usually termed polypropylene (PP) depend very significantly on small proportions of copolymerized polyethylene. Alternatively, two polymers may be blended, or a homopolymer may be mixed with inert or active fillers like chalk, wood, glass fibre or carbon black. Most elastomers (rubbers) are processed in compound form, with the amount and type of carbon black filler exerting a profound (and advantageous) influence on their rheological, and hence processing, behaviour.

Many polymers are susceptible to chemical degradation at temperatures near those at which they can be thought of as substantially fluid. Therefore, they are processed in what some have called a high-elastic state, i.e. they are maintained as far as possible at temperatures for which the relaxation times are very long. In consequence, they may never, at any stage in their existence in bulk form, be in thermodynamic equilibrium—they may never be in a homogeneous isotropic state. Consequently, the influence of past times on the present rheological behaviour may extend over the entire history of the material in question, including its formation in a polymerization reactor.

To take a simple example, it has been found that the viscous properties of some commercial PP samples, as measured by capillary rheometry, can be critically affected by the thermal history, in the (bulk) rest state, of the material while in the reservoir. Material held above the crystalline melting temperature for several days will differ significantly in behaviour, at high rates of shear, from material that has been rapidly melted from solid form.

Many raw elastomers suffer chain scission during milling and processing and so the basic rheological properties of the material change during processing. On attempting to measure viscosities (and elasticities) of raw rubber by capillary extrusion at room temperature and high shear rate, it is found that their values can change with number of passes through the extrusion rheometer.

For all of these reasons, it is not surprising that there exists no readily available compendium of rheological properties of polymers.

There are, however, certain rheological properties which are, broadly speaking, possessed by all long-chain polymer systems above their glass transition or crystalline melting points, and these can best be exemplified in diagrammatic form for various material parameters.

5.1.1 Steady Shear Viscosity

This is the property $\eta(D)$ defined by eqn (4.2.2) above, which can be measured by concentric-cylinder, cone-and-plate or parallel-plate viscometers over a moderate range of shear rates, and by capillary viscometry at high rates of shear. Almost all polymer melts exhibit a noticeable and, in most cases, a dominant amount of pseudo-plasticity. This decrease in viscosity with increasing shear rate D is shown in Fig.

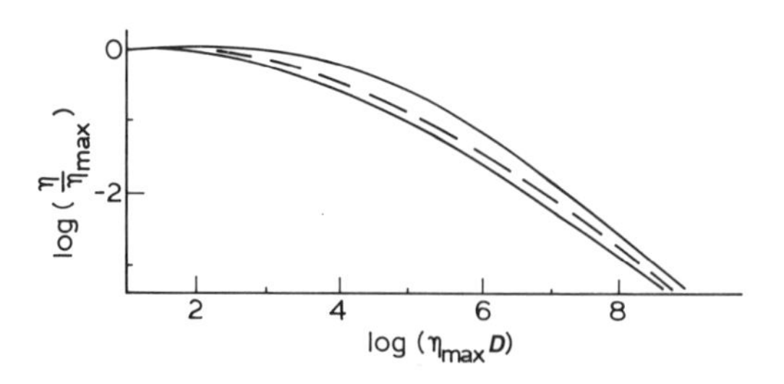

Fig. 5.1. Universal viscosity function characteristic of polymer melts ($\eta_{max}D$ in Pa). --- eqn (5.1.1); —— range of polymers studied. (From Vinogradov & Malkin, 1966.)

5.1, which is taken from Vinogradov & Malkin (1966), who collected data for a range of polymers, including HDPE, LDPE, PP, PS, PMMA, PIB, NR and SBR. The, admittedly rather broad band, curve depends upon a single material parameter characteristic of any particular material at any given temperature within its range of fluidity, namely η_{max} $(=\eta_0)$, the limiting, low shear, 'Newtonian' viscosity. They proposed the purely empirical relation

$$\eta_{max}/\eta = 1 + A_1(D\eta_{max})^{\alpha} + A_2(D\eta_{max})^{2\alpha} \tag{5.1.1}$$

with $A_1 = 6{\cdot}12 \times 10^{-2{\cdot}645}$, $A_2 = 2{\cdot}85 \times 10^{-3{\cdot}645}$, $\alpha = 0{\cdot}355$ with η and D expressed in SI units (Pa s and s^{-1}, respectively) as typical of a polymer melt. It is worth noting that the contribution of the first term on the right-hand side (1) is 80% or over of the total for all $(\eta_{max}D) \leqslant 10^3$ Pa, the Newtonian range, and that the contribution of the last term has risen to 80% of the total for all $(\eta_{max}D) \geqslant 10^6$ Pa. Consequently, a power-law relationship

$$\eta(D) = \mathrm{K}D^{-\frac{2}{3}} \tag{5.1.2}$$

is a good approximation for nearly three decades of shear. This is precisely the approximation that was introduced in Section 3.3, eqn (3.3.3), with $n = \frac{1}{3}$.

The dependence of the Newtonian limit, η_{max}, upon material and temperature is considered in detail in Fox *et al.* (1956). For all polymers over a critical molecular weight, M_c, and this will be the case for almost all commercial high polymers, it was found empirically that

$$\ln \eta_{max} = 3{\cdot}4 \ln \bar{M}_w + \mathrm{fn}(T) \tag{5.1.3}$$

where fn(T) will be a function of the polymer type and temperature, and $\bar{M}_w$ is the weight-average molecular weight. Also empirically, it was found that

$$\text{fn}(T) - \text{fn}(T_g) \simeq \frac{-17{\cdot}44(T - T_g)}{51{\cdot}6 + (T - T_g)} \tag{5.1.4}$$

with T measured in °K, and T_g the glass transition temperature. (For further details see Tables II and IV of the article in question.) A typical value of η would be 10^4 Pa s (10^5 Poise).

If we consider the implications of (5.1.3) and (5.1.4) for Fig. 5.1, we see that the ordinate scale will be unaffected by varying $\eta_{\max}$ and that change of temperature can be accommodated by an equivalent change of shear rate on the abscissa. This equivalence is usually called the time–temperature superposition principle (D being an inverse time) and is described by the WLF equation

$$\ln a_T(T; T_s) = \frac{-C_1(T - T_s)}{C_2 + (T - T_s)} \tag{5.1.5}$$

where

$$a_T = \frac{\eta T_s \rho_s}{\eta_s T \rho} \tag{5.1.6}$$

is the shift factor, eqn (5.1.4) being a useful approximation. Further details can be found in Ferry (1980).

One interpretation of the success of this shift factor is that any polymeric material can be characterized, at least as far as steady simple-shear flow is concerned, by a single characteristic timescale τ_{char}, which varies simply with temperature. This does not necessarily mean that there will be only one relevant characteristic time for all rheological behaviour, but that all those relevant in simple shear will scale similarly. Insofar as time-dependent rheological behaviour is determined by rate processes, i.e. by irreversible thermodynamical processes, it is tempting to associate them all with the theory of absolute rate processes in terms of a basic activation energy E characteristic of the segmental motions of polymer chains. This gives rise to an Arrhenius term $\exp(E/RT)$, in which E proves to be temperature- and, later, deformation-dependent. This simple interpretation can be extended to the case of non-isothermal deformations, i.e. it can be used to convert the constitutive relations developed in Chapter 3, which do not contain T explicitly, into constitutive relations that include the effect of temperature history. This is discussed later.

The effect of pressure on viscosity is described in Cogswell (1973).

The use of cone and plate rheometers has meant that first normal-stress differences (as defined by eqn (4.2.3)) are now regularly measured as functions of shear rate and temperature (Cogswell, 1981). For the shear-rate region where the viscosity η is a power-law function of D, $\Psi_1 D^2$ is often equal to or greater than $|\eta D|$, and can be approximated by a power-law relation, with $\Psi_1 D/\eta \propto D^\beta$, $\beta > 0$.

More recently, measurements on the second normal stress difference have been made (Okobo & Hori, 1980).

5.1.2 Dynamic Mechanical Properties

The response of a polymer to a small-amplitude oscillatory shear flow—see (4.6.5)—can be expressed in terms of one of many interrelated viscoelastic functions. For comparison with the steady-shear viscosity $\eta(D)$, the complex viscosity η^*—defined by (4.6.9)—is often used, and is appropriate for fluid-like melts. It is simply related to the more common complex elastic modulus G^* by

$$G^* = \mathrm{i}w\eta^* \tag{5.1.7}$$

and to the relaxation modulus $G(s)$, defined in (3.4.3), by Fourier transformation

$$\eta^*(w) = \int_0^\infty G(s)\mathrm{e}^{-\mathrm{i}ws}\,\mathrm{d}s \tag{5.1.8}$$

Because of the simple interpretation that can be placed on $G(s)$ when it has simple negative exponential form (it represents a single Maxwell element), it is convenient to decompose $G(s)$ into a relaxation spectrum $\mathrm{H}(u)$ with respect to relaxation time u, according to

$$G(s) = \int_0^\infty \frac{\mathrm{H}(u)}{u}\,\mathrm{e}^{-s/u}\,\mathrm{d}u \tag{5.1.9}$$

This is also viewed as a relaxation frequency spectrum $\mathrm{N}(w)$ according to

$$\mathrm{N}(w) = \mathrm{H}(1/w)/w \tag{5.1.10}$$

It is also found useful to define a complex compliance $J^*(w)$ as the inverse of the complex elastic modulus,

$$J^*(w) = 1/G^*(w) \tag{5.1.11}$$

In a purely mathematical sense, full knowledge of one of the functions η^*, G^*, J^*, H or N suffices to obtain all of the rest. In practice, numerical conversion leads to a great deal of uncertainty, and so various authors use whichever of the various functions seems most reliable. It is worth noting that the transformation (5.1.8) implies that η' and η'' (or equivalently G' and G'') cannot be independent arbitrary functions: they are strongly related by virtue of their being cosine and sine transforms of $G(s)$, respectively.

A most important empirical observation has been shown to apply in a remarkable number of cases (including PE and PS melts) is the Cox–Mertz rule, whereby

$$|\eta^*(w)| \equiv \{\eta'(w)^2 + \eta''(w)^2\}^{\frac{1}{2}} = \eta(D)|_{D=w} \qquad (5.1.12)$$

The correlation with $|\eta^*|$ is much better than with η' alone.

Vinogradov & Malkin (1966) give an approximate universal form for the relaxation frequency spectrum in the form

$$\mathrm{N}(w\eta_{max})/\eta_{max} = \begin{cases} 2{\cdot}24 \times 10^{-2\cdot 6}(w\eta_{max})^{0\cdot 40} & (0 < w\eta_{max} < 4{\cdot}68 \times 10^3) \\ 0{\cdot}166 & (4{\cdot}68 \times 10^3 < w\eta_{max} < 3{\cdot}80 \times 10^4) \\ 7{\cdot}1 \times 10^{1\cdot 35}(w\eta_{max})^{-0\cdot 65} & (3{\cdot}80 \times 10^4 < w\eta_{max}) \end{cases} \qquad (5.1.13)$$

with η_{max} in Pa.

The Cox–Mertz rule leads to an apparent paradox. A strongly shear-dependent viscosity is, by all the usual criteria, a sign of non-linearity in steady-shear behaviour, and yet η^* is representative of a strictly linear response in oscillatory behaviour. This paradox can be resolved partly by supposing that some of the steady-shear non-linearity is not really a non-linearity in material response but a consequence of non-linear terms in the relevant deformation measure. This is one reason for hoping that the single-memory-function integral implied by $m(s)$ in (3.3.36) or $G(s)$ in (3.3.37) can be carried over directly from the strictly linear regime into a non-linear regime by suitable choice of strain measure $\mathbf{G}_t^t$. This is the point that was raised in Chapter 3, as for example in the discussion preceding (3.4.7). See also BA & H Section 8.2 for further discussion and Section 5.2 below.

Figure 5.2 shows a typical shape for the dynamic shear modulus. The values obtained for $G'(w)$ as $w \to \infty$ become very large (of order 10^9 Pa) and so represent an essentially rigid material. The highly elastic

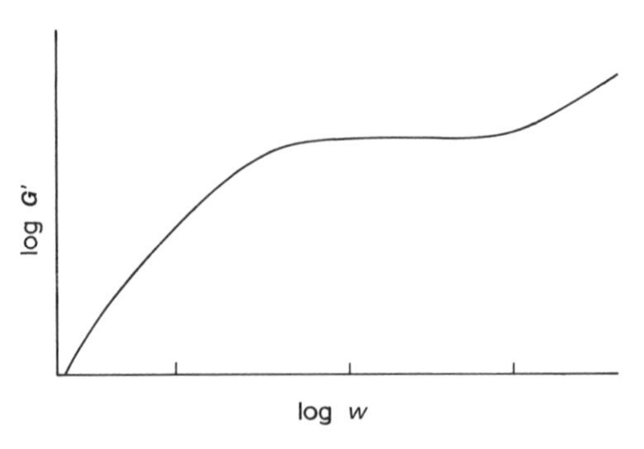

Fig. 5.2. Dynamic shear modulus for typical polymer in arbitrary units.

nature of elastomers and polymer melts implies shear moduli five orders of magnitude smaller at lower frequencies. Use of such spectra is made in Chapters 15 and 17 where analysis of fibre spinning and film blowing requires the essentially viscoelastic nature of polymer melts to be included.

5.1.3 Stress Relaxation

Here we consider the function $\eta^{-}(t; D)$, defined by (4.6.19), as a function of both t and D. For small D, it is related to the other linear viscoelastic functions by

$$\left(\frac{\partial \eta^{-}}{\partial t}\right)_{D \to 0} = -G(t) \tag{5.1.14}$$

However, it is observed that for D in the range for which $\eta(D)$ varies, $\eta^{-}(t; D)$ also varies with D. The style of the variation is shown in Fig. 5.3. It will be seen that more than a single relaxation time is required,

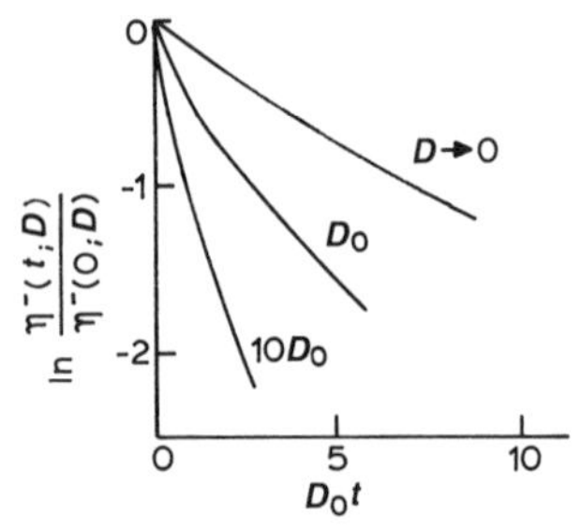

Fig. 5.3. Typical non-linear behaviour of stress relaxation function after shearing at rate D. D_0 is arbitrary (from Vinagradov & Malkin, 1966).

though, at long times, a definite largest relaxation time does appear characteristic of each shear rate D_0, this characteristic λ_{max} being a decreasing function of D_0.

5.1.4 Stress Growth

This is described by the function $\eta^+(t; D)$ defined by (4.6.14). In the linear limit we have

$$\left(\frac{\partial \eta^+}{\partial t}\right)_{D\to 0} = G(t) \tag{5.1.15}$$

so the information provided is exactly that of the stress relaxation experiment. In the non-linear regime we obtain results of the type shown in Fig. 5.4, where it is seen that the shear stress reaches a maximum for all but very small rates of shear (a phenomenon known as stress overshoot) and that it occurs at a shear strain of about 4. The

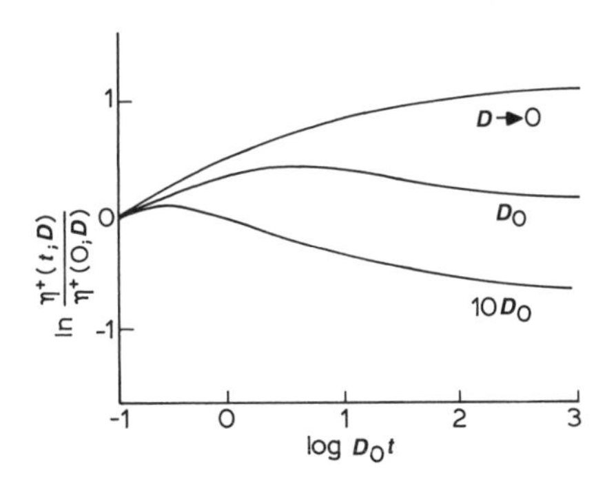

Fig. 5.4. Typical non-linear behaviour of stress growth function after sudden imposition of rate of shear D. D_0 is arbitrary (from BA & H).

steady state viscosity is usually approached when the strain is between 10 and 20.

5.1.5 Creep and Constrained Recoil

If we impose a constant shear stress t_{12}^0, at times $t \geqslant 0$, and measure the corresponding shear strain $e_{12}(t; t_{12}^0)$, we can define the creep compliance $J(t; t_{12}^0)$ as

$$J(t; t_{12}^0) = e_{12}(t; t_{12}^0)/t_{12}^0 \tag{5.1.16}$$

In the limit $t \to \infty$, we expect to recover viscometric flow, with

$$J(t;\, t_{12}^0) \to J_e^0(t_{12}^0) + t/\eta(t_{12}^0) \tag{5.1.17}$$

$\eta(t_{12}^0)$ being the simple-shear viscosity expressed as a function of shear stress. For small stresses and strains, we are again in the region of linear viscoelasticity, and we can write

$$\underset{t_{12}^0 \to 0}{\text{Lt}}\, J(t;\, t_{12}^0) = J(t) = J_e^0(t) + t/\eta_0 \tag{5.1.18}$$

where $J_e^0(t)$ is the elastic, recoverable, deformation and t/η_0 is the viscous irreversible part.

We note that a Maxwell element would involve a discontinuous jump in J at $t=0$, i.e. $J_e^0(t) = H(t)/G_e^0$, G_e^0 being the shear modulus, while $\eta_0 \to \infty$ for a viscoelastic solid.

The linear creep compliance is related to the relaxation modulus by the convolution integral

$$\int_0^t J(t')G(t-t')\,\mathrm{d}t' = t \tag{5.1.19}$$

If we consider the reverse operation, namely remove the previously constant stress t_{12}^0 at $t=0$, then a constrained elastic recoil function can be defined by

$$\frac{e_{12}(t;\, t_{12}^0) - e_{12\infty}(t_{12}^0)}{e_{12}(0;\, t_{12}^0) - e_{12\infty}(t_{12}^0)} = \varepsilon_{\text{rec}}(t;\, t_{12}^0) \tag{5.1.20}$$

Figure 5.5 shows an example of the dependence of ε_{rec} on t and t_{12}^0, from which it can be seen—as with the stress relaxation—that the recovery cannot be characterized by a single retardation time, although for long times an effective largest retardation time may become relevant. This effective maximum retardation time decreases as t_{12}^0 increases. The maximum recovery is

$$e_{\text{rec}}(t_{12}^0) = e_{12}(0;\, t_{12}^0) - e_{12\infty}(t_{12}^0) \tag{5.1.21}$$

Values of e_{rec} increase monotonically with t_{12}^0 but seem to have a limiting value between 2 and 4.

It can be shown that the behaviour described in Subsections 5.1.1–5.1.4 above (except the very high $G'(w)$ as $w \to \infty$) is all compatible with Maxwell-type models, particularly those described by (3.3.30) where a discrete sum over relaxation times u_N replaces the relaxation time spectrum $\mathrm{H}(u)$. However, if we consider creep or recoil experi-

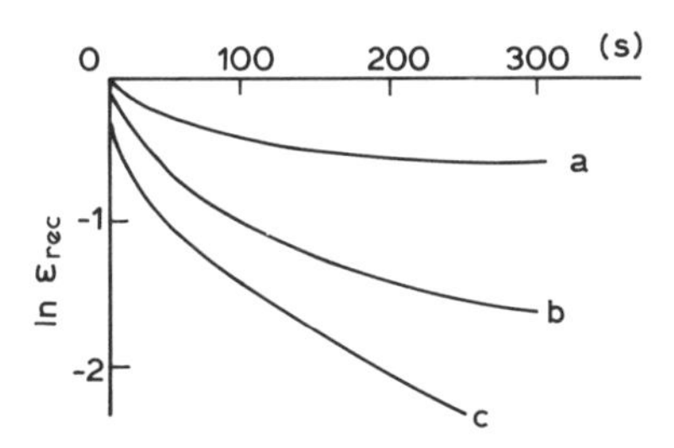

Fig. 5.5. Example of constrained elastic shear recovery after steady-state shearing at shear stress t_{12}^0. Final recovery is given by e_{rec} in absolute shear units. (From Meissner, 1975.)

	a	*b*	*c*
t_{12}^0 (Pa)	10^2	10^3	10^4
e_{rec}	0·093	0·46	1·51

ments (see below), where the stress rather than the deformation rate is prescribed, we find that real polymers do not display the capacity for instantaneous elastic deformation (e.g. $J_e^0(0)>0$) that is required by the Maxwell models. This was discussed in Chapter 3, where it was noted that the required retardation effect is achieved by putting $\alpha_M=0$ in (3.3.32), leading to a generalization of the Jeffreys fluid model.

5.2 RHEOLOGICAL MODELS

The approach taken in this book is largely based on continuum mechanics with the derivation of constitutive equations for polymer melts as one important preliminary objective. As explained earlier, proposed model relations have been based, not only on general continuum mechanical principles, but also on empirical observation and molecular theories. We now turn our attention to those few detailed investigations that have concentrated on simple well-characterized samples and have sought to test specific constitutive models for them, based on relatively simple molecular models, against as wide a range of rheological measurements as possible for the specified samples.

There are two major sets of experimental evidence that are currently relevant:

(1) Those accumulated by Vinogradov and co-workers in Moscow and published over the last twenty years (see Vinogradov & Malkin, 1980), primarily on PIB (polyisobutylene) but including BR (butyl rubber), PB, EPR (ethylene propylene rubber), and PI among traditional elastomers and PS, PE and PP among the thermoplastics.

(2) Those obtained by members of an IUPAC (Macromolecular Division) Working Party under Meissner,† and published largely by West German workers (Laun, Wagner, Münstedt) and Meissner himself, on LDPE.

The former group's work has been used as data for many proposed models, but has been most fully interpreted by Leonov, whose latest work deserves special discussion, in that it represents an unusual blend of the mechanistic and thermodynamic approaches, expressed in unexceptionable continuum mechanical form.

The latter group's work represents the most determined effort so far to provide reliable data for a thermoplastic of commercial importance and to relate the rheological observation to processing and end-use properties.

One consequence of the collaborative IUPAC investigation was to demonstrate just how difficult it is to obtain accurate and reproducible data, thereby casting considerable doubt on much published work on normal-stress differences and other viscoelastic data. Certainly, it warns against too faithful an attempt to reproduce limited rheological data exactly by lengthy mathematical adjustments to crude constitutive models.

In both cases, transient elongational flow data were obtained because these provided sensitive qualitative tests of model theories. Although in many cases of constant rate of deformation, particularly at high rates of deformation, a steady state was not achieved, large recoverable elastic strains (usually quoted in the Hencky measure, which is the natural logarithm of the linear engineering strain) were obtained. As for the case of simple shear, these recoverable strains appear to be monotonically dependent upon time and deformation rate and to rise to a maximum value between 2 and 8 (in engineering strain). Further details can be obtained from a recent research monograph by Petrie (1979). More recently, a desire to select a general

† See Meissner (1975).

constitutive equation suitable for polymer melts under all conditions of deformation history has led to a wide range of different experiments (transient and steady shear, transient elongational, oscillatory shear, large transient strains) being conducted on a single polymer sample by various workers. See for example, Tsang & Dealy (1981), Dashner & Van Arsdale (1981).

5.2.1 The Leonov–Vinogradov Model

The formal mathematical model is described in Leonov (1976) and applied in Leonov *et al.* (1976). A key concept is that of recoverable elastic strain, which embodies the structural, anisotropic, or elastic departure of the material from an equilibrium, isotropic, undeformed state. The dependence of instantaneous rheological response on past history of deformation, which is the characteristic feature of a simple fluid, is here described completely by its (equivalent) instantaneous state of elastic strain. For the most general form chosen by Leonov, the elastic strain refers to a discrete set of N separate internal structural elements. These are shown diagrammatically in a reduced one-dimensional form in Fig. 5.6 as N Maxwell elements; deformations of the spring-like components are retarded by the single viscous element shown as 0 in parallel with them. The total elastic strain energy

$$W = \sum_{k=1}^{N} W_k(I_{\mathbf{c}_k}, I_{\mathbf{c}_k^{-1}}) \tag{5.2.1}$$

is the sum of N components. Each (kth) element is described in terms

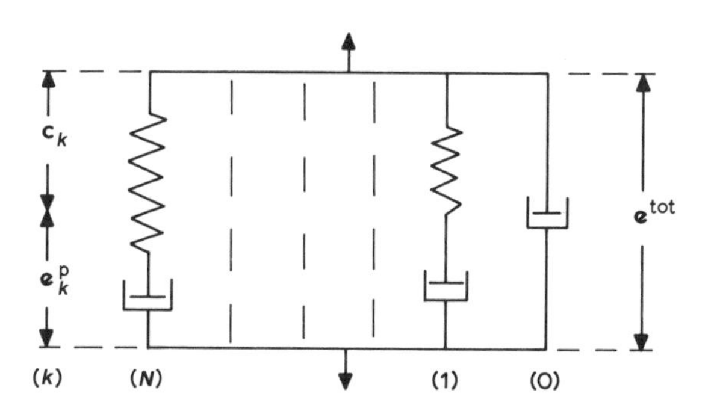

Fig. 5.6. Simple diagrammatic representation of Leonov model in mechanistic terms, showing total deformation $\int \dot{\mathbf{e}} = \mathbf{e}^{tot}$ as the sum of an elastic deformation $\mathbf{c}_k$ and an irreversible deformation $\int \dot{\mathbf{e}}_k^p = \mathbf{e}_k^p$ for each element k.

of a strain tensor $\mathbf{c}_k$ ($\mathbf{B}_R(t) = \mathbf{F}_R^T(t) \cdot \mathbf{F}_R(t)$ in the notation of A & M) for its incompressible elastic deformation relative to a virtual relaxed undeformed state, and the associated strain energy function W_k is taken to be a function of the invariants of $\mathbf{c}_k$. Since

$$\det \mathbf{c}_k = 1 \tag{5.2.2}$$

$$I_1 = I_{\mathbf{c}_k} = \operatorname{tr} \mathbf{c}_k \quad \text{and} \quad I_{-1} = I_{\mathbf{c}_k^{-1}} = \operatorname{tr} \mathbf{c}_k^{-1} \tag{5.2.3}$$

are a suitable pair.

The bulk rate of deformation, $\mathbf{D}$, is supposed to be relevant for each element, and can be split into an elastic portion, $\mathbf{D} - \mathbf{D}_k^{\mathrm{p}}$, and a dissipative, irrecoverable portion $\mathbf{D}_k^{\mathrm{p}}$, defined by

$$\begin{aligned}\mathbf{D}_k^{\mathrm{p}} = \frac{1}{2\eta_k}\Bigg[&(\mathbf{c}_k - \tfrac{1}{3}I_1\mathbf{1})\left\{\frac{\partial W_k(I_1, I_{-1})}{\partial I_1} + \frac{\partial W_k(I_{-1}, I_1)}{\partial I_1}\right\} \\ &- (\mathbf{c}_k^{-1} - \tfrac{1}{3}I_{-1}\mathbf{1})\left\{\frac{\partial W_k(I_{-1}, I_1)}{\partial I_{-1}} + \frac{\partial W_k(I_1, I_{-1})}{\partial I_{-1}}\right\}\Bigg]\end{aligned} \tag{5.2.4}$$

The strain tensor $\mathbf{c}_k$ obeys the evolution equation

$$\mathring{\mathbf{c}}_k = \mathbf{c}_k \cdot (\mathbf{D} - \mathbf{D}_k^{\mathrm{p}}) + (\mathbf{D} - \mathbf{D}_k^{\mathrm{p}}) \cdot \mathbf{c}_k \tag{5.2.5}$$

while the stress is given by

$$\mathbf{T}^{\mathrm{E}} = 2\left(\eta_0 - \sum_{k=1}^{N} \eta_k\right)\mathbf{D} + 2\sum_{k=1}^{N}\left[\frac{\partial W_k(I_1, I_{-1})}{\partial I_1}\mathbf{c}_k - \frac{\partial W_k(I_1, I_{-1})}{\partial I_{-1}}\mathbf{c}_k^{-1}\right] \tag{5.2.6}$$

η_k is the viscosity relevant in the kth relaxation mechanism, with which will be associated a relaxation time λ_k, while η_0 is the 'Newtonian' viscosity as $D \to 0$.

For small deformation, we suppose

$$W_k = \mu_k(I_1 - 3), \qquad 2\mu_k = \eta_k/\theta_k \tag{5.2.7}$$

in which case the system of equations (5.2.1)–(5.2.7) is a simple extension of a linear viscoelastic model similar to those defined by (3.3.32) with $\alpha_M = 0$. However, by letting W_k take other forms, non-linear elastic behaviour of the type well investigated for lightly cross-linked elastomers can be introduced. Temperature variations can be treated by making W_k linear in T. An interaction between the high orientation characteristic of large elastic strains and the temperature-dependent relaxation viscosities η_k can be added by letting

$$\eta_k = \eta_k^0(T)\exp\frac{\beta(\partial W/\partial I_1 + \partial W/\partial I_{-1})}{2\mu_1(T)} \tag{5.2.8}$$

where β is a constant and $\mu_1(T)$ the high elastic modulus defined by (5.2.7).

The arguments used to reach these results are best sought in the original paper. They have been developed in such a way that the basic parameters η_k, μ_k, η_0 can be obtained from linear viscoelastic measurements. The relation (5.2.5) carries the model forward into the quasi-linear region while observations of stress relaxation, elastic recoil and creep in the fully non-linear region provide forms for W_k and allow (5.2.8) to be tested.

The model is used (Leonov *et al.*, 1976) to match observed data on 40% butyl rubber solution in transformer oil (a concentrated solution that displays most of the behaviour of polymer melts), and fair agreement is obtained using a model with $N=2$. The authors show that even $N=1$ yields reasonable predictions. It may be shown that, if the approximation

$$\mathbf{1} - \mathbf{c}^{-1} \approx \mathbf{c} - \mathbf{1} \tag{5.2.9}$$

is made, then the model with $N=1$ becomes the Jeffreys model

$$\mathring{\mathbf{T}}^{\mathrm{E}} + \frac{2\mu_1}{\eta_1}\mathbf{T}^{\mathrm{E}} = 2(\eta - \eta_1)\left[\mathring{\mathbf{D}} + \frac{2\mu_1\eta}{\eta_1(\eta - \eta_1)}\mathbf{D}\right] \tag{5.2.10}$$

It is not obvious what correspondences can be obtained, for the non-linear situation, with other models we have described in Chapter 3.

For the case of steady state shear, as $D \to \infty$, it can readily be seen that

$$\begin{gathered} c_{12}^{(k)} \to 1, \qquad c_{11}^{(k)} \to 2\theta_k D, \qquad c_{22}^{(k)} \to (\theta_k D)^{-1} \\ \eta \to \left(\eta_0 - \sum_{k=1}^{N} \eta_k\right), \qquad t_{11} - t_{12} \to \text{const} \end{gathered} \tag{5.2.11}$$

while for the case of steady elongation

$$c_{11}^{(k)} = (c_{22}^{(k)})^{-2} = (c_{33}^{(k)})^{-2} = 6\theta_k\dot{e}, \qquad t_{11} - t_{22} \to 3\eta_0\dot{e} \tag{5.2.12}$$

The significance of these results is, in simple terms, that the elastic elements contribute fully to extensional viscosity but not at all, in the limit, to shear viscosity.

An earlier treatment (Vinogradov, 1972) was based on the idea that rapid steady shear effectively truncates the relaxation time spectrum $H(u)$. In order to demonstrate this suggestion, oscillatory shear flows of progressively greater amplitude, but constant frequency, were examined experimentally. These can, of course, be analysed also in terms of constant maximum shear rate and varying amplitude. In the latter case, one might expect the limiting situation of large amplitude and low frequency to correspond to steady shear flow. Figure 5.7 shows results for PIB at various frequencies marked 1 to 4. The portions of the curves parallel to the abscissa (log D_{max} axis) represent the range of linear viscoelasticity. The limit is seen to arise at a maximum shear of about unity after which the flow behaves in large measure as if it were in steady shear flow. Vinogradov and Leonov realized that each of the values of D_{max} yields an effective $\eta^*(w, D_{max})$ which can be decomposed into a relaxation time spectrum $H(u)$ on the assumption of linear viscoelasticity relationships. The result is shown in Fig. 5.8.

On the basis that

$$\eta = \int_0^\infty H(u)\,du \tag{5.2.13}$$

in linear viscoelasticity, the connection between increasing shear rate and decreasing viscosity (Cox–Mertz rule) can be explained in terms of

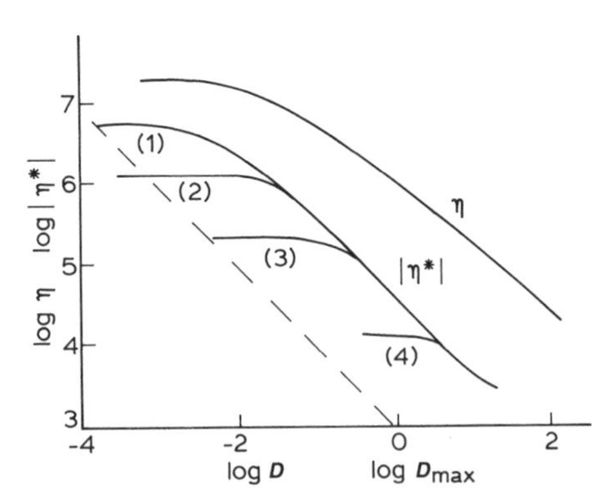

Fig. 5.7. Variation of complex viscosity modulus η^* with maximum shear rate D_{max} for fixed frequency w (i.e. variation with maximum shear for each numbered curve) compared to variation of steady shear viscosity η with shear rate D. SI units are used. log w = (1) −1·2 (2) −0·2 (3) 0·9 (4) 2·6. – – –, line of constant maximum shear. (From Vinogradov, 1971.)

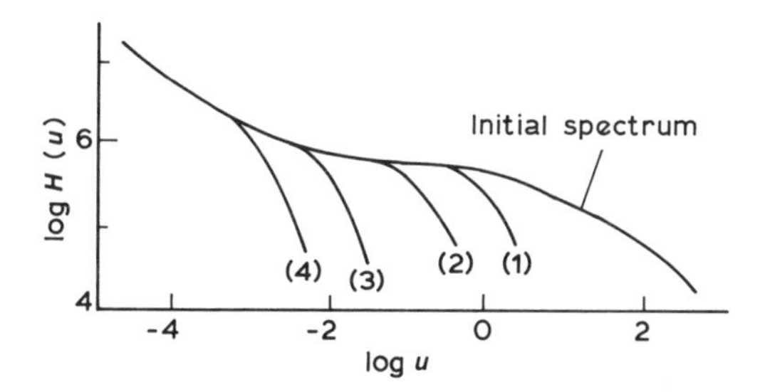

Fig. 5.8. Effect on relaxation-time spectrum $H(u)$ of increasing maximum shear rate D_{max} in oscillatory shear flow. SI units are used. $\log D_{max} =$ (1) $-1{\cdot}5$ (2) $-0{\cdot}5$ (3) $0{\cdot}5$ (4) $1{\cdot}5$.

an erosion of the high-relaxation-time end of the relaxation-time spectrum with rapid shearing. In Fig. 5.8 the actual spectra are truncated at u_{max}, that satisfies

$$D_{max}^{-1} = u_{max} = 2{\cdot}8\ \mathrm{s}^{-1} \tag{5.2.14}$$

which is chosen to give the right shift factor to superpose the $|\eta^*|$ and η curves in Fig. 5.7.

Although the explicit models are not compared with experiment at very high deformation rates, Vinogradov (1981) argues that at such rates, elastic fluids become brittle; qualitatively, they lose fluidity because of a network that forms and which raises both the characteristic relaxation time and the viscosity. This appears to be related to the W-dependent term in (5.2.8) introduced by Leonov. This phenomenon is said to be associated with melt fracture, though no convincing explanation is given.

One peculiarity of Vinogradov's presentation should be noted by the unwary. In transient shear and elongational flows, when the recoverable elastic strain is changing, the term viscosity (analogous to η^+) is applied to the dissipative part of the rate of strain only and not to the full rate of strain. This does not have an obvious analogue in the theory represented by (5.2.1)–(5.2.8).

5.2.2 Network Models

A familiar starting point for many models of polymer melts, as explained in Section 3.4 under integral models, is the theory of rubber-like elasticity, which envisages a network of polymer chains connected by a set of point junctions. In general four subchains, each a segment of a long-chain polymer molecule, meet at a junction. Under steady-state, isotropic conditions, these junctions will be distributed randomly and the chain orientations will be distributed isotropically. When such a material is deformed, there will be, in some mean sense, a homogeneous (affine) deformation of the whole structure leading to non-isotropic (rheological) stresses. If the material is to be fluid, then there cannot be a single permanent network spanning the whole bulk, as there is in a cured rubber; indeed, if we are dealing with uncross-linked linear polymers, then we suppose that all junction points are temporary and that they form, break and reform (or else move along chains) under the influence of thermal fluctuations. The basic principle of the Lodge–Yamomoto theory is that, during their lifetime, the networks based on temporary junctions deform affinely, become anisotropic and so lead to a strain-induced extra stress. The networks are strained, but display stress relaxation, and so steady uniform stresses can only result from continuous deformation; it is thus possible to associate stresses with bulk rates of deformation. This duality is best illustrated by the alternative representations we employed in describing linear viscoelasticity, i.e. eqns (3.4.1) and (3.4.3). Strain retardation, namely an inability to achieve a sudden deformation without a stress singularity, is an additional feature that must be built into the theory for either elastic solids or liquids.

This cardinal point seems to be neglected by many authors, possibly because the kinetics of junction formation and destruction lead even in the simplest model to a perfectly satisfactory steady-shear viscosity. Despite pragmatic experimental evidence to the contrary, most memory function or Maxwell-type fluid theories do allow for sudden imposition of strain at a significant level; thus, though all our experience of polymers in linear viscoelastic experiments shows them to display a transition to glassy (i.e. effectively rigid) behaviour at a low enough temperature, or using the WLF transformation, at a high enough frequency, much of the formal treatment of simple fluids relies on thermodynamic argument involving instantaneous deformation. The matter is discussed in Astarita (1975, Chapter 5) and briefly in A & M (Section 4.4 and the end of Section 6.4). The issue is a quantitative one for our purposes: whether or not 0(1) recoverable elastic strains can be instantaneously imposed on polymer melts, not whether instantaneous 0(1) strains are permissible. Models that allow the former are potentially grossly unrealistic; models that exclude the latter are not. Retardation times are shorter, perhaps by a factor of 5, than relaxation times, but both are significant, measurable and in many cases obvious.

The simplest elastic fluid model of Lodge yields a constant simple shear stress and a critical elongation rate above which a steady flow is not possible. This led to theories (De Kee & Carreau (1979), Murayama (1981)) in which the rate of formation and/or destruction of temporary junctions (cross-links or entanglements) is stress-dependent (or strain or even strain-rate dependent). This additional complication, needed to achieve closer correspondence between model predictions and observation, can be introduced directly into a structural model which deals with cross-links explicitly or indirectly by making the memory function dependent on stress, strain, or strain rate. The latter procedure, typified by the Kaye–BKZ model (3.4.9), is the more common and is usually carried out empirically.

Thus, in connection with the IUPAC investigations on LDPE, Laun (1978), Wagner (1979*a*) proposed a model

$$\mathbf{T}^{\mathrm{E}} = \int_0^{\infty} \mathring{m}(s) h(I_{\mathbf{c}_t^t}, I_{\mathbf{c}_t^{t-1}}) \mathbf{c}_t^{t-1}(s)\, \mathrm{d}s \tag{5.2.15}$$

with

$$\mathring{m}(s) = \sum_i a_i \exp(-s/\tau_i) \tag{5.2.16}$$

and $h(\,)$ a strain-dependent scalar function. $\mathring{m}(s)$ can be obtained by fitting linear viscoelastic measurements—Laun used $\tau_{i+1} = 10\tau_i$—and h from non-linear measurements. Laun sought the degenerate form of h, $\bar{h}(e_{12})$, relevant in simple shear flow, with e_{12} the shear strain. He expressed $\bar{h}$ in the form

$$\bar{h}(e_{12}) = f_1 \exp(-n_1 e_{12}) + f_2 \exp(-n_2 e_{12}) \tag{5.2.17}$$

An analogous form for extensional flow was provided by Wagner (1978). This decoupling of time and strain dependence had earlier been tested by Tschoegl and co-workers (Blatz *et al.*, 1973; Chang *et al.*, 1976*a*, *b*) using a strain measure first proposed by Seth. Many other authors have developed this theme; see particularly BA & H (Example 9.4.1 and Tables 9.4.2 and 9.4.3), who use a strain rate-dependent form for the memory function.

Marrucci Model

One of the most strongly-favoured models dealing explicitly with cross-link densities is that of Marrucci and co-workers (see e.g. Acierno *et al.*, 1976). It describes non-linear effects (for example, shear-rate reduction of viscosity) in terms of a stress-dependent increase in the rate of cross-link destruction. Thus, a non-dimensional structural scalar parameter x_i representing cross-link density is used for each of a set of elastic structures characterized by a relaxation time λ_i and a modulus G_i. The model, rather like the Leonov model, is based on a series of non-linear Maxwell elements, but has no viscous element in parallel with them. Formally, it is given by

$$\mathbf{T}^{\mathrm{E}} = \sum_{i=1}^{N} \mathbf{T}_i^{\mathrm{E}} \tag{5.2.18}$$

$$\frac{\mathbf{T}_i^{\mathrm{E}}}{G_i} + \lambda_i \overset{\nabla}{\left(\frac{\mathbf{T}_i^{\mathrm{E}}}{G_i}\right)} = 2\lambda_i \mathbf{D} \tag{5.2.19}$$

$$G_i = G_{0i} x_i, \qquad \lambda_i = \lambda_{0i} x_i^{1\cdot 4} \tag{5.2.20}$$

$$\frac{\mathrm{D}x_i}{\mathrm{D}t} = \frac{1}{\lambda_i}(1 - x_i) - a\frac{x_i}{\lambda_i}\left[\frac{\frac{1}{2}\,\mathrm{tr}\, T_i^{\mathrm{E}}}{G_i}\right]^{\frac{1}{2}} \tag{5.2.21}$$

It is assumed that, in linear viscoelastic experiments, $\mathrm{tr}\,\mathbf{T}_i^{\mathrm{E}}$ will be negligibly small, and that $x_i = 1$. Hence, G_{0i}, λ_{0i} are the relevant linear viscoelastic parameters that follow from G^*. The dependence of G_i and λ_i on x_i is chosen to match the known dependence of modulus and zero-shear rate viscosity on molecular weight.

The sharply non-linear nature of the model is introduced by the rate equation (5.2.21) for the entanglement density with the destructive term increasing with the relevant elastic stored energy $\frac{1}{2}\,\mathrm{tr}\,\mathbf{T}_i^{\mathrm{E}}$. Only the parameter a is free to vary for any particular polymer to match non-linear observations. The model is said to fit a wide range of transient observations in shear and elongation. Earlier versions (Marrucci *et al.*, 1973) used a different form for the last term on the right-hand side of (5.2.21). De Cleyn & Mewis (1981) suggest a yet different term, while the model discussed in Liu *et al.* (1981) is clearly similar.

The model has obvious similarities to the Leonov model, both of them attaching importance to elastic energy functions, which Leonov treats as given in terms of the structural strains, while Marrucci equates them to the traces of the equivalent stresses (though a more usual form would have used the second invariant). It is also related to the simpler White–Metzner fluid (3.3.35) which allows the relaxation time to be strain-rate dependent. Jongschaap (1981) has shown how the same equations can be derived using the basic Lodge–Yamomoto network theory.

Further generalization of the basic network model has been proposed by Ronca (1976*b*) on the basis of a tensor structural parameter, but it seems too early to say whether his model will have significant advantages over others.

A range of other models has been proposed (Johnson & Segalman, 1977; Phan Thien & Tanner, 1977), but not specifically for polymer melts, that invoke the notion of slip between structural elements and the mean deforming motion, i.e. the affine deformation postulate for each and every averaged structure is rejected. Their origin seems to lie more in theories of loosely inter-penetrating polymer molecules (in solution) than in dense network theories. The model of Phan Thien and Tanner with its small number of arbitrary constants is currently being tested against experiment. Bird (1982) has surveyed the role of kinetic theory in providing constitutive equations for polymer melts and in particular in a series of papers with Curtiss and other coworkers (Curtiss & Bird, 1981) has investigated an equation based on a bead-rod model that is a generalization of the Doi–Edwards model. Use of this model is in its early stages.

5.2.3 Non-isothermal Models

We discussed earlier the proposal (in connection with the WLF equation) that the viscoelastic behaviour of polymers could be described by their having a single temperature-dependent timescale, so that all rate processes could be made time-independent by a simple scaling of time with the relevant temperature-dependent material parameter. Such materials are called thermorheologically simple. Let $a(T)$ be the scaling parameter. Then, if the material history in real time t is associated with a varying temperature history $T(t)$, we define a reduced time (or pseudo time) z_t with respect to a constant temperature T_0 by the simple transformation

$$\frac{\mathrm{d}z_t}{\mathrm{d}t}=\frac{a[T(t)]}{a[T_0]}=a^*(T;\,T_0) \tag{5.2.22}$$

Clearly, all of the differential and integral constitutive models given in Chapter 3 can be simply converted into non-isothermal models by using z_t instead of t. For example,

$$\begin{gathered}\mathbf{G}_t(t') \to \mathbf{G}_{z_t}(z_t') \\ \mathbf{D}(t), \mathbf{W}(t) \to a^*\mathbf{D}(z_t), \mathbf{W}(z_t) \\ \frac{\mathrm{D}}{\mathrm{D}t} \to a^* \frac{\mathrm{D}}{\mathrm{D}z_t}\end{gathered} \tag{5.2.23}$$

and so eqn (3.3.25) can be written

$$\mathbf{T}^{\mathrm{E}}(z_t) + \lambda_1 \overset{\square}{\mathbf{T}}{}^{\mathrm{E}}(z_t) = 2\eta_0\{\mathbf{D}(z_t) + \lambda_2 \overset{\square}{\mathbf{D}}(z_t)\} \tag{5.2.24}$$

where

$$\begin{aligned}\overset{\square}{\mathbf{J}}(z_t) = {} & \frac{\mathrm{D}\mathbf{J}}{\mathrm{D}z_t} + \mathbf{J} \cdot \mathbf{W}(z_t) - \mathbf{W}(z_t) \cdot \mathbf{J} \\ & + a\{\mathbf{J} \cdot \mathbf{D}(z_t) + \mathbf{D}(z_t) \cdot \mathbf{J}\} + b\{\mathbf{J} : \mathbf{D}(z_t)\}\mathbf{1} + c\mathbf{D}(z_t)\operatorname{tr}\mathbf{J}\end{aligned} \tag{5.2.25}$$

or, by the substitution of (5.2.23) as

$$\mathbf{T}^{\mathrm{E}}(t) + \frac{\lambda_1}{a^*} \overset{\square}{\mathbf{T}}{}^{\mathrm{E}}(t) = \frac{2\eta_0}{a^*}\left\{\mathbf{D}(t) + \frac{\lambda_2}{a^*} \overset{\square}{\mathbf{D}}(t)\right\} \tag{5.2.26}$$

when $\overset{\square}{\mathbf{J}}(t)$ has the value given by (3.3.23) and (3.3.24). We see that the form (5.2.26) could have been obtained from (5.2.24) by simply reducing each of λ_1, λ_2 and η_0 by the factor a^*, which is physically what the original postulate of thermorheological simplicity means.

Finally, for the integral model (3.4.6) we would obtain:

$$\begin{aligned}\mathbf{T}^{\mathrm{E}}(z_t) = {} & \frac{1}{a^*}\int_0^\infty m(u) G_{z_t}^{z_t}(u)\,\mathrm{d}u + \frac{1}{(a^*)^2}\int_0^\infty\int_0^\infty \{\alpha(u_1, u_2)\mathbf{G}_{z_t}^{z_t}(u_1) \cdot \mathbf{G}_{z_t}^{z_t}(u_2) \\ & + \beta(u_1, u_2)\operatorname{tr}[\mathbf{G}_{z_t}^{z_t}(u_1)]\mathbf{G}_{z_t}^{z_t}(u_2)\}\,\mathrm{d}u_1\,\mathrm{d}u_2\end{aligned} \tag{5.2.27}$$

where u_1 and u_2 are dummy time variables in the z_t scale.

Some of the not very abundant experimental work undertaken to test non-isothermal theories has been reported by Bogue and co-workers (Matsui & Bogue, 1976, 1977; Matsumoto & Bogue, 1977; Carey *et al.*, 1980) who stretched polystyrene samples at a constant elongation rate. They used a strain-rate-dependent memory function, rather like those mentioned in BA & H (referred to above), in the form

$$\mathbf{T}^{\mathrm{E}} = \sum_{n=1}^{N} G_n^0 \frac{T(t)}{T_0} \int_{-\infty}^{t} \frac{\exp - \int_{t'}^{t} \left[\frac{\mathrm{d}t''}{\tau_n^*(t'')} \right]}{\tau_n^*(t')} \mathbf{C}_t^{-1}(t')\, \mathrm{d}t' \tag{5.2.28}$$

where

$$\tau_n = a^*(T; T_0)\tau_n^0 \tag{5.2.29}$$

and

$$\tau_n^* = \frac{\tau_n}{1 + bII_{\mathbf{D}}^{\frac{1}{2}}\tau_n} \tag{5.2.30}$$

(5.2.29) is the same as (5.2.22) and implies that the material is thermorheologically simple. Equation (5.2.30) makes the timescale shear-rate dependent and is reminiscent of the Vinogradov–Leonov arguments about relaxation times being cut off (or reduced) as a result of orientation. It is worth noting that the elastic moduli G_n of the components of the stress tensor are made temperature-dependent according to

$$G_n = G_n^0 T/T_0 \tag{5.2.31}$$

which is a prediction of the simplest theories of rubber-like elasticity. This would mean that the viscosity η would scale as T/a^*T_0 in a differential model (see, for example (5.2.25)) rather than as $1/a^*$. In practice, since T and T_0 in (5.2.31) are measured on an absolute scale of temperature, the variation due to T/T_0 will be small compared to that due to a^*.

Bogue and Matsumoto concluded that the observations showed a dependence on $\mathrm{D}T/\mathrm{D}t$ as well as on T alone. This possibility is effectively included in the more general theory of Crochet & Naghdi (1972, 1974) who suppose that the relationship between pseudo-time z_t and true time t is dependent on the temperature history $T(t-u)$. Thus, they write, as the relevant time-like variable z_s, for past times s, t being the current time,

$$z_s = \mathop{\mathscr{X}}_{u=0}^{\infty} [T(t-u); s] \tag{5.2.32}$$

with the constraint that, in the isothermal case,

$$\mathop{\mathscr{X}}_{u=0}^{\infty} [T_0; s] = s \tag{5.2.33}$$

They also take account of thermal strain by supposing a situation in which the material had experienced the same temperature history under stress-free conditions. Thus, from a general rheological equation of state (of the form (3.2.14))

$$\mathbf{T}^{\mathrm{E}} = \rho \mathbf{F}_{\mathrm{R}} \cdot \mathop{\mathscr{T}}_{s=0}^{\infty} [\mathbf{C}_{\mathrm{R}}(t-s), T(t-s); \mathbf{C}_{\mathrm{R}}(t), T(t)] \cdot \mathbf{F}_{\mathrm{R}}^{\mathrm{T}} \quad (5.2.34)$$

the subscript R defining a reference state and $\mathscr{T}$ being a constitutive functional, they define the thermal strain $\bar{\mathbf{C}}_{\mathrm{R}}$ according to

$$\mathop{\mathscr{T}}_{s=0}^{\infty} [\bar{\mathbf{C}}_{\mathrm{R}}(t-s), T(t-s); \bar{\mathbf{C}}_{\mathrm{R}}(t), T(t)] = 0 \quad (5.2.35)$$

which, formally will have solutions

$$\bar{\mathbf{C}}_{\mathrm{R}}(t) = \mathop{\mathscr{C}}_{s=0}^{\infty} [T(t-s); T(t)] = \{\rho_0/\bar{\rho}(t)\}^{\frac{2}{3}}\mathbf{1} \quad (5.2.36)$$

which can obviously be described in terms of the density $\bar{\rho}(t)$. This leads to a reduced (isothermal) strain tensor $\mathbf{C}_{\mathrm{R}}^{*}$ defined by

$$\mathbf{C}_{\mathrm{R}}^{*} = (\bar{\rho}/\rho_0)^{\frac{2}{3}}\mathbf{C}_{\mathrm{R}}$$

with (5.2.37)

$$\mathbf{F}_{\mathrm{R}}^{*} = (\bar{\rho}/\rho_0)^{\frac{1}{3}}\mathbf{F}_{\mathrm{R}}$$

and so (5.2.34) can be written successively as

$$\mathbf{T}^{\mathrm{E}} = \rho \mathbf{F}_{\mathrm{R}}^{*} \cdot \mathop{\mathscr{T}'}_{s=0}^{\infty} [\mathbf{C}_{\mathrm{R}}^{*}(t-s), T(t-s); \mathbf{C}_{\mathrm{R}}^{*}(t), T(t)] \cdot \mathbf{F}_{\mathrm{R}}^{*\mathrm{T}} \quad (5.2.38)$$

where $\mathscr{T}'$ is a new functional derived from $\mathscr{T}$, and then

$$\mathbf{T}^{\mathrm{E}} = \rho \bar{\mathbf{F}}_{\mathrm{R}}^{*} \cdot \mathop{\mathscr{T}^{*}}_{s=0}^{\infty} [\mathbf{C}_{\mathrm{R}}^{*}(t-z_s); \mathbf{C}_{\mathrm{R}}^{*}(t)] \cdot \mathbf{F}_{\mathrm{R}}^{*T} \quad (5.2.39)$$

where the functional $\mathscr{T}^{*}$ embodies the equivalent of a thermorheological simple fluid principle. Crochet (1975) applied this theory in the special form

$$\mathrm{d}s = \phi[T(t-z_s)]\,\mathrm{d}z_s \quad (5.2.40)$$

which is essentially (5.2.22), to show that an additional term arises, in the non-isothermal extension of (5.2.24), namely $\dfrac{2\eta_0\lambda_2}{a^{*3}}\dfrac{\mathrm{d}a^{*}}{\mathrm{d}T}\dfrac{\mathrm{D}T}{\mathrm{D}t}\mathbf{D}$.

A more specific treatment based on dumb-bell models by Gupta & Metzner (1982) introduces terms in $\mathrm{D}(\ln T)/\mathrm{D}t$ into the Maxwell model (3.3.26). This brings us back to the conclusion of Bogue and Matsumoto.

If we consider the physical meaning of the Crochet–Naghdi theory, we are led to compare it with the much older free-volume interpretation of pressure and temperature dependence of viscoelastic parameters (see Ferry, 1980, Chapter 11). This suggests that the functional dependence (5.2.32) of z_s upon the history of $T(t-u)$ can be interpreted in terms of a slow approach to thermodynamic equilibrium (defined by density) and that the difference between (5.2.22) and (5.2.32) can be ascribed to using an equilibrium and a non-equilibrium free-volume, respectively. Thus, we might combine them in the form

$$\mathrm{d}z_0 = b(T, \bar{\rho}; T_0, \rho_0)\,\mathrm{d}t \tag{5.2.41}$$

where

$$\frac{\mathrm{D}\bar{\rho}}{\mathrm{D}t} = \dot{\bar{R}}(\bar{\rho}, \rho_0, T, p) \tag{5.2.42}$$

Here we have taken $\dot{\bar{R}}$ to be strain or strain-rate independent for convenience. It need not be so. It should be remembered that ρ/ρ_0 differs little from unity in most cases of interest in polymer processing, the maximum changes being of order 5% within the fluid state. Thus, in many parts of the mechanical theory, the incompressibility assumption is perfectly adequate. However, as far as free-volume effects on rheological parameters are concerned, changes of the order of fractions of 1% may have substantial consequences. This may be compared with the Boussinesq approximation in gravitational convection of dense non-isothermal fluid systems, where the only allowed effect of density variations is on the body force (gravitational). An extension of the Marrucci model of Subsection 5.2.2 to include such free-volume change is given in Lamantia & Titomanlio (1979). An interpretation of temperature-jump experiments based on apparent activation energies is given in Hooley & Cohen (1978).

5.2.4 Thermodynamic Treatment

We conclude this section with a mention of entropic constitutive equations for fluids with fading memory. A brief treatment is given in A & M (Section 4.4) following work of Coleman (1964*a*, *b*) and has been elaborated in various articles (e.g. Astarita, 1975, Chapter 6; Astarita & Sarti, 1976).

Briefly, a single thermodynamic function is proposed for the Helmholtz free energy

$$A(t) = \underset{s=0}{\overset{\infty}{a}} [T^t(s), \mathbf{F}_R^t(s); T(t), \mathbf{F}_R(t)] \tag{5.2.43}$$

where some reference configuration R is assumed for each fluid element. It is important in the theory that at this stage A be an explicit function of the instantaneous values $T(t)$ and $\mathbf{F}_R(t)$ as well as being a functional over past time of the history $T^t(s)$, $\mathbf{F}_R^t$. By an extension of the thermodynamics of rubbery materials, the other thermodynamic and mechanical functions are obtained as

entropy $$S = -\frac{\partial a}{\partial T} \tag{5.2.44}$$

internal energy $$U = a - T\frac{\partial a}{\partial T} \tag{5.2.45}$$

stress $$\mathbf{T} = \rho \mathbf{F}_R \cdot \left(\frac{\partial a}{\partial \mathbf{F}_R}\right)^T \tag{5.2.46}$$

It should be emphasized that purely viscous fluids are not included in this treatment because the relation (5.2.46) presupposes the imposition of an instantaneous jump in strain at time t, without change in the history $\mathbf{F}^t(s)$; in the same way (5.2.44) and (5.2.45) presuppose an instantaneous change in temperature. This need not be regarded as too serious a constraint; a similar situation arises in the theory of linear viscoelasticity, where both Hookean solids and Newtonian liquids can be described as perfectly regular limiting states of a general material. What is required is care when moving to the limit in a formal mathematical sense.

It should also be noted that in many of our earlier constitutive equations for fluids, it was argued that the obvious reference state should be the instantaneous state, in which case

$$\mathbf{F}_R(t) = \mathbf{1} \tag{5.2.47}$$

and so an apparent difficulty in interpreting (5.2.46) would arise. This matter is succinctly dealt with by Sarti (1977) when reconciling the Crochet–Naghdi theory (partly described above) and the Astarita–Sarti approach. For an incompressible elastic fluid, displaying the time–temperature superposition principle implied by (5.2.32) and (5.2.39) and purely conformational entropic elasticity, the relevant thermodynamic function can be written

$$A = A_0(T) - T\mathcal{S}^*_{s>0}[\mathbf{C}^t_t(z_s)] \tag{5.2.48}$$

$$U = U(T) \tag{5.2.49}$$

$$S = S_0(T) + \mathcal{S}^*_{s>0}[\mathbf{C}^t_t(z_s)] \tag{5.2.50}$$

$$\mathbf{T}^{\mathrm{E}} = -\rho T \mathbf{F}_{\mathrm{R}}(t) \cdot \frac{\partial}{\partial \mathbf{F}^T_{\mathrm{R}}(t)} \mathcal{S}^*_{s>0}[\mathbf{F}_{\mathrm{R}}^{-1T}(t) \cdot \mathbf{C}^t_{\mathrm{R}}(z_s) \cdot \mathbf{F}_{\mathrm{R}}^{-1}(t)] \tag{5.2.51}$$

Here A_0, U, S_0 are functions which are analogous to those for an unstructured liquid, while fading memory effects are described by a single entropy functional

$$\mathcal{S}^*_{s>0}[\mathbf{C}^t_t(z_s)]$$

with z_s related to t, s by a second functional

$$\mathcal{Z}_{s>0}[T^t(s); s]$$

What becomes clear is that the partial derivative involved in (5.2.51) is effectively a Fréchet derivative as far as $\mathcal{S}^*$ is concerned.

Part II

COMPLEX FLOW FIELDS

Chapter 6

General Procedures

So far we have only considered prescribed uniform flow fields with a simple element of unsteadiness. We now consider flows of greater complexity, for which the conservation equations of mass, momentum, and energy, together with specified constitutive equations relating stress, heat flux and free energy to the other variables (reference density, temperature and deformation histories) provide a sufficient set of field equations. The actual flows that arise will depend upon the imposed boundary conditions and can be sought as solutions of the relevant field equations.

The general procedure will be as follows:

(1) Choose a reference frame that best suits the boundary conditions.

(2) Choose, within that reference frame, a coordinate system that is best suited to the geometrical symmetries of the boundaries. In almost all cases, this means an orthogonal coordinate system in terms of which the various field equations can be written as scalar equations for the (scalar) physical components of the field variables. For any of the explicit constitutive equations given earlier, this will lead to coupled sets of algebraic, differential and integro-differential equations, usually non-linear. The symmetries of the problem are normally used to impose the maximum symmetry on the sought solution with consequential simplification of the relevant equations.

(3) Make the equations dimensionless in such a fashion that the various dimensionless variables are, or are expected to be, of order unity within the field of interest. This leads to a finite set of independent dimensionless groups that include:

(a) geometrical parameters—arising from the configuration of the boundaries;

(b) material parameters—arising from the constitutive equations used for the type of material in question;
(c) operating parameters—arising from the velocities, stresses, temperatures and heat fluxes imposed at the boundaries;
(d) universal parameters—like the gravitational acceleration.

If the original problem were specified algebraically, in the sense that the actual values of the various parameters 3(a)–3(c) were arbitrary, in a numerical sense, then the full set of K independent dimensionless groups ($\mathcal{N}_k$) could be regarded as coordinates of a K-parameter space in which mathematical solutions are sought. The advantage of treating the problem in terms of the $\mathcal{N}_k$ ($k=1, \ldots, K$) instead of the original dimensional parameters P_l ($l=1, \ldots, L$) is twofold. First, $K<L$ and presentation of results is simpler. Secondly, the order of magnitude (e.g. $\ll 1$, $0(1)$ or $\gg 1$) of any $\mathcal{N}_k$ can often be readily interpreted in both mathematical and physical terms.

It should be noted that the dimensionless groups obtained in this process are really inseparable from: (a) the basic (parametrized) geometry imposed in the problem, (b) the form of constitutive equation adopted for the material and (c) the form of the imposed boundary conditions. However, it is obvious that similar ratios of parameters will arise in various problems, differing in basic geometry, material behaviour and boundary conditions, and so the same name will be given to members of such a class of dimensionless groups. In many cases, this is convenient and helpful because similar physical and mathematical consequences are associated with varying magnitudes of such a dimensionless group in differing situations; this is not always so, and therefore care should be taken in generalizing results, proved for one simple system, to related but more elaborate systems.

It proves helpful in our treatment of polymer-processing flows to classify our systems in terms of a limited number of these dimensionless groups. Some involve only geometrical parameters, and so are angles or ratios of lengths; some can be interpreted as ratios of time intervals, and need not necessarily involve material parameters; but most involve material parameters as well as operating and geometrical parameters. The main advantage of this approach is that it tells us when to neglect various physical effects (and thus the related terms in our mathematical equations).

(4) Seek physically relevant approximate or asymptotically-limiting solutions of these dimensionless equations with their dimensionless

boundary conditions. In practice, simplifications arise when a significant number of the dimensionless groups (or their inverses) become very small.

Before the advent of high-speed digital computers, there was seldom any alternative to this step. Exact analytical solutions are in general only available for linear systems, and asymptotic expansions involving small parameters are still the preferred means for tackling non-linear systems. However, there is an alternative school of thought that advocates large-scale computational (numerical) solution for engineering problems. Particularly strong claims are made for finite-element methods which allow great flexibility in the specification of boundary geometry and which can be combined with graphical display facilities. (Aspects of this are discussed in Pearson & Richardson, 1983 in the context of polymer processing.)

The most symmetrical solutions consistent with the boundary conditions have already, according to (2) above, been assumed, unless there is clear evidence to the contrary. Whether this leads to possible solutions depends not only on the boundary conditions but also on the field equations; non-linearities in the latter can lead to loss or change of symmetry in their solutions. Examples of this will be mentioned in what follows. In some cases the less symmetrical solutions are unique; in others they are the result of having multiple solutions due to bifurcation at critical values of various of the dimensionless groups appearing in the field equations and boundary conditions.

If the scalar subset of equations used to obtain solutions already contains an implied symmetry in the solution, then neither analytical nor numerical methods will necessarily yield a relevant solution because alternative more complex solutions that would arise in practice cannot be generated.

(5) Consider the stability of the solutions so generated. If steady-state solutions have been obtained, stability of the solutions to small temporal and spatial disturbances can often be examined. This last step is sometimes a first stage in the study of a fully-bifurcated steady-state solution.

The procedure (1)–(5) is quite standard in traditional fluid mechanics though it is rarely enunciated formally in any texts on the subject. Geometrical considerations are seldom stressed, non-dimensionalization is often done haphazardly, while stability or bifurcation effects are regarded as unnecessarily esoteric for routine consideration. Most attention is usually paid to the manipulation of restricted sets of equations and to elaborate solution schemes for them. This text will attempt to give equal importance to all aspects and to illustrate the basic procedure. For convenience, the sequence (1)–(5) will not always be followed exactly, many consequences of the approximating step (4) being employed before carrying out the non-dimensionalization step (3). The next section provides an illustrative example of the procedure.

6.1 THE COUETTE VISCOMETER

We have already described viscometric, or steady simple-shear, flow in Section 4.2 in terms the flow field (4.2.1). End, though not edge, effects can be avoided by confining the fluid to the annular gap between two rigid coaxial circular cylinders, mounted with their axes vertical, one of which is capable of rotating about its axis. The system and associated flow field are illustrated in Fig. 6.1, and are the basis of the Couette viscometer.

(1) We take as reference frame one fixed in the motionless cylinder, say the inner one for definiteness.

(2) We use a cylindrical polar (r, ϕ, z) coordinate system such that $r = 0$ is the common vertical axis, the two cylindrical surfaces bounding the fluid being at $r = a$ and $r = b$. The upper and lower surfaces of the fluid are taken to be at or near $z = l$ and $z = 0$. We shall not write out the full equations of motion and energy in these coordinates, though they can be found in many standard texts. For example, the momentum equation is given in BA & H, Table B1, as three scalar equations involving the following dependent variables: ρ, p (scalars), v_r, v_ϕ, v_z (vector components), t^E_{rr}, $t^E_{r\phi}$, t^E_{rz}, $t^E_{\phi\phi}$, $t^E_{\phi z}$, t^E_{zz} (tensor components). The differential-type constitutive equations, i.e. those from (3.3.1)–(3.3.35), can all be written down explicitly in terms of the v and t^E components. It is less easy to do so for the integral forms. However, we do not need to do this at this stage because we shall start by supposing the only non-zero component of velocity to be $v_\phi(r)$, i.e. the flow to be steady, and axially and cylindrically symmetric. The boundary conditions on $v_{\phi\phi}$ are

$$v_\phi(a) = 0, \qquad v_\phi(b) = Wb \tag{6.1.1}$$

and these we can regard as the primary driving force for the motion. Obviously, this flow field will be an approximation: unless we can apply exactly compatible end boundary conditions at $z = 0$ and $z = l$, the flow will lose its symmetry near those planes; furthermore, at this stage we cannot be sure that the coupled equations for $p, \mathbf{T}^{\mathrm{E}}, \rho$ and T will allow of a solution with $v_r = v_z = 0$, even with neglect of end boundary conditions. If we regard the fluid as incompressible and rheologically insensitive to temperature variations, then the energy equation is immediately decoupled from the momentum equations, while the mass conservation requirement $\nabla . \mathbf{v} = 0$ (eqn (2.2.7)) is immediately seen to be satisfied. The rate of strain and vorticity tensors associated with

$$\mathbf{v} = \{0, v_\phi(r), 0\} \tag{6.1.2}$$

are

$$\mathbf{D} = \begin{bmatrix} d_{rr} & d_{r\phi} & d_{rz} \\ d_{r\phi} & d_{\phi\phi} & d_{\phi z} \\ d_{rz} & d_{\phi z} & d_{zz} \end{bmatrix} = \begin{bmatrix} 0 & \frac{1}{2}D(r) & 0 \\ \frac{1}{2}D(r) & 0 & 0 \\ 0 & 0 & 0 \end{bmatrix} \tag{6.1.3}$$

$$\mathbf{W} = \begin{bmatrix} 0 & -w_{r\phi} & -w_{rz} \\ w_{r\phi} & 0 & -w_{\phi z} \\ w_{rz} & w_{\phi z} & 0 \end{bmatrix} = \begin{bmatrix} 0 & \frac{1}{2}D(r) & 0 \\ -\frac{1}{2}D(r) & 0 & 0 \\ 0 & 0 & 0 \end{bmatrix} \tag{6.1.4}$$

where

$$D(r) = \frac{\partial v_\phi}{\partial r} - \frac{v_\phi}{r} \tag{6.1.5}$$

from which it can be shown that the flow field is viscometric (Walters, 1975, BA & H 1977) and so the stress field will be, using (4.4.2)–(4.4.4),†

$$\mathbf{T} = \begin{bmatrix} t_{rr} & t_{r\phi} & t_{rz} \\ t_{r\phi} & t_{\phi\phi} & t_{\phi z} \\ t_{rz} & t_{\phi z} & t_{zz} \end{bmatrix} = \begin{bmatrix} -p + \Psi_1(D)D^2 & \eta(D)D & 0 \\ \eta(D)D & -p & 0 \\ 0 & 0 & -p - \Psi_2(D)D^2 \end{bmatrix} \tag{6.1.6}$$

† This usual definition of $\mathbf{T}^{\mathrm{E}}$ for a viscometric flow leaves an element of arbitrariness in p because only the differences of normal stresses and not the normal stresses $t_{11}^{\mathrm{E}}, t_{22}^{\mathrm{E}}, t_{33}^{\mathrm{E}}$, themselves are defined.

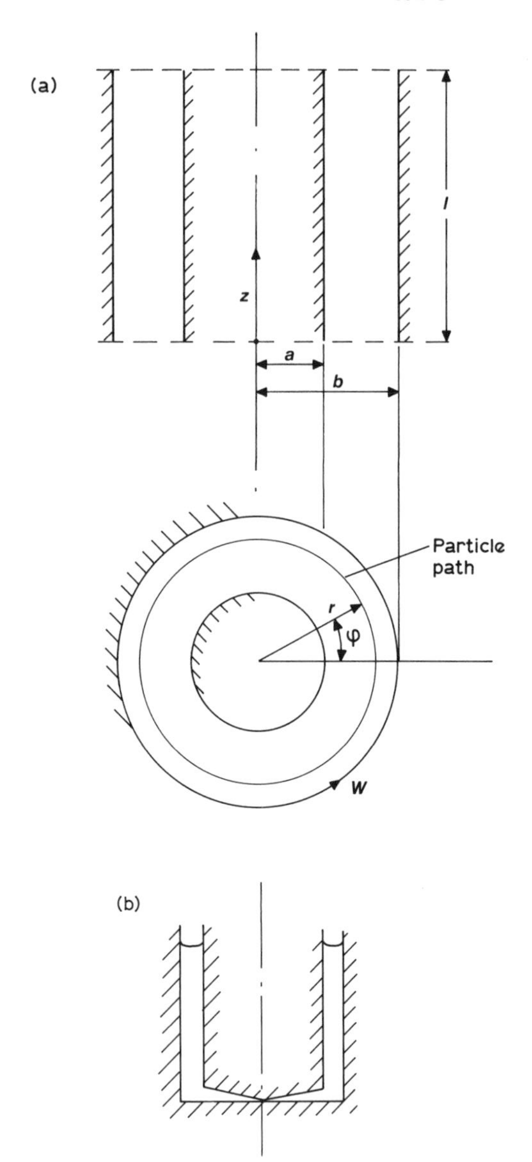

Fig. 6.1. Couette viscometer. (a) Basic cylindrical section and plan view showing (r, ϕ, z) coordinate system and boundaries at $r = a, b$. (b) Narrow-gap version with cone-and-plate end unit. (c) Taylor vortices in (r, z) section showing secondary flow (v_r, v_z).

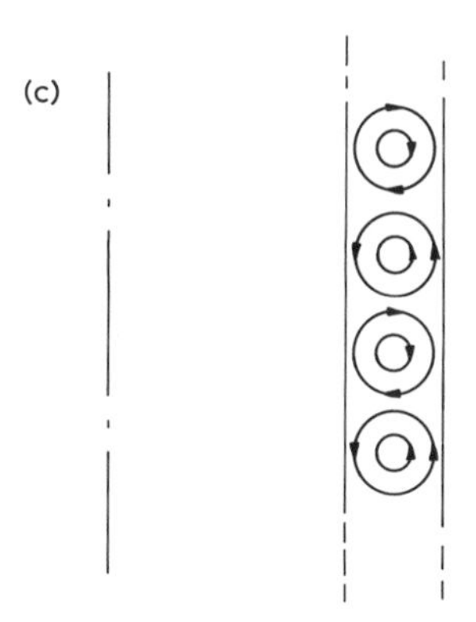

Fig. 6.1—*contd.*

If we now substitute into the momentum equations, we find that

$$r: \quad \frac{\rho v_\phi^2}{r} = -\frac{\partial p}{\partial r} + \frac{1}{r}\frac{\partial}{\partial r}(r\Psi_1(D)D^2) \tag{6.1.7r}$$

$$\phi: \quad 0 = -\frac{1}{r^2}\frac{\partial}{\partial r}(r^2\eta(D)D) \tag{6.1.7ϕ}$$

$$z: \quad 0 = -\frac{\partial p}{\partial z} + \rho g \tag{6.1.7z}$$

The r-equation represents the balance between inertial (centrifugal) force, radial pressure gradient, and normal-force gradient while the z-equation represents the hydrostatic balance between body force (gravity) and vertical pressure gradient. The ϕ-equation expresses the constancy of torque,

$$r^2\eta(D)D = L, \text{ constant} \tag{6.1.8}$$

Clearly if η is a known function of D, then (6.1.8) and (6.1.5) given (6.1.1) in principle yield v_ϕ and L whence, from (6.1.7r,z), p can be obtained. Alternatively, if L is specified, then integration of (6.1.8) will yield W in (6.1.1).

Usually, however, $\eta(D)$ is not known and experiments are undertaken at a variety of rotation rates W, with simultaneous measurement of the torque, $2\pi lL$, in order to determine the viscosity function $\eta(D)$.

The difficulty is that D is not constant in any one experiment, and so determination of D is not a trivial matter (see Walters 1975, Section 4.4). We do not pursue these details here.

(3) We can non-dimensionalize the equations we have employed in this very simple representation using b as a length scale, Wb as a velocity scale, and $\eta W(b-a)/b$ as a viscosity scale. Thus writing

$$\begin{aligned} \tilde{r} = r/b, \qquad \tilde{z} = z\mathrm{G}_2/b \\ \tilde{v}_\phi = v_\phi/Wb, \qquad \tilde{D} = D\mathrm{G}_1/W \\ \tilde{\eta}(\tilde{D}) = \eta(D)/\eta(W\mathrm{G}_1^{-1}) \end{aligned} \tag{6.1.9}$$

where

$$\mathrm{G}_1 = (b-a)/b, \qquad \mathrm{G}_2 = b/l \tag{6.1.10}$$

the relevant equations (6.1.5), (6.1.7ϕ) and (6.1.1) become

$$\tilde{D} = \left(\frac{\partial v_\phi}{\partial \tilde{r}} - \frac{\tilde{v}_\phi}{\tilde{r}}\right)\mathrm{G}_1 \tag{6.1.11}$$

$$\frac{\partial}{\partial r}(\tilde{r}^2\tilde{\eta}(\tilde{D})\tilde{D}) = 0 \tag{6.1.12}$$

subject to

$$\tilde{v}_\phi(1-\mathrm{G}_1) = 0, \; \tilde{v}_\phi(1) = 1 \tag{6.1.13}$$

in the region

$$0 \leqslant \tilde{z} \leqslant 1$$

We note that at this level only the two geometrical ratios G_1 and G_2 have arisen. We can subsequently show that $\tilde{D}$, $\tilde{v}_\phi$ and η will all be 0(1), $1-\mathrm{G}_1<\tilde{r}<1$. If we make the other equations (6.1.7r,z) dimensionless in terms of

$$\begin{aligned} \tilde{p} = p/\eta(W\mathrm{G}_1^{-1})W\mathrm{G}_1^{-1} \\ \tilde{\Psi}_{1,2}(\tilde{D}) = \Psi_{1,2}(\tilde{D}W\mathrm{G}_1^{-1})W\mathrm{G}_1^{-1}/\eta(W\mathrm{G}_1^{-1}) \end{aligned} \tag{6.1.14}$$

we get

$$-\frac{\partial \tilde{p}}{\partial \tilde{r}} + \frac{1}{\tilde{r}}\frac{\partial}{\partial \tilde{r}}(\tilde{r}\tilde{\Psi}_1(\tilde{D})\tilde{D}^2) = -\mathrm{Re}\frac{\tilde{v}_\phi^2}{\tilde{r}} \tag{6.1.15}$$

$$-\frac{\partial \tilde{p}}{\partial \tilde{z}} + \frac{\mathrm{Re}}{\mathrm{Fr}} = 0 \tag{6.1.16}$$

where

$$\mathrm{Re} = \rho W \mathrm{G}_1 b^2 / \eta(W\mathrm{G}_1^{-1}) \tag{6.1.17}$$

and

$$\mathrm{Fr} = W^2 b \mathrm{G}_2 / g \tag{6.1.18}$$

are the relevant Reynolds and Froude numbers.

Note that both geometrical parameters G_1 and G_2 appear in the definitions of Re and Fr and so a unique non-dimensionalization of the governing equations, without reference to the flow field geometry, is not necessarily a useful exercise. This illustrates why the steps (2), (3) and (4) of the basic procedure are often interrelated in practice.

If $\mathrm{G}_1 \ll 1$, then it is fairly clear to see that $\tilde{D} \simeq 1$, $\tilde{\eta}(\tilde{D}) \simeq 1$ throughout the flow field, and that, in (6.1.8)

$$L \approx b^2 \eta(W\mathrm{G}_1^{-1}) W \mathrm{G}_1^{-1} \tag{6.1.19}$$

so the viscosity is trivially related to the torque. Thus, if a narrow gap viscometer is used, interpretation of observations is relatively simple.

Neglect of the end boundary conditions is obviously justified in the limit $\mathrm{G}_2 \to 0$, though a more careful analysis shows that the product $\mathrm{G}_1\mathrm{G}_2 = (b-a)/l$ is the relevant parameter.

(4) We can now go back to the unapproximated situation and study the nature of the effects and terms neglected.

Our solution based on (6.1.2)–(6.1.6) would satisfy free-surface boundary conditions exactly at $z = 0, l$ provided p_a could be made to vary with r according to

$$\frac{\mathrm{d}p_\mathrm{a}}{\mathrm{d}r} = -\frac{1}{r}\frac{\mathrm{d}}{\mathrm{d}r}\{r\Psi_1(D)D^2\} + \frac{\mathrm{d}}{\mathrm{d}r}\{\Psi_2(D)D^2\} + \frac{\rho v_\phi^2}{r} \tag{6.1.20}$$

In general, this will not be possible and so a local flow variation will arise in practice. If the test fluid lies on top of a much heavier but relatively inviscid fluid at $z = 0$ and is open to the atmosphere at $z = 1$, then gravity forces caused by a small disturbance to the free surface can almost balance (6.1.20). This will be the case provided

$$W^2\mathrm{G}_1^{-2}\,|\Psi_1(W\mathrm{G}_1^{-1})|/\Delta\rho g b \ll 1 \quad \text{and} \quad \rho W^2 b/\Delta\rho g \ll 1 \tag{6.1.21}$$

where $\Delta\rho$ is the density difference between the two fluids at either interface.

A fuller discussion of the secondary-flow problem that arises at such interfaces is given in Joseph (1976), Joseph and Beavers (1977) and

Yoo *et al.* (1979). They show how the problem can be solved asymptotically for slow flows, when the retarded-motion expansion of Coleman and Noll is relevant, and hence that measurement of free-surface shape will lead to information about the constants β_1, γ_1 arising in the third-order fluid model (3.3.11). Constraints (6.1.21) are necessarily satisfied in Joseph's expansion because the numerators go to zero in the slow-flow limit while the denominators are constant. The elegant solution of Joseph & Fosdick (1973) shows that the influence of the free surface falls off exponentially-rapidly with $\tilde{z}$, and that the error in the torque calculated according to (6.1.8) will be of order G_1G_2, assuming that $\Psi_1(D)D$ is of the same order as $\eta(D)$.

The relationship between (6.1.21) and the dimensionless groups (6.1.17) and (6.1.18) introduces yet another material parameter, namely

$$\mathrm{Ws} = D\Psi_1(D)/\eta(D) \tag{6.1.22}$$

the Weissenberg number, which is seen to be a function of shear rate D. (6.1.21) becomes

$$\mathrm{Ws\,Fr/G_2\,Re} \ll 1, \qquad \mathrm{Fr/G_2} \ll 1 \tag{6.1.23}$$

$\mathrm{Fr/G_2}$ is the local Froude number W^2b/g. We note that if $\mathrm{Ws}=0$, i.e. if there are no difference of normal stresses, then only the centrifugal effect is relevant: the ratio Ws/Re compares difference of normal stress to inertial effects.

Let us now consider thermal effects. The energy equation, to the same approximation as (6.1.7) becomes

$$\frac{\alpha}{r}\frac{\partial}{\partial r}\left(r\frac{\partial T}{\partial r}\right) = \eta(D)D^2 \tag{6.1.24}$$

Since the right-hand side is known, a solution for T can be found by repeated integration, assuming either T or $\alpha\partial T/\partial r$ is specified at $r = a, b$ (see eqns (2.3.5, 2.3.6)). If we are to make (6.1.24) dimensionless, then we must choose a temperature scale. Because our application is here to viscometry, we choose the base temperature to be that of both cylinders, i.e.

$$T(a) = T(b) = T_\mathrm{W} \tag{6.1.25}$$

and the temperature scale to be $\eta(\partial\eta/\partial T)^{-1}$. Thus we write

$$\tilde{T} = \left[\frac{1}{\eta}\left(\frac{\partial \eta}{\partial T}\right)\right]_{T=T_\mathrm{W}, D=\mathrm{WG_1^{-1}}} T - T_\mathrm{W} \tag{6.1.26}$$

and (6.1.24) becomes

$$\frac{\partial}{\partial \tilde{r}}\left(\tilde{r}\frac{\partial \tilde{T}}{\partial \tilde{r}}\right)=\tilde{\eta}(\tilde{T},\tilde{D})\tilde{D}^2\,\mathrm{Gn} \tag{6.1.27}$$

where

$$\mathrm{Gn}=\frac{W^2b^2}{\alpha \mathrm{G}_1^2}\left[\frac{(\partial\eta)}{\partial T}\right]_{T=T_w,D=W\mathrm{G}_1^{-1}}=\frac{\mathrm{Na}}{\mathrm{G}_1^2} \tag{6.1.28}$$

is a generation number.

Its more usual form is the Nahme number Na, also defined by (6.1.28), the extra factor G_1^{-2} arising because r was made dimensionless with b, instead of with $(b-a)$, which is the natural length scale when G_1 becomes small.

If $\mathrm{Gn} \ll 1$ then thermal effects are indeed negligible in (6.1.12). If, however, $\mathrm{Gn} \geqslant 0(1)$ then the decoupling of eqns (6.1.12) and (6.1.27) is no longer possible, though the flow symmetry can be preserved to the same extent as before.

For completeness we consider departures from the incompressibility assumption. These will be of order

$$\mathrm{Co}=\max\left(\frac{gl}{\rho}\frac{\mathrm{d}\rho}{\mathrm{d}p},\eta\frac{W^2b^2}{\alpha\rho}\left|\frac{\mathrm{d}\rho}{\mathrm{d}r}\right|\right) \tag{6.1.29}$$

since gravity forces (gl) will be the dominant ones in determining the maximum variations in pressure, and $\eta W^2b^2/\alpha$ represents the scale temperature variation caused by heat generation.

In practice G_2 will be $0(1)$; a typical value is 1/3 . G_1, particularly in view of the simplifications achieved when it is small, is typically in the range 0·2–0·05. $\mathrm{G}_1\mathrm{G}_2$ is thus about 1/30 and so free-surface end effects are likely to be small.

Although the use of a much heavier, relatively inviscid, fluid such as mercury in the bottom portion of the viscometer is not conventionally adopted, other methods of reducing or calculating bottom end effects have been developed. Figure 6.1(b) shows a narrow-gap viscometer with a cone-and-plate (constant shear rate) end unit; this is one way in which end effects can be calculated. For pressurized versions, an upper cone-and-plate is also used. The theory of cone-and-plate viscometry is very similar to that of Couette viscometry and can be found in Walters (1975), Coleman *et al.* (1966), Lodge (1964), or BA & H (1977).

The dimensionless groups that arose in our dynamical analysis and which we required to be small were Fr/G_2, $\mathrm{Fr\,Ws}/\mathrm{G}_2\mathrm{Re}$, Na and Co. All of these increase with W: we therefore expect difficulties with a Couette viscometer at high shear rates. Processing operations certainly involve shear rates of $100\,\mathrm{s}^{-1}$ at viscosities of $10^3\,\mathrm{N\,s\,m}^{-2}$, with Ws as high as 10. If we put $b=0{\cdot}04$ m and $(b-a)=0{\cdot}002$ m, i.e. $\mathrm{G}_1=0{\cdot}05$, then $W=5\,\mathrm{rad\,s}^{-1}$ yields $W\mathrm{G}_1^{-1}=100$. Taking $\rho=1000\,\mathrm{kg\,m}^{-3}$, $\alpha=0{\cdot}2\,\mathrm{W\ mK}^{-1}$, $(\mathrm{d}\eta/\partial T)=20\,\mathrm{N\,s\,m}^{-2}\,\mathrm{K}$, and $\rho^{-1}\,\mathrm{d}\rho/\mathrm{d}T=5\times10^{-4}$, we find that

$$\begin{aligned} \mathrm{Fr}/\mathrm{G}_2 &\lesssim 0{\cdot}1 \\ \mathrm{Fr\,Ws}/\mathrm{G}_2\,\mathrm{Re} &\approx 10^3 \\ \mathrm{Na} &= 4 \\ \mathrm{Co} &= 0{\cdot}1 \end{aligned} \qquad (6.1.30)$$

the pressure variation of density being negligible.

It is clear from these that inertia forces are negligible ($\mathrm{Re}\ll1$) and that, for stiff polymer melts, gravity forces are totally inadequate to prevent rod climbing even at moderate shear rates ($\mathrm{Fr\,Ws}/\mathrm{Gr}_2\,\mathrm{Re}\gg1$). Thus, ideas relevant for Newtonian fluids or dilute solutions are misleading for polymer melts. Furthermore, temperature generation can be important ($\mathrm{Na}>1$). In practice, any rise in temperature due to shear heating takes place very slowly, so the result $\mathrm{Na}=4$ which refers to a steady-state temperature distribution is not so severe a restriction as might be thought the case. Similarly the criterion $\mathrm{Co}\ll1$ will apply during transient operation far better than (6.1.30) suggests. For transient operation it is obviously necessary that enough time elapse for η^+—defined by eqn (4.6.14)—to have reached its stationary value η.

(5) To complete our mechanical analysis, we should study the stability of the symmetric flow field to disturbances not so far permitted. In particular, we consider the possibility that, even with $\mathrm{G}_1\mathrm{G}_2\to0$, cellular secondary flows periodic in the z-direction could arise. Such secondary flows were observed for a rotating inner and a stationary outer cylinder, and correctly ascribed to inertial instability, by G. I. Taylor, in circumstances that are totally outside our range of interest. For $\mathrm{G}_1\ll1$, with W relating to the inner cylinder, his criterion for instability is (see Lin 1955, Section 2.4)

$$\mathrm{Re}^2/\mathrm{G}_1\simeq1700, \quad \text{i.e.} \quad \mathrm{Re}=0(10) \qquad (6.1.31)$$

well above anything we can expect. Variations on Taylor's analysis have been carried out by several authors for special equations of state, including Joseph (1976), but none of these really related to the circumstances likely to be relevant for polymer melts where non-linear rheological forces dominate the flow. The question is therefore an open one, worthy of attention.

The above discussion has been given as a simple example (based on a flow situation familiar to most readers) of the type of analysis that can be carried out for polymer processes. The objective has been to illustrate the features of the procedure that are not usually emphasized and to concentrate on the difficulties rather than to reproduce detailed or elegant solution schemes. For those interested in Couette viscometry, it is naturally vital to obtain relations between $L(W, t, T_W)$, and (W, T_W) but that is not our task here. A variety of further difficulties associated with Couette instruments have in practice led to greater use of cone-and-plate rheometers for low shear rates (since they provide a good means for measuring $\Psi_1(D)$ also) and capillary extrusion rheometers, to which we now turn our attention, for high shear rates.

EXAMPLE 6.1.1. Derivation of inequalities (6.1.21).

Put (6.1.20) into dimensionless form, using (6.1.9), (6.1.10) and (6.1.14), and equate $\mathrm{d}p_a/\mathrm{d}r$ to $\mathrm{d}(g\Delta\rho\Delta z)/\mathrm{d}r$ where Δz is the vertical displacement of the free surface necessary to produce the necessary pressure gradient. This yields

$$\frac{\mathrm{d}}{\mathrm{d}\tilde{r}}\left(\frac{\Delta z}{b}\right)=\frac{WG_1^{-1}\eta(WG_1^{-1})}{gb\Delta\rho}\left[\frac{1}{\tilde{r}}\frac{\mathrm{d}}{\mathrm{d}\tilde{r}}\{\tilde{r}\tilde{\Psi}_1(\tilde{D})\tilde{D}^2\}+\frac{\mathrm{d}}{\mathrm{d}\tilde{r}}\{\tilde{\Psi}_2(\tilde{D})\tilde{D}^2\}\right]$$
$$+\frac{\rho W^2 b}{g\Delta\rho}\frac{\tilde{v}_\phi^2}{\tilde{r}}$$

Clearly if the total displacement is to be small

$$\Delta z/b \ll 1$$

Because of the non-dimensional variables chosen, $\tilde{v}_\phi^2/\tilde{r}=0(1)$ and so directly this requires

$$W^2 b/g\Delta\rho \ll 1$$

As discussed in Section 5.1.1, $|\tilde{\Psi}_1(\tilde{D})|$ will in general be larger than $|\tilde{\Psi}_2(\tilde{D})|$, but $|\tilde{\Psi}_1(\tilde{D})\tilde{D}|$ may be $\gg 1$, i.e. $\mathrm{Ws} \gg 1$ using definition (6.1.22).

It is therefore better for order-of-magnitude arguments to introduce the non-dimensional quantities

$$\hat{\Psi}_{1,2}(\tilde{D}) = \Psi_{1,2}(\tilde{D}WG_1^{-1})/\Psi_{1,2}(WG_1^{-1})$$

in which case the dimensionless quantity

$$|\Psi_1(WG_1^{-1})|\ W^2G_1^{-2}/gb\Delta\rho$$

arises above instead of $\dfrac{WG_1^{-1}\eta(WG_1^{-1})}{gb\Delta\rho}$

$$\left[\frac{1}{\tilde{r}}\frac{\mathrm{d}}{\mathrm{d}\tilde{r}}\{\tilde{r}\hat{\Psi}_1(\tilde{D})\tilde{D}^2\} + \frac{\mathrm{d}}{\mathrm{d}\tilde{r}}\{\hat{\Psi}_2(\tilde{D})\tilde{D}^2\}\right]$$

will be of order unity and so for $\Delta z/b$ to be small

$$|\Psi_1(WG_1^{-1})|\ W^2G_1^{-2}/gb\Delta\rho \ll 1$$

EXERCISE 6.1.1. Show that, in a Couette viscometer with a stator of variable diameter, the relation (6.1.8) allows the viscosity function η to be obtained from

$$\eta(-a\,\partial W/\partial a) = -Lb^2/a^3(\partial W/\partial a)_{L,b}$$

To do this, regard the shear rate

$$\dot{\gamma} = r\frac{\mathrm{d}}{\mathrm{d}r}\left(\frac{v_\phi}{r}\right)$$

as a function of shear stress

$$\tau = L/r^2$$

Derive the relation $W = \displaystyle\int_{a/b}^{1} \frac{1}{r}\dot{\gamma}\left(\frac{L}{r^2}\right)\mathrm{d}r$ and hence $\left(\dfrac{\partial W}{\partial a}\right)_{L,b}$. The required result follows directly.

EXERCISE 6.1.2. Derive equation (6.1.20) by requiring that

$$p_a = p - t^{\mathrm{E}}_{zz} \quad \text{at} \quad z = l.$$

Chapter 7

The Capillary Rheometer

The capillary rheometer is the most commonly used experimental tool for measuring the rheological properties of polymer melts. Interpretation of the observations that can be made is a much more subtle and difficult matter than is usually admitted (Kamal & Nyun, 1980). Many complex effects are relevant for elastic fluids and so a detailed account of the relevant flow fields is given here. This exercise has the further great advantage for our purposes of introducing almost all of the important dimensionless groups and associated physical effects that will be relevant in our later discussions of confined flow processes, with the minimum of geometrical complexity.

Figure 7.1(a) shows diagrammatically the most important parts of the flow field in a capillary rheometer. Fluid material is forced, either by the controlled motion of a piston or by gas pressure, from a reservoir into a capillary tube of constant small circular cross-section, whence it emerges into the ambient air, a heated environment or a low-viscosity liquid. The reservoir is shown as having a diameter d_r which is usually much larger than d_c, the capillary diameter. The entry to the capillary is usually sharp-edged and axisymmetric: the capillary is shown to be flush-mounted with (i) a θ_e-half-angle conical entry in the upper half of the diagram and (ii) a flat base-plate entry in the lower half of the diagram. These are alternatives. In any actual apparatus, because of the requirements of mechanical design and the need to exchange the capillary used, the geometry of the entry region may be less simple in the near-neighbourhood of the capillary end than described here. In Fig. 7.1(b) a conical piston is shown near the bottom of its travel in a conically-ended reservoir, to show that in that position ($h_m \ll d_r$) the entry flow field can obviously be totally different from that arising when the piston end is far ($>d_r$) from the capillary entry.

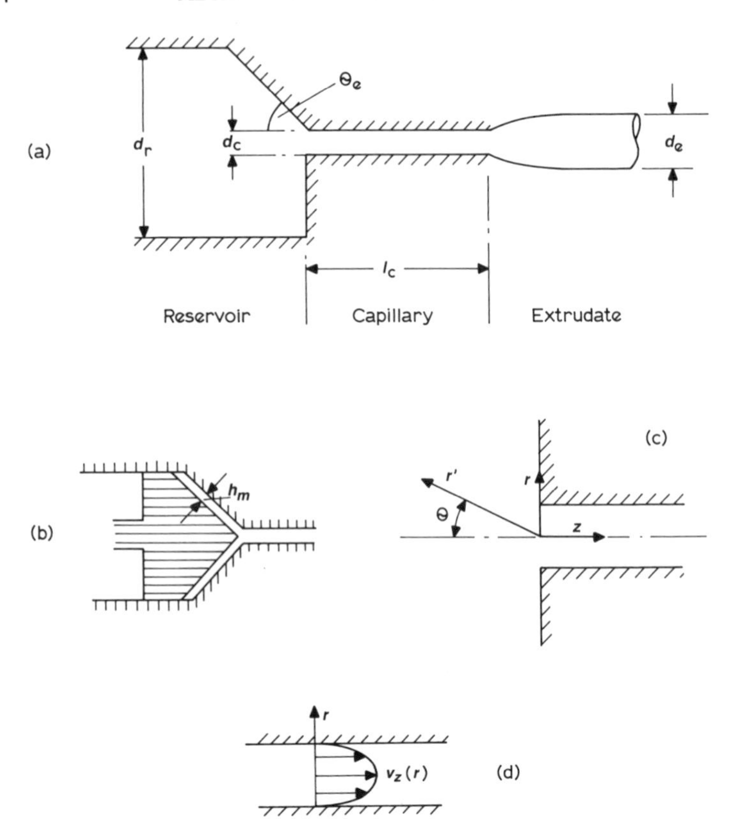

Fig. 7.1. Capillary rheometer. (a) Geometry of entire flow field (not to scale) showing alternative entry shapes. (b) Geometry of reservoir when a conical piston approaches the base. (c) Coordinate systems for reservoir (r', θ) and capillary (r, z). (d) Fully developed velocity profile.

We shall not analyse the situation (b), and assume that the reservoir extends a long way ($\gg d_r$) in the axial upstream direction. Some erratic observations can be ascribed to situation (b).

The extrudate is shown to swell as it emerges, deliberately. Although the extrudate (or die) swell associated with $d_e > d_c$ is irrelevant for viscometric measurement, it is a very important indication of highly elastic behaviour and so measurements of the observed swelling ratio

$$\mathrm{Es} = d_e / d_c \tag{7.1}$$

are regarded as important quantitative rheological information.

For long narrow capillaries, it is assumed that the force (or pressure) applied to the fluid in the reservoir can be related to the steady rate of output by considering the viscous dissipation in the capillary, and that this assumption can be checked by using capillaries of different lengths and diameters. It is found that in practice, for polymer melts, significant entry losses (pressure drop within the reservoir entry region) arise and that these also have rheological significance. In this section, we shall discuss the mechanics of the whole flow field, not only because it provides necessary information for the interpretation of experimental measurements, but also because it illustrates many effects that are of very wide significance in polymer processing generally, and introduces analytical techniques that will be used repeatedly later.

Recollecting the value we obtained for the Reynolds number in Section 6.1 we shall henceforth neglect inertial terms, and so use

$$\boldsymbol{\nabla} \cdot \mathbf{T} = \mathbf{0} \tag{7.2}$$

as the relevant form of (2.2.8).

7.1 CAPILLARY FLOW

We take a frame of reference fixed in the apparatus and a cylindrical-polar coordinate system (Fig. 7.1(c)) based on the capillary axis with $z = 0$ at the entry, and $z = l_c$ at the exit. The tube wall is at $r = \frac{1}{2}d_c$. Steady flow is implied. The flow field geometry is axisymmetric and so we assume the velocity field to be axisymmetric also. Thus,

$$\mathbf{v} = \{v_r(r, z), 0, v_z(r, z)\} \tag{7.1.1}$$

with p and T functions of r and z only. The stress components $t_{r\phi}$ and $t_{\phi z}$ are zero by symmetry.

Note that this implies that the entry flow field is axisymmetrical also.

We can write the conservation equations (2.2.5), (2.2.8) and (2.2.9), neglecting inertia terms, and using (2.2.10) (see Bird *et al.*, 1960, Table 3.4.1(B), Table 3.4.3(A), (B) and (C) and various results in Chapter

10) as

$$\frac{1}{r}\frac{\partial(\rho r v_r)}{\partial r}+\frac{\partial(\rho v_z)}{\partial z}=\frac{\partial \rho}{\partial t} \tag{7.1.2}$$

$$\frac{1}{r}\frac{\partial}{\partial r}(rt_{rr}^{\mathrm{E}})+\frac{\partial}{\partial z}t_{rz}^{\mathrm{E}}-\frac{t_{\phi\phi}^{\mathrm{E}}}{r}-\frac{\partial p}{\partial r}=-\rho g_r \tag{7.1.3}$$

$$\frac{1}{r}\frac{\partial}{\partial r}(rt_{rz}^{\mathrm{E}})+\frac{\partial}{\partial z}t_{zz}^{\mathrm{E}}-\frac{\partial p}{\partial z}=-\rho g_z \tag{7.1.4}$$

$$\begin{aligned}\rho\left(v_r\frac{\partial U}{\partial r}+v_z\frac{\partial U}{\partial z}\right)=&-\rho\frac{\partial U}{\partial t}+\frac{1}{r}\frac{\partial}{\partial r}\left(\alpha r\frac{\partial T}{\partial r}\right)+\frac{\partial}{\partial z}\left(\alpha\frac{\partial T}{\partial z}\right)\\&-p\left\{\frac{1}{r}\frac{\partial(rv_r)}{\partial r}+\frac{\partial v_z}{\partial z}\right\}+t_{zz}^{\mathrm{E}}\frac{\partial v_r}{\partial r}+t_{rz}^{\mathrm{E}}\left(\frac{\partial v_r}{\partial z}+\frac{\partial v_z}{\partial r}\right)\\&+t_{zz}^{\mathrm{E}}\frac{\partial v_z}{\partial z}+t_{\phi\phi}^{\mathrm{E}}\frac{v_r}{r}\end{aligned} \tag{7.1.5}$$

where U, ρ and α are assumed known functions of p and T.

The wall boundary conditions are, using (2.3.2) and (2.3.4); (2.3.5), (2.3.6) or (2.3.7),

$$v_r(\tfrac{1}{2}d_c, z)=0 \tag{7.1.6}$$

$$\left.\begin{aligned}v_z(\tfrac{1}{2}d_c, z)&=\mathrm{fn}(t_{rr}, t_{rz})t_{rz}\\&=0 \text{ for adhesion at the wall}\end{aligned}\right\} \tag{7.1.7}$$

and either

$$T(\tfrac{1}{2}d_c, z)=T_{\mathrm{W}} \quad \text{or} \tag{7.1.8i}$$

$$\alpha\frac{\partial T}{\partial r}(\tfrac{1}{2}d_c, z)=q_{\mathrm{W}} \quad \text{or} \tag{7.1.8ii}$$

$$\alpha\frac{\partial T}{\partial r}(\tfrac{1}{2}d_c, z)=h_{\mathrm{W}}\{T_{\mathrm{W}}-T(\tfrac{1}{2}d_c, z)\} \tag{7.1.8iii}$$

where T_{W}, q_{W} and h_{W} are known functions of z.

The initial conditions at $z=0$ have to be chosen to match the flow in the region $z>0$. There are certain constraints at $z=l_c$.

In the above we have made use of the fact that the velocity gradient tensor can be written

$$\nabla\mathbf{v} = \begin{bmatrix} d_{rr} & d_{r\phi} - w_{r\phi} & d_{rz} - w_{zz} \\ d_{r\phi} + w_{r\phi} & d_{\phi\phi} & d_{\phi z} - w_{\phi z} \\ d_{rz} + w_{rz} & d_{\phi z} + w_{\phi z} & d_{zz} \end{bmatrix} = \begin{bmatrix} \dfrac{\partial v_r}{\partial r} & 0 & \dfrac{\partial v_r}{\partial z} \\ 0 & \dfrac{v_r}{r} & 0 \\ \dfrac{\partial v_z}{\partial r} & 0 & \dfrac{\partial v_z}{\partial z} \end{bmatrix} \quad (7.1.9)$$

Further progress can only be made by making approximations which result in (a) simplification of the governing set of equations and/or (b) more precise specification of the constitutive equation and of the entry- and end-boundary conditions. As explained in the previous sections, consideration of suitably non-dimensional forms of the governing equations can help in determining what approximations are relevant because such non-dimensionalization gives rise to dimensionless groups whose numerical values can often be calculated for any given polymer melt before solution of the equations themselves. The relevant groups are dependent on the geometry of the flow field and on the imposed boundary conditions (i.e. the operating conditions).

Thus, in the reservoir flow, the purely geometrical ratio

$$\mathrm{G}_3 = d_c/d_r \quad (7.1.10)$$

and θ_e will determine how reservoir wall effects will affect the entry flow field. This in turn will affect the capillary flow through the entry conditions to the latter. The length-to-diameter ratio

$$\mathrm{L/D} = l_c/d_c \quad (7.1.11)$$

will clearly be important in determining whether an asymptotic flow field independent of z is effectively established in the capillary.

The obvious length scale is $\frac{1}{2}d_c$, giving dimensionless variables

$$\tilde{r} = 2r/d_c, \qquad \tilde{z} = 2z/d_c \quad (7.1.12)$$

A velocity scale can be obtained from the volume flow rate $\dot{V}$; thus

$$\tilde{v}_{r,z} = d_c^2 v_{r,z}/\dot{V} = v_{r,z}/v^* \quad (7.1.13)$$

These lead to a rate-of-deformation scale

$$D^* = \dot{V}/d_c^3 \quad (7.1.14)$$

If the viscosity function is assumed known, then a scale stress—to make $\mathbf{T}^{\mathrm{E}}$ and p dimensionless—is given by $\eta(D^*)D^*$ and we can write

$$\tilde{t}^{\mathrm{E}}_{rr} = t^{\mathrm{E}}_{rr}/\eta(D^*)D^* \quad \text{etc.} \tag{7.1.15}$$

and

$$\tilde{p} = p/\eta(D^*)D^* \tag{7.1.16}$$

In taking $\eta(D^*)D^*$ as a scale stress for capillary flow, we are implicitly assuming that the flow is essentially one of uniform steady simple shear (viscometric flow). As yet we have not established that this will be so. We could argue that the supposition is supported by visual observation and by analogy with the flow of strictly Newtonian fluids, or alternatively we can merely require that the consequences of the assumption be consistent with the original assumption. This latter method of *a posteriori* justification using tests for self-consistency is not so satisfying, but it is forced upon us by the complexity of many of the rheological equations of state that are relevant for polymer melts, reflecting the complexity of their observed rheological behaviour. The point is not a trivial one for $\eta(D^*)$ may well be only 5% of the zero-shear-rate viscosity $\eta(0)$, while the Weissenberg number may be as high as 10. This, the dimensionless value of the difference of normal stress $\tilde{t}^{\mathrm{E}}_{rr}$ may be as large as 10 even in viscometric flow, while the elongational viscosity η_{e} at an extension rate of D^* may be 20 times as large as $\eta(D^*)$, and so slight departures from exactly circular-cylindrical flow may, in practice, have profound consequences, as we shall discuss later on.

If, however, the capillary pressure drop, P, were given instead of $\dot{V}$, then the scale stress becomes $Pd_{\mathrm{c}}/l_{\mathrm{c}}$ and the scale velocity v^* can be chosen such that

$$P/l_{\mathrm{c}} = \eta(2v^*/d_{\mathrm{c}})v^*/d_{\mathrm{c}}^2 \tag{7.1.17}$$

Various numerical factors could have been introduced in equations (7.1.13)–(7.1.17) in order to lead to elegant forms in the case of a Newtonian fluid, but these have been dropped here to make clear the important factors.

$\mathbf{g}$ can be absorbed into p for confined flows, since it has no dynamical consequences. Several different temperature scales may be relevant. A rheological temperature scale

$$T^*_{\mathrm{rheol}} = \eta(\partial\eta/\partial T)^{-1} \tag{7.1.18}$$

was introduced in the last section. A generation temperature scale

$$T^*_{\text{gen}} = \eta v^{*2}/\alpha \tag{7.1.19}$$

can be obtained from arguments that led to the Nahme number

$$\text{Na} = \frac{v^{*2}}{\alpha}\frac{\partial \eta}{\partial T} \tag{7.1.20}$$

Finally there may be an operating temperature difference

$$T^*_{\text{op}} = |T_{\text{w}} - T_0| \tag{7.1.21}$$

if the fluid entering at $z = 0$ is at a mean temperature T_0, which leads to a further dimensionless group

$$\text{Br} = \frac{\eta^* v^{*2}}{\alpha\,|T_{\text{w}} - T_0|} \tag{7.1.22}$$

Any one of the three T^*'s defined above can be used to make the energy equation dimensionless, noting that a corresponding scale for U is $\rho^*\Gamma_{\text{p}}{}^*T^*$ where Γ_{p} is the specific heat at constant pressure/unit mass, and

$$\rho^* = \rho(T_{\text{w}}, p_{\text{a}}), \qquad \Gamma^*_{\text{p}} = \Gamma_{\text{p}}(T_{\text{w}}, p_{\text{a}}) \tag{7.1.23}$$

This choice is suggested by the following form (see BA & H equation (1B.15.4)) for the energy equation

$$\rho\Gamma_{\text{p}}\frac{\text{D}T}{\text{D}t} = -(\boldsymbol{\nabla}\cdot\mathbf{q}) + (\mathbf{T}:\boldsymbol{\nabla}\mathbf{v}) - \left(\frac{\partial \ln \rho}{\partial \ln T}\right)_{\text{p}}\frac{\text{D}p}{\text{D}t} \tag{7.1.24}$$

which follows from the assumptions (2.2.2) and (2.2.4).

We write

$$\tilde{\rho} = \rho/\rho^*, \qquad \tilde{\alpha} = \alpha/\alpha^* = \alpha/\alpha(T_{\text{w}}, p_{\text{a}}) \tag{7.1.25}$$

and

$$\tilde{t} = 2tv^*/d_{\text{c}}, \qquad \tilde{T} = (T - T_{\text{w}})/T^* \tag{7.1.26}$$

where v^* is given either by $\dot{V}/d_{\text{c}}^2$ or eqn (7.1.17).

Equations (7.1.2)–(7.1.5) now become

$$\frac{1}{\tilde{r}}\frac{\partial}{\partial \tilde{r}}(\tilde{\rho}\tilde{r}\tilde{v}_z) + \frac{\partial}{\partial \tilde{z}}(\tilde{\rho}\tilde{v}_z) = -\frac{\partial \tilde{\rho}}{\partial t} \tag{7.1.27}$$

$$\frac{1}{\tilde{r}}\frac{\partial}{\partial \tilde{r}}(\tilde{r}\tilde{t}^{\text{E}}_{rr}) + \frac{\partial}{\partial \tilde{z}}\,\tilde{t}^{\text{E}}_{rz} - \frac{\tilde{t}^{\text{E}}_{\phi\phi}}{\tilde{r}} - \frac{\partial \tilde{p}}{\partial \tilde{r}} = 0 \tag{7.1.28}$$

$$\frac{1}{r}\frac{\partial}{\partial \tilde{r}}(\tilde{r}\tilde{t}^{\mathrm{E}}_{rz})+\frac{\partial}{\partial \tilde{z}}\tilde{t}^{\mathrm{E}}_{zz}-\frac{\partial \tilde{p}}{\partial \tilde{z}}=0 \tag{7.1.29}$$

$$\begin{aligned}\tilde{\rho}\tilde{\Gamma}_{\mathrm{p}}\left(\frac{\partial \tilde{T}}{\partial \tilde{t}}+\tilde{v}_r\frac{\partial \tilde{T}}{\partial \tilde{r}}+\tilde{v}_z\frac{\partial \tilde{T}}{\partial \tilde{z}}\right)\mathrm{Pe}&=\frac{1}{\tilde{r}}\frac{\partial}{\partial \tilde{r}}\left(\tilde{\alpha}\tilde{r}\frac{\partial \tilde{T}}{\partial \tilde{r}}\right)+\frac{\partial}{\partial \tilde{z}}\left(\tilde{\alpha}\frac{\partial \tilde{T}}{\partial \tilde{z}}\right)\\&+\mathrm{Gn}\left[-\tilde{p}\left(\frac{1}{r}\frac{\partial(\tilde{r}\tilde{v}_r)}{\partial \tilde{r}}+\frac{\partial \tilde{v}_z}{\partial \tilde{z}}\right)+\tilde{t}^{\mathrm{E}}_{rr}\frac{\partial \tilde{v}_r}{\partial \tilde{r}}+\tilde{t}^{\mathrm{E}}_{rz}\left(\frac{\partial \tilde{v}_r}{\partial \tilde{z}}+\frac{\partial \tilde{v}_z}{\partial \tilde{r}}\right)\right.\\&\left.+\tilde{t}^{\mathrm{E}}_{zz}\frac{\partial \tilde{v}_z}{\partial \tilde{z}}+\tilde{t}^{\mathrm{E}}_{\phi\phi}\frac{\tilde{v}_z}{\tilde{r}}-\left(\frac{\partial \ln \tilde{\rho}}{\partial \ln \tilde{T}}\right)_{\tilde{p}}\left(\frac{\partial \tilde{p}}{\partial \tilde{t}}+\tilde{v}_z\frac{\partial \tilde{p}}{\partial \tilde{r}}+\tilde{v}_z\frac{\partial \tilde{p}}{\partial \tilde{z}}\right)^{\dagger}\right]\end{aligned} \tag{7.1.30}$$

where

$$\mathrm{Pe}=\frac{\rho^*\Gamma^*_{\mathrm{p}}v^*d_{\mathrm{c}}}{2\alpha^*} \tag{7.1.31}$$

and

$$\mathrm{Gn}=\eta^*v^{*2}/2\alpha^*T^* \tag{7.1.32}$$

with boundary conditions (7.1.6)–(7.1.8) becoming

$$\tilde{v}_r=\tilde{v}_z=\tilde{T}=0 \quad \text{at} \quad \tilde{r}=1 \tag{7.1.33}$$

for the case of no-slip and specified wall temperature.

As in the previous section, we take

$$\begin{aligned}&D^*=100\ \mathrm{s}^{-1}, \qquad \eta^*=10^3\ \mathrm{N\,s\,m^{-2}}, \qquad \rho^*=10^3\ \mathrm{kg\,m^{-3}},\\&\alpha^*=0{\cdot}2\ \mathrm{W\,mK^{-1}}, \qquad \Gamma^*_{\rho}=2\times10^3\ \mathrm{J\,kg^{-1}\,K^{-1}}, \qquad T^*_{\mathrm{rheol}}=50\ \mathrm{K}\end{aligned} \tag{7.1.34}$$

as being typical of polymer melts under operating conditions. For

$$\tfrac{1}{2}d_{\mathrm{c}}=0{\cdot}5\times10^{-3}\ \mathrm{m} \tag{7.1.35}$$

this gives

$$v^*=0{\cdot}1\ \mathrm{m\,s^{-1}} \tag{7.1.36}$$

and so from (7.1.31)

$$\mathrm{Pe}=500 \tag{7.1.37}$$

which, because of the relatively small values of d_{c} and v^* used in this example, is smaller in magnitude than it is for most polymer processes. Nevertheless $\mathrm{Pe}\gg1$, and so convective transport of heat energy is very

† Here $\ln\tilde{T}$ has to be interpreted as $\ln(T_{\mathrm{abs}}/T^*)$.

much more effective than conduction of heat; also, from (7.1.32) with $T^* = T^*_{\mathrm{rheol}}$

$$\mathrm{Gn} = 0{\cdot}5 \tag{7.1.38}$$

which, again, is relatively small for processing operations. $\mathrm{Gn} \ll 1$ implies that viscosity variation due to 'generated' temperature differences will be small, although even in the case chosen these actual temperature differences will be of the order of 25 K.

Strictly speaking (7.1.38) refers to a balance between generation and conduction alone. However, as we shall explain later, (7.1.37) suggests that most heat generated will be convected towards the end of the capillary rather than conducted towards the walls. This suggests that an alternative way of looking at the process is as a locally adiabatic one. We then use the simple relationship between pressure drop and temperature rise

$$\Delta P = \rho \Gamma_{\mathrm{p}} \Delta T \tag{7.1.39}$$

to give the average temperature changes within the capillary. This can be written

$$T^*_{\mathrm{adiab}} = \frac{\eta^* v^*}{\rho^* \Gamma^*_{\mathrm{p}}} \frac{l_{\mathrm{c}}}{d^2_{\mathrm{c}}} \tag{7.1.40}$$

from (7.1.17). In our case, using (7.1.34), this predicts

$$\frac{T^*_{\mathrm{adiab}}}{T^*_{\mathrm{rheol}}} = 10^{-3}\,\mathrm{L/D} \tag{7.1.41}$$

which will be equal to 0·5 (the value (7.1.38) gives for $T^*_{\mathrm{gen}}/T^*_{\mathrm{rheol}} \equiv \mathrm{Gn}$) when

$$\mathrm{L/D} = 500 \tag{7.1.42}$$

This is equal to Pe, an order relation that might be expected. If we write

$$\mathrm{Gz} = 2\mathrm{Pe}(\mathrm{L/D})^{-1} \tag{7.1.43}$$

the criterion for fully developed flow ($T^*_{\mathrm{adiab}} \gg T^*_{\mathrm{gen}}$) becomes

$$\mathrm{Gz} \ll 1 \tag{7.1.44}$$

Actual values of L/D are significantly less than 100, and so we would never expect fully developed flow in a capillary rheometer, at least as far as temperature development is concerned.

Having established the orders of magnitude of expected variations in temperature and pressure, we can readily show that in most cases

$$\mathrm{Co} < 1 \tag{7.1.45}$$

where Co is defined analogously to (6.1.24) and thus a first approximation to flow in a capillary can be obtained by assuming constant values for the density and other thermophysical parameters, and regarding the momentum (now a stress-equilibrium) equation and the energy equation to be decoupled.† However, there are cases, involving long capillary tubes and very viscous materials, where large pressure drops, of order 1000 bar, occur. In these cases, the pressure dependence of viscosity becomes an important factor and so the dimensionless group $\mathrm{Vp} = P(\mathrm{d}\eta/\mathrm{d}p)^*/\eta^*$, defined in (19.3.14) and used later, can be of order unity. Under these circumstances, the temperature-induced changes in viscosity, measured by Na, can be $\gg 1$, while even the density changes induced by pressure and temperature measured by

$$\mathrm{Cm} = \frac{P^*}{\rho^*}\left(\frac{\partial \rho}{\partial p}\right)^*, \quad \text{and} \quad \mathrm{Ex} = \frac{\eta^* v^{*2}}{\alpha^* \rho^*}\left(\frac{\partial \rho}{\partial T}\right)^*$$

can be significant in transient situations.

There is relatively little difficulty in accommodating variations of η or ρ with z; it is only where η becomes a sharply varying function of r that difficulties in calculations arise; this is all discussed at length in Chapter 19.

We therefore consider here the reduced set of equations (7.1.28) and (7.1.29) and the incompressibility condition

$$\frac{1}{\tilde{r}}\frac{\partial(\tilde{r}\tilde{v}_r)}{\partial \tilde{r}} + \frac{\partial \tilde{v}_z}{\partial \tilde{z}} = 0 \tag{7.1.46}$$

together with an as-yet-unspecified isothermal rheological equation of state.

Some insight is provided by changing the length scale for z from $\frac{1}{2}d_c$ to l_c and writing

$$\hat{z} = \frac{z}{l_c} = \frac{\tilde{z}}{2\mathrm{L/D}}, \quad \hat{p} = \frac{\tilde{p}}{2\mathrm{L/D}} \tag{7.1.47}$$

† Solutions for the uncoupled equations are given in BA & H Section 5.4, and will be developed where necessary in Chapter 19.

Entry and exit boundary conditions will now be at $\hat{z}=0,1$, while $\hat{p}\sim 1$ when $p=P$.

Our governing equations become

$$\frac{1}{\tilde{r}}\frac{\partial}{\partial \tilde{r}}(\tilde{r}\tilde{v}_r) = -\frac{1}{2\mathrm{L/D}}\frac{\partial \tilde{v}_z}{\partial \hat{z}} \tag{7.1.48}$$

$$\frac{1}{\tilde{r}}\frac{\partial}{\partial \tilde{r}}(\tilde{r}\tilde{t}^{\mathrm{E}}_{zz}) - \frac{\tilde{t}^{\mathrm{E}}_{\phi\phi}}{\tilde{r}} + 2\mathrm{L/D}\frac{\partial \hat{p}}{\partial \tilde{r}} = -\frac{1}{2\mathrm{L/D}}\frac{\partial \tilde{t}^{\mathrm{E}}_{rz}}{\partial \hat{z}} \tag{7.1.49}$$

$$\frac{1}{\tilde{r}}\frac{\partial}{\partial \tilde{r}}(\tilde{r}\tilde{t}^{\mathrm{E}}_{rz}) + \frac{\partial \hat{p}}{\partial \hat{z}} = -\frac{1}{2\mathrm{L/D}}\frac{\partial \tilde{t}^{\mathrm{E}}_{zz}}{\partial \tilde{z}} \tag{7.1.50}$$

From (7.1.48) we can deduce that, if $\tilde{v}_z$ and $\partial\tilde{v}_z/\partial\tilde{z}$ are both 0(1), which is consistent with the boundary conditions, then $\tilde{v}_r = 0[(\mathrm{L/D})^{-1}]$ almost everywhere and can be neglected for most of the flow field to that order of approximation, namely $0[(\mathrm{L/D})^{-1}]$.

From (7.1.49) we deduce that $\partial\hat{p}/\partial\tilde{r} = 0[(\mathrm{L/D})^{-1}]$ assuming that $\tilde{t}^{\mathrm{E}}_{zz}$ and $\tilde{t}^{\mathrm{E}}_{\phi\phi}$ are of smaller order than L/D, and then from (7.1.50) we deduce that $\partial\hat{p}/\partial\hat{z} = 0(1)$, so that finally we get

$$\frac{1}{\tilde{r}}\frac{\partial}{\partial \tilde{r}}(\tilde{r}\tilde{t}^{\mathrm{E}}_{rz}) = \mathrm{Const} + 0[(\mathrm{L/D})^{-1}] \tag{7.1.51}$$

provided $\tilde{t}^{\mathrm{E}}_{rz} = 0(1)$ and is a function of $\tilde{r}$ only, to order $(\mathrm{L/D})^{-1}$, given by the flow field

$$\tilde{\mathbf{v}} = \{0, 0, \tilde{v}_z(\tilde{r})\} + 0[(\mathrm{L/D})^{-1}] \tag{7.1.52}$$

This flow field shown in Fig. 7.1(d) is necessarily viscometric in that (7.1.52) gives rise by substitution in (7.1.9) to only one component in $\nabla\mathbf{v}$, and this will be constant along streamlines. (This is proved in Lodge, 1964; BA & H 1977; A & M, 1974.) This is the assumption usually made *ab initio*: the flow is regarded as the fully developed symmetric flow over most of the flow field.

For Newtonian fluids, it can be shown rigorously that any 0(1) variation from (7.1.52) at the ends $z=0, l_c$ will fall off exponentially rapidly far from the ends with an exponent of order unity.

We have assumed that the rheological equation of state for the material in question will be consistent with $\tilde{t}^{\mathrm{E}}_{rz}$ being a function of $\tilde{v}_z(r)$ only. This requires that the memory of the fluid does not extend back to the time it entered the capillary. Note that we cannot merely take the limit $\mathrm{L/D}\rightarrow\infty$, because that would destroy the argument that

$T^*_{\text{adiab}}/T^*_{\text{rheol}}$ is in practice small and that $\text{Co} \ll 1$, both conditions that we have already used.

One way to test the matter is to consider the ramp flow of Subsection 4.6.2 as providing the most severe discontinuity at the entry $z = 0$, which we identify as $t = 0$ in the Lagrangian sense of Subsection 4.6.2. The development of $\eta^+(t; D)$ with tD for typical polymer melts is shown in Fig. 5.4. From there and what is said in the text, it is clear that the steady-state viscosity is not reached until the strain is larger than 10, and so at first sight L/D must be much larger than 10 for entry effects to be negligible.

An alternative approach is to study the stress relaxation behaviour implied by the function $\eta^-(t; D)$, which provides a relaxation time t_{rel}. Again, we find that $t_{\text{rel}} \sim 10\, D^{-1}$ and so L/D must be larger than 10 for entry effects to be negligible.

What we have now determined, on rough order-of-magnitude grounds, is that typical extrusion rheometers, which have proved excellent and accurate devices for measuring fluids, probably suffer from significant weaknesses when used to measure the steady-shear viscosity of polymer melts at high rates of shear (see also Kamal & Nyun 1980). In particular, we can expect the details of the entry flow to affect the capillary flow unless the capillary is so long that pressure and temperature variations can no longer be neglected.

For the shear rates for which flow is rheologically fully developed, and for which viscometric flow is relevant, we write the dimensional equation

$$\frac{1}{r}\frac{\mathrm{d}}{\mathrm{d}r}\left\{r\eta\left(\left|\frac{\mathrm{d}v_z}{\mathrm{d}r}\right|\right)\frac{\mathrm{d}v_z}{\mathrm{d}r}\right\} = \frac{\partial p}{\partial z} = -\frac{P}{l_c} \tag{7.1.53}$$

with its integral

$$t_{rz}\left(\frac{\mathrm{d}v_z}{\mathrm{d}r}\right) = \eta\left(\left|\frac{\mathrm{d}v_z}{\mathrm{d}r}\right|\right)\frac{\mathrm{d}v_z}{\mathrm{d}r} = -\frac{1}{2}\frac{Pr}{l_c} \tag{7.1.54}$$

which satisfies

$$\frac{\mathrm{d}v_z}{\mathrm{d}r} = 0 \quad \text{at} \quad r = 0 \tag{7.1.55}$$

For stable shearing we require that t_{rz} be a monotonic function of $(\mathrm{d}v_z/\mathrm{d}r)$ with an inverse

$$\mathrm{d}v_z/\mathrm{d}r = \dot{\gamma}(t_{rz}) \tag{7.1.56}$$

such that $\dot{\gamma}(0)=0$. $\dot{\gamma}$ and t_{rz} are assumed to take the same sign. Since

$$v_z = -\int_r^{\frac{1}{2}d_c} \frac{dv_z}{dr}\,dr = \frac{2l_c}{P}\int_\tau^{\tau_w} \dot{\gamma}(u)\,du \tag{7.1.57}$$

where

$$\tau = -t_{rz} = \frac{Pr}{2l_c} \tag{7.1.58}$$

we can write the volume flux as

$$\dot{V} = 2\pi\int_0^{\frac{1}{2}d_c} rv_z\,dr = \frac{16\pi l_c^3}{P^3}\int_0^{\tau_w}\left(\int_\tau^{\tau_w}\dot{\gamma}(u)\,du\right)\tau\,d\tau \tag{7.1.59}$$

or integrating by parts

$$P^3\dot{V} = 8\pi l_c^3\int_0^{\tau_w}\dot{\gamma}(u)u^2\,du \tag{7.1.60}$$

If we now differentiate with respect to P, we get

$$\left(3\dot{V}+P\frac{d\dot{V}}{dP}\right)\frac{8}{\pi d_c^3} = \dot{\gamma}(\tau_w) = \dot{\gamma}_w \tag{7.1.61}$$

whence

$$\eta(\dot{\gamma}_w) = \frac{\pi P d_c^4}{32 l_c(P\,d\dot{V}/dP+3\dot{V})} \tag{7.1.62}$$

where $\dot{\gamma}_w$ is the wall shear rate given by (7.1.61). In the Newtonian range, where $P \propto \dot{V}$, eqn (7.1.62) becomes the familiar Poseuille law $\eta = \dfrac{\pi P(\frac{1}{2}d_c)^4}{8l_c\dot{V}}$. More generally it is known as the Mooney–Rabinowitz relation.

For a Newtonian fluid, the wall shear rate (7.1.61) is simply $(32\dot{V}/\pi d_c^3)$. More generally this is called the apparent wall shear rate, which when plotted against the wall shear stress yields a flow curve. The apparent viscosity is then given by the ratio of the wall shear stress to the apparent wall shear rate. A double logarithmic plot of the apparent wall shear rate against the wall shear stress yields a straight line of slope $1/\nu$ for a power-law fluid, for, as shown below in Example 7.1.1, the apparent wall shear rate is a simple multiple of the true wall shear rate.

7.1.1 Temperature Effects

The effect of temperature variations can be analysed by rescaling the energy equation (7.1.30) using the incompressibility approximation (7.1.46) and the variables given in (7.1.47) along with a rescaled radial velocity

$$\hat{v}_r = 2(L/D)\tilde{v}_r \tag{7.1.63}$$

Neglecting terms of order $(L/D)^{-1}$ or higher, we obtain in the steady case, with $\tilde{\rho}$, $\tilde{\Gamma}_p$ and $\tilde{\alpha}$ all put equal to unity,

$$\mathrm{Gz}\left(\hat{v}_r \frac{\partial \tilde{T}}{\partial \tilde{r}} + \tilde{v}_z \frac{\partial \tilde{T}}{\partial \tilde{z}}\right) = \frac{1}{\tilde{r}} \frac{\partial}{\partial \tilde{r}}\left(\tilde{r} \frac{\partial \tilde{T}}{\partial \tilde{r}}\right) + \mathrm{Gn}\, \tilde{t}^{E}_{rz} \frac{\partial \hat{v}_z}{\partial \tilde{r}} \tag{7.1.64}$$

It may readily be seen from (7.1.64) what the criterion (7.1.44) means analytically. The factor involving Gz is the convection term, the factor involving Gn the generation term and the middle term is the conduction term.

EXAMPLE 7.1.1. Derive the flow rate as a function of the pressure gradient in the case of a power-law fluid flowing in a uniform capillary.

From (3.3.5), (7.1.54) and (7.1.58) we have

$$\tau = \eta(\dot{\gamma})\dot{\gamma} = K\dot{\gamma}^{\nu} \tag{7.1.65}$$

so

$$\dot{\gamma} = (\tau/K)^{1/\nu} \tag{7.1.66}$$

which is the relevant form of (7.1.56). Substituting in (7.1.59) we obtain

$$\dot{V} = 16\pi\left(-\frac{\partial p}{\partial z}\right)^{-3} \int_0^{\tau_w} \left[\int_{\tau}^{\tau_w} \left(\frac{u}{K}\right)^{1/\nu} du\right] \tau \, d\tau \tag{7.1.67}$$

Carrying out the integration, putting

$$\tau_w = -\frac{\partial p}{\partial z}\frac{d_c}{4} \tag{7.1.68}$$

and treating $(\partial p/\partial z)$ as negative, yields

$$\dot{V} = \left(\frac{\nu}{3\nu+1}\right)\frac{\pi d_c^3}{8}\left[-\left(\frac{dp}{dz}\right)\frac{d_c}{4K}\right]^{1/\nu} \tag{7.1.69}$$

Simple rearrangement gives

$$-\left(\frac{\mathrm{d}p}{\partial z}\right)=\frac{4\mathrm{K}}{d_{\mathrm{c}}}\left[\frac{8(3\nu+1)\dot{V}}{\nu\pi d_{\mathrm{c}}^{3}}\right]^{\nu} \tag{7.1.70}$$

Writing (7.1.70) using (7.1.68) in the form

$$\tau_{\mathrm{w}}=\mathrm{K}\left[\frac{8(3\nu+1)\dot{V}}{\nu\pi d_{\mathrm{c}}^{3}}\right]^{\nu}$$

and comparing this with (7.1.66) shows that the true wall shear rate is $8(3\nu+1)\dot{V}/\nu\pi d_{\mathrm{c}}^{3}$ or $(3\nu+1)/4\nu$ times the apparent wall shear rate.

EXERCISE 7.1.1. Derive eqns (7.1.2)–(7.1.5) paying particular attention to the last.

EXERCISE 7.1.2. Verify the meaning of the statement '(7.1.38) refers to a balance between generation and conduction alone' by considering (7.1.30) for the case $\mathrm{Pe}\to 0$; Gn fixed.

Show that if $\mathrm{Pe}\gg\mathrm{Gn}\gg 1$, then an adiabatic approach is relevant, as is implied by (7.1.39).

EXERCISE 7.1.3. Show that the velocity profile in capillary flow of a power-law fluid is

$$v_{z}(r)=\frac{\partial p/\partial z}{|\partial p/\partial z|}\left[\frac{|\partial p/\partial z|}{2\mathrm{K}}\right]^{1/\nu}\left(\frac{\nu}{1+\nu}\right)\{(\tfrac{1}{2}d)^{(1+\nu)/\nu}-r^{(1+\nu)/\nu}\} \tag{7.1.71}$$

and hence that the maximum velocity

$$v_{z}(0)=\left(\frac{3\nu+1}{\nu+1}\right)\bar{v}_{z} \tag{7.1.72}$$

where $\bar{v}_{z}$ is the mean velocity $4\dot{V}/\pi d_{\mathrm{c}}^{2}$.

7.2 ENTRY FLOW

We can use either the (r, ϕ, z) coordinate system already defined for the capillary flow, or the spherical polar system (r', θ, ϕ) shown at the left-hand side of Fig. 7.1(c). We suppose the entry to the capillary to be sharp-edged. Figure 7.2 displays two possible types of entry flow,

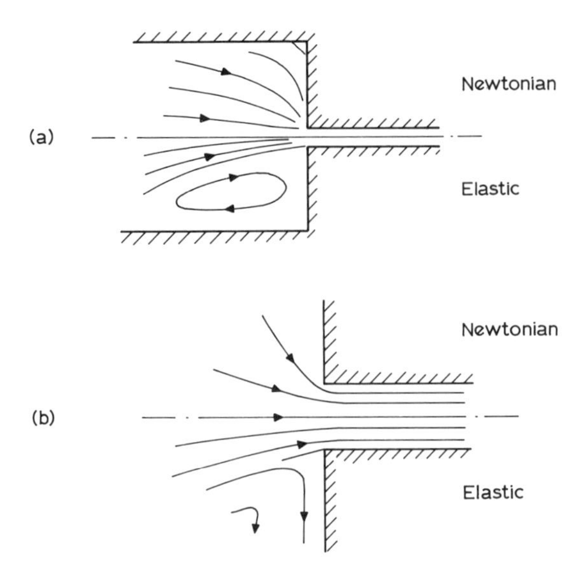

Fig. 7.2. Entry flow field (streamlines). (a) Overall view. (b) Expanded view near capillary. Upper and lower halves in each diagram are alternative flow fields typical of Newtonian and elastic fluid behaviour, respectively.

shown for convenience on the same diagrams, with $\theta_e = \frac{1}{2}\pi$. The upper halves, typical of Newtonian fluid flow, have streamlines entering the capillary from the entire half space to the left; the lower halves show a dividing streamline, which separates an inner core flow (that enters the capillary) from an outer (recirculating) vortex flow. The latter flow is more typical of highly elastic fluids. Any one elastic fluid can show both types of behaviour as the flow rate is increased.

The boundary conditions on the flow are simply that

$$\mathbf{v} = \mathbf{0}, \qquad T = T_w \quad \text{at} \quad \theta = \theta_e \tag{7.2.1}$$

and on the cylindrical part of the reservoir. In cases where $G_3 \gg 1$,† we can hope that entry conditions, which would be applied at some radius $r' \simeq \frac{1}{2}d_e$, can suitably be replaced by an entry flux condition

$$2\pi r'^2 \int_0^{\theta_e} v_{r'}(\theta, \phi) \sin\theta \, d\theta = -\dot{V} \tag{7.2.2}$$

where $\dot{V}$ is the same as used in (7.1.13) above, with $T = T_w$. If G_3 turns

† G_3 is defined in eqn (7.1.10) as d_c/d_r.

out not to be sufficiently large, then a matching condition, with the flow in the cylindrical portion of the reservoir, would be needed. In this account, we shall suppose that $G_3 \gg 1$.

We first assume the flow to be axisymmetric and therefore look for a solution with

$$\mathbf{v} = \{v_{r'}(r', \theta), v_\theta(r', \theta), 0\}, \qquad t^E_{r'\phi} = t^E_{\theta\phi} \equiv 0 \tag{7.2.3}$$

and all other components of $\mathbf{T}$, namely $t^E_{r'r'}$, $t^E_{r'}$, $t^E_{\theta\theta}$, $t^E_{\phi\phi}$ and p, functions of r' and θ only. On the basis of the quantitative arguments given for the capillary flow region, we suppose the energy equation to be decoupled from the other flow and constitutive equations.

The velocity gradient tensor $\nabla\mathbf{v}$ becomes (see BA & H Tables B.3 and B.4)

$$\nabla\mathbf{v} = \begin{bmatrix} \dfrac{\partial v_{r'}}{\partial r'} & \dfrac{1}{r'}\dfrac{\partial v_{r'}}{\partial \theta} - \dfrac{v_\theta}{r'} & 0 \\ \dfrac{\partial v_\theta}{\partial r'} & \dfrac{1}{r'}\dfrac{\partial v_\theta}{\partial \theta} + \dfrac{v_{r'}}{r'} & 0 \\ 0 & 0 & \dfrac{v_{r'}}{r'} + \dfrac{v_\theta}{r'}\cot\theta \end{bmatrix} = (\nabla\mathbf{v})^T \tag{7.2.4}$$

with obvious symmetric and antisymmetric parts for $\mathbf{D}$ and $\mathbf{W}$.

We can define a stream function $Q(r', \theta)$ such that

$$v_{r'} = \frac{1}{r'^2 \sin\theta}\frac{\partial Q}{\partial \theta}, \qquad v_\theta = -\frac{1}{r'\sin\theta}\frac{\partial Q}{\partial r'} \tag{7.2.5}$$

which automatically satisfies the incompressibility condition (which we shall apply on the basis of calculations done for the capillary flow).

It is less easy to define characteristic length and velocity scales than in the capillary flow. Clearly, it will be convenient to use d_c as a scale length and $v^*_c = \dot{V}/d^2_c$ as a scale velocity in order to match the entry flow with the capillary flow, but $d_r = G_3 d_c$ and $v^*_r = v^*_c/G^2_3$ are really the more appropriate over most of the flow field. The range of relevant rate-of-deformation scales is from $D^*_c = \dot{V}/d^3_c$ to $D^*_r = D^*_c/G^3_3$ and so with $G_3 \gg 1$, it is unlikely that the same approximations will necessarily be relevant over the whole of the flow field. In particular we may easily have circumstances where

$$D^*_c \Lambda \gg 1 \tag{7.2.6c}$$

yet

$$D_{\mathrm{r}}^{*}\Lambda \ll 1 \tag{7.2.6r}$$

Λ being a characteristic relaxation time for the fluid. This would suggest a Newtonian approximation for $r' = 0(d_{\mathrm{r}})$ but a highly elastic regime (whatever that proves to mean) for $r' = 0(d_{\mathrm{c}})$. Elastic effects would become relevant at values of r' for which

$$D(r', \theta)\Lambda = 0(1) \tag{7.2.7}$$

The dimensionless group

$$\mathrm{De} = D\Lambda \tag{7.2.8}$$

can be regarded as one example of the widely used dimensionless group known as Deborah number. This is generally intended to provide an estimate of elastic effects as compared with viscous effects and is most easily described as a ratio of time scales. The flow field provides one and the material's constitutive equation the other. For example, if we go back to the definition of weak and strong flows in Section 4.1, $(\mathrm{tr}\,\mathbf{L}^2)^{-\frac{1}{2}}$ defines the characteristic time of the flow field, while Λ, the relaxation time of the fluid, defines the characteristic material time. Other examples will arise later.

Apparently successful progress in analysing the flow field has been made in three ways:†

(1) By using computational methods. This has meant choosing a particular constitutive equation, expressing it and the flow equation $\boldsymbol{\nabla} . \mathbf{T} = 0$ in component form, and then using finite-difference or finite-element methods to evaluate the field variables given by (7.2.3). The consensus of opinion (Mendelson *et al.*, 1982; Black *et al.*, 1975; Pearson & Richardson, 1983) is that such methods only converge satisfactorily so long as $D_{\mathrm{c}}^{*}\Lambda \leqslant 0(1)$, though this seems to be associated with the way iterative schemes (necessary to deal with the non-linear nature of the constitutive equation) are constructed.

(2) By using analytical expansion schemes based on the Newtonian limit. These should be valid for $D_{\mathrm{c}}^{*}\Lambda \ll 1$ and might be useful for

$$D_{\mathrm{c}}^{*}\Lambda \rightarrow 0(1)$$

† Much of the relevant work has been carried out for plane wedge flows for which most of the arguments used here for conical flow apply.

(3) By making further kinematic assumptions for the case $D_c^*\Lambda \gg 1$, based partly on observation and partly on the known rheological behaviour of polymer melts at high shear and elongation rates (Hull, 1981*a*, *b*). in particular, it has been suggested that the entry flow can be split into an essentially elongational core flow (for which $|\mathrm{tr}\, \mathbf{W}^2| \ll |\mathrm{tr}\, \mathbf{D}^2|$ and a toroidal lubricating recirculating flow (for which $|\mathrm{tr}\, \mathbf{W}^2| \simeq |\mathrm{tr}\, \mathbf{D}^2|$). The material entering the capillary takes part in a 'strong' flow while the vortices represent 'weak' flow—where strong and weak flow are used in the general sense of Section 4.1.

For simplicity we start with the Newtonian limit. We find (see, for example, Happel and Brenner, 1973, Section 4.23) that the stream function Q_N satisfies the equation

$$\left[\frac{\partial^2}{\partial r'^2} + \frac{\sin\theta}{r'^2}\frac{\partial}{\partial\theta}\left(\frac{1}{\sin\theta}\frac{\partial}{\partial\theta}\right)\right]^2 Q_N = 0 \tag{7.2.9}$$

with

$$Q_N(r', 0) = 0 \quad \text{and} \quad Q_N(r', \theta_e) = -\dot{V}/2\pi \tag{7.2.10}$$

This has an obvious solution†

$$Q_N = -\frac{\dot{V}}{2\pi}\left[\frac{(1-3\cos^2\theta_e)(1-\cos\theta)+\sin^2\theta\cos\theta}{(1+2\cos\theta_e)(1-\cos\theta_e)^2}\right] \tag{7.2.11}$$

leading to

$$v_{r'N} = -\frac{3\dot{V}}{2\pi r'^2}\frac{(\cos^2\theta-\cos^2\theta_e)}{(1+2\cos\theta_e)(1-\cos\theta_e)^2}, \qquad v_{\theta N} = 0 \tag{7.2.12}$$

and

$$p_N = p_{\infty N} + \frac{\mu\dot{V}}{\pi r'^3}\frac{1-3\cos^2\theta}{(1+2\cos\theta_e)(1-\cos\theta_e)^2} \tag{7.2.13}$$

Thus, we note that a basic radial flow field can exist. In practice, the need to match the solution (7.2.11) to the cylindrical flows in the regions $r' \sim \frac{1}{2}d_c$ and $r' \sim \frac{1}{2}d_r$ will introduce extra terms in Q dependent on r' and θ, but this provides no difficulties in principle. Separable solutions of (7.2.9) in ascending powers of r' would be used for the large r' matching and in descending powers (inverse powers) of r' for the small r' matching.

† The negative sign arises because we use the convention that $\dot{V}$ is positive for inflow, i.e. when V_r is negative.

For small values of the Deborah number (see (7.2.8) above)

$$\mathrm{De} = \Lambda \dot{V}/d_{\mathrm{c}}^3 \tag{7.2.14}$$

we can expect the flow to be expressible in terms of an expansion

$$Q(r', \theta) = Q_{\mathrm{N}}(\theta) + \mathrm{De}\, Q^{(1)}(r', \theta) + \mathrm{De}^2\, Q^{(2)}(r', \theta) \tag{7.2.15}$$

about the Newtonian solution Q_{N}, with

$$Q^{(1)}(r', \theta_{\mathrm{e}}) = Q^{(2)}(r', \theta_{\mathrm{e}}) = \ldots = 0 \tag{7.2.16}$$

and with the $Q^{(J)}(r', \theta)$ obeying an inhomogeneous form of (7.2.9), the inhomogeneous terms being expressible in terms of Q_{N}, $Q^{(1)}, \ldots, Q^{(J-1)}$. To show this requires the constitutive equation to be written in the retarded motion form (see Subsection 3.3.3)

$$\mathbf{T}^{\mathrm{E}} = \mathbf{T}_{\mathrm{N}}^{\mathrm{E}} + \mathrm{De}\, \mathbf{T}^{(1)\mathrm{E}} + \mathrm{De}^2\, \mathbf{T}^{(2)\mathrm{E}} + \ldots \tag{7.2.17}$$

A very general procedure is advocated in Joseph (1979). The limit $\mathrm{De} \rightarrow 0$ can be viewed as one obtained either by letting $\Lambda \rightarrow 0$, or, which is physically more understandable, letting $\dot{V} \rightarrow 0$. Briefly, we see that the flow field (7.2.12) leads to the ordering

$$\begin{aligned} \mathbf{A}_{1\mathrm{N}} &= 0(\mathrm{De}) \\ \mathbf{A}_{2\mathrm{N}}, \mathbf{A}_{1\mathrm{N}}^2 &= 0(\mathrm{De}^2) \end{aligned} \tag{7.2.18}$$

or, more generally

$$\mathbf{A}_{J_1\mathrm{N}}^{n_1} \cdot \mathbf{A}_{J_2\mathrm{N}}^{n_2} \ldots \mathbf{A}_{J_K\mathrm{N}}^{n_K} = 0(\mathrm{De}^{\sum_1^K J_i n_i})$$

which is consistent with (7.2.17) in a very obvious way. The method has been used by Langlois & Rivlin (1959), Schümmer (1967) and Strauss (1975), and represents approach (2) mentioned above. No non-trivial matching for large or small r' has been attempted (see Hull, 1981*a*, *b*).

The 'strong'-core, 'weak'-vortex approach discussed under (3) above was introduced by Metzner (1971) and Pearson & Pickup (1973) as a means of estimating elongational or pure shear viscosities in polymer solutions. The approach has been adopted by Cogswell (1981) for polymer melts and examined carefully by Hull (1981*b*, see also Hull, *et al.* 1981). In all cases, it appears that at high elongational rates, the ratio of elongational stress to rate of elongation is one or two orders of magnitude larger than the equivalent ratio for simple shear flow at the same deformation rate. Put crudely

$$\frac{\eta_e(D)}{\eta(D)}, \frac{\eta_{PS}(D)}{\eta(D)} \gg 1 \tag{7.2.19}$$

for high D, where for Newtonian fluids the ratios would be 3 and 4, respectively.†

For dilute polymer solutions, η is almost constant, and the effect is due to an increase in η_e with D, while for polymer melts, η_e is more nearly constant so the effect is largely due to the decrease of η with D. Note that justification for the approach does not depend on a constant-stretch-history flow having been reached; the inequalities (7.2.19) are merely meant to be illustrative of the trends observed.

What concerns us most here is whether the heuristic approach of Metzner and others can be adapted as a formal asymptotic starting point for analysing the case (7.2.6c), i.e. large values of the Deborah number. No successful work along these lines has so far been published, so we can only make comments at this stage.

Some insight can be gained by considering spherically symmetrical sink flow. This flow would be strictly relevant if a zero wall shear stress rather than a zero wall shear velocity were applied at $\theta = \theta_e$ in (7.2.1), and appears to be approached in practice, at least for small values of θ, when $\theta_e \leqslant \pi/6$ and $\mathrm{De} \gg 1$. In this latter case, a thin highly-sheared lubricating layer close to the wall achieves the sharp change in radial velocity needed to satisfy the first part of (7.2.1).

In this case the flow field can be specified as

$$v_{r'} = -\dot{V}/4\pi r'^2, \qquad v_\theta = v_\phi = 0 \tag{7.2.20}$$

The vorticity $\mathbf{W}$ is identically zero, and the rate of deformation tensor becomes diagonal, with

$$d_{r'r'} = \dot{V}/2\pi r'^3 = -2d_{\theta\theta} = -2d_{\phi\phi} \tag{7.2.21}$$

The Jaumann derivative $\mathscr{D}/\mathscr{D}t$ or (°) becomes, in our steady flow situation,

$$\frac{\mathscr{D}}{\mathscr{D}t} = -\frac{\dot{V}}{4\pi r'^2}\frac{\mathrm{d}}{\mathrm{d}r'} \tag{7.2.22}$$

and all the Rivlin–Ericksen tensors $\mathbf{A}_N$ can be readily evaluated. The

† See (4.3.7), (4.3.9) and (4.2.2) for the definition of η_e, η_{PS} and η, and (4.3.10) for the Newtonian result.

position of a particle at all past times $0<s<\infty$ can be obtained simply by integrating

$$\frac{dr''}{ds} = -v_{r'} = \frac{\dot{V}}{4\pi r''^2} \tag{7.2.23}$$

to give $r''(r', s)$ as

$$r''^3 = r'^3 + 3\dot{V}s/4\pi \tag{7.2.24}$$

From this the deformation gradient tensor $\mathbf{C}_t^t(s)$, defined after (3.2.9) can be written as a function of r' and s. Like $\mathbf{D}$, it is diagonal with

$$C_{tr'r'}^t = (C_{t\theta\theta}^t)^{-2} = (C_{t\phi\phi}^t)^{-2} = \left(\frac{dr''}{dr'}\right)^2 = \left(1+\frac{3\dot{V}s}{4\pi r'^3}\right)^{-4/3} \tag{7.2.25}$$

Thus, the stress tensor can be evaluated as a function of r' for all the constitutive equations discussed in Chapters 3 and 5. Furthermore, only one quantity is of interest,

$$\sigma_1 - \sigma_2 = t_{r'r'}^E - t_{\theta\theta}^E \tag{7.2.26}$$

because $t_{\phi\phi}^E = t_{\theta\theta}^E$ and all other components of $\mathbf{T}^E$ are zero by symmetry.

As a useful and illustrative example, we take the Oldroyd 8-constant model (3.3.22) which leads to the equations

$$\sigma_1 - \frac{\dot{V}\lambda_1}{4\pi r'^2}\frac{d\sigma_1}{dr'} + \frac{\mu_0\dot{V}}{2\pi r'^3}(\sigma_1+2\sigma_2) - \frac{\mu_1\dot{V}}{\pi r'^3}\sigma_1 + \frac{\nu_1\dot{V}}{2\pi r'^3}(\sigma_1-\sigma_2)$$
$$= \frac{\eta_0\dot{V}}{\pi r'^3}\left\{1+\frac{\dot{V}}{4\pi r'^3}(3\lambda_2-4\mu_2+3\nu_2)\right\} \tag{7.2.27}$$

$$\sigma_2 - \frac{\dot{V}\lambda_1}{4\pi r'^2}\frac{d\sigma_2}{dr'} - \frac{\mu_0\dot{V}}{4\pi r'^3}(\sigma_1+2\sigma_2) + \frac{\mu_1\dot{V}}{2\pi r'^3}\sigma_2 + \frac{\nu_1\dot{V}}{2\pi r'^3}(\sigma_1-\sigma_2)$$
$$= -\frac{\eta_0\dot{V}}{2\pi r'^3}\left\{1+\frac{\dot{V}}{4\pi r'^3}(3\lambda_2+2\mu_2-6\nu_2)\right\} \tag{7.2.28}$$

It is convenient to define the new variables

$$\tilde{x} = 4\pi r'^3/3\dot{V}\lambda_1, \qquad \tilde{\sigma}_i = \sigma_i\lambda_1/\eta_0 \tag{7.2.29}$$

and the dimensionless material parameters

$$\tilde{\mu}_i = \mu_i/\lambda_1, \qquad \tilde{\nu}_i = \nu_i/\lambda_1, \qquad \tilde{\lambda}_2 = \lambda_2/\lambda_1 \tag{7.2.30}$$

in terms of which (7.2.27) and (7.2.28) become

$$-\frac{d\tilde{\sigma}_1}{d\tilde{x}}+\tilde{\sigma}_1\left[1+2\frac{(\tilde{\mu}_0-2\tilde{\mu}_1+\tilde{\nu}_1)}{3\tilde{x}}\right]+\tilde{\sigma}_2\left[\frac{2(2\tilde{\mu}_0-\tilde{\nu}_1)}{3\tilde{x}}\right]$$
$$=\frac{4}{3\tilde{x}}\left\{1+\frac{1}{3\tilde{x}}(3\tilde{\lambda}_2-4\tilde{\mu}_2+3\tilde{\nu}_2)\right\} \quad (7.2.31)$$

$$\tilde{\sigma}_1\left[\frac{(-\tilde{\mu}_0+2\tilde{\nu}_1)}{3\tilde{x}}\right]-\frac{d\tilde{\sigma}_2}{d\tilde{x}}+\tilde{\sigma}_2\left[1+\frac{2(-\tilde{\mu}_0+\tilde{\mu}_1-\tilde{\nu}_1)}{3\tilde{x}}\right]$$
$$=-\frac{2}{3\tilde{x}}\left[1+\frac{1}{3\tilde{x}}(3\tilde{\lambda}_2+2\tilde{\mu}_2-6\tilde{\nu}_2)\right] \quad (7.2.32)$$

We note that $\tilde{x}^{-1}$ is a local Deborah number, so our interest lies in small values of $\tilde{x}$. Formally, either $\tilde{\sigma}_1$ or $\tilde{\sigma}_2$ can be eliminated between (7.2.31) and (7.2.32), leading to a second-order ordinary differential equation for the other. We consider the special cases.

I. $\tilde{\mu}_1=1$; $\tilde{\mu}_2=\tilde{\lambda}_2$; $\tilde{\nu}_1=\tilde{\mu}_0=\tilde{\nu}_2=0$: This is the upper-convected Jeffreys fluid of (3.3.25). The $\tilde{\sigma}_1$ and $\tilde{\sigma}_2$ equations decouple, yielding

$$\frac{d\tilde{\sigma}_1}{d\tilde{x}}+\tilde{\sigma}_1\left(\frac{4}{3\tilde{x}}-1\right)=-\frac{4}{3\tilde{x}}\left(1-\frac{\tilde{\lambda}_2}{3\tilde{x}}\right) \quad (7.2.33)$$

and

$$\frac{d\tilde{\sigma}_2}{d\tilde{x}}-\tilde{\sigma}_2\left(\frac{2}{3\tilde{x}}+1\right)=\frac{2}{3\tilde{x}}\left(1+\frac{5}{3}\frac{\tilde{\lambda}_2}{\tilde{x}}\right) \quad (7.2.34)$$

A particular integral of (7.2.33) and (7.2.34) can be found in the form of a series expansion

$$\tilde{\sigma}_{1,2}=\frac{1}{\tilde{x}}(\sigma_{1,2}^{(0)}+\sigma_{1,2}^{(1)}\tilde{x}+\ldots) \quad (7.2.35)$$

with the leading term determined by $\tilde{\lambda}_2$. This we note is of exactly the same form as the leading term of the Newtonian solution, and allows us in a simple sense to say that the $\tilde{\lambda}_2$ term represents a high-extensional-rate viscosity. ($\tilde{\lambda}_2$, when relevant, has a value near 1/5.) If we take the Maxwell model-fluid approximation, then $\tilde{\lambda}_2=0$, and the dominant term in the particular integral expansion becomes, surprisingly, a constant isotropic stress: $\tilde{\sigma}_1=\tilde{\sigma}_2=\tilde{\sigma}_3=1$. However, we cannot expect the particular integral alone to match the outer Newtonian solution so we must consider the complementary functions given by

$$\frac{d\tilde{\sigma}_1}{d\tilde{x}}+\tilde{\sigma}_1\left(\frac{4}{3\tilde{x}}-1\right)=0, \qquad \frac{d\tilde{\sigma}_2}{d\tilde{x}}-\tilde{\sigma}_2\left(\frac{2}{3\tilde{x}}+1\right)=0 \qquad (7.2.36)$$

Written as series solutions, these are dominated by $\tilde{x}^{-\frac{4}{3}}$ and $\tilde{x}^{\frac{2}{3}}$, respectively, and so $\tilde{\sigma}_1-\tilde{\sigma}_2$ will be $0(\tilde{x}^{-\frac{4}{3}})$ for $\tilde{x}\rightarrow 0$. This is a more singular solution than the particular integral (7.2.35) dominated by $\tilde{\lambda}_2$. Thus, going back to the original Oldroyd model (3.3.22) we deduce that the far field $\tilde{x}\geqslant 0(1)$ matching together with the degenerate form

$$\overset{\nabla}{\mathbf{T}}^{E}=\mathbf{0} \qquad (7.2.37)$$

dominate the inner field expansion in this particular 'upper-convected' form of the Oldroyd equation. The conclusion remains true for all $\tilde{\mu}_1$ such that

$$1\geqslant\tilde{\mu}_1>3/4 \qquad (7.2.38)$$

and for the Maxwell model for all $|\tilde{\mu}_1|>0$ if we use the appropriate $\overset{\Box}{\mathbf{T}}^{E}=0$. In the F_{abc} notation, $\tilde{\mu}_1=-a$, while $b=c=0$ for this case I.

II. $\tilde{\mu}_1=\tilde{\nu}_1=\tilde{\mu}_0=\tilde{\mu}_2=\tilde{\nu}_2=0$. Again the equations for σ_1 and σ_2 decouple to give

$$\frac{d(\tilde{\sigma}_1-\tilde{\sigma}_2)}{d\tilde{x}}-(\tilde{\sigma}_1-\tilde{\sigma}_2)=-\frac{2}{\tilde{x}}\left(1+\frac{\tilde{\lambda}_2}{\tilde{x}}\right) \qquad (7.2.39)$$

The particular-integral solutions

$$\tilde{\sigma}_1-\tilde{\sigma}_2\sim\begin{cases}\dfrac{2\tilde{\lambda}_2}{\tilde{x}} & \text{for} \quad \tilde{\lambda}_2\geqslant 0\\ -2\ln\tilde{x} & \text{for} \quad \lambda_2=0\end{cases} \qquad (7.2.40)$$

now dominate over the complementary function $e^{\tilde{x}}$ and so the leading term is given by

$$\lambda_1\overset{\circ}{\mathbf{T}}^{E}=2\eta_0(\mathbf{D}+\lambda_2\overset{\circ}{\mathbf{D}}) \qquad (7.2.41)$$

in the 'highly elastic' limit. It can readily be seen that the conclusions can be modified to cover the case

$$\tfrac{3}{4}>\tilde{\mu}_1>1 \qquad (7.2.42)$$

Petrie (1979), Chapter 2, discusses in some detail the relation between Maxwell models and various extensions of the linear viscoelastic model. The exact correspondence between the upper- and lower-convected Maxwell models ($a = \pm 1$, $b = c = 0$ in F_{abc}) and the Lodge integral models can be checked directly by substituting (7.2.25) into (3.4.1) or into its analogue using (3.4.7) with $m(s) = \mathrm{e}^{-s/\lambda_1}$ for this particular sink flow. This latter approach has the advantage that the necessary matching between Newtonian and highly elastic regions is carried out automatically; it is implicit in the integral formulation.

Armed with our exact solutions, or expansions thereof, for the radial sink flow, we can now in principle return to the more complex problem posed by (7.2.1) and (7.2.2). We may hope to be guided by the asymptotic limits (7.2.37), (7.2.41) or whatever proved appropriate for sink flow, in establishing an expansion scheme encompassing θ-dependence of $\mathbf{D}$ and $\mathbf{T}^{\mathrm{E}}$. No successful treatment along these lines has been given.

In practice, experimental rheologists undertake $(\dot{V}, P)$ experiments using a variety of capillary lengths l_c at fixed $\dot{V}$ in order to distinguish between the entry pressure drop P_{entry} and the capillary drop P_{cap}. The former can always be expressed in terms of an apparent entry length connection

$$l_e = \tilde{l}_e d_c \tag{7.2.43}$$

where $\tilde{l}_e$ is a function of $\dot{V}$, or equivalently of D_c.

EXAMPLE 7.2.1. Estimate the entry pressure drop for a Newtonian fluid in a capillary rheometer, and hence the entry length correction.

We start with the known solution given by (7.2.11)–(7.2.13) and note at once that $p_{\mathrm{N}} - p_{\infty\mathrm{N}}$ is not independent of θ. Furthermore, we note that $t^{\mathrm{E}}_{r'r'} = 2\eta_0 d_{r'r'} = 2\eta_0\, \partial v_{r'}/\partial r'$ is also of order r'^3 and dependent upon it. Some approximation is therefore required in estimating the entry pressure drop.

We therefore consider the total force $F_{z\text{ entry}}$ in the z-direction (see Fig. 7.3) obtained by integrating the total normal stress component $t_{rr}(\theta)$ over the hemisphere $r' = \frac{1}{2}d_c$, where

$$t_{rr}(\theta) = (-p_{\mathrm{N}} + 2\eta_0\, \partial v_{r'}/\partial r')_{r' = \frac{1}{2}d_c}$$

This can be converted into an equivalent entry pressure drop by taking

$$\begin{aligned} P_{\text{entry}} &= \frac{4F_{z\text{ entry}}}{\pi d_c^2} + p_{\infty\mathrm{N}} \\ &= 2\int_0^{\pi/2} (p_{\infty\mathrm{N}} - p_{\mathrm{N}} + 2\eta_0\, \partial v_{r'}/\partial r')\cos\theta\, \sin\theta\, \mathrm{d}\theta \end{aligned}$$

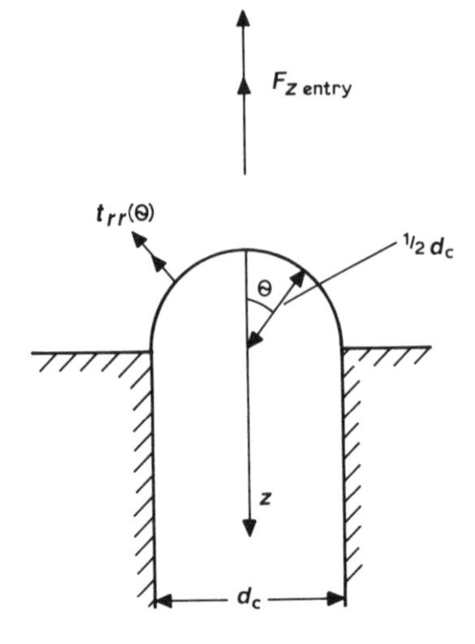

Fig. 7.3. Idealized geometry for entry-pressure-drop analysis.

Using (7.2.12) and (7.2.13) with $\theta_e = \frac{1}{2}\pi$, this gives

$$P_{\text{entry}} = -\frac{16\eta_0 \dot{V}}{\pi d_c^3} \int_0^{\pi/2} \{(1-3\cos^2\theta) - 6\cos^2\theta\}\cos\theta \sin\theta \, d\theta$$

$$= \frac{7\eta_0 \dot{V}}{2\pi(\frac{1}{2}d_c)^3}$$

If we write

$$l_e = \frac{\pi P_{\text{entry}}(\frac{1}{2}d_c)^4}{8\eta_0 \dot{V}}$$

and substitute for P_{entry}, we obtain

$$l_e = \frac{7}{32} d_c$$

If we had used instead the spherically symmetrical sink flow solution given by

$$v_{r'} = -\dot{V}/2\pi r'^2$$

then the r' component of the stress equilibrium equation

$$\frac{\partial p}{\partial r} = \eta_0 \left\{ \frac{1}{r'^2} \frac{\partial}{\partial r'} \left(r'^2 \frac{\partial v_{r'}}{\partial r'} \right) - \frac{2v_{r'}}{r'^2} \right\}$$

yields

$$p^{\infty}-p=\int_{r}^{\infty}\frac{\partial p}{\partial r}\,\mathrm{d}r=\frac{2\eta_0\dot{V}}{3\pi(\frac{1}{2}d_c)^3}$$

Adding on the component $2\eta_0\,\partial v_{r'}/\partial r'$ gives a stress independent of θ, and hence equivalent to

$$P_{\text{entry}}=\frac{8}{3}\frac{\eta_0\dot{V}}{\pi(\frac{1}{2}d_c)^3}$$

and so

$$l_e=\frac{1}{6}d_c$$

In practice both of these solutions will underestimate the P_{entry} pressure drop because of the viscous losses involved in the velocity vector rearrangement between $r'=\frac{1}{2}d_c$ and $z=0$. A reasonable estimate would therefore be in the range

$$0{\cdot}8d_c>l_e>0{\cdot}4d_c$$

EXERCISE 7.2.1. Show that (7.2.5) satisfies the relation $\boldsymbol{\nabla}\cdot\mathbf{v}=\mathrm{tr}(\boldsymbol{\nabla}\mathbf{v})=0$ where $\boldsymbol{\nabla}\mathbf{v}$ is given by (7.2.4).

EXERCISE 7.2.2. Show that the equations of motion for a Newtonian incompressible fluid with velocity field given by (7.2.3) become (Bird *et al.*, 1960, Table 3.4.4) in spherical polar coordinates when inertia is neglected

$$\frac{1}{\eta_0}\frac{\partial p}{\partial r'}=\nabla^2 v_{r'}-\frac{2}{r'^2}v_{r'}-\frac{2}{r'^2}\frac{\partial v_{r'}}{\partial\theta}-\frac{2}{r'^2}v_\theta\cot\theta$$

$$\frac{1}{\eta_0 r'}\frac{\partial p}{\partial\theta}=\nabla^2 v_\theta+\frac{2}{r'^2}\frac{\partial v_{r'}}{\partial\theta}-\frac{v_\theta}{r'^2\sin^2\theta}$$

Show that (7.2.9) follows by use of (7.2.5) and elimination of p.

EXERCISE 7.2.3. Show that

$$\sigma_1^{(0)}=\frac{4}{3}\tilde{\lambda}_2,\qquad \sigma_2^{(0)}=-\frac{2}{3}\tilde{\lambda}_2$$

where $\sigma_{1,2}^{(0)}$ are defined by (7.2.35) satisfying (7.2.33) and (7.2.34).

Hence show that the leading term in the particular integral solution of $\sigma_1-\sigma_2$ is of the form $3\eta_0(\lambda_2/\lambda_1)d_{rr}$ which is analogous to a Newtonian viscosity $\frac{1}{2}\eta_0\lambda_2/\lambda_1$.

Show also that the terms in (7.2.36) that yield the complementary function solution $\tilde{x}^{-\frac{4}{3}}$ and $\tilde{x}^{\frac{2}{3}}$ arise from the $\overset{\nabla}{\mathbf{T}}^{\mathrm{E}}$ term in the original Oldroyd model, and hence are solutions of (7.2.57).

Finally deduce that in a simple spherical sink-flow given by (7.2.20), elastic effects are dominant in the sense that an infinite memory is exhibited.

EXERCISE 7.2.4. Obtain the complementary-function solutions for the analogue of (7.2.36) in case I with $|\tilde{\mu}_1|\leqslant 1$ in the form

$$\tilde{x}^{-4\tilde{\mu}/3}, \qquad \tilde{x}^{2\tilde{\mu}/3}$$

and hence justify the conclusion associated with (7.2.38).

EXERCISE 7.2.5. Carry out in detail the analysis leading to (7.2.39) for case II and the result (7.2.41).

Show that the extension of the argument to the cases (7.2.42) can be obtained from the result of Exercise 7.2.4.

EXERCISE 7.2.6. Show that the upper-convected Maxwell model is equivalent to using $\mathbf{H}_t^t(s)$ (or $\mathbf{C}_t^{t-1}(s)$), as defined by (3.4.7), in (3.4.1) instead of $\mathbf{G}_t^t(s)$, with $m(s)$ given by (3.4.13). Hence express $\sigma_1-\sigma_2$ as defined in (7.2.26) for the velocity field (7.2.20) in terms of $3\dot{V}\pi/4\lambda r'^3$ as an integral. This is a result used in Section 7.4.

Obtain expansions for $\sigma_1-\sigma_2$ for the cases of $r'\gg 1$ and $r'\ll 1$.

EXERCISE 7.2.7. Show that if a material exhibits power-law behaviour in simple shear and Newtonian behaviour in extensional (sink) flow then the entry lengths increases as $\dot{V}^{1-\nu}$. Derive an exact relation as in Example 7.2.1.

7.3 CORNER FLOW

One feature of the matching between the entry and capillary flows deserves mention: the singularity in stress and rate-of-strain that arises at a sharp-edged entry, such as shown in Fig. 7.4. If we take a local planar coordinate system $(\hat{r}, \hat{\theta})$ for the region $\hat{r}\ll d_c$, then we look for

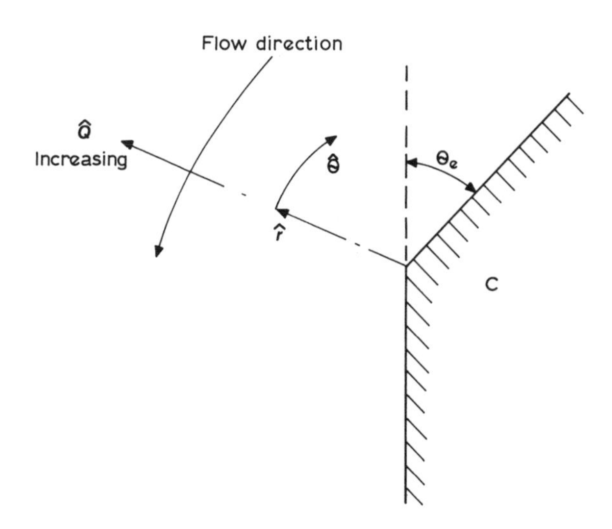

Fig. 7.4. Flow near a corner C. The planar polar coordinate system $(\hat{r}, \hat{\theta})$ will be relevant for the near neighbourhood of the corner, when $\hat{r} \ll d_c$.

flows such that

$$\mathbf{v} = \mathbf{0} \quad \text{on} \quad \hat{\theta} = \pm\tfrac{1}{2}(\pi + \theta_e) \tag{7.3.1}$$

Along the line $\hat{\theta} = 0$, we expect the stream function $\hat{q}$ to increase, where

$$v_{\hat{r}} = \frac{1}{\hat{r}}\frac{\partial \hat{q}}{\partial \hat{\theta}}, \qquad v_{\hat{\theta}} = -\frac{\partial \hat{q}}{\partial \hat{r}} \tag{7.3.2}$$

so as to give flow in the direction indicated in the figure. The Lagrange stream function $\hat{q}$ allows the incompressibility condition to be automatically satisfied.

Once again we can take the Newtonian limit for illustrative purposes, in which case it can be shown (see Exercise 7.3.1) that $\hat{q}$ satisfies

$$\nabla^2\nabla^2\hat{q} \equiv \left(\frac{\partial^2}{\partial \hat{r}^2} + \frac{1}{\hat{r}}\frac{\partial}{\partial \hat{r}} + \frac{1}{\hat{r}^2}\frac{\partial^2}{\partial \hat{\theta}^2}\right)^2 \hat{q} = 0 \tag{7.3.3}$$

We look for solutions of the form (see Moffatt, 1964)

$$\hat{q} = \hat{r}^c X_c(\hat{\theta}) \tag{7.3.4}$$

whence by substitution we find that

$$X_c = X_{c1} \cos c\,\hat{\theta} + X_{c2} \cos(c-2)\hat{\theta} \tag{7.3.5}$$

if we assume symmetry about $\hat{\theta}=0$. The vector boundary condition (7.3.1) leads to the two scalar conditions

$$X_c = \frac{\partial X_c}{\partial \hat{\theta}} = 0, \qquad \hat{\theta} = \pm\tfrac{1}{2}(\pi + \theta_e) \tag{7.3.6}$$

which in turn require c to satisfy an eigenvalue equation involving θ_e,

$$\sin\{c\pi + (c-1)\theta_e\} = -(c-1)\sin\theta_e \tag{7.3.7}$$

For $\theta_e > 0$, this has at least one real solution. We take as relevant the solution $\bar{c}(\theta_e)$ that satisfies

$$\bar{c}(\theta_e) \to 2 \quad \text{as} \quad \theta_e \to 0 \tag{7.3.8}$$

$$\bar{c}(\theta_e) \to \frac{3}{2} \quad \text{as} \quad \theta \to \pi \tag{7.3.9}$$

The limit (7.3.8) is consistent with Poiseuille flow through a circular-cylindrical tube, while (7.3.9) has the same singularity as is given for flow through a circular hole in a plate (Happel & Brenner 1973), so the solution is *a priori* the one we want. It can be shown that $\bar{c}$ is monotonic in the range $0 < \theta_e < \pi$, so that $\bar{c} < 2$ for $\theta_e > 0$. This means that the rate of deformation, D, defined by (2.3.19) behaves as $\hat{r}^{\bar{c}-2}$, which is singular as $\hat{r} \to 0$. The stresses are also singular but integrable in the sense that $\int_0^1 \hat{r}^{\bar{c}-2}\,dr = \dfrac{1}{\bar{c}-1}$ is bounded, so that the total force exerted on the corner is bounded. The details are not so important as the general conclusion that a retarded-motion expansion can never justify the use of a Newtonian model as the asymptotic limit in the whole of a corner flow: however slow the flow, i.e. however small $|X_{\bar{c}1}|$, there will always be a region $\hat{r} < \hat{r}_{\text{crit}}$ such that non-linear effects will dominate in the sense that the Deborah number

$$\text{De} = |X_{\bar{c}1}|\, \hat{r}_{\text{crit}}^{\bar{c}-2}\, \Lambda \gg 1 \tag{7.3.10}$$

The only way of avoiding this conclusion is to suppose that there exists a length $\hat{r}_{\min}$ below which either the corner cannot be sharp, or the continuum hypothesis cannot be maintained.

Thus, formally, for any given fluid whose constitutive equation is supposed known, we might seek its characteristic corner flow to which all outer flows should be matched. No such non-linear solutions have been obtained. To show that such aspects must be important, we need only point out that no non-linear viscosity effect can be expected in a fully-developed capillary flow without such non-linearities having been relevant in the corner region at lower throughputs. These difficulties are conveniently disregarded when computational methods are used;† the integrability conditions mean that the linear or quadratic approximating functions used in finite difference or finite element methods are non-singular.‡ It must be admitted that comparison of these solutions with experiment largely justifies the conclusion that corner singularities are not of critical importance in smooth flows of the type shown in the upper half of Fig. 7.2(b). It is an open question whether they exert a significant effect on the outer flow in cases like that shown in the lower half of the figure where a dividing streamline meets the corner.

It must now be clear that reasonable predictions of entry pressure losses (ΔP_e) as a function of volume flow rate $\dot{V}$ cannot be made on the basis of fluid mechanical calculations for any but strictly Newtonian fluids. For the latter the extra pressure drop is equivalent to rather less than one diameter's length (see Example 7.2.1) of capillary and can be accurately checked experimentally by using capillaries of different length and equal diameter (see Pearson, 1966, Section 2.6). In elastic fluids, the same experimental observations show end-corrections equivalent to many diameters, where these corrections are themselves increasing functions of shear rate. From the point of view of processing predictions, it is probably preferable to use the raw capillary-extrusion data for entry losses than to interpret the latter first in terms of constitutive relations and assumed flow fields.

EXERCISE 7.3.1. Show that, in the slow-flow limit, $\nabla \cdot \mathbf{T} = 0$, for a Newtonian fluid given by (3.3.1), the equations of motion in cylindrical-polar coordinates (Fig. 7.4) become, when

$$\mathbf{v} = [v_{\hat{r}}(\hat{r}, \hat{\theta}), v_{\hat{\theta}}(\hat{r}, \hat{\theta}), 0],$$

$$\frac{1}{\eta}\frac{\partial p}{\partial \hat{r}} = \frac{\partial}{\partial \hat{r}}\left(\frac{1}{\hat{r}}\frac{\partial}{\partial \hat{r}}(\hat{r}v_{\hat{r}})\right) + \frac{1}{\hat{r}^2}\frac{\partial^2 v_{\hat{r}}}{\partial \hat{\theta}^2} - \frac{2}{\hat{r}^2}\frac{\partial v_{\hat{\theta}}}{\partial \hat{\theta}}$$

$$\frac{1}{\eta \hat{r}}\frac{\partial p}{\partial \hat{\theta}} = \frac{\partial}{\partial \hat{r}}\left(\frac{1}{\hat{r}}\frac{\partial}{\partial \hat{r}}(\hat{r}v_{\hat{\theta}})\right) + \frac{1}{\hat{r}^2}\frac{\partial^2 v_{\hat{\theta}}}{\partial \hat{\theta}^2} + \frac{2}{\hat{r}^2}\frac{\partial v_{\hat{r}}}{\partial \hat{\theta}}$$

† But see Holstein (1981) who uses the Stokesian, i.e. Newtonian, limit at the corner.

‡ For further details see Pearson & Richardson (1983).

Derive (7.3.3) using the substitution (7.3.2) and eliminating p from the above two equations.

The relevant equations of motion are given in Table 3.4.3 of Bird *et al.* (1960).

7.4 EXIT FLOW (EXTRUDATE SWELL)

The flow at the exit from the capillary differs from that at the entry in that a free surface forms in almost all cases. This is shown in Fig. 7.5 where streamlines lying near the wall for upstream ($\hat{\theta} \ll \frac{1}{2}\pi$) regions emerge into the ambient environment ($\hat{\theta} > \frac{1}{2}\pi$) and converge towards the free surface. In the analogous inlet flow with $\theta_e = \frac{1}{2}\pi$, the fluid remains in contact with the wall on both sides of the corner.

In practice, even for strictly Newtonian fluids, a small swell ratio Es, (7.1), is observed when the Reynolds number becomes inappreciable, provided gravity and surface tension effects are also negligible, while for strongly elastic polymer melts such as LDPE or PP swell ratios of

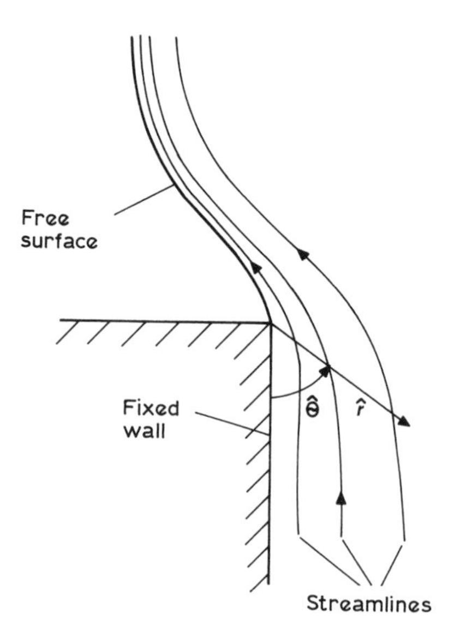

Fig. 7.5. Flow near a corner when a free surface forms. A local polar coordinate system ($\hat{r}$, $\hat{\theta}$) is shown with streamlines that converge towards the free surface, for $\hat{r} \ll d_c$.

2·0 are common (Pearson & Richardson, 1983, Chapter 3; Huang & White, 1980). A careful visual investigation of the free surface in the neighbourhood of the corner suggests that the tangent plane at the corner does not coincide with $\hat{\theta} = \pi$ and that a sharp change in direction occurs at the corner $\hat{r} = 0$; however, it is also clear that a considerable curvature exists at the corner, so estimates of the angle that the tangent plane makes with the axis $\hat{\theta} = 0$ are unlikely to be accurate. The axisymmetric free surface settles down to a fairly uniform cross-sectional diameter d_e within a few capillary diameters d_c in most cases, though there are reports of a variable stand-off distance before a sudden rearrangement of flow takes place. The rapidity of swelling is at least as great with elastic melts as with Newtonian fluids.

In a trivial sense we can see that the final velocity v_∞ will be given by

$$\tfrac{1}{4}\pi d_e^2 v_\infty = \dot{V} \tag{7.4.1}$$

$\dot{V}$ being defined earlier, and that if $\mathrm{Es} > 1$, v_∞ will correspond to some value $v_z(r)$ in the fully-developed velocity profile for capillary flow, with $0 < r < \tfrac{1}{2}d_c$. 'Core' fluid will be decelerated, while fluid near the capillary walls will be accelerated during the exit flow. Thus, streamlines near $r = 0$ will move apart, while those near $r = \tfrac{1}{2}d_c$ will move together, during the exit flow. The relevant (r, ϕ, z) coordinate system is shown in Fig. 7.6 and is chosen to coincide with that used for capillary flow in Fig. 7.1(c). In simple terms, we can say that fluid near the edges will have been stretched axially and circumferentially while fluid near the centre will have been compressed axially and stretched circumferentially.

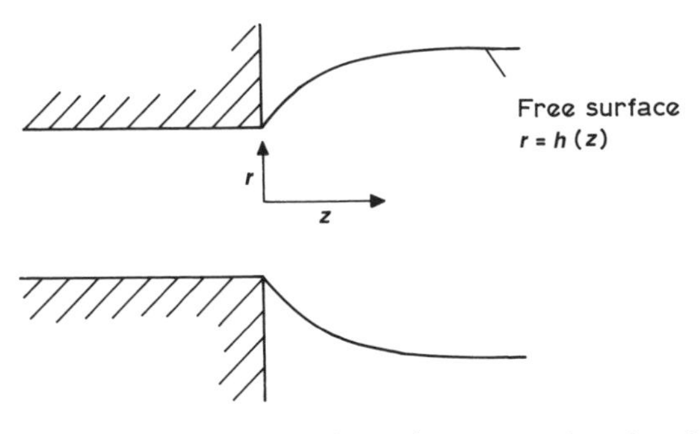

Fig. 7.6. Polar coordinate system (r, ϕ, z) relevant for description of free surface position in extrudate swell.

As on previous occasions, we can hope to be guided in a mechanical analysis of the phenomenon by a treatment of the Newtonian case. However, on this occasion, we find that no complete and satisfactory treatment of the steady situation has been given so far.

Although the governing equations for the flow are the same as for the corner flow, the boundary conditions are mixed:

$$\mathbf{v}(\tfrac{1}{2}d_c, z<0) = \mathbf{0} \tag{7.4.2}$$

$$\left.\begin{aligned} \mathbf{v}\cdot\hat{\mathbf{n}}_{FB} &= \mathbf{0} \\ \mathbf{T}\cdot\hat{\mathbf{n}}_{FB} &= \mathbf{0} \end{aligned}\right\} \quad \text{for} \quad r = Y(z>0) \tag{7.4.3}$$

(7.4.3) represents the conditions that the free surface has no component of velocity normal to itself and be stress-free if, as explained above, surface tension and gravity forces are neglected. The free surface is assumed to be at $r = Y(z)$ for $z > 0$, where obviously

$$2Y(0) = d_c, \qquad 2Y(\infty) = d_e \tag{7.4.4}$$

and

$$\hat{\mathbf{n}}_{FB} = \left[\frac{1}{(1+Y'^2)^{\frac{1}{2}}}, 0, -\frac{Y'}{(1+Y'^2)^{\frac{1}{2}}}\right] \tag{7.4.5}$$

The ambient pressure p_a has been set equal to zero without loss of generality.

Joseph (1974) gives the full stress boundary conditions when $\sigma, \mathbf{g}$ and inertia are included.

The position of the free surface is unknown *ab initio* and its determination is part of the problem. This introduces many difficulties, whatever method of solution is attempted.

We cannot even argue that a unique solution exists (as is normal for linear problems), in the sense that absence of a free surface ($Y = \infty$) is a possible alternative solution of the problem, while the free-surface boundary conditions are equivalent to non-linearities (see Jean & Pritchard, 1980).

Various analytical contributions to a full solution have been made. Joseph (1974) following earlier work of Richardson (1967) has shown how the ultimate steady uniform flow $2Y = d_e$, $v = v_\infty$ can be perturbed for $z \gg 1$, yielding eigen solutions of the form $\exp\{-k_n z\}\ R_n(r)$, and that an expansion in terms of these eigen functions could, in principle, be matched to a local flow field in the region $z = 0(1)$. From this a value for $\mathrm{d}Y/\mathrm{d}z$ as $z \rightarrow 0$, consistent with $Y(0) = \frac{1}{2}d_c$, should be derivable. However, this work is not immediately compatible with other results for flow in the immediate neighbourhood of the edge where the free surface forms.

Michael (1958) has shown that if the free surface is locally plane near the edge, i.e. lies at some fixed $\hat{\theta}_{FS}$ in the notation of Fig. 7.5, then $\hat{\theta}_{FS} = \pi$. He reaches this result by considering separable solutions of the form (7.3.4) and satisfying the boundary conditions, derived from (7.4.2) and (7.4.3).

$$\hat{q}(\hat{r}, 0) = \hat{q}_{\hat{\theta}}(\hat{r}, 0) = 0 \tag{7.4.6}$$

$$\left.\begin{aligned} &\hat{q}(\hat{r}, \hat{\theta}_{FS}) = \hat{q}_{\hat{\theta}\hat{\theta}}(\hat{r}, \hat{\theta}_{FS}) = 0 \\ &3\hat{r}\hat{q}_{\hat{r}\hat{\theta}} + 3\hat{r}^2\hat{q}_{\hat{r}\hat{r}\hat{\theta}} - \hat{q}_{\hat{\theta}\hat{\theta}\hat{\theta}} - 4\hat{q}_{\hat{\theta}} = 0 \end{aligned}\right. \tag{7.4.7}$$

The solution admits of the single eigen value

$$\hat{\theta}_{FS} = \pi \tag{7.4.8}$$

which implies a stream function dominated by $\hat{r}^{\frac{3}{2}}$ near the corner. This, in turn, leads to an $\hat{r}^{-\frac{1}{2}}$ singularity in the shear stress as $\hat{r} \to 0$ for $\hat{\theta} = 0$, as was the case for confined flow through an orifice in a flat plate, and so as before casts doubt on the suitability of a Newtonian solution as the leading term of an uniformly-valid retarded-motion expansion. Furthermore, the perturbation expansion mentioned above will not yield the value of zero for $\mathrm{d}Y/\mathrm{d}r$ as $z \to 0$, which is necessary to match (7.4.8).

Tayler (1972) and Richardson (1970) have suggested that the function $Y(z)$ need not be analytic at $z = 0$ and Tayler has proposed a free surface shape

$$Y(z) = z^{\nu}$$

with ν non-integral. Some unpublished work of Millard Johnson (University of Wisconsin, private communication), based on an integral-equation representation for (7.3.3), (7.4.6) and (7.4.7) has suggested that the curvature of the free surface at $z = 0$ will not be bounded although the slope can remain so.

Sturges (1981) has given a low-De solution using a domain-perturbation theory based on Richardson's (1970) stick-slip solution.

Several numerical solutions (see Pearson & Richardson, 1983, Chapter 3) have been given for the Newtonian-fluid extrudate-swell problem which correctly predict the observed value of swell (about 13%) using methods that are insensitive to the edge singularity. These suggest that the edge flow does not exert a significant influence on the outer flow field.

7.4.1 Tanner's Theory

We now turn back to a consideration of exit flow for an elastic fluid. Crochet & Keunings (1980, 1982) and Tanner (1980*b*) have undertaken numerical solutions for various rheological equations of state, which appear to be qualitatively consistent with observations. In principle the finite-element methods used can be applied to the complete entry/capillary/exit-flow field, this reducing the global fluid-mechanical problem to one of numerical computation, once the constitutive equation is specified. As before, practical difficulties arise when large Deborah numbers are attained, and asymptotic solutions valid in those cases would be desirable.

Attempts that have been made to predict extrudate swell analytically as a function of the parameters θ_e, L/D and De have been based on grossly unrealistic assumptions. In physical terms, they imply that Es for large De is determined largely by the elastic energy stored in the fluid reaching the end of the capillary. If $\mathrm{De}(\mathrm{L/D})^{-1}$ is small, then a fully-developed viscometric flow will have been obtained and the potential for swelling will be governed in a loose sense by the shear creep-recovery curve (Fig. 5.5). If, however, $\mathrm{De}(\mathrm{L/D})^{-1}$ is not small, then the core fluid will remember the elongational flow field it experienced in the entry region, and an elongational creep-recovery curve will be relevant in the same loose sense. Figure 7.7 shows diagrammatically the ideas involved. However, the kinematic constraints involved in going from the upstream velocity profile (Fig. 7.1(d)) to the flat velocity profile given by v_∞ of (7.4.1) mean that the actual recovery of individual fluid elements will be subject to varying constraints, different from those relevant for the classical shear and elongational creep-recovery curves. We cannot know what the true time-dependent constraints are until we have solved completely for the flow field in the neighbourhood of the exit. Progress can only be made by making crude assumptions about these constraints.

The earliest apparently-successful predictive theory was suggested by Lodge and developed by Tanner (1970). It supposed:

(a) The flow is fully developed up to the exit, at which point a uniform cylindrical length is suddenly released from all external stress on the cylindrical surface and allowed to alter shape subject to continuity (and incompressibility) conditions, and to zero axial tension.

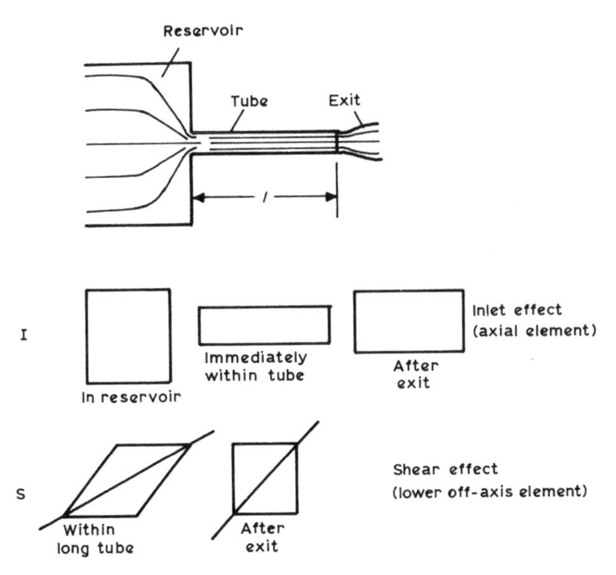

Fig. 7.7. Diagrammatic representation of 'elastic' process leading to die swell. I, irrotational distortion and recovery; S, shear distortion and recovery.

(b) The fluid is a Kaye–BKZ fluid with $\phi_1 = 0$ (eqn (3.4.9)).

(c) The observed extrudate swell Es will be the sum of the Newtonian extrudate swell, 0·13, and that predicted by (a) and (b).

The kinematics of this process are very simple because it implies that the axial length of every annulus of the test length in the die is reduced by the same ratio Es_e^{-2}. This yields the stretch ratio Es_e in each of the r and ϕ directions by continuity. Only the amount of shear recovery is free to vary with radius r. The relative deformation gradient tensor is given by a sudden jump

$$\mathbf{F}_0 = \begin{bmatrix} \mathrm{Es}_e^2 & g(r) & 0 \\ 0 & \mathrm{Es}_e^{-1} & 0 \\ 0 & 0 & \mathrm{Es}_e^{-1} \end{bmatrix} \tag{7.4.9}$$

while the previous viscometric flow is given by

$$\mathbf{F}^{(v)}(s) = \begin{bmatrix} 1 & -\dfrac{\partial v_z}{\partial r} s & 0 \\ 0 & 1 & 0 \\ 0 & 0 & 1 \end{bmatrix} \tag{7.4.10}$$

in the (z, r, ϕ) representation† with the final state being the reference state in (7.4.9) and the exit state being the reference state in (7.4.10). The history of the deformation is then

$$\mathbf{F}_{t=0}(\tau)=\mathbf{F}=\mathbf{1} \quad \text{for} \quad \tau>0; \qquad \mathbf{F}=\mathbf{F}^{(v)}\cdot\mathbf{F}_0 \quad \text{for} \quad \tau<0 \tag{7.4.11}$$

(see (3.2.2)–(3.2.6) for definitions of the various **F**'s).

The functions $g(r)$ and Es are determined by the requirement that the stress tensor **T** be diagonal in the (z, r, ϕ) representation and that

$$\int_0^{\frac{1}{2}d_e} rt_{zz}\,\mathrm{d}r=0 \tag{7.4.12}$$

$$t_{rr}(\tfrac{1}{2}d_e)=0 \tag{7.4.13}$$

The final result is that

$$\mathrm{Es}_e=\left[1+\frac{\Psi_1^2(D_w)D_w^2}{8\eta^2(D_w)}\right]^{\frac{1}{6}}=(1+\tfrac{1}{2}\mathrm{Ws}^2)^{\frac{1}{6}} \tag{7.4.14}$$

or

$$\mathrm{Es}=0{\cdot}13+\mathrm{Es}_e \tag{7.4.15}$$

where Ψ_1 and η are the viscometric functions, eqns (4.2.3), (4.2.2), and D_w is the wall shear rate in the viscometric flow. The mean Weissenberg number Ws is defined by (7.4.11).

The Tanner theory has one weakness: it does not correspond to reality in that what comes out of the exit in any given time is not a cylindrical element whose length is independent of cross-sectional position, but one whose length is proportional to the fully-developed velocity profile. This is easily appreciated by recognizing that fluid at the die wall never leaves the die, and so the fluid forming the outer skin of the extrudate has been infinitely extended axially.

An explanation based upon viscous resistance to elongation of the fluid near the free surface has recently been given by Tanner (1980a).

7.4.2 Pearson & Trottnow's Theories

An alternative long-capillary theory taking this into account has been given by Pearson & Trottnow (1978) whose assumptions are:

(i) The flow is fully developed right up to the exit.

† This order of coordinates corresponds to the (1, 2, 3) terminology of viscometric flow introduced in Section 4.2.

(ii) At the exit a sudden deformation takes place for each differential fluid element.
(iii) Continuity is preserved, the fluid being incompressible.
(iv) The final flow is that of an undeforming rod.
(v) The stress-equilibrium equation is satisfied in the final undeforming state, with zero boundary stress and zero net axial tension.

Because they discovered that applying (v) immediately after (ii) led to an infinite extrudate swell B_e, they added (vi) a variable delay time (function of radial position r) for each fluid annulus before (v) was applied.

This was intended to take some account of the fact that the outer layers of fluid actually take much longer to move from the die exit to the almost fully swollen region than the inner layers. Once again the kinematics of the initial (fully-developed-capillary) and final (undeforming-rod) flow fields specifies the nature of the sudden deformation.

$$\mathbf{F}_s = \begin{bmatrix} v_z'/v_\infty & k(r) & 0 \\ 0 & \mathrm{d}r'/\mathrm{d}r & 0 \\ 0 & 0 & r'/r \end{bmatrix} \tag{7.4.16}$$

where r', v_z' refer to radial position and velocity in the die and r and v to the final swollen extrudate, $k(r)$ being an initially-unspecified shear-recovery factor.

Pearson & Trottnow used a Maxwell fluid model (3.3.26) in integral form ((3.4.1) using (3.4.7) and (3.4.13)). This is a simple example of the restricted Kaye–BKZ model used by Tanner, with a single-exponential function of s for ϕ_{-1} in (3.4.9). It also corresponds to the upper-convective form of the differential relation (3.3.26) and so is the same as the model used in I above with $\tilde{\lambda}_2 = 0$, leading to the result (7.2.37). The Deborah and Weissenberg numbers are essentially equivalent for this model and flow field. The latter is defined as

$$\mathrm{Ws} = \frac{32\dot{V}\lambda_1}{\pi d_c^3} \tag{7.4.17}$$

where λ_1 is the single relaxation time, which is $32/\pi$ times the De defined by (7.2.14).

Their results are shown in Fig. 7.8 for various values of E, where the

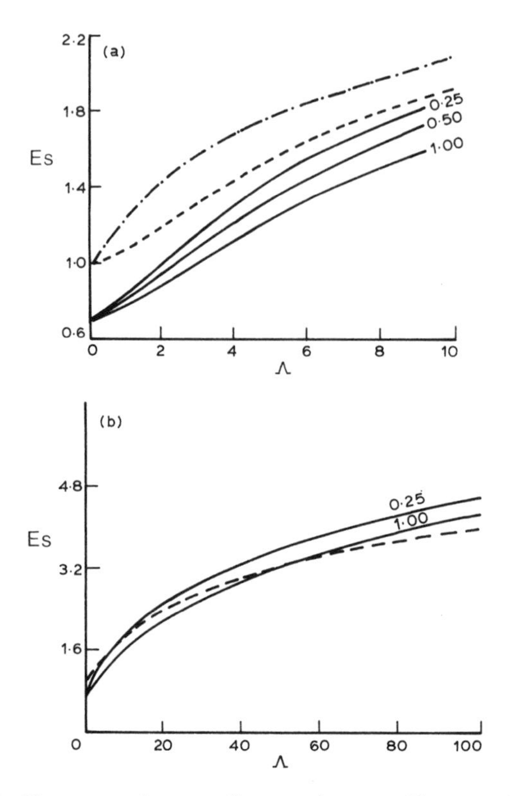

Fig. 7.8. Capillary extrudate swell. ---, long-capillary model (values of E shown against curves) of Pearson and Trottnow. ---, long-capillary model of Tanner. -·-·-, short-capillary model, L/D = 0.

delay time t_+, referred to in (vi) above, is given by

$$t_+(r') = Ed_c/2v_z'(r') \tag{7.4.18}$$

For values of Ws between 10 and 50, corresponding to Es between 1·5 and 3·0, which is the interesting range of extrudate swell for many elastic melts, there is little to choose between the two theories, given the crudeness of the models.

However, both fail to take account of the observed fact that extrudate swell is also dependent upon the die (capillary) L/D ratio. Indeed our earlier discussion of capillary flow suggested that for Ws(De) $\geqslant$ 0(1), capillary flows would not be fully developed in a rheological sense. Pearson & Trottnow give a very elementary treatment for extrudate swell in a short capillary, which is based entirely on the idea of recovery from the elastic strain resulting from the entry flow alone. The extreme case arises when $L/D=0$, and their assumptions then become:

(i) The flow is spherically-symmetric sink flow up to the exit hole (this is the flow discussed in Section 7.2 above, particularly Exercise 7.2.6) and is uniform for each element of fluid.
(ii) At the exit hole a sudden deformation takes place, uniformly for each element of fluid using a diagonal form for $\mathbf{F}_s$ similar to (7.4.9) with $g(r)=0$, such that
(iii) the subsequent flow is that of a moving undeforming stress-free rod,
(iv) continuity having been preserved.

Using the same constitutive equation as for the long-capillary model and defining Ws as in (7.4.17), an explicit result was obtained for Es which is also plotted in Fig. 7.8. For large values of Ws its asymptotic form is

$$\mathrm{Es}_0=(\tfrac{3}{2}\mathrm{Ws}^2)^{\frac{1}{6}} \tag{7.4.19}$$

which is very close to that given for Es_e in (7.4.14).

The model is extended to cover short dies by supposing plug flow to take place in the die with consequent relaxation of stress (and hence loss of stored elastic energy) leading to a reduction in swell ratio with L/D. For a simple Maxwell model, exhibiting a single relaxation time λ_1, the decay is simply exponential, although the relation for Es_e is a little more complex

$$\mathrm{Es}^6=\frac{\int_{\mathrm{Ws}^{-1}}^{\infty} \mathrm{e}^{-x}x^{\frac{2}{3}}\,\mathrm{d}x+[\exp\{4(\mathrm{L/D})\mathrm{Ws}^{-1}\}-1]}{\mathrm{Ws}^{-2}\int_{\mathrm{Ws}^{-1}}^{\infty} \mathrm{e}^{-x}x^{-\frac{2}{3}}\,\mathrm{d}x+[\exp\{4(\mathrm{L/D})\mathrm{Ws}^{-1}\}-1]} \tag{7.4.20}$$

This result is not incompatible with the empirical proposal of Locati (1976), which written in our terminology becomes:

$$(\mathrm{Es}^2-1)=(\mathrm{G}_3^{-2}-1)\exp\{4(\mathrm{L/D})\mathrm{Ws}^{-1}\} \tag{7.4.21}$$

where

$$\Delta P_{\mathrm{entry}}=3G(\mathrm{G}_3^{-2}-1) \tag{7.4.22}$$

G being the elastic shear modulus and ΔP_{entry} the entry pressure drop (attributed wholly to a sudden inlet elastic strain).

The effect of temperature variations on die-swell is analysed in Ben-Sabar & Caswell (1981).

To complete our discussion of capillary-rheometer flow, we must consider the stability of the steady flow fields we have assumed to arise (this is step 5 of the procedure described at the beginning of Chapter 6). We have every reason to do so, because experiments show that steady symmetric flow almost invariably breaks down as Ws becomes large compared with unity. Because of the difficulty we have had in obtaining either analytical or numerical solutions to all but the fully-developed (viscometric) capillary flow, we cannot expect to carry out stability analyses in the precise fashion that is possible for many Newtonian flows. However, the phenomenon of melt-flow instability, elastic turbulence or the extreme form melt fracture, is of such importance in processing operations that we shall discuss it as fully as possible in a later section, Section 9.1.

Chapter 8

Approximate Methods

Many of the flows encountered in polymer processes are characterized by a narrow dimension in one direction. This is true both of confined flows where the channel geometry is determined by rigid boundaries and of free flows where the boundaries are usually with the ambient air. In both these cases, analyses of the flow field can be split up into a local and a global problem, involving significantly different length scales. As a result, feasible methods of solution can be developed.

8.1 THE LUBRICATION APPROXIMATION. FULLY DEVELOPED FLOW

Our discussion of capillary flow above introduced the notion of the asymptotic fully-developed flow in an infinitely long capillary, and the separate development lengths for momentum, $0(d_c)$, rheological history, $0(10d_c)$, and temperature, $0(\mathrm{Pe}\, d_c)$, in ascending order of magnitude. The geometry involved there was particularly simple, with a defined entry to a cylindrically and otherwise axially-symmetric flow field. What we deduced was that, for a significant and useful range of L/D (say 10–200), provided Gn was not too large, the fully-developed flow solution was a useful approximation to flow in the capillary, e.g. it would predict $(P, \dot{V})$ relations adequately.

We also noted that, in the entry-cone flow for a Newtonian fluid, the limit $\theta_e \to 0$ corresponded to entry flow having capillary flow as its asymptotic limit. This can be seen by writing

$$\sin\theta = r/r' \ll 1, \qquad \cos\theta = 1 - \frac{1}{2}\frac{r^2}{r'^2} + \frac{1}{8}\frac{r^4}{r'^4} + \text{h.o.t.} \quad (8.1.1)$$

for $\theta_e \ll 1$, and substituting in (7.2.11) to give

$$Q_N = -\frac{\dot{V}\tilde{r}}{\pi}\left\{1 - \tfrac{1}{2}\tilde{r}^2 + 0\left(\frac{r}{r'}\right)^2\right\} \tag{8.1.2}$$

where $\tilde{r} = r/r_e$. Hence, the difference in stream function (and similarly in the pressure field) between flow through a capillary of radius r_e and a cone of small angle θ_e is of order θ_e^2. In other words, we can get a result correct to order θ_e^2 by replacing the smoothly converging cone by a series of circular cylindrical tubes having a cross-sectional area locally equal to that of the cone. This is shown in Fig. 8.1, where four approximating tubes (1–4) are used to approximate to the cone C between A and B. No entry or exit corrections need be used because the real conical flow is smooth. In an analytical calculation the number of approximating tubes can become indefinitely large because (8.1.2) can be used as a purely local value for continuous integration along r', from r'_A to r'_B say. This procedure can obviously be generalized for axisymmetric tubes of slowly varying cross-section; thus, if the radius

$$X = X(z) \tag{8.1.3}$$

we merely require that

$$\mathrm{d}X/\mathrm{d}z \ll 1 \tag{8.1.4}$$

for flow fields based on the assumption that the local flow is that in a tube of radius $X(z)$ to give results correct to order $(\mathrm{d}X/\mathrm{d}z)^2$. We do not give a formal proof of this here. A better approximation useful up to larger values of $\mathrm{d}X/\mathrm{d}z$ is obtained by assuming the flow to be that given for a cone of angle θ_e corresponding to the local value of $\mathrm{d}X/\mathrm{d}z$,

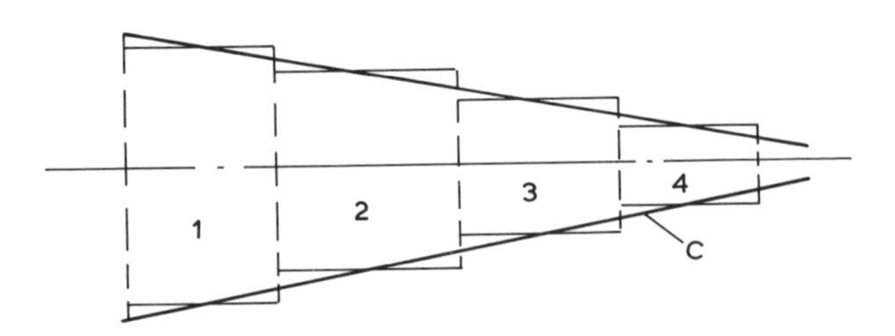

Fig. 8.1. The cone C is replaced by the series (1, 2, 3, 4, in diagram) of circular cylindrical tubes for calculating $(P, \dot{V})$ approximately.

i.e. with

$$\sin\theta_e = \mathrm{d}X/\mathrm{d}z \tag{8.1.5}$$

So far we have based our proposed approximation on conical flow of an isothermal Newtonian fluid, for which we have an exact solution anyway.

The utility of the approximating technique becomes much greater when the rheology of the fluid is sufficiently complex for only the capillary-flow solution to be obtainable at all easily. Indeed, if direct experimental evidence for $(P, \dot{V})$ is available from capillary rheometry for the fluid in question, the procedure proposed above avoids any discussion of other than viscometric flow.

If we consider the effect of elasticity, we must require that significant changes in cross-section, and hence in deformation rates of a given particle, take place in times large compared with the relaxation time Λ. This implies that we must have the Deborah number

$$\mathrm{De}_{\mathrm{lub}} = \frac{\Lambda\dot{V}}{X^3}\frac{\mathrm{d}X}{\mathrm{d}z} \ll 1 \tag{8.1.6}$$

where $\dot{V}/X^2$ is a characteristic velocity and $X(\mathrm{d}X/\mathrm{d}z)^{-1}$ a characteristic length. This is usually a more stringent condition than (8.1.4) because the Weissenberg number

$$\mathrm{Ws}_{\mathrm{lub}} = \Lambda\dot{V}/X^3 \quad \text{is} \quad \geqslant 0(1) \tag{8.1.7}$$

for elastic fluids. The criterion for a fully-developed thermal field is even more stringent and is seldom met; if it is, the effects of temperature variations are relatively unimportant anyway.

The approximation implied by Fig. 8.1 based on (8.1.4) is the essence of lubrication theory, as applied to nearly unidirectional axisymmetric flow.

Lubrication theory, as usually understood, is, however, applied mainly to flow fields that are narrow in one direction only, compared with their extent in all other directions. The flow field thus consists of the region between two neighbouring surfaces. This rather vague statement can be given precise geometrical significance. A formal general account based on asymptotic analysis, given in Pearson (1967), will be summarized here.

We consider a surface S_0, in which it is convenient to select an orthogonal coordinate system (y^1, y^2) such that it is equidistant from the prescribed solid (but not necessarily motionless) surface boundaries of the flow field, S_1 and S_2, say. At every point P on S_0, the width $h(y^1, y^2)$ of the channel between S_1 and S_2 can be defined as the distance along the normal to S_0 through P. We then define a coordinate y^3 along this line, such that the axes of (y^1, y^2, y^3) coincide locally with the axes (x_1, x_2, x_3) of a rectangular Cartesian system. It is often convenient to arrange for the surfaces S_1 and S_2 to coincide with the coordinate surfaces $y^3 = \pm 1$; if y^3 is then identified with x_3, this fixes the metric tensor for the local x_i-system.† There will be a metric tensor with diagonal components $g_1(y^i)$, $g_2(y^i)$, and $g_3 = \frac{1}{4}h^2$, associated with the y^i-system. Locally the maximum rate of change in width of the channel will be determined by

$$|\nabla_{S_0} h| = \left| \frac{1}{(g_1)^{\frac{1}{2}}} \frac{\partial h}{\partial y^1}, \frac{1}{(g_2)^{\frac{1}{2}}} \frac{\partial h}{\partial y^2} \right| \tag{8.1.8}$$

and so a local length scale in the S_0 plane is given by

$$l_{S_0} = h / |\nabla_{S_0} h| \tag{8.1.9}$$

The requirements for lubrication theory to apply are that

(i) $$h/l_{S_0} = |\nabla_{S_0} h| \ll 1 \tag{8.1.10i}$$

(ii) $$Hh \ll 1 \tag{8.1.10ii}$$

where H is the first principal curvature of the surface S_0. This can be written as

$$H = |\nabla_{S_0} \cdot \hat{\mathbf{n}}| \tag{8.1.11}$$

where $\hat{\mathbf{n}}$ is the unit normal to S_0.

(iii) that the extent of S_0 is at least l_{S_0}.

These have the consequence that, to first-order in $\nabla_{S_0} h$, using h as the length scale, and for values of x_i of order unity,

(a) the axes of the x_i-system coincide with the coordinate lines y^i, and
(b) the surfaces S_0, S_1 and S_2 are plane.

We can then consider an idealized-flow-field solution, for each and

† This can in one sense be regarded as leading to a dimensionless coordinate system $\tilde{x}_i$, the length scale being $\frac{1}{2}h$, but it is here really intended to do no more than define the unit of length, purely for arithmetic convenience.

every point $P \equiv (y^1, y^2)$ on S_0, which is determined by flow between exactly parallel planes S_1', S_2', whose local motion is the same as that of the surfaces S_1 and S_2 at P, under a pressure gradient $\nabla p'$ whose value at P is the same as $\nabla_{S_0} p$ in the true flow field. The solution to the idealized problem will yield a volume flux/unit length

$$\mathbf{q}' = (q_1', q_2') \tag{8.1.12}$$

in the (x_1, x_2) plane, which to first order in H/L (the local value of ∇h), will be the same as the two-dimensional flux

$$\mathbf{q}(y^1, y^2) = (q_1, q_2) \tag{8.1.13}$$

in the true flow field. Here

$$q_i = \tfrac{1}{2} h \int_{-1}^{1} v_{y^i}\, dy^3 \tag{8.1.14}$$

where v_{y^i} $(i = 1, 2)$ are the physical components of the velocity $\mathbf{v}$.

We consider two cases for the local idealized flow.

8.1.1 The Steady Uniform Problem

Here h and the velocity boundary conditions on S_1' and S_2', and hence the entire solution, are independent of t and (x_1, x_2). We have the analogue of fully-developed flow in a pipe; we seek a solution in the form

$$\begin{aligned} \mathbf{v}' &= \{v_1'(x_3), v_2'(x_3), 0\}, \qquad T = T'(x_3) \\ p' &= p_{,1}x_1 + p_{,2}x_2 + p_0'(x_3) \end{aligned} \tag{8.1.15}$$

where $p_{,1}$ and $p_{,2}$ are constants to be determined later, subject to boundary conditions

$$\begin{aligned} \mathbf{v}(1) &= \mathbf{u}_{S1} = (v_{11}, v_{12}, 0), \qquad & T(1) &= T_{S1} \\ \mathbf{v}(-1) &= \mathbf{u}_{S2} = (v_{-11}, v_{-12}, 0), \qquad & T(-1) &= T_{S2} \end{aligned} \tag{8.1.16}$$

It can be shown fairly readily (Example 8.1.1) that (8.1.15) represents a viscometric flow, and so the flow equations become

$$\frac{d}{dx_3}\left(\eta(D, T)\frac{dv_1'}{dx_3}\right) = p_{,1} \tag{8.1.17}$$

$$\frac{d}{dx_3}\left(\eta(D, T)\frac{dv_2'}{dx_3}\right) = p_{,2} \tag{8.1.18}$$

where

$$D^2 = \left(\frac{dv_1'}{dx_3}\right)^2 + \left(\frac{dv_2'}{dx_3}\right)^2 \tag{8.1.19}$$

and $\eta(D, T)$ is the viscosity, which is dependent on both shear rate and temperature,

$$p_0' = 0\{D^2\Psi_1(D)\} \tag{8.1.20}$$

and

$$\frac{d}{dx_3}\left(\alpha \frac{dT}{dx_3}\right) = -\eta(D)D^2 \tag{8.1.21}$$

The relation (8.1.20) arises because the viscometric functions do not specify $\mathbf{T}^E$ but merely its deviatoric part. In any case p_0' plays no dynamical role in the solution for the idealized flow problem.

From (8.1.17)–(8.1.19), (8.1.21) and the boundary conditions (8.1.16), knowing $\eta(D, T)$ and $\alpha(T)$, $\mathbf{v}'$ and T' can in principle be found in terms of $p_{,1}$ and $p_{,2}$. Thus we can calculate the two-dimensional volume flux/unit length

$$\mathbf{q}' = \mathbf{q}'(\nabla_{S'}p'; T_{S1}, T_{S2}; \mathbf{u}_{S1}, \mathbf{u}_{S2}; h) \tag{8.1.22}$$

In principle this function can be inverted to give

$$\nabla_{S'}p' = \nabla_{S'}p'(\mathbf{q}'; T_{S1}, T_{S2}; \mathbf{u}_{S1}, \mathbf{u}_{S2}; h) \tag{8.1.23}$$

where $\nabla_{S'}$ is the two-dimensional operator $\left(\frac{\partial}{\partial x_1}, \frac{\partial}{\partial x_2}\right)$, equivalent to (8.1.8) when (y^1, y^2) is a Cartesian coordinate system.

One example of this procedure is given in Example 8.1.2 and a range of such solutions is used in Chapters 10, 11, 13, 14 and 19.

EXAMPLE 8.1.1. Proof that (8.1.15) represents a viscometric flow satisfying (8.1.17)–(8.1.21).

The flow field (8.1.15) leads to a rate-of-deformation tensor

$$\mathbf{D} = \begin{bmatrix} 0 & 0 & \dot\gamma_{13} \\ 0 & 0 & \dot\gamma_{23} \\ \dot\gamma_{13} & \dot\gamma_{23} & 0 \end{bmatrix}, \qquad \dot\gamma_{13} = \frac{dv_1'}{dx_3} \qquad \dot\gamma_{23} = \frac{dv_2'}{dx_3}$$

where the components are functions of x_3. However, for any particular value of x_3 there will be a coordinate system (x_1^*, x_2^*, x_3) obtained by rotation about the x_3-axis in which

$$\mathbf{D}^* = \begin{bmatrix} 0 & 0 & \frac{1}{2}\dot{\gamma}_{13}^* \\ 0 & 0 & 0 \\ \frac{1}{2}\dot{\gamma}_{13}^* & 0 & 0 \end{bmatrix}$$

where

$$\dot{\gamma}_{13}^{*2} = \dot{\gamma}_{13}^2 + \dot{\gamma}_{23}^2$$

it having been assumed that $(\dot{\gamma}_{13}, \dot{\gamma}_{23})$ is a two-dimensional vector function of x_3. Any particle of fluid that is initially at that value of x_3 remains at the same value of x_3 because $v_3'(x_3) \equiv 0$. Hence the flow for any particle is a motion of constant stretch history (Section 4.1) and because of the specific form of $\mathbf{D}^*$ will be a viscometric flow yielding a local form for the total stress tensor (cf. eqn (6.1.6))

$$\mathbf{T}^* = \begin{bmatrix} -p + \Psi_1(\dot{\gamma}_{13}^*)\dot{\gamma}_{13}^{*2} & 0 & \eta(\dot{\gamma}_{13}^*)\dot{\gamma}_{13}^* \\ 0 & -p - \Psi_2(\dot{\gamma}_{13}^*)\dot{\gamma}_{13}^{*2} & 0 \\ \eta(\dot{\gamma}_{13}^*)\dot{\gamma}_{13}^* & 0 & -p \end{bmatrix}$$

Transforming back into the original coordinates yields

$$\mathbf{T} = \begin{bmatrix} -p + \dot{\gamma}_{13}^2\Psi_1(\dot{\gamma}_{13}^*) - \dot{\gamma}_{23}^2\Psi_2(\dot{\gamma}_{13}^*) & 0 & \eta(\dot{\gamma}_{13}^*)\dot{\gamma}_{13} \\ 0 & -p + \dot{\gamma}_{23}^2\Psi_1(\dot{\gamma}_{13}^*) - \dot{\gamma}_{13}^2\Psi_2(\dot{\gamma}_{13}^*) & \eta(\dot{\gamma}_{13}^*)\dot{\gamma}_{23} \\ \eta(\dot{\gamma}_{13}^*)\dot{\gamma}_{13} & \eta(\dot{\gamma}_{13}^*)\dot{\gamma}_{23} & -p \end{bmatrix}$$

If this is substituted into the equation $\nabla \cdot \mathbf{T} = 0$ then eqns (8.1.17) and (8.1.18) are obtained directly. (8.1.20) follows when account is taken of the fact that p_0' is typically taken as $\frac{1}{3}\,\mathrm{tr}(\mathbf{T})$.

EXAMPLE 8.1.2. Derive the volume flowrate $\mathbf{q}$ of a power-law fluid for pressure flow between 2 stationary parallel plates h apart in terms of the pressure gradient ∇p, where η is independent of T.

We use the stress equation (8.1.17) with x_1 chosen to lie along the direction of ∇p. Since $\mathbf{u}_{S1} = \mathbf{u}_{S2} = 0$ in (8.1.15), and $p_{,2} = 0$ in (8.1.18), it follows that $v_2 \equiv 0$.

A first integral of (8.1.17) becomes

$$\eta(\mathrm{d}v_1/\mathrm{d}x_3)\,\mathrm{d}v_1/\mathrm{d}x_3 = p_{,1}x_3$$

if symmetry about the plane $x_3=0$ is assumed. Using relations (7.1.63) and (7.1.64) for a power-law fluid we obtain

$$\frac{dv_1}{dx_3}=\left|\frac{p_{,1}x_3}{K}\right|^{1/\nu}\frac{p_{,1}x_3}{|p_{,1}x_3|}$$

(the modulus signs are included to avoid fractional or other non-integral powers of negative numbers).

Integration yields successively, when use is made of $\mathbf{u}_{S1}=\mathbf{u}_{S2}=0$,

$$\begin{aligned}v_1&=\frac{-p_{,1}|p_{,1}|^{1/\nu-1}}{K^{1/\nu}}\int_{-\frac{1}{2}h}^{x_3}|x_3|^{1/\nu}\,dx_3 \quad \text{for} \quad x_3<0\\ &=\frac{-p_{,1}|p_{,1}|^{1/\nu-1}}{K^{1/\nu}}\left(\frac{\nu}{\nu+1}\right)((\tfrac{1}{2}h)^{1+1/\nu}-|x_3|^{1+1/\nu})\end{aligned}$$

for all x_3, by symmetry, and

$$\begin{aligned}q&=\int_{-\frac{1}{2}h}^{\frac{1}{2}h}v_1(x_3)\,dx_3\\ &=\frac{-p_{,1}|p_{,1}|^{1/\nu-1}}{K^{1/\nu}}\left(\frac{2\nu}{2\nu+1}\right)(\tfrac{1}{2}h)^{2+1/\nu}\end{aligned}$$

Re-adopting the vector formulation yields

$$\mathbf{q}=-\frac{\nabla p}{|\nabla p|}\frac{h^2\nu}{2(2\nu+1)}\left(\frac{h|\nabla p|}{2K}\right)^{1/\nu}$$

EXERCISE 8.1.1. Show that the total pressure drop P for two-dimensional flow of a Newtonian fluid of constant viscosity η_0 in the channel

$$h(x_1)=h_0+(h_1-h_0)x_1/l \qquad 0\leqslant x_1\leqslant l$$

is given by

$$P=\frac{6\eta_0 lq}{(h_1-h_0)}\left|\frac{1}{h_1^2}-\frac{1}{h_0^2}\right|$$

where q is the flow rate per unit width.

Carry out a similar analysis for a power-law fluid, using the results of Example 8.1.2.

8.1.2 The Unsteady or Squeeze-film Problem

Here h and possible $\mathbf{u}_{S1}$, $\mathbf{u}_{S2}$ are given functions of t but not of (x_1, x_2). We show that

$$v_3(1) - v_3(-1) = \mathrm{d}h/\mathrm{d}t \tag{8.1.24}$$

will be small compared with $|\mathbf{q}'|/h$ almost everywhere. By integrating the continuity equation (2.2.7) over x_3

$$\frac{\mathrm{d}h}{\mathrm{d}t} = \nabla_{S'} \cdot \mathbf{q}' \tag{8.1.25}$$

where $\mathbf{q}'$ is now $\mathbf{q}'(x_1, x_2, t)$.

We next integrate (8.1.25) over an area of order $l_{S_0}^2$, where l_{S_0} is given by (8.1.9), and obtain, using the divergence theorem,

$$\frac{\mathrm{d}h}{\mathrm{d}t} = 0\left(\frac{|\mathbf{q}_n'|}{l_{S_0}}\right) \tag{8.1.26}$$

at most, where $|\mathbf{q}_n'|$ is the maximum value of the component of $\mathbf{q}'$ normal to the boundary of the selected area, whose perimeter will be of order l_{S_0} in length.

Thus,

$$\frac{\mathrm{d}h}{\mathrm{d}t} = 0\left(\frac{|\mathbf{q}'|}{h}\,\mathrm{H/L}\right) \ll 0\left(\frac{|\mathbf{q}'|}{h}\right) \tag{8.1.27}$$

over an area for which the idealized flow $\mathbf{q}'$ is a lubrication approximation to the actual flow $\mathbf{q}$.

We now argue that the flow field will still be dominated by the viscometric flow field given by $\mathbf{u}_{S1}$, $\mathbf{u}_{S2}$ and $\nabla_{S'}p'$. The requirement (8.1.25) is only significant at a global level, i.e. over length scales of order l_{S_0} and greater.

8.1.3 The Global Flow Problem

Our idealized local flow solution (8.1.22) contains the parameter $\nabla_{S'}p'$ (or alternatively the parameter $\mathbf{q}'$ if (8.1.23) is adopted). We can now use this solution to reduce the full three-dimensional global flow problem to a two-dimensional one in the S_0 surface.

The equivalent of (8.1.25) provides one relation

$$\frac{\mathrm{d}h}{\mathrm{d}t} = \nabla_{S_0} \cdot \mathbf{q} \tag{8.1.28}$$

in which of course $\mathrm{d}h/\mathrm{d}t$ may be zero, while the requirement that $p(y^1, y^2, t)$ be single valued on S_0 provides another, i.e.

$$\nabla_{S_0} \wedge \nabla_{S_0} p = 0 \tag{8.1.29}$$

where we employ the relation (8.1.23) to give $\nabla_{S_0} p$ in terms of $\mathbf{q}$.

Equations (8.1.28) and (8.1.29) are the differential equations governing flow in the narrow gap between S_1 and S_2. A particular solution is prescribed when boundary conditions are given on the periphery $\mathcal{B}$ of the defining surface S_0. The length scale of $\mathcal{B}$ must clearly be of order l_{S_0} at least.

The analogue of the wall boundary condition (2.3.1) is now

$$\mathbf{q} \cdot \hat{\mathbf{n}}_B = h \mathbf{v}_B \cdot \hat{\mathbf{n}}_B \tag{8.1.30}$$

where $\hat{\mathbf{n}}_B$ is the normal to the boundary $\mathcal{B}$ in the surface S_0 and $\mathbf{v}_B$ is the velocity of the boundary 'wall', if such a lateral constraint, with $\mathbf{v}_B$ independent of y^3, is relevant.

Note that the no-slip condition, at this global level, has had to be replaced by a volume conservation requirement. The actual local flow field will, because of the true no-slip condition, differ substantially from the idealized solution for a distance of order h from the boundary $\mathcal{B}$, but this will not introduce errors any larger than $0(|\nabla_{S_0} h|)$.

Parts of the boundary $\mathcal{B}$ will represent places where fluid can flow in and out, i.e. entries from and exits into reservoirs (cf. the ends of a capillary). On these we can suppose that either p or $\mathbf{q} \cdot \hat{\mathbf{n}}_B$ (as in (8.1.30)) is prescribed. Formally, then, we write:

$$p = p_B, \qquad \mathbf{q} \cdot \hat{\mathbf{n}}_B = q_B \tag{8.1.31}$$

Examples of the use of lubrication theory arise frequently in the fluid mechanical analysis of polymer processes. These account for the majority of published solutions to engineering flow problems, and form the basis for much of the analyses given in Chapters 10–14 and 18–19.

8.1.4 Thermal Effects

It is useful to develop criteria for the complete neglect of thermal effects or for the use of the fully-developed thermal flow field given by (8.1.21). To do this, we make (8.1.17), (8.1.18) and (8.1.21) dimensionless as we did for capillary flow in Section 7.1.

There is no trouble in choosing a length scale, for this will be the local value of h, regarded as constant. Choice of a velocity scale will be less simple because there are several velocities that are relevant. The wall boundary conditions yield a mean convected velocity

$$\mathbf{u}_S = \tfrac{1}{2}(\mathbf{u}_{S_1} + \mathbf{u}_{S_2}) \tag{8.1.32}$$

and a mean shearing velocity

$$\mathbf{u}_A = \tfrac{1}{2}(\mathbf{u}_{S_1} - \mathbf{u}_{S_2}) \tag{8.1.33}$$

The total flux provides an overall mean velocity

$$\bar{\mathbf{v}}_c = \mathbf{q}'/h \tag{8.1.34}$$

which has a differential component due to pressure

$$\bar{\mathbf{v}}_p = \bar{\mathbf{v}}_c - \mathbf{u}_S \tag{8.1.35}$$

For assessing heat-generation effects we are concerned with mean shear rates; this suggests using

$$v^*_{gen} = |\bar{\mathbf{v}}_p| + |\mathbf{u}_A| \tag{8.1.36}$$

as a velocity scale.

For assessing convective effects, we should more obviously use

$$v^*_{conv} = |\bar{\mathbf{v}}_c| \tag{8.1.37}$$

Thus, by analogy with (7.1.31) and (7.1.32) the relevant Peclet number will be

$$\mathrm{Pe} = \rho^* \Gamma^*_p v^*_{conv} h/\alpha^* \tag{8.1.38}$$

and the generation number, which in this case is the Nahme number

$$\mathrm{Gn} = \mathrm{Na} = \eta^* v^{*2}_{gen}/\alpha^* T^*_{rheol} \tag{8.1.39}$$

Here η^* should be evaluated at the shear rate given by (v^*_{gen}/h) and at a suitable characteristic temperature. Our boundary conditions yield T_{S_1} and T_{S_2}, which will obviously be relevant in fully-developed flow. If

$$\mathrm{B}_{lub} = |T_{S_1} - T_{S_2}|/T^*_{rheol} \not\ll 1 \tag{8.1.40}$$

then the difference in wall temperature will be rheologically significant, and the fully-developed flow solution will involve a coupling between (8.1.17), (8.1.18) and (8.1.21). If $\mathrm{Na} \not\ll 1$, then rheologically-significant temperature gradients will arise and again coupling of the velocity and temperature fields will occur. Going back to the definition of Na, we can use either T_{S_1} or T_{S_2} in evaluating η^* provided B_{lub} and Na so calculated are $\ll 1$. This then provides a suitable criterion for treating

$\eta(D, T)$ in (8.1.17) and (8.1.18) as independent of T; namely

$$\mathrm{B}_{\mathrm{lub}}, \mathrm{Na} \ll 1 \tag{8.1.41}$$

If, however, the group

$$|\boldsymbol{\nabla} h \,.\, \bar{\mathbf{v}}_c|\, \rho^* \Gamma_p^* h / \alpha^* \not\ll 1 \tag{8.1.42}$$

where we have taken the component of $\bar{\mathbf{v}}_c$ in the direction in which h is varying as an improvement on the simple product $|\boldsymbol{\nabla} h|\, v^*_{\mathrm{conv}}$, then the flow is a developing flow thermally and so (8.1.21) is not a suitable approximation of the thermal energy equation. Provided (8.1.41) is still obeyed, together with

$$\mathrm{B} = |T_{\mathrm{S}_1} - T_0| / T^*_{\mathrm{rheol}} \ll 1 \tag{8.1.43}$$

where T_0 is an entry temperature for the fluid, then decoupling of the velocity equation from the thermal energy equation can still be assumed.

If not, then the local relations (8.1.22) are not applicable, and the lubrication approximation has to be amended to allow for a rheologically-significant developing temperature field. In cases where

$$|\mathbf{u}_A| \gg |\bar{\mathbf{v}}_c|, \quad \text{or} \quad v^*_{\mathrm{gen}} \gg v^*_{\mathrm{conv}} \tag{8.1.44}$$

we are probably wiser to use v^*_{gen} rather than v^*_{conv} in the calculation of Pe and hence of the left-hand side of (8.1.42) because of the large counter-current flow within the melt that is implied by (8.1.44) and which will dominate convection.

The equations used in Subsections 8.1.1–8.1.3 have some of the characteristics of dimensionless equations, in that the local length scale h has been used to ensure that the flow boundaries are given by $y^3 = \pm 1$, and that relations (8.1.22) and (8.1.23) imply a dependence upon h. When developing flow equations are used, this dependence has to be correctly included in the energy equation.

Thus locally, for steady flow, we can write

$$\rho \Gamma_p \left(v_1 \frac{\partial T}{\partial x_1} + v_2 \frac{\partial T}{\partial x_2} \right) = \alpha \frac{\partial^2 T}{\partial x_3^2} + \eta(D, T) D^2 \tag{8.1.45}$$

where D^2 is given by (8.1.19), and T is a rapidly-varying function of x_3 and a more slowly varying function of x_1 and x_2. The various terms in eqn (8.1.45) can be brought into balance by using the stretched coordinates

$$x_1^* = x_1/\mathrm{Pe}, \qquad x_2^* = x_1/\mathrm{Pe} \tag{8.1.46}$$

where Pe is given by (8.1.38) and must be regarded as a local variable, particularly so if (8.1.42) holds. This yields

$$\left(v_1^* \frac{\partial T}{\partial x_1^*} + v_2^* \frac{\partial T}{\partial x_2^*}\right) = \frac{\partial^2 T}{\partial x_3^2} + \frac{\eta(D, T)D^2}{\alpha} \tag{8.1.47}$$

where now $\mathbf{v}^*$ is a physical velocity, whose scale is independent of local position. The way in which (8.1.47) is solved as part of the global flow problem will be considered in detail in later chapters.

8.2 THE THIN SHEET APPROXIMATION. PLANE STRESS

In lubrication theory, the lateral dimension $h(y^1, y^2)$ of the flow channel is part of the problem specification, and the velocity is prescribed on the boundaries. If, however, the flow field is supposed bounded by closely-spaced free surfaces S_1 and S_2, then we have to impose a stress boundary condition on S_1 and S_2, i.e. eqn (2.3.8), but no constraint on the velocity field. For many purposes we can neglect surface tension and external drag, and regard the ambient pressure as zero. We then get a zero boundary-stress condition

$$\mathbf{T} \cdot \hat{\mathbf{n}}_{\mathrm{FB}} = \mathbf{0} \tag{8.2.1}$$

The geometrical arguments involving a coordinate system (y^1, y^2) based on S_0 and the subsequent definition of $h(y^1, y^2)$, y^3 and then (x_1, x_2, x_3) together with the requirements (8.1.10i, ii), can be taken over to define a thin sheet approximation.† The relevant idealized flow field at each point $P \equiv (y^1, y^2)$ is provided by a plane sheet (S_0') of uniform thickness $h' = h(y^1, y^2, t)$ which is tangential to the surface S_0 at P, in which the flow field is given instantaneously by

$$\mathbf{v}' = (\boldsymbol{\nabla}\mathbf{v}')_{\mathrm{P}} \cdot \mathbf{x} \tag{8.2.2}$$

where

$$(\boldsymbol{\nabla}\mathbf{v}')_{\mathrm{P}} = \begin{vmatrix} \partial_1 v_1' & \partial_2 v_1' & 0 \\ \partial_1 v_2' & \partial_2 v_2' & 0 \\ 0 & 0 & -(\partial_1 v_1' + \partial_2 v_2') \end{vmatrix} \tag{8.2.3}$$

† The formal mathematical treatment leading to the required asymptotic result for the steady axisymmetric case is given in Pearson & Petrie (1970*a*).

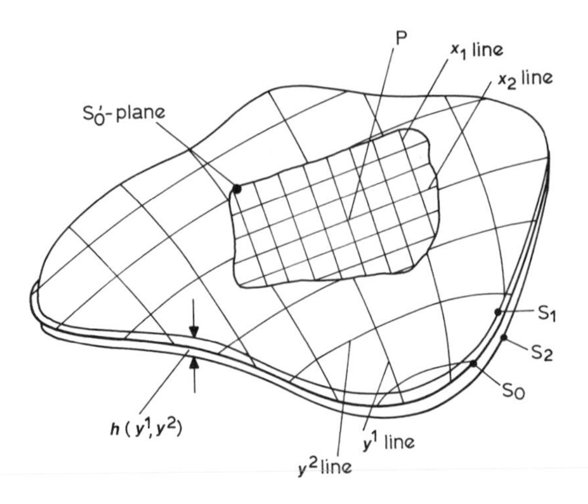

Fig. 8.2. Diagrammatic representation of thin-sheet approximation. The plane S_0', is tangential at P to the surface S_0, which is itself midway between the outer surfaces S_1 and S_2 of the deforming sheet. Orthogonal coordinate lines y^1 and y^2 are shown on S_0 and the corresponding (locally coincident) rectangular cartesian coordinate lines x_1 and x_2 on S.

is a function of (y^1, y^2) and is independent of $\mathbf{x}$ (here $\partial_1 v_1' \equiv \partial v_1'/\partial x_1$ etc.). It satisfies the incompressibility condition automatically. The local idealized flow is such that the point $\mathbf{x}=\mathbf{0}$ moves with the fluid at point P in S_0, and the plane S_0' remains tangential to S_0: the frame of reference for $\mathbf{x}$ and $\mathbf{v}'$ is thus a frame of reference which is associated with the moving and deforming surface S_0 (see Fig. 8.2). A similar frame of reference is used in the lubrication approximation, but there the fluid at P is moving relative to the frame of reference. Here it is not, $\mathbf{v}'(\mathbf{0}) \equiv 0$, and $(\nabla \mathbf{v}')_P$ is the local velocity-gradient tensor. What the criteria (8.1.10) have allowed us to do is to consider a local approximating flow in which $\partial_3 v_1'$, $\partial_3 v_2'$ and v_3 $(x_3=0)$ are zero everywhere, and in which the non-zero components of $(\nabla \mathbf{v}')_P$ are instantaneously those of $\nabla \mathbf{v}$ for the true flow, evaluated at $(y^1, y^2, y^3=0)$, i.e. on S_0 at P.

In the idealized S_0' flow, we must have an unsteady situation in the Lagrangian sense that

$$\frac{\mathrm{d}h'}{\mathrm{d}t} = -h(\partial_1 v_1' + \partial_2 v_2') \tag{8.2.4}$$

by continuity. If in the real flow $h = h(y^1, y^2, t)$, where we are now allowing for Eulerian unsteadiness also, the relationship between $h'(t)$ and $h(y^1, y^2, t)$ is given by

$$\frac{dh'}{dt} = \frac{Dh}{Dt} \tag{8.2.5}$$

The global flow field $\mathbf{v}$ has thus been approximated by the motion $\mathbf{v}_P$ of a surface S_0 together with a local deformational field $(\nabla \mathbf{v}')_P$ chosen to match the deformation of the surface $(\nabla \mathbf{v}_P)$ at each point $P \equiv (y^1, y^2)$ and time t. $(\nabla \mathbf{v}')_P$ regarded as a function of t provides all the information necessary for whatever constitutive equation we use for the fluid, and so allows $\mathbf{T}^E$ to be obtained for each point P at each time t. Indeed, the free-surface condition (8.2.1) allows us to eliminate the pressure giving

$$\mathbf{T} = \begin{bmatrix} t_{11}^E - t_{33}^E & t_{12}^E & 0 \\ t_{12}^E & t_{22}^E - t_{33}^E & 0 \\ 0 & 0 & 0 \end{bmatrix} \tag{8.2.6}$$

If the fluid is Newtonian, then the relation for $\mathbf{T}$ in terms of $\nabla \mathbf{v}_P$ is relatively simple. For other models, the relationship will be more complex, but is simpler than that for a fully-general flow field because we now have a condition of plane stress, with three different non-zero components instead of six, functions of the history of four different non-zero components of $(\nabla \mathbf{v}_P)$ instead of eight.

As we see later, the problem is much further simplified if $(\nabla \mathbf{v}_P)$ is diagonal and we can employ the functional relations

$$t_{11} = \underset{S=0}{\overset{\infty}{\mathcal{N}}} [\partial_1 v_1'(s), \partial_2 v_2'(s)] \tag{8.2.7}$$

$$t_{22} = \underset{S=0}{\overset{\infty}{\mathcal{N}}} [\partial_2 v_2'(s), \partial_1 v_1'(s)] \tag{8.2.8}$$

which are obvious extensions of (4.3.3) and (4.3.4) based on (3.2.13). If the flow field is such that

$$\Lambda \dot{h}/h \geqslant 0(1) \tag{8.2.9}$$

where Λ is a characteristic relaxation time, then elastic effects will be relevant. If further,

$$\Lambda \frac{d}{dt}\left(\frac{\dot{h}}{h}\right)\frac{h}{\dot{h}} \ll 1 \tag{8.2.10}$$

following a particle, then a MWCSH flow approximation, given by (4.1.1), will be appropriate.

Equation (8.2.9) can be viewed as a condition on the Weissenberg number, (8.2.10) on the Deborah number. The analogues in lubrication flow are given by (8.1.7) and (8.1.6). It is worth noting that a useful definition of Ws and De for fluid mechanical purposes in general flow fields is provided by

$$\mathrm{Ws} = \Lambda\,\|\mathbf{L}\|, \qquad \mathrm{De} = \Lambda\,\|\dot{\mathbf{L}}\|/\|\mathbf{L}\| \tag{8.2.11}$$

where $\|\mathbf{L}\|$ is some suitable measure of $\mathbf{L}$ and $\|\dot{\mathbf{L}}\|$ of the Lagrangian derivative $\dfrac{\mathrm{D}\mathbf{L}}{\mathrm{D}t}$.

The solution of the global problem, now that the constitutive equation has been employed in the idealized flow (S_0'), requires the use of suitably selected forms of the stress-equilibrium equations. In the tangent plane (y^1, y^2) we work with the two-dimensional symmetric tension tensor

$$\mathbf{T} = h\begin{vmatrix} t_{11} & t_{12} \\ t_{12} & t_{22} \end{vmatrix} \tag{8.2.12}$$

which, to our given order of approximation,† satisfies

$$\nabla_{S_0} \cdot \mathbf{T} = 0 \tag{8.2.13}$$

There is also a condition for the balance of forces in the direction perpendicular to S_0'. This involves terms an order of magnitude smaller than the terms in (8.2.13) but the result is part of a consistent approximation scheme. The resultant of the tensions in $\mathbf{T}$ acting over the surface S_0, now considered curved, is balanced by any small pressure differential between the two sides of the surface S_0. Geometrically, the surface S_0 instantaneously defines two principal radii of curvature, R_1 and R_2, at each point P. With these are associated locally-orthogonal coordinates ($\hat{y}^1$, $\hat{y}^2$) and tensions

$$\hat{t}_1 = \mathbf{T}:\hat{\boldsymbol{\delta}}_1\hat{\boldsymbol{\delta}}_1, \qquad \hat{t}_2 = \mathbf{T}:\hat{\boldsymbol{\delta}}_2\hat{\boldsymbol{\delta}}_2 \tag{8.2.14}$$

† The meaning of this surface divergence based on the stress equilibrium equation can be followed up in detail in Aris (1962, Chapter 10, particularly in his eqn (10.43.5) with inertia terms and body forces neglected). The divergence operator involves covariant differentiation of a contravariant stress tensor. Examples of the use of eqn (8.2.13) will arise in Chapters 16, 17 and 20.

where $\hat{\boldsymbol{\delta}}_1$, $\hat{\boldsymbol{\delta}}_2$ are unit vectors in the $\hat{y}^1$, $\hat{y}^2$ directions. The relevant balance of forces in the y^3 direction is

$$\frac{\hat{t}_1}{R_1}+\frac{\hat{t}_2}{R_2}=\Delta p_a = p_a(\text{inside}) - p_a(\text{outside}) \tag{8.2.15}$$

In arriving at this result it is important to remember that the sign of Δp_a will depend upon the choice of an inside and an outside for the surface S_0 and that R_1 and R_2 will have a positive sign if the relevant centre of curvature is inside S_0 and negative if outside. Also by writing $\dfrac{\hat{t}_{1,2}}{R_{1,2}}$ as $\dfrac{ht_{1,2}}{R_{1,2}}$, and remembering that, using (8.1.10ii),

$$Hh = \frac{h}{R_1}+\frac{h}{R_2} \ll 1 \tag{8.2.16}$$

we see that $\Delta p_a \ll |\mathbf{T}|$ and so the assumption that $p_a = 0$ in the approximating idealized flow is consistent within the order of approximation considered. This particular point is considered in detail in Pearson & Petrie (1970*a*) where it is likened to the well-known thin-shell approximation in shell theory for elastic structures.

Clearly at the boundary $\mathcal{B}$ to the surface S_0 we have to apply suitable conditions on $\mathbf{v}$ or $\mathbf{T}$. Formally then we have either

(i) $$\mathbf{v} = \mathbf{v}_B \tag{8.2.17}$$

or

(ii) $$\mathbf{T} = \mathbf{T}_B \tag{8.2.18}$$

The set of equations (8.2.15),

$$\frac{Dh}{Dt} = h(\boldsymbol{\nabla}\mathbf{v}_P)_{33} \tag{8.2.19}$$

obtained from (8.2.4) and (8.2.5),

$$\boldsymbol{\nabla} \cdot (h\mathbf{T}) = 0 \tag{8.2.20}$$

obtained from (8.2.6), (8.2.12) and (8.2.13), together with a constitutive equation given symbolically by

$$\mathbf{T} = \mathcal{T}(\boldsymbol{\nabla}\mathbf{v}_0) \tag{8.2.21}$$

and boundary conditions (8.2.17) and (8.2.18), constitute a preliminary statement of the mathematical problem to be solved in order to predict the flow. However, this formulation is deceptively simple, and does not immediately lead to an obvious procedure for solution. Embedded in the eqns (8.2.15)–(8.2.21) is an unspecified and largely arbitrary coordinate system (y^1, y^2), which is time-dependent if the surface S_0 moves in space. The evaluation of Dh/Dt in (8.2.19) is thus not a trivial matter. The asymptotic analysis of Pearson & Petrie (1970*a*) was restricted to a steady axisymmetric flow, which suggested an obvious fixed orthogonal coordinate system in S_0. Yeow (1972, 1976) has extended the procedure, without developing the full analysis, to cover small perturbation to the steady axisymmetric flow, but no completely general approach has yet been developed.

EXAMPLE 8.2.1. Continuous extension of a tubular sheet by stretching between 2 coplanar rings.

Material is supplied at a steady speed v_i and thickness h_i in the plane $z=0$ at the circumference of a circle $r=r_i$ and is drawn axisymmetrically over a coaxial coplanar ring at $r=r_0$ at steady speed v_0. It may be assumed that r_0-r_i, $r_i \gg h_i$ and that $v_0>v_i$. We examine the kinematics and dynamics of the motion. We use the obvious (r, ϕ, z) coordinate system in a fixed frame of reference. The local $\mathbf{x}$ frame will be that given by the local unit vectors in the (r, ϕ, z) orthogonal system. We suppose that the velocity field is given by

$$\mathbf{v}=\{v(r), 0, 0\}$$

whence

$$(\nabla \mathbf{v}')_P=\begin{bmatrix} dv/dr & 0 & 0 \\ 0 & v/r & 0 \\ 0 & 0 & -(v/r+dv/dr) \end{bmatrix}$$

as defined in (8.2.3). Hence, using (8.2.19),

$$\frac{Dh}{Dt}=v\frac{dh}{dr}=-h\left(\frac{v}{r}+\frac{dv}{dr}\right)$$

From this, or directly, it follows that $vhr=v_i h_i r_i$.

The stress-equilibrium relation (8.2.20) becomes

$$\frac{d}{dr}(rht_{rr})-ht_{\phi\phi}=0, \qquad t_{zz}=0$$

If a particular form for (8.2.21) can be given, then in principle the problem can be solved.

For a Newtonian fluid,

$$t_{rr} = t^{\mathrm{E}}_{rr} - t^{\mathrm{E}}_{zz} = 2\eta_0\left(2\frac{\mathrm{d}v}{\mathrm{d}r} + \frac{v}{r}\right)$$

$$t_{\phi\phi} = t^{\mathrm{E}}_{\phi\phi} - t^{\mathrm{E}}_{zz} = 2\eta_0\left(\frac{\mathrm{d}v}{\mathrm{d}r} + \frac{2v}{r}\right)$$

and so

$$2\frac{\mathrm{d}^2}{\mathrm{d}r^2}(\ln v) = \frac{3}{r^2} + \frac{1}{r}\frac{\mathrm{d}}{\mathrm{d}r}(\ln v)$$

It follows that $v = \frac{\mathrm{B}}{r^2}\mathrm{e}^{\mathrm{C}r^{\frac{3}{2}}}$ in general, and, using the boundary conditions, that

$$\frac{v}{v_i} = \frac{r_i^2}{r_0^2}\exp\left\{\ln\left(\frac{v_0}{v_i}\frac{r_0^2}{r_i^2}\right)\left(\frac{r^{\frac{3}{2}} - r_i^{\frac{3}{2}}}{r_0^{\frac{3}{2}} - r_i^{\frac{3}{2}}}\right)\right\}$$

Chapter 9

Instability

9.1 MELT FLOW INSTABILITY

This phenomenon has been exhaustively investigated for the past 30 years but has not yet been properly and fully explained. It remains the most interesting and challenging problem in the mechanics of polymeric fluids. The publications of many authors (Clegg, 1958; Howells & Benbow, 1962; Benbow & Lamb, 1963; Lupton, 1963; Dennison, 1967; Bialas & White, 1969; Vinogradov, 1977) show how varied, yet reproducible and ubiquitous, is the observed breakdown of axisymmetric capillary-flow extrusion. The polymer melts investigated include HDPE, LDPE, PP, PS, PMMA, PVC, PDMS (poly dimethyl-siloxane), PIB, SBR and other rubbers. This phenomenon restricts very significantly the output rates that are possible (in terms of acceptable product) in extrusion operations and introduces, even at relatively low rates of output, mild surface imperfections that are a constant source of dissatisfaction to processors and clients.

Numerous review articles (e.g. Dennison, 1967; Pearson, 1969) have summarized the findings of previous workers. The best and most recent is that of Petrie & Denn (1976), whose comments on melt fracture may be regarded as essential reading for serious students of the subject, and whose reference list may be regarded as needing very few, if any, additions.

In what follows we shall be referring to an extrusion system of the type shown in Fig. 7.1 and to breakdown of flows arising in it as discussed briefly at the end of Chapter 7. In the various observations that have been reported, differences have arisen as between polymers, between low- and high-molecular-weight samples, between broad and narrow molecular-weight distributions, between low and high L/D ratios, between large- and small-angle entry cones, and between constant-pressure and constant-rate extrusion. Some special experiments have been carried out with very large reservoirs, some with no reservoirs, while the material forming the capillary has been varied. Finally, the temperature of extrusion has been varied. In this account we shall try to select certain obvious and important features that have been regularly described.

9.1.1 Observations

The empirical observations that require explanation include the following:

M.1. The output rate $\dot{V}$ as a function of P sometimes shows:

(i) A discontinuity (over a range of pressure) usually resulting in hysteresis as P is increased and then decreased over the relevant range in a constant-pressure experiment. This is shown in Fig. 9.1(a)—note the large jump in output rate at a given pressure and the occurrence of an apparently inaccessible range of output rate. This has been observed chiefly with HDPE, and is reproducible for a given polymer under given conditions.

(ii) A change in slope over a relatively narrow range of pressure drop. This often coincides with the onset of asymmetric and unsteady flow, and may be more or less pronounced depending on the polymer. It is usually observed in constant-rate extrusion experiments (see Fig. 9.1(b)).

M.2. A periodic fluctuation in reservoir pressure is often noted in constant-output-rate experiments when flow is unsteady and/or asymmetric. The magnitude and frequency of this fluctuation varies considerably, but as expected correlates well with the periodicities observed in the extrudate. It can be 25% or more of the mean pressure.

M.3. The distortions observed on the extrudate can be:

(i) Surface irregularities, called matte or sharkskin, that vary from imperceptible rippling to deep fissures—see Fig. 9.2—arising, on an otherwise almost uniform flow, at the die exit.

(ii) Periodic pulsations leading to unsteady but largely axisymmetric extrudates particularly for HDPE—see Fig. 9.3—corresponding in many cases to pulsations in flowrate. The periodicity may be extremely regular or somewhat erratic, and surface distortions may be superimposed (Okubo & Hori, 1980).

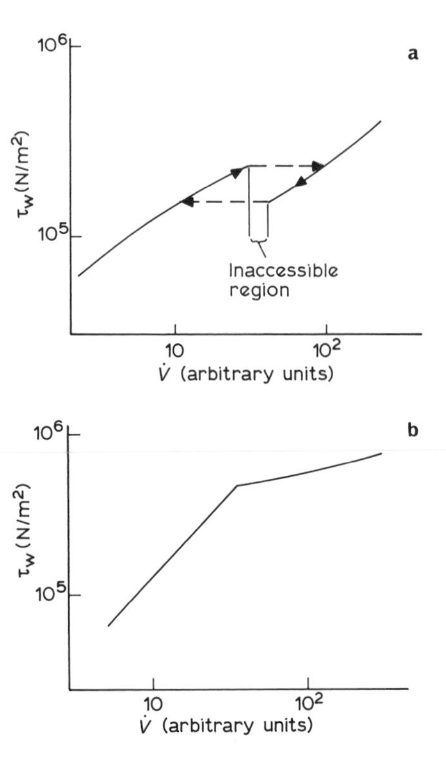

Fig. 9.1. (a) Typical flow curve for HOPE under constant pressure conditions. (b) Typical flow curve for various polymers at constant output rate conditions. Diameter not specified and hence $\dot{V}$ is in arbitrary units.

(iii) Wave-like distortions, usually of a helical or double helical kind, with extremely regular and reproducible periodicity, and often a smooth surface (see Fig. 9.4).

(iv) So pronounced that the extrudate becomes incoherently broken up (see Fig. 9.5).

M.4. The severity of the distortions of type (iii) can be reduced in most cases either by increasing L/D or by reducing the entry-cone angle. However, type (ii) distortions can sometimes be made worse by increasing L/D. Type (i) distortions are largely independent of L/D. In all cases, flow is smooth at sufficiently low flowrates, and unacceptably distorted above a certain critical flowrate. Reducing the severity of the

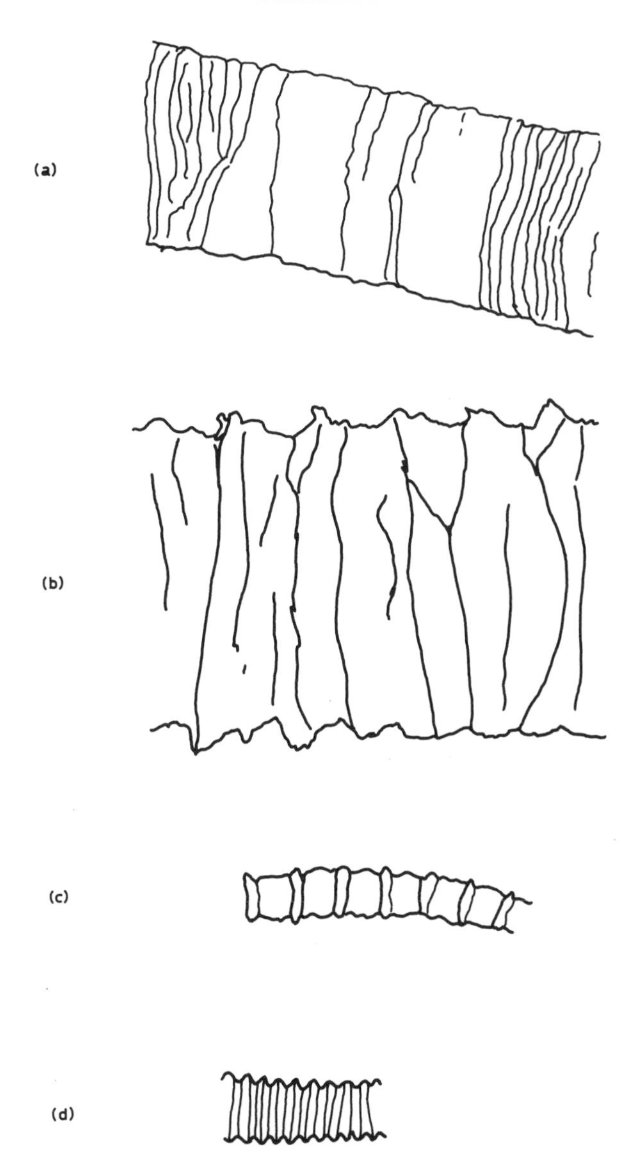

Fig. 9.2. (a) From Dennison (1967, Fig. 1); (b) from Howells & Benbow (1962, Fig. 5(c)); (c), (d) from Lupton (1963).

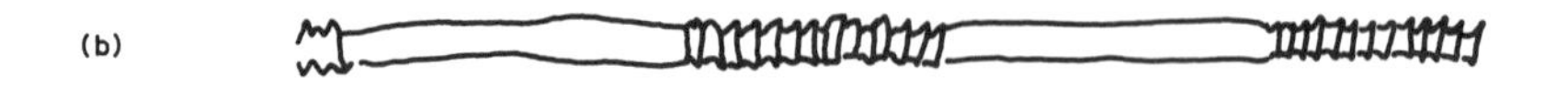

Fig. 9.3. (a) From Lupton (1963); (b) from Vinogradov (1977, Fig. 3).

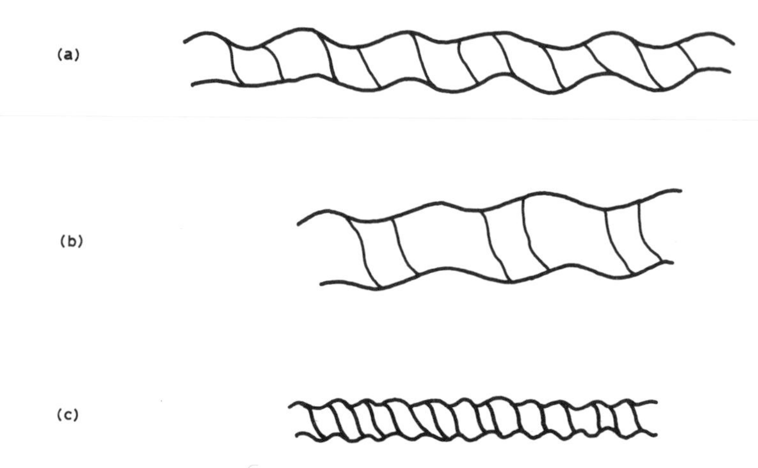

Fig. 9.4. (a) From Dennison (1967, Fig. 2(a)); (b) from Bialas & White (1969, Fig. 3B); (c) from Benbow & Lamb (1963, Fig. 1.3).

Fig. 9.5. From Dennison (1967, Fig. 2(d)).

distortion at a given flowrate goes hand-in-hand with increasing the critical flowrate. In certain cases, supercritical stable flow has been demonstrated.

M.5. The best correlating parameters found so far for determining the onset of distortion are the capillary-wall shear-stress, the associated Weissenberg number Ws evaluated in terms of viscometric flow at the corresponding shear stress, or a less precisely defined recoverable elastic-shear.

9.1.2 Comments

We can make various comments on the basis of our fluid-mechanical analysis of the three flow regions—entry, capillary die, and exit—and of the rheology of polymer melts.

C.1. As flow rate $\dot{V}$ increases, so does the magnitude of the rate of deformation D (in the mean and probably almost everywhere), and this will be true whether the capillary flow is fully developed or not. Thus all the Deborah and Weissenberg numbers defined in Chapter 7 will increase. In physical terms, so will elastic effects; the amount of stored elastic energy and the importance of non-linear rheological terms in the flow equations will in general correspondingly increase. Hence, all mathematical experience shows that the chance of instability, or of bifurcation of solutions, will increase. Critical flowrates are to be expected, with De or Ws as relevant parameters, as noted in M.5.

C.2. If sharp edges occur at the entry or exit to the die, we can expect local regions of very high deformation-rate there, as could be argued convincingly in the near-Newtonian case discussed in Sections 7.3 or 7.4, and hence we have every reason to expect sharp changes in flow behaviour to arise in these regions. Thus, we have, at a very simple level, an explanation† for the exit effects M.3(i).

† Different workers seek different levels of understanding and so are satisfied with explanations of various degrees of precision. To some it is enough to say that elasticity is the cause of elastic turbulence, and so a suitable technological criterion for the stable/unstable flow transition need be no more than a rough value of an easily measured Weissenberg number. For others, some causal chain of effects must be added, which introduces the effect of L/D ratio and die-entry geometry and the observed differences between polymers; unfortunately many of the mechanisms proposed are often little more than a restatement of observed phenomena in terms of imprecisely measured secondary or local variables. For students of classical mechanics, a full explanation can be nothing less than a successful quantitative comparison between well-characterized experimental observations and mathematically rigorous predictions based on the conservation and constitutive relations discussed at length here and elsewhere; lesser levels of understanding are usually achieved through qualitatively successful comparisons between slightly inadequate experiments and the solutions to approximate model equations. In a textbook on the mechanics of polymer processing, it is natural that a formal mechanical explanation be sought; in the practical field of processing, it is to be expected that simpler and more immediately suggestive explanations will hold sway.

The critical elastic regime is reached preferentially near these sharp edges. Our kinematical analysis of flow near the die edge shows a convergence of the streamlines (Fig. 7.5) near the free surface with maximally-high rates of extension there as well as high rates of shear just inside the die. Flow fields that can avoid or relieve the sharp growth in stress associated with these high deformation rates can be expected to arise spontaneously, particularly in view of the unsteadiness that can be accommodated by the presence of a free surface. What we are not able to show is why the unsteady flow fields have a particular form and a particular periodicity.

We would not expect exit effects to be propagated far upstream since the dissipative nature of the flow makes the governing equations, if expressed in differential form, elliptic or parabolic rather than hyperbolic in nature. This agrees with the third observation of M.4, that type (i) distortions are largely independent of L/D.

C.3. Similarly, we would expect upstream disturbances to arise first at a sharp-edged entry to the die. If the entry flow field had a recirculating portion as shown at the bottom of Fig. 7.2(a) or (b), then some upstream influence on the die entry flow could arise and even induce a large-scale unsteadiness (or instability) in it, which would be missing in the more uniformly convergent flow shown at the top of Fig. 7.2(a) and (b). In both cases some downstream influence would be expected; this, on the most naive assumptions, would be damped because the critical stress level would be reached at the entry edge for smaller $\dot{V}$ than in a developed die flow. This provides some explanation for the first observation of M.4. It is also consistent with the view that periodic wave-like instabilities of type M.3(iii) do correlate with large fluctuations in a recirculating entry flow pattern.

In our discussion of flow in the capillary, we pointed out that L/D ratios of much more than 10 may be necessary to give fully-developed flow. Indeed the necessary value of L/D will be proportional to the entry Deborah number, suitably calculated to take account of any rate dependence of the relaxation time, and so increasing apparent severity of extrudate distortion (arising in the entry region) as a function of $\dot{V}$ may be expected on that ground alone.

C.4. Both C.2 and C.3 refer to fluid-mechanical instabilities that need reflect no catastrophic change in rheological behaviour of the fluid, nor breakdown in the boundary conditions and physical assumptions applied so far, nor mechanical resonance with the extrusion apparatus. If we had a satisfactory analytical description of the entire flow field for steady symmetric flow, then we could expect to investigate and perhaps verify the suggested mechanisms by means of linear (or non-linear) perturbation analysis of the relevant mathematical equations about the steady solution.

The only part of the flow field for which this has been attempted in any systematic fashion is fully developed capillary flow, with as yet unfortunately no definitive results. See *inter alia* Vanderborck and Platten (1977), Vanderborck *et al.* (1979) and Ho *et al.* (1977).

For reasons of mathematical simplicity most stability analyses have concentrated on plane channel flow, for which there are fewer experimental observations, while most of those have restricted in one way or another the class of disturbances that have been permitted. Confidence in any of the results quoted cannot be very high because of the contradictory results obtained in more than one case (see Petrie & Denn, 1976, for details).

We give a sketch of the procedure for two specific models, the Jeffreys fluid and the Kaye–BKZ fluid, in Appendix 1, showing in general terms how an eigenvalue problem results. The analytical and subsequent computational difficulties associated with the search for

$$(P/l_c)_{crit}, \text{ with associated } k_{crit}, c_{crit}$$

has so far prevented any satisfactory results being obtained for constitutive equations even reasonably representative of polymer melts.

C.5. Some experiments have suggested that slip takes place at the capillary walls, a process that is permitted by the boundary condition (2.3.4). An analogy with static and dynamic friction may then locate the source of instability at the wall; if the function $\mathrm{fn}(\mathbf{T}\cdot\hat{\mathbf{n}}_w)$ is multivalued in $\mathbf{v}-\mathbf{v}_w$ for values of the argument above some critical value, as suggested by Fig. 9.6, then the output $\dot{V}$ as a function of P may be correspondingly multivalued as shown in Fig. 9.7. This is precisely the phenomenon noted as M.1 and shown in Fig. 9.1(a).

C.6. Many of the simple rheological equations of state discussed above, including the very appealing model of Doi and Edwards, predict an instability in viscometric flow in the sense that the shear stress $\eta(D)D$ is not a monotonically increasing function of D for sufficiently large values of D.

It has been customary to argue that constitutive equations showing such maxima in stress are unrealistic. Certainly it is true that, where the models predict zero or negative values of shear stress for large enough positive values of shear rate, they must be totally false, because

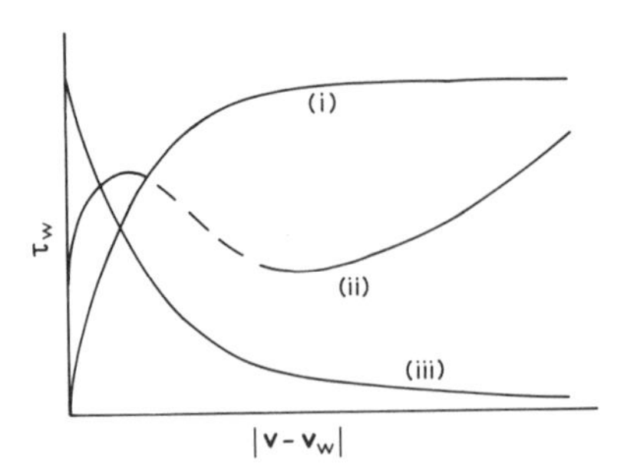

Fig. 9.6. Three possible forms of slip velocity function. (i) Smoothly failing but stable. (ii) Triple-valued curve with two stable (——) and one unstable (– – –) portions. (iii) Wholly unstable. Note that the wall shear stress is used as the relevant stress. It is quite possible that curves (i)–(ii) should really be plotted as surfaces with the tangential and normal components of $\mathbf{T} \cdot \hat{\mathbf{n}}_w$ as coordinates. Units are arbitrary.

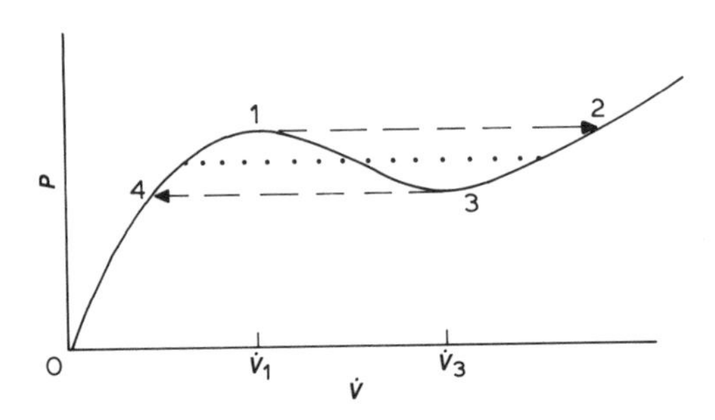

Fig. 9.7. Multivalued pressure-output curves based on slip or apparent-slip effects. The curve from 0 to 1 is stable, but increasing pressure beyond that at 1 would necessarily cause a jump to 2, which lies on another stable portion of the curve. On reducing the pressure from 2 to 3 and below, a sudden fall in output would take place from 3 to 4. This is very similar to the hysteresis noted in Fig. 9.1 for constant pressure extrusion. Note that transitions (. . .) from points between 4 and 1 to corresponding points between 3 and 2 are possible for large enough disturbances.

they then disobey the second law of thermodynamics. However, there is no basic physical objection to multivalued curves like that in Fig. 9.7 as solutions of the coupled heat and flow equations for temperature sensitive Newtonian fluids in Couette or pressure flow (Vanderborck *et al.*, 1979). If we accept the possibility of such maxima in viscometric shear stress, then the predicted flow curves (i.e. pressure–output curves), even assuming axisymmetry, present a non-uniqueness.

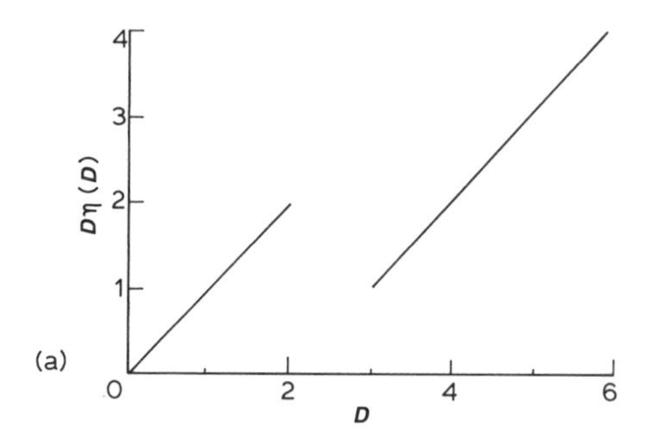

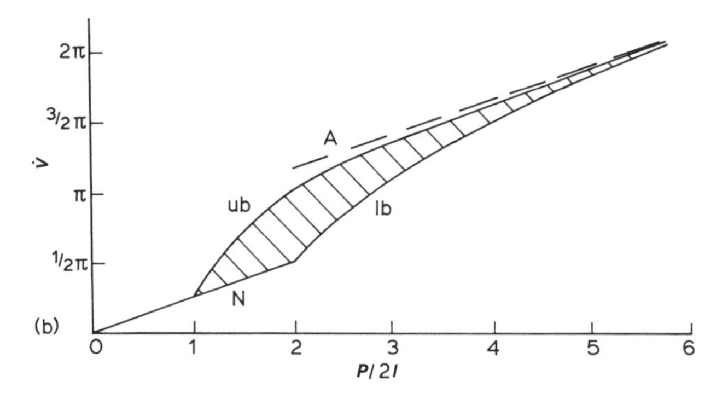

Fig. 9.8. (a) Simple form of discontinuous shear-stress/rate-of-shear relation with unstable region between (2, 2) and (3, 1) not shown. (b) Corresponding flow curve showing: N Newtonian linear region, eqn (9.1.4); ub upper bound, eqn (9.1.2); lb lower bound, eqn (9.1.3); A high-flowrate asymptote, eqn (9.1.5).

A very simple example is provided in Fig. 9.8 for the artificially simplified viscosity function

$$\eta(D)=\begin{cases}1 & D<2\\ 1-2D^{-1} & D>3\end{cases} \tag{9.1.1}$$

which shows an overlap in $D\eta(D)$ but a gap in D. The volume flow rate in a tube of unit radius has an upper bound beyond $\tau_w \equiv P/2l_c = 1$ given by

$$\dot{V}_{ub}=\pi\left(\frac{2}{3}+\frac{\tau_w}{4}-\frac{2}{3\tau_w^3}\right) \tag{9.1.2}$$

and a lower bound beyond $\tau_w = 2$ given by

$$\dot{V}_{lb} = \pi\left(\frac{2}{3} + \frac{\tau_w}{4} - \frac{16}{3\tau_w^3}\right) \tag{9.1.3}$$

with the $\eta = 1$ line relevant only up to $\tau_w = 2$ being given by

$$\dot{V}_N = \tfrac{1}{4}\pi\tau_w \tag{9.1.4}$$

Both bounds asymptote for large τ_w to

$$\dot{V}_A = \tfrac{1}{4}\pi\tau_w + \tfrac{2}{3}\pi \tag{9.1.5}$$

What would happen in practice has not been fully investigated but it is clear that the behaviour shown in Fig. 9.1(b) could appear to result. It is also possible that behaviour closer to that of Fig. 9.1(a) could ensue.

C.7. Both of the comments C.5 and C.6 have supposed that steady axisymmetric flow could result from slip or an internal discontinuity in shear rate. If the latter were very large, it would arise first at the wall and so give the impression of boundary slip.

An alternative possible consequence of such slip or slip-like effects in a very long capillary would be to lead to instabilities of the type examined in Appendix 1. This possibility has not been exhaustively examined, but it might resolve the uncertainty shown in Fig. 9.8(b) and replace the hatched region by one or more stable curves (or possibly stable and unstable regions). There are, of course, profound difficulties in carrying out such analyses because a properly-invariant tensor relation for extra stress as a function of rate-of-deformation would be required, valid for unsteady flow fields. What Fig. 9.8(a) shows is a hypothetical viscometric function only, and not a constitutive equation.

Catastrophic, i.e. sudden, changes in either shear rate or shear stress or both would be exhibited if portions of fluid in a flow field attempted to jump from the low to the high viscosity state. Coleman & Gurtin (1968) have considered the possibility that melt fracture as extreme as M.3(iv) might involve steep shear-rate waves involving acceleration forces; although the mean Reynolds number of the flow may be very small, local accelerations represented by the term $\rho\, \partial\mathbf{v}/\partial t$ might be of the same order as stress gradients. Though this is a possibility that cannot be ruled out, it certainly cannot explain the instabilities M.3(ii) and (iii). Vinogradov (1977) gives physical arguments based on critical changes in physical state.

C.8. The crudest 'explanations' of melt flow instability often avoid or neglect consideration of the kinematic requirements of continuity.

In a real experiment, unless voids are to form in the flow field, a possibility that is not ruled out in flow defects M.3(i) and (iv), the conservation equation (2.2.5) must be obeyed everywhere in the fluid. This couples flow in the capillary die to flow in the reservoir and leads to very significant effects that control the observed form of flow breakdown.

A particularly clear and important example is provided by the interaction of reservoir flow and the multivalued die flow described in C.5 and C.6. If the extrusion device is a constant pressure device, then step changes (or in practice very rapid changes) in volume flowrate are permissible and jumps in output as described in Fig. 9.1(a) and predicted in Fig. 9.7 can be accommodated as pressure is slowly increased, with steady axisymmetric flow being possible on either side of the jump. This can be seen in portions of the extrudate in the figure.

If, however, the extrusion device employs a piston which reduces the reservoir volume at a constant rate, then sudden jumps in output are not permissible as piston speed is increased. Indeed, if a steady output rate $\dot{V}$ lying between 1 and 3 in Fig. 9.3 is imposed, then the corresponding steady axisymmetric flow in the die would be a basically unstable regime. If the fluid is strictly incompressible, then the rate $\dot{V}$ imposed by the piston will be exactly balanced by the same steady flow rate $\dot{V}$ in the die. However, the flow field in both capillary and die need not be asymmetric or steady. Observations on the die entry and exit flow fields have shown that secondary flows (perturbations superposed on a steady symmetric flow) having a character reasonably well described by $\exp\{ik(z-ct)+im\theta\}$, as described in Appendix 1, can arise in a suitably matched fashion throughout the flow field. These are the distortions M.3(iii).

We are left with the pulsations described in M.3(ii). These are usually correlated with large pressure fluctuations in the reservoir, and a convincing explanation for these in terms of bulk compressibility of the reservoir fluid has been given by Lupton & Register (1965) and tested by other workers. We give here the elementary explanation from Pearson (1966, p. 58).

Let M be the mass, ρ the density, β the isothermal compressibility and V the volume of the melt within the reservoir. Then

$$M = V\rho = V\rho_0(1+\beta P) = (V_0 - \dot{V}t)\rho_0(1+\beta P) \qquad (9.1.6)$$

where V_0 is the initial volume (at $t=0$) and ρ_0 is the density at zero pressure. We have taken the compressibility to be constant over the range of pressure involved. $\dot{V}$ is fixed. From this we can deduce that the mass flowrate in the capillary will be

$$-\frac{dM}{dt}=\rho_0\dot{V}(1+\beta P)-\rho_0(V_0-\dot{V}t)\beta\dot{P} \qquad (9.1.7)$$

which will be constant, as expected, if $\beta=0$ or if $\dot{P}=0$. We take the curve for steady flows given by Fig. 9.3 in the form

$$\dot{V}=\begin{cases}f_1(P) & \dot{V}<\dot{V}_1\\ f_2(P) & \dot{V}>\dot{V}_3\end{cases} \qquad (9.1.8)$$

where f_1 and f_2 refer to the two branches of the curve between 0 and 1 and from 3 upwards beyond 2 respectively, and assume that we can interpret $\dot{V}$ as $-\dot{M}/\bar{\rho}$ where $\bar{\rho}$ is that density which effectively makes the relation (9.1.8) hold over the relevant range of density in the capillary.

On substitution we find that

$$(t^*-t)\beta\dot{P}=(1+\beta P)-f(P)\bar{\rho}/\dot{V}\rho_0 \qquad (9.1.9)$$

where

$$t^*=V_0/\dot{V} \qquad (9.1.10)$$

represents the time to empty the reservoir at output rate $\dot{V}$. Clearly if $\dot{V}$, as imposed by the ram, lies within the range $(0, \dot{V}_1)$ or is $>\dot{V}_3$, then we would expect $\dot{P}$ to approach zero rapidly with $\bar{\rho}=\rho_0(1+\beta P)$ to yield $\dot{V}=f(P)$ as required.

However, if $\dot{V}$ lies between $\dot{V}_1$ and $\dot{V}_3$, then

$$f_2(P)>\dot{V}>f_1(P) \qquad (9.1.11)$$

and we would expect P to vary periodically around the circuit 1234. The timescale of the periodicity

$$t_p=0\left(\frac{t^*\beta P}{1-\dot{V}_1/\dot{V}_3}\right) \qquad (9.1.12)$$

from which we draw the conclusion that it is linearly dependent upon t^* and β. A typical value of βP is 0·05 and so $\beta(P_3-P_1)\sim 0\cdot 01$. In extreme cases $\dot{V}_3=2\dot{V}_1$ and so we might expect $t_p=0\cdot 02t^*$ and of the order of 1 s.

C.9. The one characteristic of extrudate deformations that has been little reported on is the frequency (or wave number) of the distortions. A careful experimental analysis of these as various geometrical and operating parameters were varied could give significant insight into the mechanics of melt-flow instability.

Thus, in the compressibility/multivalued flow-curve mechanism described in C.8, the wavelength of the extrudate variations, at fixed output rate, should be a linear function of reservoir volume. That this has been observed (Lupton, 1963) provides a strong, though not necessarily unassailable, confirmation of the proposed mechanism in the cases described.

Alternatively, if extrudate periodicity were governed entirely by the timescale of flow recirculation in the reservoir eddies, its temporal frequency would be independent of die diameter and die length but would be noticeably dependent on reservoir geometry, at fixed output rate. If, however, the extrudate periodicity were found to be dependent only on capillary diameter and flowrate, and not on reservoir shape, or die length, then arguments based on the breakdown of a fully-developed die flow would be relevant.

Periodicities of exit-induced distortions would clearly be affected by changes in local exit geometry, or by heating the die lips, for example, rather than by entry or die geometry, except in so far as extrudate (die) swell were determined by upstream conditions. In general, this has been verified.

9.2 STEADY SECONDARY FLOWS AND OTHER FLOW INSTABILITIES

We met a secondary flow (usually called rod-climbing) caused primarily by normal-force (elastic) effects in our consideration of flow near the free surface in a Couette viscometer (Chapter 6). We also met an instability (Taylor instability) associated with the primary viscometric flow that leads in practice to secondary flow vortices, which are incidentally much altered in shape from the Newtonian form for strongly elastic liquids.

In our discussion of conical entry flow, we found that the purely radial (and self-similar) flow field possible for a Newtonian fluid was altered, for slightly elastic fluids, so as to have steady non-zero angular components of velocity. The flow field lost its self-similar structure and, for very elastic fluids, a reversal in the velocity field near the walls is known to arise (Ballenger & White, 1971). In the literature, most analytical solutions are perturbation expansions about the Newtonian solution, and the secondary flows referred to are the perturbation velocity field rather than the full flow field. Comparison with observation is often made by using a limited number of terms in the expansion beyond the range (in Deborah number) for which they are, strictly speaking, valid. In this way the perturbation flow field—$Q^{(1)}$ say in (7.2.15)—can dominate the Newtonian field Q_N, at least for very small r', and gross departures in the velocity pattern are successfully, though necessarily qualitatively, reproduced.

Such secondary flows also arise in plane converging flows, which have been more exhaustively analysed (Langlois & Rivlin, 1959; Strauss, 1975) even though they are a basically weaker phenomenon in plane flow, being third-order effects rather than second-order effects as they are in conical flow. Their stability has been investigated (Strauss, 1975).

Similarly, flow in non-circular cylindrical tubes of viscoelastic fluids (Langlois & Rivlin, 1959) displays qualitative differences (see Fig. 9.9) from Newtonian flow. It can be shown quite simply that the latter, in slow isothermal flow, have stable unidirectional solutions given by

$$\nabla^2 v_{1N} = \frac{dp}{dx_1} \tag{9.2.1}$$

where

$$\mathbf{v}_N\{v_{1N}(x_2, x_3), 0, 0\} \tag{9.2.2}$$

x_1 being the axial coordinate, ∇^2 being a two-dimensional Laplacian,

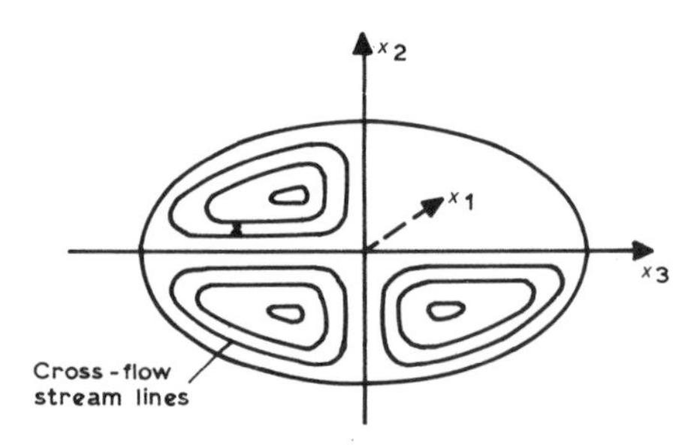

Fig. 9.9. Cross-section of a cylindrical tube of elliptic (non-circular) cross-section. Typical cross-flow streamlines are shown for an elastic fluid.

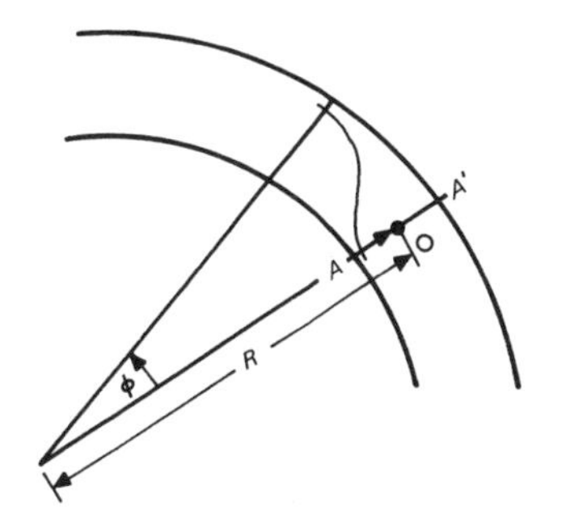

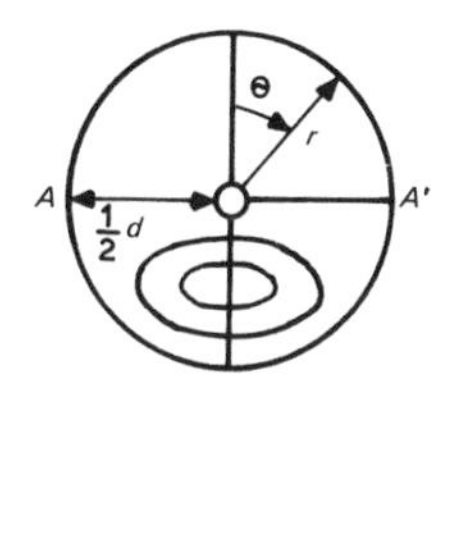

Fig. 9.10. Diagram of coordinate system and geometry for flow in a curved tube of circular cross-section. Typical streamlines are shown for a highly elastic fluid.

and dp/dx_1 a constant, subject to

$$v_{1N} = 0 \quad \text{on the boundary} \quad b(x_2, x_3) = 0 \tag{9.2.3}$$

For slightly non-Newtonian fluids, a regular perturbation expansion in terms of the Weissenberg number shows that, at fourth order, a steady secondary flow field involving non-zero v_2 and v_3 velocity components must arise.

Flow in curved tubes also shows (Thomas & Walters, 1963) secondary flow for elastic liquids, where the creeping flow of Newtonian fluids does not. The nature of this flow is shown in Fig. 9.10. The unperturbed Newtonian solution has the form

$$\mathbf{v}_N = \{v_{\phi N}(r, \theta), 0, 0\} \tag{9.2.4}$$

with

$$v_{\phi N}(\tfrac{1}{2}d, \theta) = 0$$

where $v_{\phi N}$ is related to the pressure gradient $R^{-1}\, dp/d\phi$ according to an equation very like (9.2.1). Elastic liquids lead to flow equations that can only be satisfied with non-zero values of v_r and v_θ.

This secondary flow will have little significance in terms of $(\dot{V}, P)$ relations but, for very high Pe flows, can enhance heat transfer very significantly.

In all these examples, a distinction must be drawn between the steady secondary flows that can be predicted from regular perturbations about the Newtonian solution and instabilities that might be predicted by linearized analyses of the type sketched in Appendix 1.

There is no critical value of the Weissenberg or Deborah number for the secondary flow effect: these secondary flows increase smoothly from zero according to some power of the relevant dimensionless group. They are part of the only feasible basic flow field. Instabilities that occur at critical values of the relevant Ws or De lead to steady, oscillatory or conceivably aperiodic, flow fields that are usually qualitatively different from the basic unperturbed flow. Instabilities are associated, in a mathematical sense, with bifurcations of the steady solution as Ws or De is increased. Because both effects are caused by the non-linear nature of the constitutive equation—and mathematically can be traced to the same terms—secondary flows will tend to dominate the basic flow at about the same (order unity) values of Ws or De at which bifurcations will arise. In other words, in any given processing flow, for which Ws or De is about 10, the steady flow that is observed may either be a smooth modification of the creeping, Newtonian, flow or be the result of a major change at some critical value of Ws or De. This distinction could be of importance when numerical solutions are being sought because techniques used to make the chosen iterative procedure converge might, unfortunately, be so devised that they unsuccessfully attempted to attain unstable solutions.

Particular instabilities will be discussed later in connection with particular processes.

Part III

CONTINUOUS PROCESSES

A—EXTRUSION

Extrusion refers to a process whereby polymer is melted and pumped through fixed channels and orifices to form cylindrical shapes of various cross-section. The melting and pumping device is the extruder and is, in most cases, a single-screw plasticating extruder (but it could also include a ram or gear-pump), while the forming device is the extruder die. On emerging from the confining channels of the die, the extrudate can either:

(i) be maintained and frozen in its extruded form—subject perhaps to fine controls imposed by sizing mandrels, sizing rolls and tensioning haul-offs, or
(ii) be further deformed substantially as in fibre spinning, film blowing and film casting, or
(iii) be used to coat other substrates as in wire coating, cable covering and sheet coating.

In almost all cases† the objective of the operation is to provide a steady stream of polymer at a given temperature with a given cross-section. In general, a homogeneous extrudate in terms of both composition and temperature will be aimed at. (Exceptions arise in the case of deliberate co-extrusion of different polymers, while a completely uniform temperature profile is seldom obtained in practice.) The object of die and extruder design is to ensure these objectives. In simple terms the extruder provides the homogeneous melt, while the die forms it. We shall therefore consider the two operations separately.

† Unsteady extrusion arises in the case of injection moulding or blow moulding; its peculiarities are deferred until Chapters 19 and 20.

Chapter 10

Extrusion Dies

For our purposes, a die consists of a rigid metal boundary to a flow channel joining an orifice (usually circular) from an extruder to a die orifice of chosen shape. The pressure at the die orifice is that of the ambient atmosphere, while that at the extruder orifice is supposed known, as a function of volume flowrate, in terms of the extruder characteristics and operating conditions. The temperature of the incoming material is also supposed known, while the thermal boundary conditions at the die-channel walls can be specified.

If we prescribe:

(a) die-channel geometry,
(b) either flowrate or pressure drop in the die,
(c) entering material temperature and thermal boundary conditions at the die walls,
(d) material properties through constitutive equations,

then we can in principle attempt to solve for the velocity and temperature field that ensues. This is the procedure for flow analysis that has been explained in Chapter 6.

However, from an engineering point of view, we are often interested in less detailed information about the flow. Thus, we may only wish to answer specific questions such as:

(a) What will be the relation between flowrate and pressure drop within the die to an accuracy of 10%?
(b) What temperature gradients will exist in the extrudate?
(c) How long will it take to attain a steady state after changes are made in the operating conditions?
(d) What departures from axisymmetry will arise in the extrudate from a particular axisymmetric die orifice?

(e) How sensitive will the extrudate flow be to small changes in properties of the material supplied by the extruder?

Having provided answers for a given system, we then often wish to answer further questions such as:

(f) How do we alter the design (i.e. shape of the die) to make the extrudate symmetric within 1%, or to reduce temperature differences to 3°C, or to reduce the pressure drop by 40%, or to restrict cross-sectional area variations to 6% when we vary the material properties within certain specified limits?

To most polymer-processing engineers, die design is an art founded on experience. To a great extent, this will always be so. The aim here is to provide design aids on the basis of approximate analyses and to provide simple criteria for basic design choices. We concentrate attention on a restricted number of extrudate shapes and die designs. Because dies are relatively expensive to fabricate, many designs provide means for on-line adjustment, i.e. for the operator to change the geometry of the flow channels in the light of the observed extrudate shape.

10.1 PLATE DIES

These are the simplest devices, used primarily in the rubber industry to produce various profiled extrudates such as tyre-tread blanks. The shape of the extrudate is determined by the shape of the hole cut in the plate as shown in Fig. 10.1. The plate thickness, l_p, is usually a small multiple or large fraction of the typical hole dimension $\frac{1}{2}d_p$. The pressure drop and hence temperature changes through the entire die system will be relatively small (a few MN/m^2 at most), and the hole shape is arrived at by trial and error. The mechanics of the flow can best be understood in terms of an entry flow followed by an exit (swelling) flow, as in an extrusion rheometer with a short and broad capillary.

Thus, by comparison with Fig. 7.1(a) and the analysis of Chapter 7, we put

$$d_r = d_e, \qquad d_c = d_p, \qquad l_c = l_p \tag{10.1.1}$$

where d_p may be calculated in various ways, two of which are men-

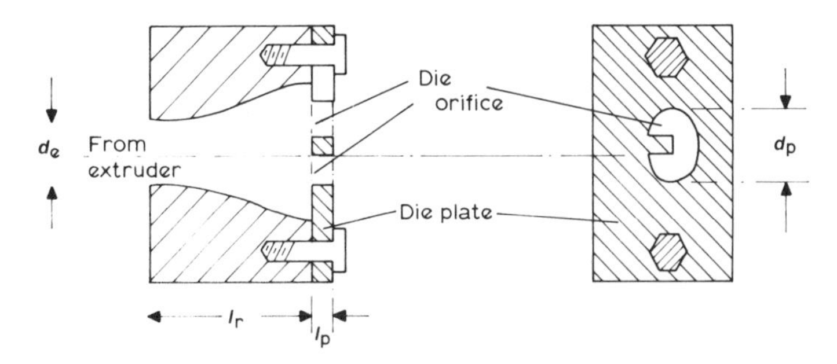

Fig. 10.1. A simple plate die. d_p is shown as the diameter of the circumscribing circle for the orifice; it could have been calculated alternatively as 4(hole area)/hole perimeter.

tioned in Fig. 10.1. If

$$d_p \ll d_e \tag{10.1.2}$$

then entrance losses associated with extensional (sink-type) flow are to be expected and can be estimated from entrance pressure drop information from capillary rheometry. Shear-flow pressure drop can be estimated on the basis of developed tube flow in a tube with the cross-section of the die orifice, and if l_r is significantly greater than d_e, of a tube with a cross-section representative of the 'reservoir' behind the plate.

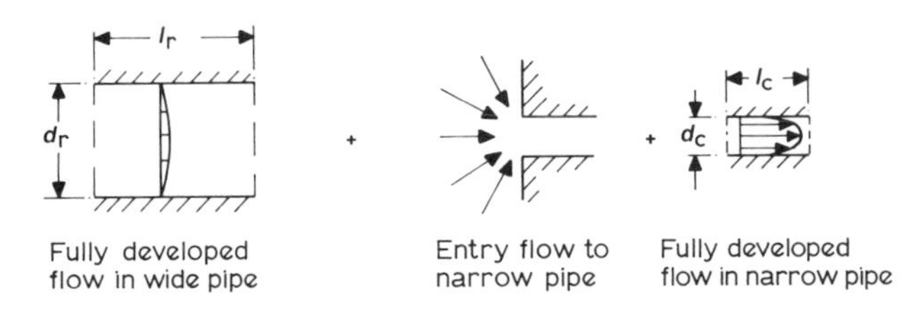

Fig. 10.2. Crude representation of pressure drops in plate die.

Thus, as shown in Fig. 10.2,

$$\Delta P_{\text{total}} = \Delta P_{\text{res}} + \Delta P_{\text{entry}} + \Delta P_{\text{orifice}} \tag{10.1.3}$$

Using (3.3.5), the power-law approximation to the fluid viscosity

$$\eta(\dot{\gamma}_{\text{w}}) = \text{K}\dot{\gamma}_{\text{w}}^{\nu-1}$$

and the capillary-flow results (7.1.61)–(7.1.63) in the form

$$\dot{\gamma}_w = 8(3\nu+1)\dot{V}/\pi\nu d_c^3$$

$$\Delta P = 4l_c\eta(\dot{\gamma}_w)\dot{\gamma}_w/d_c = 4l_c K\dot{\gamma}_w^\nu/d_c$$

We can write, as in (7.1.67) or Example 7.1.1,

$$\Delta P_{res} \approx \frac{4l_r K}{d_r}\left[\frac{(3\nu+1)8\dot{V}}{\nu\pi d_r^3}\right]^\nu \tag{10.1.4}$$

$$\Delta P_{orifice} \approx \frac{4l_p K}{d_p}\left[\frac{(3\nu+1)8\dot{V}}{\nu\pi d_p^3}\right]^\nu \tag{10.1.5}$$

while ΔP_{entry} can be written as the right-hand side of (10.1.5) with l_p replaced by l_e the entry-length correction appropriate to the apparent wall shear rate $32\dot{V}/\pi d_c^3$ (see Example 7.2.1).

As an example, we take $\nu = 0{\cdot}25$ and $l_e = 3d_p$. For $d_r = 3d_p$ we find that

$$\Delta P_{total} = 8K\left(\frac{7\dot{V}}{2\pi}\right)^{0\cdot25}\frac{l_p}{d_p^{1\cdot75}}\left(1+\frac{l_r}{3^{1\cdot75}l_p}+\frac{3d_p}{l_p}\right) \tag{10.1.6}$$

It is clear that since $3d_p$ can be greater than l_p, the dominant term could easily be the elastic entry pressure drop. Accurate predictions cannot therefore be expected, but reasonable upper and lower bounds can be given. There is little advantage here in trying to get better approximations to the pressure drops for fully-developed flow in pipes of non-circular cross-section. The practical limitations on output rate are set by exit flow instabilities, not by pressure drop or shear heating in the die. The extrudate shape will be determined largely by Deborah-number effects, associated with strong flow into the die orifice, and not by weak viscometric flow within the orifice. Degradation or cross-linking in the dead spaces behind the die plate can be a disadvantage.

Swelling (in the absence of positive draw-down of the extrudate) can be reduced by using a longer, smoother, entry to the die orifice itself; this can be achieved in a crude but simple fashion if the single die plate is replaced by a sheaf of plates which reduce the main channel cross-section to that of the final orifice cross-section by small discrete steps, thus preserving cheapness of manufacture and ease of adjustment.

Many plate dies have multiple orifices, a typical example being the spinneret used in fibre spinning. This consists of tens or hundreds of circular holes, sometimes countersunk, arranged in some pattern suited to the draw-down operation that follows extrusion. The diameter of these orifices is usually of the order of hundreds of microns. When making polymer pellets (chips or granules) by die-face cutting or by later chopping of quenched rod, the diameter is a few millimetres. The die design problem is largely one of ensuring that the output is equal from each hole: if the separation of the holes is not large compared with the diameter, then there can be fluid-dynamical interference between the separate entry flows with possible starvation of some holes. This will be particularly true if the $(P, \dot{V})$ curves for developed flow in the die orifices are multivalued as shown in Fig. 9.3. As output rates are increased by using pressure drops—across the die plate—greater than P_4, then it is in principle quite possible for some holes to carry a flow corresponding to the P_4P_1 arm of the curve while others carry the much larger flow corresponding to the P_3P_2 arm. The flow need not be unstable in the sense of varying with time; it will merely be grossly non-uniform from hole to hole. An alternative cause of variation could be boundary slip as shown in Fig. 9.2(ii) or (iii) or even thermal effects.

The subject has not been exhaustively analysed.

10.2 SLOT DIES

These are used to produce film ($\leqslant$1 mm thick) and sheet (>1 mm thick) of uniform thickness. Once extruded, film is chilled by letting it fall onto a cooled metal chill-roll or into a bath of cooling water and then wound onto shipment rolls (after drying, if necessary). Sheet is usually passed between two chill-rolls and then cut or wound up. The same system is used for film coating.

Figure 10.3 shows three typical slot dies. The first is end-fed and of the simplest construction. The long tapering manifold feeds the shallow die channel between roughly adjustable lands. The second is centre-fed. Besides the cylindrical manifold, it has an adjustable choke bar and adjustable but massive die-lips. The lip lands are in practice set as parallel as possible, while the choke bar, being more flexible, is used for correcting crosswise variations in gauge (thickness) by acting as a restrictor. The third is centre-fed but has a coat-hanger-shaped tapering manifold intended to eliminate the need for an adjustable restrictor bar. The die-lips are adjustable. The wedge-like cross-section is to allow the die to penetrate close to the nip between two chill-rolls.

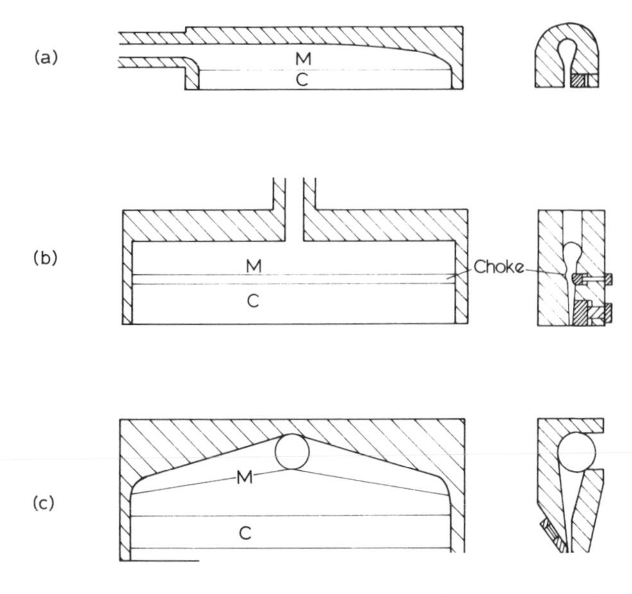

Fig. 10.3. Slot dies. (a) End-fed. (b) Centre-fed with choke bar. (c) Centre-fed, coat-hanger type. M, manifold; C, narrow channel to die-lips.

In all cases the dies can be provided with heaters top and bottom, and if necessary can be insulated from excessive heat loss to the environment. Experience shows that at the highest coating temperatures, 350°C say, variations of up to 15°C in the metal temperature can arise near the die-lips unless care is taken, these variations being caused more by losses to the environment than by heat transfer to the flowing polymer.

The width of such dies can vary from 0·5 m to 4 m. The length of the narrow channel, from manifold to die-lips, will be in proportion, and of the order of 1 m. The channel depth will vary from about 100 mm at die entry to about 1 mm at the die-lips for most film and correspondingly more for sheet.

The obvious intention in all slot-die designs is to provide a uniform plane flow through the die-lips and as far back towards the manifold as possible. The basic idea for ensuring this is to have relatively little pressure drop in the manifold and to have significant pressure drop through the flat channel C leading from the manifold M to the die exit. The geometry is idealized slightly in Fig. 10.4, where the manifold M

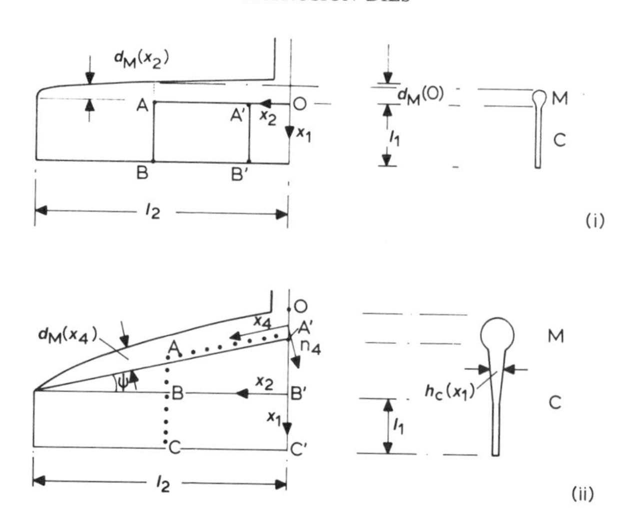

Fig. 10.4. Idealized geometry for slot dies. (i) Constant land length. (ii) Coat-hanger type.

is taken to be a tube of slowly-varying circular cross-section, feeding (i) a plane rectangular (case (a) and (b) of Fig. 10.3), or (ii) a plane trapezoidal (case (c) of Fig. 10.3) flow-channel C of slowly-varying depth.

A Cartesian coordinate system is introduced, so that x_1 and x_2 are surface coordinates in C, with the channel depth being $h_c(x_1, x_2)$. The manifold lies along $x_1 = 0$ in (i) and along $x_1 = (|x_2| - l_2)\tan\psi$ in (ii) for $0 \leqslant |x_2| \leqslant l_2$. The die exit is at $x_1 = l_1$. From symmetry we need only consider positive values of x_2. We introduce the coordinate

$$x_4 = x_2 \sec\psi \tag{10.2.1}$$

to describe flow in the manifold. We assume

$$d_M, h_c \ll \min(l_1, l_2) \tag{10.2.2}$$

We are thus able to apply the lubrication approximation described in Section 8.1, both to the manifold and the flat channel.

An estimate of $|\nabla h_c|$ based on the values given above is $10\ \text{mm}/1\ \text{m} = 0{\cdot}01$ and of $\mathrm{d}d_M/\mathrm{d}x_2$ is $40\ \text{mm}/2\ \text{m} = 0{\cdot}02$, both well within the range of applicability.

The flow field can then be described by a flow rate $Q_M(x_4)$ in M and $q_c(x_1, x_2)$ in C with associated pressures $p_M(x_4)$ and $p_c(x_1, x_2)$. If we assume for the present that the temperature field does not significantly affect the viscosity (what we may loosely term the isothermal approximation), then we can write

$$Q_M = Q_M(dp_M/dx_4, d_M) \tag{10.2.3}$$

$$\mathbf{q}_c = q_c(|\nabla p_c|, h_c)\nabla p_c/|\nabla p_c| \tag{10.2.4}$$

where Q_M and q_c are universal functions for any given material at the specified temperature, representing the flowrate in a uniform tube of diameter d_M under constant pressure gradient dp_M/dx_4, and the plane flowrate/unit width in a channel of depth h_c under uniform pressure gradient $|\nabla p_c|$, respectively. Note that because the walls at $x_3 = \pm\frac{1}{2}h$ are motionless in C, $\mathbf{q}_c$ is parallel to ∇p_c.

In C

$$\nabla \cdot \mathbf{q}_c = 0 \tag{10.2.5}$$

hence we can write

$$q_{c1} = \frac{\partial Q_c}{\partial x_2}, \qquad q_{c2} = -\frac{\partial Q_c}{\partial x_1} \tag{10.2.6}$$

where $Q_c(x_1, x_2)$ is a surface stream function. Continuity between the Q_M and Q_c flows is given by

$$\sec\psi \frac{d}{dx_4}(Q_M + Q_c) \equiv \frac{dQ_M}{dx_2} + \frac{\partial Q_c}{\partial x_2} + \frac{\partial Q_c}{\partial x_1}\tan\psi = 0$$

along

$$x_1 = (x_2 - l_2)\tan\psi = x_{1M} \tag{10.2.7}$$

which can trivially be integrated to give

$$[Q_M + Q_c]_{x_{1M}} = \dot{V} \tag{10.2.8}$$

where $2\dot{V}$ is the total flowrate into the die regarded as symmetric about $x_2 = 0$. At the entry to the die, $x_2 = 0$,

$$Q_M = \dot{V}, \qquad Q_c = 0 \tag{10.2.9}$$

and at the end of the die, $x_2 = l_2$, $x_4 = l_2 \sec\psi$

$$Q_M = 0, \qquad Q_c = \dot{V} \tag{10.2.10}$$

while at the outlet to the die, $x_1 = l_1$,

$$p_c = 0 \tag{10.2.11}$$

taking the ambient pressure p_a to be zero. (There is a singularity in pressure gradient at the origin $x_4 = 0$ in that it is a source of zero extent, but no singularity in the pressure.) The desired uniform flow distribution at the outlet is satisfied when

$$Q_c(x_1 = l_1) = \dot{V}x_2/l_2 \tag{10.2.12}$$

The driving pressure

$$P = p_M(x_4 = 0) \tag{10.2.13}$$

while continuity of pressure requires that

$$p_M(x_4) = p_c(x_{1M}, x_2) \tag{10.2.14}$$

Once the relations Q_M and q_c are known in terms of the viscosity $\eta(D)$, a solution to the flow problem can be obtained by requiring the pressure to be single-valued in C. This is the relation (8.1.29). Note that if d_M and h_c are prescribed, then the boundary condition (10.2.12) cannot be applied: $Q_c(x_1 = l_1)$ is determined by the solution. This is the straightforward flow analysis problem.

If, however, we insist that (10.2.12) be satisfied, then d_M and h_c cannot be completely prescribed, and we have a die-design problem to which, in this as in most other cases, there is no unique solution.

10.2.1 Power-law Approximation

Using the rheological model of (3.3.5)

$$\eta = \mathrm{K}D^{\nu-1}$$

it follows (see Examples 7.1.1 and 8.1.1) that

$$Q_M = \frac{\pi d_M^3 \nu}{8(3\nu+1)} \left(\frac{d_M}{4\mathrm{K}} \left| \frac{\mathrm{d}p_M}{\mathrm{d}x_4} \right| \right)^{1/\nu} \left(-\mathrm{sn} \frac{\mathrm{d}p_M}{\mathrm{d}x_4} \right) \tag{10.2.15}$$

and

$$q_c = -\frac{\nu h_c^2}{2(2\nu+1)} \left(\frac{h_c}{2\mathrm{K}} |\boldsymbol{\nabla} p_c| \right)^{1/\nu} \tag{10.2.16}$$

For flow analysis, given h_c and d_M we use (10.2.15) and (10.2.16) in their inverted form

$$\frac{\mathrm{d}p_M}{\mathrm{d}x_4} = -\frac{4\mathrm{K}}{d_M} \left(\frac{8(3\nu+1)}{\pi \nu d_M^3} \right)^{\nu} Q_M |Q_M|^{\nu-1} \tag{10.2.17}$$

$$\boldsymbol{\nabla} p_c = -\frac{2\mathrm{K}}{h_c} \left(\frac{2(2\nu+1)}{\nu h_c^2} \right)^{\nu} |\mathbf{q}|^{\nu-1} \mathbf{q} \tag{10.2.18}$$

Using (10.2.6) and the simple relation

$$\frac{\partial p_c^2}{\partial x_1\,\partial x_2}=\frac{\partial^2 p_c}{\partial x_2\,\partial x_1} \tag{10.2.19}$$

we obtain from (10.2.18)

$$\frac{\partial}{\partial x_2}\left[\frac{1}{h_c^{2\nu+1}}\left\{\left(\frac{\partial Q_c}{\partial x_1}\right)^2+\left(\frac{\partial Q_c}{\partial x_2}\right)^2\right\}^{\frac{1}{2}(\nu-1)}\frac{\partial Q_c}{\partial x_2}\right] + \frac{\partial}{\partial x_1}\left[\frac{1}{h_c^{2\nu+1}}\left\{\left(\frac{\partial Q_c}{\partial x_1}\right)^2+\left(\frac{\partial Q_c}{\partial x_2}\right)^2\right\}^{\frac{1}{2}(\nu-1)}\frac{\partial Q_c}{\partial x_1}\right]=0 \tag{10.2.20}$$

The boundary conditions are to be given on $x_1=\tan\psi(x_2-l_2)$, l_1; $x_2=0$, l_2. From (10.2.9, 10.2.10) we get

$$Q_c(x_1,0)=0 \qquad Q_c(x_1,l_2)=\dot{V} \tag{10.2.21}$$

From (10.2.11) we get

$$\frac{\partial Q_c}{\partial x_1}(l_1,x_2)=0 \tag{10.2.22}$$

From (10.2.14), (10.2.17), (10.2.18) and (10.2.8) we get

$$\frac{\partial Q_c}{\partial n_4}=-(\dot{V}-Q_c)\frac{(3\nu+1)}{(2\nu+1)d_M}\left(\frac{2h_c}{d_M}\right)^{2+1/\nu} \tag{10.2.23}$$

on x_{1M}, where n_4 is the inward normal. The pressure P can be calculated from

$$P=\int_{-l_2\tan\psi}^{l_1}\frac{2K}{h_c}\left(\frac{(2\nu+1)2}{\nu h_c^2}\left|\frac{\partial Q_c}{\partial x_2}\right|\right)^{\nu}\mathrm{d}x_1 \tag{10.2.24}$$

evaluated along $x_2=0$. We note that if we use the non-dimensional variables

$$\tilde{Q}_c=Q_c/\dot{V},\qquad \tilde{x}_1=x_1/l_1,\qquad \tilde{x}_2=x_2/l_1,\qquad \tilde{x}_4=x_4/l_1 \tag{10.2.25}$$

$$\tilde{d}_M=d_M/d_M(0),\qquad \tilde{h}_c=h_c/h_c(l_1,0) \tag{10.2.26}$$

$$\tilde{p}=ph_1^{1+2\nu}/Kl_1^{1-\nu}\dot{V}^{\nu} \tag{10.2.27}$$

then (10.2.20) is unchanged, (10.2.21) and (10.2.22) are only trivially altered, while (10.2.23) becomes

$$-\frac{\partial\tilde{Q}_c}{\partial\tilde{n}_4}=(1-\tilde{Q}_c)\frac{(3\nu+1)2^{2+1/\nu}}{\pi(2\nu+1)}\frac{\mathrm{H}^{2+1/\nu}}{\mathrm{D}^{3+1/\nu}}\frac{\tilde{h}_c^{2+1/\nu}}{\tilde{d}_M^{3+1/\nu}} \tag{10.2.28}$$

for all $0<\tilde{x}_2<L$, x_{1M} where the dimensionless groups L, H, and D are defined by

$$L=\frac{l_2}{l_1},\qquad H=\frac{h_c(l_1,0)}{l_1},\qquad D=\frac{d_M(0)}{l_1} \tag{10.2.29}$$

(10.2.20) is an elliptic partial-differential equation for which (10.2.21), (10.2.22) and (10.2.28)—or their equivalent dimensionless forms—form a suitable set of boundary conditions defining a unique solution for $\tilde{Q}_c$ and hence $\tilde{p}$.

A most important characteristic of the dimensionless solution for our purposes is that it is influenced only by the geometrical parameters ψ, L, H and D and the power-law index ν and not by the rheological constant K or the operating constant $\dot{V}$. Thus, the streamlines are uninfluenced by flow rate or changes in operating temperature, provided ν is essentially constant. Thus, any design that satisfies (10.2.12) for given ν will be a suitable design for all polymers having that power-law index.

We have already seen that many polymer melts can be reasonably represented, as far as viscometric flows are concerned, by a power-law relation, with ν between $\frac{1}{2}$ and $\frac{1}{3}$, for shear rates between 1 and 1000, say. If we take the values

$$\left.\begin{aligned}&\dot{V}=7\times10^4\,\mathrm{mm^3\,s^{-1}}\ (250\,\mathrm{kg\,h^{-1}})\\ &l_2=2l_1=2\,\mathrm{m},\ d_c(0)=50\,\mathrm{mm},\ h_c(l_1,0)=1\,\mathrm{mm},\ \nu=1/3\end{aligned}\right\} \tag{10.2.30}$$

to get upper and lower limits to the shear rate D, we find that

$$D_M\simeq 8,\qquad D_c\simeq 350 \tag{10.2.31}$$

which is what we had hoped. For these values we see that

$$H=10^{-3},\qquad D=0{\cdot}05 \tag{10.2.32}$$

and the coefficient of $(1-\tilde{Q}_c)\tilde{h}_c^{2+1/\nu}/\tilde{d}_M^{3+1/\nu}$ in (10.2.28) is of order 10^{-7}. However, the factor $\tilde{h}_c^5/\tilde{d}_M^6$ will usually be large because $\tilde{h}_c>1$ and $\tilde{d}_M<1$, a typical value being $3^5/\frac{1}{2}^{6}\sim 1{\cdot}5\times10^5$. This still makes the coefficient of $(1-\tilde{Q}_c)$ of order 10^{-2}, which is consistent with having a low pressure drop in the manifold.

Further progress can be made in special cases, where particular forms for $\tilde{h}_c$ and $\tilde{d}_M$ are specified, and where the coefficient of $(1-\tilde{Q}_c)$ in (10.2.28) is small.

Case (i): Rectangular Channel Plan. Analysis
Thus, for example, if the solution $\tilde{Q}_c^{(0)}$ to (10.2.20) subject to (10.2.21), (10.2.22) and

$$\left(\frac{\partial \tilde{Q}_c^{(0)}}{\partial \tilde{x}_1}\right)_{x_1=0} = 0 \tag{10.2.33}$$

is relatively easy to obtain, then we can obtain a better approximation

$$\tilde{Q}_c = \tilde{Q}_c^{(0)} + \tilde{Q}_c^{(1)} \tag{10.2.34}$$

where $\tilde{Q}_c^{(1)}$ is chosen to satisfy the linearized equation

$$\sum_{i=1}^{2} \frac{\partial}{\partial \tilde{x}_i}\left[\frac{1}{\tilde{h}_c^{2\nu+1}}\left\{\left(\frac{\partial \tilde{Q}_c^{(0)}}{\partial \tilde{x}_1}\right)^2 + \left(\frac{\partial \tilde{Q}_c^{(0)}}{\partial \tilde{x}_2}\right)^2\right\}^{\frac{1}{2}(\nu-3)} \left\{\sum_{j=1}^{2}\left(\frac{\partial \tilde{Q}_c^{(0)}}{\partial \tilde{x}_j}\right)^2 \frac{\partial \tilde{Q}_c^{(1)}}{\partial \tilde{x}_i} + (\nu-1)\frac{\partial \tilde{Q}_c^{(0)}}{\partial \tilde{x}_j}\frac{\partial \tilde{Q}_c^{(0)}}{\partial \tilde{x}_i}\frac{\partial \tilde{Q}_c^{(1)}}{\partial \tilde{x}_j}\right\}\right] = 0 \tag{10.2.35}$$

together with the boundary conditions

$$\tilde{Q}_c^{(1)}(\tilde{x}_1, 0) = \tilde{Q}_c^{(1)}(\tilde{x}_1, \mathrm{L}) = \frac{\partial \tilde{Q}^{(1)}}{\partial \tilde{x}_1}(1, \tilde{x}_2) = 0 \tag{10.2.36}$$

and

$$\frac{\partial \tilde{Q}_c^{(1)}}{\partial \tilde{x}_1} = (1 - \tilde{Q}_c^{(0)})\frac{(3\nu+1)2^{2+1/\nu}}{(2\nu+1)}\frac{\mathrm{H}^{2+1/\nu}}{\mathrm{D}^{3+1/\nu}}\frac{h_c^{2+1/\nu}}{D_M^{3+1/\nu}} \text{ on } \tilde{x}_1 = 0 \tag{10.2.37}$$

Although (10.2.35) may seem to have no advantages over the original equation, its linearity makes numerical solution much more straightforward once $\tilde{Q}_c^{(0)}$ is known.

If, further, $\tilde{d}_M = 1$ and $\tilde{h}_c$ is a function of $\tilde{x}_1$ only, then the solution

$$\tilde{Q}_c^{(0)} = \tilde{x}_2/\mathrm{L} \tag{10.2.38}$$

satisfies (10.2.20) with $\tilde{p}$ given by

$$\tilde{p}(\tilde{x}_1) = \mathrm{L}^{-\nu}\int_{\tilde{x}_1}^{1} \tilde{h}_c^{-(1+2\nu)}\, d\tilde{x}_1 \tag{10.2.39}$$

Equation (10.2.35) then simplifies to

$$\frac{\partial}{\partial \tilde{x}_1}\left(\frac{1}{\tilde{h}_c^{2\nu+1}}\frac{\partial \tilde{Q}_c^{(1)}}{\partial \tilde{x}_1}\right) + \frac{1}{\tilde{h}_c^{2\nu+1}}\frac{\partial^2 \tilde{Q}_c^{(1)}}{\partial \tilde{x}_2^2} = 0 \tag{10.2.40}$$

with (10.2.37) becoming

$$\frac{\partial \tilde{Q}_c^{(1)}}{\partial \tilde{x}_1}=\frac{(\mathrm{L}-\tilde{x}_2)A}{\mathrm{L}} \text{ on } \tilde{x}_1=0 \tag{10.2.41}$$

where A is a small constant.

At $\tilde{x}_1=0$, $\tilde{x}_2=0$, we have the apparently incompatible boundary conditions

$$\frac{\partial \tilde{Q}_c^{(1)}}{\partial \tilde{x}_1}=A \text{ from (10.2.41) and } \frac{\partial \tilde{Q}_c^{(1)}}{\partial \tilde{x}_1}=0 \text{ from (10.2.36)}$$

However, this problem is a familiar one in harmonic analysis and leads to no difficulty. It has no physical significance because the die entry will have finite cross-section and so the manifold approximation will not be meaningful where $\tilde{x}_1$ and $\tilde{x}_2$ are of order D.

If $\mathrm{L}^2 \gg 1$, which corresponds to a wide die, a different non-dimensionalization is applied using

$$\tilde{x}_2^* = \tilde{x}_2/\mathrm{L} \tag{10.2.42}$$

instead of $\tilde{x}_2$. This makes the governing equation

$$\frac{\partial}{\partial \tilde{x}_1}\left[\frac{1}{h_c^{2\nu+1}}\left|\frac{\partial \tilde{Q}_c}{\partial \tilde{x}_1}\right|^{\nu-1}\frac{\partial \tilde{Q}_c}{\partial \tilde{x}_1}\right]+\frac{1}{\mathrm{L}^2}\left\{\frac{\partial}{\partial \tilde{x}_1}\left[\frac{(\nu-1)}{2h_c^{2\nu+1}}\left|\frac{\partial \tilde{Q}_c}{\partial \tilde{x}_1}\right|^{\nu-3}\right.\right.$$

$$\left.\left.\times\left(\frac{\partial \tilde{Q}_c}{\partial \tilde{x}_2^*}\right)^2\frac{\partial \tilde{Q}_c}{\partial \tilde{x}_1}\right]+\frac{\partial}{\partial \tilde{x}_2^*}\left[\frac{1}{h_c^{2\nu+1}}\left|\frac{\partial \tilde{Q}_c}{\partial \tilde{x}_1}\right|^{\nu-1}\frac{\partial \tilde{Q}_c}{\partial \tilde{x}_2^*}\right]\right\}=0\left(\frac{1}{\mathrm{L}^4}\right) \tag{10.2.43}$$

with boundary conditions on $\tilde{x}_1=0$, 1 and $\tilde{x}_2^*=0$, 1. A solution can be written formally as

$$\tilde{Q}_c=\tilde{Q}_c^{*(0)}+\mathrm{L}^{-1}\tilde{Q}_c^{*(1)}+\ldots \tag{10.2.44}$$

with $\tilde{Q}_c^{*(0)}$ satisfying

$$\frac{\partial}{\partial \tilde{x}_1}\left[\frac{1}{h_c^{2\nu+1}}\left(\frac{\partial \tilde{Q}_c^*(0)}{\partial \tilde{x}_1}\right)^{\nu}\right]=0 \tag{10.2.45}$$

and so

$$\tilde{Q}_c^{*(0)}=\tilde{Q}_c^{*(0)}(\tilde{x}_2^*) \text{ because of (10.2.22)} \tag{10.2.46}$$

$\tilde{Q}_c^{*(0)}$ is then obtained by matching the pressure drop along the pathline OAB to that along OA′B′ in Fig. 10.4(i). Thus, from

(10.2.28),

$$\frac{\partial}{\partial \tilde{x}_2^*}\left[\left(\frac{\partial \tilde{Q}_c^{*(0)}}{\partial \tilde{x}_2^*}\right)^{\nu}\int_0^1 \tilde{h}_c^{-(2\nu+1)}\,d\tilde{x}_1\right]-\left(\frac{3\nu+1}{(2\nu+1)\pi\tilde{d}_M}\right)^{\nu}$$

$$\times\left(\frac{2\tilde{h}_c(x_1=0)}{\tilde{d}_M}\right)^{2\nu+1}\frac{\mathrm{L}^{\nu+1}\mathrm{H}^{2\nu+1}}{\mathrm{D}^{3\nu+1}}(1-\tilde{Q}_c^{*(0)})=0 \qquad (10.2.47)$$

This will lead to a consistent approximation provided

$$\mathrm{L}^{\nu+1}\mathrm{H}^{2\nu+1}/\mathrm{D}^{3\nu+1}\leqslant 0(1) \qquad (10.2.48)$$

and can be treated as independent of L.

The equation for $\tilde{Q}_c^{*(1)}$ proves to be unpleasantly non-linear, so there is really no advantage to the expansion (10.2.44) unless it is truncated after the first term. It has been used (Pearson, 1963, 1966, p. 105) to select a profile for h_c that will lead $\tilde{Q}_c^{*(0)}$ to satisfy the uniform exit condition (10.2.12).

Case (ii): Coat-hanger Channel Plan. Design

In this case we can choose $h_c(x_1)$ and $d_M(x_4)$ so as to give, within the lubrication approximation used here, an exact solution satisfying the uniform exit condition (10.2.12) with streamlines OA′B′C′ and OABC as shown in Fig. 10.4(ii). If

$$\tilde{Q}_c\equiv\tilde{Q}_c(x_2)=\tilde{x}_2^* \qquad (10.2.49)$$

then

$$\tilde{p}=\tilde{p}(\tilde{x}_1) \qquad (10.2.50)$$

directly from (10.2.18). Moreover (10.2.28) becomes simply

$$(\tfrac{1}{2}d_M)^{3+1/\nu}=\frac{h_c^{2+1/\nu}(3\nu+1)(-x_1)}{2\pi(2\nu+1)\sin\psi\tan\psi} \qquad (10.2.51)$$

for all $-l_2\tan\psi<x_1<0$.

10.2.2 Oldroyd Fluid Approximation

Using the rheological model (3.3.22), it can be shown (see Table 4.1, or Williams & Bird, 1962) that

$$\eta(D)=\eta_0\frac{1+\lambda_3^2D^2}{1+\lambda_4^2D^2} \qquad (10.2.52)$$

where

$$\left.\begin{aligned}\lambda_3^2 &= \lambda_1\lambda_2 + \mu_0(\mu_2 - \tfrac{3}{2}\nu_2) - \mu_1(\mu_2 - \nu_2)\\ \lambda_4^2 &= \lambda_1^2 + \mu_0(\mu_1 - \tfrac{3}{2}\nu_1) - \mu_1(\mu_1 - \nu_1)\end{aligned}\right\} \tag{10.2.53}$$

The corresponding functions Q_M, q_c as defined in (10.2.3) and (10.2.4) are given in dimensionless form in Williams & Bird (1962, Figs 2 and 3). It is readily seen that for all reasonable values of λ_3/λ_4, the relevant curves show a fairly sharp change in slope for a value of apparent wall shear rate equal to about $1/2\lambda_4$, or equivalently a wall shear stress of about $\eta_0/2\lambda_4$.

Without attempting to solve the flow problem in any detail, it is clear that unless the relevant wall shear rates are everywhere either much less than $1/2\lambda_4$ or much greater than $1/2\lambda_4$ (corresponding to limiting Newtonian viscosities of η_0 and $\lambda_3^2\eta_0/\lambda_4^2$, respectively), the stream function Q_c will be significantly dependent on the magnitude of $\dot{V}$. Thus, for example, a type (ii)—coat-hanger—die designed to operate at a given output rate would show non-uniform flow distribution at the die exit as either flowrate or material (as parametrized by λ_3 and λ_4) were altered significantly from the design conditions. Adjustable die-lips or choke bars have been introduced to cover this situation.

10.2.3 Thermal Effects

Following the development given for thermal effects in Section 8.1.3 (the lubrication approximation), we start by calculating typical values of Gz, Br (or B) and Na.

Taking the values (10.2.30) used to obtain (10.2.31), we obtain using the definition (8.1.37)

$$v^*_{\text{conv}} \leqslant 35 \text{ mm s}^{-1} \text{ with } h \geqslant 1 \text{ mm} \tag{10.2.54}$$

Using the values for ρ^*, Γ_p and α^* given in (7.1.34), we obtain a value for the thermal diffusivity

$$\kappa^* = \alpha^*/\rho^*\Gamma_p^* = 10^{-7} \text{ m}^2 \text{ s}^{-1} \tag{10.2.55}$$

and hence a Peclet number, as defined in (8.1.38),

$$\text{Pe} = v^*_{\text{conv}} h/\kappa^* \approx 350 \tag{10.2.56}$$

Note that for a given output rate $\dot{V}$ and fixed l_2, the Peclet number is given by $\dot{V}/2l_2\alpha^*$ and is independent of h.

The Graetz number will therefore be determined by the relevant value of H, as defined in (10.2.29), since analogously to (7.1.43), we write

$$\mathrm{Gz} = \mathrm{PeH} \tag{10.2.57}$$

Because in many cases h varies along the 'streamlines', it is useful to define an integrated Graetz number, by

$$(\mathrm{Gz_I})^{-1} = \int_0^{l_1} (\mathrm{Pe})^{-1} \frac{\mathrm{d}x_1}{h(x_1)} \tag{10.2.58}$$

If we suppose that the die-lip lands are 20 mm long, we obtain a local Graetz number for heated die-lips of

$$\mathrm{Gz} \approx 350/20 = 17{\cdot}5 = 0(10) \tag{10.2.59}$$

For $h_c = 5$ mm and $l_1 = 0{\cdot}5$ m, the Graetz number falls to 3·5.

Thus, we see that the temperature of the extrudate can be altered by changing the die body temperature and even by preferentially heating or cooling the die-lips.

Using, as in (7.1.34), $T^*_{\mathrm{rheol}} = 50$ K, and $\eta^* = 10^3$ N s m^{-2}, we obtain using the definition (8.1.39) with $v^*_{\mathrm{gen}} = v^*_{\mathrm{conv}}$

$$\mathrm{Na} \leqslant 10 \tag{10.2.60}$$

which shows that generated temperature difference may be very important, though since Na is proportional to the square of the scale velocity, this is unlikely to be the case except at the die-lips for thin film production. A typical value of Na is about $\frac{1}{2}$.

In practice, a temperature differential of about 20 K is about as large as is imposed between die body and extruder barrel. Whether the difference in wall temperature is likely to lead to larger or smaller temperature gradients within the flowing melt and hence within the extrudate will depend on many factors:

(i) On whether the bulk mean temperature of the melt leaving the extruder is greater or less than the barrel wall temperature. As will be seen later, extruders operated slowly can yield a melt that is colder than the imposed wall temperatures, but large extruders operated near maximum capacity often yield a melt that is much hotter than the barrel walls—and indeed barrel cooling rather than barrel heating is the norm. Under these circumstances, additional heating of the die body may tend to even out the temperature profile, provided $\mathrm{Gz_I} > 10$.

(ii) On whether the pressure drop within the die is large or not. A significant pressure drop is necessary to ensure stability of operation. The energy so dissipated will appear as heat preferentially in the layers near the wall (where the shear stress and shear rates are highest).

If $Gz_I \ll 10$, then a thermally fully-developed flow is obtained, and so a generation number will serve to compare the final temperature differences with whatever scale temperature difference is important.

If $Gz_I \gg 10$, then the ratio of the adiabatic temperature difference (see (7.1.40)) and the chosen scale temperature difference will be a better measure of generation effects than the generation number. A reasonable estimate of this ratio is given by $10\,Gn/Gz_I$, the factor 10 being included to make the two measures equal when $Gz_I = 10$. If this ratio is large, then generation effects will be dominant; if small, negligible.

(iii) On whether the viscosity of the melt is significantly altered by the temperature differences involved. This is measured by

$$\begin{array}{ll} Na & \text{for } Gz_I \ll 10 \\ 10\,Na/Gz_I & \text{for } Gz_I \gg 10 \end{array}$$

as far as generation is concerned, and by (see (8.1.40))

$$B = Na/Br \text{ where Br is defined in (7.1.22)}$$

as far as wall temperature differences are concerned.

For the case discussed above, leading to the estimates (10.2.55)–(10.2.60) we find a minor but non-trivial effect ($B < 0{\cdot}4$, $Na \sim 0{\cdot}5$) except at the die-lips. Even there, if we use Na/Gz as the relevant parameter, we obtain $Na/Gz \sim 0{\cdot}5$, and so we do not expect the effect to be overwhelming.

To say anything more definite in any particular case, we have to return to the original model equations given by (10.2.3), (10.2.4) and re-introduce the temperature field. Because the lubrication approximation still applies to the momentum equation, the actual relationship between $\boldsymbol{\nabla} p$ and $\mathbf{q}$ (or dp/dx_4 and Q_M) now becomes a relatively complicated function of the local temperature field $T_c(x_3)$ (or $T_M(r)$). Furthermore, the energy equation that yields $T_c(x_3)$ (or $T_M(r)$) involves the velocity distribution $\mathbf{v}_c(x_1, x_2, x_3)$ (or $\mathbf{v}_M(r, x_4)$) and not just the local stream function $\mathbf{q}_c$ or $\mathbf{Q}_M$.

Some simplification can be achieved by supposing the isothermal approximation to be the leading term of an expansion in terms of Na. In this way, the solution already obtained can be used to provide a detailed first-order temperature distribution, on the assumption that the relations (10.2.3) and (10.2.4) actually imply particular (isothermal) velocity distributions that can be used in the thermal energy equation. This first-order temperature field can then be substituted back into the (local) lubrication equation for the velocity profile to yield the first-order perturbations to it and hence to (10.2.3) and (10.2.4). A very simple example of this procedure is given in Appendix 2. It requires that the Brinkman number, Br, be no larger than order unity (or rather that B be small enough to be used as an alternative expansion parameter).

Alternatively, fully numerical solutions obtained by digital computational methods can be sought. Winter (1977) has provided an extensive review of the subject.

EXERCISE 10.2.1. Derive the boundary condition (10.2.23), paying attention to the direction of n_4.

EXERCISE 10.2.2. Derive the eqns (10.2.35) and (10.2.40) from those given earlier.

EXERCISE 10.2.3. Derive eqns (10.2.43) and (10.2.47) from those given previously.

EXERCISE 10.2.4. Derive eqn (10.2.51).

10.3 ANNULAR DIES

These are used to produce pipes and film. The former is extruded at or very near the desired diameter and is not drawn down, whereas the latter is reduced in thickness by a factor of 10 or more by drawing down in the axial direction and blowing outwards in the radial direction after extrusion. The channel thickness at the die-lips will therefore range from about 1 mm to 10 mm. The internal structure of the die, providing a flow channel leading from the nose of the extruder screw to the annular die lips, varies with the type of die used. Conventionally, three types are distinguished:

(a) cross-head dies;
(b) spiral dies; and
(c) spider dies.

Typical examples are shown diagrammatically in Figs. 10.5, 10.6 and 10.7. The actual dimensions have been chosen for ease of visualization

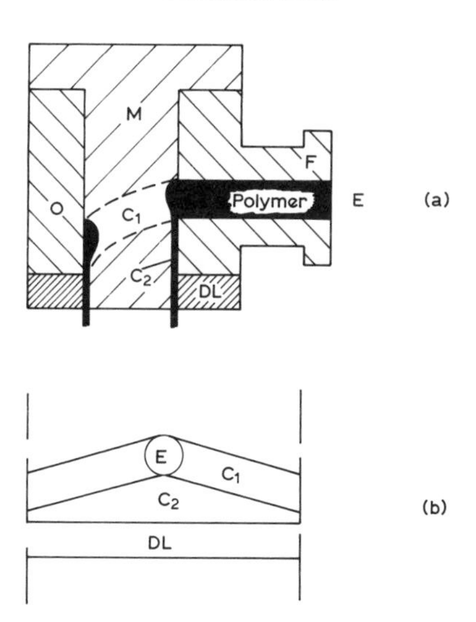

Fig. 10.5. Diagrammatic representation of simple cross-head die. (a) Cross-section taken along axis of extruder (E) showing flange (F), outer die body (O), inner die body, or mandrel, (M) and adjustable die-lips (DL). (b) Channel (C_1, C_2) shown opened out.

rather than as actual examples. In all three cases, much of the channel has been shown as lying on a circular cylinder, which is in any case consistent with ease of manufacture. Provided the channel width is small compared to the cylindrical radius, then the channels can be unrolled and shown as lying on a plane.

The cross-head die is then seen (Fig. 10.5(b)) to be essentially equivalent to the coat-hanger slot die analysed in the previous subsection, and so we can take over all the arguments used there. A cross-head die design is given in Pearson (1963), Pearson & Devine (1963) and is discussed in Pearson (1966, p. 103).

10.3.1 Spiral Die

The unrolled form of the spiral die shown in Fig. 10.6(b) is similar to that considered for the slot die. Using analogous coordinates (x_1, x_2) for the leakage areas L, and x_4 for the spiral distribution channel, we have the same relations (10.2.3) and (10.2.4) to relate Q_S, $\mathbf{q}_L$ to dp_S/dx_4 and ∇p_L. Equations (10.2.5) and (10.2.6) still apply although

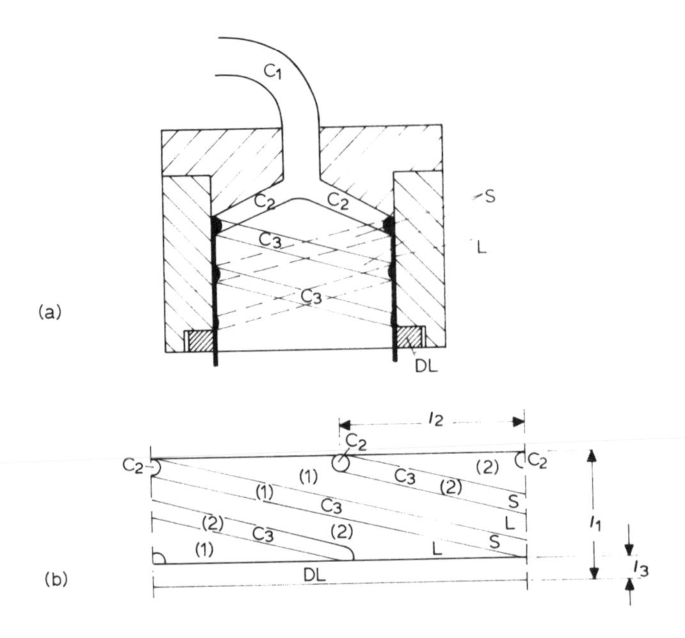

Fig. 10.6. Diagrammatic representation of two-start spiral die. (a) Cross-section showing internal channels C_1 and C_2 from the extruder and spiral channels C_3 wound around the inner die body down to adjustable die-lips DL. A leakage channel L between the inner and outer die bodies extends right around the die increasing in thickness towards the die-lips. (b) Unrolled form showing two (one split) entries C_2 and helical channels C_3 leading to die lips DL.

$\mathbf{q}_L$ will show discontinuities along the lines

$$x_2 = 2l_2x_1/(l_1 - l_3) \quad \text{and} \quad l_2\{1 + 2x_1/(l_1 - l_3)\} \tag{10.3.1}$$

Analogues of equation (10.2.7) can be written if L is supposed to be continuous beneath S.

$$\sec\psi \frac{d(Q_{S1} + Q_{L2} - Q_{L1})}{dx_4} \equiv \frac{dQ_{S1}}{dx_2} + \frac{\partial Q_{L2}}{\partial x_2} - \frac{\partial Q_{L1}}{\partial x_2} + \left(\frac{\partial Q_{L2}}{\partial x_1} - \frac{\partial Q_{L1}}{\partial x_1}\right)\tan\psi = 0$$

along

$$x_1 = (l_1 - l_3)x_2/2l_2 \tag{10.3.2}$$

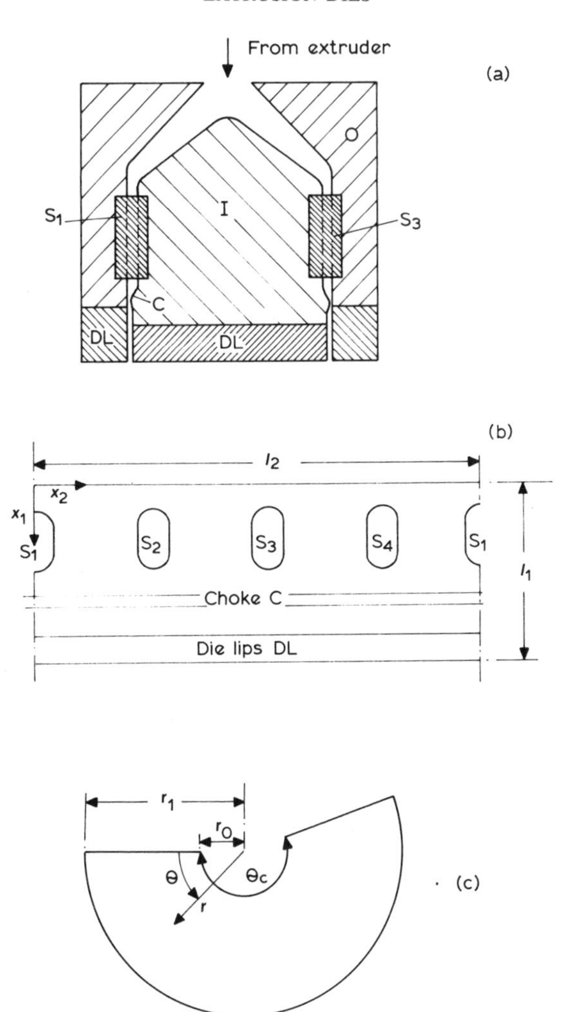

Fig. 10.7. Diagrammatic representation of four-legged spider-die. (a) Cross-section showing outer die body O and inner die body I fixed relative to one another by two of four spider legs S_1 and S_3. The flow channel is choked (C) just before the die lips DL. (b) Unrolled form of cylindrical portion of channel showing four symmetrical obstacles to flow caused by spider legs, and (x_1, x_2) coordinate system. (c) Unrolled form of conical portion of channel showing (r, θ) coordinate system, where $r = r_1$, $\theta = (0, \theta_c)$ are equivalent to $x_1 = 0$, $x_2 = (0, r_1\theta_c)$ in (b) above, and $r = r_0$, $\theta = (0, \theta_c)$ are taken to be the entry point from the extruder proper.

and

$$\sec\psi \frac{d(Q_{S2}+Q_{L1}-Q_{L2})}{dx_4} \equiv \frac{dQ_{S2}}{dx_2}+\frac{\partial Q_{L1}}{\partial x_2}-\frac{\partial Q_{L2}}{\partial x_2}+\left(\frac{\partial Q_{L1}}{\partial x_1}-\frac{\partial Q_{Ls}}{\partial x_1}\right)\tan\psi \equiv 0$$

along

$$x_1=(l_1-l_3)(x_2-l_2)/2l_2 \tag{10.3.3}$$

where

$$\tan\psi=(l_1-l_3)/2l_2$$

with Q_{S1} and Q_{S2} defined along the channels (1) and (2) shown in the Figure and Q_{L1} and Q_{L2} defined in the regions marked (1) and (2) in the same Fig. 10.6(b).

Obvious integral forms follow when we put

$$\left.\begin{aligned} Q_{L1}=Q_{L2}=0 \quad \text{at} \quad x_1=0 \\ Q_{S1}(0)=Q_{S2}(0)=\tfrac{1}{2}\dot{V} \end{aligned}\right\} \tag{10.3.4}$$

For convenience we write

$$Q_{S1}(l_4)=Q_{S2}(l_4)=0; \qquad l_4^2=(l_1-l_3)^2+4l_2^2 \tag{10.3.5}$$

In practice, this will be the case because the cross-section of the spiral channel is chosen to decrease from a maximum at the inlets C_2 to 'zero' at $x_1=l_1-l_3$, when allowance is made for the depth of the leakage channel which usually increases from zero to a maximum at $x_1=l_1-l_3$.

The pressure condition (10.2.11) still holds, while a uniform distribution condition (10.2.12) can still be applied. If the channel depth between $x_1=l_1-l_3$ and $x_1=l_1$ is constant, then (10.2.12) is applied to Q_{L1} and Q_{L2} at $x_1=l_1-l_3$. A continuity of pressure condition (10.2.14) will apply to p_{S1}, p_{S2}, p_{L1} and p_{L2}.

As in the case of the slot die, the problem can be regarded either as one of analysing the flow pattern if d_S and h_L are given, or as a design problem constraining d_S and h_L if the analogue of (10.2.12) is imposed as a boundary condition.

10.3.2 Spider Die

The spider die (a typical example of which is shown in Fig. 10.7) presents a slightly different flow problem as can be seen from the unrolled form when an otherwise completely symmetric plane flow field (corresponding to the original circular cylindrical field) is disturbed by four obstructions S_1–S_4 and has to be matched to a radial flow field (corresponding to the original conical flow field) along the edge $x_1 = 0$. Within the circular segment $\theta = (0, \theta_c)$, $r = (r_0, r_1)$ we assume a channel depth $h_c(r, \theta)$, a pressure $p_c(r, \theta)$ and a stream function $Q_c(r, \theta)$; within the plane rectangular field $x_1 = (0, l_1)$, $x_2 = (0, l_2)$, we assume a channel depth $h_s(x_1, x_2)$, a pressure $p_s(x_1, x_2)$ and a stream function $Q_s(x_1, x_2)$. There is fourfold symmetry so we can impose the conditions

$$\left.\begin{aligned} Q_c(r, 0) &= Q_s(x_1, 0) = 0 \\ Q_c(r, \tfrac{1}{4}\theta_c') &= Q_s(x_1, \tfrac{1}{4}l_2) = \tfrac{1}{4}\dot{V} \end{aligned}\right\} \tag{10.3.6}$$

except where $(x_1, 0)$ or $(x_1, \frac{1}{4}l_2)$ lies within the obstructions,

$$p_c(r_0, \theta) = P,\ p_s(l_1, x_2) = 0 \tag{10.3.7}$$

If h_s and h_c are both prescribed, then clearly P and $\dot{V}$ are not independent, one implying the other. On the half boundaries of S_1 and S_2 having $\frac{1}{4}l_2 > x_2 > 0$, we have

$$Q_s = 0, \tfrac{1}{4}\dot{V} \text{ respectively} \tag{10.3.8}$$

This neglects any viscous effects at the spider boundaries but is consistent with the lubrication approximation adopted and can in any case be dealt with by extending the effective area of the spiders.

The matching at $r = r_1$, $x_1 = 0$ requires that

$$\left.\begin{aligned} Q_c(r_1, \theta) &= Q_s(0, r_1\theta) \\ p_c(r_1, \theta) &= p_s(0, r_1\theta) \end{aligned}\right\} \tag{10.3.9}$$

Within the flow fields C and S, the eqns (10.2.4), (10.2.5) and (10.2.6) developed for slot-flow hold, with the obvious change that the stream function Q_c is defined by

$$q_r = -\frac{1}{r}\frac{\partial Q_c}{\partial \theta}, \qquad q_\theta = \frac{\partial Q_c}{\partial r} \tag{10.3.10}$$

Clearly the design problem is to increase h_s in the neighbourhood of the spiders S_1–S_4 so as to counter the weakening of the flow that would otherwise arise along the lines $x_2 = kl_2/4$, $k = 0, 1, 2, 3$.

This approach has been used by Pearson (1962) to estimate the effect of the spiders when h_s is taken to be constant. Using the power-law model leading to eqn (10.2.20), it was shown that, as ν decreases from 1, the effect of the spiders becomes more pronounced.

10.3.3 Solutions for Deep Annular Channels

The unrolling technique used to give the planar forms of the flow channels was based on the assumption that

$$G^* = h^*/R^* \ll 1 \tag{10.3.11}$$

where h^* is a representative depth, and R^* a representative radius. Some check on the accuracy of this assumption, or more properly, some guide as to the best choice of radius to use for determining the area of the planar surface, can be obtained by solving for the exact viscometric flow field in an annulus.

Let the flow channel be between $r = R_i$ and $r = R_0$, in terms of an (r, θ, z) coordinate system. In a fully-developed flow, we have

$$\mathbf{v} = \{0, 0, v_z(r)\} \tag{10.3.12}$$

with

$$v_z(R_i) = v_z(R_0) = 0 \tag{10.3.13}$$

The equation governing the flow is the same as for capillary flow, eqn (7.1.53), and so we write

$$\frac{1}{r}\frac{\mathrm{d}}{\mathrm{d}r}\left\{r\eta\left(\left|\frac{\mathrm{d}v_z}{\mathrm{d}r}\right|\right)\frac{\mathrm{d}v_z}{\mathrm{d}r}\right\} = \frac{\partial p}{\partial z} \tag{10.3.14}$$

Integration of (10.3.14) yields

$$\tau(\dot{\gamma}) = t_{rz}\left(\frac{\mathrm{d}v_z}{\mathrm{d}r}\right) = \frac{1}{2}\frac{\partial p}{\partial z}\left(r - \frac{r_0^2}{r}\right) \tag{10.3.15}$$

where the constant r_0^2 can only be obtained after further integration of (10.3.15), when the constitutive function $\tau(\dot{\gamma})$ is inverted to yield $\dot{\gamma}(t_{rz})$—cf. eqn (7.1.56)—and the boundary conditions (10.3.13) are satisfied. In the case of a capillary, r_0^2 can be put equal to zero by symmetry, but not in this more general case which makes the algebra much more complex.

Solutions are given in BA & H. A curvature correction factor for a Newtonian fluid is given in their Example 5.2.5. This suggests that for $R_0 - R_i = \varepsilon R_0$ with ε small, the effective width of a plane channel of depth εR_0 is $2\pi R_0(1 - \frac{1}{2}\varepsilon)$, a fairly obvious result. For larger values of

ε, the relevant radius is

$$\tfrac{3}{4}R_0\varepsilon^{-3}\{1-(1-\varepsilon)^4-\varepsilon^2(2-\varepsilon)^2/\ln(1/1-\varepsilon)\} \qquad (10.3.16)$$

and it is suggested that this be used as a suitable value for the Ellis fluid model also. The relevant values for a power-law model are plotted in Fig. 5C.1, where it should be noted that the limit $\varepsilon \to 1$ ($x \to 0$ in BA & H's notation) is not the same as capillary flow because the velocity is here zero on the axis $r = 0$.

10.3.4 Elastic Effects

The spider die shown in Fig. 10.7 clearly involves relatively rapid changes in flow field geometry in the neighbourhood of the spider and relatively large extension rates along streamlines both there and in the cone region unless channel depths are suitably chosen to avoid this. Thus elastic effects due to large extension rates, mentioned in connection with profile dies, but so far neglected in the slot die, could be relevant in the more compact annular dies.

A simple and effective way to minimize such effects seems to be to adopt a constant overall channel cross-section. Thus, in the (r, θ) flow field of Fig. 10.7(c) we would choose

$$h_c = h_{c0}r_0/r = A_0/\theta_c r \qquad (10.3.17)$$

and arrange in Fig. 10.7(b) for

$$4\int_0^{l/4} h_s\,dx_2 = A_0 \qquad (10.3.18)$$

for all x_1 up to the line of the choke. In practice, the area $A(x_1)$ will decrease at the choke and at the die-lips, but here the flow can be expected to be planar (i.e. independent of x_2) and so elastic effects should not affect the streamline pattern, unless the plane flow is itself unstable. Recent work by Winter & Fischer (1981) uses the assumption that the die flow can be calculated by lubrication theory, even when predicting large elastic effects on exit.

10.3.5 Instabilities in Plane Channel or Narrow Annulus Flow

The most conclusive experimental evidence of instability of plane flow has been provided by Giesekus (1972) and Pickup (1970), using highly elastic polymer solutions flowing through a slot die. In the latter case, the instability arose in the wedge-shaped entry to the relatively short plane slot die, for which the Weissenberg and Deborah numbers were both high. Bhatnagar & Giesekus (1970) have considered theoretically the stability of fully-developed parallel plane flow to roll-like disturbances with axes parallel to the flow direction, but without obtaining any conclusive results.

Observations on molten polymers extruded through annular dies have often shown, in practice, relatively large, steady variations (around the die) in extrudate thickness which could not obviously, or even plausibly, be explained in terms of channel depth variations, die-wall temperature variations or earlier asymmetries in the die channel geometry. To be specific, the local output q_1 at $x_1 = l_1$ in Fig. 10.7(b) (or equivalently in Fig. 10.4) can in practice show relatively short wavelength† variations as a function of x_2 that are not to be expected on the basis of the lubrication flow analyses given above. The use of deformable choke bars or die-lips can eliminate most of the large-scale variations, i.e. those with a length scale large compared with h^*, but have no effect on or may even intensify the small-scale fluctuations.

Possible explanations for this phenomenon can be taken over from the discussion on melt flow instability given in Chapter 9 but must bear in mind the fact that we are not considering an unsteady phenomenon. We are naturally led to consider situations where the flowrate in a constant depth plane channel can be multivalued for a given value of pressure drop, as suggested for capillary flow in Figs. 9.3 or 9.4. Thus, if corresponding to a given pressure gradient P/l_1 in the uniform-depth channel shown in Fig. 10.8 a range of local flowrates $q_{\min} < q_1 < q_{\max}$ were possible, any given overall flowrate

$$\dot{V} = l_2 \bar{q}_1 \tag{10.3.19}$$

could be achieved provided $q_{\min} < \dot{V}/l_2 < q_{\max}$. This could, in principle, within the lubrication approximation, lead to a distribution of the stream function Q as shown by the flow lines in Fig. 10.8(a) parallel to the x_1 axis with p a linear function of x.

In any real flow we might expect terms neglected in the lubrication approximation, such as the interaction between the entry flow and the channel flow or viscous drag (t_{12}^{E}) in the x_2 direction, to play a part in

† The term wavelength is used rather loosely here to mean the distance between successive maxima in flow rate q_1.

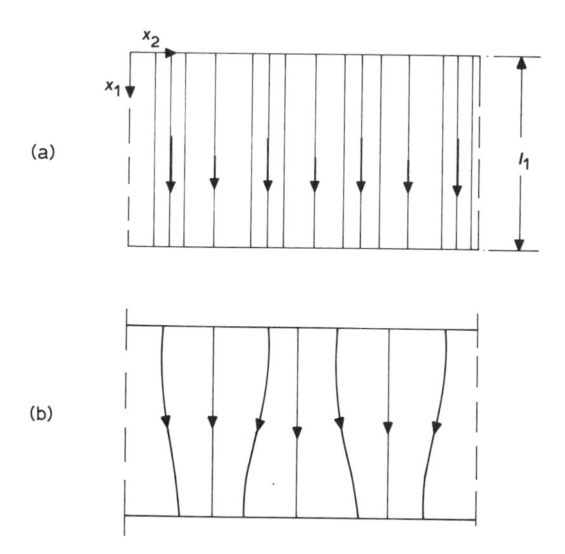

Fig. 10.8. Non-uniform flow in a constant depth channel. (a) Unidirectional but non-uniformly spaced flow lines. (b) Curved flow lines.

determining the distribution of Q. Thus, a more realistic pattern of flow lines might be as shown in Fig. 10.8(b), with p a non-trivial function of x_1 and x_2. Correspondences between melt flow instability in capillaries and slot dies have not been exhaustively analysed.

A crude explanation in terms of a variable heat-transfer mechanism has been given by Shah & Pearson (1974), whereby hot melt entering a cold die can flow either fast (because it then cools little) or slowly (because it then cools down a lot); this, however, depends on a large value of B or Na being relevant in the channel, with a suitable value of Gz, and is unlikely to be directly applicable to extrusion dies.

It is probable that many of the observed anomalies in output distribution at the die-lips have arisen because upstream effects involving thermal or elastic relaxation times have not been effectively enough damped within the symmetrical part of the die. It is well-known in the industry that, in general, longer dies give smoother extrudates, the limitations being weight (and hence cost) of the die and pressure demand. From equation (10.2.58), it can be seen that large $(\mathrm{Gz})^{-1}$, which is desirable to eliminate upstream temperature effects, demands large l and small h. This is also consistent with large shear and large shear rate and so a relaxation of upstream elastic effects. However, the pressure drop will increase at least in proportion to $(\mathrm{Gz})^{-1}$, while the onset of melt fracture (an unsteady phenomenon) will necessarily be approached as h decreases at fixed q.

EXAMPLE 10.3.1. Obtain simplified relations for a spiral die when both the spiral diameter and the leakage depth are constant and the fluid is Newtonian. Develop the symmetries of the flow field for $l_3 = 0$.

The greatest symmetry is obtained by considering, as a singly-connected region, the leakage channel lying 'below' spiral channel 2 and 'above' spiral channel 1. In the notation of Fig. 10.6(b) this can be considered to be bounded by $x_1 = 0$, $x_1 = l_1$, $x_1 = x_2 \tan \psi$ and $x_1 = (x_2 - l_2)\tan \psi$. The flow equation (10.2.20), with $\nu = 1$ (Newtonian fluid), $h_c = h_{c0}$ (constant) becomes simply

$$\left(\frac{\partial^2}{\partial x_2^2} + \frac{\partial^2}{\partial x_1^2}\right) Q_L = 0$$

the harmonic equation. The inlet and outlet boundary conditions become, using (10.3.4) and (10.2.11),

$$Q_L(x_1 = 0) = 0, \qquad Q_{S1}(x_1 = 0) = Q_{S2}(x_1 = 0) = \tfrac{1}{2}\dot{V}$$

$$\frac{\partial Q_L}{\partial x_2}(x_1 = l_1) = 0$$

The integrated form of the side boundary conditions (10.3.2) and (10.3.3) can be simplified if the periodicity conditions

$$\left[p_{L1}, \frac{\partial Q_{L1}}{\partial x_2}, \frac{\partial Q_{L1}}{\partial x_1}\right](x_1, x_2) = \left[p_{L2}, \frac{\partial Q_{L2}}{\partial x_2}, \frac{\partial Q_{L2}}{\partial x_1}\right](x_1, x_2 \pm l_2)$$

are employed. Thus they can be written

$$[Q_{S1} - Q_{L1}](x_2 = x_1 \cot \psi) + Q_{L1}(x_2 = l_2 + x_1 \cot \psi) = \tfrac{1}{2}\dot{V}$$

$$[Q_{S2} + Q_{L1}](x_2 = l_2 + x_1 \cot \psi) - Q_{L1}(x_2 = x_1 \cot \psi) = \tfrac{1}{2}\dot{V}$$

where x_1 is regarded as the independent variable of position. By symmetry

$$Q_{S1}(x_2 = x_1 \cot \psi) = Q_{S2}(x_2 = l_2 + x_1 \cot \psi)$$

and so the two integrated equations degenerate into a single relation applicable on both side boundaries.

The flow equations (10.2.15) and (10.2.16) allow Q_S to be expressed in terms of ∇Q_L; writing

$$\bar{d} = 3\pi d_s^4/32h_{c0}^3$$

the side boundary equations become

$$\bar{d}\frac{\partial Q_L^\pm}{\partial n^\pm} = Q_L^\pm - Q_L^\pm + \tfrac{1}{2}\dot{V} \quad \text{for all} \quad 0 < x_1 < l_1$$

where the $\pm$ refers to the lines

$$x_2 = (x_1 \cot \psi + \tfrac{1}{2}l_2) \pm \tfrac{1}{2}l_2$$

and $n^\pm$ refers to the inward facing normal. These are somewhat unusual mixed boundary conditions for a harmonic equation, and no standard method of solution can be quoted.

EXERCISE 10.3.1. Show that, by taking the spiral diameter in Example 10.3.1 as

$$d_s = d_{s0}(1 - x_1^4/l_1^4)$$

the singularities at $(l_1, l_1 \cot \psi + \tfrac{1}{2}l_2 \pm \tfrac{1}{2}l_2)$ can be eliminated. Derive the relevant 'side' boundary conditions.

EXERCISE 10.3.2. Solve (10.3.15) subject to (10.3.13) for a Newtonian fluid to derive the relation

$$\dot{V} = -\frac{\pi}{8\eta_0}\frac{\partial p}{\partial z}\left[(R_0^4 - R_i^4) - \frac{(R_0^2 - R_i^2)^2}{\ln(R_0/R_i)}\right]$$

and show that the correct channel-flow result is obtained as $R_0 - R_i \to 0$, i.e.

$$\dot{V} = -\frac{\pi R_i}{6\eta_0}\frac{\partial p}{\partial z}(R_0 - R_i)^3$$

10.4 WIRE-COATING DIES

In these a rapidly-moving circular-cylindrical wire moves rapidly through the centre of an essentially conical annular die usually of the cross-head variety. A typical example is shown in Fig. 10.9 with relevant dimensions labelled and a polar coordinate system shown in

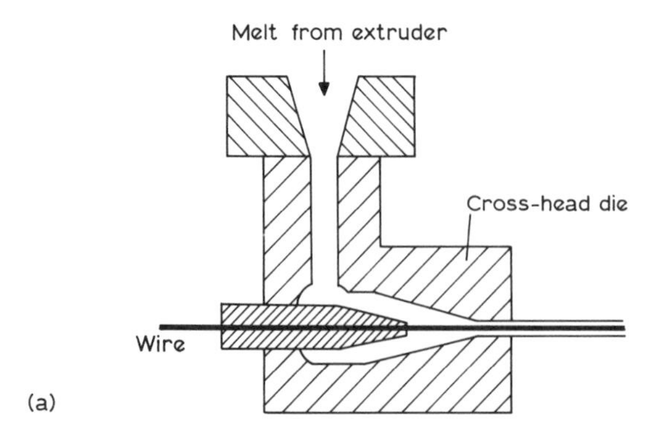

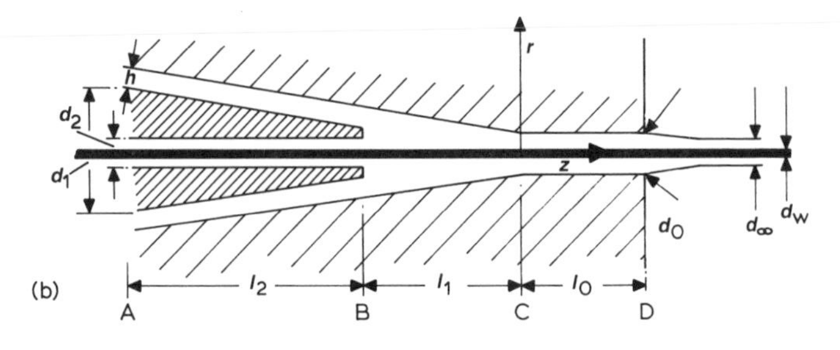

Fig. 10.9. Diagrammatic view of wire-coating die. (a) Typical cross-head design allowing wire to move rapidly across the front of the extruder acquiring a coating of plastic under pressure. (b) Expanded view of flow field where coating operation takes place, showing coordinate system.

Fig. 10.9(b). The flow field is similar in some ways to that shown in Fig. 10.5, but the final die-exit diameter is much smaller and the shear rates far higher. The flow is no longer simply a pressure flow, because the moving wire drags molten polymer out through the die orifice.

Typical values of the relevant dimensions are:

$$\begin{array}{lll} l_0 = 10\ \text{mm} & l_1 = 10\ \text{mm} & l_2 = 60\ \text{mm} \\ d_w = 0{\cdot}5\ \text{mm} & d_\infty = 0{\cdot}9\ \text{mm} & d_0 = 1{\cdot}1\ \text{mm} \\ d_2 = 0{\cdot}7\ \text{mm} & h = 2\ \text{mm} & d_1 = 40\ \text{mm} \end{array} \tag{10.4.1}$$

Hence

$$2l_0/(d_0 - d_w) = 300, \qquad l_2/h = 25 \tag{10.4.2}$$

Thus with the exception of the region near B, we can expect the lubrication approximation to hold at least in the limited sense required to simplify the stress equilibrium equation. Some calculations made on this basis have been given by Fenner & Williams (1967), using an isothermal power-law rheological model. For the conical annular region between A and B, they adopt the solution given for the corresponding region in the die shown in Fig. 10.7 leading to the unrolled form (Fig. 10.7(c)). (The flow direction is reversed, but this does not alter the magnitude of the pressure-drop/flowrate relations.) For the region between B and D, they find a local relation between $\partial p/\partial z$ and $Q(=\dot{V})$ given the wire speed v_∞. This follows the development given above for deep annular channels, using eqns (10.3.12), (10.3.14) and (10.3.15), but with the boundary conditions

$$v_z(\tfrac{1}{2}d_w) = v_\infty, \qquad v_z(R_0) = 0 \tag{10.4.3}$$

where $R_0 = \frac{1}{2}d_0$ between C and D, instead of (10.3.13). Once again numerical solution is necessary to provide output pressure-gradient curves (see Fig. 3 of Fenner & Williams). For any given wire speed and output rate, the final coated diameter d_∞ must obey the relation

$$\pi(d_\infty^2 - d_w^2)v_\infty = 4\dot{V} \tag{10.4.4}$$

and the total pressure demand of the die can be approximated by

$$P = \int_{-(l_1+l_2)}^{l_0} \frac{\partial p}{\partial z}(z)\,dz \tag{10.4.5}$$

A feasible, if not always typical, pressure profile is shown in Fig. 10.10. In region AB there is, as there must be, a pressure-driven flow.

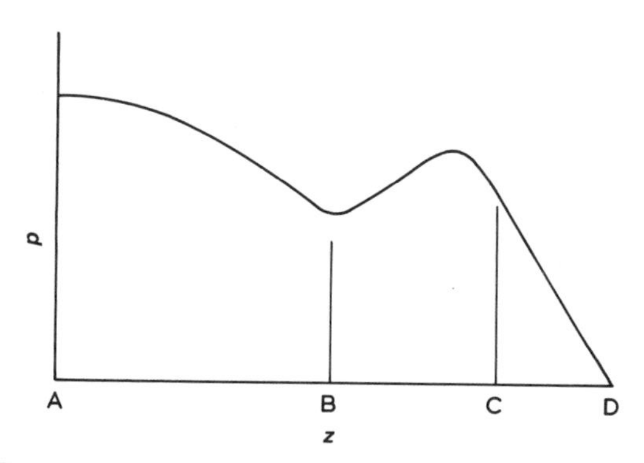

Fig. 10.10. Pressure distribution along die axis in coating die. Positions A–D correspond to those so labelled on Fig. 10.9(b). Arbitrary units.

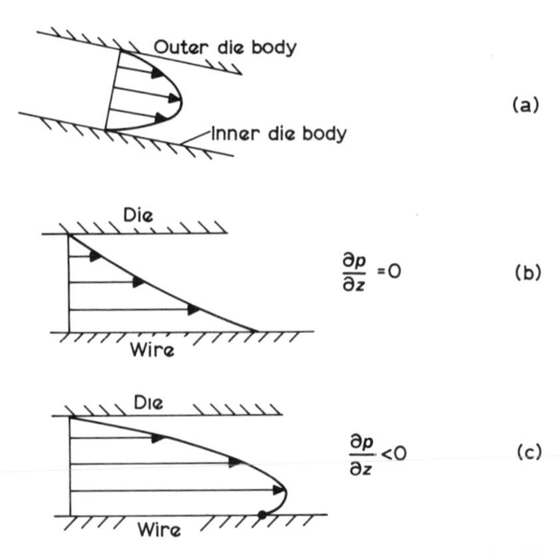

Fig. 10.11. Velocity profiles in wire-coating die shown in Fig. 10.9(b). (a) Pressure flow in conical annulus (A–B). (b) Drag flow in coating zone (B–C). (c) Drag flow plus pressure flow in coating zone (C–D).

In region BC there is mainly an adverse pressure gradient retarding the drag flow, while in region CD there is a constant pressure gradient aiding the drag flow. Velocity profiles are shown in Fig. 10.11. In practice, large positive pressure at B is necessary to obtain reliable wetting of the wire near the nose of the inner die body through which the wire is fed. Flow upstream along the axis of the die from B is discussed in Example 10.4.1.

In the neighbourhood and just downstream of B, we can expect a recirculation zone—a toroidal vortex attached to the outer conical surface driven by the adverse pressure gradient and the drag flow caused by the die as shown in Fig. 10.12. Calculations for this have

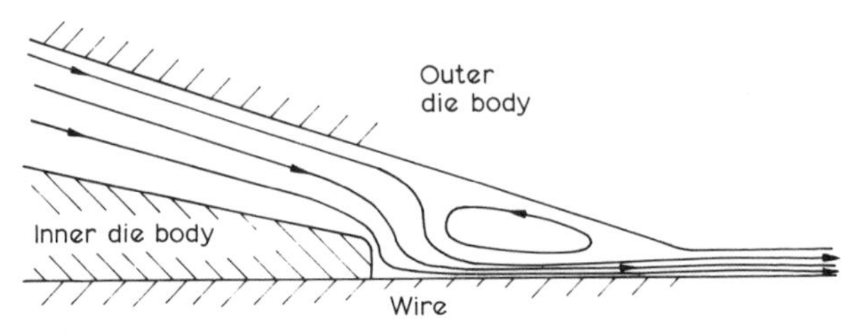

Fig. 10.12. Sketch of streamlines showing recirculating vortex.

been carried out by Caswell & Tanner (1978) for power-law fluids ($\nu = 1$ and $\frac{1}{2}$), not using the lubrication approximation, and showing how modifications to the geometry can avoid the recirculating zone for their rheological model.

However, in a typical wire-coating situation the wire speed will be about 10 m/s (2000 ft/min) and so in the upstream neighbourhood of B, we can expect extension rates of the order of $20\,\text{s}^{-1}$ and shear rates of the order of $150\,\text{s}^{-1}$. In the region CD the shear rate will be of the order of $4 \times 10^4\,\text{s}^{-1}$ (see Example 10.4.2 for detailed calculations). In the region between B and C, we can expect circumstances similar to those arising in the entry region to a capillary die at high Ws and De, unless the melt is very hot. Elastic effects can therefore be expected to dominate and so the power-law approach used by both Fenner & Williams, and Caswell & Tanner, may well be significantly in error. One of the problems to be faced in selecting a more suitable rheological equation of state for numerical calculations is that the rates of shear and extension are at the limit of experimental feasibility, using current rheological measurement techniques.

Thermal effects will also be highly significant. Using some of the figures for stress given by Fenner & Williams (who do not quote actual viscosities) we take

$$K = 5 \times 10^3\,\text{N}\,\text{s}^{\frac{1}{2}}\,\text{m}^{-2}, \qquad \nu = \tfrac{1}{2} \tag{10.4.6}$$

in (3.3.5). Assuming that

$$\kappa^* = 10^{-7}\,\text{m}^2\,\text{s}^{-1}, \qquad \alpha^* = 0{\cdot}2\,\text{W}\,\text{mK}^{-1}, \qquad \zeta = 0{\cdot}02\,\text{K}^{-1} \tag{10.4.7}$$

for the annular region around the wire we get

$$\text{Gz}_{\text{CD}} \approx 400 \gg 10 \tag{10.4.8}$$

and

$$\text{Na}_{\text{CD}} \approx 250 \tag{10.4.9}$$

with $(\text{Na}/\text{Gz})_{\text{CD}} = 0(1)$. For the annular region about the entry cone we have

$$\text{Gz}_{\text{AB}} \approx 40 \tag{10.4.10}$$

and

$$\text{Na}_{\text{AB}} \approx 0{\cdot}125 \ll 1 \tag{10.4.11}$$

where Gz and Na are defined as in Subsection 10.2.3 earlier. The 0(1) value for $(Na/Gz)_{CD}$ suggests that a large temperature rise can be expected in that zone on an adiabatic basis. This is confirmed by the figures given by Fenner & Williams for pressure drop (4000 psi) and tension build-up in the wire (equivalent to a further 12 000 psi).

All our discussion of wire-coating so far has supposed that the flow field will be axially symmetrical. However, with the very fine clearances (0·2 mm) involved in the operation it is obvious that alignment of the rapidly-moving wire will present problems. It is therefore to be hoped—if not expected in view of the practicability of the operation—that the moving wire will be self-aligning because of viscous forces. Tadmor & Bird (1974) have attempted to analyse this on the basis of an earlier solution of Jones (1965) for an eccentrically-positioned (off-centre) moving wire. Their solution uses dipolar coordinates, and a Criminale–Ericksen–Filbey model, eqn (3.3.21); it presupposes steady unidirectional flow in the z-direction. They give an expression for the lateral force acting on the wire, from which they deduce that the system is self-centring if Ψ_2, as defined in (4.2.5), is negative, which is the case for most elastic fluids investigated experimentally, although the authors are careful to say that experimental data is 'completely unavailable in the (relevant) shear-rate range' for the polymer melts used. They also carry out an analysis of the forces generated if the wire moves laterally using a perturbation technique which allows of a Newtonian model; as expected such displacements are opposed by viscous forces. Finally, they give some numerical estimates which they regard as conjectural rather than conclusive.

EXAMPLE 10.4.1. Obtain a relation for the length, l^*, of wire that is wetted upstream of B in Fig. 10.9(b) for a Newtonian fluid.

The flow can be treated by lubrication theory; within the region $-(l^*+l_1)<z<-l_1$, $\frac{1}{2}d_w \leq r \leq \frac{1}{2}d_2$, the eqn (10.3.15) will apply subject to boundary condition (10.4.3) with $R_0=\frac{1}{2}d_2$.

For a Newtonian fluid, (10.3.15) becomes

$$\frac{dv_z}{dr}=\frac{\partial p/\partial z}{2\eta_0}\left(r-\frac{r_0^2}{r}\right)$$

where r_0 is as yet arbitrary. Further integration using $v_z(\frac{1}{2}d_2)=0$ yields

$$v_z=\frac{1}{4\eta}\frac{\partial p}{\partial z}\left(r^2-2r_0^2\ln\frac{2r}{d_2}-\frac{1}{4}d_2^2\right)$$

whence

$$v_\infty = \frac{1}{16\eta}\frac{\partial p}{\partial z}\left(d_w^2 + 8r_0^2 \ln\frac{d_2}{d_w} - d_2^2\right)$$

In steady state, there is no net melt flow, so

$$\int_{\frac{1}{2}d_w}^{\frac{1}{2}d_2} 2\pi r v_z(z)\,\mathrm{d}r = 0$$

whence

$$\left[\frac{r^4}{4} - \frac{d_2^2 r^2}{8} - r_0^2 r^2 \ln\frac{2r}{d_2} - \frac{1}{2}\right]_{\frac{1}{2}d_w}^{\frac{1}{2}d_2} = 0$$

or

$$(d_2^2 - d_w^2)^2 = 8r_0^2\left\{-2d_w^2 \ln\left(\frac{d_2}{d_w}\right) + (d_2^2 - d_w^2)\right\}$$

Resubstitution yields

$$\frac{\partial p}{\partial z} = \frac{16\eta v_\infty\{2d_w^2 \ln(d_2/d_w) - (d_2^2 - d_w^2)\}}{(d_2^2 - d_w^2)\{(d_2^2 + d_w^2)\ln(d_2/d_w) - (d_2^2 - d_w^2)\}} > 0$$

l^* then follows from $l^* \partial p/\partial z = p_B$. Note that backflow will arise in the velocity profile.

EXAMPLE 10.4.2. Derive estimates for the shear and extension rate in regions AB and CD of Fig. 10.9.

In region CD we take the shear rate $\dot{\gamma}_{CD}$ to be given by $2v_\infty/(d_0 - d_w)$, assuming drag flow as in Fig. 10.11(b). Hence

$$\dot{\gamma}_{CD} \approx 2 \times 10^4\ \mathrm{mm\ s^{-1}}/(1{\cdot}1\text{–}0{\cdot}5)\ \mathrm{mm} = 33\,000\ \mathrm{s^{-1}}$$

In region AB, the shear rate varies because the cross-sectional area of the channel and hence the mean velocity of the flow decreases linearly from A to B. If we let h be the (assumed) uniform depth of the channel and d its effective diameter, then the total flowrate

$$\dot{V} = \tfrac{1}{4}\pi(d_\infty^2 - d_w^2)v_w$$

leads to a mean velocity

$$\bar{v} = \dot{V}/2\pi dh = (d_\infty^2 - d_w^2)v_w/8dh$$

and, assuming parabolic flow in AB, an apparent wall shear rate

$$\dot{\gamma}_w = 2\dot{V}/\pi dh^2 = (d_\infty^2 - d_w^2)v_w/2dh^2$$

Taking $h = 2$ mm, $d_A = 40$ mm and $d_B = 3$ mm, this gives

$$\bar{v}_A \sim 9 \text{ mm s}^{-1} \qquad \dot{\gamma}_A \sim 15 \text{ s}^{-1}$$

$$\bar{v}_B \sim 120 \text{ mm s}^{-1} \qquad \dot{\gamma}_B \sim 200 \text{ s}^{-1}$$

The rate of extension $\dot{e}_z$ can be estimated from $-\mathrm{d}\bar{v}/\mathrm{d}z$; in the case above we have assumed that

$$\bar{v} \sim -720/(4+z) \text{ mm s}^{-1}$$

and so

$$\dot{e}_z \sim 720/(4+z)^2$$

For $z = -10$, which corresponds to the point B, this yields a value of

$$\dot{e}_{z\max} \sim 20$$

EXERCISE 10.4.1. Derive $\dot{V}$ in terms of $\partial p/\partial z$ for flow of a Newtonian fluid in region CD. The procedure is very similar to Exercise 10.3.2 and Example 10.4.1.

EXERCISE 10.4.2. Derive the pressure drop $p_B - p_C$ using the result of Exercise 10.4.1 treating d_0 as a function of z, and integrating according to lubrication theory.
Derive a condition for $p_B > 0$.

EXERCISE 10.4.3. Verify the estimates for Gz and Na given in eqns (10.4.8)–(10.4.11).

10.5 CO-EXTRUSION DIES

It is often desirable to form multilayer polymer sheets that will not delaminate in use. Good bonding between two polymers can often best be achieved by holding them together under pressure in the molten state. It thus becomes advantageous for many reasons to co-extrude polymer sheet through a single die orifice. This requires two or more extruders feeding a single die. A very simple example for a sheet die that gives a symmetric coating of polymer 2 on either side of a sheet of

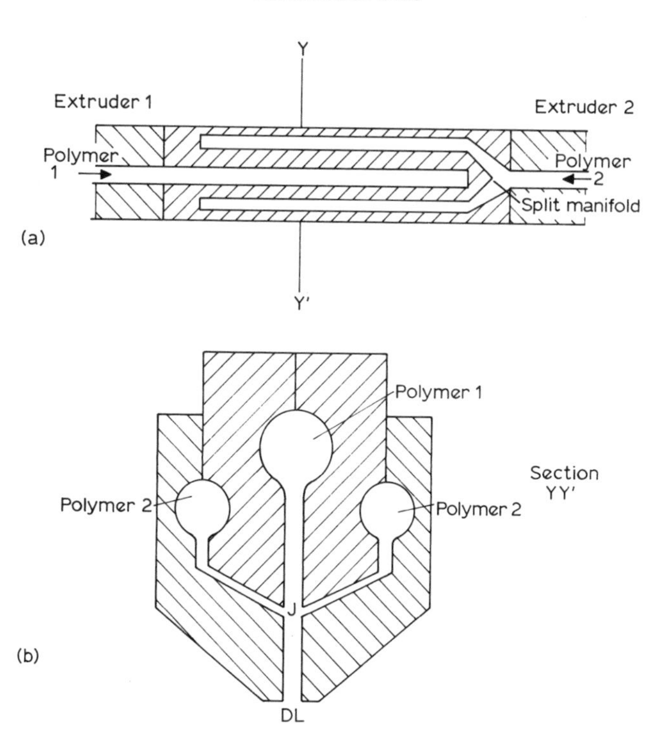

Fig. 10.13. Sketch of coextrusion sheet die. (a) Double-end feeding. (b) Cross-section showing central manifold and channel for polymer 1 and two channels for polymer 2 symmetrically arranged to join the main channel (at J) leading to die-lips DL.

polymer 1, is shown in Fig. 10.13. The die is fed from both ends. A single manifold of the type shown in Fig. 10.13(a) provides the flow of polymer 1, from extruder 1, while a split manifold of the same type provides the flow of polymer 2 from extruder 2, as shown in (a). A cross-section of the die is given diagrammatically in (b). Up to the position J, the flow of each polymer can be treated exactly as though it were a simple single polymer system ending at a virtual exit J. From J to the die-lips DL co-current planar flow within the two-dimensional channel arises. We can analyse this in terms of lubrication theory. Figure 10.14 shows a symmetric flow field $v_1(x_3)$ with polymer 1 occupying the region $|x_3|<\frac{1}{2}h_1$ and polymer 2 the region $\frac{1}{2}h_1<|x_3|<\frac{1}{2}h$.

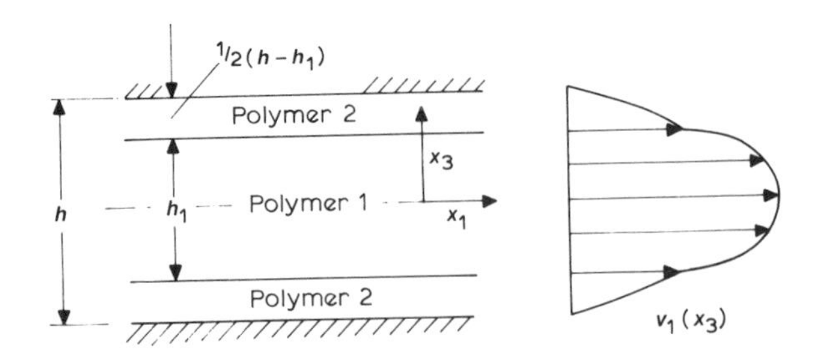

Fig. 10.14. Velocity profile for two co-extruding polymers according to the lubrication approximation.

The unidirectional velocity vector $v_1(x_3)$ is continuous but can show a discontinuity in gradient at $x_3 = \frac{1}{2}h_1$ corresponding to t_{13} continuous, but $\eta(\dot{\gamma})$ discontinuous.

As before, we seek a relation between pressure gradient and flow-rate. We can write, as in (7.1.54)

$$t_{13} = \frac{\partial p}{\partial x_1} x_3 = -\tau \tag{10.5.1}$$

and assume that the inverse stress relation

$$\frac{\partial v_1}{\partial x_3} = \dot{\gamma}(\tau_{13}) \tag{10.5.2}$$

is known for both polymers 1 and 2. As in (7.1.52), from (10.5.1) and (10.5.2),

$$v_1 = -\left(\frac{\partial p}{\partial x_1}\right)^{-1} \int_{\tau}^{\tau_w} \dot{\gamma}(u)\, \mathrm{d}u \tag{10.5.3}$$

where

$$\dot{\gamma}(u) = \begin{cases} \dot{\gamma}_2(u) & |u| \geq \frac{1}{2}h_1 \left|\dfrac{\partial p}{\partial x_1}\right| \\ \dot{\gamma}_1(u) & |u| < \frac{1}{2}h_1 \left|\dfrac{\partial p}{\partial x_1}\right| \end{cases} \tag{10.5.4}$$

and $\tau_w = \frac{1}{2}h\,\partial p/\partial x_1$ is treated as positive.

On further investigation we get for the separate flowrates

$$q_2 = 2\left(\frac{\partial p}{\partial x_1}\right)^{-2} \int_{-\frac{1}{2}h_1 \partial p/\partial x_1}^{-\frac{1}{2}h \partial p/\partial x_1} \left(u + \tfrac{1}{2}h_1 \frac{\partial p}{\partial x_1}\right) \dot{\gamma}_2(u)\, \mathrm{d}u \tag{10.5.5}$$

$$q_1 = 2\left(\frac{\partial p}{\partial x_1}\right)^{-2} \left[\int_0^{-\frac{1}{2}h_1 \partial p/\partial x_1} u\dot{\gamma}_1(u)\, \mathrm{d}u - \frac{1}{2}\left(\frac{\partial p}{\partial x_1}\right) h_1 \int_{-\frac{1}{2}h_1 \partial p/\partial x_1}^{-\frac{1}{2}h \partial p/\partial x_1} \dot{\gamma}(u)\, \mathrm{d}u\right] \tag{10.5.6}$$

If q_1 and q_2 are separately specified, then (10.5.5) and (10.5.6) can be regarded as a pair of implicit equations for h_1 and $\partial p/\partial x_1$. In general a numerical solution would be necessary, although an analytical solution is possible for the Newtonian case.

In the analysis of any but uniform flow in a channel of constant depth, the pressure gradient ∇p would be required as a function of the local flow rates $\mathbf{q}_1$ and $\mathbf{q}_2$, by analogy with (8.1.23); we could assume that $\mathbf{q}_1$ and $\mathbf{q}_2$ were parallel for a die with fixed boundaries, i.e. where $\mathbf{u}_{S1}$ and $\mathbf{u}_{S2}$ were zero, and write

$$\nabla p = -\hat{\mathbf{q}} P'(q_1, q_2, h) \tag{10.5.7}$$

where P' involves a knowledge of $\dot{\gamma}_1$ and $\dot{\gamma}_2$, and $\hat{\mathbf{q}}$ is a unit vector in the flow direction. No dependence on temperature is admitted at this stage. The continuity relation (10.2.5) has to be replaced by the separate solenoidal requirements

$$\nabla \cdot \mathbf{q}_1 = \nabla \cdot \mathbf{q}_2 = 0 \tag{10.5.8}$$

These can lead to two separate stream functions Q_1 and Q_2 defined as in (10.2.6). However, the parallelism requirement for Q_1 and Q_2 streamlines means that q_1/q_2 must be constant along streamlines, which means a unique monotonic relation between Q_1 and Q_2, determined by inlet boundary conditions. Note that h_1/h need not be constant along streamlines.

The importance of thermal effects can be determined by considerations similar to those given for single-polymer die flows.

10.5.1 Stability of Co-extrusion Flows

Experience of co-extrusion, in both cylindrical tubes and channels, has shown that steady uniform flow of the type assumed in Fig. 10.14 will not always be stable. The interfaces at $x_3 = \pm\frac{1}{2}h_1$ share many of the properties of a free surface and so waves upon them may grow either in time or in a downstream direction. A linearized stability analysis of the type given in Appendix 1 can be undertaken, although there is little reliable theoretical or experimental work in this field. Pearson & Pickup (1973, Appendix 3) have analysed the symmetrical co-current flow of two Newtonian fluids and concluded that both sinuous and varicose plane waves of arbitrary real wavelength k move downstream or upstream with a real wave velocity $c_s(k)$ or $c_v(k)$ that depends upon the relative proportions and viscosities of the two fluids. Thus, each of the periodic disturbances is neutrally stable whether the more or the less viscous fluid forms the core. The situation may be quite different for elastic fluids, although a non-zero critical Weissenberg number would then be expected; observations suggest that stability could depend upon which fluid formed the core of any given pair.

EXERCISE 10.5.1. Derive relations (10.5.5) and (10.5.6) and show that for a Newtonian fluid

$$q_2 = -\frac{p_{,1}}{12\eta_2}(h-h_1)(h^2 - \tfrac{1}{2}hh_1 - \tfrac{1}{2}h_1^2)$$

$$q_1 = -\frac{p_{,1}}{12\eta_2}\left\{\frac{\eta_2}{\eta_1}h_1^3 + \tfrac{3}{2}h_1(h^2 - h_1^2)\right\}$$

Show also that for specified ratios (q_2/q_1) and (η_2/η_1), the dimensionless ratio (h_1/h) is given by a cubic equation involving (q_2/q_1) and (η_2/η_1) only.

Chapter 11

The Single-screw Extruder

The single-screw extruder is the most important piece of equipment used in the processing of polymer melts. Because the extrusion die is essentially a passive device, the quality of its output will usually reflect the quality of its input, and so it is normally found that success in any extrusion process will be largely dependent upon the performance of the extruder. Furthermore, it is at the extrusion stage that the crucial operation of converting an intermittent feed into a steady or regularly periodic output is carried out. The screw extruder is the means by which the necessary smoothing and averaging is achieved. It performs an essentially active role and can be, simultaneously, a solids pump, a melt pump, a melting (or plasticating) device and an intensive mixer.

The history of the screw extruder in one or more of these roles is an interesting one, and the range of detailed designs employed far greater than will be considered here. Interested readers should read the fascinating textbook by Schenkel (1966), which remains the most exhaustive account of plastics extrusion technology available; its blend of scholarship, analytical insight and practical experience make it an exceptional contribution to our understanding of the process. The earlier work by Fisher (1958) recently re-edited, is similarly reliable, though briefer, while the relevant chapters in Bernhardt (1959) and Frados (1976) form excellent introductions to this chapter. Useful information, largely of an analytical nature, can be found in books by McKelvey (1962), Fenner (1970), and Tadmor and Klein (1970).

The development of screw extruders has been largely an art: there is no unique design to satisfy any given purpose, no single solution to any given extrusion problem. However, the advantages of mass production are such that standard extruders offered by big machinery manufacturers are often the cheapest and most reliable machines for processers to

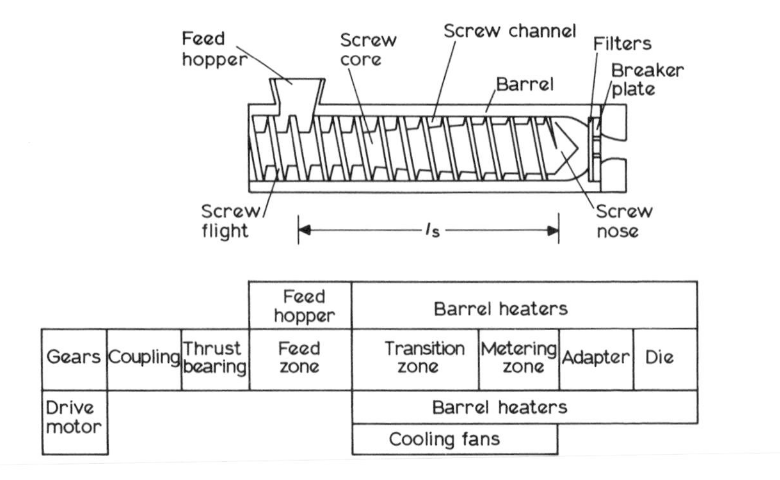

Fig. 11.1. Basic elements of single-screw extruder.

use. (A typical plasticating extruder is shown diagrammatically in Fig. 11.1.) Thus the issue is often less of design from first principles or of laborious optimization of a complex design than of selection from a relatively limited range. It is usual for any given machine to be equipped with a range of screws, and so extruder design becomes in practice a matter of screw design (the relevant channels along which the polymer passes are machined out of the screw blank). Although by no means necessary, it is convenient that the barrel of the screw should be a hollow circular cylinder forming the outer boundary of the screw channel. Where possible the screw flights are continuous helices of constant pitch (lead). Both of these simplifications are a help during manufacture. They are also a help to the theorist wishing to analyse flow in extruders!

As has been, and will be, emphasized on many occasions in this text, the geometry of flow boundaries is often far more important than all other factors in determining flow fields and hence process behaviour.† It is therefore desirable for any feasible optimization program to be

† This may seem so obvious as not to need stating: unfortunately, many processors seek to describe the behaviour of individual polymers in an all-embracing fashion, so that it is quite usual to say that one plastic compound extrudes better than another without reference to the machinery used, the manner in which it is operated, or the purpose to which the extrudate is put.

able to describe the important aspects of this geometry, quantitatively, in terms of as few parameters as possible. Such limitation represents a constraint on the art of a designer; but only after ruthless simplification can reliable quantitative analyses be undertaken.

The same argument applies to the control system for an extruder. If the operating conditions can be varied in a great many ways to give the greatest amount of flexibility any processor might desire, the problem of selecting even an economically satisfactory set of operating conditions becomes overwhelming. Better results are sometimes obtained when fewer controls are available.

In this account, most attention will be given to the simplest geometrical form of single-screw extruder, operating in a steady mode on a homogeneous material displaying typical polymer properties and rheology. No claim will be made that a full thermal and mechanical analysis of plasticating extrusion is yet available, nor that satisfactory extruder analysis and design can be undertaken by subscribing to and employing a commercially available computer program. Our understanding of extruder mechanics is still based on critical experimental observations which allow us to postulate certain basic mechanisms and flow regimes. These suggest certain approximate, albeit complex, methods of modelling methematically, and thereby analysing qualitatively, different aspects of the process.

There are two ways of proceeding with our analysis. One is to work backwards from the nose end, or die, on the grounds that the quality of the extrudate will depend most critically on what happens nearest to the exit, and the other is to work forward from the feed end, on the grounds that convection is a dominant effect and thus implies a continuing dependence on initial (i.e. feed) conditions.

Insofar as models for the basic successive zones—solids compacting and conveying, melting, melt pumping and mixing—have to be made compatible, the order of treatment is ultimately irrelevant. Nevertheless, it is useful to try to distinguish which features of a process are the critical ones and to concentrate effort on those portions, particularly if one seeks to remedy malfunctions such as irregular or unsteady operation.

It is often by no means clear, in practice, whether any given extruder is essentially feed-controlled or is operating in a metering mode; whether it is starve-fed, force-fed or neutrally fed. The design of an extruder will be determined by the anticipated answer to these questions. The need for extreme steadiness and uniformity in output for many continuous plastics processes, like fibre-spinning, film-making, pipe-extrusion, etc., has led to the development of long-barrelled extruders with dominant metering (or melt pumping) zones, whose output is relatively low so that feed and melting zones can easily perform their function. On the other hand, less demanding operations like compounding, granulation or strip production, where total output proves to be the key factor, have led to relatively short extruders, dominated by the feed and softening sections.

Fortunately, all zones share the same geometrical feature: the relevant channel lengths are large compared with the channel depth or width. Changes take place slowly in the direction of the downstream coordinate (y^1) and matching of the outputs and pressures of the separate zones can conveniently be carried out at the relevant values of y^1.

11.1 GEOMETRY OF THE SINGLE-SCREW EXTRUDER

In common with manufacturers we take the inner barrel diameter, d_b, as the primary defining parameter. Figure 11.2 shows slightly over one full turn of the flight helix, whose lead angle at the barrel is ϕ_b. The axial width of the flight is w_{fa} and of the screw channel is w_{ca}, their sum being the screw pitch (or lead) $\pi d_b \tan \phi_b$. The clearance between top

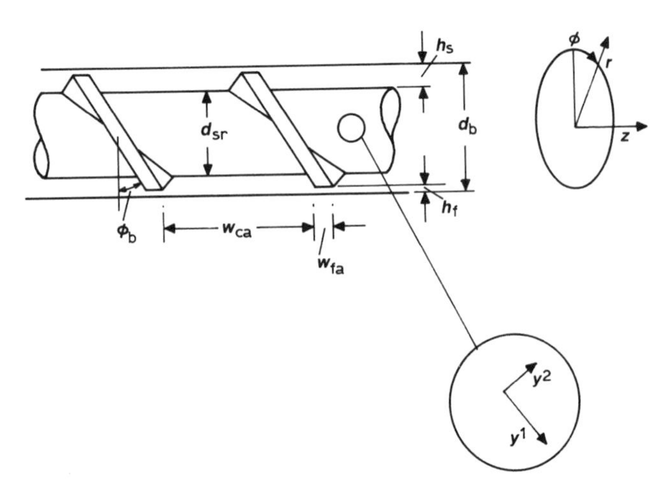

Fig. 11.2. Screw geometry.

of screw flight and inner surface of the barrel is h_f. The depth of the screw channel is h_s. These dimensions are sufficient to define the geometry of the screw channel provided:

(a) the bottom of the screw channel (the screw root) lies on a circular cylinder coaxial with the barrel;
(b) the top of the screw flight also lies on a circular cylinder coaxial with the barrel;
(c) the sides of the screw flight are defined by radii through and perpendicular to the screw/barrel axis. The (r, ϕ, z) cylindrical polar coordinate system shown in the figure defines the radial and axial directions.

In practice, chamfers and fillets may be used, while surfaces may be rounded. More important, the channel depth h_s may be a slowly varying function of the helical coordinate y^1 measured along the flight, in which case condition (a) above can be replaced by the requirement that h_s be constant between leading and trailing edges of the flight, either in the axial direction z or the direction y^2 orthogonal to $r(y^3)$ and y^1 for $y^3 = r_b$. It will be supposed here that any real screw design can be adequately represented as a first approximation by a screw of simple type specified above. Table 11.1 gives typical ratios for the various dimensions.

Table 11.1
Dimensionless Ratios for Typical Extruder Geometry

$L/D = l_s/d_b$	30
$W/D = w_{ca}/d_b$	1·0
$H/D = h_s/d_b$	0·04–0·20
w_{fa}/d_b	0·1
$F/D = h_f/d_b$	0·001–0·005

It will be seen that between the minimum flight clearance and the full helical length ($\sim \pi l_s$) of the screw channel, there is a difference of five orders of magnitude, i.e. 10^5. We see that the screw channel, in each of the three zones, can be more than ten times as long as it is broad and more than ten times as broad as it is deep. The curvature of the channel is $2/d_b$. This strongly suggests that the lubrication theory approximation (Section 8.1) can be safely employed from the point of view of the stress equations, though our earlier estimates of Peclet and

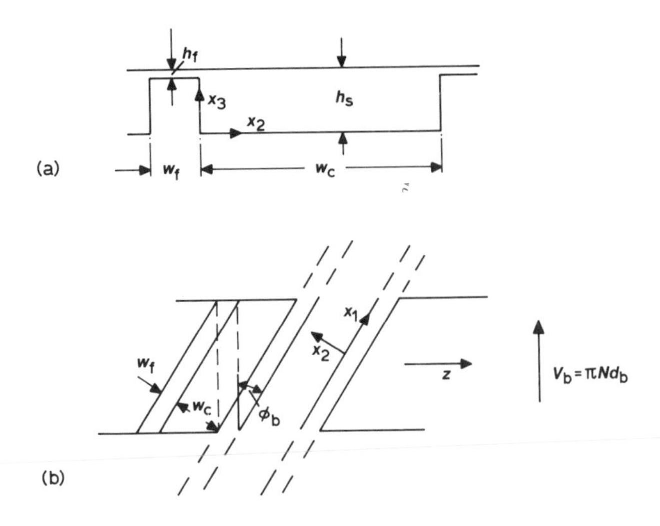

Fig. 11.3. Unrolled form of screw geometry.

Graetz numbers for polymer flows make it unlikely that fully-developed thermal equilibrium will be achieved on a local (y^1) basis.

With this in mind, we consider the unrolled form of screw geometry shown in Fig. 11.3, with the Cartesian coordinate system (x_1, x_2, x_3) based on the locally orthogonal coordinate system (y^1, y^2, y^3) adopted in Fig. 11.2. It will be noted that the geometrical conditions (a)–(c) lead to the boundaries of the flow field being given by

$$x_2 = 0,\ w_c; \qquad 0 < x_3 < h_s - h_f$$
$$x_3 = h_s; \qquad 0 < x_2 < w_c + w_f$$
$$x_3 = h_s - h_f; \qquad w_c < x_2 < w_c + w_f$$
$$x_3 = 0; \qquad 0 < x_2 < w_c$$

with entry at $x_1 = 0$ and exit at $x_2 = l_s \operatorname{cosec} \phi_b$. h_s will be a slowly-varying function of x_1. (In fact, it can be seen that no further complexity is in principle introduced by making h_f, w_c and w_f also slowly-varying functions of x_1 and x_2.) It is consistent with the lubrication approximation that the full problem involving the y^1 (and y^2) variations be treated exactly in the unrolled (x_1, x_2) coordinate system with the local problem based on the locally planar boundaries $x_3 = 0, h$.

Any improvement on the fully unrolled solution would require the solution of the local equations of motion in terms of the helical coordinates (see Appendix 3) or alternatively, but less elegantly, in terms of the cylindrical coordinates (r, ϕ, z).

Note that by choosing our frame of reference fixed in the screw, the driving force for motion down the channel is provided by an apparently-rotating barrel. In the unrolled geometry of Fig. 11.3, the upper boundary $x_3 = h$ moves relative to the screw with velocity

$$\mathbf{v}_b = (v_b \cos\phi_b, v_b \sin\phi_b, 0) \qquad (11.1.1)$$

It should be remembered that the screw thread can be either left- or right-handed, and so it is not worth being too specific about whether right- or left-handed Cartesian axes are being used.

No significant difficulties are experienced from the use of a rotating frame of reference because inertia (and hence centrifugal) forces are negligible in our application. More significant is the fact that the position of the centre of the feed pocket varies periodically and linearly with time, along the line

$$x_2 = x_1 \tan\phi_b \qquad (11.1.2)$$

11.2 FEED ZONE: SOLIDS TRANSPORT

When the feed is in solid form, there will be a portion of the screw channel that acts primarily as a solids conveyer. In most cases the material is in fairly uniform granular or powder form and enters the screw channel through a hole on the top or side of the barrel (the feed pocket) above which sits a hopper containing a substantial amount† of solid material.

It is usual to differentiate between granules (usually larger than 1 mm in diameter) and powder (usually less than 100 μm in diameter). The former have a size large compared with the flight clearance, while the latter particles are of the order of the flight clearance. Granules feed smoothly under gravity alone, but powders may have to be force-fed using an auger. It is thought that some of the difficulty experienced when feeding powders arises from the fluidizing effect of the air that they entrain and which has to be squeezed out of them, involving counter-current air flow in the feed zone.

† By substantial is meant many times the total volume of the screw channel.

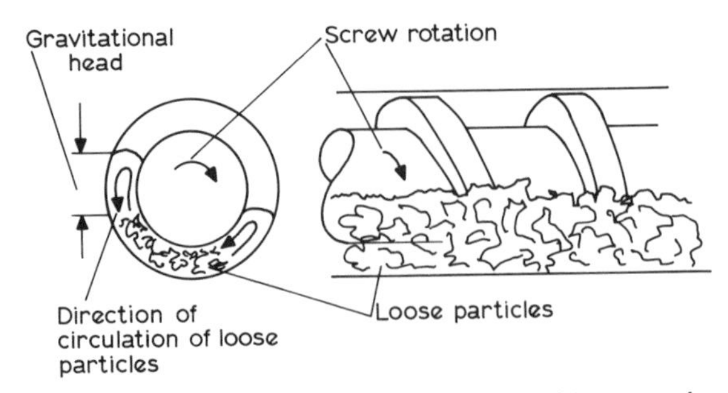

Fig. 11.4. Archimedean-screw mode of solids conveying.

There are two modes in which the screw feed can work. One is rather like an Archimedean screw pump, with the channel only partly full of loosely-packed and even freely tumbling material, moving forward axially and prevented from rotating with the screw by gravity forces. This is shown in Fig. 11.4: it is rarely observed in screw extruders, largely because of flood feeding at the hopper and because of the reduction in channel depth in the transition section which ensures that the feed channel will fill up. The natural, i.e. zero-pressure-gradient, delivery rate decreases as the depth decreases towards the nose, and so it is usual for the pressure to rise away from the feed pocket, thus constraining the delivery rate at the feed end to match the final output rate. The normal mode is therefore one in which the feed-zone channel is full of progressively more compacted material and is driven by the rotating screw against a pressure gradient. It is noticed that most granule feeds—whether of the cube-cut, cylindrically-cut, or die-face-cut (which are more nearly spherical) variety—rapidly form a well-consolidated and interlocked plug which resists internal shear and slips at both barrel and screw surfaces. The simplest model to visualize is that of an undeforming plug of polymer just filling a feed channel of constant depth, which rotates relative to the screw like a nut on a bolt.

The kinematics of the motion are relatively easily described in terms of a single helical angle, that of the motion of the outside of the plug relative to the screw axis. The true stress distribution over the surface of, and within, the plug will depend upon the deformation and rheology of the compacted medium and the laws governing the interfacial forces between plug and metal boundaries. No full analysis of the situation has been given, but considerable insight can be gained from an overall force balance on an axial segment of the plug.

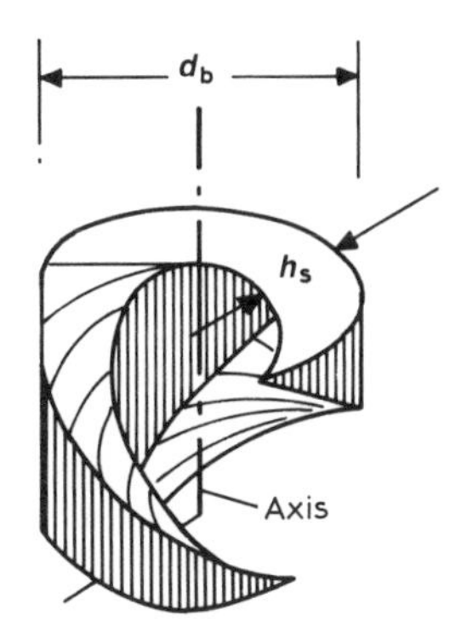

Fig. 11.5. A section of the solid plug.

Figure 11.5, following Darnell & Mol (1956), shows such a slice, in which it is quite clear that the ratio 2H/D is not so small that the unrolling proposed in Fig. 11.3 to represent the channel in Fig. 11.2 could be regarded as quantitatively satisfactory in this case. Along with various other authors, Darnell and Mol therefore retained the true helical nature of the geometry in their analysis, using a balance of axial forces and of axial torques. However, ruthless simplifications that they all then applied to the problem make their results more illustrative than quantitatively predictive. In this account, the conceptual simplification of a rectangular channel will be adopted deliberately; all the features of the true problem can be adequately illustrated thereby.

Figure 11.6 shows a rectangular element of length Δx_1, height h_s and width w_c, which moves at speed v_s relative to the screw, and speed v_{rel} relative to the barrel.† In the (x_1, x_2, x_3) coordinate system

$$\mathbf{v}_{rel} = (v_s - v_b \cos \phi_b, -v_b \sin \phi_b, 0) \tag{11.2.1}$$

where v_b is the barrel speed relative to the screw. The forces acting across three of the faces are shown in Fig. 11.6(b) in terms of the (t_{jk}) components of the continuum stress tensor $\mathbf{T}$ relevant within the plug. The overbar denotes an average over the face in question. The balance of forces involved can be obtained either by summing the forces in the

† Note that the screw is of opposite sense to that used elsewhere in the chapter, while the origin of the coordinate system is in the centre of the channel rather than at one side.

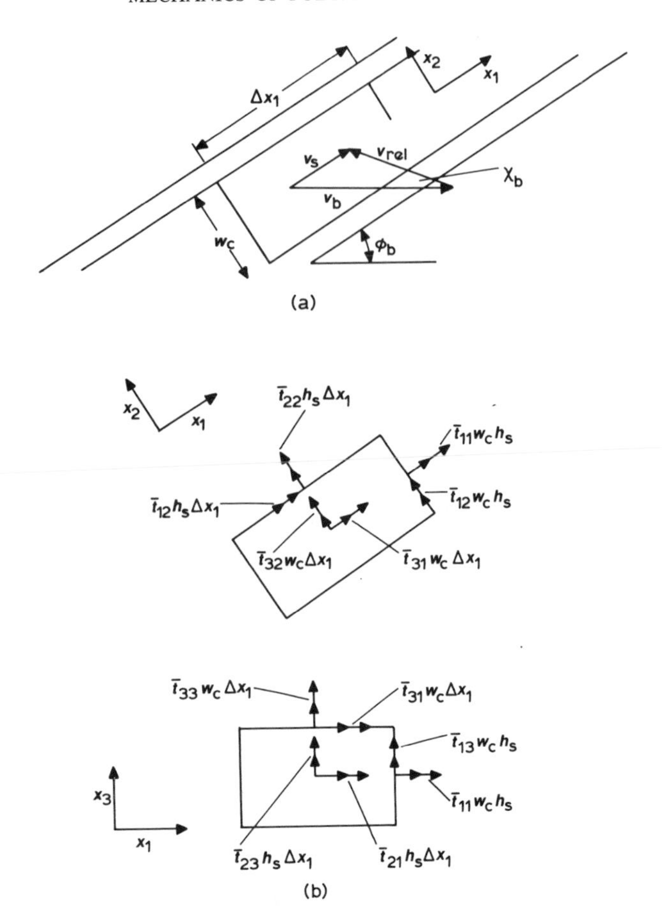

Fig. 11.6. (a) Kinematics of plug in rectangular channel: $\vec{v}_s$, velocity of plug relative to screw; $\vec{v}_b$, velocity of barrel relative to screw; $\vec{v}_{rel}$, velocity of plug relative to barrel. (b) Forces on 3 faces of plug of length Δx_1, width w_c and depth h_s in (x_1, x_2) and (x_1, x_3) planes. $\bar{t}_{12}$, $\bar{t}_{22}$, $\bar{t}_{11}$, $\bar{t}_{31}$, $\bar{t}_{32}$ are all averaged values. (c) Balance of forces on same element assuming frictional forces at interfaces.

three coordinate directions over the six faces of the element, or by integrating the stress equilibrium equation, $\nabla . \mathbf{T}=0$. Thus we get

$$\frac{\partial}{\partial x_1}\int_0^{h_s}\int_{-\frac{1}{2}w_c}^{\frac{1}{2}w_c} t_{j1}\,dx_2\,dx_3+\int_0^{h_s}\{t_{j2}(\tfrac{1}{2}w_c)-t_{j2}(-\tfrac{1}{2}w_c)\}\,dx_3$$

$$+\int_{-\frac{1}{2}w_c}^{\frac{1}{2}w_c}\{t_{j3}(h_s)-t_{j3}(0)\}\,dx_2=0; \qquad j=1,2,3 \quad (11.2.2)$$

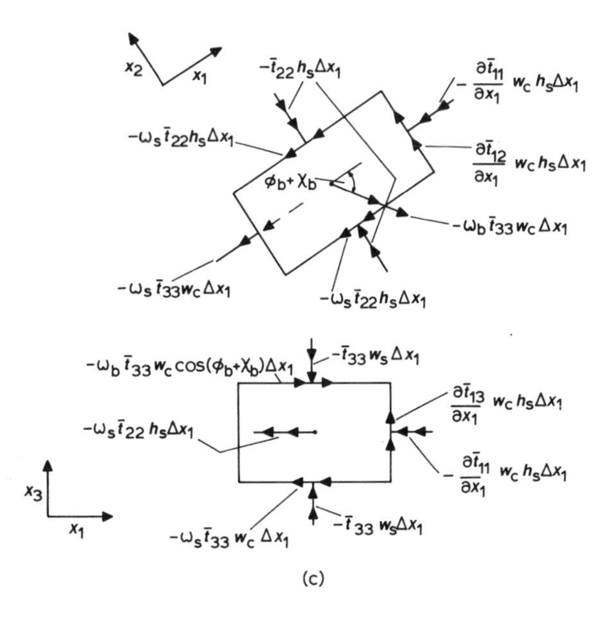

Fig. 11.6. *contd.*

or, using the definitions

$$\bar{t}_{j1} = \frac{1}{h_s w_c} \int_0^{h_s} \int_{-\frac{1}{2}w_c}^{\frac{1}{2}w_c} t_{j1}\, dx_2\, dx_3$$

$$\bar{t}_{j2} = \frac{1}{h_s} \int_0^{h} t_{j2}\, dx_3, \qquad \bar{t}_{j3} = \frac{1}{w_c} \int_{-\frac{1}{2}w_c}^{\frac{1}{2}w_c} t_{j3}\, dx_2 \tag{11.2.3}$$

this gives

$$\frac{\partial}{\partial x_1}(\bar{t}_{j1}) + \frac{1}{w_c}\{\bar{t}_{j2}(\tfrac{1}{2}w_c) - \bar{t}_{j2}(-\tfrac{1}{2}w_c)\} + \frac{1}{h_s}\{\bar{t}_{j3}(h_s) - \bar{t}_{j3}(0)\} = 0 \tag{11.2.4}$$

Here t_{jk} represents the stress over the face k in direction j. Further progress can only be made by postulating various relations between the various $\bar{t}_{jk}$.

Note that the three forces $\bar{t}_{j1}$ are functions of x_1 only, whereas the forces $\bar{t}_{j2}$ are functions of x_2 and x_1 and those of $\bar{t}_{j3}$ are functions of x_3 and x_1. Hence $\bar{t}_{13}$ is not necessarily equal to $\bar{t}_{31}$.

11.2.1 Frictional Hypothesis

Figure 11.6(c) illustrates the simplifications that ensue if we assume a linear friction law at the screw and barrel surfaces, with frictional coefficients ω_s and ω_b, respectively. Because frictional forces are parallel to and oppose the direction of relative motion, several of the components in eqn (11.2.4) become zero. Detailing those that remain yields

$$\left.\begin{aligned}
&\text{across } x_3=0\text{: } t_{33},\ t_{13}=\omega_s t_{33},\ t_{23}=0;\dagger\\
&\text{across } x_3=h_x\text{: } t_{33},\ t_{13}=+\omega_b t_{33}\cos(\phi_b+\chi_b);\ t_{23}=-\omega_b t_{33}\\
&\qquad\qquad\qquad\qquad\qquad\qquad\times\sin(\phi_b+\chi_b);\\
&\text{across } x_2=\tfrac{1}{2}w_c\text{: } t_{22},\ t_{12}=-\omega_s t_{22},\ t_{32}=0\\
&\text{across } x_2=-\tfrac{1}{2}w_c\text{: } t_{22},\ t_{12}=+\omega_s t_{22},\ t_{32}=0\\
&\text{across } x_1=x_1^* \quad\text{and}\quad x_1^*+\Delta x_1;\ t_{11},\ t_{21},\ t_{31}
\end{aligned}\right\}\qquad(11.2.5)$$

When averaged, this yields a total of seven unknown stresses or stress gradients, namely $\bar{t}_{33}(0)$, $\bar{t}_{33}(h_s)$, $\bar{t}_{22}(\frac{1}{2}w_c)$, $\bar{t}_{22}(-\frac{1}{2}w_c)$, $\partial\bar{t}_{11}/\partial x_1$, $\partial\bar{t}_{21}/\partial x_1$, and $\partial\bar{t}_{31}/\partial x_1$. Since the equations are completely homogeneous, we may hope that the solution becomes homogeneous in the simplest fashion, in which case a single length scale, X say, characterizes the x_1-dependence, and

$$\bar{t}_{jk}=\bar{t}_{jk}^{(0)}e^{x_1/X}\qquad(11.2.6)$$

To obtain a determinate set of algebraic equations, assuming (11.2.6) to hold, Darnell and Mol proposed that

$$\bar{t}_{11}^{(0)}=\bar{t}_{22}^{(0)}(-\tfrac{1}{2}w_c)=\bar{t}_{33}^{(0)}(0)=\bar{t}_{33}^{(0)}(h_s)=-p_0\qquad(11.2.7)$$

$$\bar{t}_{21}^{(0)}=\bar{t}_{31}^{(0)}=0\qquad(11.2.8)$$

with

$$\bar{t}_{22}^{(0)}(\tfrac{1}{2}w_c)=-(p_0+f_0),\qquad f_0>0\qquad(11.2.9)$$

In physical terms, this supposes that the stress state of the plug is determined primarily by an isotropic pressure p, together with an additional pushing pressure f over the leading face of the flight and the consequential frictional forces. No justification was given for this proposal, although it seems clear that if ω_s, $\omega_b \ll 1$, it could form the

† The sign of the term t_{13} depends upon the facts that t_{33} has to be negative for friction to be operative, and that $\mathbf{v}_s$ is in the positive x_1 direction.

basis for an asymptotic approximation with $X \gg h_s$. Equations (11.2.4) then become

$$h_s w_c \frac{dp}{dx_1} + \omega_s\{(2p+f)h_s + pw_c\} = \omega_b p w_c \cos(\phi_b + \chi_b) \tag{11.2.10}$$

$$fh_s = \omega_b p w_c \sin(\phi_b + \chi_b) \tag{11.2.11}$$

with the three-direction relation identically satisfied. From (11.2.10) and (11.2.11) we can eliminate f to yield, using (11.2.6),

$$\frac{1}{X} = \frac{\omega_b}{h_s}\{\cos(\phi_b + \chi_b) - \omega_s \sin(\phi_b + \chi_b)\} - \omega_s\left(\frac{2}{w_c} + \frac{1}{h_s}\right) \tag{11.2.12}$$

The volumetric delivery of the screw will be

$$\dot{V} = v_b w_c h_s \sin \chi_b / \sin(\chi_b + \phi_b) \tag{11.2.13}$$

Equations (11.2.12) and (11.2.13) thus provide a monotonic relation between X and $\dot{V}$, with ω_s, ω_b, h_c, w_c, v_b and ϕ_b as parameters by elimination of χ_b. The exponential nature of p as a function of x_1 means that p_0 becomes a crucial quantity in the theory.

Schneider (1969) noted that, for many elastoplastic solids compressed within a cylindrical (or rectangular) box, a difference arises between the normal stresses in the active and passive directions. This led him to amend the assumption (11.2.7)–(11.2.9) of Darnell and Mol to

$$\bar{\tau}_{11}^{(0)} = p_0; \qquad \bar{\tau}_{33}^{(0)}(h_s) = \bar{\tau}_{33}^{(0)}(0) = k_3 p_0; \qquad \bar{\tau}_{22}^{(0)}(\tfrac{1}{2}w_c) = k_2 p_0 \tag{11.2.14}$$

with the coefficients derived from experiment. The additional normal stress f was retained. A result similar to (11.2.12) follows.

Lovegrove and Williams (1973, 1974) sought to take account of the actual stress state within the plug, allowing it to vary in the x_2 direction. They also sought to tackle the origin of the initial pressure p_0 in the Darnell & Mol-type formulation. Tadmor & Klein (1970) had suggested that the weight of the material in the hopper was the determining factor. However, Lovegrove and Williams argue that this is not the case, and that the gravitational and even centrifugal body forces within the plug are the necessary and inevitable origin of a pressure gradient. They employed neither the true helical geometry, nor the simplified Cartesian geometry used here, but an approximating cylindrical geometry relevant when the lead angle is small. When interpreted in a form suitable for the eqns (11.2.4), the centrifugal force would yield a body force

$$F_3^{(c)} = \tfrac{1}{2}\rho_s(v_b - v_s \cos\phi_b)^2 \tag{11.2.15}$$

while the gravity force would have components

$$\left.\begin{aligned} F_3^{(g)} &= -\rho_s g \cos\{\pi x_1 \sin 2\phi_b/(w_f + w_c)\} \\ F_1^{(g)} &= \rho_s g \sin\{\pi x_1 \sin 2\phi_b/(w_f + w_c)\}\cos\phi_b \\ F_2^{(g)} &= \rho_s g \sin\{\pi x_1 \sin 2\phi_b/(w_f + w_c)\}\sin\phi_b \end{aligned}\right\} \tag{11.2.16}$$

It is immediately clear that these forces added to eqns (11.2.4) would destroy their homogeneous nature. For small values of x_1, i.e. assuming $x_1 = 0$ at the feed pocket, the inhomogeneous terms (11.2.15) and (11.2.16) would dominate the solution. However, for large values of x_1, p becomes large and the effect of the inhomogeneous terms becomes negligible. Their effect would by then have become equivalent to specifying $p_0 = p(0)$. A solution of the set of equations would still require assumptions analogous to (11.2.7)–(11.2.9) or to (11.2.14).

Lovegrove and Williams retained the assumptions of Schneider embodied in (11.2.14) when considering variations of p with x_2, although their local stress-equilibrium equation (integrated over x_3 between 0 and h_s) contains an independent, non-zero, component t_{12}. The object of their model is to account for the 'distribution' of the pushing force f, which is balanced by the frictional forces $\bar{t}_{12}$ at the barrel and screw walls, within the plug. A pair of first-order inhomogeneous partial hyperbolic differential equations for p and τ (their integrated form of t_{12}) with x_1 and x_2 as independent variables results, whose homogeneous forms have two exponential solutions of the form (11.2.6).

The details of this solution will not be given here: although it is probably closer to reality than those of Darnell and Mol or Schneider, there remain so many assumptions implicit in it that it should be regarded as still having only qualitative significance.

(1) No account is taken of compressibility, which is an important factor since the loose granular material compresses by a factor of 10–25% before it behaves like solid polymer. The primary effect of compressibility is in causing $\dot{V}$ in (11.2.13) to vary with x_1, so, if the

effective bulk density is known as a function of p, the relevant correlation to the solution could be made.

(2) No account is taken of changes in h_s, which are relevant when the solids conveying regime extends into the transition section. The main consequence of this could be to alter the relevant values of k_3 and k_2, particularly if the direction of greatest compressive stress changed from x_1 to x_3, say.

(3) No really satisfactory account of the stress state within the plug has yet been given. At one extreme, as in extrusion of lubricated PVC powder, a suitable model might be that of an incipiently yielding plastic solid, displaying internal Coulombic friction; at the other, with well-compacted and interlocking granules of nylon, an adequate model might be that of a Hookean elastic solid. Behaviour would be different in the two cases.

(4) Constant friction coefficients are used. All the evidence (Chung, 1970; Tadmor & Klein, 1970, Section 4.2; Briscoe & Tabor, 1978; Huxtable *et al.*, 1981) shows that the 'apparent' friction coefficient is stress, temperature, velocity and time dependent. Frictional mechanisms are anything but steady, and reliable experimental data are difficult to obtain. It is this unreliability of frictional data that makes the solid-feed-zone models of doubtful utility. In any case the object of plasticating extrusion is to melt (or soften) the solid feed, and so the regime of hard, dry friction is unlikely to be relevant for more than a few turns; in practice, the barrel often has to be cold-water-cooled near the feed pocket to provide such a regime at all.

11.2.2 Viscous Hypothesis

Once the metal surfaces of barrel or screw exceed the melting temperature of the polymer, then the latter can adhere to the former and form a molten lubricating layer between the solid plug and the metal. Provided the layer is very thin then the basic force balance implied by (11.2.4) can be retained; what is altered is the relation for the shear stresses t_{13}, t_{23} and t_{12} across the interfaces.

At the earliest stages of formation, in the case of relatively rigid spherical particles forming the solid bed, the mean lubricating layer thickness will be small compared to the dimensions of the particle. If the frictional contacts prior to melting had been essentially point-wise, i.e. with mean areas small compared with the mean particle cross-sections, then it is possible to visualize the type of lubrication shown in Fig. 11.7(b) as being relevant. Here the bulk of the lubricating layer

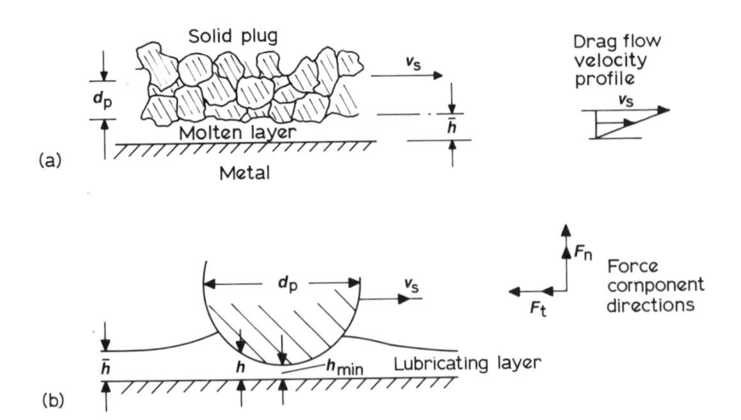

Fig. 11.7. Two interpretations of the action of a molten lubricating layer between a solid plug and a metal surface. (a) Layer thickness, $\bar{h}$, essentially uniform and continuous with velocity profile essentially independent of position in metal surface. (b) Lubricating layer significantly penetrated by spherical particles such that

$$d_p \gg \bar{h} \gg h_{min}$$

will contain fluid at the ambient air pressure. The normal stresses required to balance any normal force F_n on the particle (equivalent to the normal stresses $\bar{t}_{33}$ and $\bar{t}_{22}$ envisaged in the frictional hypothesis) will be generated by an asymmetric free surface of the lubricating layer where it meets the particle: this is the traditional mechanism of hydrodynamic lubrication and the relevant theory can be applied (Martin *et al.*, 1969, Appendix 3) provided

$$d_p \gg \bar{h} \gg h_{min} \tag{11.2.17}$$

Under these circumstances, the drag force F_t on each spherical particle will be (asymptotically) linearly dependent upon v_s (and parallel to it) and upon η_0, the viscosity, taken to be constant, and independent of F_n, the normal force, and h. The quantitative result obtained obviously depends upon the spherical nature of the particle and upon an otherwise uniform layer thickness $\bar{h}$ outside the regions of contact. This is unlikely to be relevant if the molten layer comes from the particles themselves as they melt, while the viscosity of the material will be significantly temperature and shear-rate dependent. Nevertheless, the change from frictional resistance, linearly proportional to normal pressure and independent of relative speed, to a viscous resistance, more linearly proportional to relative speed and independent of normal pressure, will be a real effect, probably occurring smoothly as x_1, and $T_s(x_1)$ and $T_b(x_1)$, increase downstream.

Once a coherent lubricating layer has formed, we may expect melting of the asperities on the compacted granular plug and chilling of the molten material forced up into the interstices of the solid bed to provide a situation close to that shown in Fig. 11.7(a). Here the normal stress $\bar{t}_{33}$ at screwroot and barrel (or $\bar{t}_{22}$ at the flight walls) is carried uniformly across the layer of essentially constant depth $\bar{h}$ as a uniform pressure p. (More strictly, what we mean is that $\partial\bar{h}/\partial x_1 \ll 1$ and $(\bar{h}/p)\partial p/\partial x_1 \ll 1$.) Since there is relative motion between solid plug and metal, we expect a sharply-sheared drag-flow velocity profile within the lubricating layer.

So long as $\bar{h}_b < h_f$, we can argue that the development of $\bar{h}_b$, the thickness of the layer between barrel surface and solid plug, will take place primarily in the axial, z, direction while the layer between screw and solid, $\bar{h}_s$, will develop in the x_1 direction.

We therefore consider the following fundamental steady problem: that of a solid bed of polymer moving over a hot metal surface held at temperature T_W, above or near the melting temperature T_m, at a speed v_W in the x_1 direction. Melting will occur at the solid–melt interface; the thickness $\bar{h}(x^1)$ of the layer cannot be specified, nor can the pressure field $p(x^1)$ within the developing layer until the rheological behaviour of the solid bed is known. Figure 11.8 shows the idealized problem with coordinate system (x_1, x_3) and an arbitrary starting point $x_1 = 0$ where $\bar{h} = 0$. The velocity field is given by

$$\mathbf{v} = \{v_1(x_1, x_3), 0, v_3(x_1, x_3)\} \quad (11.2.18)$$

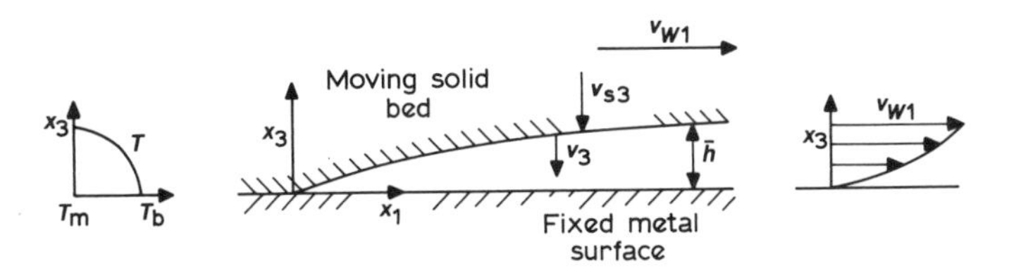

Fig. 11.8. The developing molten layer. The curvature of the temperature profile is caused by heat generation and of the velocity profile by temperature dependent viscosity.

with

$$\left.\begin{array}{ll} & v_1(x_1, 0) = 0, \qquad v_2(x_1, 0) = 0; \\ v_1(x_1, \bar{h}) = v_{W1}; & \rho_m v_3(x_1, \bar{h}) = \rho_s v_{W3} - (\rho_s - \rho_m) v_{W1}\, d\bar{h}/dx_1 \end{array}\right\} \tag{11.2.19}$$

the pressure field by $p(x_1)$ and the temperature field by $T(x_1, x_3)$ with

$$T(x_1, 0) = T_W,\ x_1 > 0; \qquad T(x_1, \bar{h}) = T_m \tag{11.2.20}$$

At the interface $x_3 = \bar{h}$, we have a balance of energy involving the latent heat of fusion ξ_s of the solid material

$$\left(\alpha_m \frac{\partial T}{\partial x_3}\right)_{x_3 = \bar{h}-} - \left(\alpha_s \frac{\partial T}{\partial x_3}\right)_{x_3 = \bar{h}+} = \rho_s \xi_s \left(v_{W3} - v_{W1} \frac{\partial \bar{h}}{\partial x_1}\right) \tag{11.2.21}$$

which involves the flux of heat into the solid bed. We expect $|v_{W3}| \ll v_{W1}$ in which case the lubrication approximation will apply to the stress equation. Thus the familiar lubrication approximation

$$\frac{\partial p}{\partial x_1} = \frac{\partial}{\partial x_3}\left(\eta \frac{\partial v_1}{\partial x_3}\right) \tag{11.2.22}$$

results with the conservation of mass, assuming uniform melt density, taking the form

$$\rho_m \frac{d}{dx_1} \int_0^{\bar{h}} v_1\, dx_3 = -\rho_s \left(v_{W3} - v_{W1} \frac{d\bar{h}}{dx_1}\right) \tag{11.2.23}$$

The energy equation, particularly in view of the melting process giving a small but non-zero normal velocity at $x_3 = \bar{h}$, will contain convective terms that may not be negligible. It will be written, by analogy with (7.1.72) or (8.1.45), as

$$\rho_m \Gamma_m \left(v_1 \frac{\partial T}{\partial x_1} + v_3 \frac{\partial T}{\partial x_3}\right) = \alpha_m \frac{\partial^2 T}{\partial x_3^2} + \eta \left(\frac{\partial v_1}{\partial x_3}\right)^2 \tag{11.2.24}$$

where we shall suppose η can be written

$$\eta = K_m e^{\zeta_m (T_m - T)} \left|\frac{\partial v_1}{\partial x_3}\right|^{\nu - 1} \tag{11.2.25}$$

if we wish to derive relatively explicit results, and where we use the continuity equation in the form

$$\frac{\partial v_1}{\partial x_1}+\frac{\partial v_3}{\partial x_3}=0 \tag{11.2.26}$$

The set of equations (11.2.22)–(11.2.26), together with the boundary conditions (11.2.19) and (11.2.20) do not provide a determinate system. If the solid bed were a completely rigid system, then for steady flow

$$v_{W3}=0 \tag{11.2.27}$$

If the bed were freely deformable, but coherent, then we could use the approximation

$$\partial p/\partial x_1=0 \tag{11.2.28}$$

In a real situation we may expect some behaviour between the two, i.e. some relation between v_{W3}, p and $d\bar{h}/dx_1$. The temperature within the solid bed can be expected to obey a heat conduction equation modified by the gross convective velocity v_{W1} but also by any minor convective terms caused by slow deformation. Within the lubrication approximation this becomes

$$v_{W1}\frac{\partial T}{\partial x_1}+v_{W3}\frac{\partial T}{\partial x_3}=\kappa_s\frac{\partial^2 T}{\partial x_3^2} \tag{11.2.29}$$

To get an idea of the importance of the various terms through their orders of magnitude, we use typical values for the various physical quantities. With $\kappa_s=\kappa_m=10^{-7}\ \mathrm{m^2\,s^{-1}}$, a development length l_1 in the x_1 direction of 1 m and a solid bed-speed relative to the wall of $100\ \mathrm{mm\,s^{-1}}$, then a 'contact' time of $l_1/v_{W1}=10$ s is obtained, with a consequential penetration depth of temperature into a fixed solid bed of order $(\kappa_s l_1/v_{W1})^{\frac{1}{2}}=1$ mm.† For $h_s=15$ mm, this means that conduction can have little effect in heating the solid bed overall. A contact length of order 100 m would be needed to ensure significant internal warming. The same argument shows that, in the absence of significant generation, the lubricating layer must have a depth $\bar{h}$ small compared to 1 mm for substantial melting to take place by conduction across the layer.

† This result follows by using an equivalent unsteady heat conduction equation to represent (11.2.29).

For $\bar{h}=0{\cdot}1$ mm, $\alpha_m=0{\cdot}2\ \mathrm{W\,mK^{-1}}$, $T_W-T_m=50$ K, then $\alpha_m(T_W-T_m)/\bar{h}$, an estimate for heat conducted across the layer, yields 10^5 W. Neglecting altogether conduction into the bulk in eqn (11.2.21) and putting $\rho_s=10^3\ \mathrm{kg\,m^{-3}}$, $\xi_s=2\times10^5\ \mathrm{J\,kg^{-1}}$ yields

$$-v_{W3}+v_{W1}\frac{d\bar{h}}{dx_1}\sim 0{\cdot}5\ \mathrm{mm\,s^{-1}} \tag{11.2.30}$$

This suggests that the melted depth can be as large as the penetrated depth. It is therefore not unreasonable to replace the effect of the terms $\alpha_s(\partial T/\partial x_3)_{x_3=\bar{h}+}$ in boundary condition (11.2.21) by an effective addition $\Gamma_s(T_m-T_0)$ to ξ_s on the right-hand side, where T_0 is the bulk temperature of the feed. Thus we write

$$\left(\alpha_m\frac{\partial T}{\partial x_3}\right)_{x_3=\bar{h}}=\rho_s\xi_s^*\left(v_{W3}-v_{W1}\frac{d\bar{h}}{dx_1}\right) \tag{11.2.31}$$

where

$$\xi_s^*=\xi_s+\Gamma_s(T_m-T_0) \tag{11.2.32}$$

as a boundary condition at $\bar{h}$ which decouples the temperature field in $x_3>\bar{h}$ from that in the molten layer. Example 11.2.2 further justifies this argument.

We now have a determinate set of equations for velocity, pressure and temperature in the molten lubricating layer, provided we adopt either (11.2.27) or (11.2.28). Computational solution is perfectly practicable. From any given value of x_1 the solution is stepped forward to $x_1+\Delta x_1$ primarily by use of the relation (11.2.23), written as

$$\frac{dq_1}{dx_1}=\frac{d}{dx_1}\int_0^h v_1\,dx_3=-\frac{\alpha_m}{\rho_m\xi_s^*}\left(\frac{\partial T}{\partial x_3}\right)_{x_3=\bar{h}} \tag{11.2.33}$$

to give the new value of q_1.

The convective term $v_1\,\partial T/\partial x_1$ in (11.2.24) carries forward information about the temperature. At the new value $x_1+\Delta x_1$, (11.2.27) will give us, using (11.2.31), the new value of $\bar{h}$ so (11.2.22) can be solved, knowing q_1 and a first approximation to T, the new values of v_1 and dp/dx_1. If an implicit representation is used for (11.2.24), then iteration around (11.2.22) and (11.2.24) yields improved values for v_1 and dp/dx_1. If (11.2.28) is used, then (11.2.22) together with q_1, and a first approximation to T, gives new values of v_1 and $\bar{h}$; iteration around (11.2.22) and an implicit form for (11.2.24) again yields improved values for v_1 and $\bar{h}$. A step-by-step method is thus defined provided an initial situation at $x_1=0$ can be prescribed.

11.2.3 Similarity Solutions

Analytical insight can best be gained by seeking a similarity solution.

We take a characteristic length scale h^* for the layer thickness and a separate—later to be determined—length scale l^* in the x_1 direction. v_{W1} is the relevant velocity scale for the x_1 direction while $T^* = T_w - T_m$ is a suitable temperature scale. We now define

$$\left.\begin{array}{lll} \tilde{h} = \bar{h}/h^*, & \tilde{x}_1 = x_1/l^*, & \hat{x}_3 = x_3/\bar{h} \\ \tilde{T} = (T - T_m)/T^*, & \tilde{v}_1 = v_1/v_{W1}, & \tilde{v}_3 = v_3 l^*/v_{W1} h^* \end{array}\right\} \quad (11.2.34)$$

The boundary conditions (11.2.19), (11.2.20) and (11.2.31) now become

$$\left.\begin{array}{ll} \tilde{v}_1(\tilde{x}_1, 0) = 0, & \tilde{v}_3(\tilde{x}_1, 0) = 0 \\ \tilde{v}_1(\tilde{x}_1, 1) = 1, & \rho_m \tilde{v}_3(\tilde{x}_1, 1) = \rho_s \tilde{v}_{W3} - (\rho_s - \rho_m)\dfrac{d\tilde{h}}{d\tilde{x}_1} \end{array}\right\} \quad (11.2.35)$$

$$\tilde{T}(\tilde{x}_1, 0) = 1, \qquad \tilde{x}_1 > 0; \qquad T(\tilde{x}_1, 1) = 0$$

and

$$\frac{\partial T}{\partial \hat{x}_3}(\tilde{x}_1, 1) = \tilde{h}\left(\tilde{v}_{W3} - \frac{d\tilde{h}}{d\tilde{x}_1}\right) \quad (11.2.36)$$

provided

$$l^* = \rho_s \xi_s^* v_{W1} h^{*2}/\alpha_m T^* \quad (11.2.37)$$

Equations (11.2.22), (11.2.23), (11.2.24) and (11.2.26) become

$$\frac{\partial}{\partial \hat{x}_3}\left\{\rho^{-B\tilde{T}} \left|\frac{\partial \tilde{v}_1}{\partial \hat{x}_3}\right|^{\nu-1} \frac{\partial \tilde{v}_1}{\partial \hat{x}_3}\right\} = \tilde{h}^{\nu+1} \frac{\partial \hat{p}}{\partial \tilde{x}_1} \quad (11.2.38)$$

$$\frac{d}{d\tilde{x}_1}\int_0^1 \tilde{v}_1 \, d\hat{x}_3 = -\frac{\rho_s}{\rho_m \tilde{h}}\left(\tilde{v}_{W3} - \frac{d\tilde{h}}{d\tilde{x}_1}\right) \quad (11.2.39)$$

$$\left.\begin{array}{l} \mathrm{Sf}\left(\tilde{v}_1 \dfrac{\partial \tilde{T}}{\partial \tilde{x}_1} - \dfrac{1}{\tilde{h}}\dfrac{d\tilde{h}}{d\tilde{x}_1}\tilde{v}_1 \hat{x}_3 \dfrac{\partial \tilde{T}}{\partial \hat{x}_3} + \dfrac{1}{\tilde{h}}\tilde{v}_3 \dfrac{\partial \tilde{T}}{\partial \hat{x}_3}\right) \\ \quad = \dfrac{1}{\tilde{h}^2}\dfrac{\partial^2 \tilde{T}}{\partial \hat{x}_3^2} + \dfrac{\mathrm{Gn}}{\tilde{h}^{\nu+1}}\left|\dfrac{\partial \tilde{v}_1}{\partial \hat{x}_3}\right|^{\nu+1} e^{-B\tilde{T}} \end{array}\right\} \quad (11.2.40)$$

$$\frac{\partial \tilde{v}_1}{\partial \tilde{x}_1} - \frac{1}{\tilde{h}}\frac{d\tilde{h}}{d\tilde{x}_1}\hat{x}_3 \frac{\partial \tilde{v}_1}{\partial \hat{x}_3} + \frac{1}{\tilde{h}}\frac{\partial \tilde{v}_3}{\partial \hat{x}_3} = 0 \quad (11.2.41)$$

where

$$\hat{p} = p\alpha_m T^*/K_m \rho_s \xi_s^* v_{W1}^{\nu+1} h^{*-\nu+1} \tag{11.2.42}$$

$$B = \zeta_m T^* \tag{11.2.43}$$

$$Sf = \rho_m \Gamma_m T^*/\rho_s \xi_s^* \tag{11.2.44}$$

$$Gn = K_m v_{W1}^{\nu+1} h^{*1-\nu}/T^* \alpha_m \tag{11.2.45}$$

Here Gn is the Brinkman number based on $T_{op}^* = T^*$ and Sf a Stefan number. B is the ratio of the Nahme number to the Brinkman number.

If $\nu = 1$, which is the Newtonian fluid approximation or if $Gn \ll 1$, in which case conduction dominates, then (11.2.35)–(11.2.45) admit of the similarity solution

$$\tilde{v}_1 = \tilde{v}_1^s(\hat{x}_3), \qquad \tilde{v}_3 = \tilde{v}_3^s(\hat{x}_3)\frac{d\tilde{h}}{d\tilde{x}_1}(\tilde{x}_1), \qquad \tilde{T} = \tilde{T}^s(\hat{x}_3) \tag{11.2.46}$$

satisfying

$$\hat{x}_3 \frac{d\tilde{v}_1^s}{d\hat{x}_3} - \frac{d\tilde{v}_3^s}{d\hat{x}_3} = 0 \tag{11.2.47}$$

$$\frac{d}{d\hat{x}_3}\left\{e^{-B\tilde{T}^s}\left|\frac{d\tilde{v}_1^s}{d\hat{x}_3}\right|^{\nu-1}\frac{d\tilde{v}_1}{d\hat{x}_3}\right\} = \tilde{P}_{,1}^s \tag{11.2.48}$$

$$Sf\, A(\hat{x}_3 \tilde{v}_1^s - \tilde{v}_3^s)\frac{d\tilde{T}^s}{d\hat{x}_3} + \frac{d^2\tilde{T}^s}{d\hat{x}_3^2} + Gn\left|\frac{d\tilde{v}_1^s}{d\hat{x}_3}\right|^{\nu+1}\tilde{h}^{1-\nu}e^{-B\tilde{T}^s} = 0 \tag{11.2.49}$$

$$\tilde{h}\frac{d\tilde{h}}{d\tilde{x}_1} = A \tag{11.2.50}$$

with boundary conditions

$$\tilde{v}_1^s(0) = \tilde{v}_3^s(0); \qquad \tilde{T}^s(0) = 1 \tag{11.2.51}$$

$$\tilde{v}_1^s(1) = 1, \qquad \tilde{v}_3^s(1) = \left(1 + \frac{\rho_s}{\rho_m A}\frac{d\tilde{T}^s}{d\hat{x}_3}(1)\right), \qquad \tilde{T}^{(s)}(1) = 0 \tag{11.2.52}$$

where either $\nu = 1$ or $Gn = 0$, and A and $\tilde{P}_{,1}^s$ are constants to be determined. The rigid body approximation (11.2.27) yields

$$\tilde{v}_3^s(1) = -\frac{\rho_s - \rho_m}{\rho_m} \tag{11.2.53}$$

while the drag flow approximation (11.2.28)

$$\tilde{P}^{s}_{,1} = 0 \tag{11.2.54}$$

In the first case (11.2.51)–(11.2.53) represent 7 boundary conditions for the 5th-order system of equations (11.2.47)–(11.2.49), and so A and $\tilde{P}^{s}_{,1}$ act as joint eigenvalues, while in the second case there is one fewer boundary condition and A is the single eigenvalue. In both cases we find on integration of (11.2.50) that

$$\tilde{h} = (2A\tilde{x}_1)^{\frac{1}{2}} \tag{11.2.55}$$

which is consistent with our boundary condition at $x_1 = 0$. In dimensional form, this becomes

$$\bar{h} = \left(\frac{2A\alpha_m T^* x_1}{\rho_s \xi_s^* v_{W1}}\right)^{\frac{1}{2}} = \left(\frac{2\,\mathrm{Sf}\,A\kappa_m x_1}{v_{W1}}\right)^{\frac{1}{2}} \tag{11.2.56}$$

and the arbitrary scale h^* has, as expected, disappeared.

Using various values for plasticating extruders and various feeds, we find that B, Sf are usually less than unity, while Gn can be significantly smaller than or of order unity. In very large extruders run fast, Gn becomes large compared with unity. At the high shear rates involved in lubricating layers ν is rarely unity and is most likely to be as small as 0·3. Thus the similarity solution with Gn$\ll$1 is likely to be the most useful. Although A is not available as a function of Sf, B and Gn or ν, it is likely to be of order 1 for real situations. However, the important fact that follows from this analysis is that the molten layer will reach the thickness of the flight clearance in very few turns of the screw flight and so will be swept up by the flight moving over the barrel surface at a relatively early stage downstream of the point at which the barrel rises substantially above the melting temperature. Our first approximation from (11.2.56) is that this will occur at an axial distance $\bar{z}$ given by

$$\bar{z} = \frac{v_{\mathrm{rel}} \sin \chi_b h_f^2}{2A\kappa_m\,\mathrm{Sf}} = \frac{v_s \sin \phi_b h_f^2}{2A\kappa_m\,\mathrm{Sf}} \tag{11.2.57}$$

where χ_b is the angle between $\mathbf{v}_{\mathrm{rel}}$ and $\mathbf{v}_b$ as shown in Fig. 11.6, and used in (11.2.12).

It is possible that the melt being sheared between solid bed and barrel is not swept up until the amount left behind is thicker than the flight clearance h_f. This condition can be written in the form

$$\bar{h}_b \int_0^1 (1 - v_1)\,d\hat{x}_3 = h_f \tag{11.2.58}$$

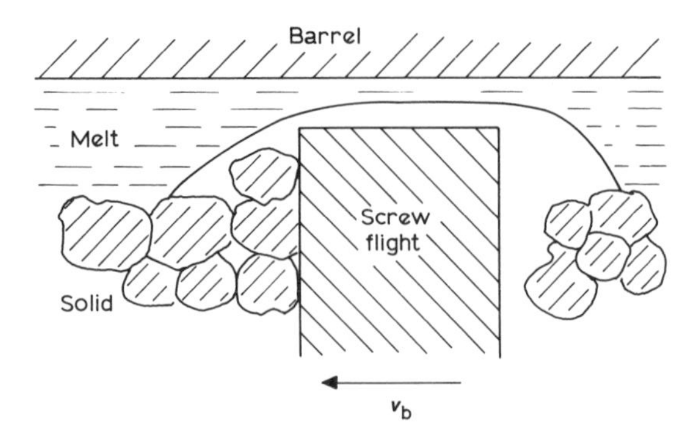

Fig. 11.9. Diagram showing molten layer pulling away from solid bed enabling it to pass through flight clearance.

where $\bar{h}_b$ refers to the thickness of the lubricating layer on the barrel. For a constant viscosity fluid and $\text{Sf} \gg 1$, this will give $\bar{h}_b = 2h_f$, and if $\text{B}_b \ll 1$, with $\nu < 1$, the value could be as high as $4h_f$. The only evidence that this might be the case is provided by Tadmor and Klein (1970) whose experiments suggest a ratio as high as 5. A diagrammatic indication of free surface forming is shown in Fig. 11.9.

Alternatively, it can be argued that much of the solid material being melted at $x_3 = \bar{h}$ percolates back into the solid bed. If this is the case, then the analysis carried out in this section (11.2) based on a single velocity vector $\mathbf{v}$ defined in (11.2.18) becomes inadequate to describe the full physical situation. Instead, for $x_3 > h$, two velocity vectors would be required, $\mathbf{v}_s$ and $\mathbf{v}_m$, to refer to solid and melt respectively, together with effective densities $\rho_s(1-\varepsilon_s)$ and $\rho_m\varepsilon_m$ where $\varepsilon_m < \varepsilon_s$, and ε_s is taken to be the voidage in the solid bed. The relevant governing equations would be difficult to determine because of the complicated nature of the solid melt interface within the percolation region.

Our analysis of the viscous lubricating layers allows us to derive values for the forces in (11.2.2) that are tangential to the metal surfaces. For example, for the face $x_3 = 0$, the viscous stress will be

$$t_{13}(x_1) = \eta\left(\frac{\partial v_1}{\partial x_3}\right)_{x_3=0} = K_m\left(\frac{v_s^3}{2A_s\kappa_m x_1\,\text{Sf}_s}\right)^{\frac{1}{2}\nu} e^{-B_s\tilde{T}_s}\left|\frac{d\tilde{v}_{1s}}{d\hat{x}_3}\right|^{\nu-1}\frac{d\tilde{v}_{1s}}{d\hat{x}_3} \tag{11.2.59}$$

Carrying out a balance of forces in the x_1 direction, yields

$$h_s w_c \frac{dp}{dx_1} + (2h_s + w_c)K_m \left(\frac{v_s^3}{2A_s\kappa_m x_1 \, \mathrm{Sf}_s}\right)^{\frac{1}{2}\nu} e^{-B_s\tilde{T}_s} \left|\frac{d\tilde{v}_{1s}}{d\hat{x}_3}\right|^{\nu-1} \frac{d\tilde{v}_{1s}}{d\hat{x}_3}$$

$$= \cos(\chi_b + \phi_b) w_c K_m \left(\frac{v_{rel}^3 \sin \chi_b}{2A_b\kappa_m \sin \phi_b x_1 \, \mathrm{Sf}_b}\right)^{\frac{1}{2}\nu} e^{-B_b\tilde{T}_b} \left|\frac{d\tilde{v}_{1b}}{d\hat{x}_3}\right| \frac{d\tilde{v}_{1b}}{d\hat{x}_3} \qquad (11.2.60)$$

where Sf_s, v_s, A_s, B_s refer to the screw plug interface near $x_3 = 0$ and Sf_b, v_{rel}, A_b and B_b refer to the barrel plug interface near $x_3 = h_s$ and $\tilde{T}_{s,b}$, $d\tilde{v}_{1s,b}/d\hat{x}_3$ are evaluated at the surface of the plug. Equation (11.2.13) still holds and so (11.2.60) and (11.2.13) represent an explicit relationship for dp/dx_1 in terms of χ_b, i.e. of $\dot{V}$.

We now return to the question of the relation between the pressure within the lubricating layers and the deformation of the solid plug. Because $\bar{h}_s$, $\bar{h}_b \ll h_s$ in this case very small differences in the forward velocity v_s of the plug, i.e. very small compressions in the x_2 and x_3 directions and compensating extension in the x_1 direction, would be sufficient to change from the rigid boundary condition $v_{W3} = 0$, (11.2.27), to the constant pressure conditions (11.2.28). As one might guess the pressure difference scale derived from (11.2.60) is $\bar{h}/h_s$ smaller than that given by (11.2.42); put differently, the ratio between the scale length for change of pressure in the channel as a whole is very small. So unless the solid bed is really rigid, the approximation $\tilde{p}^s_{,1} = 0$, (11.2.54), can be used in the lubricating layer analysis. This successfully decouples the rheology of the bed from the lubricating melt flow. It also justifies neglect of any pressure-driven flow tending to move fluid normal to the relative direction of motion of solid bed and metal surface. This is equivalent to there being no coupling between the overall force balances in the x_1, x_2 and x_3 directions and so the quantities $e^{-B_s\tilde{T}_s}|d\tilde{v}_{1s}/d\hat{x}_3|^\nu$ and $e^{-B_b\tilde{T}_b}|d\tilde{v}_{1b}/d\hat{x}_3|^\nu$ are functions of Sf_s, v_s, B_s and Sf_b, v_{rel}, B_b respectively, only.

Various approximate solutions for A can be obtained by methods similar to those given in Pearson (1976*a*) for a related problem that will be met and discussed in the next subsection.

We note that integration of (11.2.60) leads to a form

$$p = p_0 + Y(\mathrm{Sf}_s, B_s, \mathrm{Sf}_b, B_b, v_b, v_s) x_1^{1-\frac{1}{2}\nu} \qquad (11.2.61)$$

For $\nu = \frac{1}{3}$, this is not very different from linear behaviour, as proposed by Tadmor & Klein (1970).

Finally we observe that if $T_b > T_s$ for any given value of x_1, or vice versa, then there may be a region in which frictional forces are relevant across some interfaces and viscous forces over the others.

EXAMPLE 11.2.1. Derive the relation for F_t in the circumstances illustrated in Fig. 11.7(b).

We take a coordinate system based on the point of the wall that is closest to the sphere of diameter d_p representing a polymer pellet, with x_1 in the direction of motion of the pellet and x_3 perpendicular to the plane. It follows that the local thickness of the fluid layer

$$h \approx h_{min} + \tfrac{1}{2}(x_1^2 + x_2^2)/d_p$$

According to lubrication theory there will be a flow field between the sphere and wall given by

$$\mathbf{v} = [v_1(x_1, x_2), v_2(x_1, x_2), 0]$$

The symmetry of the problem suggests that any forces due directly to v_2 will cancel out. Furthermore, it can be argued that, in the Newtonian case where pressure and drag contributions can be added linearly, flow due to pressure gradients will lead to equal but *not* opposite tangential (in directions 1 and 2) forces on wall and sphere. These will therefore cancel out on integration over the entire surfaces of wall or sphere, and so the drag force F_t will therefore be the result of the drag force in the x_1 direction integrated over the surface of the sphere.

The drag flow component will everywhere be given by

$$v_{1\ \mathrm{drag}} = v_s(h - x_3)/h$$

and the local tangential stress by

$$t_{13}(x_1, x_2) = -\eta_0 v_s/h$$

The total drag

$$F_t = -\int_{\substack{\text{melted}\\ \text{area } \mathcal{A}}} \eta_0 v_s/h \, d\mathcal{A}$$

If, for the sake of simplicity, we let the melted area tend to infinity in both the x_1 and x_2 directions, and write

$$\xi = x_1/(h_{min}\, d_p)^{\frac{1}{2}}, \qquad \zeta = x_2/(h_{min}\, d_p)^{\frac{1}{2}}$$

then

$$F_t = -d_p \mu V_s \int_{-\infty}^{\infty} d\zeta \int_{-\infty}^{\infty} d\xi \left[\frac{1}{1 + (\xi^2 + \zeta^2)}\right]$$

This shows at once that F_t is independent of both F_n and h_{min} (h_{min} will be determined by F_n), but is linearly proportional to μV_s.

For the approximation given above to be relevant, the diameter of the wetted area has to be large compared with $(h_{min} d_p)^{\frac{1}{2}}$, i.e. the 'flooded' condition, and some degree of asymmetry in the wetted area is necessary in order to support the load F_n.

EXAMPLE 11.2.2. Obtain the solution to (11.2.29) for $x_3 > \bar{h}$ when $\bar{h}$ and T are independent of x_1, and a heat flux $\dot{q}$ is specified at $x_3 = \bar{h}$, corresponding to $(\alpha_m \partial T/\partial x_3)_{x_3=\bar{h}-}$ in (11.2.21). The object of this simple analysis is to get further insight into the penetration depth and justification for (11.2.31).

Integrating (11.2.29) with respect to x_3 when $v_{W1}\, \partial T/\partial x_1 = 0$ yields, using $T \to T_0$ as $x_3 \to \infty$

$$v_{W3}(T - T_0) = \kappa_s \, dT/dx_3$$

Further integration using $T(\bar{h}) = T_W$ gives

$$T - T_0 = (T_W - T_0)\exp\{v_{W3}(x_3 - \bar{h})/\kappa_s\}$$

where v_{W3} is negative.

Relation (11.2.21) with $v_{W1} = 0$ then provides a relation between v_{W3} and $\dot{q}$, namely

$$\dot{q} = -v_{W3}\{\rho_s \xi_s + \rho_s \Gamma_s (T_W - T_0)\}$$

Relations (11.2.31) and (11.2.32) are precisely the above when $-v_{W3}$ is replaced by $v_{W1}\, d\bar{h}/dx_1 - v_{W3}$.

Note that the penetration length scale is now $\kappa_s/|v_{W3}|$ which is smaller than the value obtained earlier $(\kappa_s l_1/v_{W1})^{\frac{1}{2}}$, provided

$$\left|\frac{v_{W1}}{v_{W3}}\right| < \frac{l_1}{\kappa_s} |v_{W3}|$$

EXERCISE 11.2.1. Show how the boundary conditions (11.2.19) for $x_3 = \bar{h}$ follow from the lubrication approximation, and that the terms neglected are of order $d\bar{h}/dx_1$ smaller than those included.

EXERCISE 11.2.2. Derive the boundary condition (11.2.21) by using a coordinate system moving with the solid bed.

EXERCISE 11.2.3. Show the dimensionless group $\rho_s \xi_s \bar{h} v_{W1}/\alpha_m(T_w - T_m)$ can be written as the ratio of a Peclet number and a Stefan number, the latter being defined here as in (11.2.44).

EXERCISE 11.2.4. Show that l^*/h^* is the ratio of a Peclet number and a Stefan number.

EXERCISE 11.2.5. Derive eqn (11.2.60), noting that the lubricating layer flow is taken to be separately uniform over the screw/plug and barrel/plug interfaces respectively.

EXERCISE 11.2.6. Consider feed zone dynamics in the situation where melting takes place at the barrel ($T_b > T_m$) but not at the screw ($T_s < T_m$).

11.3 MELTING OR PLASTICATING ZONE

Strictly speaking, the lubricated-layer solution given in the discussion of feed-zone mechanics represents the earliest, and indeed most rapid, part of the melting process. However, the most important part of the melting process usually arises once the screw flight effectively scrapes the molten layer on the barrel into a melt pool which accumulates between the solid bed and the forward face of the screw flight. This is shown diagrammatically in Fig. 11.10.

This type of melting behaviour has been observed, or more strictly inferred, by Maddock (1959), Street (1961) and many later workers in conventional single-screw extrusion of most thermoplastics. However,

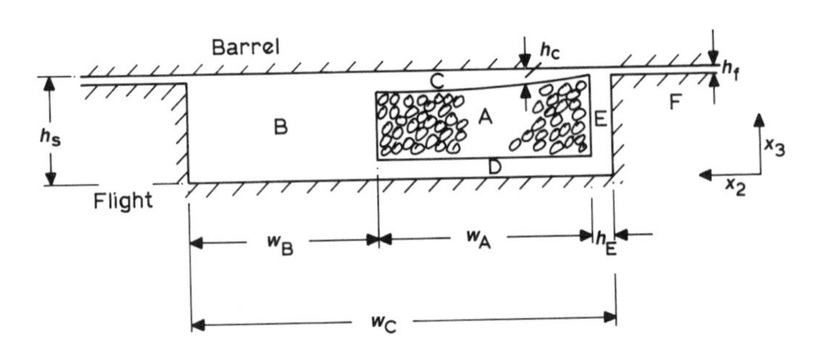

Fig. 11.10. The six regions of the six-zone model. A, solid bed; B, melt pool; C, melting layer at barrel surface; D, E, lubricating layers at screw root and screw flight; F, flight clearance.

it is not universal. Klenk (1968) (see also Gale, 1970; Cox *et al.*, 1981) has reported a different mechanism in the case of powders, e.g. PVC, sufficiently fine to pass between screw flight and barrel, in which case the melt pool accumulates on the training face of the screw flight, and the bulk of the melting process takes place within the flight clearance. Lindt (1976) has observed and modelled a further—almost certainly undesirable—mode of behaviour in PP extrusion whereby the melt pool distributes itself as a thick layer between solid bed and hot metal surfaces. This can in some sense be regarded as a continuation of the process begun in the feed zone, and is clearly less efficient at melting the solid bed than the melt pool mode. In the Lindt mode the 'lubricating' layers become so thick that they effectively insulate the solid bed thermally from heat conducted from the barrel and screw. In the scraped-surface mode, the solid bed is always brought close to the barrel surface at the trailing edge of the flight and so there is always a region of rapid melting across a gap that starts no wider than the flight clearance.

No successful analysis has yet been given to predict the mode of melting. Many designs for single-screw extruders contain additional features designed to facilitate and stabilize melting, suggesting that the scraped-surface/melt-pool mode cannot always be relied upon. The most straightforward of the modified designs is that due to Maillefer (1963), in which a helical barrier flight divides the main screw channel into a feed channel that decreases in width and a melt channel that increases in width towards the nose of the screw. The barrier has a relieved flight clearance and is intended to separate solid from melt.

Several authors (Tadmor, 1966; Tadmor & Klein, 1970; Shapiro *et al.*, 1976; Edmondson & Fenner, 1975) have presented model analyses of the melt-pool melting process. All of them have started with very specific kinematical assumptions based on observations. That due to Shapiro, Halmos, Pearson and Trottnow is emphasized here, partly because its predictions are as good as any others, and partly because the authors were able to justify *a posteriori* some of the approximations made. Lindt's analysis will be discussed later.

11.3.1 The Six-zone Model

Figure 11.10 shows six separate regions within the channel cross-section, taken to be slowly varying in the x_1-direction. The unrolled approximation is employed. We take a frame of reference and coordinate system $\mathbf{x}$ fixed in the screw.

Region A

This is the remnant of the solid bed formed in the feed zone, which is here shown to be of nearly rectangular cross-section. Melting takes place at its four interfaces, AB–AE, but the most important is that with the lubricating layer C, which separates region A from the barrel wall, while that at AB is usually negligible. The same basic assumptions are made for the solid in region A as were made for the solid plug in the feed zone. Its downstream velocity component is taken to be a (slowly-varying) function of x_1 only, while its x_2 and x_3 velocity components are regarded as small, and are neglected except for their contribution to melting at the interfaces AB–AE. The rheological behaviour of the solid bed is not specified, and only an integrated force balance in the x_1-direction is satisfied. The contributions to this force balance are the pressure gradient in the x_1-direction and the shear stresses in the x_1-direction at the four interfaces of AB–AE of which that at AB is usually neglected. The density of the solid bed is permitted to be a slowly-varying function of x_1, through dependence on the mean pressure $p(x_1)$. In practice, the bulk compressive modulus of the bed will be an increasing function of p, as the interstitial air is forced out (back towards the feed pocket) and ρ_s may be taken as constant if the bed is well-compacted before melting starts.

Region B

This is a fully molten region which moves co-currently downstream with the solid bed under the influence of drag forces from the barrel and solid bed and the, usually opposing, pressure force due to the pressure gradient in the x_1-direction. Over most of the melting region, the width-to-depth ratio w_B/h_s is not very large and so a full melt-flow analysis would involve all three components of velocity. The Graetz number for the flow is rarely small and so convective effects will be important in the thermal energy equation making the flow far from fully developed. Methods of solution will be discussed in the section on melt pumping.

Region C

This is the most important region in the melting zone. It is observed† that the layer thickness h_c is an increasing function of x_2 and that its

† These 'observations' are difficult to make, and depend upon rapid chilling of the barrel and screw after a sudden stop followed by careful extraction of the screw and analysis of the polymer unwound from the screw channel.

value near the trailing flight ($x_2 = 0$) appears to be determined by the flight clearance h_f; it does not appear to change with x_1 as much as do the lubricating layers D and E (between the screw and solid bed). The drag flow of the barrel surface relative to the moving bed acts to keep the layer thickness small: material swept from the layer C at its downstream end $x_2 = w_A + h_E$ enters the pool B, with only part of it passing through the (further) flight clearance F. Thus the intervention of the flight causes the development of region C to be different from that of regions D, E or the initial lubricating layers described for the viscous mode of solid transport in the feed zone. The basic mechanism of melting is, however, the same and so the analysis given earlier will be applicable with few changes.

Regions D and E

These two regions can be treated together insofar as the model used here will predict h_E and h_D to be given by the same slowly-varying function of x_1 as used in connection with $\bar{h}$ in Subsection 11.2.2 above. Indeed, h_E and h_D are merely a continuation of $\bar{h}$, there being no change in its behaviour when the melt pool B begins to form, except insofar as v_{A1}, i.e. v_s, changes slowly.

Region F

The mechanics of flow in this region is very simply understood because it is a confined flow between two metal surfaces moving parallel to one another. It is treated separately here because the 'output' from region F at $x_2 = 0$ is the 'input' to region C.

In our model analysis, we shall satisfy overall conservation of mass and force balances in the x_1-direction in regions A and B. We shall not formally satisfy any force balances in the x_2- and x_3-directions. Energy balances will be carefully considered in the drag-flow regions C–F but will not be explicitly relevant in the A or B regions. The flow will be taken as steady.

Figure 11.11 provides a diagrammatic illustration of the various mass fluxes that will be considered below.

Mass Balances

The overall mass balance is given by

$$\sum_{K=A}^{F} \dot{M}_{K1} = \rho_s \dot{V}_{A1} + \sum_{K=B}^{F} \rho_m \dot{V}_{K1} = \dot{M}_T \qquad (11.3.1)$$

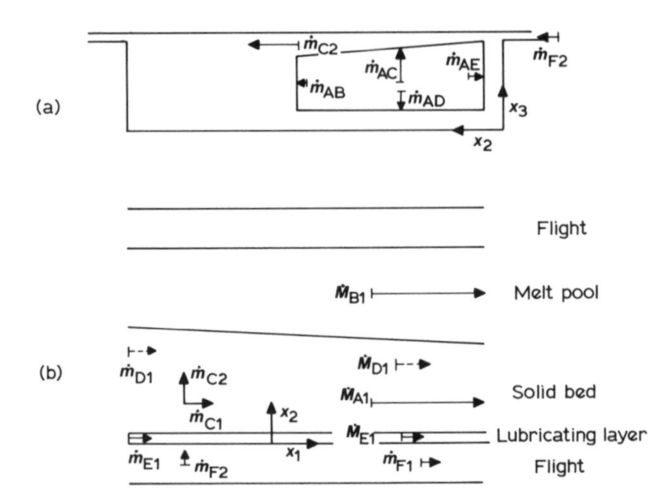

Fig. 11.11. Components of mass balance. (a) In (x_2, x_3) plane. Note that there is no component in the x_2-direction from region D, and none in the x_3-direction from region E. (b) In (x_1, x_2) plane. The length of arrows is indicative of the relative importance of the various flow rates/unit width, $\dot{M}$'s mass flowrates.

in the 1-direction, where $\dot{M}_{K1}$ refers to the mass flowrate and $\dot{V}_{K1}$ to the volume flowrate in region K, ρ_s is the solid density, ρ_m is the melt density at any cross-section $x_1 = \text{constant}$. $\dot{M}_T$ is the constant output of the extruder.

Region A. The volume flowrate in this region will be given by

$$\dot{V}_{A1} = v_{A1} w_A \bar{h}_A \tag{11.3.2}$$

where

$$\bar{h}_A = h_s - h_D - \bar{h}_C \tag{11.3.3}$$

and

$$w_A = w_c - w_B - h_E \tag{11.3.4}$$

Writing the melting rates at the four interfaces AK (K = B, C, D, E) as $\dot{m}_{AK}$ we can write a mass conservation equation for region A alone as

$$\frac{d\dot{M}_{A1}}{dx_1} = -\sum_{K=B}^{E} \dot{m}_{AK} \tag{11.3.5}$$

Writing the shear forces in the x_1-direction at the interfaces AK as F_{AK1} we can write a force balance in the x_1-direction for region A alone as

$$\bar{h}_A w_A \frac{dp}{dx_1} = \sum_{K=B}^{E} F_{AK1} \tag{11.3.6}$$

Here $\bar{h}_c$ is the mean (over x_2) thickness of layer C. The quantities $\dot{m}_{AK}$ and F_{AK1} will be obtained by separate consideration of the other regions.

Region B. Following our usual procedure for channel flow, we shall seek a relation between volume flow rate $\dot{V}_{B1}$ and pressure drop dp/dx_1 in terms of the geometry and velocities of the boundaries and the viscous properties of the material. This can be written symbolically as

$$\dot{V}_{B1} = Q_B[T(x_2, x_3), h_s, w_B, dp/dx_1, v_{b1}, v_{b2}, v_{A1}, h_f, h_C] \tag{11.3.7}$$

where dependence upon h_f and h_C ($x_2 = w_A + h_E$) will be very minor, and where the greatest complication arises through the temperature field $T(x_2, x_3)$. Here $\dot{V}_{B1}$, w_B, dp/dx_1, v_{A1} and h_C are expected to be slowly-varying functions of x_1. A full treatment of the problem requires solution for the three-dimensional velocity field $\mathbf{v}_B(\mathbf{x})$ and the temperature field $T(\mathbf{x})$ in terms of the full conservation equations (2.2.5), (2.2.8) and (2.2.9) using appropriate approximations, such as (2.2.7), $\nabla . \mathbf{T} = 0$, and (2.2.10). We can expect $\mathbf{v}_B$ and T to be slowly-varying functions of x_1 too, so that the stress equations of motion and the continuity equations are solved locally for a given value of x_1. However, the convective terms in the thermal energy equation will be such that a local solution will usually not be directly relevant. This matter has been discussed by Martin (1969), who shows that a quantitatively accurate numerical solution would involve a dauntingly large amount of computing even for a purely viscous fluid. All detailed applications of the 6-zone model have therefore made very severe approximations for this zone, usually employing a temperature-independent power-law viscous fluid model in unidirectional flow, which avoid any discussion of the thermal energy equation.

For the present, we suppose that a relation (11.3.7) can be specified and that approximate values, usually zero, can be given for F_{AB1} and $\dot{m}_{AB}$.

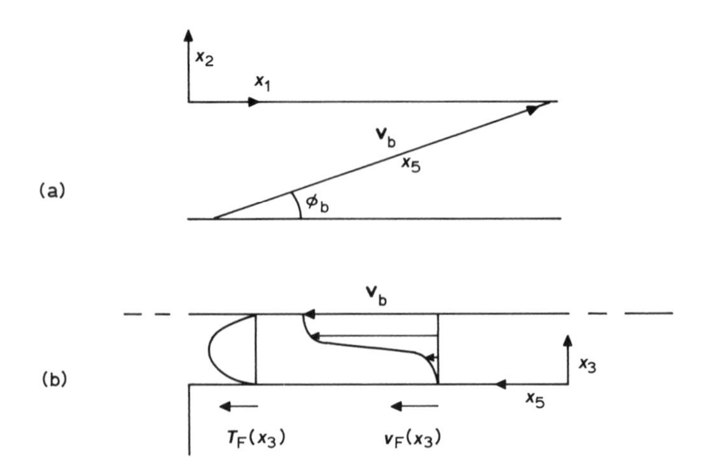

Fig. 11.12. (a) Diagram showing direction of flow in the flight clearance: coordinate x_5 is in the direction of $\mathbf{v}_b$, the velocity of the barrel relative to the screw. (b) Fully-developed velocity and temperature profiles within the flight clearance.

Region F. This is the simplest flow field to analyse, because it obeys the requirements of lubrication flow discussed in Section 8.1. Figure 11.12 refers to this.

The boundaries are parallel plates and the flow is dominated by unidirectional drag flow. The overriding geometrical ratio is

$$(\mathrm{H/L})_\mathrm{F} = h_\mathrm{f}/w_\mathrm{f}\,\mathrm{cosec}\,\phi_\mathrm{b} \tag{11.3.8}$$

which is usually no more than 0·005.† The Graetz number is given by

$$\mathrm{Gz_F} = \mathrm{Pe_F}(\mathrm{H/L})_\mathrm{F} = v_\mathrm{b} h_\mathrm{f}^2/\kappa_\mathrm{m} w_\mathrm{f}\,\mathrm{cosec}\,\phi_\mathrm{b} \tag{11.3.9}$$

Taking $d_\mathrm{b} = 100$ mm, $W_\mathrm{s} = 10$ rad s^{-1}, $v_\mathrm{b} = 0{\cdot}5$ m s^{-1}, $\kappa_\mathrm{m} = 10^{-7}$ m^2 s^{-1}, $w_\mathrm{f}\,\mathrm{cosec}\,\phi_\mathrm{b} = 30$ mm, $h_\mathrm{f} = 0{\cdot}15$ mm, we get $\mathrm{Gz_F} = 3{\cdot}7$, which is marginally fully developed. The relevant generation number will be based on $T^*_{\mathrm{rheol}}, = 30$ K say, and η $(v_\mathrm{b}/h_\mathrm{f}, T_\mathrm{b})$. For the values given above $v_\mathrm{b}/h_\mathrm{f} = 3\times10^3$ s^{-1}, which means a relatively low viscosity for a strongly shear-rate dependent material. If we take $\eta = 300$ N s m^{-2}, $\alpha = 0{\cdot}2$ W mK^{-1}, then $\mathrm{Na_F} = \eta v_\mathrm{b}^2/\alpha T^* = 12$, which is still large enough to ensure significant heating of the fluid.

† A simple rule of thumb is that

$$h_\mathrm{f}/d_\mathrm{b} = 1{\cdot}5\times10^{-3}$$

A very useful approximation is provided by the thermally fully developed drag flow, given by the equations

$$\eta \frac{dv_{F5}}{dx_3} = t_{F35} \tag{11.3.10}$$

$$\frac{d^2T_F}{dx_3^2} = -\frac{1}{\alpha_m} t_{F35} \frac{dv_{F5}}{dx_3} \tag{11.3.11}$$

with boundary conditions

$$\left.\begin{array}{ll} v_{F5}(0) = 0, & v_{F5}(h_f) = v_b \\ T_F(0) = T_s, & T_F(h_f) = T_b \end{array}\right\} \tag{11.3.12}$$

where t_{F35} is a constant to be determined and η is supposed given by (11.2.25). The plane $x_3 = 0$ is the flight tip of the screw and the plane $x_3 = h_f$ is the barrel surface. This problem has an analytical solution, whose parameters cannot, however, be explicitly given in terms of the problem parameters. Numerical solution of transcendental equations is therefore required. Details can be found in Martin (1967) or Gavis & Laurence (1968*a*, *b*). Both make the point that a maximum shear stress is achieved at a finite value of v_b. Table 11.2, recalculated from Table 1 of Gavis & Laurence (1968*b*) gives the maximum values of temperature reached, i.e. the centre-line temperature, when $T_s = T_b$, for representative values of Na_F and ν.

Table 11.2
Maximum Dimensionless Temperatures as Functions of ν and Na_F in Fully Developed Couette Flow

ν	Na_F	$(T_{F\,max} - T_b)/T^*_{rheol}$
0·5	0·0125	0·02
1	0·025	0·04
0·5	2·0	0·24
1	4·0	0·42
0·5	100	2·6
1	200	3·25

Because the velocity profile is antisymmetric about $x_3=\frac{1}{2}h_f$, the flowrates can easily be written

$$\dot{M}_{F1}\sim\tfrac{1}{2}v_b\cos\phi_b h_f w_f\rho_m \tag{11.3.13}$$

$$\dot{m}_{F2}=\tfrac{1}{2}v_b\sin\phi_b h_f\rho_m \tag{11.3.14}$$

The precise solution for T_F as a function of q_F where

$$\frac{\partial q_F}{\partial x_3}=-v_{F5}, \qquad q_F(h_f)=0 \tag{11.3.15}$$

would have to be obtained from the full solution. However, a look at Figs. 3 and 4 of Gavis & Laurence (1968*a*) shows that, for large values of Na_F ($\geqslant$10, say), a good approximation is given by

$$\tilde{T}=\tilde{T}_{max}2q_F/v_b h_f \tag{11.3.16}$$

If we associate $q_F\sin\phi_b$ with q_C defined by (11.3.37) where $q_F(0)=\frac{1}{2}v_b h_f$, then (11.3.16) provides initial conditions for the full calculation described next in region C.

Region C. To investigate this region we take a frame of reference $\mathbf{x}_A$ fixed at its origin in the moving solid bed so that $\mathbf{x}_A=0$ coincides with $\mathbf{x}=(x_1, h_E, 0)$ at the instant in question. Next we define the coordinate x_{A4} in the direction of $\mathbf{v}_{rel}$, the velocity of the barrel relative to the solid bed, so that $x_{A4}=0$ coincides with $x_{A2}=0$, the leading edge of the swept surface AC. Within the lubrication approximation, i.e. neglecting slow changes with x_{A1}, we can consider the flow field in region C to be given by

$$\left.\begin{aligned} \mathbf{v}_{C\,rel}&=\{v_{C\,rel\,4}(x_{A4}, x_{A3}), v_{C\,rel\,3}(x_{A4}, x_{A3})\} \\ T_C&=T(x_{A4}, x_{A3}), \qquad h_C=h_C(x_{A4}) \end{aligned}\right\} \tag{11.3.17}$$

where the two components of $\mathbf{v}_{C\,rel}$ are in the x_{A4} and x_{A3} directions, and $x_{A4}\geqslant 0$.

The relevant equations and boundary conditions are very similar to those given in Subsection 11.2.2 following eqn (11.2.18); in particular (11.2.28) rather than (11.2.27). Thus we have

$$\left.\begin{aligned} v_{C\,rel\,4}(x_{A4}, h_s)&=v_{rel}; \qquad v_{C\,rel\,3}(x_{A4}, h_s)=0 \\ v_{C\,rel\,4}(x_{A4}, h_s-h_C)&=0; \qquad \rho_m v_{C\,rel\,3}(x_{A4}, h_s-h_C)=\rho_s v_{A3} \end{aligned}\right\} \tag{11.3.18}$$

$$T(x_{A4}, h_s)=T_b; \qquad T(x_{A4}, h_s-h_C)=T_m \tag{11.3.19}$$

$$\left(\alpha_m \frac{\partial T}{\partial x_{A3}}\right)_{x_{A3}=h_s-h_C+} - \left(\alpha_s \frac{\partial T}{\partial x_{A3}}\right)_{x_{A3}=h_s-h_C-} = \rho_s \xi_s v_{A3} \quad (11.3.20)$$

as boundary conditions and

$$\frac{\partial}{\partial x_{A3}}\left(\eta \frac{\partial v_{C\,rel\,4}}{\partial x_{A3}}\right) = 0 \quad (11.3.21)$$

$$\rho_m \Gamma_m \left(v_{C\,rel\,4} \frac{\partial T}{\partial x_{A4}} + v_{C\,rel\,3} \frac{\partial T}{\partial x_{A3}}\right) = \alpha_m \frac{\partial^2 T}{\partial x_{A3}^2} + \eta \left(\frac{\partial v_{C\,rel\,4}}{\partial x_{A3}}\right)^2 \quad (11.3.22)$$

$$\frac{\partial v_{C\,rel\,4}}{\partial x_{A4}} + \frac{\partial v_{C\,rel\,3}}{\partial x_{A3}} = 0 \quad (11.3.23)$$

as defining equations if we assume (11.2.28) to be relevant, i.e. pressure gradients to be negligible. A first integral of (11.3.23) using the last of (11.3.18) gives

$$\frac{d\dot{m}_{C4}}{dx_{A4}} = \rho_m \frac{d}{dx_{A4}} \int_{h_s-h_C}^{h_s} v_{C\,rel\,4}\, dx_{A3} = \rho_s v_{A3} \quad (11.3.24)$$

We take (11.2.25) to be relevant.

This is the problem considered in Pearson (1976*a*; see also Griffin, 1977) where many special cases are worked out. The similarity variables defined by (11.2.34), (11.2.42)–(11.2.45) can be used in somewhat modified form. We now have a relevant length scale in the x_{A4} direction, i.e. the width of the swept layer,

$$l_C^* = w_A \operatorname{cosec}(\chi_b + \phi_b) \quad (11.3.25)$$

where χ_b and ϕ_b are as defined in Figs. 11.6 and 11.13. And so from (11.2.37)

$$h_C^* = |\alpha_m T^* w_A / \sin(\chi_b + \phi_b) \rho_s \xi_s^* v_{rel}|^{\frac{1}{2}} \quad (11.3.26)$$

As before, we write

$$\left.\begin{aligned} \tilde{h}_C = h_C/h_C^*, \quad \tilde{x}_{C4} = x_{A4}/l_C^*, \quad \hat{x}_{C3} = (h_s - x_{A3})/h_C \\ \tilde{T}_C = (T - T_m)/(T_b - T_m), \quad \tilde{v}_{C4} = v_{C\,rel\,4}/v_{rel}, \\ \tilde{v}_{C3} = -v_{C\,rel\,3} l^*/v_{rel} h^* \end{aligned}\right\} \quad (11.3.27)$$

$$\mathrm{B}_C = \zeta_m (T_b - T_m) = \zeta_m T_C^* \quad (11.3.28)$$

$$\mathrm{Sf}_C = \rho_m \Gamma_m (T_b - T_m)/\rho_s \xi_s^* \quad (11.3.29)$$

$$\mathrm{Br}_C = K_m v_{rel}^{\nu+1} h_C^{*1-\nu} / T_C^* \alpha_m \quad (11.3.30)$$

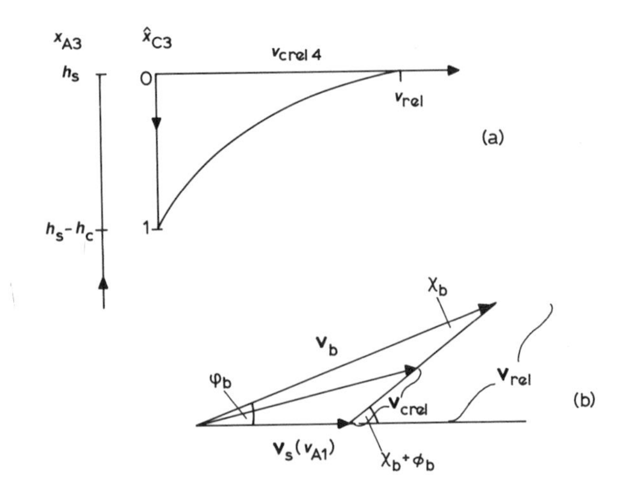

Fig. 11.13. (a) Velocity profile in region C showing $x\,A_3$ and $\hat{x}C_3$ coordinates. (b) Velocity vectors: $\mathbf{v}_b$, barrel relative to screw; $\mathbf{v}_s$, solid bed relative to screw; $\mathbf{v}_{rel}$, barrel relative to solid bed—in x_{A4} direction. Diagram also shows the angles χ_b and ϕ_b.

to give dimensionless equations

$$\frac{\partial \tilde{v}_{C4}}{\partial \tilde{x}_{C4}} - \frac{\tilde{h}'_C}{\tilde{h}_C}\hat{x}_{C3}\frac{\partial \tilde{v}_{C4}}{\partial \hat{x}_{C3}} + \frac{1}{\tilde{h}_C}\frac{\partial \tilde{v}_{C3}}{\partial \hat{x}_{C3}} = 0 \tag{11.3.31}$$

$$\frac{\partial}{\partial \hat{x}_{C3}}\left\{ e^{-B_C\tilde{T}_C}\left|\frac{\partial \tilde{v}_{C4}}{\partial \hat{x}_{C3}}\right|^{\nu-1}\frac{\partial \tilde{v}_{C4}}{\partial \hat{x}_{C3}}\right\} = 0 \tag{11.3.32}$$

$$\mathrm{Sf}_C\left(\tilde{v}_{C4}\frac{\partial \tilde{T}_C}{\partial \tilde{x}_{C4}} - \frac{\tilde{h}'_C}{\tilde{h}_C}\tilde{v}_{C4}\hat{x}_{C3}\frac{\partial \tilde{T}_C}{\partial \hat{x}_{C3}} + \frac{\tilde{v}_{C3}}{\tilde{h}_C}\frac{\partial \tilde{T}_C}{\partial \hat{x}_{C3}}\right)$$
$$= \frac{1}{\tilde{h}_C^2}\frac{\partial^2 \tilde{T}_C}{\partial \hat{x}_{C3}} + \frac{\mathrm{Br}_C}{\tilde{h}_C^{\nu+1}}\left|\frac{\partial \tilde{v}_{C4}}{\partial \hat{x}_{C3}}\right|^{\nu+1} e^{-B_C\tilde{T}_C} \tag{11.3.33}$$

$$\frac{d}{d\tilde{x}_{C4}}\int_0^1 \tilde{v}_{C4}\,d\hat{x}_{C3} = -\frac{\rho_s \tilde{v}_{A3}}{\rho_m \tilde{h}} \tag{11.3.34}$$

and dimensionless boundary conditions

$$\left.\begin{array}{llll} \hat{x}_{C3}=0: & \tilde{v}_{C4}=1, & \tilde{v}_{C3}=0, & \tilde{T}_C=1 \\ \hat{x}_{C3}=1: & \tilde{v}_{C4}=0, & \dfrac{\partial \tilde{T}_C}{\partial \hat{x}_{C3}}=\tilde{h}_c\tilde{v}_{A3}, & \tilde{T}_C=0 \end{array}\right\} \tag{11.3.35}$$

The intial condition at $\tilde{x}_{C4}=0$ will, however, be given by the flow over the flight, which provides

$$\dot{m}_{F2}=\dot{m}_{C2}(x_{A4}=0)=\dot{m}_{C4}(x_{A4}=0)\sin(\phi_b+\chi_b) \qquad (11.3.36)$$

where $\dot{m}_{C4}$ is defined as the integral in (11.3.24), and the temperature field T_C as a function of the stream function q_C, defined by

$$\frac{\partial q_C}{\partial x_{A3}}=v_{C\,rel\,4}, \qquad q_C(h_s-h_C)=0 \qquad (11.3.37)$$

It follows that

$$\rho_m q_C(h_s)=\dot{m}_{C4} \qquad (11.3.38)$$

The initial conditions lead to an initial value $\tilde{h}_C(0)$ which in the form

$$h_{C0}=h_C^*\tilde{h}_C(0) \qquad (11.3.39)$$

could have been used as the basic length scale in the x_{A3} direction, giving

$$l_{C0}=\rho_s\xi_s^*V_{rel}h_{C0}^2/\alpha_m T_C^* \qquad (11.3.40)$$

as the length scale in the x_{A4} direction.

A step-by-step solution of the parabolic set of differential equations (11.3.31)–(11.3.33) yields $\dot{m}_{C4}(x_{A4})$ which defines, in the frame of reference fixed in the screw,

$$\dot{m}_{C2}(x_{A2})=\dot{m}_{C4}\{x_{A4}\sin(\phi_b+\chi_b)\}\sin(\phi_b+\chi_b) \qquad (11.3.41)$$

$$\dot{m}_{C1}(x_{A2})=\rho_m h_C\{x_{A4}\sin(\phi_b+\chi_b)\}v_{A1} + \dot{m}_{C4}\{x_{A4}\sin(\phi_b+\chi_b)\}\cos(\phi_b+\chi_b) \qquad (11.3.42)$$

Integration yields

$$\dot{M}_{C1}=\int_0^{w_A}\dot{m}_{C1}\,dx_{A2} \qquad (11.3.43)$$

while we can write

$$F_{AC1}=\frac{1}{2}\int_0^{l_C^*}\eta\frac{\partial v_{C\,rel\,4}}{\partial x_{A3}}\,dx_{A4}\sin 2(\phi_b+\chi_b) \qquad (11.3.44)$$

$$\bar{h}_C=\frac{1}{l^*}\int_0^{l_C^*}h_C\,dx_{A4} \qquad (11.3.45)$$

and

$$\dot{m}_{AC} = \dot{m}_{C2}(w_A) - \dot{m}_{F2} \tag{11.3.46}$$

This provides all the quantities we need from region C for the overall mass and force balance equations.

Regions D and E. As mentioned earlier, these are merely continuations of the viscous lubricating layers considered in connection with feed zone behaviour. If T_s is constant over the screw, then in our approximate solution

$$h_D = h_E, \qquad q_{D1} = q_{E1}, \quad \text{and} \quad t_{D13} = t_{E12} \tag{11.3.47}$$

everywhere, where $q_{D1} = q_{E1}$ are the volume flow rates (in the x_1-direction)/unit width in regions D and E given by (11.2.33), t_{D13} is the shear stress at the AD interface and t_{E12} the shear stress at the AE interface, both given by (11.2.59). Thus regions D and E can be treated together. In the solution discussed for the feed zone, v_s, which now becomes v_{A1}, was treated as constant. For the similarity solution (11.2.46) to be valid in the full melting region, we require that the development length for v_{A1}, namely

$$l_A = \dot{M}_{A1}/(\mathrm{d}\dot{M}_{A1}/\mathrm{d}x_1) \tag{11.3.48}$$

be large enough to make the Graetz number based on l_A and h_D smaller than Sf. This ensures (see Example 11.3.2) that even a full convective solution is locally applicable, with $\bar{h} = h_D$. If this is not the case, then eqns (11.2.22)–(11.2.24) have to be solved with v_{A1} a slowly-varying function of x_1 in boundary conditions (11.2.19). No difficulty in principle is involved.

From these solutions we can derive approximations for $\dot{M}_{D1}$, $\dot{M}_{E1}$, F_{AD1}, F_{AE1}, which are required for the overall mass and force balance equations. Providing

$$h_f, h_D \ll h_s \tag{11.3.49}$$

it is consistent to write

$$\dot{M}_{D1} = \rho_m w_A q_{D1} = w_A \dot{m}_{D1}, \qquad \dot{m}_{AD} = w_A\, \mathrm{d}\dot{m}_{D1}/\mathrm{d}x_1 \tag{11.3.50}$$

$$\dot{M}_{E1} = \rho_m h_s q_{E1} = h_s \dot{m}_{E1}, \qquad \dot{m}_{AE} = h_s\, \mathrm{d}\dot{m}_{E1}/\mathrm{d}x_1 \tag{11.3.51}$$

$$F_{AD1} = w_A t_{D31} \tag{11.3.52}$$

$$F_{AE1} = h_s t_{E21} \tag{11.3.53}$$

This completes the terms we need for the overall equations.

11.3.2 Step-by-step and Iterative Solution of the Six-zone Model

The sum total of relations developed above constitute a highly non-linear system of equations that can only be tackled conveniently by digital computation. In practice, a step-by-step process based on a suitable increment Δx_1 in the x_1 coordinate is adopted. At any given value of x_1, we suppose $\dot{M}_{A1}$ and $\dot{m}_{D1} = \dot{m}_{E1}$ to be prescribed, having been calculated from their values at the previous value $x_1 - \Delta x_1$, using (11.3.5), (11.3.50) and (11.3.51). The dependent variables are v_{A1}, w_A and dp/dx_1, which are to be calculated from the three equations (11.3.2), (11.3.6) and (11.3.7). The F_{AK1}'s that appear in (11.3.6) are dependent upon v_{A1} and w_A but not upon dp/dx_1, while $\dot{V}_{B1}$ in (11.3.7), which is prescribed via (11.3.1) is very weakly dependent upon v_{A1} and w_A. Thus if a reasonable guess for v_{A1} is carried forward, based on its values at $x_1 - 2\Delta x_1$ and $x_1 - \Delta x_1$, then the x_1 value for $\dot{M}_{A1}$ yields, using the $x_1 - \Delta x_1$ value for h_D and $\bar{h}_C$, a good approximation for w_A. Equation (11.3.14) yields $\dot{m}_{F2}$ and (11.3.16) the distribution of T with q_F. Turning to region C, using the initial conditions $\dot{m}_{F2}$ and $T(q_F)$ and the estimated first approximations for v_{A1} and w_A, we obtain first estimates for $\dot{M}_{C1}$, F_{AC1} and $\bar{h}_C$ from (11.3.43)–(11.3.45). Equations (11.3.50)–(11.3.53) provide values for $\dot{M}_{D1}$, $\dot{M}_{E1}$, F_{AD1} and F_{AE1}. h_D ($= h_E$) is similarly obtained. Having good estimates for $\dot{M}_{K1}$ gives us a reliable value for $\dot{V}_{B1}$. We suppose F_{AB1} to be approximated at this stage by its value at $x_1 - \Delta x_1$. We now use all these estimates in (11.3.2), (11.3.6) and (11.3.7) to derive, by iteration around the three equations, better estimates for w_A, v_{A1} and a value for dp/dx_1. If v_{A1}, w_A are significantly different from their guessed values then F_{AC1}, F_{AD1} and F_{AE1} can be recalculated and the three equations (11.3.2), (11.3.6) and (11.3.7) used iteratively again to provide still better estimates for w_A, v_{A1}, and dp/dx_1. Once these are seen to have converged, (11.3.46), (11.3.50) and (11.3.51) can be used to provide values for $\dot{m}_{AC}$, $\dot{m}_{AD}$ and $\dot{m}_{AE}$ and hence new values for $\dot{M}_{A1}$, $\dot{m}_{D1}$ and $\dot{m}_{E1}$ at $x_1 + \Delta x_1$. At every stage the relevant geometrical variables, especially h_s, and operating temperatures T_b and T_s, entered earlier into the program, are used. This is shown in Fig. 11.14. Initiating the process at x_{10} and carrying out the first few steps presents some problems. When the end of the lubricated plug flow in the feed zone is reached,

$$w_B = h_B \; (= \bar{\bar{h}}_{screw} \quad \text{from Subsection 11.2.3}) \qquad (11.3.54)$$

while $\dot{M}_{C1}$ is provided by a uniform layer of thickness $\bar{h}_b$. Although v_s

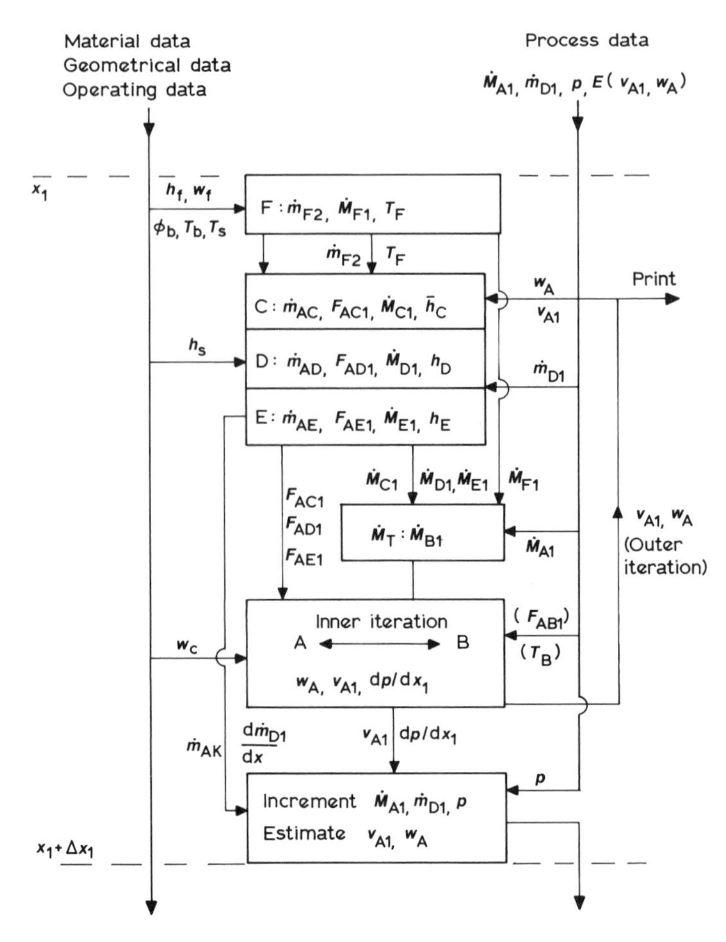

Fig. 11.14. Flow diagram for step-by-step computation of six-zone model. E( ) indicates estimated value. (:) prefaces quantities to be calculated by routine.

could have been adjusted to take into account the effect of x_1-varying mass flux in the lubricating layers, the correction is negligible in the pre-melt-pool zone. What is obvious is that $\dot{m}_{AB}$ and F_{AB1} will be important in the early stages of the formation of the melt pool and so although $\dot{M}_{B1}$ will be very small, a sudden relative change in $\dot{M}_{B1}$ of order one which occurs if $\dot{M}_{C1}$ is calculated according to a constant layer thickness h_f will lead to a significant discontinuity in dp/dz unless $h_s/w_c \ll 1$.

This can be patched up by making the criterion for the change-over the relation

$$\bar{h}_b = \bar{h}_C(w_A = w_c, v_{A1} = v_s) \qquad (11.3.55)$$

or even continuity of $\dot{M}_{C1}$.

The importance of $\dot{m}_{AB}$ and F_{AB1} at the early stages of melting brings up again the question of flow in region B. While w_B is small, eqn (11.3.7) is best represented by a lubrication formula such as those used in regions D and E, although the temperature of the material joining region B from the flux $\dot{m}_{C2}-\dot{m}_{F2}$ will be significantly different (usually higher) in mean from that generated by $\dot{m}_{AB}$. As $\dot{M}_{B1}$ increases so will w_B, the effective Graetz number Gz_B based on w_B and l_A will increase, and the coupling between eqns (11.3.2), (11.3.6) and (11.3.7) will increase through dp/dx_1 leading to more rapid variations in v_{A1} and w_A with x_1. The terms $v_{B1}\,\partial T_B/\partial x_1$ in the energy equation for region B will be important if Na_B is significant, and so (11.3.10) should really be written as a functional equation

$$\frac{\partial p}{\partial x_1} = \mathcal{P}^{x_1}_{x=x_{10}}(\dot{V}_{B1}, v_{b1}, v_{b2}, v_{A1}, w_B, h_s, T_B, T_s; T_m) \qquad (11.3.56)$$

Calculation routines based on the six-zone or similar models are given and discussed extensively in Tadmor & Klein (1970), Fenner & Edmondson (1975) and in Shapiro *et al.* (1976), Halmos *et al.* (1978). The skills required to get rapidly converging routines are to a large extent those of the numerical analyst. Ths most recent calculations that have been carefully compared with experiment are those of Cox & Fenner (1980); these cover LDPE, PS and PP. A variety of simple solutions are given in Ingen-Housz & Meijer (1981).

11.3.3 The Three-layer Model

Dekker (1976) and others have observed a melting mechanism which is noticeably different from the melt pool model in that the solid bed remains almost as wide as the screw channel and is separated from the screw and barrel by relatively thick layers of melt whose thicknesses are largely independent of the cross channel coordinates x_2, and which develop slowly in a downstream x_1-direction (see Fig. 11.15). As explained in connection with the melt pool model, the analysis to be given here will be based on kinematical assumptions that are taken over directly from observation. It cannot be argued conclusively at this

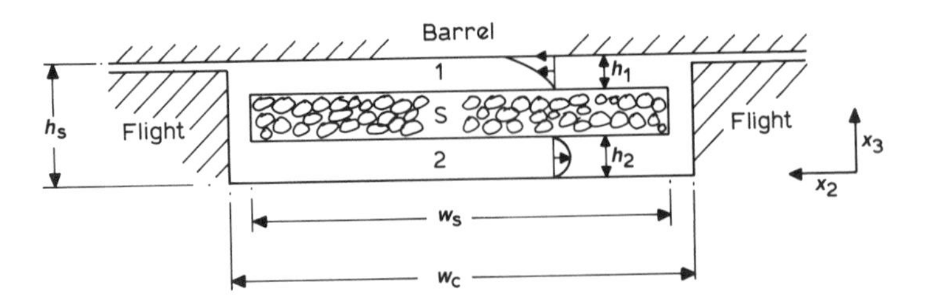

Fig. 11.15. The three regions of the three-layer model. S, solid bed; 1, upper melt layer; 2, lower melt layer.

stage why or when either model becomes relevant, although Lindt (1976) has suggested that the determining factor is the force balance in the x_2-direction (he used the 'unrolled' approximate geometry shown in Fig. 11.15). The primary difference between the three-layer and the six-zone model is that the melting mechanism is not determined by the flow over the flight in the former in the way that it is in the latter. The molten material that accumulates in the melt pool in the six-zone model distributes itself instead between the layers 1 and 2 above and below the advancing solid bed.

A general theory, based on approximations very similar to those used for the melt pool model, can be given for the three-layer model. If we take values from Lindt (1976), namely $v_{b2}=0{\cdot}085\ \mathrm{m\,s^{-1}}$, $h_1=0{\cdot}5\text{–}1{\cdot}0$ mm, $h_f=0{\cdot}2$ mm, $w_c=77$ mm, $\rho_m=720\ \mathrm{kg\,m^{-3}}$, $\Gamma_m=2{\cdot}3\times10^3\ \mathrm{J\,kg\,K}$, $\alpha_m=0{\cdot}25\ \mathrm{W\,mK^{-1}}$, $\zeta_m=0{\cdot}009\ \mathrm{K^{-1}}$, $\xi_s^*\rho_s=0{\cdot}35\times10^6\ \mathrm{J\,kg^{-1}}$, $T_b-T_m=80$ K, $K_0=5{\cdot}6\times10^3\ \mathrm{N\,s\,m^{-2}}$, $\nu=0{\cdot}372$, then

$$\mathrm{Gz}_1\sim2\text{–}8 \tag{11.3.57}$$

$$\mathrm{Na}_1\sim0{\cdot}1\text{–}0{\cdot}4\sim\mathrm{Br}_1 \tag{11.3.58}$$

$$\mathrm{Sf}_1\sim\tfrac{1}{2} \tag{11.3.59}$$

This suggests that thermal effects will be small as far as viscosity is concerned, though neither conductive nor convective effects will be dominant in the cross-channel direction. It therefore becomes rather crucial to decide whether the cross-channel circulation is essentially a circulatory path involving one-way motion in each of layers 1 and 2 as is implied by Fig. 5(a) of Lindt (1976) when $h_f\ll h_1$, or is wholly within layer 1 including the flight clearance, with layer 2 essentially stagnant, as is implied by the mathematical model employed by Lindt. (Lindt specifically separates the down-channel flows in layers 1 and 2 whose magnitudes are separately determined in his model according to the melting at interfaces S1 and S2, which is only justifiable if there is virtually no exchange of melt between layers 1 and 2.) In the linked circulatory path, the streamlines would be largely closed (in the (x_2, x_3)-plane) and the effective Graetz number would be relatively small, thus ensuring that a fully-developed solution would be relevant, with streamlines treated as isotherms. (It should be noted that both the very large and the very small Graetz number situations involve streamlines as isotherms, a fact that will be made use of when providing models for metering zone flows.) The slow development due to melting at interfaces S1 and S2 will be considered part of the x_1-flow, while the x_2-flow will be treated as a virtually decoupled circulatory flow driven by the moving barrel surface between four parallel surfaces at $x_3 = 0$, h_2 and $x_3 = h_s - h_1$, h_s.

Region S. Using nomenclature analogous to that used for the six-zone model we can write

$$\dot{M}_{S1} = v_{S1} w_s (h_s - h_1 - h_2) \tag{11.3.60}$$

$$\frac{d\dot{M}_{S1}}{dx_1} = -(\dot{m}_{S1} + \dot{m}_{S2}) \tag{11.3.61}$$

$$(h_s - h_1 - h_2) w_s \frac{\partial p}{\partial x_1} = F_{S11} + F_{S21} \tag{11.3.62}$$

$$(h_s - h_1 - h_2) w_s \frac{\partial p}{\partial x_2} = F_{S12} + F_{S22} \tag{11.3.63}$$

Here v_{S1} is the x_1-component of the solid bed taken to be independent of x_2 and x_3, w_s is the width of the bed, $\dot{M}_{S1}$ is the mass flowrate in the solid bed, F_{AKj} (K = 1, 2; j = 1, 2) is the force on the solid bed/unit downstream distance in the j-direction caused by the viscous flow in the (Kth) layer, $(h_s - h_1 - h_2)$ is the thickness of the solid bed, and $\dot{m}_{S1}$ and $\dot{m}_{S2}$ are the rates of melting at the S_1 and S_2 interfaces. Equation (11.3.63) represents a force balance in the 2-direction. (11.3.62) and (11.3.63) are consistent with

$$p = p_0 + p_1(x_1) + p_2(x_2) \tag{11.3.64}$$

where

$$p_0 = p(0, \tfrac{1}{2} w_c, \tfrac{1}{2} h_s) \tag{11.3.65}$$

and dp_1/dx_1, dp_2/dx_2 are slowly-varying functions of x_1 only. Again as before, we can consider two extreme approximations:

(1) Rigid solid bed (cf. (11.2.27))

$$v_{S1} = v_s, \quad \text{constant} \tag{11.3.66}$$

(2) Deformable solid bed, which in this case can be equated to continuity of p across the S_1 and S_2 interfaces. We shall not neglect $\partial p/\partial x_1$ and $\partial p/\partial x_2$ in the equations for layers 1 and 2 because h_1 and h_2 are of the same order as h_s. It can be argued that the same approximation (continuity of p across interface AB) is really what is used in the six-zone model, and that the drag flow approximation in regions C, D and E is a natural consequence of $h_{C,D,E} \ll h_s$.

Lindt uses (11.3.66); the equations relevant for approximation (2) are given here.

Regions 1 *and* 2. The equations of motion in these regions can be written ($K = 1, 2$)

$$\frac{\partial}{\partial x_3}\left(\eta \frac{\partial v_{K1}}{\partial x_3}\right) = \frac{\partial p_1}{\partial x_1} \tag{11.3.67}$$

$$\frac{\partial}{\partial x_3}\left(\eta \frac{\partial v_{K2}}{\partial x_3}\right) = \frac{\partial p_2}{\partial x_2} \tag{11.3.68}$$

where

$$\eta = K_m D^{\nu-1} e^{\zeta_m(T_m - T)} \tag{11.3.69}$$

$$D = \left\{\left(\frac{\partial v_{K1}}{\partial x_3}\right)^2 + \left(\frac{\partial v_{K2}}{\partial x_3}\right)^2\right\}^{\frac{1}{2}} \tag{11.3.70}$$

with boundary conditions

$$x_3 = 0\colon v_{21} = v_{22} = 0 \tag{11.3.71}$$

$$x_3 = h_2\colon v_{21} = v_{S1}, \qquad v_{22} = 0 \tag{11.3.72}$$

$$x_3 = h_s - h_1\colon v_{11} = v_{S1}, \qquad v_{12} = 0 \tag{11.3.73}$$

$$x_3 = h_s\colon v_{11} = v_{b1}, \qquad v_{12} = v_{b2} \tag{11.3.74}$$

where $\mathbf{v}_K = \mathbf{v}_K\,(x_3)$ are slowly-varying functions of x_1. If we write the mass flux through the flight $\dot{m}_{F2}$ as before, and for $j = 1, 2$,

$$\dot{m}_{1j} = \rho_m \int_{h_s - h_1}^{h_s} v_{1j}(x_3)\, dx_3 \tag{11.3.75}$$

$$\dot{m}_{2j} = \rho_m \int_0^{h_2} v_{2j}(x_3)\, dx_3 \tag{11.3.76}$$

then

$$\dot{m}_{F2} = \dot{m}_{12} + \dot{m}_{22} \tag{11.3.77}$$

is the integrated mass balance equation in the 2-direction. The overall mass balance equation in the 1-direction is given by

$$\dot{M}_{S1} + (\dot{m}_{11}w_1 + \dot{m}_{21}w_2) = \dot{M}_T \tag{11.3.78}$$

where w_s, w_1 and w_2 can all be approximated by w_c for simplicity, assuming $h_s \ll w_c$.

Finally, the energy equations can be written as

$$\rho_m \Gamma_m \left(v_{K1} \frac{\partial T_K}{\partial x_1} + v_{K2} \frac{\partial T_K}{\partial x_2} \right) = \alpha_m \frac{\partial^2 T_K}{\partial x_3^2} + \eta D^2 \tag{11.3.79}$$

with

$$T_1(h_s - h_1) = T(h_2) = T_m; \qquad T_1(h_s) = T_{b1}, \qquad T_2(0) = T_s \tag{11.3.80}$$

and

$$\alpha_m \left[-\left(\frac{\partial T_1}{\partial x_3} \right)_{x_3 = h_s - h_1} + \left(\frac{\partial T_2}{\partial x_3} \right)_{x_3 = h_2} \right] = -\rho_m \xi_s^* \frac{\partial}{\partial x_1} (\dot{m}_{11} + \dot{m}_{21}) \tag{11.3.81}$$

where $w_1 = w_2 = w_s$ has been assumed, and individual boundary conditions of the form (11.2.31) are implied.

The solution to the pair of energy equations (11.3.79) subject to (11.3.80) yield $\dfrac{d}{dx_1}(\dot{m}_{11} + \dot{m}_{21})$ at any station x_1 whence $\dot{M}_{S1}$, $\dot{m}_{11} + \dot{m}_{21}$ are prescribed at the next station $x_1 + \Delta x_1$. The dependent variables then to be determined are h_1, h_2, v_{S1}, dp_1/dx_1 and dp_2/dx_2. There are five independent equations to be satisfied, namely (11.3.60), (11.3.62), (11.3.63), (11.3.77) and one expressing $\dot{m}_{11} + \dot{m}_{21}$ from (11.3.78) in terms of (11.3.75) and (11.3.76), assuming that (11.3.67) and (11.3.68) have been used to evaluate F_{SKj} and $\dot{m}_{Kj}$ in terms of the dependent variables. An iterative solution can obviously be obtained as described in more detail for the more complex six-zone model provided a suitable approximation scheme can be obtained for (11.3.79) based on the solutions given in Section 11.4.

11.3.4 Comparison of Various Models for the Melting Process

Despite the elaborate structure of the models described above for the melting process, it is by no means clear whether a truly satisfactory predictive model exists. All of those described rely in the first place on observation-based kinematic assumptions and so confidence in them can likewise only come from comparison with experiment. This has proved difficult to arrange except insofar as overall extruder response is concerned, which involves feed-zone, metering-zone and die-zone performance also. It is relatively easy to measure flowrate and mean output temperature as a function of screw speed for a given extruder, given homogeneous feed and set barrel temperatures. It is not difficult to measure screw temperature, and normal stresses at metal/melt interfaces, which latter are essentially equivalent to pressure in the lubrication approximation. It is less easy to measure 'pressures' at metal/solid polymer interfaces, and very difficult to measure temperature profiles in the flowing melt. No attempt has been made to measure velocity profiles. All experimental observations are subject to error.

Calculations that have been made have all involved very significant approximations to the model, usually going beyond those introduced above. One that has simplified many calculation schemes has involved either neglect or replacement of the convective terms in the energy equation for thin melt layers, thus making calculations for regions C, D and E, or 1 and 2, into purely local ones. Pearson (1976*a*) provides a heuristic argument for increasing the value of ξ_s^* by the amount $F\Gamma_m(\bar{T}-T_m)$ as an alternative to the convective terms, where $\bar{T}$ is a mean temperature given by

$$\bar{T}=\int_0^{\bar{h}} v_1 T\,\mathrm{d}x_3 \bigg/ \int_0^{\bar{h}} v_1\,\mathrm{d}x_3 \tag{11.3.82}$$

and

$$0<F<\tfrac{1}{2} \tag{11.3.83}$$

with $F=0{\cdot}3$ as a representative value.

Compressibility, melt permeability and frictional behaviour of the solid bed remain little understood, but important, variables.

Screw temperature has to be guessed or measured: its value is not an operating variable that can be prescribed in practice; it is determined by the process and should therefore be predicted by the model solutions. To do so would require coupling a heat transfer analysis within the screw to the energy equations in the melt layers near the screw flight and screw root. An order of magnitude argument is developed in Appendix 4 to show that, in typical extruders with a narrow flight clearance wetted with polymer melt, the screw temperature at any axial position z tends to follow the barrel temperature fairly closely. This proves, in any case, to be a very convenient approximation for calculation purposes; sensitivity of predictions to changes of order 10 K in screw surface temperature can be undertaken to investigate the likely effects of screw cooling.

It can be argued that the distinction between solid bed and melt is unnecessary for glassy polymers and that a formal solution for flow in the plasticating zone could be obtained by treating the whole flow as that of a continuum with properties (ρ, η, α, Γ) that change smoothly with temperature, even if the changes are rapid near the glass-transition temperature. However, even the largest of modern digital computers prove to be overtaxed by the requirements of solving the fully three-dimensional coupled sets of equations that ensue. On economic grounds, it is likely that most use will continue to be made of relatively simple models that exploit any analytical solutions that can be derived on the basis of linearity or weak coupling.

EXAMPLE 11.3.1. Derive (11.3.6) by considering a short length Δx_1 of the solid bed A.

The pressure p will contribute forces in the downstream (1) direction over the faces $x_1 = x_{1c}$, $x_1 = x_{1c} + \Delta x_1$ and AB between x_{1c} and $x_{1c} + \Delta x_1$. These will be respectively

$$(\bar{h}_A w_A p)_{x_{1c}}, -(\bar{h}_A w_A p)_{x_{1c}+\Delta x_1} \quad \text{and} \quad \left(\bar{h}_A p \frac{dw_A}{dx_1} \Delta x_1\right)_{x_{1c}}$$

The first two can be combined to yield

$$\bar{h}_A \frac{d}{dx_1}(w_A p)\Delta x_1$$

which is the differential force component usually employed. However, adding the small (third) contribution given by the slight angle between the face AB and the coordinate direction x_1 yields finally the right-hand side of (11.3.6).

This must obviously be balanced by the ΣF_{AK1}; note that the inclination of the AB surface yields a factor $[1-(dw_A/dx_1)^2]^{\frac{1}{2}}$ in F_{AB1}, which can be neglected.

EXAMPLE 11.3.2. Verify that Sf will be 0(1) but not large for most processing situations and that $\mathrm{Gz_D}$ will, in general, be larger than Sf. Explain the significance of $\mathrm{Gz_D}/\mathrm{Sf}$.

By definition, (11.2.44) and (11.2.32),

$$\mathrm{Sf} = \frac{\rho_m \Gamma_m T^*}{\rho_s \xi_s^*} = \frac{\rho_m \Gamma_m T^*}{\rho_s\{\xi_s + \Gamma_s(T_m - T_0)\}}$$

with (11.2.51) and (11.2.52) implying that $T^* = (T_s - T_m)$. Typical values give $\rho_s \sim \rho_m$, $\Gamma_s \sim \Gamma_m = 2\times10^3$ J/kg K, $T_m - T_0 \sim 100$ K, $T^* \sim 50$ K. ξ_s can vary from close to zero to 2×10^5 J kg^{-1}. Thus Sf will vary from about $\frac{1}{2}$ to $\frac{1}{4}$.

The development length l_A defined by (11.3.48) is of the order of one-half of the total helical length of the extruder. For $h_s = 15 \rightarrow 5$ mm assuming a compression ratio of 3, $d_b = 100$ mm, this means that l_A can be conservatively taken as 1 m. A reasonable value for v_{A1} is $0{\cdot}4$ m s^{-1}. The Graetz number

$$\mathrm{Gz_D} = \frac{v_{A1} h_D^2}{\kappa_m l_A} \quad \text{is therefore} \leqslant 4h_D^2$$

where h_D is expressed in mm.

The development length for the similarity solution given in Subsection 11.2.3 is l^* given by (11.2.37). This leads to an apparent $\mathrm{Gz}^* = \mathrm{Sf}$ from eqn (11.2.40). From the above, $\mathrm{Gz}^* \sim \frac{1}{3}$ is significantly less than the value (4) thought necessary in channel flow to ensure it being fully developed. Thus a criterion for the use of the similarity solution would be

$$l_A \gg 0{\cdot}1 l^*$$

or

$$\mathrm{Gz_D} \ll 10\,\mathrm{Sf} \sim 3$$

This requires h_D to be significantly less than 1 mm which will only be strictly true for the very early stages of melting; however, it is consistent with (11.3.49).

EXERCISE 11.3.1. Verify that eqns (11.3.10)–(11.3.12) correspond to the case of zero pressure gradient in both x_1 and x_2, and hence the x_5, directions.

Justify the neglect of the pressure difference across and along the flight tip in order-of-magnitude terms. (The argument will be similar to that given at the end of Subsection 11.2.2 leading to condition (11.2.54).)

EXERCISE 11.3.2. Include the downstream pressure gradient in eqns (11.3.21) and (11.3.32). From the dimensionless formulation, estimate the error involved in neglecting it. (This exercise continues Exercise 11.3.1.)

EXERCISE 11.3.3. Use eqns (11.2.46)–(11.2.52) and (11.2.54) with $\mathrm{B} = \mathrm{Gn} = 0$ to obtain an analytic solution for $\tilde{v}_1$, $\tilde{v}_3$ and $\tilde{T}$ in terms of A and Sf. Show that Sf obeys the transcendental equation

$$2\rho_s e^{-\mathrm{A\,Sf}/6} = \rho_m \mathrm{A} \int_0^1 \exp(-\mathrm{A\,Sf}\,x^3/6)\,\mathrm{d}x$$

whatever the value of ν, and that for small Sf, $\mathrm{A} \rightarrow 2\rho_s/\rho_m$. Develop the large-Sf result directly.

EXERCISE 11.3.4. Develop a similar solution for zone C to that given in Exercise 11.3.3 for zones D and E. Evaluate $\bar{h}_C$, and calculate F_{AC1}.

EXERCISE 11.3.5. Obtain an approximate form for (11.3.7) when $w_B \gg h_s$, $\mathrm{B} = \mathrm{Gn} = 0$, $\nu = 1$. (This will be simple channel flow of a constant viscosity fluid with one moving boundary and a pressure gradient.)

EXERCISE 11.3.6. Using the results of Exercises 11.3.3–11.3.5, product a set of algebraic equations that have to be solved to simulate the six-zone melting process.

EXERCISE 11.3.7. Put $\eta = \text{constant}$ and $\dot{m}_{F2} = 0$ into the equations for the three-layer model to obtain three algebraic equations for v_{S1}, h_2 and h_1 in terms of v_{b1}, $\dot{M}_{S1}$, $\dot{M}_T$ and h_s.

11.4 METERING ZONE: MELT PUMPING

Once the solid bed has been substantially melted, which normally takes place towards the end of, or just after, the transition zone, the screw channel is full of a homogeneous viscous melt. If the shallow metering section is relatively long, say 1/3 of the channel length, then the mechanics and heat transfer within this section will be major factors in determining the quantity and quality of melt leaving the extruder. An accurate analysis of this part of the flow field has been the objective of many authors.

The unrolled approximation implied by Figs. 11.3 and 11.10 will apply even more strongly in this zone than in the feed or melting zones, because h_s will be significantly smaller in most cases. Thus the aspect ratio

$$\mathrm{A} = w_c/h_s \tag{11.4.1}$$

will usually be greater than 5 and often greater than 10.† Furthermore, the ratio L/H where 'l_M' is measured along the screw channel in the metering zone will usually lie between 10^2 and 10^3 (cf. Table 11.1), so that the lubrication approximation will hold for most situations, while fully developed flow (in the x_1-direction) may be reached in the smaller diameter extruders. The most general description of the system we shall consider is given by:

$$\left.\begin{aligned} \text{Velocity field: } \mathbf{v}(\mathbf{x}) &= \{v_1(x_1, x_2, x_3), \\ &\quad\; v_2(x_1, x_2, x_3), \\ &\quad\; v_3(x_1, x_2, x_3)\} \\ \text{Pressure field: } p(\mathbf{x}) &= p(x_1, x_2, x_3) \\ \text{Temperature field: } T(\mathbf{x}) &= T(x_1, x_2, x_3) \end{aligned}\right\} \tag{11.4.2}$$

with boundary conditions

$$\left.\begin{aligned} &\text{On barrel:} \quad \mathbf{v} = \mathbf{v}_b, \text{ constant}; \; T = T_b(x_1) \\ &\text{On screw:} \quad \mathbf{v} = \mathbf{0}, \qquad T = T_s(x_1) \end{aligned}\right\} \tag{11.4.3}$$

Previous approaches have considered the more general condition

$$h_s(\partial T/\partial n) = \mathrm{Bi}\{T - T_s(x_1)\} \tag{11.4.4}$$

at the screw, with the normal direction at the screw surface facing into the fluid, and Bi a Biot number. However, T_s still has to be specified as

† Exceptions arise in some melt fed extruders and in older forms of rubber extruders, and as we shall see later, in twin screw extruders. A = 1 can be considered a lower limit in all cases.

a function of x_1 and Bi as a function of peripheral position in the (x_2, x_3) plane if this approach is to be realistic, so little has been gained over a specification of T_s in (11.4.3), which if necessary can be regarded as a function of x_3 on the screw flight surfaces, and x_2 on the screw root as well. Arguments given in Appendix 4 suggest that $\mathrm{Bi}=0$ is the best alternative to the last equation in (11.4.3).

Formally we can include the flight clearance within the channel, in which case we have to be careful either to suppose that $w_f \to \infty$ with q_{f2} and dp_f/dx_2 $(x_2 \to \infty)$ specified or to prescribe a periodicity condition with 'wavelength' $(w_c + w_f)$.

The quantities of most interest for overall analysis of the extruder will be

$$\dot{M}_{T1} = \rho_m Q_{T1} = \int_{\mathscr{A}}\int \rho_m v_1(x_1, x_2, x_3)\, d\mathscr{A} \tag{11.4.5}$$

$$\frac{\partial p}{\partial x_1}(x_1, \tfrac{1}{2}w_c, \tfrac{1}{2}h_s),\ q_{f2}$$

and

$$\bar{T}(x_1) = \int_{\mathscr{A}}\int T(x_1, x_2, x_3) v_1(x_1, x_2, x_3)\, d\mathscr{A}/Q_{T1} \tag{11.4.6}$$

where ρ_m has been taken as at most a function of x_1, and $\mathscr{A}$ is the cross-sectional area of the channel.

If the dependence of ρ_m on T or p treated as a function of (x_2, x_3) is to be included, then $\bar{T}$ loses much of its significance in a strict sense and should be replaced by $\bar{U}$.

11.4.1 The Shallow Channel Approximation: A ≫ 1

Locally we adopt the approximation provided by eqns (8.1.15)–(8.1.19) as far as solving for the velocity profile at any 'point' (x_1, x_2) is concerned. This assumes that we are sufficiently far from flight walls. The relevant form of energy equation that is to be used to prescribe $T(x_3)$ locally depends upon the values of the down-channel and cross-channel Graetz numbers

$$\mathrm{Gz}_{\mathrm{d.c.}} = v^*_{\mathrm{conv\ d.c.}} h_s^2/\kappa_m l_M \tag{11.4.7}$$

and

$$\mathrm{Gz}_{\mathrm{c.c.}} = v^*_{\mathrm{conv\ c.c.}} h_s^2/\kappa_m w_c \tag{11.4.8}$$

whose ratio

$$\mathrm{Gz}_{\mathrm{d.c.}}/\mathrm{Gz}_{\mathrm{c.c.}} = \mathrm{A(L/H)}^{-1} \qquad (11.4.9)$$

can, according to the values mentioned after (11.4.1) above, be as small as 0·07. In practice, $\mathrm{Gz}_{\mathrm{c.c.}}$ will always be large while $\mathrm{Gz}_{\mathrm{d.c.}}$ can range from small to large, usually being an increasing function of extruder size.

If T_b, T_s or h_s vary with x_1, within the melt pumping zone, then l_M should be taken as the least of the development lengths

$$l_Z = Z/(\mathrm{d}Z/\mathrm{d}x_1) \qquad (11.4.10)$$

where Z stands for T_b, T_s or h_s. In practice, these will rarely be smaller than l_M, and so the inequality criterion (8.1.42) will usually not be satisfied provided $\mathrm{Gz}_{\mathrm{d.c.}}$ is small.

$\mathrm{Gz}_{\mathrm{d.c.}}$ *small.* The flow can be regarded as fully developed in the x_1-direction and so the temperature field will be given by

$$\rho_m \Gamma_m v_2 \frac{\partial T}{\partial x_2} = \frac{\partial}{\partial x_3}\left(\alpha_m \frac{\partial T}{\partial x_3}\right) + \eta(D, T)D^2 \qquad (11.4.11)$$

instead of by (8.1.21). An important feature of this equation is that v_2 changes sign within the flow channel, as shown in Fig. 11.16(a), and so (11.4.11) is not a parabolic partial differential equation in the normal sense. Two-point boundary conditions have to be provided for T (at $x_2 = 0$ and $x_2 = w_c$ respectively) depending on whether v_2 is >0 or <0 at any given value of x_3. If account is taken of the essentially recirculatory nature of the cross-channel (v_2) flow, then the calculated exit

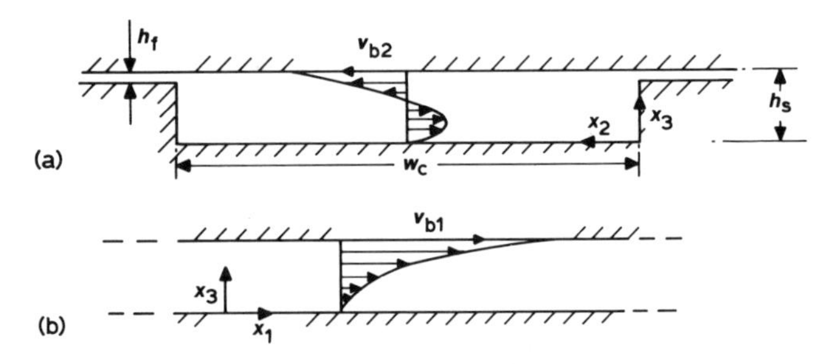

Fig. 11.16. (a) Cross-channel geometry and velocity profile $v_2(x_3)$. (b) Down channel velocity profile $v_1(x_3)$.

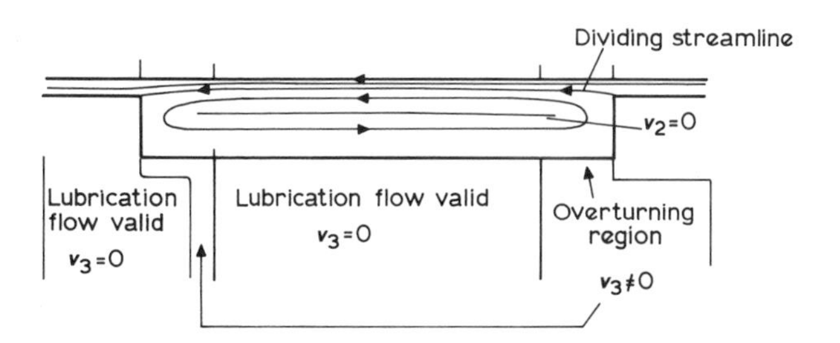

Fig. 11.17. Cross-channel streamline pattern for large A. For $Gz_{c.c.} \gg 1$, streamlines are isotherms, the dividing streamline having temperature T_s.

temperatures for those x_3 that make $v_2 > 0$ at $x_2 = w_c$ will provide, in some sense, the entry (imposed boundary condition) temperatures for those x_3 that make $v_2 < 0$ at $x_2 = w_c$—and vice versa at $x_2 = 0$. This is shown in Fig. 11.17.

Quasi-adiabatic cross-flow. If $Gz_{c.c.}$ is large, then we can suppose the cross-channel streamlines to be essentially isotherms,† except those that pass close to the metal boundaries. An asymptotic approximation is obtained by considering the temperature to be given by

$$T = T(q_2) \tag{11.4.12}$$

where

$$q_2 = \int_0^{x_3} v_2(x_3)\, dx_3 \tag{11.4.13}$$

Figure 11.18 shows the cross-channel stream function q_2, whence it is clear that x_3 is a double-valued function of q_2 for all $q_{2\min} < q_2 < 0$. In the (doubly) asymptotic limit of $A \to \infty$, $Gz_{c.c.} \to \infty$, we suppose that the velocity profile $v_2(x_3)$ shown in Fig. 11.16(a) will apply for the whole channel width w_c. Thus (11.4.11) can be integrated over $0 \leqslant x_2 \leqslant$

† The first author to make this point clear was Griffith (1962) who provided numerical solutions to the problem. However, his eqn (7) differs from eqn (11.4.14) given below, by a factor v_2^{-1}, which should be included when the averaging procedure he proposed is carried out. This has very considerable effect on the nature of the solution and introduces singularities that he appeared to avoid. The formulation given here was first discussed in Pearson (1972).

w_c to give, assuming ρ_m, Γ_m and α_m are constant,

$$-\alpha_m\left\{\left(\frac{1}{v_2}\frac{d^2T}{dx_3^2}\right)_{x_3^+(q_2)}-\left(\frac{1}{v_2}\frac{d^2T}{dx_3^2}\right)_{x_3^-(q_2)}\right\}=\left(\eta\frac{D^2}{v_2}\right)_{x_3^+(q_2)}-\left(\eta\frac{D^2}{v_2}\right)_{x_3^-(q_2)} \tag{11.4.14}$$

where

$$D^2=\left(\frac{\partial v_2}{\partial x_3}\right)^2+\left(\frac{\partial v_1}{\partial x_3}\right)^2, \qquad \eta=\eta(D, T) \tag{11.4.15}$$

and $x_3^+(q_2)$ and $x_3^-(q_2)$ refer to those values of x_3 corresponding to some given negative value of q for which v_2 is positive and negative respectively. This is illustrated in Fig. 11.18. For the streamlines for which $0<q<q_{2\,\max}$, the integration would have to be taken over $-w_f \leqslant x_2 \leqslant w_c$ to give

$$-\alpha_m\left\{\left(\frac{w_c}{v_2}\frac{d^2T}{dx_3^2}\right)_{x_{3c}(q_2)}+\left(\frac{w_f}{v_2}\frac{d^2T}{dx_3^2}\right)_{x_{3f}(q_2)}\right\}=w_c\left(\eta\frac{D^2}{v_2}\right)_{x_{3c}(q_2)}+w_f\left(\eta\frac{D^2}{v_2}\right)_{x_{3f}(q_2)} \tag{11.4.16}$$

where now x_{3c} and x_{3f} refer to the values of x_3 corresponding to any given positive value of q in screw channel and flight clearance respectively. Clearly both (11.4.14) and (11.4.16) will fail when $v_2=0$, and so the streamline $q_2=0$ will have to be treated differently.

The failure of the isothermal streamline approximation at or near no-slip bounding surfaces can be understood in terms of the local

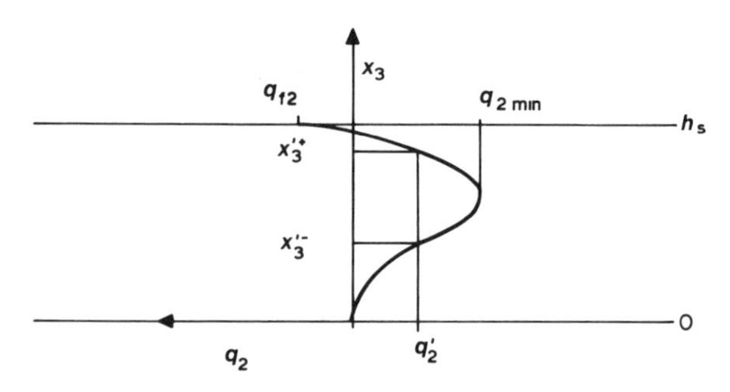

Fig. 11.18. Stream function for profiles shown in Fig. 11.16(a). The values $x_3'^+$ and $x_3'^-$ refer to the values of x for which $q_2=q_2'$.

Graetz number near the screw root,

$$\mathrm{Gz}(x_2) = v_2(x_3)x_3^2/\kappa_m x_2 \tag{11.4.17}$$

which will be small for sufficiently small x_3 for any given x_2. Thus temperature development will arise along any streamline close enough to a fixed boundary provided some change of thermal boundary condition arises at some point along the bounding streamline. In the case we are considering this will be at the points of attachment and detachment, namely the leading and trailing edges of the screw flight ($x_3 = h_s - h_f$). If we had taken the extreme case of $h_f = 0$, then such a change would have arisen at $x_3 = h_s$ provided $T_b \neq T_s$.

Singular behaviour at the streamline $q_2 = 0$ can be included in the asymptotic model by allowing T_s, $T(q_2 = 0-)$, and $T(q_2 = 0+)$ if it is relevant, to have different values, with the requirement that the heat fluxes calculated for $q_2 = 0+$, $q_2 = 0-$ within the fluid, on the basis of $T = T(q_2)$, should be balanced by the heat fluxes across the surfaces of temperature discontinuity; the latter fluxes have to be calculated on the basis of inner approximations, for which the local Graetz number is effectively of order unity. Thus, in our case, we can consider 3 inner approximations:

Flight wall $\quad x_3 = h_s - h_f; \quad 0 > x_2 > -w_f$

For flow in the flight clearance, the relevant dimensionless group is

$$\mathrm{Gz}_f = v_{b2}h_f^2/\kappa_m w_f \tag{11.4.18}$$

which is the same as Gz_F defined by (11.3.9). This can range in practice, as was explained in Subsection 11.3.1, from small (fully-developed) to a value of order unity (marginally developed). Certainly we expect $\mathrm{Gz}_f < \mathrm{Gz}_{c.c.}$. The fully-developed case leads to $T(q_2 > 0)$ being specified at $x_2 = 0$, as a solution of (11.3.10)–(11.3.12); this is not consistent with the relation (11.4.16), which means that the whole of the region $q_2 > 0$, $0 \leqslant x_2 \leqslant w_c$ has to be absorbed into an inner approximation to be used for matching at $q_2 = 0$ to the asymptotic solution given by (11.4.14), (11.4.15). In practice this is perfectly reasonable because

$$h_s - x_3^+(q_2 = 0) \ll h_s \tag{11.4.19}$$

However, if we suppose that Gz_f is large, then an inner approximation will be relevant at the flight wall. We suppose that

$$\zeta_m\{T(q_2 = 0+) - T_s\} \ll 1 \tag{11.4.20}$$

so that, for the inner region concerned, the velocity v_2 can be represented by a uniform shear flow with constant gradient $(dv_{f2}/dx_3)_{x_3=h_s-h_f}$. The important terms in (11.4.11) will be the convection and diffusion terms; generation can be neglected. The relevant energy equation becomes

$$(x_3-h_s+h_f)\left(\frac{dv_{f2}}{dx_3}\right)_{x_3=h_s-h_f}\frac{\partial T}{\partial x_2}=\kappa_m\frac{\partial^2 T}{\partial x_3^2} \qquad (11.4.21)$$

with boundary and initial conditions

$$x_3=h_s-h_f: T=T_s \qquad (11.4.22)$$

$$x_3\to+\infty: T\to T(q_2=0+) \qquad (11.4.23)$$

$$x_2=-w_f: T=T(q_2=0+)\equiv T_0^+ \qquad (11.4.24)$$

This is to be integrated from $x_2=-w_f$ to $x_2=0$ to provide the heat flux into the screw flight

$$\dot{Y}_f=\alpha_m\int_{-w_f}^{0}\frac{\partial T}{\partial x_3}\,dx_2 \qquad (11.4.25)$$

The solution to (11.4.21)–(11.4.24) is standard (Bird *et al.* (1960), p. 349) and we can write

$$\dot{Y}_f=0{\cdot}8(v'_{f20}\alpha_m^2\rho_m\Gamma_m w_f^2)^{\frac{1}{3}}(T_0^+-T_s) \qquad (11.4.26)$$

where

$$v'_{f20}=\left(\frac{dv_{f2}}{dx_3}\right)_{x_3=h_s-h_f}$$

is given by the solution of

$$\frac{d}{dx_3}\left(\eta(D,T)\frac{dv_{f2}}{dx_3}\right)=0 \qquad (11.4.27)$$

that applies within the flight clearance; here η is given by (11.4.15) and the pressure gradient has been neglected as it was for region F in Subsection 11.3.1 above.

If we cannot make assumption (11.4.20) then the solution is complicated because v_{f2} is no longer linear in the relevant energy equation. Instead we have to use (11.4.27) coupled to

$$v_{f2}\frac{\partial T}{\partial x_2}=\kappa_m\frac{\partial^2 T}{\partial x_3^2}$$

Screw root $\qquad x_3=0;\ w_c>x_2>0$

The same arguments apply as for the screw flight, with no question of fully-developed flow. The relevant quantity is now the heat flux to the screw, given by

$$\dot{Y}_s = \alpha_m \int_0^{w_c} \frac{\partial T}{\partial x_3} \mathrm{d}x_2 = 0{\cdot}8(v'_{c20}\alpha_m^2\rho_m\Gamma_m w_c^2)^{\frac{1}{3}}(T_0^- - T_s) \tag{11.4.28}$$

where

$$T_0^- = T(q_2 = 0-) \tag{11.4.29}$$

and

$$v'_{c20} = \left(\frac{\mathrm{d}v_{c2}}{\mathrm{d}x_3}\right)_{x_3=0}$$

is given by the solution of

$$\frac{\mathrm{d}}{\mathrm{d}x_3}\left(\eta(D, T)\frac{\mathrm{d}v_{c2}}{\mathrm{d}x_3}\right) = \frac{\partial p_c}{\partial x_2} \tag{11.4.30}$$

The temperature in (11.4.30) is given by the coupled equation (11.4.14).

Dividing streamline $\quad x_3 = x_3^+(0);\ w_c > x_2 > 0$

In view of (11.4.19), which implies $q_{f2} \ll |q_{2\,\mathrm{min}}|$, an obvious simplification for heat transfer across the dividing streamline is to put

$$v_{c2}(x_3) = v_2(q_2, x_2 > 0) = v_{b2} \tag{11.4.31}$$

for an inner approximation near $x_3 = x_3^+(0)$, and to use (11.4.11) in the form of the heat conduction equation

$$v_{b2}\frac{\partial T}{\partial x_2} = \kappa_m \frac{\partial^2 T}{\partial x_3^2} \tag{11.4.32}$$

One obvious boundary condition is provided by matching to the core flow, i.e.

$$x_3 \to -\infty : T \to T_0^- \tag{11.4.33}$$

If we have used the approximation $\mathrm{Gz_f} \gg 1$, then another boundary condition is given by

$$x_3 \to \infty : T \to T_0^+ \tag{11.4.34}$$

The initial condition will be

$$x_2 = 0: x_3 < x_3^+(0), \qquad T = T_0^-$$
$$x_3 > x_3^+(0), \qquad T = T_0^+ \qquad (11.4.35)$$

If, however, $\mathrm{Gz_f} \leqslant 0(1)$, then a fully-developed solution at $x_2 = 0$ will apply and the relevant boundary condition will be

$$x_3 = h_s : T = T_b \qquad (11.4.36)$$

The initial condition is now no longer so clear cut; for $h_s > x_3 > x_3^+(0)$, we take the fully-developed flow solution, (11.3.10)–(11.3.12); for $x_3 < x_3^+(0)$ to be consistent, we should take the solution of (11.4.21) as used to obtain (11.4.28), evaluated at $x_2 = w_c$, namely

$$T(x_3) = T_0^- + (T_s - T_0^-) \times \left\{1 - 1{\cdot}12 \int_0^{(x_3^+(0)-x_3)(v'_{c20}/\kappa_m w_c)^{\frac{1}{3}}} \exp(-u^3/9)\, du\right\} \qquad (11.4.37)$$

A simple alternative to the rather elaborate problems provided by these initial conditions, which would necessarily involve digital computation, is to absorb the fluid layer $x_3^+(0) \leqslant x_3 \leqslant h_s$ into the boundary condition. Thus (11.4.33) is retained; (11.4.36) is replaced by

$$x = x_3^+(0): \frac{\partial T}{\partial x_3} = \frac{(T_b - T)}{h_s - x_3^+(0)} \qquad (11.4.38)$$

while the initial condition becomes (11.4.35) for $x_3 < x_3^+(0)$. Clearly, in the limit where $h_s - x_3^+(0) \to 0$, then

$$\dot{Y}_b = -\alpha_m \int_0^{w_2} \left(\frac{\partial T}{\partial x_3}\right)_{x_3^+(0)}$$
$$= 2(T_0^- - T_b)\rho_m \Gamma_m (\kappa_m v_{b2} w_c/\pi)^{\frac{1}{2}} \qquad (11.4.39)$$

represents the heat flux to the barrel from the recirculating zone. A similar result obtains when conditions (11.4.33)–(11.4.35) are used, to give an exchange of heat

$$\dot{Y}_e = (T_0^- - T_0^+)\rho_m \Gamma_m (\kappa_m v_{b2} w_c/\pi)^{\frac{1}{2}} \qquad (11.4.40)$$

The quantities $\dot{Y}_s$, $\dot{Y}_f$ and $\dot{Y}_e$, or $\dot{Y}_s$ and $\dot{Y}_b$, provide effective boundary conditions to be applied to the ordinary second-order differential equations (11.4.14) and (11.4.16). These can be written

$$w_c\alpha_m\left\{-\left(\frac{dT}{dx_3}\right)_{x_3=0+}+\left(\frac{dT}{dx_3}\right)_{x_3=x_3^+(0)-}\right\}=\dot{Y}_s+\dot{Y}_e \quad (11.4.41)$$

$$\alpha_m\left\{w_c\left(\frac{\partial T}{\partial x_3}\right)_{x_3=x_3^+(0)+}+w_f\left(\frac{dT}{dx_3}\right)_{x_3=h_s-h_f+}\right\}=\dot{Y}_f-\dot{Y}_e \quad (11.4.42)$$

if $Gz_f \gg 1$, or more simply, as (11.4.41) alone with $\dot{Y}_b$ replacing $\dot{Y}_e$ if $Gz_f \leqslant 0(1)$.

Fully-developed cross-flow. Most of the published analyses for melt flow in shallow channels (see, for example, Fenner, 1977) have assumed that $Gz_{c.c.}$ is small. The first solution of the fully-coupled energy and flow equations was given by Zamodits & Pearson (1969; see also Zamodits, 1964; Pearson, 1966, Section 3.2) who solved eqns (8.1.17)–(8.1.21) subject to appropriate boundary conditions and using the power-law fluid (11.2.25), i.e. (3.3.5) with

$$\mathrm{K}=K_m e^{-\zeta_m(T-T_b)} \quad (11.4.43)$$

They took the leakage flow in $\dot{m}_{f2}$ to be zero and expressed their results in dimensionless form. Writing

$$\tilde{x}_3=x_3/h_s, \qquad \tilde{v}_1=v_1/v_b, \qquad \tilde{v}_2=v_2/v_b$$
$$\tilde{T}=\zeta_m(T-T_b) \quad (11.4.44)$$

with

$$\tilde{p}_{,\tilde{x}_1}=\frac{\partial p}{\partial x_1}\frac{h_s^{1+\nu}}{K_m v_b^{\nu}}, \qquad \tilde{p}_{,\tilde{x}_2}=\frac{\partial p}{\partial x_2}\frac{h_s^{1+\nu}}{K_m v_b^{\nu}} \quad (11.4.45)$$

$$\mathrm{B}=\zeta_m(T_s-T_b), \qquad \mathrm{Na}=K_m v_b^{1+\nu_h} h_s^{1-\nu}\zeta_m/\alpha_m \quad (11.4.46)$$

and integrating the flow equations once gives

$$\tilde{p}_{,\tilde{x}_1}(\tilde{x}_3-\tilde{X}_1)=e^{-\tilde{T}}\left\{\left(\frac{\partial\tilde{v}_1}{\partial\tilde{x}_3}\right)^2+\left(\frac{\partial\tilde{v}_2}{\partial\tilde{x}_3}\right)^2\right\}^{\frac{1}{2}(\nu-1)}\frac{\partial\tilde{v}_1}{\partial\tilde{x}_3} \quad (11.4.47)$$

$$\tilde{p}_{,\tilde{x}_2}(\tilde{x}_3-\tilde{X}_2)=e^{-\tilde{T}}\left\{\left(\frac{\partial\tilde{v}_1}{\partial\tilde{x}_3}\right)^2+\left(\frac{\partial\tilde{v}_2}{\partial\tilde{x}_3}\right)^2\right\}^{\frac{1}{2}(\nu-1)}\frac{\partial\tilde{v}_2}{\partial\tilde{x}_3} \quad (11.4.48)$$

where $\tilde{X}_1$ and $\tilde{X}_2$ are as yet undetermined constants. The energy equation becomes

$$\frac{\mathrm{d}^2\tilde{T}}{\mathrm{d}\tilde{x}_3^2}=-\mathrm{Na}\,\mathrm{e}^{-\tilde{T}}\left\{\left(\frac{\partial\tilde{v}_1}{\partial\tilde{x}_3}\right)^2+\left(\frac{\partial\tilde{v}_2}{\partial\tilde{x}_3}\right)^2\right\}^{\frac{1}{2}(\nu+1)} \tag{11.4.49}$$

while the boundary conditions are

$$\tilde{v}_1(0)=\tilde{v}_2(0)=0 \tag{11.4.50}$$

$$\tilde{v}_1(1)=\cos\phi_\mathrm{b}; \qquad \tilde{v}_2(1)=\sin\phi_\mathrm{b} \tag{11.4.51}$$

$$\tilde{T}(1)=0; \qquad \tilde{T}(0)-\mathrm{B}=\mathrm{Bi}^{-1}\frac{\partial\tilde{T}(0)}{\partial\tilde{x}_3} \tag{11.4.52}$$

where $\mathrm{Bi}^{-1}=0$ is the assumption made in (11.4.3). Zero net flux in the 2 direction ($\dot{m}_{\mathrm{f}2}=0$) implies

$$\tilde{q}_2=\int_0^1 \tilde{v}_2\,\mathrm{d}\tilde{x}_3=0 \tag{11.4.53}$$

while we write

$$\tilde{q}_1=\int_0^1 \tilde{v}_1\,\mathrm{d}\tilde{x}_3 \tag{11.4.54}$$

The solution of (11.4.47)–(11.4.53) provides

$$\tilde{q}_1=\tilde{q}_1(\tilde{p}_{,\tilde{x}_1};\phi_\mathrm{b};\mathrm{Na},\mathrm{B},\mathrm{Bi}^{-1},\nu) \tag{11.4.55}$$

For later purposes, we shall need

$$p_{,\tilde{x}_1}=\tilde{p}_{,\tilde{x}_1}(\tilde{q}_1;\phi_\mathrm{b};\mathrm{Na},\mathrm{B},\mathrm{Bi}^{-1},\nu) \tag{11.4.56}$$

No analytical forms can be given in general for (11.4.55) or (11.4.56). Some typical plots of $\tilde{p}_{,\tilde{x}_1}$ vs $\tilde{q}_1$ for fixed ϕ_b, Na, B, Bi^{-1}, ν are shown in Fig. 11.19. These refer to the case $\mathrm{Bi}=0$, which means that T_s is greater than T_b ($\mathrm{B}>0$) for the ensuing solution. The effect of increasing Bi is seen in Fig. 11.20 where $\mathrm{d}\tilde{T}/\mathrm{d}\tilde{x}_3$ is specified. Figure 11.21 shows some typical temperature profiles for $\mathrm{Bi}=0$, which suggests that $\mathrm{B}=0$ will imply $\mathrm{d}\tilde{T}/\mathrm{d}\tilde{x}_3=0(\mathrm{Na})$. It is worth noting that, for $\mathrm{Na}>2$, $\tilde{q}_1$ is not a single-valued function of $\tilde{p}_{,\tilde{x}_1}$ in Fig. 11.19(d). Although the magnitudes of $\tilde{q}_1$ for which a maximum in $\tilde{p}_{,\tilde{x}_1}$ is reached are impracticably low, the strength of the non-linearity introduced by thermal coupling is clearly a factor that cannot be neglected in such circumstances. Only small changes arise in the screw characteristics when small positive values of $\tilde{q}_2$ are used, based on a fully-developed flight clearance solution ($\mathrm{Gz}_\mathrm{f}\ll 1$).

As explained earlier, the $Gz_{c.c.} \gg 1$ solution is more likely to be relevant than the $Gz_{c.c.} \ll 1$ solution, although in many cases $Gz_{c.c.} \sim 0(1)$. It is therefore interesting to compare the predictions of the two asymptotic limits. Unpublished work by J. Gluza has shown that differences in both the screw characteristics and the mean temperature $\bar{T}$, defined by (11.4.6), are small in practical terms when the two predictions are compared, using realistic values of the parameters in (11.4.55), provided $B = 0$.

$Gz_{d.c.} = 0(1)$. We must now consider temperature development of the flow in the x_1-direction. To be rigorous, we must assume that $Gz_{c.c.} \gg 1$, and so we should look for a solution in which most cross-channel streamlines are isotherms, although the only published work (Yates, 1968) on temperature development considers the case of $A \to \infty$, $Gz_{c.c.} \ll 1$.

The flow (stress equilibrium) equations will be unchanged, and so the local dimensionless forms (11.4.47), (11.4.48) with boundary conditions (11.4.50) and (11.4.51) will still apply. In the rigorous $Gz_{c.c.} \gg 1$ case, we have to amend eqn (11.4.14) to read

$$\left\{\left(\frac{v_1}{v_2}\right)_{x_3^+(q_2)} - \left(\frac{v_1}{v_2}\right)_{x_3^-(q_2)}\right\}\frac{\partial T(q_2)}{\partial x_1} = \kappa_m\left\{\left(\frac{1}{v_2}\frac{d^2T}{dx_3^2}\right)_{x_3^+(q_2)} - \left(\frac{1}{v_2}\frac{d^2T}{dx_3^2}\right)_{x_3^-(q_2)}\right\} + \left(\frac{\eta D^2}{\rho_m \Gamma_m v_2}\right)_{x_3^+(q_2)} - \left(\frac{\eta D^2}{\rho_m \Gamma_m v_2}\right)_{x_3^-(q_2)} \qquad (11.4.57)$$

Since we can assume $Gz_f \leqslant 0(1)$, and $h_f \ll h_s$, we can retain all the rest of the eqns (11.4.16) to (11.4.42) as valid in a local (x_1) sense. Matching at the singular streamline $q_2 = 0$ is a little more delicate than in the fully-developed case. No published solutions of this case have been given, but unpublished work of C. Miller has shown that the procedure is feasible for a Newtonian fluid approximation. This method can only work if $v_1 > 0$ for all q_2.

The work of Yates (1968) and subsequent extensions of Martin and Gluza was based on eqns (11.4.47), (11.4.48), boundary conditions (11.4.50)–(11.4.52) and the constraint (11.4.53), with (11.4.49) replaced by

$$\mathrm{Pe_D}\, \tilde{v}_\mathrm{D} \frac{\partial \tilde{T}}{\partial \tilde{x}_\mathrm{D}} = \frac{\partial^2 \tilde{T}}{\partial \hat{x}_3^2} + \mathrm{Na}\, e^{-\tilde{T}}\left\{\left(\frac{\partial \tilde{v}_1}{\partial \tilde{x}_3}\right)^2 + \left(\frac{\partial \tilde{v}_2}{\partial \tilde{x}_3}\right)^2\right\}^{\frac{1}{2}(\nu+1)} \qquad (11.4.58)$$

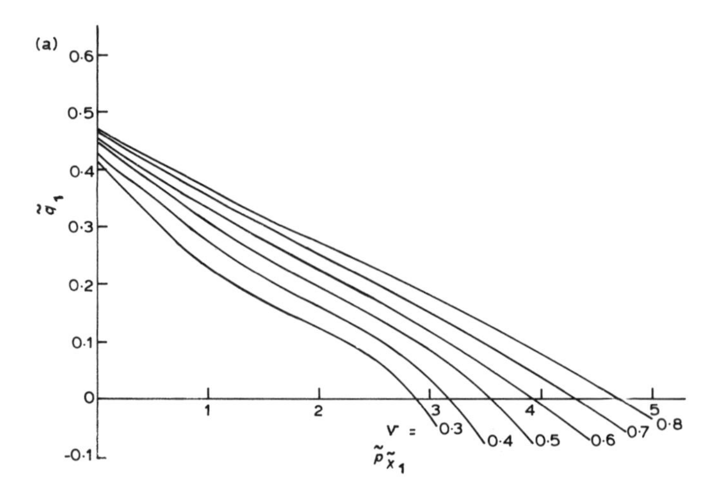

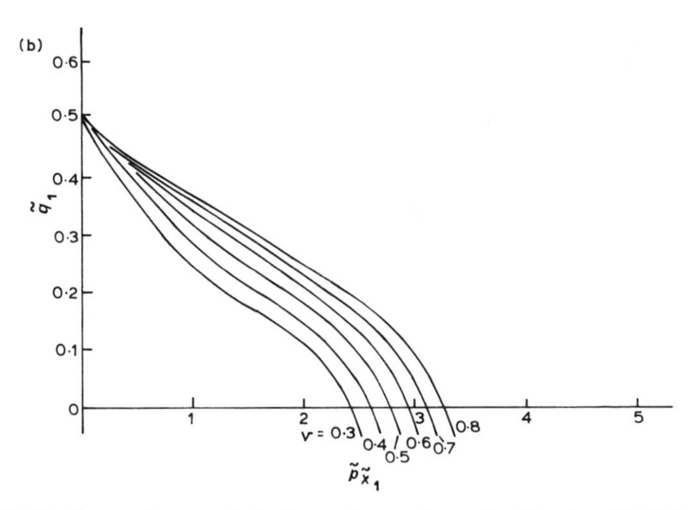

Fig. 11.19. Screw characteristics for various values of ν, Na. $\phi_b = 17{\cdot}7°$ (square pitch); Bi = 0 and so B not specified. (a) Na = 0. (b) Na = 1. (c) Na = 2. (d) $\nu = 0{\cdot}3$.

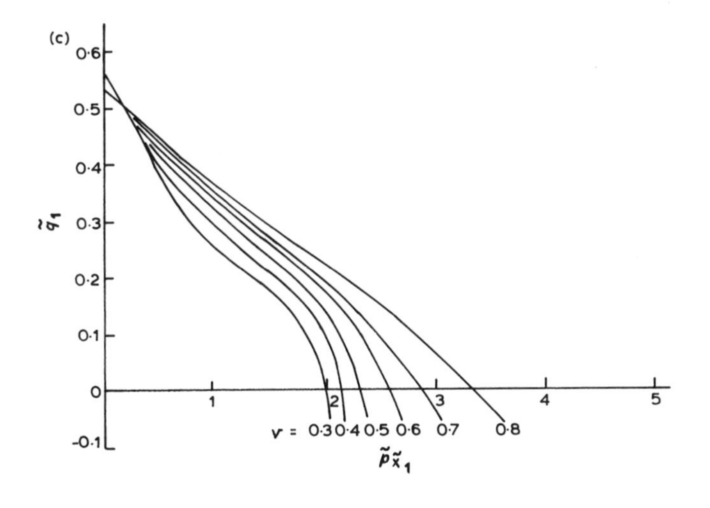

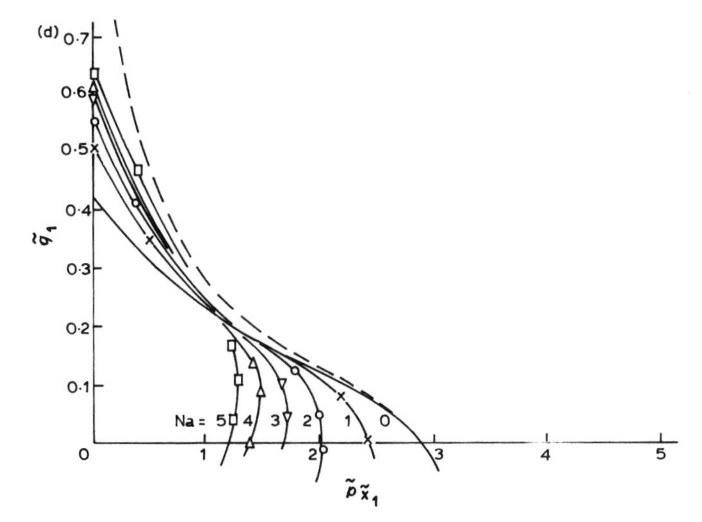

Fig. 11.19. *contd.*

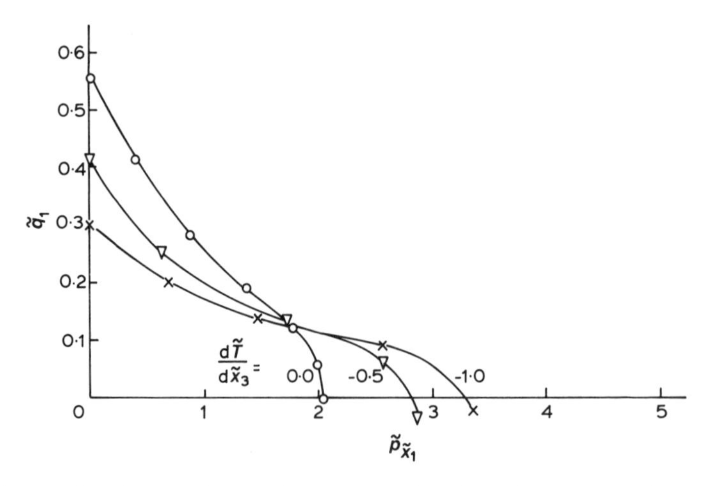

Fig. 11.20. Screw characteristics as a function of screw cooling. $\nu = 0{\cdot}3$, $\mathrm{Na} = 2$, $\phi_b = 17{\cdot}7°$.

where $\tilde{v}_D$, $\tilde{x}_D$, are taken to be $\tilde{v}_1$, $\tilde{x}_1$, provided $\tilde{v}_1 > 0$ for all $0 < \tilde{x}_3 < 1$, but are replaced by $\tilde{v}_z$, $\tilde{z}$, where $\tilde{z}$ is the dimensionless axial coordinate if $v_1 \not> 0$ for all $0 < \tilde{x}_3 < 1$. It was found in practice that $\tilde{v}_z > 0$ for every case studied and so (11.4.58) is suitably parabolic for integration in the positive $\tilde{z}$-direction. Although the formulation is almost perversely internally inconsistent, experience of the fully-developed case suggested that the results might be predictively useful. Indeed, a long and careful comparison† with experimental results proved that this was the case. Yates recast many of his results in terms of a dimensionless development length $\tilde{X}_{1D}$, defined as the distance downstream required for an initial mean temperature $\bar{T} = 0$ (given by $\tilde{T}(\tilde{x}_3) \equiv 0$ for $\tilde{x}_1 = 0$) to achieve 95% of its asymptotic value $\bar{T}_\infty$. A typical case is shown in Fig. 11.22, for which the approximation

$$\tilde{X}_{1D} = 2\tilde{q}_1^2 \tag{11.4.59}$$

seems appropriate.

† These results were not published in the open literature but were reported at various meetings of the European Federation of Chemical Engineers Working Party on non-Newtonian liquid processing held at Delft in the period 1968–1971. The theoretical results were provided by an exhaustively-tested computer program, while the experimental results were provided by various European companies.

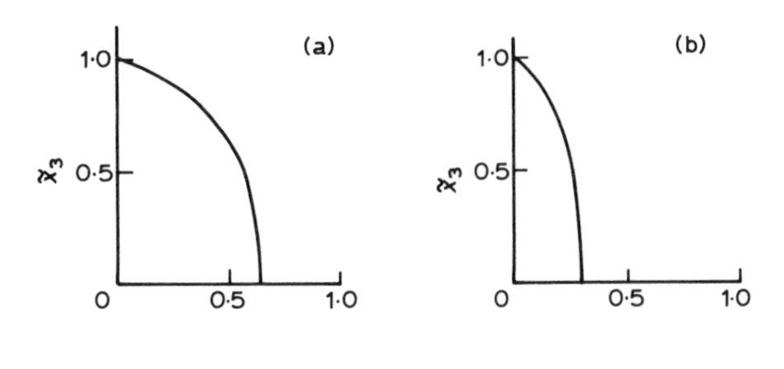

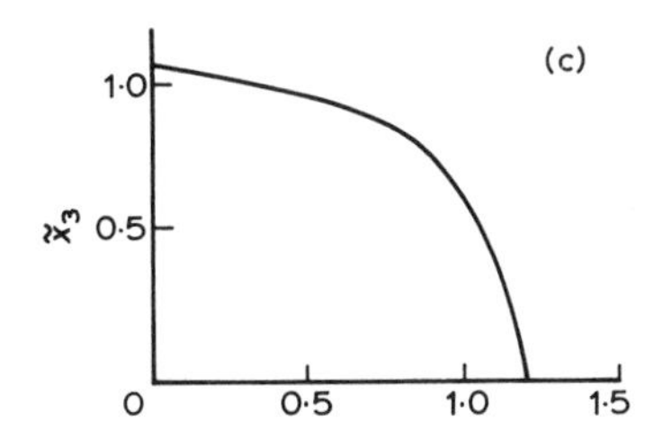

Fig. 11.21. Typical temperature profiles in fully-developed flow in a shallow-channel extruder.

	ν	Na	$\tilde{p}_{,\tilde{x}_1}$	$\tilde{q}_1$
(a)	0·3	2·25	0·35	0·43
(b)	0·5	0·875	2·55	0·097
(c)	0·7	4·70	1·45	0·21

The Graetz number corresponding to $\tilde{X}_{1D}$ is thus $1/2\tilde{q}_1$ where $\tilde{q}_1 v_b$ is taken as the relevant convective velocity v^*_{conv}. In practical cases $1/2\tilde{q}_1$ will not be far from unity, thus justifying $\text{Gz}_{\text{d.c.}} \ll 1$ as the criterion for fully-developed flow. Yates also represented his partial-differential-equation solutions in terms of a single second-order non-linear ordinary differential equation for $\tilde{T}$ involving a Taylor (1953, 1954) apparent axial diffusion coefficient and a Nusselt number as well as the convective velocity $\tilde{q}_1$. The hope was to provide vastly simpler computational techniques. Lastly he undertook a regular perturbation solution for $\tilde{v}_1$, $\tilde{v}_2$, $\tilde{T}$ and $\tilde{p}$ in powers of Na for Na $\leqslant$ about 4, as discussed in Appendix 2.

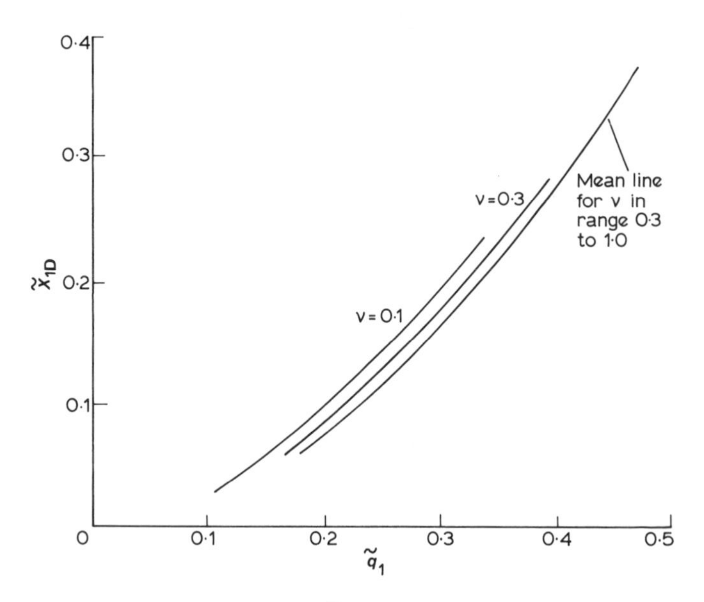

Fig. 11.22. Development length $\tilde{X}_{1D}$ as a function of $\tilde{q}_1$ for various values of ν. $\phi_b = 17{\cdot}7°$, Na = 0, Bi = 0.

11.4.2 The Deep Channel Case: A = 0(1)

There are circumstances in which the effect of the flight walls are felt directly, through viscous action, over the whole of the flow channel. This is the case for melt-fed extruders having a deep channel chosen to maximize output, for extruders having a very small helix angle (as is usually the case in twin-screw extruders), or, as we saw in Subsection 11.3.1, for much of the length of the melt pool in the melting region of a plasticating extruder. We can define

$$\mathrm{Gz}_{\mathrm{d.c.}} = \frac{Q_{T1} h_s}{\kappa_m w_c l_M} \tag{11.4.60}$$

as the relevant Graetz number for down-channel flow. In most of the cases considered, $\mathrm{Gz}_{\mathrm{d.c.}}$ will be significantly greater than 1 and so the development length as defined in the test section will be greater than l_M. It is worth observing that the generation number will also be high enough for temperature effects on viscosity to be relevant for most practical situations.

A direct numerical attack on the problem of obtaining theoretical predictions, even given the power of modern computers, is impracticable. Two partial approaches have been attempted.

$Gz_{d.c.} \ll 1$. Martin (1969) has sought finite-difference numerical solutions to the dimensionless set of equations

$$\frac{\partial \tilde{p}}{\partial \tilde{x}_1} = \frac{\partial}{\partial \tilde{x}_2}\left(\tilde{\eta}\frac{\partial \tilde{v}_1}{\partial \tilde{x}_2}\right) + \frac{\partial}{\partial \tilde{x}_3}\left(\tilde{\eta}\frac{\partial \tilde{v}_1}{\partial \tilde{x}_3}\right) \tag{11.4.61}$$

$$2\frac{\partial^2}{\partial \tilde{x}_3\,\partial \tilde{x}_2}\left\{\tilde{\eta}\left(\frac{\partial \tilde{v}_2}{\partial \tilde{x}_2} - \frac{\partial \tilde{v}_3}{\partial \tilde{x}_3}\right)\right\} + \left(\frac{\partial^2}{\partial \tilde{x}_3^2} - \frac{\partial^2}{\partial \tilde{x}_2^2}\right)\left\{\tilde{\eta}\left(\frac{\partial \tilde{v}_2}{\partial \tilde{x}_3} + \frac{\partial \tilde{v}_3}{\partial \tilde{x}_2}\right)\right\} = 0 \tag{11.4.62}$$

$$\frac{\partial \tilde{v}_2}{\partial \tilde{x}_2} + \frac{\partial \tilde{v}_3}{\partial \tilde{x}_3} = 0 \tag{11.4.63}$$

$$\mathrm{Pe}_{c.c.}\left(\tilde{v}_2\frac{\partial \tilde{T}}{\partial \tilde{x}_2} + \tilde{v}_3\frac{\partial \tilde{T}}{\partial \tilde{x}_3}\right) = \left(\frac{\partial^2 \tilde{T}}{\partial \tilde{x}_2^2} + \frac{\partial^2 \tilde{T}}{\partial \tilde{x}_3^2}\right) + \mathrm{Na}\,\tilde{\eta}\tilde{D}^2 \tag{11.4.64}$$

where

$$\mathrm{Pe}_{c.c.} = v_b h_s/\kappa_m \tag{11.4.65}$$

$$\tilde{\eta} = e^{-\tilde{T}}(\tilde{D}^2)^{\frac{1}{2}(\nu-1)} \tag{11.4.66}$$

and

$$\tilde{D}^2 = \left(\frac{\partial \tilde{v}_1}{\partial \tilde{x}_2}\right)^2 + \left(\frac{\partial \tilde{v}_1}{\partial \tilde{x}_3}\right)^2 + 2\left(\frac{\partial \tilde{v}_2}{\partial \tilde{x}_2}\right)^2 + \left(\frac{\partial \tilde{v}_2}{\partial \tilde{x}_3} + \frac{\partial \tilde{v}_3}{\partial \tilde{x}_2}\right)^2 + 2\left(\frac{\partial \tilde{v}_3}{\partial \tilde{x}_3}\right)^2 \tag{11.4.67}$$

subject to the boundary conditions

$$\left.\begin{array}{ll}\tilde{x}_2 = 0, \mathrm{A};\ 0 \leqslant x_3 < 1 \\ \tilde{x}_3 = 0, \quad 0 < x_2 < \mathrm{A}\end{array}\right\}\tilde{v}_1 = \tilde{v}_2 = \tilde{v}_3 = \tilde{T} = 0 \tag{11.4.68}$$

$$\tilde{x}_3 = 1, \qquad 0 \leqslant x_2 \leqslant \mathrm{A}:\ \tilde{v}_1 = \cos\phi_b, \qquad \tilde{v}_2 = \sin\phi_b, \qquad \tilde{v}_3 = 0, \qquad \tilde{T} = \mathrm{B} \tag{11.4.69}$$

He chose two values for A: 1 as a lower bound and 5, which proved large enough to exhibit within its central portion $1 < x_2 < 4$ most of the characteristics of the solution for $\mathrm{A} \gg 1$. By writing $\hat{T} = \tilde{T}/\mathrm{Na}$, eqns (11.4.61)–(11.4.69) can be rewritten in a form that allows $\mathrm{Na} \to 0$,

$\hat{T} \neq 0$ even when $B \to 0$, i.e. the decoupled solution can be obtained. Martin found difficulties arose at $\tilde{x}_2 = 0$, A, $\tilde{x}_3 = 1$ with a singularity (in the local velocity and pressure fields) that is concealed in the numerical solution. To study the local behaviour of the temperature field he included a finite flight clearance, with boundary and periodicity conditions like those used in the $A \gg 1$ solutions given above. As expected, he found the solutions for $Pe = 0$ to be totally different from those for $Pe = 0(1000)$, the latter being a not unrealistic value in many cases of practical interest.

Analytical solutions, in the form of infinite series, can be given for $\tilde{v}$ and $\tilde{p}$ when $\nu = 1$, $Na = B = 0$, from which screw characteristics can be obtained, because the equations decouple. Such a solution is useful for the melt-pool flow in region B of our six-zone model, for which we use, instead of part of (11.4.68), the moving side wall boundary condition

$$\tilde{x}_2 = 0, \qquad 0 \leqslant \tilde{x}_3 < 1 : \tilde{v}_2 = \tilde{v}_3 = 0, \qquad \tilde{v}_1 = \tilde{v}_{A1} \tag{11.4.70}$$

The relevant value for Q_B is given, in dimensionless form, by

$$\begin{aligned} \tilde{Q}_B = \frac{Q_B}{h_s^2 A v_b} = \cos\phi_b \frac{8A}{\pi^3} \sum_{k=1}^{\infty} \frac{1}{k^3} \tanh \frac{k\pi}{2A} \\ + \tilde{v}_{A1} \frac{8}{\pi^3 A} \sum_{k=1}^{\infty} \frac{1}{k^3} \tanh \frac{k\pi A}{2} \\ - \tilde{p}_{,\tilde{x}_1} \left(\frac{A^2}{12} - \frac{16A^3}{\pi^5} \sum_{k=1}^{\infty} \frac{1}{k^3} \tanh \frac{k\pi}{2A} \right) \end{aligned} \tag{11.4.71}$$

Shapiro (1971) has used solutions of (11.4.61) using (11.4.66) with $\tilde{T} \equiv 0$, and assuming $\tilde{v}_2, \tilde{v}_3 \approx 0$.

We shall not dwell on these solutions here, because the formulation (11.4.61)–(11.4.69) already involves several weaknesses:

(i) $Gz_{d.c.}$ is rarely $\ll 1$.

(ii) The zero-curvature assumption implicit in the unrolling procedure will not be satisfactorily met for deep channels.

(iii) The rheological model (11.4.66) will be unrealistic for the strongly non-viscometric flows predicted by the numerical solutions obtained. The main significance of the work is that it provides some criterion for deciding what the quantitative effect of the flight walls on Q_{T1}/A will be, compared with the value that would be obtained for a very wide channel ($A \to \infty$). Some of the results are given diagrammatically in Appendix 6.

Recently Choo *et al.* (1981) have obtained a numerical solution when curvature effects are taken into account using helical coordinates.

$\mathrm{Gz}_{\mathrm{d.c.}} \geqslant O(1)$. We have seen earlier that in this situation $\mathrm{Pe}_{\mathrm{c.c.}}$ (which is now equivalent to $\mathrm{Gz}_{\mathrm{c.c.}}$) will necessarily be large, and so we can expect much of the heat transfer to the metal walls to take place in a thin layer near to the barrel wall, comparable to the region $q_2<0$ in the $\mathrm{A} \gg 1$ case. Jepson (1953; see also Janeschitz-Kriegl & Schijf, 1969) presented a simple analysis based on the same arguments as used in the 'dividing streamline' analysis in Subsection 11.4.1 above. The bulk of the flow is approximated by a Newtonian flow field whose viscosity is that relevant to the mean temperature $\bar{T}(x_1)$, for which the total rate of heat generation in the cross-section per unit distance downstream,

$$\mathscr{G} = \int_0^{w_c} \int_0^{h_s} \eta(\bar{T}) D^2 \, \mathrm{d}x_3 \, \mathrm{d}x_2 \tag{11.4.72}$$

can be evaluated (cf. Example 11.4.1 and Exercise 11.4.4). The rate of heat transfer/unit distance downstream is given by

$$\mathscr{H} = \frac{\alpha_{\mathrm{m}} w_{\mathrm{c}}}{h_{\mathrm{f}}} (\bar{T} - T_{\mathrm{b}}) \left\{ \frac{2\mathrm{y}}{\sqrt{\pi}} \exp(-\mathrm{y}^2) - 2\mathrm{y}^2[1 - \mathrm{erf}(\mathrm{y})] + \mathrm{erf}(\mathrm{y}) \right\} \tag{11.4.73}$$

where

$$\mathrm{y} = \frac{h_{\mathrm{f}}}{2} \left(\frac{v_{\mathrm{b}} \sin \phi_{\mathrm{b}}}{\kappa_{\mathrm{m}} w_{\mathrm{c}}} \right)^{\frac{1}{2}} = \frac{\sin^2 \phi_{\mathrm{b}}}{2} \mathrm{Gz}_{\mathrm{F}}^{\frac{1}{2}} \tag{11.4.74}$$

For

$$h_{\mathrm{f}} \to 0, \qquad \mathscr{H} \to \frac{2\alpha_{\mathrm{m}}}{\sqrt{\pi}} \left(\frac{v_{\mathrm{b}} w_{\mathrm{c}} \sin \phi_{\mathrm{b}}}{\kappa_{\mathrm{m}}} \right)^{\frac{1}{2}} (\bar{T} - T_{\mathrm{b}}) \tag{11.4.75}$$

which is relation (11.4.39). The rate of change of internal energy carried by the melt is

$$\dot{\mathscr{U}} = \alpha_{\mathrm{m}} \Gamma_{\mathrm{m}} Q_{\mathrm{T1}} \frac{\mathrm{d}\bar{T}}{\mathrm{d}x_1} \tag{11.4.76}$$

Equating the various contributions we get

$$\dot{\mathscr{U}} = \mathscr{G} - \mathscr{H} \tag{11.4.77}$$

and so $\tilde{T}^* = \tilde{\bar{T}} - \tilde{T}_b$ obeys an equation of the form

$$\frac{d\tilde{T}^*}{d\tilde{x}_1} = g\, e^{-\tilde{T}^*} - \frac{\tilde{T}^*}{\tilde{X}_1} \tag{11.4.78}$$

where $\tilde{X}_1 = \frac{1}{2} Q_{T1}\, (\pi/\kappa_m w_c v_b \sin\phi_b)^{\frac{1}{2}}/h_s$ for the case $h_f = 0$, and g is a dimensionless generation number.

This approach can be used in dealing with the melt-pool, when appropriate adjustments are made to deal with the fact that Q_B increases with x_1, so that

$$\frac{d}{dx_1}(\rho_m \Gamma_m Q_B \bar{T}_B) = \dot{m}_{C2} \Gamma_m \bar{T}_C + \mathscr{G}_B - \mathscr{H}_B \tag{11.4.79}$$

Elastic effects. When $A = 0(1)$ and $\sin\phi_b \approx 1/3$, the flow field in the screw channel is not a lubrication flow. Consequently, the assumption that the rheology of the polymer can be represented by a viscosity alone is unsatisfactory. No solutions exist for any other rheological models. The Deborah number relevant to the cross-channel flow is given by

$$\mathrm{De}_{\mathrm{c.c.}} = \frac{v_b \sin\phi_b \Lambda}{A h_s} \tag{11.4.80}$$

since $A h_s / v_b$ is the least time required to cross the channel, while the relevant Weissenberg number is

$$\mathrm{Ws}_{\mathrm{c.c.}} = \frac{v_b \Lambda}{h_s} = A \operatorname{cosec}\phi_b\, \mathrm{De}_{\mathrm{c.c.}} \tag{11.4.81}$$

In practice, $\mathrm{De}_{\mathrm{c.c.}}$ will be small for hot melts of crystalline polymers like PA and PET, but at least of order one for elastomers like SBR or PI. $\mathrm{Ws}_{\mathrm{c.c.}}$ will be significantly larger than $\mathrm{De}_{\mathrm{c.c.}}$ and may be large compared with unity for rubbery thermoplastics and elastomers. Some idea of the effect of large Ws can be gained by considering the limiting case of $\phi_b = 0$, for which $v_{b2} = 0$. For any generalized Newtonian fluid, defined by $\eta(D, T)$, a fully-developed unidirectional solution $\mathbf{v} = (v_1(x_2, x_3), 0, 0)$, $T(x_2, x_3)$, can be found for the mixed drag (v_b) and pressure flow (dp/dx_1) in the rectangular screw channel, satisfying $\boldsymbol{\nabla}\cdot\mathbf{T} = 0$. If however, the CEF-fluid approximation (3.3.21) is used, for which

$$\left|\frac{\Psi_1(D, T) + \Psi_2(D, T)}{\eta(D, T) D}\right| \propto \mathrm{Ws}$$

then it is in general not true that such a unidirectional solution can be found. If it did exist, it would be that given for $\Psi_1 = \Psi_2 = 0$; however, the contributions of $\nabla \cdot (\Psi_1 + \Psi)\mathbf{A}_1^2$, in the x_2- and x_3-directions, could not in general be balanced by a pressure gradient field. The true solution would include a circulation in the (x_2, x_3) plane driven by the difference of normal stresses, which would itself alter the v_1 flow.

Perturbation expansions in Ws have been carried out for related flows (Thomas & Walters, 1963; Langlois & Rivlin, 1959) displaying circulation in pressure-driven pipe flows.

These elastic secondary flows will not, however, be strong flows, in the sense that the long-term deformation rates of fluid elements constrained by closed streamlines in the (x_2, x_3) plane will be those corresponding to viscometric flow and not to elongational flow. We shall still expect $\mathbf{v}$ to be independent—in a lubrication sense—of x_1, and so there is no possibility of elongation or shear in the x_1-direction. Thus although the distribution of $\mathbf{v}$ as a function of (x_2, x_3) may be very different from that based on a purely viscous model, the estimates of $\mathcal{G}$ so obtained, for example, may be reasonably accurate. Indeed when evaluating $\eta(\bar{T})$ in (11.4.72) for a melt known to exhibit power-law behaviour in simple shear with $1-\nu = 0(1)$, it is worth evaluating η at a shear rate corresponding to the mean shear rate for the flow field, a simple approximation for D being given by v_b/h_s.

EXAMPLE 11.4.1. Examine the case when $h_f = 0$, $T_s = T_b$ and $\eta = \eta_0$ constant, for $\mathrm{Gz}_{\mathrm{d.c.}} \ll 1$, $\mathrm{Gz}_{\mathrm{c.c.}} \gg 1$.

The velocity field can readily be shown to be

$$\frac{v_2}{v_{b2}} = \frac{3x_3^2}{h_s^2} - \frac{2x_3}{h_s}, \qquad \frac{v_1}{v_{b2}} = \cot\phi_b \frac{x_3}{h_s} + \frac{p_{,x_1}}{2\eta_0} \operatorname{cosec}\phi_b \left(\frac{x_3^2}{h_s^2} - \frac{x_3}{h_s}\right)$$

(from balance and sum, respectively, of the Newtonian pressure and drag flows) and hence

$$\frac{q_2}{v_{b2}} = \frac{x_3^3}{h_s^2} - \frac{x_3^2}{h_s}$$

The minimum value of q_2 is $-\frac{4}{27}h_s v_{b2}$ at $x_3 = \frac{2}{3}h_s$, where $v_2 = 0$. Equation (11.4.14) can be written as

$$\frac{\mathrm{d}}{\mathrm{d}q_2}\left\{\left(\frac{\mathrm{d}T}{\mathrm{d}x_3}\right)_+ - \left(\frac{\mathrm{d}T}{\mathrm{d}x_3}\right)_-\right\} = \frac{\eta_0}{\alpha_m}\left(\left[\frac{1}{v_2}\left\{\left(\frac{\mathrm{d}v_2}{\mathrm{d}x_3}\right)^2 + \left(\frac{\mathrm{d}v_1}{\mathrm{d}x_3}\right)^2\right\}\right]_- - \left[\frac{1}{v_2}\left\{\left(\frac{\mathrm{d}v_2}{\mathrm{d}x_3}\right)^2 + \left(\frac{\mathrm{d}v_1}{\mathrm{d}x_3}\right)^2\right\}\right]_+\right)$$

Integrating, when use is made of $(\mathrm{d}T/\mathrm{d}x_3)_- = (\mathrm{d}T/\mathrm{d}x_3)_- = 0$ at $q_2 = q_{2\min}$

$$\left(\frac{\mathrm{d}T}{\mathrm{d}x_3}\right)_{h_s} - \left(\frac{\mathrm{d}T}{\mathrm{d}x_3}\right)_0 = -\frac{\eta_0}{\alpha_m}\int_{-4h_s/27}^{0} \mathrm{d}q_2\left(\left[\frac{1}{v_2}\left\{\left(\frac{\mathrm{d}v_2}{\mathrm{d}x_3}\right)^2 + \left(\frac{\mathrm{d}v_1}{\mathrm{d}x_3}\right)^2\right\}\right]_{x^-(q_2)}\right.$$
$$\left. - \left[\frac{1}{v_2}\left\{\left(\frac{\mathrm{d}v_2}{\mathrm{d}x_3}\right)^2 + \left(\frac{\mathrm{d}v_1}{\mathrm{d}x_3}\right)^2\right\}\right]_{x^+(q_2)}\right) = \frac{\eta_0 v_{b2}^2}{\alpha_m h_s^2}\int_0^{h_s} \mathrm{d}x_3\left[4\left(\frac{3x_3}{h_s} - 1\right)^2\right.$$
$$\left. + \left\{\cot\phi_b + \tfrac{1}{2}\frac{p_{,1}}{\eta_0}\left(\frac{2x_3}{h_s} - 1\right)\right\}^2\right] = \frac{\eta_0 v_{b2}^2}{\alpha_m h_s}\left[4 + \cot^2\phi_b + \tfrac{1}{12}\frac{p_{,1}^2}{\eta_0^2}\right] = -\frac{(\dot{Y}_b + \dot{Y}_s)}{\alpha_m W_c}$$

The last result follows directly from an energy balance. Using (11.4.28) and (11.4.39) the above relation can be written

$$\frac{\eta_0 v_{b2}^2}{\alpha_m h_s}\left[4 + \cot^2\phi_b + \tfrac{1}{12}\frac{p_{,1}^2}{\eta_0^2}\right] = \frac{2}{\sqrt{\pi}}\frac{\mathrm{A}}{W_c}\,\mathrm{Gz}_{\mathrm{c.c.}}^{\frac{1}{2}}(1 + 1{\cdot}4\,\mathrm{Gz}_{\mathrm{c.c.}}^{\frac{1}{6}})(T_0 - T_b)$$

or

$$(T_0 - T_b) = \frac{\sqrt{\pi}\,\eta_0 v_{b2}^2[4 + \cot^2\phi_b + \frac{1}{12}p_{,1}^2/\eta_0^2]}{2\alpha_m\,\mathrm{Gz}_{\mathrm{c.c.}}^{\frac{1}{2}}[1 + 1{\cdot}4\,\mathrm{Gz}_{\mathrm{c.c.}}^{\frac{1}{6}}]}$$

where T_0 is to be interpreted as $T(q_2)$ when $q_2 \to 0$.

Further progress in solving for $T(q_2)$ as a function of q_2 and hence for $T(x_3)$ requires that $\mathrm{d}T/\mathrm{d}x_3$ be expressed as $v_2(\mathrm{d}T/\mathrm{d}q_2)$, with v_2 a function of q_2, and that (11.4.14) be integrated a second time. Thus writing the first integral as

$$\left(\frac{\mathrm{d}T}{\mathrm{d}x_3}\right)_{x_3^+(q_2)} - \left(\frac{\mathrm{d}T}{\mathrm{d}x_3}\right)_{x_3^-(q_2)} = \frac{\mathrm{d}T}{\mathrm{d}q_2}\{v_2^+(q_2) - v_2^-(q_2)\}$$
$$= \frac{\eta_0}{\alpha_m}\int_{q_2}^{0}\mathrm{d}p\left(\left[\frac{1}{v_2}\left\{\left(\frac{\mathrm{d}v_2}{\mathrm{d}x_3}\right)^2 + \left(\frac{\mathrm{d}v_1}{\mathrm{d}x_3}\right)^2\right\}\right]_{x_3^+(p)} - \left[\frac{1}{v_2}\left\{\left(\frac{\mathrm{d}v_2}{\mathrm{d}x_3}\right)^2 + \left(\frac{\mathrm{d}v_1}{\mathrm{d}x_3}\right)^2\right\}\right]_{x_3^-(p)}\right)$$

the second integral becomes

$$T(q_2) - T_b = \int_{q_2}^{0}\frac{\eta_0\,\mathrm{d}s}{\alpha_m\{v_2^+(s) - v_2^-(s)\}}\int_s^0 \mathrm{d}p$$
$$\times\left(\left[\frac{1}{v_2}\left\{\left(\frac{\mathrm{d}v_2}{\mathrm{d}x_3}\right)^2 + \left(\frac{\mathrm{d}v_1}{\mathrm{d}x_3}\right)^2\right\}\right]_{x_3^+(p)} - \left[\frac{1}{v_2}\left\{\left(\frac{\mathrm{d}v_2}{\mathrm{d}x_3}\right)^2 + \left(\frac{\mathrm{d}v_1}{\mathrm{d}x_3}\right)^2\right\}\right]_{x_3^-(p)}\right)$$

EXAMPLE 11.4.2. Examine the case $\mathrm{Gz}_{\mathrm{c.c.}} \ll 1$ when $h_f = 0$, $T_s = T_b$ and $\eta = \eta_0$ constant (the analogue of Example 11.4.1 which relates to $\mathrm{Gz}_{\mathrm{c.c.}} \gg 1$).

The velocity fields can be written down almost directly as

$$\tilde{v}_2 = \sin\phi_b(3\tilde{x}_3^2 - 2\tilde{x}_3)$$
$$\tilde{v}_1 = \tfrac{1}{2}\tilde{p}_{,\tilde{x}_1}(\tilde{x}_3^2 - \tilde{x}_3) + \cos\phi_b\,\tilde{x}_3$$

whence

$$\tilde{q}_1 = -\tfrac{1}{12}\tilde{p}_{,\tilde{x}_1} + \tfrac{1}{2}\cos\phi_b$$

Equation (11.4.49) becomes

$$\frac{d^2\tilde{T}}{d\tilde{x}_3^2} = -\text{Na}\{(\tilde{p}_{,\tilde{x}_1}^2 + 36\sin^2\phi_b)\tilde{x}_3^2 + [(2\cos\phi_b - \tilde{p}_{,\tilde{x}_1})\tilde{p}_{,\tilde{x}_1} - 24\sin^2\phi_b]\tilde{x}_3$$
$$+ (\cos\phi_b - \tfrac{1}{2}\tilde{p}_{,\tilde{x}_1})^2 + 4\sin^2\phi_b\}$$

which may be integrated twice to give

$$\frac{T - T_b}{\text{Na}} = \{\tfrac{1}{24}\tilde{p}_{,\tilde{x}_1}^2 - \tfrac{1}{6}\tilde{p}_{,\tilde{x}_1}\cos\phi_b + \tfrac{1}{2}\sin^2\phi_b + \tfrac{1}{2}\}\tilde{x}_3$$
$$-\tfrac{1}{2}\{(\cos\phi_b - \tfrac{1}{2}\tilde{p}_{,\tilde{x}_1})^2 + 4\sin^2\phi_b\}\tilde{x}_3^2$$
$$-\tfrac{1}{6}\{\tilde{p}_{,\tilde{x}_1}(2\cos\phi_b - \tilde{p}_{,\tilde{x}_1}) - 24\sin^2\phi_b\}\tilde{x}_3^3$$
$$-\tfrac{1}{12}\{\tilde{p}_{,\tilde{x}_1}^2 + 36\sin^2\phi_b\}\tilde{x}_3^4$$

EXERCISE 11.4.1. Derive relation (11.4.14) from (11.4.11) on the assumption that

$$T(w_c, x_3^+) = T(w_c, x_3^-) \neq T(0, x_3^+) = T(0, x_3^-)$$

for every $q_2 < 0$.

EXERCISE 11.4.2. Show that (11.4.37) follows from (11.4.21) where $T(x_2 = 0)$ is taken to be T_0^-.

EXERCISE 11.4.3. Examine the case where $h_f = 0$, $T_s = T_b$ and $\eta = K_m\dot{\gamma}^\nu$ instead of η = constant in Example 11.4.1.

EXERCISE 11.4.4. Examine the case where $h_f \neq 0$, $\eta = \eta_0$ constant and $T_b \neq T_s$ in Example 11.4.1.

EXERCISE 11.4.5. Examine the cases described in Exercises 11.4.3 and 11.4.4 for $\text{Gz}_{\text{c.c.}} \ll 1$ rather than $\text{Gz}_{\text{c.c.}} \gg 1$.

EXERCISE 11.4.6. Show that $\tilde{X}_1 = \frac{1}{2}\pi^{\frac{1}{2}}\cot\phi_b\,\text{Pe}_{\text{c.c.}}\,\tilde{q}_1$ in (11.4.78).

11.5 CALCULATIONS OF EXTRUDER AND DIE CHARACTERISTICS: OPERATING POINTS

Earlier sections have provided mechanical models for separate zones of extruder flow. To describe any complete extruder process, these models must be matched, in sequence, so that the geometrical, material and operating parameters can be used to predict the performance of the entire system, i.e. the quantity and state of its output.

Figure 11.23 shows, in block diagrammatic form, the sequence of zones that are dynamically and thermally interconnected in a typical steady single-screw extrusion process. The nature and state of the feed falling onto the extruder screw through the feed pocket is assumed to be part of the initial data: these provide the entry boundary conditions. In their simplest form these are simply a bulk temperature $\bar{T}_0$ and a zero gauge pressure, $p_0 = 0$. Unless a force-feed device is employed, the mass flux $\dot{M}_T$ is unspecified. For steady operation, $\dot{M}_T$ is constant throughout the system. The frictional model (Subsection 11.2.1) based on a compacted plug will apply from the entry point $z = z_0 = 0$ to the axial position z_1 at which either the barrel or screw temperature (T_b or T_s) reaches the melting temperature T_m (or approaches it sufficiently closely for local frictional heat to cause melting at the metal/solid interface). From then on a viscous (Subsection 11.2.2), or frictional/viscous model (Exercise 11.2.6) will apply until the molten layer on the barrel reaches a thickness equal to the flight clearance

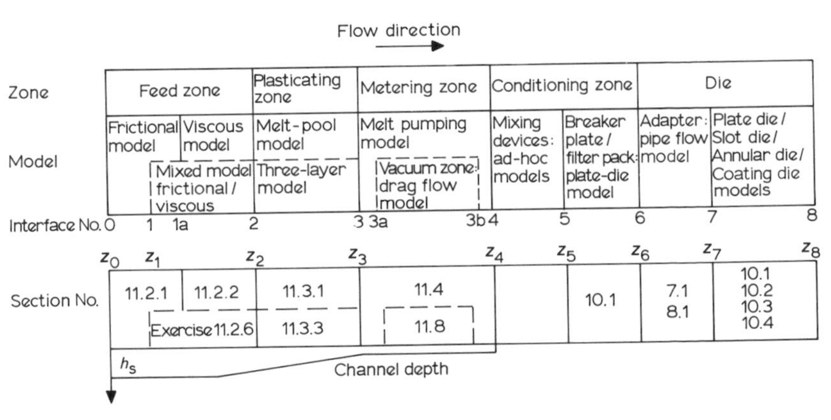

Fig. 11.23. Block diagram for dynamic and kinematic analysis of the single-screw extrusion process.

(the precise meaning of this statement has been discussed earlier). Up to this point, z_2, most of the mass flow is assumed to be in the form of a very slowly-deforming plug, much of the deformation being a compression in the direction of the helical channel caused by the increasing pressure $p(z)$. The pressure is obtained by integration from $x_1 = 0$ $(z = 0)$ of the pressure gradient dp/dx, calculated according to the relevant model equations for the chosen value of $\dot{M}_T$. (As we shall see, $\dot{M}_T$ cannot be prescribed arbitrarily: we seek, by trial and error, the value that makes $p(z_8) = 0$.) Unless there is good evidence that the three-layer model (Subsection 11.3.3) will be relevant, a melt-pool model will apply up to the point (z_3) at which the solid bed melts completely or disintegrates. Most of the information that is carried across the interface 2 is contained within the shallow lubricating layers at the barrel and screw (assuming that melting has by then taken place at the screw): the penetration depth for heat transfer will not in practice be significantly greater than the depth of the layers themselves and, according to the approximations used, the entire rise in internal energy of the feed is assumed to be carried by the molten layers alone. As these layers and the solid bed continue into the plasticating zone, the initial conditions for the melt-pool model are provided very satisfactorily by the final state in the lubricated-feed-zone model, except for the discontinuity going into zone C which has already been mentioned. Interface 2 at z_2 therefore represents a very smooth transition for the full field variables $\mathbf{v}$, T and p. If the melting model can be taken to the point z_3 at which $w_A = 0$ (or equivalently $h_s = h_1 + h_2$ in the three-layer model), then region B will have grown to occupy most of the channel. The information carried across the interface 3 will therefore be as detailed as the model used to represent the flow in region B. In its weakest form, only the pressure $p(z_3)$ will have been calculated. The temperature will have been approximated by some mean value, $\bar{T}_4$, for which T_b probably provides a reasonable approximation. If, however, the aspect ratio at z_3 is already relatively large $(\geqslant 10)$ then a shallow channel approximation could have been used for region B and so $\mathbf{v}$ and T would be known as functions of x_3. In either case, a smooth transition can be achieved at z_4. It is thus clear that a significant discontinuity in the distribution of the energy flux, $\mathcal{U}$, will only arise when going from an $\mathrm{A} = 0(1)$ to an $\mathrm{A} \gg 1$ model for the melt pumping process (whether this transition occurs within the plasticating or the metering zone). A discontinuity in the total energy flux will only arise if a fully-developed model for $\mathrm{A} \gg 1$ is matched directly to a different model for $A = 0(1)$. However, little harm will be done to the final predictions in that case, because a fully-developed approximation implies that upstream behaviour and conditions do not affect the state of the flowing material at the end of the metering zone.

The melt pumping zone splits into three sections if a vented screw is used. At $z = z_{3a}$, the pressure falls to something below atmospheric, i.e. $p < 0$, and drag flow takes place until at $z = z_{3b}$ the channel fills again and the melt pumping model applies; $p(z_{3a}) = p(z_{3b})$ is specified.

The information contained in the solution at the end of the metering zone can rarely be applied directly as input data for extruder-die calculations of the type described in Chapter 10. In any real extrusion system, the nose end of the screw channel is separated from the die entry by one or more devices, such as:

(i) a mixing head (see Schenkel, 1966, Subsections 3.5.6 and 6.3.8; Tadmor & Gogos, 1979, Section 11.2, for examples);
(ii) a breaker plate and filter pack;
(iii) an adaptor pipe or manifold.

In all of these, and particularly in (i) and (ii), significant pressure drop and heat exchange take place. The range and geometrical complexity of designs used make it virtually impossible to give any general theoretical treatment of particular devices here. However, the modelling principles outlined above should suffice in most cases to undertake an analysis of the flow and heat transfer taking place in them. Much insight can be gained by evaluating the relevant Gz, Br and Na's, and by subdividing the flow field into zones characterized by flows already analysed exhaustively above.

Although it is usually argued that mixing heads are introduced simply to make the melt more homogeneous or to disperse pigment more evenly, and that filter packs are used to catch foreign particles that would otherwise weaken or blemish the extrudate, experience suggests that their main function may often be to prevent unmolten material from reaching the die. In other words neither the plasticating section alone, nor the entire helical channel of the single screw extruder being used in some specific application is adequate to provide complete melting, and so some additional device is required to complete the process (mixing head) or at worst, to hold back solid material at the screw extruder exit (filter pack). If the analysis given above were reasonably accurate, then such a situation would be predicted, and the true significance of post-metering devices made clear.

At the various interfaces, z_4, z_5 and z_6, it is unlikely that much more can be matched than $\bar{T}$ and p, although if solid granules penetrate interface 4, the solid flow rate $\dot{M}_{solid}$ ($\dot{M}_A$) would also have to be conserved. Our analysis of extrusion dies has shown that the inlet melt temperature at z_7 can be important in determining die pressure drop, flow pattern and extrudate temperature; it was also argued that a good die should, if possible, make die exit flow independent of the entry conditions, and thus stabilize the process, but that in practice this requirement has to be balanced against the need to minimize die pressure drop.

It is usual to distinguish between the active, pumping, role of the extruder and the passive, resistive behaviour of the die. Two sets of characteristic curves are used as shown in Fig. 11.24, one for the extruder and one for the die. Both relate $p(z_6)$ to Q_T (or, more strictly, $\dot{M}_T$) and are considered to be essentially independent. The intersection of the relevant curves provides the operating point. The curves will depend upon geometrical, material and operating data.

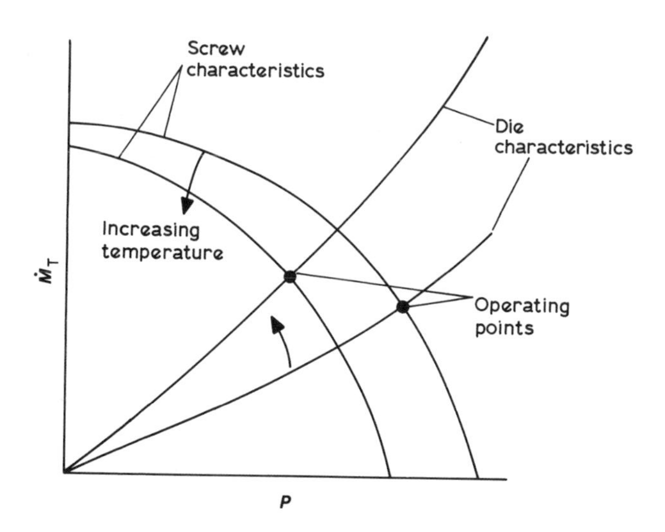

Fig. 11.24. Screw and die characteristics in generalized form for a given polymer feed. Two representative curves are shown to indicate the general effect of increasing the temperature of the metal boundaries. The screw characteristics presuppose a fixed operating speed. Two operating points are shown.

The simplest situation arises when the extruder is melt-fed and the rheological model for the melt is temperature independent. For given geometry there is a unique die characteristic, and a one-parameter (v_b) set of extruder characteristics. If the rheological model is a power-law viscous model then by suitable non-dimensionalization, a single extruder characteristic can be calculated for any given ν. Any heat transfer calculations are decoupled from the dynamical calculations. The operating point is unique. However, all our estimates of Gz, Br and Na suggest that this will at best be a poor representation.

If the extruder is solid fed, then its characteristics will inevitably involve heat transfer and heat generation effects, because the process of melting is involved. At low speeds, the process will be dominated by conduction from the barrel (the relevant dimensionless group being Sf_c, Br_c being negligible) while at high speeds, generation may dominate ($\mathrm{Br}_c \gg 1$). This is true whether the rheological equation is strongly temperature dependent or not, i.e. whether B_c and Na_c are $\geqslant 0(1)$ or not. It is clear, then, that the extruder characteristics will involve Gz, Sf, Br and Na as parameters as well as the purely geometrical parameters and ν, provided $\dot{M}_T$ and $p(z_b)$ are expressed in dimensionless form.

If Br and Na are significant then the die characteristic will also be a function of Gz, Br and Na. The actual values of these parameters will depend upon the precise region of the flow field that is relevant but as $\dot{M}_T$ changes, they will all change similarly, so it will still be relevant to talk about characteristic die and extruder curves and operating points, once geometrical and operating factors have been defined. This aspect is discussed further in the section on scale-up.

The operating point gives a throughput $\dot{M}_T$ and a pressure at the end of the extruder, $p(z_6)$. The model calculations involved provide much more detailed information about p and T as a function of z. For comparison with experiment, the pressure p is the most useful. Numerous examples of such pressure profiles (or of $\mathrm{d}p/\mathrm{d}z$) are given in Tadmor & Klein (1970, Section 8.4), Edmondson & Fenner (1975), or Halmos *et al.* (1978). The most interesting result is that $p(z_4)$ is much more sensitive to $\dot{M}_T$ than is $p(z_3)$, which accounts for the intervening region being called the metering zone in high-compression-ratio (h_{smax}/h_{smin}) screws. Figure 11.25 shows how the metering zone can act either as a true pumping (pressure raising) region for low enough $\dot{M}_T$ or as the reverse for high enough $\dot{M}_T$. In all cases, there is a significant contribution to pumping from the melting zone. If the screw is starve

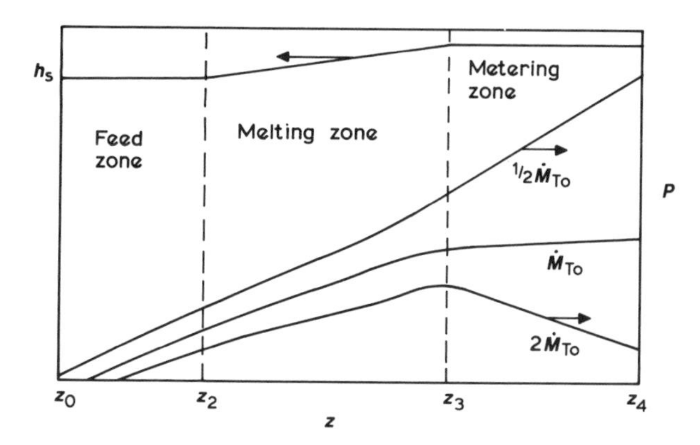

Fig. 11.25. Schematic illustration of role of metering zone. The pressure profile in the feed and melting zones may be far less smooth in practice.

fed,† then the point at which the screw fills and pressure begins to rise moves towards the nose. This is suggested by the change in intersection point of the pressure profile with the z-axis, between z_0 and z_2, as $\dot{M}_T$ is increased from $\frac{1}{2}\dot{M}_{T0}$ to $2\dot{M}_{T0}$.

Other useful information that can be obtained includes:

(1) *Mechanical power consumption.* This is best obtained by calculating the torque at the barrel, since power = torque × screw speed in suitable units. The torque is simply

$$L = \tfrac{1}{2} d_b \int_{A_b} t_{r\theta} \, dA_b \qquad (11.5.1)$$

where $t_{r\theta}$ is the circumferential wall shear stress and A_b is the entire barrel area in contact with solid feed or melt. It can easily be shown that the region opposite the flights contributes very significantly to this, provided the flight clearance is full of melt, so that zone F, as it has been called earlier, must be included in the calculations. The torque is provided by the screw, so the distribution of L as a function of z is useful if a mechanical strength analysis of the screw is being undertaken.

† This aspect, which involves flow through the feed pocket and the compacting mechanism just below it, has not been treated in this account.

(2) *Mean degree of mixing.* The screw extruder is a very effective laminar mixer, and this quantity can be a useful measure of extrudate quality. It is defined as

$$\overline{Mx} = \int_V D \, \mathrm{d}V/Q_T \tag{11.5.2}$$

where V is the total melt volume being sheared, D is the local rate of deformation, as defined in (3.3.4) and Q_T is the volume flowrate. If dispersive mixing is involved, then the maximum values of D will also be relevant.

(3) *Mean residence time.* This is given by

$$\bar{t}_{res} = \int_V \mathrm{d}V/Q_T \tag{11.5.3}$$

and can be written as $\overline{Mx}/\bar{D}$, where $\bar{D}$ is the mean rate of deformation (mean shear rate).

(4) *Heat transfer at barrel and screw surfaces.* These are important in that they determine whether or not the thermal boundary conditions applied at screw and barrel are realistic in terms of the flows of heat that have to be provided by the external heat exchange mechanisms to screw and barrel. Formally we can write the total fluxes as

$$\int_{\mathcal{A}_b} \alpha_m \frac{\partial T}{\partial n_b} \mathrm{d}\mathcal{A}_b \quad \text{and} \quad \int_{\mathcal{A}_s} \alpha_m \frac{\partial T}{\partial n_s} \mathrm{d}\mathcal{A}_s$$

where $\mathbf{n}_b$ and $\mathbf{n}_s$ are the inward normals at barrel and screw respectively, and $\mathcal{A}_b$ and $\mathcal{A}_s$ are the barrel and screw melt interface areas. However, it is the local distribution of the integrands that is often of most importance: what ancillary calculations have to determine is whether the predicted heat fluxes into screw and barrel are consistent with the assumed screw and barrel temperature z-profiles. This is discussed in Appendix 2.

EXAMPLE 11.5.1. Calculate the torque on the barrel for some of the various dynamical models described earlier.

Frictional model for feed zone: Using eqn (11.2.5) with $t_{33} = -p$ at $x_3 = h_s$, and noting that

$$t_{r\theta} = t_{13} \cos \phi_b - t_{23} \sin \phi_b$$

eqn (11.5.1) yields

$$\mathrm{d}L = \tfrac{1}{2}\pi \frac{d_b^2 w_c}{w_c + w_f} \omega_b p \cos \chi_b \, \mathrm{d}z$$

as would be expected from Fig. 11.6(a), the $\cos \chi_b$ term arising because v_{rel} is inclined to v_b at an angle χ_b, and $w_c/(w_c + w_f)$ is the areal ratio between deep channel end screw flight regions opposite the inner barrel surface. p is an exponentially increasing function of $x_1 = z \sin \phi_b$ according to (11.2.6).

Viscous model for feed zone: Here we use eqns (11.2.56) and (11.2.59) for v_{rel} and resolve in the θ direction to obtain

$$\mathrm{d}L = \tfrac{1}{2}\pi \frac{d_b^2 w_c \cos \chi_b}{w_c + w_f} K_m \left(\frac{\mathrm{Sf}_b \, v_{rel}^3 \sin \chi_b}{2A_b \kappa_m z} \right)^{\frac{1}{2}\nu} \mathrm{e}^{-\mathrm{B}_s \tilde{T}} \left| \frac{\mathrm{d}\tilde{v}_{1b}}{\mathrm{d}\hat{x}_3} \right|^{\nu} \mathrm{d}z$$

Six-zone model for melting zone: Because the flight clearance is now filled, there will be a contribution from each of the zones F, C and B. By comparison with the contributions from C and F, that from B can be neglected.

Here, using (11.3.10)

$$\mathrm{d}L_f = \tfrac{1}{2}\pi \frac{d_b^2 w_f}{w_c + w_f} |t_{F35}| \, \mathrm{d}z$$

and using (11.3.44) and Fig. 11.13,

$$\mathrm{d}L_C = \tfrac{1}{2}\pi d_b \, |F_{AC1} \cos \chi_b / \sin 2(\phi_b + \chi_b)| \, \mathrm{d}z$$

Fully molten zone: For $\mathrm{A} \gg 1$, the velocity vector is given by $\mathbf{v}(\mathbf{x}) \approx \{v_1(x_3), v_2(x_3), 0\}$ and

$$\mathrm{d}L \approx \tfrac{1}{2}\pi \frac{d_b^2 w_c}{w_c + w_f} \eta \left| \left(\frac{\mathrm{d}v_1}{\mathrm{d}x_3} \right)_{x=h_s} \cos \phi_b - \left(\frac{\mathrm{d}v_2}{\mathrm{d}x_3} \right)_{x=h_s} \sin \phi_b \right| \mathrm{d}z$$

The total torque is obtained by integration over z. This can be carried out analytically for the first two models for $\mathrm{d}L_F$ and in simple cases for the fully molten zone. However, no simple analytic form can be given by $\mathrm{d}L_C$, because F_{AC1} varies with z in a non-analytic fashion.

EXERCISE 11.5.1. Show that, for all z for which the barrel wall is completely wetted,

$$\frac{\mathrm{d}\overline{Mx}}{\mathrm{d}z} \geqslant \frac{\pi d_b v_b}{Q_T}$$

What causes the inequality rather than the equality to hold? ($\overline{Mx}$ is defined by (11.5.2)).

EXERCISE 11.5.2. Show that

$$\overline{Mx} \geqslant \bar{t}_{\mathrm{res}} v_{\mathrm{b}} / \bar{h}_{\mathrm{s}}$$

where $\bar{h}_{\mathrm{s}}$ is the mean value of the channel depth over the wetted area. Estimate $\overline{Mx}$ for some of the typical values given earlier, thus verifying that

$$\overline{Mx} > l_{\mathrm{s}} / h_{\mathrm{s}}$$

11.6 DESIGN AND SCALE-UP OF EXTRUDER SYSTEMS

The objective of much engineering analysis of processes is to evolve rational design procedures. The first stage is to evolve a mathematical model for the process; this is what has been done in the previous sections.† The next stage is to specify the requirements of the system to be designed. If the system to be designed includes the die, then the requirement is likely to be:

(a) a given output rate of a given polymer within
(b) a given temperature range, in
(c) a given shape.

Constraints (tolerances) will usually be placed on:

(d) the allowable dimensions of the extrudate;
(e) the temperature variations about the mean in the extrudate;
(f) the minimum degree of mixing subject to a maximum shear rate;
(g) the residence time; and
(h) the maximum temperatures attained by the polymer within the system—or combinations of (d)–(h). If the die is provided then the requirements no longer include (c) or (d) except indirectly, but will include:
(i) a pressure requirement.

† A lumped parameter method which may prove to be the simplest approach, based on the detailed ideas described above, is discussed in Parnaby *et al.* (1981).

The list (a)–(h) neither specifies a unique design nor, in most cases, does it present difficult design problems, except perhaps for the die. Complications arise when an optimizing criterion is added. This will, in most cases, be based on a minimum cost criterion: the ultimate cost will be calculated in terms of an objective function involving capital cost of the machinery and operating costs including space, power, maintenance and labour. For our purposes, it can be supposed that a sub-optimal problem connected with a particular type of extruder design is being investigated, in which case the objective function reduces to a monotonic function of extruder diameter, extruder length and power supplied.

At this point, it is clear that the problem can be inverted to that of maximizing $\dot{M}_T$ for given barrel diameter d_b, if it is assumed that d_b is the most important variable in the objective function (see Fenner, 1975, for an example).

The advantage of this approach is that it also covers the requirements of a processor who wishes to expand his business by maximizing the output of his existing installed extruders. The most easily replaced component is the screw, and so, on all counts, the most important factor in engineering design of extruders is screw design.

Figure 11.26 shows in very simplified diagrammatic fashion† what is involved in an optimization procedure. A single scalar variable, h_s, is taken to represent the geometry of the screw design, though in practice this should include the lengths l_F, l_T and l_M of the feed, transition and metering zones, the depths h_{sF}, h_{sM} of the feed and metering zones, the flight width w_f and the clearance h_f, and the helix angle ϕ_b. A single scalar variable, v_b, is taken to represent the operating conditions, though these should include temperature or heat-transfer boundary conditions on barrel and screw. Mechanical constraints, like the strength of the screw, restrict the possible range of h_s and v_b. Within the region thus allowed, mixing, temperature and melting constraints will further restrict the region over which an optimum design can be sought to maximize output $\dot{M}_T$. To each point of the allowed region will correspond a particular value of $\dot{M}_T$. For simplicity, a single maximum is shown.

The actual dimensionality of the space to be searched is so large, and the calculation routines required to calculate $(\dot{M}_T, \bar{T})$ and to decide

† No importance should be attached to the fact that all boundaries of constrained regions are straight lines, nor that two of them are parallel to the axes.

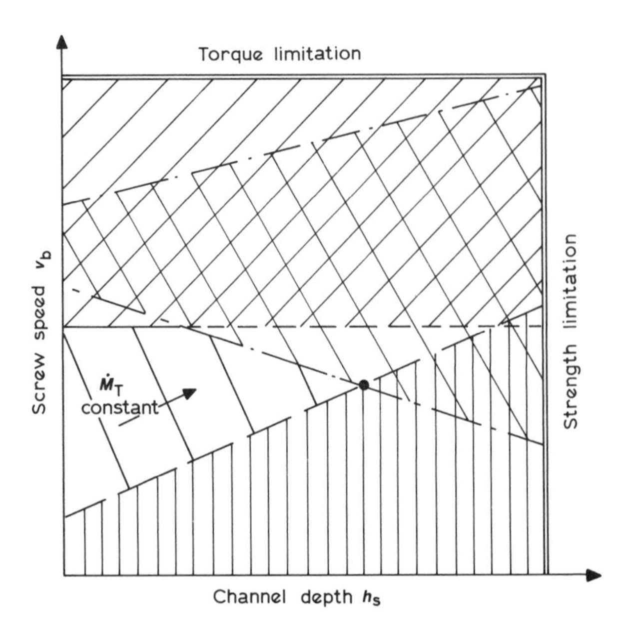

Fig. 11.26. Simple illustration of optimizing procedure for screw extruder design. Constraints are given by: Insufficient mixing (— — —). Excess temperature and/or temperature variations (- - -). Inadequate melting (—·—·—). Mechanical limitations. For visual simplicity, the boundaries of the allowable region (═══) and of the constraints are shown as straight lines. Curves for $\dot{M}_T$ constant (——) are shown as straight lines also. The arrow indicates the direction in which $\dot{M}_T$ increases. (●) Optimum design and operating point.

whether the constraints are being violated so complex, that a rigorous solution of the optimal problem cannot be regarded as feasible. The history of the industry is such that pragmatic adjustments have been made experimentally to overcome the limitations of various constraints on the basis of intuitive physical arguments. In particular, the elementary single-screw design discussed above has been amended in a variety of ways to improve performance (Schenkel, 1966, gives many examples). These include:

(i) grooving the barrel in the feed zone (Kosel, 1971);
(ii) interrupting the screw flight in the transition zone, with or without lengths of a reverse-angle helix interposed;
(iii) placing rows of pins across the screw channel at a series of axial positions;

(iv) introducing a barrier or mixing head before or after the metering zone;

(v) using a two-stage screw with a vacuum venting section;

(vi) introducing a second channel on the screw separated from the fed channel by a relieved barrier flight; the intention behind this approach, pioneered by Maillefer (1963) is to achieve a separation between solid and melt, holding back the former at the nose end of the solid-fed channel. We consider this device in more detail in Appendix 5.

It is obvious that many of the mixing devices act to break up the solid bed; their efficacy may possibly be explained by the argument that when the solid ratio, $\dot{M}_s/\dot{M}_T$, in the channel flow becomes sufficiently low, melting can be accelerated by maximizing the interfacial area between solid granules and melt, particularly if the bulk of the melt is well above the melting temperature. Ultimate melting is assured once the total internal energy flux in the mixture of solid and melt is high enough to melt all of the flow on the basis of thermal equilibrium being attained. This mechanism has a particular advantage if a cool melt is required at the die exit.

Once a satisfactory design (optimized by calculation or experiment over the most important range of variables) has been achieved for a given purpose, the question of scaling-up (or down) the design arises when a different output range is required for the same purpose. How can this best be done? In this context, scale-up is intended to mean the alteration of geometrical and operating parameters according to simple arithmetic laws, keeping the basic design of the system unaltered. The maximum degree of geometrical similarity is sought. It is implied, though rarely stated, that the basic physical regimes—the basic mechanisms—governing the process will be unaltered by scale-up. The quantitative measure of scale-up is conveniently taken to be the diameter ratio

$$\mathrm{Sr} = d_b/d_{b0} \tag{11.6.1}$$

Possible scale-up laws can best be investigated in terms of a dimensionless formulation of the model equations for the process. Such a formulation introduces a basic set of dimensionless groups characterizing the satisfactory design: these groups involve material, operating and geometrical parameters. In the case of the single-channel single-screw plasticating extruder being considered here, the relevant groups consist of the Graetz, Brinkman and Nahme numbers defined for each of the flow regions in each of the operating zones. The material parameters, describing the polymer in question, will be unaltered by scaling-up. What can change are the geometrical parameters, h, w, ϕ and l, and the operating parameters, v_w and T_w. The model equations describe the dimensionless velocity, pressure and temperature fields, from which follow the quantities that determine the quality of the output, like $\bar{T}$, T_{max}, $\overline{Mx}$, $\bar{t}_{res}$, $\bar{D}$, the efficiency of the process, like power consumption and heat flux, and the operating point $(\dot{M}_T, P)$.

We must now decide which features of the original design are to be invariant after scale-up. Going back to our discussion of the design problem we are likely to retain (b), (e), (f), (g), (h) and (i). Some of these, (e), (f), (g) and (h) are constraints that may not have limited the original design. Our scale-up laws will therefore be chosen to satisfy (b), (i) and those of (e)–(h) that were limiting. At our disposal are the ratios

$$\left(\frac{v_b}{v_{b0}}\right), \left(\frac{\phi_b}{\phi_{b0}}\right), \left(\frac{h_{sF}}{h_{sF0}}\right), \left(\frac{h_{sM}}{h_{sM0}}\right), \left(\frac{l_F}{l_{F0}}\right),$$
$$\left(\frac{l_T}{l_{T0}}\right), \left(\frac{l_M}{l_{M0}}\right), \left(\frac{h_f}{h_{f0}}\right), \left(\frac{w_f}{w_{f0}}\right)$$

and the barrel temperature T_b.

It is shown in Pearson (1976*b*) that if ϕ_b and T_b are held constant and w_f scales as d_b, then (b), (i), (e) and (h) are held invariant if all the Gz, Br and Na are held invariant. This implies a unique set of scale-up laws

$$\frac{v_b}{v_{b0}} = \mathrm{Sr}^{v}, \qquad \frac{h_s}{h_{s0}} = \mathrm{Sr}^{h}, \qquad \frac{l_F}{l_{F0}} = \mathrm{Sr}^{l_F}$$
$$\frac{l_T}{l_{T0}} = \frac{l_M}{l_{M0}} = \mathrm{Sr}^{l_M}, \qquad \frac{\dot{M}_T}{\dot{M}_{T0}} = \mathrm{Sr}^{m} \tag{11.6.2}$$

involving the indices, v, h (for all zones), l_F, l_M and m as functions of the rheological power law index. Table 11.3 gives their values. It is also shown that $\bar{t}_{res}$ and $\overline{Mx}$ increase with Sr while D_{max} (and indeed $\bar{D}$) decrease with Sr. It is thus possible that constraints (f) and (g) could be violated. To hold $\bar{t}_{res}$ or $\overline{Mx}$ or D_{max} invariant, some other requirement or constraint would have to be relaxed.

Alternative scale-up procedures are given in Yi & Fenner (1976) and Potente & Fischer (1977). Yi & Fenner are concerned only with the melting process, and use numerical calculations to obtain an

Table 11.3
Scale-up Indices for Single-screw Extruders

ν	*1·0*	*0·8*	*0·6*	*0·4*	*0·2*
v	−1	−1·06	−1·14	−1·27	−1·5
h	0·5	0·53	0·57	0·64	0·75
l_F	0·5	0·59	0·71	0·92	1·25
l_M	1	1	1	1	1
m	1·5	1·47	1·43	1·36	1·25

approximate value of 0·3 for h given $m=2$, $l=1$. Their melting model, though similar to the six-zone model described above, is not identical and it is not clear that in dimensionless form it would lead to the same scaling laws. Yi & Fenner do not apply the same constraints on the output however. The value of 2 for m is closer to what industry expects than the 1·3 predicted by Table 11.3. Nevertheless, it can be argued that if industry can get such scale-up ratios in practice, then it is running its small extruders too slowly, a comment that has been made, and proved, on many occasions. Similar ideas are contained in Pawlowski (1967).

11.7 UNSTEADINESS: SURGING

It has been tacitly assumed that the extrusion process under investigation is steady. One of the major constraints, in practice, on increasing output from a given extruder has been unsteadiness in output, often called surging. This manifests itself as cyclic or random fluctuations in output rate, temperature and pressure at the die. The matter is discussed in Tadmor & Klein (1970, Chapter 9) where observations on plasticating extruders are classified according to their apparent periodicities.

(1) Fluctuations caused by the helical symmetry, having the same frequency as the screw rotation rate. These can be traced back to the effect of the flight passing across the feed pocket and to the associated change in solid feed rate. Similarly, cross-channel variations in the output from the screw channel at the nose end can lead to residual helical, i.e. non-axisymmetric, variations in the output, which persist up to the die-lips in an annular die, for example. These are ultimately inseparable from the screw design, but can be made tolerable.

(2) Long-term variations due to changes in the feed, the cooling water, the electrical supply or ambient air conditions. These will interact with the control systems used to hold various operating parameters constant, such as screw speed and barrel and die temperatures. If these fluctuations are slow enough, they can be thought of as leading to a continuous set of quasi-steady states. It is the province of the control engineer (e.g. Hassan & Parnaby, 1981) to seek control systems that will nullify the effect of external variations. To select the optimal control system, it can be useful to study the response of the model equations described above to small perturbations of the material and operating parameters about any given steady solution. Solution of the linearized equations provides the necessary 'transfer' matrices, $\|Z_T\|$, which give the output fluctuations in terms of the input perturbations. Thus we can write, for example,

$$\begin{pmatrix} \Delta\dot{M} \\ \Delta\bar{T} \\ \Delta P \end{pmatrix} = \|Z_T(w)\| \begin{pmatrix} \Delta T_b \\ \Delta v_b \\ \Delta K_m \\ \Delta\zeta_m \end{pmatrix} \tag{11.7.1}$$

where the components of Z_T are complex functions of the frequency w, and of all the steady-state parameters of the motion, it being assumed that all the perturbations are of the form $\mathrm{Re}\{\Delta J \exp(\mathrm{i}wt)\}$. Feed-forward and feed-back control systems are currently used on commercial extruders.

(3) Intermediate frequency fluctuations apparently unconnected with either 1 or 2. These must be intrinsic to the internal mechanics of the process. Because the model equations are intrinsically non-linear we have no *a priori* grounds for assuming that stable steady solutions will exist, even when the inlet and boundary conditions are steady. We must be prepared for limit cycles in certain circumstances. Thus, although our present model equations might predict a unique steady solution, extension of the model to include unsteady terms—such as $\partial w_A/\partial t$, for example, in region A of the six-zone model—might also show that solution to be unstable to small disturbances.

Observations show that gross fluctuations in output, temperature and pressure often correlate with periodic break-up of the solid bed, as shown in Fig. 11.27. It is generally agreed (Fenner *et al.*, 1979) that the position of break-up, $x_{1\mathrm{Bmin}}$, coincides with the position of relatively rapid increase in v_{A1}, i.e. $\mathrm{d}v_{A1}/\mathrm{d}x_1 > v_{A1}/w_A$ say. The six-zone model in its present form is unable to provide a mechanical explana-

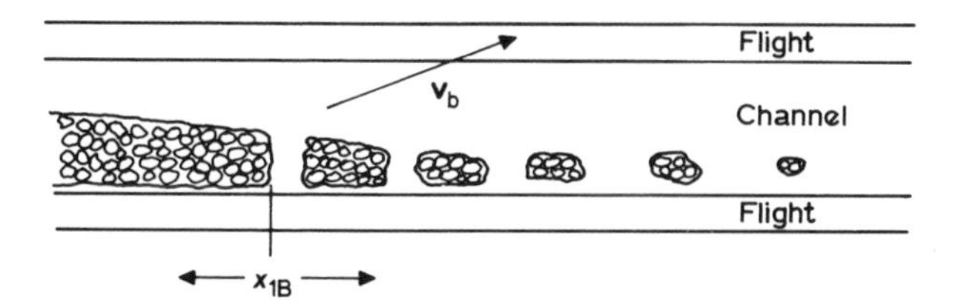

Fig. 11.27. Diagrammatic example of bed break-up. The view shown is an instantaneous one: the position of the broken edge of the solid bed, x_{1B}, moves forward in time, jumping back as a new fracture occurs.

tion for the break-up process. Any successful attempt to do so would have to consider more closely the internal rheology of the solid bed.

11.8 TWO-STAGE VENTED SCREWS: VACUUM EXTRACTION

In a traditional plasticating single-screw extruder, all gases and vapours are assumed to be expelled back towards the feed pocket through the interstices of the continuous solid bed. This is not always adequate either for plasticating or melt-fed screws and so a venting pocket is sometimes added about half-way down the barrel opposite a deepened section of the screw channel. In effect, a two-stage screw is used with feed, transition and metering zones followed by a rapid reverse transition to another deep channel and subsequent transition and metering zones. The design of the screw is such that at any given screw speed, the maximum delivery of the first stage to the reverse transition is less than the amount that would be delivered by the second stage against the die pressure if all the channels in the second stage were full. Therefore steady operation is only possible if some of the second-stage channel runs partly empty, and in this way a vacuum can be drawn in the neighbourhood of the vacuum vent.

A modification of the melt-pumping model of Section 11.4 can be used to cover this situation, as suggested by Berlis *et al.* (1973). We suppose that Q_{T1} is known and that $A \gg 1$. The channel is supposed filled to a width (see Fig. 11.28)

$$w_m(x_1) < w_c \tag{11.8.1}$$

and that

$$\partial p/\partial x_1 = 0 \tag{11.8.2}$$

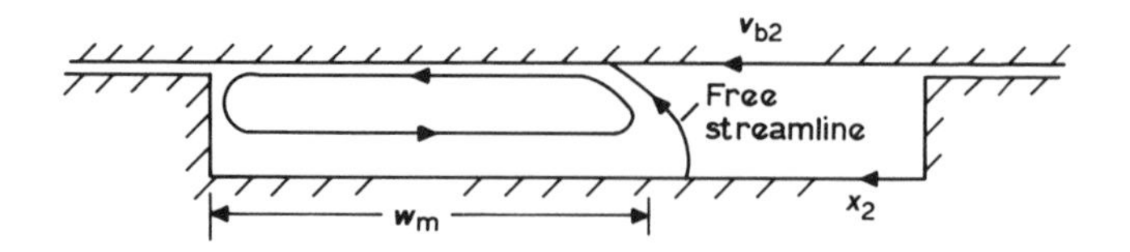

Fig. 11.28. Partially-filled screw showing a free streamline with a void region on the trailing wall of the flight.

For the simplest approximation we suppose that w_m changes very slowly with x_1 because h_s does, and so the local stress equilibrium relations will be satisfied on the basis of w_m constant. Thus we can take over eqns (11.4.47)–(11.4.48) with boundary conditions (11.4.50)–(11.4.51) and constraint (11.4.53) to yield $\tilde{q}_1$ for $\tilde{p}_{,\tilde{x}_1} \equiv 0$. The width w_m is then given by

$$w_m = Q_{T1}/v_b h_s \tilde{q}_1 \tag{11.8.3}$$

If Na$\not\ll 1$ then either (11.4.49) and (11.4.52) or (11.4.14) with (11.4.28) and (11.4.39) can be used to give the temperature field. For simplicity flow over the flight has been disregarded.

This model does not explain what happens in the neighbourhood of $x_2 = w_c - w_m$ where a free streamline forms. This aspect is neglected on the basis that $w_m \gg h_s$.

At some point x_{1D}, the channel must fill, for otherwise no increase in pressure towards the die would occur. This cannot arise smoothly, in a region where $dh_s/dx_1 < 0$ (which corresponds to $dw_m/dx_1 > 0$) because for values of $x_1 > x_{1D}$ a negative value of dp/dx_1 would be required to achieve the necessary Q_{T1}, and this would certainly not correspond to an increase in pressure. It must therefore do so suddenly: in our model this means a discontinuity in w_m, which is consistent with our local approximation, and a corresponding discontinuity in dp/dx_1 with $dp/dx_1 > 0$ for $x_1 > x_{1D}$.

The position of x_{1D} would therefore be determined by back calculation from the die for a Q_{T1}, given by solution of the model equations for the first stage extruding into a vacuum. The second stage is necessarily starve-fed if the vacuum port is not to be filled with melt.

Chapter 12

The Twin-screw Extruder

The twin-screw extruder is sometimes used as an alternative to a single-screw extruder. To some extent, they are interchangeable devices, with some processors preferring one to the other. However, it is generally agreed that:

(a) twin-screw extruders are more expensive to manufacture than their single-screw equivalents and therefore carry a cost penalty;
(b) twin-screw extruders can be designed to provide far better mixing than basic single-screw machines, i.e. those without separate mixing devices;
(c) the output of most twin-screw extruders is less sensitive to feed fluctuations and back pressure than that of single-screw extruders.

These comments should not be taken as invariably true, since the behaviour of any given type of screw in the single-screw extruder, or screw and barrel system in the twin-screw extruder, depends on its particular design. They can best be interpreted as relating to general purpose machines.

As in the case of the single-screw extruder, a full understanding of the geometry of the device is an essential prerequisite of any satisfactory mechanical analysis. This geometry is necessarily more complex for twin-screw devices than for the basic single-screw device. Although a cylindrical form for the barrels, and helical screws of equal diameter and constant pitch, can be assumed for simplicity (as in the single-screw case), two new additional degrees of freedom are necessarily present: the first is that the two screws can be either co-rotating or contra-rotating—and both types are used—and the second is that the separation of the screw axes can be varied to range from limitingly close intermeshing to no overlapping of the screws. In the latter extreme case, the twin-screw extruder degenerates effectively into two independent side-by-side single-screw machines; while in the former extreme case, the intermeshing screws act like a pair of positive displacement pumps, whose behaviour is radically different from that of a traditional single-screw extruder. The range of possible regimes of behaviour is thus much wider than in the basic single-screw case, and it is hardly surprising that theories for predicting performance are not well-developed for twin-screw systems. A careful account has been given recently by Janssen (1978) which forms the basis of the present development while much useful information can be found in Schenkel (1966).

12.1 GEOMETRY OF THE TWIN-SCREW EXTRUDER

Figure 12.1 provides a simple illustration of the four extreme types of twin-screw extruder discussed above. Figure 12.2 shows plan and cross-sectional views for intermeshing screws with a Cartesian coordinate system **z** based upon the plane defined by the two parallel screw axes, $z_3=0$, $z_2=\pm\frac{1}{2}d_s \cos\frac{1}{2}\psi_1$. Here d_s is the diameter of the smallest coaxial circular cylinder that would enclose either of the screws. For complete intermeshing as shown in Fig. 12.1(a, b), the channel depth

$$h_s = d_s(1-\cos\tfrac{1}{2}\psi_1) \tag{12.1.1}$$

There is no intermeshing when $\psi_1=0$ as shown in Fig. 12.1(c, d). Each screw will have a helical screw angle ϕ_b defined by the axial pitch

$$w_a = \pi d_s \tan\phi_b \tag{12.1.2}$$

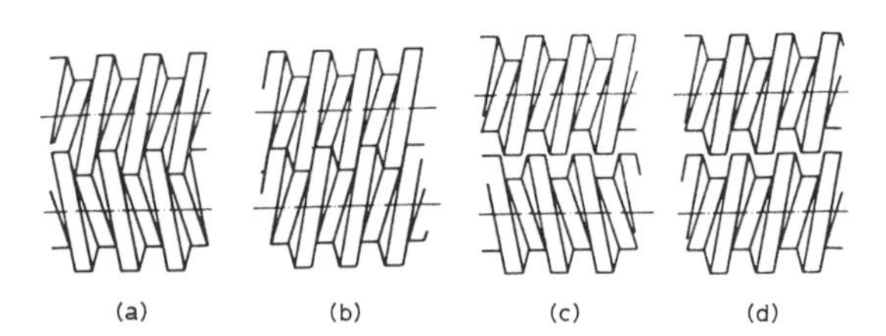

Fig. 12.1. The four basic types of twin-screw extruder. (a) Contra-rotating, intermeshing; (b) co-rotating, intermeshing; (c) contra-rotating, non-intermeshing; (d) co-rotating, non-intermeshing.

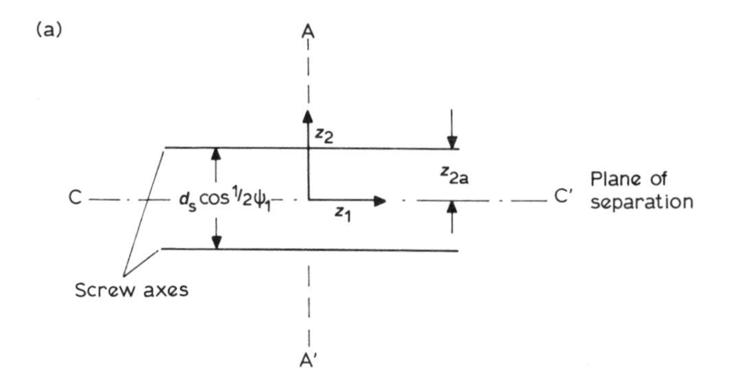

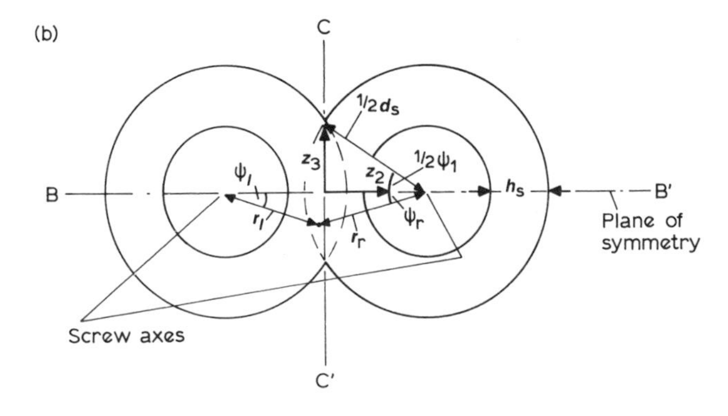

Fig. 12.2. (a) Plan view BB′ of screw axes and coordinate system. (b) Cross-sectional view AA′ showing coordinate system and geometrical nomenclature. Note that ψ_l and ψ_r are negative in this diagram according to definition (12.1.5).

In general ϕ_b will be small compared with values common in single-screw extruders. The maximum degree of intermeshing arises in the plane $z_3=0$; a suitably general description of the screw-channel and screw-flight geometry, consistent with engineering practice, is illustrated in Fig. 12.3, where the flight wall angle ψ_2 and the basic minimum axial clearance between flight walls, c_{00}, are shown in (a), and a small radial clearance c_{r0} is shown in (b), in which case the axial clearance for $z_3=0$ is

$$c_0 = c_{00} + c_{r0} \tan \psi_2 \tag{12.1.3}$$

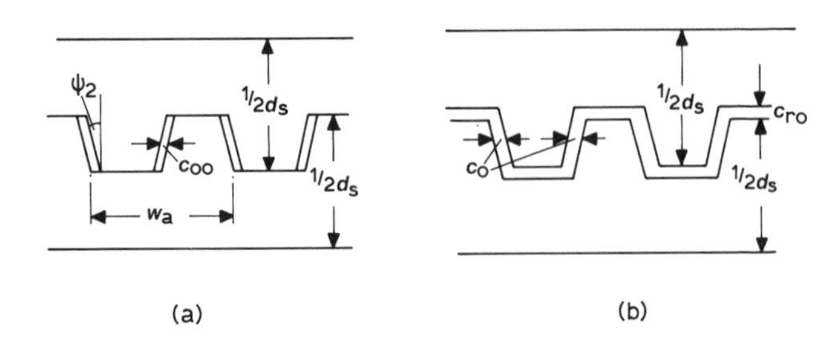

Fig. 12.3. Plan view BB′ (see Fig. 12.2) showing geometry of clearances in twin screw extruder: (a) close meshing, (b) with radial clearance c_{r0}.

It is worth noting that the mean flight width

$$\bar{w}_f = \tfrac{1}{2}w_a - c_{00}$$

is little different from the mean channel width $\bar{w}_c = \frac{1}{2}w_a + c_{00}$ for closely-fitting screws; the channel depth h_s will usually be of the same order as the channel width and a substantial fraction of the screw radius $\frac{1}{2}d_s$. Consequently, neither the small-curvature nor the wide-channel approximations will be strictly relevant, as they proved to be for the typical single-screw extruder.

The intermeshing of the two screws has the effect of breaking up the helical channels on each screw into a series of largely independent C-shaped volumes (or chambers) as illustrated in Fig. 12.4. As the screws rotate, so these volumes progress axially towards the delivery end of the extruder; to the extent that material within one of these

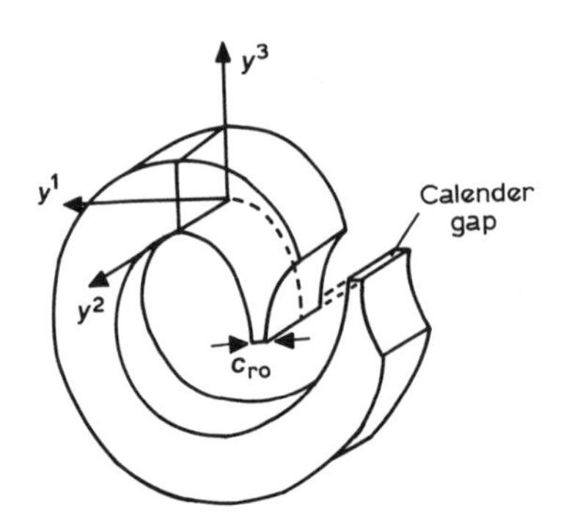

Fig. 12.4. A C-shaped chamber. c_{00} and $\psi_2 \approx 0$ in this diagram.

volumes is constrained to remain within that volume, the extruder acts as a positive displacement pump, with a delivery rate that is independent of back pressure and linearly proportional to screw speed. Any departure from this simple relation will depend upon flows through the clearances in the intermeshing regions. It is convenient at this stage to distinguish between four separate clearance flow regions.

(1) *The flight gap* between barrel and flight land which is exactly analogous to that which arises in a single-screw extruder. This is shown in Fig. 12.5 as c_f, and is responsible for melt leaking backwards from chamber to chamber on the same screw.

(2) *The calender gap* between screw root of one screw and flight land of the other. It is shown in Fig. 12.3 and Fig. 12.4 as c_{r0}. This also allows flow between adjacent chambers on the same screw.

(3) *The side gap* between the pushing (or trailing) face of the flight on one screw and the corresponding trailing (or pushing) face of the flight on the other screw. These are the gaps c_0 shown in Fig. 12.3(b), and the relevant flows add to that through the calender gap.

(4) *The tetrahedron gap* which is connected with the side gap, but which relates to flow between opposite chambers in the two screws. This gap can best be visualized from a cross-section in the plane of separation ($z_2 = 0$ in Fig. 12.2), as shown in Fig. 12.6. At this point a clear difference arises between contra- and co-rotating screws. The intermeshing can be much tighter for the contra-rotating configuration than for the co-rotating; the plane $z_2 = 0$ represents a plane of symmetry for the former but not for the latter. The axial clearance c between the faces for all points P within the intermeshing region (see

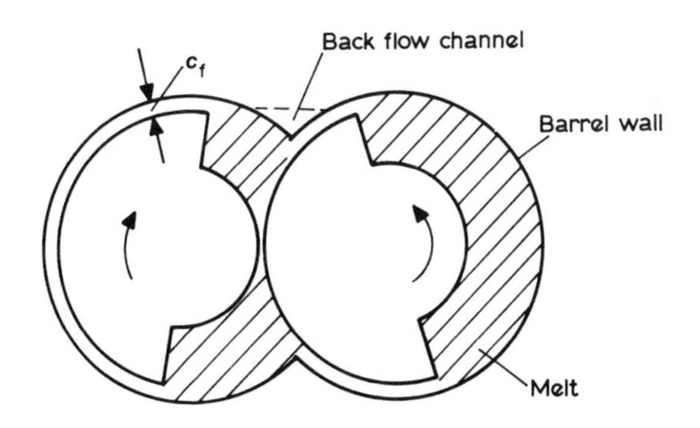

Fig. 12.5. Cross-sectional view AA′ (see Fig. 12.2) showing flight clearance c_f and presence of melt (contra-rotating case).

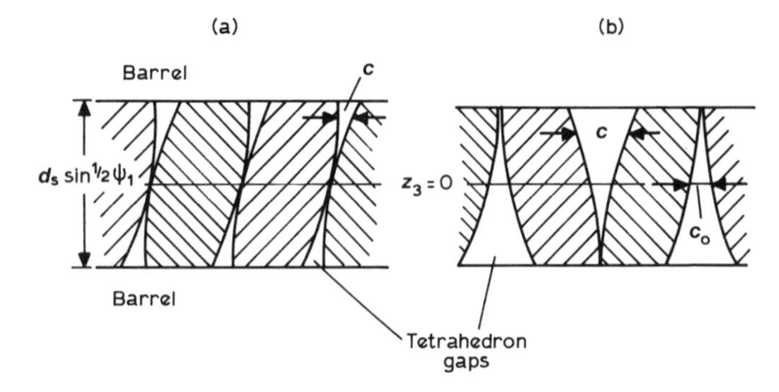

Fig. 12.6. Sectional view CC′ (see Fig. 12.2) of intermeshing zone for (a) contra-rotating screws, (b) co-rotating screws.

Fig. 12.2(b)) can be written

$$c = w_a(\psi_l \pm \psi_r)/2\pi + (r_r + r_l - d_s + h_s - c_{r0})\tan \psi_2 + c_{00} \quad (12.1.4)$$

where the + sign refers to co-rotating and the − sign to contra-rotating screws. Figure 12.6 corresponds to $\tan \psi_2 \neq 0$, with

$$\left.\begin{aligned} \psi_l = \psi_r &= \tan^{-1}(z_3/\tfrac{1}{2}d_s \cos \tfrac{1}{2}\psi_1) \\ r_r = r_l &= (z_3^2 + \tfrac{1}{4}d_s^2 \cos^2 \tfrac{1}{2}\psi_1) \end{aligned}\right\} \quad (12.1.5)$$

The position of the minimum value for c will depend upon ψ_2, ϕ_b, c_{r0}/d_s and h_s/d_s. It will not necessarily arise when $r_r = r_l$, i.e. when $z_2 = 0$. This can be seen by taking $\psi_2 = 0$ in which case

$$\frac{2\pi(c - c_{00})}{w_a} = \psi_l \pm \psi_r = \tan^{-1}\frac{z_3}{(z_{2a} - z_2)} \pm \tan^{-1}\frac{z_3}{(z_{2a} + z_2)} \quad (12.1.6)$$

$$= \begin{cases} \tan^{-1}\dfrac{2z_3 z_{2a}}{z_{2a}^2 - (z_2^2 + z_3^2)} & \text{for co-rotating case} \quad (12.1.7) \\[2ex] \tan^{-1}\dfrac{2z_3 z_2}{z_{2a}^2 - z_2^2 + z_3^2} & \text{for contra-rotating case} \quad (12.1.8) \end{cases}$$

where

$$z_{2a} = \tfrac{1}{2}(d_s - h_s + c_{r0}) = \tfrac{1}{2}d_s \cos \tfrac{1}{2}\psi_1 \quad (12.1.9)$$

It can be shown that the least value of $(\psi_l + \psi_r)$ for the co-rotating case arises when $z_2 = 0$, $z_3 < 0$ and so the flight can just touch in the manner shown in Fig. 12.6(b). Analysis shows that the least value of $\psi_l - \psi_r$

arises when $z_2 \approx \frac{1}{3}h_s$, $z_3^2 \approx \frac{1}{3}h_s z_{2a}$ for small h_s/d_s. The corresponding value of Δc will be, approximately,

$$\frac{w_a}{2\pi}(\psi_l - \psi_r)_{min} \approx -d_s \tan \phi_b \left(\frac{h_s}{3z_{2a}}\right)^{\frac{3}{2}} \tag{12.1.10}$$

This must be balanced by the other terms in (12.1.4), c_{00} in this case. Thus for $\psi_2 = 0$, the screws cannot actually touch at $z_2 = z_3 = 0$, as can be verified directly by consideration of (12.1.8).

We may compare the value (12.1.10) with the maximum variation in c shown in Fig. 12.6(a). This is given approximately by

$$|\Delta(r_r + r_l)|_{max} \tan \psi_2 \approx d_s\left(\frac{h_s}{d_s}\right)\tan \psi_2 \tag{12.1.11}$$

The ratio of the former to the latter is in magnitude less than $\tan \phi_b h_s^{\frac{1}{2}}/3\sqrt{3} \tan \psi_2 z_a^{\frac{1}{2}}$. Thus unless

$$\tan \psi_2 < \left(\frac{h_s}{z_a}\right)^{\frac{1}{2}} \frac{\tan \phi_b}{3\sqrt{3}} \ll 1 \tag{12.1.12}$$

then the appearance of near close contact at $z_3 = 0$ for the contra-rotating case is perfectly realistic.

Indeed the term $(r_r + r_l) - 2z_{2a}$ is always positive for $z_3 \neq 0$ and so the contribution from the second term on the right-hand side of (12.1.4) may more than counterbalance the negative value obtained from the first term, i.e. (12.1.10), almost everywhere if $\tan \psi_2$ is large enough. It cannot for very small z_3 since the second term is $0(z_3^2)$ and the first is $0(z_3)$.

EXERCISE 12.1.1. Describe the geometry of the boundaries of one of the two screws shown in Figs. 12.1–12.3 in terms of the helical coordinates introduced in Appendix 3.

EXERCISE 12.1.2. Show, from first principles, that

$$\tilde{c} = 2c/d_s = \tilde{w}_a \tan^{-1}\left(\frac{2\tilde{z}_3\tilde{z}_2}{\tilde{z}_{2a}^2 - \tilde{z}_2^2 + \tilde{z}_3^2}\right) + \tilde{c}_{00}$$
$$+ [\{\tilde{z}_3^2 + (\tilde{z}_{2a} - \tilde{z}_2)^2\}^{\frac{1}{2}} + \{\tilde{z}_3^2 + (\tilde{z}_{2a} + \tilde{z}_2)^2\}^{\frac{1}{2}} - 2\tilde{z}_{2a}]\tan \psi_2$$

where

$$\tilde{w}_a = w_a/\pi d_s = \tan \phi_b, \qquad \tilde{z}_{2a} = (1 - h_s/d_s + c_{r0}/d_s)$$
$$\tilde{c}_{00} = 2c_{00}/d_s, \qquad \tilde{z}_2 = 2z_2/d_s, \qquad \tilde{z}_3 = 2z_3/d_s$$

for the contra-rotating intermeshing screws illustrated in Figs. 12.1–12.6.

Show also that the intermeshing region for which $\tilde{c}$ can be defined is

$$|\tilde{z}_2|<1-\tilde{z}_{2a}, \qquad \tilde{z}_3^2<1-(|\tilde{z}_2|+\tilde{z}_{2a})^2$$

Investigate the least value for $\tilde{c}$ within this region. (This is a non-trivial exercise in constrained minima.) In particular, verify the approximate results given in Section 12.1 for $h_s/d_s \ll 1$.

12.2 MECHANICS OF FLOW IN THE TWIN-SCREW EXTRUDER

It is helpful to consider the flow in a twin-screw extruder as consisting primarily of the flow within independent C-shaped chambers: this determines a first approximation to the pressure field and to the delivery rate $\dot{M}_T$. The leakage flows through the four gaps described above can then be regarded as perturbations, and can be calculated on the basis of the known kinematics of the boundaries, and of the first approximation to the pressure field calculated for the primary field. These leakage flows will reduce $\dot{M}_T$ and, after recalculation, will alter the pressure field. This procedure can be iterated.

If the extruder is solid fed, then the full flow field can be expected to involve solids-conveying, melting and fully-molten zones, as in the single-screw case, and provided the leakage flows are small, we can expect the primary flow itself to exhibit these separate regions.

We shall analyse the primary flow first, and then consider the effect of leakage.

12.2.1 Flow in the C-shaped Chambers

We note that the length to depth (or length to width) ratio will be at least 10, and so there will be some justification for seeking solutions that depend only slowly, i.e. not locally, on the y^1 coordinate (see Fig. 12.4) measuring distance in the helical channel direction. A true lubrication approximation cannot be employed, even for fully molten flow, because the aspect ratio $A = 0(1)$ (cf. Subsection 11.4.2). It is not possible, because of the intermeshing zones, to look for steady solutions in a frame of reference fixed in the screw. It is more convenient to take a frame of reference fixed in any C-chamber volume itself, i.e. one that moves relative to both barrel and screw; this volume moves, without rotation, steadily along the screw axis, and the flow field within it, particularly if a melting process is involved, will in general be unsteady. Formally then, we shall look for a solution of the form

$$\mathbf{v}(\mathbf{y}, t) = [v_{(1)}, v_{(2)}, v_{(3)}](y^1, y^2, y^3, t),\ T(\mathbf{y}, t),\ p(\mathbf{y}, t) \qquad (12.2.1)$$

where dependence of $\mathbf{v}$ upon y^1 and t may be weak over most of any one chamber flow field but cannot be neglected if a melting process is to be modelled. The conservation and constitutive equations will be those relevant for the single-screw extruder given earlier. It is the boundary conditions that will be different. For purposes of comparison with the single-screw analysis of Chapter 11, a crude approximating (x_1, x_2, x_3) Cartesian coordinate system oriented as shown in Fig. 11.3 will be employed in this situation also, with h_s/d_s supposed $\ll 1$ and $\psi_2 = 0$. The wall velocities will be given by†

$$\left.\begin{array}{l} v_{1W} = -\pi W_s d_s \sec\phi_b, \qquad v_{2W} = v_{3W} = 0 \\ \text{on } x_3 = 0, \text{ on } x_2 = 0 \text{ and on } x_2 = w_c \cos\phi_b \end{array}\right\} \qquad (12.2.2)$$

$$\left.\begin{array}{l} v_{1W} = -\pi W_s d_s \sin\phi_b \tan\phi_b, \qquad v_{2W} = -\pi W_s d_s \sin\phi_b \\ v_{3W} = 0 \text{ on } x_3 = h_s \end{array}\right\} \qquad (12.2.3)$$

where W_s is the screw rotation rate. An adequate approximation is provided by

$$\cos\phi_b \approx 1, \qquad \sin\phi_b \approx \phi_b, \qquad \sin\phi_b \tan\phi_b \approx 0 \qquad (12.2.4)$$

We note that these are steady. However, the temperature boundary conditions, may, because of the motion of the frame of reference relative to the barrel, be unsteady. A variation of T_b or T_s with the axial coordinate z can be converted into a time dependence by the transformation

$$\pi W_s d_s \tan\phi_b \, dt \leftrightarrow dz$$

The mass conservation integral, assuming negligible leakage at this level of approximation, can be written

$$\int_0^{w_c}\int_0^{h_s} \rho v_1(x_1, x_2, x_3, t)\, dx_3\, dx_2 \approx 0 \qquad (12.2.5)$$

for all t and $0 \leqslant x_1 < d_s(\pi - \psi_1)$. The relevant range of x_1 corresponds

† Here $\pi W_s d_s$ is the same as v_b in Chapter 11.

(see Fig. 12.2) to the length of the C-chamber volume unaffected by the intermeshing of the screw flights.

For $T_W(t) < T_m$, a frictional plug flow model of the type described in Subsection 11.2.1 will be relevant. The analogue of (11.2.5) would be

$$\left.\begin{array}{l} \text{across } x_3 = 0\text{: } t_{33},\ t_{13} = \omega_s t_{33},\ t_{23} = 0 \\ \text{across } x_3 = h_s\text{: } t_{33},\ t_{13} \approx 0,\ t_{23} = -\omega_b t_{33} \\ \text{across } x_2 = 0\text{: } t_{22},\ t_{12} = \omega_s t_{22},\ t_{23} = 0 \\ \text{across } x_1\text{: } t_{11},\ t_{21},\ t_{31} \\ \text{across } x_2 = w_c\text{: } t_{22},\ t_{12} = -\omega_s t_{22},\ t_{23} = 0 \end{array}\right\} \tag{12.2.6}$$

with (11.2.10) and (11.2.11) replaced by

$$h_s w_c \frac{dp}{dx_1} + \omega_s\{(2p + f)h_s + pw_c\} = 0 \tag{12.2.7}$$

and

$$fh_s = \omega_b p w_c \tag{12.2.8}$$

On eliminating f we get

$$\frac{1}{X} = -\frac{\omega_s}{h_s w_c}\{2h_s + w_c(1 + \omega_b)\} \tag{12.2.9}$$

Although the result (12.2.9) is a very crude one it shows as expected that $X < 0$. The problem of deciding on $p_0 = p(0)$ is more difficult to resolve even than before, and so we shall not discuss this frictional mechanism any further.

For $T_W(t) > T_m$, a lubricated plug flow of the type described in Subsection 11.2.2 will arise. At the screw plug interfaces, the molten layer will develop in the $-x_1$-direction, while at the plug barrel interface, it will develop in the z_1 (essentially the x_2-direction). However, we cannot follow this process beyond the point at which these lubricated layers reach the thickness of the various gaps (c_f, c_0, c_{r0}) because the observational evidence does not provide a clean picture of the kinematic regime that then ensues in the melting zone. Janssen (1978, Section VII.2) presents a limited amount of information for PP powder in a contra-rotating system, which is compatible with neither of the models (Maddock/Street/Tadmor/6-zone and Dekker/Klenk/3-layer) analysed above in Subsections 11.3.1 and 11.3.3; if anything, it is closest to the Menges–Klenk (1967; see also Klenk, 1968) observations on PVC powder which have not so far been satisfactorily analysed.

However, whichever model is relevant, it is clear that the melting mechanism associated with the deep chamber flow is insufficient to account for the relatively rapid melting rate that was observed by Janssen (1978, pp. 1–3): '... in the twin-screw extruder the melting length is surprisingly short compared with similar experiments carried out with single-screw extruders'. We can conclude tentatively that flow in the intermeshing region plays a significant part in the melting process, particularly for powder feed.

Once melting is complete, and the chamber can be taken to be full of melt, an analysis very similar to that given for $A=0(1)$ in Subsection 11.4.2 will be applicable. If thermal effects can be neglected, then a good approximation to the flow everywhere is given by

$$\mathbf{v} \approx \{v_1(x_2, x_3), 0, 0\} \tag{12.2.10}$$

since $\phi_b \ll 1$ in most cases, subject to

$$\int_0^{w_c} \int_0^{h_s} v_1 \, dx_3 \, dx_2 \approx 0 \tag{12.2.11}$$

This can be converted into the situation studied in Subsection 11.4.2 by a simple change in frame of reference to one fixed in the screw, so that the boundary conditions now become

$$\left.\begin{aligned} v_1^* = v_2^* = v_3^* = 0 \quad \text{on} \quad x_3 = 0;\ x_2 = 0,\ w_c \\ v_1^* = \pi W_s d_2,\ v_2^* = v_3^* = 0 \quad \text{on} \quad x_3 = h_s \end{aligned}\right\} \tag{12.2.12}$$

with (12.2.11) replaced by

$$\int_0^{w_c} \int_0^{h_s} v_1^* \, dx_3 \, dx_2 = \pi W_s d_s h_s w_c = \dot{M}_{T1}/\rho_m \tag{12.2.13}$$

Solution of the relevant equations, as carried out by Martin (1969), yields the desired value of dp/dx_1, which we can anticipate to be <0. Integrated along the chamber length it yields

$$P_c \approx |dp/dx_1| \, (\pi - \tfrac{1}{2}\psi_1)(d_s - h_s) \tag{12.2.14}$$

So far, because the leakage flows have been neglected, the analysis is the same for both co- and contra-rotating systems. A difference will only arise when we take account of flow in the intermeshing region. This we consider in the following subsections, where we shall necessarily be concerned with the interactions between the flows in adjacent or opposite chambers.

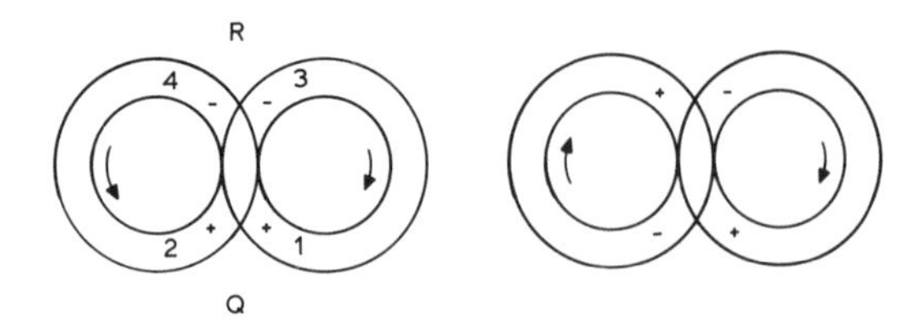

Fig. 12.7. Pressure fields in (a) contra-rotating, (b) co-rotating twin screws.

Figure 12.7 shows the basic difference between the pressure fields in the chambers for contra- and co-rotating systems that follows from our analysis above. In the former the pressure field is (in a crude sense) symmetrical and so there will be little exchange of material between screws. In the latter, however, it is asymmetrical about the dividing plane and so we can expect a certain amount of exchange. Indeed, in the case of poorly-fitting screws, we expect this to be large.

The significant differences in meshing region geometry illustrated in Fig. 12.6, in primary flow pressure distribution illustrated in Fig. 12.7, and in relative boundary motions lead to quite different behaviour in the co- and contra-rotating cases. They will therefore be dealt with separately.

12.2.2 Flow in the Intermeshing or Leakage Zones: Contra-rotating Case

The relation between the various C-shaped chambers is illustrated in the unrolled 'plan' shown in Fig. 12.8. This view is obtained by supposing the inner surface of the figure-of-eight shaped barrel, as shown in Fig. 12.5, for example, to be cut along the axial line R, and then flattened symmetrically about the axial line through Q onto the plane of Fig. 12.8. The inclined shaded regions then correspond to regions where the flight lands lie close to the barrel, while the unshaded regions correspond to regions bounding the deep screw channel. The arrows marked v_s indicate the direction of motion of the screws relative to the barrel surface viewed from outside the barrel looking in. The pattern shown in Fig. 12.8 therefore moves from left to right as the screws rotate, this being the direction of bulk flow.

The particular realization shown is that for which R′ and Q′ represent the nose end of the screws. The plus and minus signs represent the pressure differentials (P_c from (12.2.14)) shown in Fig. 12.7(a), where Q and R are again shown. Material moving into the calender gaps

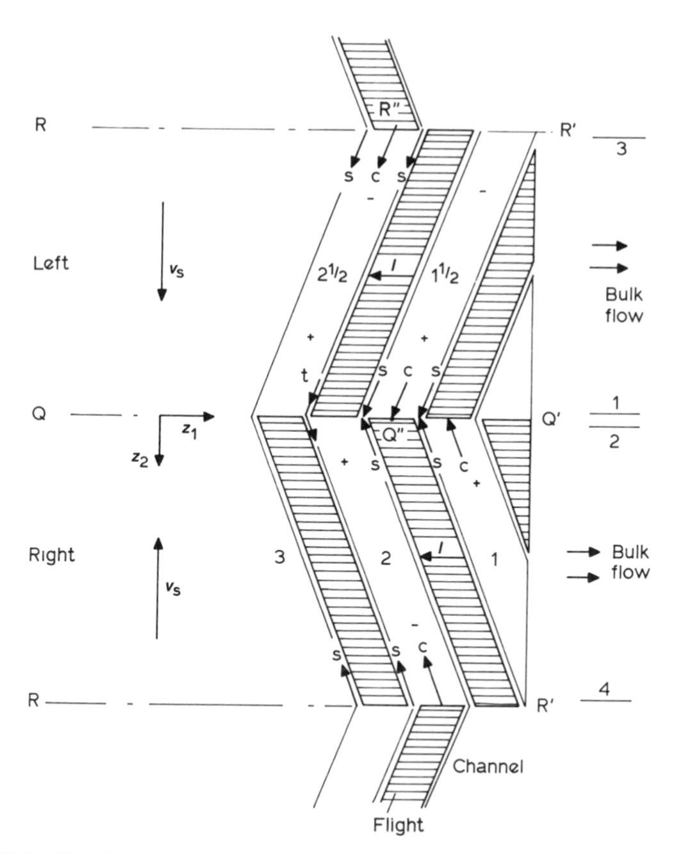

Fig. 12.8. Unrolled geometry for contra-rotating twin-screw extruder screws. The line QQ′ represents the intermeshing line looking down (in the z_3-direction, Fig. 12.2), while RR′ is that obtained looking up (in the z_3-direction). The pattern is periodic in the z_2 direction. The arrows v_s represent the direction of movement of the screw relative to the barrel viewed from below. Calender (c) and side (s) flows are primarily drag flows (large). Leakage (l) flow is primarily a pressure-driven flow (small).

along QQ″, as exemplified by the arrow labelled c at Q″ must by continuity reappear in the same direction at R″. Similarly for the side gaps where the flow of material corresponds to the arrows labelled s. These are driven largely by the motion of the screw-channel boundaries. Material moving through the tetrahedral gap, shown by the arrow labelled t, crosses the lines QQ′ or RR′ and is driven by pressure differences. Pressure profiles along the lines QQ′ and RR′ for the left

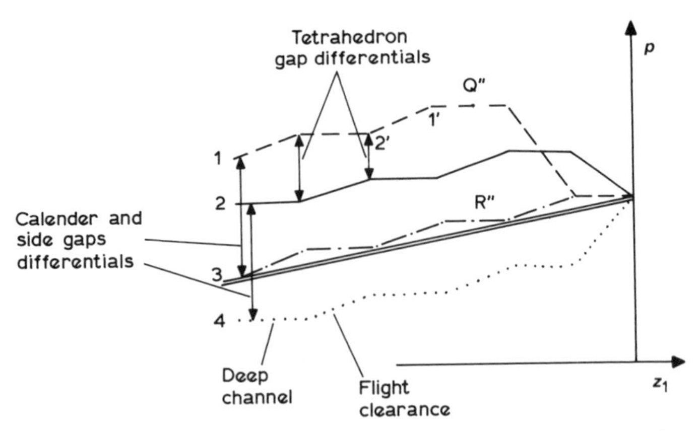

Fig. 12.9. Example of instantaneous pressure distribution along various axes (1–4) shown in Figs. 12.7 and 12.8 for contra-rotating twin screw extruder. ═══ Mean pressure profile, with slope yielding the mean pressure gradient $d\bar{p}/dz_1$. Pressure profiles along: --- axis 1; —·—·— axis 3; —— axis 2; ····· axis 4.

and right screw chambers are shown in Fig. 12.9, for the fully molten region near the nose end of the screws. It is supposed that the pressure p is uniform—at the delivery pressure P—along the line R′Q′R′. Assuming that the fully molten region extends back for many chambers, i.e.

$$\Delta z_{1\ \text{fully molten}} \gg w_a \tag{12.2.15}$$

then the delivery pressure P can be large compared with the pressure differential P_c, although its contribution, through $d\bar{P}/dz_1$, to the pressure differences in the various gaps can be small. The relative positions of the various pressure profiles in Fig. 12.9 depend on the assumptions that are made concerning the pressure profiles within an axial distance w_a of the nose, for which the theory given in Subsection 12.2.1 is not wholly relevant. However, the diagram shows that for all but the right-most C-chambers the pressure differences across the various gaps will be independent of the chamber concerned and are, in principle, calculable at any given instant, i.e. screw configuration. We can now turn to a consideration of the flows within each of the gaps.

(1) *Flight gap.* The region concerned is that shown in Figs. 12.5 and 12.10(a) as a narrow gap c_f between flight land and barrel. The axial

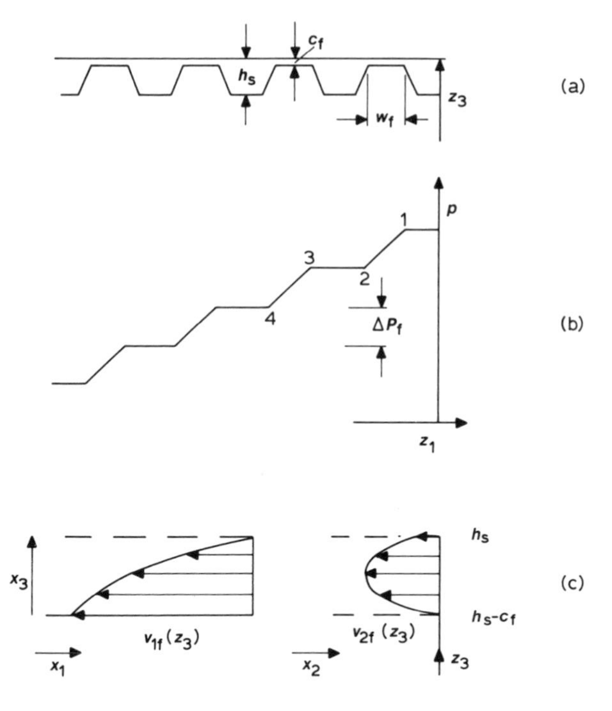

Fig. 12.10. Leakage flow in the flight gap. (a) Geometry of axial section. (b) Pressure profile corresponding to (a). The adverse gradients (1, 2) or (3, 4) represent the effect of back pressure and correspond to (1′, 2′) on Fig. 12.9. (c) Typical velocity profiles in the leakage gap.

pressure drop ΔP_f is that between equivalent positions in two adjacent chambers on one screw and is due largely to back pressure (at the nose) in the fully molten zone; any other contribution will be due to differences in dp/dx_1, i.e. to P_c in the neighbouring chambers. This is shown in Fig. 12.10(b). The velocity profile within the clearance will be largely drag flow $v_{1f}(x_3)$ in the x_1-direction, with

$$v_{1f}(h_s - c_f) \approx -v_s; \qquad v_{1f}(h_s) = 0 \tag{12.2.16}$$

and largely pressure-driven flow $v_{2f}(x_3)$ in the x_2-direction, with

$$v_{2f}(h_s - c_f) = 0; \qquad v_{2f}(h_s) = -v_s \sin \phi_b \tag{12.2.17}$$

where $\partial p/\partial x_2 \approx \partial p/\partial z_1$. These are shown in Fig. 12.10(c). Because w_f is often significantly larger than it would be in the single-screw-extruder case, the actual leakage flow will be smaller in the twin-screw-extruder case. Calculation procedures described earlier in Subsection 11.4.1 for $\mathrm{Gz_f} \ll 1$ will be directly relevant and a leakage flow q_{2f} obtained. This is unlikely to be important.

(2) *Calender gap.* The geometry of this gap is best understood by considering the planes $z_1 = \text{const}$. We then have the calender gap as shown in Fig. 12.11 given by

$$h_{cal} = c_{r0}(1 + z_3^2/Z_{cal}^2) \tag{12.2.18}$$

where

$$Z_{cal}^2 = \frac{c_{r0} d_s (d_s - 2h_s)}{2(d_s - h_s)} \approx \tfrac{1}{2} c_{r0}(d_s - h_s) \tag{12.2.19}$$

This is precisely the geometry studied in Chapter 13 in connection with a calender or 2-roll mill. The boundary conditions are given by

$$v_1(0) = \pi d_s W_s, \qquad v_1(h_{cal}) = \pi(d_s - 2h_s) W_s \tag{12.2.20}$$

where the lubrication approximation has been employed, the x_1-direction being along the centreline of the calender gap (i.e. along z_3), and the x_3-direction being perpendicular to the bounding planes; $v_1(x_3)$ is the only component of velocity considered. Equation (12.2.20) describes the strong calendering action of the contra-rotating screws. The velocities differ by the ratio $(1 - 2h_s/d_s)$ because the diameters of the 2 'rolls' are those of the screw roots and the flight lands respectively; thus there is some frictioning action. The inlet

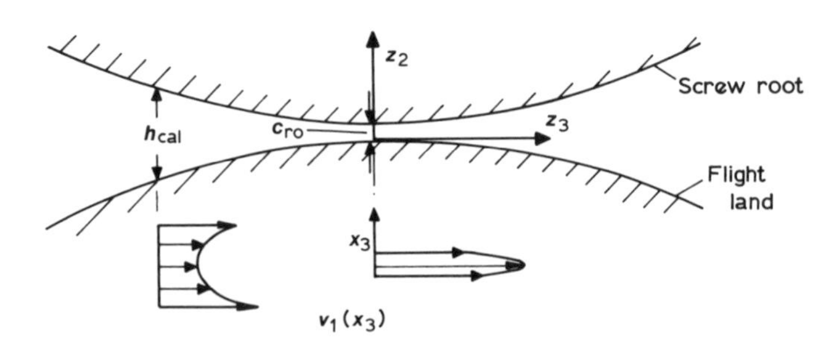

Fig. 12.11. The calender gap, showing typical velocity profiles at different stations z_3 (or x_1). The lower 'roll' moves with speed $\pi d_s W_s$ and the upper roll with speed $(d_s - 3h_s) W_s$.

$(x_1 \to \infty)$ pressure will be greater than the outlet $(x_1 \to +\infty)$ pressure by the sum of the chamber pressure-drop P_c and the back pressure ΔP_f already discussed in connection with the flight gap flow. This can be seen by reference to Fig. 12.9. We write

$$\Delta P_{cal} = P_c + \Delta P_f \tag{12.2.21}$$

From our calender analysis (see Subsection 13.1.1, in particular (13.1.23)) we can derive a characteristic pressure difference p^*_{cal}. Its value will depend—as will ΔP_f and P_c also—upon the rheology of the material, the operating conditions and the extruder geometry. If we assume that

$$\mathrm{Na}_{cal} \ll 1, \qquad \mathrm{Gz}_{cal} \ll 1 \tag{12.2.22}$$

for simplicity at this stage, and that a power-law model (3.3.5) applies, then

$$p^*_{cal} = K_m \left(\frac{d_s}{c_{r0}}\right)^{\nu+\frac{1}{2}} W_s^{\nu} \tag{12.2.23}$$

A similar very crude calculation for P_c yields

$$P_c \sim K_m \left(\frac{d_s}{h_s}\right)^{\nu+1} W_s^{\nu} \tag{12.2.24}$$

If we assume that ΔP_f is at most of order P_c, then ΔP_{cal} will only affect the magnitude of the calender gap flow when

$$c_{r0}^{\nu+\frac{1}{2}} d_s^{\frac{1}{2}} / h_s^{\nu+1} \geqslant 0(1) \tag{12.2.25}$$

For $\nu = 1$ this is when

$$\frac{c_{r0}}{h_s} \sim \left(\frac{h_s}{d_s}\right)^{\frac{1}{3}} \tag{12.2.26}$$

and, for the typical polymer value $\nu = 1/3$, when

$$\frac{c_{r0}}{h_s} \sim \left(\frac{h_s}{d_s}\right)^{\frac{3}{5}} \tag{12.2.27}$$

For typical twin-screw extruders

$$0{\cdot}1 < h_s/d_s < 0{\cdot}25 \tag{12.2.28}$$

At one extreme limit, with $\nu = 1$ and large h_s/d_s, c_{r0} has to be a substantial fraction of h_s before the pressure is effective; at the other extreme limit, with $\nu \to 0$, say, and h_s/d_s small, the chamber pressure drop will be significant even for relatively closely meshing ($c_{r0} \sim h_s/20$) screws. The ratio of the actual magnitude of the calender leakage flux to the volumetric delivery will be given by

$$\frac{q_{1cal} w_c \rho_m}{\dot{M}_{T1}} = \frac{q_{1cal}}{\pi W_s d_s h_s} = \frac{(d_s - h_s) c_{r0}}{d_s h_s} \tilde{h}_{e.c.} \tag{12.2.29}$$

using (12.2.13) and defining $\tilde{h}_{e.c.}$ as the dimensionless calender flow parameter. $\tilde{h}_{e.c.}$ is the quantity $\tilde{q}(\pm 1)$ defined in (13.1.43) below and is related to H/H_0 in Middleman (1977, Section 7.2); it will usually be between 1·25 and 1·50. Thus provided $c_{r0} \ll h_s$, the leakage flow will be negligibly small. If the leakage flow is not negligibly small, then it will be sensitive to the pressure drop across the calender gap, and so the contra-rotating twin-screw extruder will behave more like a traditional single-screw extruder, and less like a positive displacement pump.

(3) *Side gap.* Here, once again, we are concerned with flow in the x_1, i.e. z_3, direction (near $z_3 = 0$). However, whereas the channel profile for the calender gap (Fig. 12.11), represents the plane x_2 (or z_1) = constant, here we take channel profiles in the planes x_3 (or z_2) = constant. One such is provided by Fig. 12.6(a) ($z_2 = 0$) for which

$$h_{\text{side}}^{(0)} = c_{00}(1 + z_3^2/Z_{\text{side}}^{(0)2}) \tag{12.2.30}$$

$$Z_{\text{side}}^{(0)2} = \tfrac{1}{2} c_{00} d_s \cos \tfrac{1}{2}\psi_1 / \tan \psi_2 \tag{12.2.31}$$

with wall boundary conditions

$$v_1(x_2 = 0) = v_1(x_2 = h_{\text{side}}^{(0)}) \approx \pi(d_s - h_s) W_s \tag{12.2.32}$$

for the case $c_{00}, c_{r0} \ll h_s$. For this particular case we can use the calender analysis given earlier if we neglect v_3 effects. However, for other values of z_2, h_{side} has a term proportional to z_3 and so the channel is asymmetrical about $z_3 = 0$. It would be possible to undertake a lubrication analysis on the assumption that

$$c \ll h_s \tag{12.2.33}$$

everywhere, c being defined by (12.1.4). However, as soon as relatively wide gaps are permitted, and the effective length of the nip region, e.g. $Z_{\text{side}}^{(0)}$ in (12.2.31), becomes comparable with h_s, then the component v_3 cannot be neglected compared with v_1, and interactive effects would be relevant when $\nu \neq 1$. An estimate of $q_{1\text{side}}$ is provided by the pseudo-calender flow (12.2.30)–(12.2.32), and so the relative importance of the side and calender leakage flows is provided by the ratio c_0/c_{r0} where c_0 is defined by (12.1.3).

From a design point of view, it is clearly better to increase the leakage flux between adjacent chambers by increasing c_{r0} than by increasing c_0. The calender-gap flow is more controllable in a design sense than the side-gap flow. Their effect is essentially equivalent, and for an overall mass balance their fluxes can be added. It should be noted that the side gap relating to one screw ($z_2<0$ in Fig. 12.2 say) is adjacent to that relating to the other ($z_2>0$ in Fig. 12.2), i.e. each inclined gap c_0 shown in Fig. 12.3(b) consists of two adjacent side gaps, one for each screw.

(4) *Tetrahedron gap.* Although in an overall sense the tetrahedron gap is geometrically the same as the side gap—they are both defined by c—it is distinguished from the latter by relating to flow in the z_2, i.e. x_3-direction, between a C-shaped chamber on one screw and an opposing chamber on the other. Thus we are concerned with material passing perpendicular to the cross-sectional view shown in Fig. 12.6(b) through the symmetrical gaps above and below the line $z_3 = z_2 = 0$. For any given $z_3 \neq 0$ we find that the dominant terms in the expression for the tetrahedral gap width are

$$c \approx h_{\text{side}}^{(0)} + \left[\frac{4 \tan \phi_b z_3 (d_s - h_s)}{(d_s - h_s)^2 + 4z_3^3}\right] z_2 \qquad (12.2.34)$$

The boundary conditions on $v_3(x_2)$, the velocity component in the x_3, i.e. z_3, direction, will be

$$v_3(0) = -v_3(c) \approx \pi W_s z_3 \qquad (12.2.35)$$

which represents no more than a weak shearing effect. The only mass flux will thus be that due to the pressure differences between the opposed chambers. These are shown in Fig. 12.9 as being such as to drive material from the left screw to the right screw chambers (see Fig. 12.8) through all the tetrahedral gaps. This asymmetrical situation (dependent on the assumptions made in deriving the pressure profiles near Q′) cannot be true for all relative positions of the screws and must on symmetry grounds have a zero time mean. What will be left is a leakage flow driven by the pressure gradient term ΔP_f as shown in Fig. 12.12.

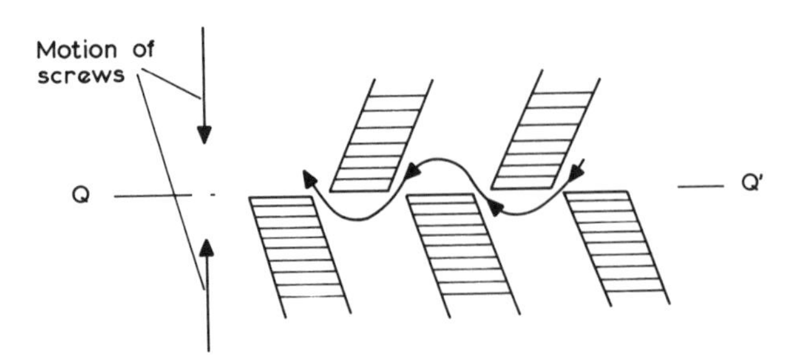

Fig. 12.12. Curved arrows show the mean effect of tetrahedral gap pressure flows in a contra-rotating twin screw extruder, leading to a sinuous leakage involving all C-shaped chambers. The diagram is similar to that shown in Fig. 12.8.

Thus the whole of the leakage flow in the intermeshing region for contra-rotating screws acts effectively independently for each screw. The flow follows the sequence of chambers numbered $1\frac{1}{2} \rightarrow 2\frac{1}{2} \rightarrow 3\frac{1}{2}$ for the upper screw and $1 \rightarrow 2 \rightarrow 3$ for the lower screw. That part of the flow which is driven by the back pressure gradient ΔP_{f} (which is related to $\mathrm{d}\bar{p}/\mathrm{d}z_1$ as discussed in Exercise 12.2.1) represents the departure of the system from pure 'volumetric' delivery, and is probably dominated by the tetrahedral gap flow; that part which is dominated by the relative motion of the screw channel and flight boundaries (in the calender and side-gap flows) represents an unavoidable back flow that reduces what would otherwise be the maximum volumetric delivery rate of the twin-screw extruder viewed as a positive displacement pump. If a temperature independent power-law relation like (3.3.5) applies then the latter back flow will be linearly proportional to the screw speed and to the net delivery rate of the screw pump. The back flow has been interpreted by Janssen (1978, Section V.3) as back mixing. It also represents for small c_{r0} and c_{00}, a region of intense shearing, and will in many cases act like the corresponding region in an internal (dispersive) mixer or a 2-roll mill. Figure 12.13 gives a simple diagrammatic view of the unrolled geometry to compare with that shown in Fig. 11.3 for a single-screw extruder.

Contribution to Melting

The analysis given above for flow in the intermeshing zone relates to fully molten flow both in the gap regions and in the chambers associated with them. However, it was earlier stated (Subsection 12.2.1

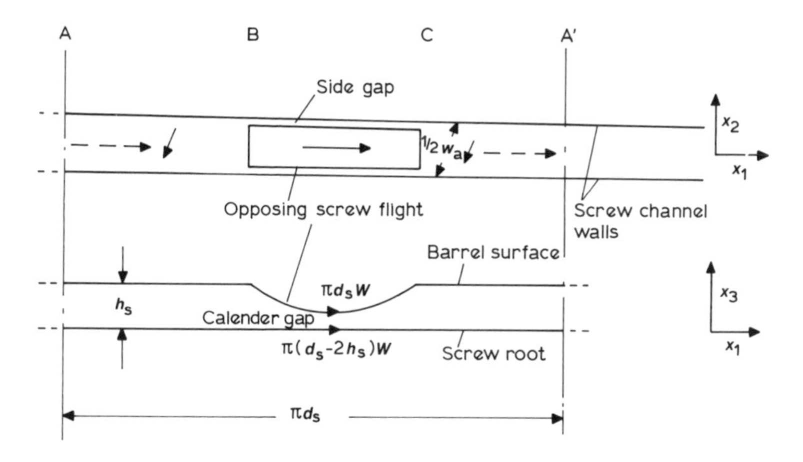

Fig. 12.13. Unrolled view of the periodically-repeating channel for one screw in a contra-rotating twin-screw extruder. The frame of reference is a slowly moving C-shaped chamber. The region BC is the intermeshing region. The region CAA′B is the uniform part of the chamber in contact with the barrel. ---→ motion of the screw relative to the chamber; ↙ motion of the barrel relative to the chamber; ⟶ motion of the opposing screw flight relative to the chamber. The calender leakage flow is from B to C in the lower diagram; the side gap leakage flow is from B to C in the upper diagram along the screw channel walls. The tetrahedron gap flow lies outside the upper diagram and goes from BC to B′C′ (not shown but lying to the right of A′) through the opposing screw channel.

and Exercise 12.2.3) that observed melting rates could not be explained in terms of the chamber flow above and that a substantial, indeed in some cases even a dominant, contribution would have to come from the intermeshing regions. So far neither a quantitative theory nor clear experimental evidence has been advanced to describe such a melting process, although it has long been known that solid material can be completely melted by passing through a narrow calender gap, even when the metal rolls are not significantly above, or even quite at, the melting temperature. A significant factor is thought to be internal heating by viscous dissipation, even in a plastic, rather than viscous, flow range. Such a process would clearly be much more critically dependent on c_{r0} than would the overall output rate. Indeed if we were to argue that passage through a calender nip were the sole requirement for melting, then we could obtain a rough estimate of the number of turns n_m required for melting from the crude relation

$$n_m = kh_s/c_{r0}$$

by comparing the delivery rate per screw, $\frac{1}{2}k_1\phi_b d_s h_s(\pi - \frac{1}{2}\psi_1)v_s$, with the rate of flow per calender nip, $\frac{1}{2}k_2\phi_b d_s c_{r0}\,\pi v_s$, where the ratio of the parameters k_1 and k_2 will be of order unity and k will probably be near $\frac{1}{2}$. For $n_m = 10$ say, the ratio in (12.2.29) will represent a back flow of only 5%.

12.2.3 Flow in the Intermeshing or Leakage Zones: Co-rotating Case

The unrolled form of the C-shaped chambers is shown in Fig. 12.14, and is to be compared with Fig. 12.8. Once again, the arrows marked

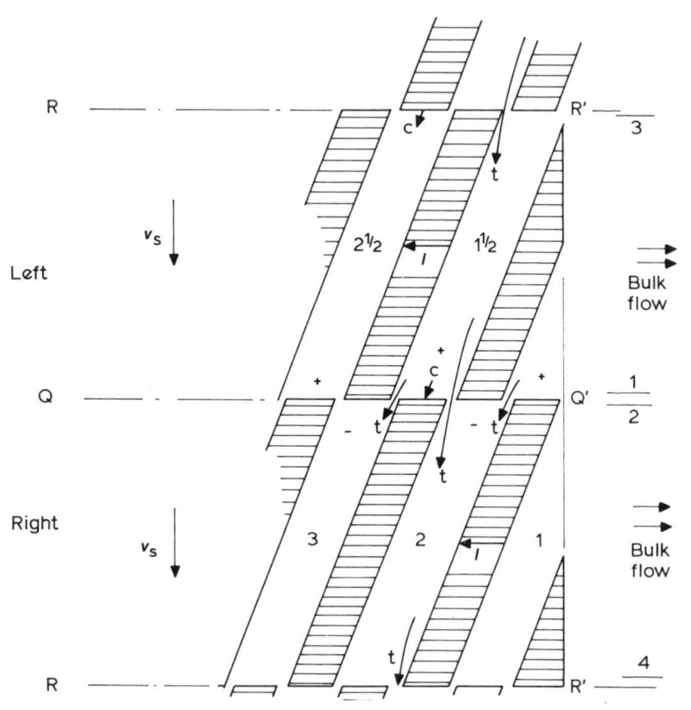

Fig. 12.14. Unrolled geometry for co-rotating twin screw extruder screws. QQ′, RR′ and v_s are defined as in Fig. 12.8. Calender (c) and side (s) flows are now primarily pressure flows (small). Leakage flow (l) is again primarily a pressure-driven flow (small). The tetrahedron flow (t) is driven by both pressure and drag (large).

v_s indicate the direction of motion of the screws, and the pattern moves from left to right. An important difference arises when the effect of the pressure differential P_c is considered, which is here represented again by + and − signs at either end of each chamber. This is such that there is a mean pressure difference forcing fluid from the 'upper' to the 'lower' screw through the tetrahedral gaps on QQ′ and from the 'lower' to the 'upper' screw through the tetrahedral gaps on RR′. This will be assisted by the motion of the screw flight surfaces. On the other hand, only the pressure differential acts to force material through the calender and side gaps. The motion of the screw channel boundaries leads largely to frictioning. The relevant pressure profiles to compare with Fig. 12.9 are given in Fig. 12.15. In the realization chosen, the only difference lies in the reversal of the profiles for the top (QQ′) and bottom (RR′) of the lower screw, with the consequences just mentioned.

(1) *Flight gap.* The flows in this region will be just as for the contra-rotating case.

(2) *Calender gap.* The geometry will be as given by (12.2.18) and (12.2.19) but the velocity boundary conditions become

$$v_1(0) = \pi d_s W, \qquad v_1(h_{cal}) = -\pi(d_s - 2h_s)W_s \tag{12.2.36}$$

The same general arguments will apply as for the contra-rotating case,

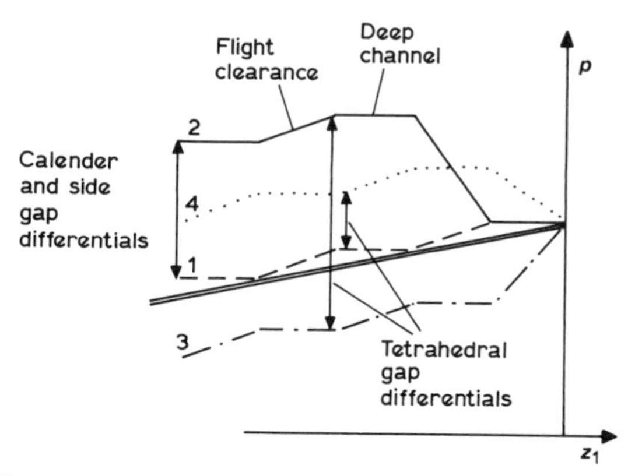

Fig. 12.15. Example of instantaneous pressure distribution along various axes (1–4) shown in Fig. 12.14 for a co-rotating screw extruder. See Fig. 12.9 for line legends.

except that now

$$\frac{q_{1\text{cal}} W_c \rho_m}{\dot{M}_{T1}} \approx \frac{c_{r0}}{d_s} \tilde{h}_{\text{e.c.}} \qquad (12.2.37)$$

with $\tilde{h}_{\text{e.c.}}$ still close to unity. The leakage flow is even less likely to be significant than in the earlier case.

(3) *Side gap.* The same argument applies as for the calender gap in that the velocity boundary condition (12.2.32) is replaced by

$$v_1(x_2=0) = -v_1(x_2=h_{\text{side}}^{(0)}) \approx \pi(d_s - h_s) W_s \qquad (12.2.38)$$

with similar consequences.

(4) *Tetrahedron gap.* There is a geometrical difference between the contra-rotating case and the present, as illustrated in Fig. 12.6. As explained in Section 12.1 the two screws can touch in the plane $z_2=0$ and the tetrahedron gap is dominated in that plane by the term (12.1.7). For any given value of z_3, the dependence of c on z_2 is $0(z_2^2)$ and so (12.2.34) is replaced by a calender gap

$$c \approx h_{\text{side}}^{(0)}(1 + z_2^2/Z_{\text{tet}}^2) \qquad (12.2.39)$$

The flow is much greater through the wide gaps than the narrow ones as is shown by the arrows marked t in Fig. 12.14; thus the tetrahedral leakage flow carries material along the sequence of chambers numbered $1 \rightarrow 1\frac{1}{2} \rightarrow 2 \rightarrow 2\frac{1}{2}$ in the figure. It is difficult to provide an adequate analogue of Fig. 12.13. However, if the tetrahedral gap is large enough, the result is to connect the two screws in such a way that they act like a single-screw extruder, with the screw channel adopting a figure-of-eight form around two parallel adjoining, but otherwise equal, screws. The back flow through the intermeshing zone from chamber to chamber on the same screw will be less than in the corresponding contra-rotating case. One is forced to conclude therefore, that unless a twin-screw extruder is designed to have very little leakage flow at all, in which case there is little difference in performance between the co- or contra-rotating systems, the contra-rotating system is the only one that can offer a substantial improvement in performance over a traditional single-screw system.

EXAMPLE 12.2.1. Estimate the relative contribution to melting in a C-shaped region from barrel and screw surfaces.

Barrel surface: Suppose that a uniform width $w_s \leqslant w_a - w_f$ is available for melting as in region C of the 6-zone model of Subsection 11.3.1. Then the rate of melting will be $\dot{m}_{C4}$ of (11.3.24) at $x_{A4} = w_s$ with $v_{c\,rel\,4} = v_{2W}$ of (12.2.3).

The special case B = Gn = 0 was set as Exercise 11.3.4. Using the similarity form introduced in Subsection 11.2.3, and with obvious simplification in notation, we have to solve the equation

$$\mathrm{Sf}\frac{d^2\tilde{T}}{d\hat{x}_3^3} = A(\hat{x}_3 - \tfrac{1}{2}\hat{x}_3^2)\frac{d\tilde{T}}{d\hat{x}_3}$$

subject to

$$\tilde{T}(0) = 1, \qquad \tilde{T}(1) = 0, \qquad \frac{d\tilde{T}}{d\hat{x}_3}(1) = -\tfrac{1}{2}A$$

where

$$\tilde{v}_4 = (1 - \hat{x}_3), \qquad \tilde{v}_3 = -\frac{1}{2}\left(\frac{A}{2\tilde{x}_4}\right)^{\frac{1}{2}}\hat{x}_3^2, \qquad \tilde{h} = (2A\tilde{x}_4)^{\frac{1}{2}}$$

provided $\rho_m = \rho_s$, with simple solution

$$\tilde{T} = \frac{\displaystyle\int_{\tilde{x}_3}^{1} \exp\{-A(3x^2 - x^3)/6\mathrm{Sf}\}\,dx}{\displaystyle\int_{0}^{1} \exp\{-A(3x^2 - x^3)/6\mathrm{Sf}\}\,dx}$$

A is given by

$$e^{-A/3\mathrm{Sf}} = \tfrac{1}{2}A\int_0^1 \exp\{-A(3x^2 - x^3)/6\mathrm{Sf}\}\,dx$$

For large Sf, $A \to 2$ as in Exercise 11.3.3.

The total flow rate of molten polymer/unit length in the y^1-direction

$$\dot{m}_2 = \rho_m\left(\frac{1}{2}\frac{A\kappa_m w_s v_{2W}}{\mathrm{Sf}}\right)^{\frac{1}{2}} \sim \rho_m\left(\frac{\kappa_m w_s v_{2W}}{\mathrm{Sf_b}}\right)^{\frac{1}{2}} \quad \text{for large Sf}$$

Screw surfaces: Suppose that a uniform length $l < (\pi - \frac{1}{2}\psi_1)d_s$ is available for melting by a process equivalent to that described for the barrel surface above, with v_{2W} replaced by v_{1W}. The relevant melting rate

$$\dot{m}_1 \sim \rho_m\left(\frac{\kappa_m l v_{1W}}{\mathrm{Sf_s}}\right)^{\frac{1}{2}}$$

Let us now suppose that the solid bed remains in contact with the screw around the whole periphery and that the melt accumulates at one end of the C-shaped channel. In this case, l becomes a function of t, while w_s can be taken for $\psi_2 \approx 0$, as roughly $\frac{1}{2}w_a$. Thus the total volumetric rate of melting becomes

$$\dot{V}_m \approx \rho_m^{-1}[\dot{m}_2 l + \dot{m}_1(2h_s + w_s)]$$

Putting $\mathrm{Sf_b} = \mathrm{Sf_s}$ and $\sec \phi_b \approx 1$ gives

$$\dot{V}_m \approx \left(\frac{\pi \kappa_m W_s d_s w_s l}{\mathrm{Sf}}\right)^{\frac{1}{2}} \left\{ l^{\frac{1}{2}} \sin^{\frac{1}{2}} \phi_b + w_s^{\frac{1}{2}}\left(1 + \frac{2h_s}{w_s}\right) \right\}$$

But $\dot{V}_m \approx -h_s w_s \dfrac{dl}{dt}$ and $\pi d_s \sin \phi_b = w_a \approx 2w_s$ and so writing $l = \pi d_s \tilde{l}$, $\tilde{t} = W_s t$

$$\frac{d\tilde{l}}{d\tilde{t}} = -\left(\frac{2\kappa_m}{\mathrm{Sf}\, W_s h_s^2}\right)^{\frac{1}{2}} \left\{ \tilde{l} + \left(\frac{1}{\sqrt{2}} + \frac{\sqrt{2}h_s}{w_s}\right) \tilde{l}^{\frac{1}{2}} \right\}$$

This has solution

$$\frac{\left(\dfrac{1}{\sqrt{2}} + \dfrac{\sqrt{2}h_s}{w_s}\right)^2 \tilde{l} + 1}{\left(\dfrac{1}{\sqrt{2}} + \dfrac{\sqrt{2}h_s}{w_s}\right)^2 \tilde{l}_0 + 1} = \exp\left\{ -\left(\frac{\kappa_m W_s}{2\,\mathrm{Sf}\, h_s^2}\right)^{\frac{1}{2}} t \right\}$$

Melting is complete when $\tilde{l} = 0$, i.e. when

$$t = \left(\frac{2\,\mathrm{Sf}\, h_s^2}{\kappa_m W_s}\right)^{\frac{1}{2}} \ln\left[1 + \left(\frac{1}{\sqrt{2}} + \frac{\sqrt{2}h_s}{w_s}\right)^2 \tilde{l}_0\right]$$

$\tilde{l}_0 \approx (1 - \frac{1}{2}\psi_1/\pi)$ and for $h_s \approx w_s$, the logarithmic factor will be approximately ln 5.

The C-shaped volume advances axially at a velocity $w_a W_s$ and so the axial length needed for complete melting will be approximately

$$\left(\frac{2\,\mathrm{Sf}\, h_s^2 W_s}{\kappa_m}\right)^{\frac{1}{2}} w_a \ln 5$$

Returning to the original question of relative efficiencies of melting, the model chosen leads to the screw contribution being dominant, starting at about 80% and rising to 100%.

EXERCISE 12.2.1. Verify that $\partial p/\partial x_2$ for the flight leakage flow satisfying (12.2.17) will be given approximately by $2\,\mathrm{d}\bar{p}/\mathrm{d}z_1$, where $\mathrm{d}\bar{p}/\mathrm{d}z_1$ is the mean pressure gradient given by the thick full line in Fig. 12.9.

EXERCISE 12.2.2. Carry out the analysis used in Example 12.2.1 for the case when the solid material retains a C-shaped length of $(\pi - \frac{1}{2}\psi_1)d_s$ but slowly decreases in width as a melt pool (similar to Region B in the 6-zone model) forms alongside it insulating one face from the screw flight.

EXERCISE 12.2.3. Show that for $w_a = 0{\cdot}1$ m, $h_s = 25$ mm, $W_s = 2\,\mathrm{s}^{-1}$, $\kappa_m = 10^{-7}\,\mathrm{m}^2\,\mathrm{s}^{-1}$, $\mathrm{Sf} = 1$, the melting length predicted by Example 12.2.1 is approximately $300 w_a$, which is (as argued in the text) unrealistic.

EXERCISE 12.2.4. Derive the relation (12.2.24). Show that if $\Delta P_f \gg P_c$, then the back pressure on the twin-screw extruder as a whole must be very much larger than could be developed by a single-screw extruder of equivalent dimensions. Deduce that in practice $\Delta P_f \not\gg P_c$.

EXERCISE 12.2.5. Verify relations (12.2.30)–(12.2.32) using the result for $\tilde{c}$ given in Exercise 12.1.2.

EXERCISE 12.2.6. Verify relations (12.2.34), (12.2.35) using the result for $\tilde{c}$ given in Exercise 12.1.2. Note that z_3^2 is taken to be more significant than z_2^2 in much of the tetrahedral gap.

EXERCISE 12.2.7. Obtain an estimate for Z_{tet} in (12.2.39).

B—ROLLING

Rolling refers to a process whereby polymeric material, usually in molten form, passes through the nip between two almost touching circular-cylindrical rolls with parallel axes. Both rollers are usually rotating and the forces required to carry the material through the nip are generated at the surfaces of the rollers in contact with it. Except in the case of embossed rolls, the result of the process is a sheet of more or less even thickness, and so it is much less versatile than screw extrusion discussed earlier.† The objective is the same in that a homogeneous output is aimed at.

The technique is used for a variety of purposes, among which it is worth distinguishing three:

(1) *Calendering.* This is used for the production of high quality sheet of from 25 to 250 μm in thickness, usually of highly viscous coherent materials, like PVC and various elastomers, that are readily degraded. The equipment is often large, and correspondingly expensive, consisting of a train of 3 or 4 polished metal rolls forming 3 sequential nips through which previously mixed and heated polymeric material passes. A useful description of the process can be found in Bernhardt (1959, Chapter 6). Close control of temperature, of bearing alignment and of roll speeds is an essential part of the process. Take-off by a smaller diameter roll often involves a 25% or so element of stretching of the calendered sheet.

(2) *Roll coating.* This involves the simultaneous passage through a

† A roller system has been used as an extruder to feed a die, e.g. Schenkel (1966), but this is not a common technique.

nip of a solid substrate material (metal sheet, polymer film or woven fabric) together with a relatively-low-viscosity liquid with which the substrate is to be coated. Coating on any given pass is usually one-sided, with the horizontal substrate below the coating liquid, gravity thus acting in a helpful manner. The coating material can be low viscosity (hot) wax, paint, PVC plastisol† or some special interfacial conditioner. Its thickness will usually be in the 5–50 μm range. Stoving, to drive off solvent or to gel polymer and/or subsequent cooling usually follows the rolling operation.

(3) *Roll milling.* This is the crudest of the three operations, and involves passing a mixture of rubber, fillers, pigments and possibly curing agents through a relatively wide nip (1–10 mm) to disperse and warm the mixture, and to yield, often on a batch basis, slabs of material for later use.

Each of these separate uses introduces its own characteristics and problems. Some of these are matters of mechanical engineering and will not be discussed in detail here. For our purposes we suppose that the geometry of the system is given and consists of a parallel pair of rigid rotating circular cylindrical rolls whose length is large compared to their diameter, and whose minimum separation is very small compared to either diameter. The confined flow in the nip can be treated according to unidirectional lubrication theory; special questions arise concerned with the rolling bank (excess material) on the entry side and the separation of material from one or both rolls on the exit side of the nip; the no-slip boundary condition will not always apply; heating and non-isothermal effects will often be significant; fluid-mechanical instabilities can be dominant features.

Calendering and coating are dealt with in the next two chapters. Milling is considered in a later chapter as an example of mixing.

† This consists of a suspension of PVC powder in a plasticizing agent. On subsequent heating, the plastisol gels into a homogeneous tough adherent sheet.

Chapter 13

Calendering

The geometry of the relevant plane flow field is shown in Fig. 13.1. A frame of reference is chosen fixed in the roller axes, and a Cartesian coordinate system has its origin at the mid-plane ($x_3 = 0$) of the nip, with the $x_1 = 0$ plane passing through the roller axes. The diameters of the rolls are d_1 and d_2 and the distance parallel to x_3 between their nearest surfaces is $h(x_1)$. The length of the rolls is supposed infinite. They rotate steadily with angular speeds W_1 and W_2, where a positive value of W_i means a positive value of the x_1 component of the peripheral velocity

$$v_{ri} = W_i d_i \tag{13.1}$$

of either roll surface in the nip region. The v_{ri} need not be equal, nor even of the same sign. The temperature of the roll surfaces is supposed held steady at T_1 and T_2 respectively. The minimum nip-gap

$$h_0 = h(0) \ll d_1, d_2 \tag{13.2}$$

Within the region for which

$$h(x_1) \ll d_1, d_2 \tag{13.3}$$

a lubrication flow dominated by

$$\mathbf{v} = \{v_1([x_1], x_3), 0, 0\} \tag{13.4}$$

will be dynamically relevant. The notation $[x_1]$ indicates that v_1 is a slowly-varying function of x_1, and so v_2 can be neglected in the stress equilibrium equation. When the energy equation becomes relevant then a more suitable relation can be obtained by using the stream function q, given by

$$v_1 = \frac{\partial q([x_1], x_3)}{\partial x_3} \tag{13.5}$$

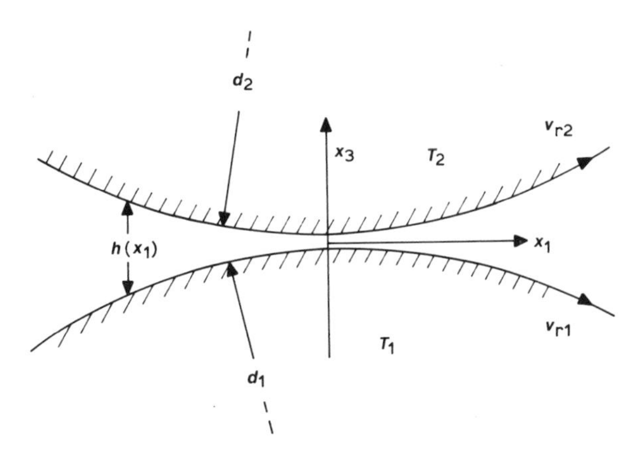

Fig. 13.1. Nip geometry.

However, when $h(x_1) = 0(d_1, d_2)$, i.e. when

$$dh/dx_1 = 0(1) \tag{13.6}$$

the lubrication approximation will not be directly relevant, if material fills the gap. This happens when material accumulates on the entry side, to form what is often called a rolling bank, as shown in Fig. 13.2. Flow patterns within the rolling bank are very complex and often unsteady, all three components (v_1, v_2, v_3) of velocity being relevant. The rolling bank acts as an intermediate reservoir between entry and exit flows; slight variations in the former are damped out by corresponding fluctuations in the content of the 'reservoir'; any slight mismatch in the flowrates (per unit length in the x_2-direction) between entry and exit flow can be eliminated in the mean by transverse (x_2) mean flow within the rolling bank. For this to be a thoroughly satisfactory mechanism, the nip-region flow has to be essentially independent of the rolling-bank flow. The rolling-bank flow is usually asymmetric about $x_3 = 0$, even when $d_1 = d_2$, $W_1 = W_2$, because material enters on only one of the two rolls.

On the exit side, it is essential that the calendered sheet separates smoothly from one, and ultimately both, roll surfaces. This means that the internal cohesive strength of the material has to be significantly greater than its adhesive strength at the roll surface. Using highly polished rolls, this means that the no-slip condition may not apply when the normal stress across the interface is of the order of 1 bar; this

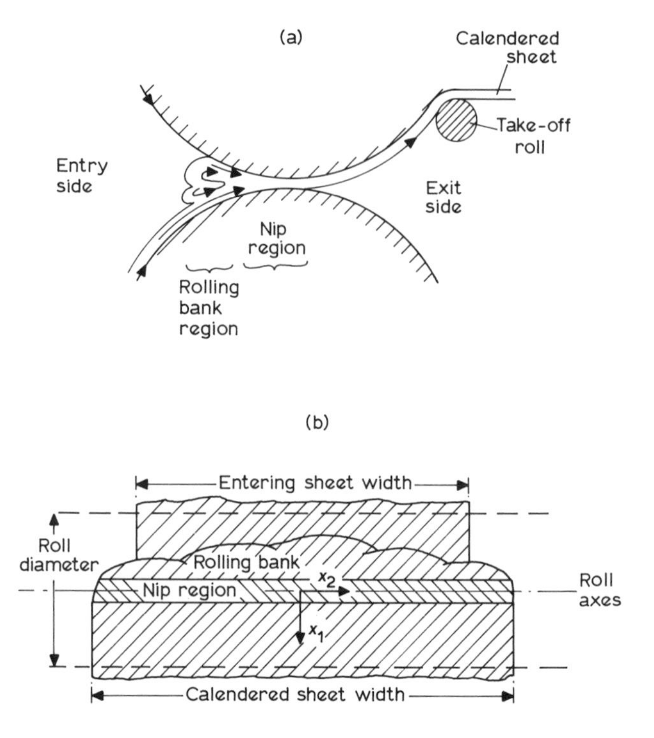

Fig. 13.2. Rolling bank. (a) Side view showing flow lines. (b) Plan view showing distribution effect.

in turn can lead to unstable behaviour on the inlet side, where the rotating rolls have to 'grip' the incoming material and force it through the rolling bank and nip. In extreme cases—which are much more common in roll milling than in calendering—the flow through the nip can be interrupted.

13.1 NIP FLOW MECHANICS. LUBRICATION APPROXIMATION

The geometry of the nip gap is rigorously defined by

$$h(x_1) = h_0 + \tfrac{1}{2}(d_1 + d_2) - (\tfrac{1}{4}d_1^2 - x_1^2)^{\frac{1}{2}} - (\tfrac{1}{4}d_2^2 - x_1^2)^{\frac{1}{2}},\ 4x_1^2 \leqslant \min(d_1^2, d_2^2) \tag{13.1.1}$$

For small values of $|x_1/d_1| \geqslant |x_1/d_2|$, this can be written

$$h(x_1) = h_0(1 + x_1^2/X^2) + 0(x_1^4/d_1^4) \tag{13.1.2}$$

where

$$X^2 = h_0 d_1 d_2/(d_1 + d_2) \tag{13.1.3}$$

A characteristic length in the x_1-direction is thus given by X. The lubrication approximation, according to Section 8.1, will apply when

$$\frac{dh}{dx_1} \approx \frac{2h_0 x_1}{X^2} = \frac{2(d_1 + d_2)x_1}{d_1 d_2} \ll 1 \tag{13.1.4}$$

This will be true when

$$x_1 \ll d_1, d_2 \tag{13.1.5}$$

Using as representative values $d_1 = d_2 = 1$ m, $h_0 = 0{\cdot}2$ mm, we find $X = 10$ mm. When $x_1 = X$, $h(x) = 2h_0$, and $dh/dx_1 = 4X/d_1 = 0{\cdot}04 \ll 1$, which is within the lubrication-approximation range. However, when $x_1 = 7X$, which is still such that $x_1/\frac{1}{2}d_1 < 0{\cdot}15$, $h = 50h_0 = X = 0{\cdot}01d_1$, and $dh/dx_1 = 0{\cdot}28 = 0(1)$. At this stage the lubrication approximation breaks down and we are within the rolling bank region.

Within the nip region we can suppose the boundaries to the flow to be given by $x_3 = \pm\frac{1}{2}h$ without loss of accuracy. The boundary conditions to the flow field (13.4) are then given by

$$v_1(\tfrac{1}{2}h) = v_{r2}, \qquad v_1(-\tfrac{1}{2}h) = v_{r1} \tag{13.1.6}$$

The mass conservation equation can be integrated to give the mass flowrate/unit length,

$$\left.\begin{aligned} \dot{m}_1 &= \int_{-\frac{1}{2}h}^{\frac{1}{2}h} \rho_m v_1(x_3)\,dx_3 \text{ (constant)} \\ &= \bar{\rho}_m q_1 \text{ if the material is treated as incompressible} \end{aligned}\right\} \tag{13.1.7}$$

As we shall see later, q_1 will always be close to $\frac{1}{2}(v_{r1} + v_{r2})h_0$ in calendering flows.

The importance of thermal effects can be estimated by considering the various dimensionless groups introduced in Subsection 8.1.3. From (8.1.37) it follows that

$$v^*_{conv} = q_1/h_0 \tag{13.1.8}$$

From (8.1.36) we can write

$$v^*_{gen} = \left.\begin{matrix} |v_{r1} - v_{r2}| & \text{if} & |v_{r1} - v_{r2}| \geqslant 0(v_{r1} + v_{r2}) \\ v^*_{conv} & \text{if} & |v_{r1} - v_{r2}| < 0(v_{r1} + v_{r2}) \end{matrix}\right\} \tag{13.1.9}$$

T^*_{rheol} is a material property. We suppose that there exists a temperature T_0 characteristic of the rolling bank. We can thus evaluate

$$\mathrm{B}=\max\left(\frac{|T_1-T_2|}{T^*_{\text{rheol}}},\frac{|T_1-T_0|}{T^*_{\text{rheol}}},\frac{|T_2-T_0|}{T^*_{\text{rheol}}}\right) \tag{13.1.10}$$

to determine whether the operating temperature conditions will be important rheologically.

In practice T_0 may not be independent of T_1 and T_2: for example, if the rolling bank is small and the material enters the nip region in contact with roll 2, then $T_0 \approx T_2$ may be a reasonable appreciation.

From (8.1.39) we can write

$$\mathrm{Na}_0=\eta^*(v^*_{\text{gen}}/h_0, T_0)v^{*2}_{\text{gen}}/2\alpha^* T^*_{\text{rheol}} \tag{13.1.11}$$

α^* and T^*_{rheol} having their T_0 values. Because we now have a characteristic length scale for the flow direction, X, we can evaluate a Graetz number

$$\mathrm{Gz}_0=\rho^*\Gamma^*_{\text{p}}\mathrm{v}^*_{\text{conv}}h_0^2/\alpha^* X=\rho^*\Gamma^*_{\text{p}}q_1\{h_0(d_1+d_2)/d_1d_2\}^{\frac{1}{2}}/\alpha^* \tag{13.1.12}$$

We choose as representative values

$$\left.\begin{array}{lll} \rho^*=10^3\,\text{kg}\,\text{m}^{-3}, & \Gamma^*_{\text{p}}=2\,\text{kJ}\,\text{kg}^{-1}\,\text{K}^{-1}, & \alpha^*=0{\cdot}2\,\text{W}\,\text{mK}^{-1} \\ v^*_{\text{gen}}=v^*_{\text{conv}}=1\,\text{m}\,\text{s}^{-1}, & d_1=d_2=1\,\text{m}, & h_0=0{\cdot}2\,\text{mm}, \\ T^*_{\text{rheol}}=50\,\text{K}; & & \end{array}\right\} \tag{13.1.13}$$

as before

$$X=10\,\text{mm}, \quad \text{and} \quad D^*=v^*_{\text{gen}}/h_0=5000\,\text{s}^{-1} \tag{13.1.14}$$

which represents a very high shear rate; a reasonable value for η^* is then $1000\,\text{N}\,\text{s}\,\text{m}^{-2}$. Any difference between T_1 and T_2 will be chosen merely to provide preferential adhesion to one roll, so we can assume

$$\mathrm{B}<\tfrac{1}{2} \tag{13.1.15}$$

Substitution of (13.1.13) into (13.1.12) gives

$$\mathrm{Gz}_0=40 \tag{13.1.16}$$

which means that flow through the nip is not fully developed. Finally, from (13.1.11)

$$\mathrm{Na}_0=50 \tag{13.1.17}$$

From (13.1.17) and (13.1.15) it is clear that generation is likely to be much more important than imposed temperature differences, particularly, if v^*_{gen} is determined by $|v_{r1}-v_{r2}|$ rather than $|v_{r1}+v_{r2}|/2$. However, from earlier arguments, we can estimate the joint effect of (13.1.16) and (13.1.17) partly in terms of (see eqn (7.1.40))

$$T^*_{adiab}/T^*_{rheol}=\mathrm{Na}_0/\mathrm{Gz}_0=1{\cdot}25 \qquad (13.1.18)$$

However this does not fully describe the influence of large Na_0, because it refers to a mean temperature rise. In practice, a very non-uniform temperature profile may occur as shown by the example given in Tadmor & Gogos (1979, Fig. 16.14); a diagrammatic indication is given in Fig. 13.3.

If, as mentioned earlier, calendering is used to deal with temperature-sensitive materials, then a detailed solution for the temperature field may be necessary in order to predict limiting velocities W_1 and W_2 at which the process can be run.

The importance of elastic effects can be estimated from an analogue of (8.1.6),

$$\mathrm{De}=\frac{\Lambda q_1}{h^{*2}}\left(\frac{\mathrm{d}h}{\mathrm{d}x_1}\right)^* \qquad (13.1.19)$$

which using the value $x_1=X$, i.e. $h=2h_0$, gives a maximum value

$$\mathrm{De}_{max}=\Lambda v^*_{conv}/X \qquad (13.1.20)$$

Using the values given above, we get

$$\mathrm{De}_{max}=(10^2\ \mathrm{s}^{-1})\Lambda \qquad (13.1.21)$$

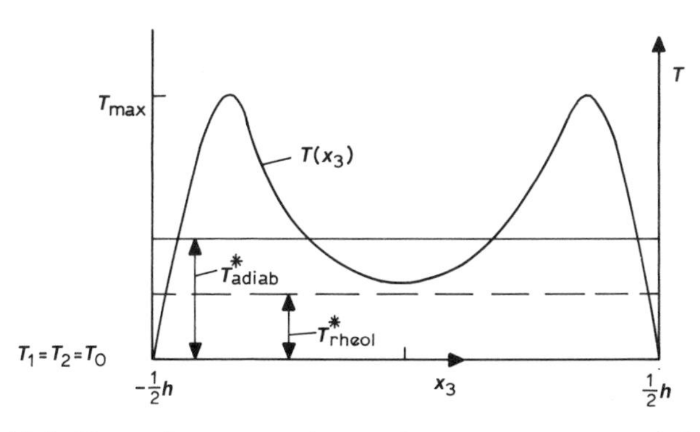

Fig. 13.3. Illustrative symmetrical profile for temperature $T(x_3)$ when $h=h_{exit}$ in calender nip.

Clearly if we take $\Lambda = 0(D^{*-1})$ from (13.1.14) then $\mathrm{De}_{\mathrm{max}} \ll 1$, as will always happen in lubrication theory. This would be consistent with the Leonov–Vinogradov (Subsection 5.2.1) notion of high shear rates 'destroying' the long relaxation times in the viscoelastic spectrum of relaxation times. However, if we look in greater detail at the flow field when $v_{r1} = v_{r2}$, we see that the central portion of the flow (near $x_3 = 0$) is not highly sheared and approximates to pure shear. The arguments given in Section 7.2 regarding entry flow to a capillary may therefore become relevant, with a very-highly-sheared boundary flow near the roll surfaces, and an essentially irrotational core flow. For this type of flow we shall expect $\Lambda \gg 10^{-2}$ s, so giving $\mathrm{De}_{\mathrm{max}} \gg 1$, according to (13.1.21).

If we take $x_1 = 7X$, which we earlier showed to be typical of the rolling-bank region, then (13.1.19) yields

$$\mathrm{De} \approx \tfrac{1}{4}\Lambda v^*_{\mathrm{conv}}/X \tag{13.1.22}$$

where now $h = 10$ mm and $D = 100\ \mathrm{s}^{-1}$. Thus even allowing Λ to be $0(D^{-1})$ still makes $\mathrm{De} = 0(1)$. From this we can deduce that an elastic analysis will be required for the rolling-bank region. Because the characteristic value of h has increased by a factor of 50 while that for l^* has increased by a factor of 10 at most, we find that the contribution to $T^*_{\mathrm{adiab}}/T^*_{\mathrm{rheol}}$ in the rolling bank, even allowing for an increase in η^*, will be less than the 1·25 given in (13.1.18). However, the relevant Graetz and Nahme numbers will still be high, and so the hope that the nip flow will be uninfluenced by the rolling-bank flow will not be realized in all cases.

13.1.1 B, Na≪1. Decoupled System

It is convenient to work as far as possible in dimensionless variables. We thus define

$$\left.\begin{aligned} \tilde{x}_1 &= x_1/X, \qquad \tilde{x}_3 = 2x_3/h_0, \qquad \hat{x}_3 = 2x_3/h \\ \tilde{\mathbf{v}} &= \mathbf{v}/v^*, \qquad \tilde{p} = ph_0^2/4Xv^*\eta^*, \qquad \tilde{h} = h/h_0 \end{aligned}\right\} \tag{13.1.23}$$

where η^* is a characteristic viscosity, say $\eta(v_{r1}/h_0)$, and v^* is a suitable velocity scale, say $\tfrac{1}{2}|v_{r1} - v_{r2}|$ or q_1/h_0. The usual stream function q is defined by

$$v_1 = \partial q/\partial x_3, \qquad v_3 = -\partial q/\partial x_1 \tag{13.1.24}$$

which leads to a dimensionless form

$$\tilde{q} = 2q/v^* h_0 \tag{13.1.25}$$

so that

$$\tilde{v}_1 = \partial\tilde{q}/\partial\tilde{x}_3, \qquad \hat{v}_3 = (2X/h_0)\tilde{v}_3 = -\partial\tilde{q}/\partial\tilde{x}_1 \tag{13.1.26}$$

The geometry of the channel, (13.1.2), becomes

$$\tilde{h} = 1 + \tilde{x}_1^2 \tag{13.1.27}$$

(note that $\partial\tilde{h}/\partial\tilde{x}_1 = 2\tilde{x}_1$ is 0(1) unlike (13.1.4)) with the roll surfaces at $\hat{x}_3 = \pm 1$. The velocity boundary conditions become

$$\tilde{v}_1(-1) = \tilde{v}_{r1}, \tilde{v}_1(1) = \tilde{v}_{r2} \tag{13.1.28}$$

where v_{r1} and v_{r2} are made dimensionless with v^* and the error in taking (13.1.28) to be equalities is of order $(h_0/2x_1)^2$, which being very small is the basis of the lubrication theory.

The use of the stream function (13.1.24) embodies the incompressible continuity equation (2.2.7) in a local sense. However, the lubrication approximation proper neglects the term v_3, replacing (2.2.7) by the integrated form (13.1.7). These approaches can be combined in a way which is particularly helpful where the (decoupled) energy equation is to be solved by making q and x_1 the independent variables, instead of x_3 and x_1. In these new coordinates, $v_q \equiv 0$, and so only the term $v_1\partial(\ )/\partial x_1$ survives in the convective term $\mathbf{v}\,.\,\boldsymbol{\nabla}$, while temperature gradients in the q-direction dominate those in the x_1-direction as far as generation is concerned. Thus, almost everywhere within the lubrication approximation, lines of constant q are effectively parallel to lines of constant $\hat{x}_3$. It matters little which representation is used for the dynamical equation, because the separation between the local solution and the global problem is complete ($\mathrm{d}p/\mathrm{d}x_1$ is treated as constant in the local stress-equilibrium equation and is trivially integrated to give the global solution), but it does simplify the solution of the energy equation to use (q, x_1) coordinates, because the latter make the equation parabolic in an obvious way that asymptotes smoothly to both the Graetz–Nusselt type of solution (for uncoupled flows in channels of constant depth) and the more complex solutions for melting interfaces.

The kinematical field will be defined by $q(x_1, x_3)$ or equivalently $\tilde{q}(\tilde{x}_1, \hat{x}_3)$. Equation (13.1.7) becomes

$$\begin{aligned} q_1 &= \tfrac{1}{2}h_0 v^* \int_{-1}^{1} \frac{\partial\tilde{q}}{\partial\hat{x}_3}\,\mathrm{d}\hat{x}_3 = \tfrac{1}{2}h_0 v^*[\tilde{q}(\tilde{x}_1, 1) - \tilde{q}(\tilde{x}_1, -1)] \\ &= q(x_1, \tfrac{1}{2}h) - q(x_1, -\tfrac{1}{2}h) \end{aligned} \tag{13.1.29}$$

Thus if we put

$$v^* = q_1/h_0 \tag{13.1.30}$$

then we can conveniently define

$$\tilde{q}(x_1, \pm 1) = \pm 1 \tag{13.1.31}$$

This has the consequence that $\tilde{q}_1 = \tilde{q}_1(+1) - \tilde{q}_1(-1) = 2$ rather than 1, but proves convenient in the symmetrical case $v_{r1} = v_{r2}$.

The local dynamical equation follows from (8.1.17) as

$$\frac{\partial}{\partial \hat{x}_3}\left(\tilde{\eta}(\tilde{D}, \tilde{T})\frac{\partial^2 \tilde{q}}{\partial \hat{x}_3^2}\right) = \tilde{h}^3 \frac{\mathrm{d}\tilde{p}}{\mathrm{d}\tilde{x}_1} \tag{13.1.32}$$

where $\tilde{h}(\tilde{x}_1)$ and $\mathrm{d}\tilde{p}/\mathrm{d}\tilde{x}_1$ are treated as constants and so the left-hand side becomes an ordinary differential equation. The boundary conditions are (13.1.31) together with

$$\frac{\partial \tilde{q}}{\partial \hat{x}_3}(-1) = \tilde{h}\tilde{v}_{r1}, \qquad \frac{\partial \tilde{q}}{\partial \hat{x}_3}(1) = \tilde{h}\tilde{v}_{r2} \tag{13.1.33}$$

from (13.1.28). Within the lubrication approximation proper,

$$\tilde{D} = \left|\frac{\partial \tilde{v}_1}{\partial \tilde{x}_3}\right| = \frac{1}{\tilde{h}^2}\left|\frac{\partial^2 \tilde{q}}{\partial \hat{x}_3^2}\right| \tag{13.1.34}$$

For present purposes we neglect dependence upon $\tilde{T}$, and so it is clear that

$$\frac{\mathrm{d}\tilde{p}}{\mathrm{d}\tilde{x}_1} = \frac{\mathrm{d}\tilde{p}}{\mathrm{d}\tilde{x}_1}(\tilde{v}_{r1}, \tilde{v}_{r2}, \tilde{h}) \tag{13.1.35}$$

which is the relevant form of (8.1.23). Solutions can be obtained for various $\tilde{\eta}$.

For engineering design purposes it is useful to know the torque and normal load/unit length on the rolls. The former is obtained, to the approximation considered, from

$$L_i = \tfrac{1}{2}d_i \int t_{13}^{(i)}(x_1)\,\mathrm{d}x_1 \tag{13.1.36}$$

where

$$t_{13}^{(1)} = -\left[\eta\frac{\partial v_1}{\partial x_3}\right]_{x_3=-\frac{1}{2}h}, \qquad t_{13}^{(2)} = \left[\eta\frac{\partial v_1}{\partial x_3}\right]_{x_3=\frac{1}{2}h}$$

and the integral is taken over the length of the nip that is full of material. The rate of working/unit width (on the material) is then

$$\dot{W} = L_1 W_1 + L_2 W_2 \tag{13.1.37}$$

The work done/unit volume of material passing through the nip is

$$\rho c T^*_{\text{adiab}} = \dot{W}/q_1 \tag{13.1.38}$$

The normal load on the rolls is given by

$$F_1 = F_2 = \int (p - p_a)\, dx_1 \tag{13.1.39}$$

again with the integral taken over the filled portion of the nip; p_a is the ambient pressure.

The integrals (13.1.36) and (13.1.39) introduce the non-trivial matter of the entry and exit conditions.

One mathematically acceptable approach is to suppose that the flow extends from $-\infty$ to $+\infty$; this is possible within the geometrical approximation (13.1.2), without leading to singularities in pressure, although the lubrication approximation itself fails for sufficiently large $|x_1|$. The obvious boundary conditions to apply to p are then

$$\tilde{p}(-\infty) = \tilde{p}(\infty) = 0 \tag{13.1.40}$$

which overdetermines the solution to (13.1.35). This is because q_1 cannot be chosen arbitrarily; in any given physical situation for which $\mathbf{v}_1, \mathbf{v}_2, h_0$ and X are specified there is a unique solution for q_1. This situation can more easily be understood if

$$v^* = \tfrac{1}{2}(v_{r1} + v_{r2}); \qquad \tfrac{1}{2}(v_{r1} - v_{r2}) = f v^* \tag{13.1.41}$$

in which case

$$\tilde{v}_{r1} = 1 + f,\ \tilde{v}_{r2} = 1 - f \tag{13.1.42}$$

and (13.1.31) is replaced by

$$\tilde{q}(\pm 1) = \pm 2q_1/(v_{r1} + v_{r2})h_0 \tag{13.1.43}$$

and so $\tilde{q}_1$ is varied until both of (13.1.40) can be satisfied.

It may be verified that $\tilde{q}_1$ will lie somewhat above 1. For a Newtonian fluid the value is 4/3 rising to 1·5 or so for a very shear-rate dependent fluid.

If the material is being calendered between two rolls with $v_{r1} = v_{r2}$, and a single sheet is formed on the exit side, adhering preferentially to one roll, then it is reasonable to suppose that $\tilde{x}_{1\ \text{exit}}$ will be given by

$$\tilde{h}(\tilde{x}_{1\ \text{exit}}) = q_1/v_{r1}h_0 \tag{13.1.44}$$

At that point, the solution to (13.1.32) is given by

$$\tilde{q} = \hat{x}_3,\ \mathrm{d}\tilde{p}/\mathrm{d}\tilde{x}_1 = 0 \tag{13.1.45}$$

This satisfies (13.1.33) since $\tilde{v}_{r1} = 1/\tilde{h}$ according to (13.1.44) using (13.1.23) and (13.1.30), and is the solution for the undeforming flow of a layer of thickness $\tilde{h}$ on a surface moving with 'velocity' $\partial\tilde{q}/\partial\tilde{x}_3 = 1$. On the upstream side, it is reasonable to assume that a rolling bank will form which effectively fills the whole inlet-side volume. The analogue of (13.1.40) is then

$$\tilde{p}(-\infty) = \tilde{p}(\tilde{x}_{1\ \text{exit}}) = \tilde{p}_a \tag{13.1.46}$$

where $\tilde{p}_a$ is the dimensionless ambient (air) pressure.

If $v_{r2} \neq v_{r1}$ then the argument given above does not hold. Neither reliable experimental evidence, nor any convincing theory, has been advanced to cover this case. It is probable that the condition $\mathrm{d}\tilde{p}/\mathrm{d}\tilde{x}_1 = 0$ would be closer to reality than (13.1.44). This gives a unique solution provided the obvious physical requirement, $\partial\tilde{h}/\mathrm{d}\tilde{x}_1 > 0$ at $\tilde{x}_{1\ \text{exit}}$, holds.

Newtonian case

As an illustrative example, which will enable us to draw further useful kinematical conclusions, we consider the constant-viscosity case, $\tilde{\eta} = 1$. From (13.3.32) using (13.1.31) and (13.1.33) we obtain the local solution

$$\begin{aligned}\tilde{q} = {} & \tfrac{1}{4}[\tilde{h}(\tilde{v}_{r1} + \tilde{v}_{r2}) - 2]\hat{x}_3^3 + \tfrac{1}{4}\tilde{h}(\tilde{v}_{r1} - \tilde{v}_{r2})\hat{x}_3^2 \\ & + \tfrac{1}{4}[6 - \tilde{h}(\tilde{v}_{r1} + \tilde{v}_{r2})]\hat{x}_3 + \tfrac{1}{4}\tilde{h}(\tilde{v}_{r1} - \tilde{v}_{r2})\end{aligned} \tag{13.1.47}$$

where

$$\frac{\mathrm{d}\tilde{p}}{\mathrm{d}\tilde{x}_1} = \frac{3\{\tilde{h}(\tilde{v}_{r1} + \tilde{v}_{r2}) - 2\}}{2\tilde{h}^3} \tag{13.1.48}$$

We note, as would be expected for a Newtonian fluid, that the pressure distribution and $\tilde{q}_1$ are independent of $\tilde{v}_{r1} - \tilde{v}_{r2}$; if non-zero, this only leads to a uniform shear, i.e. 'frictioning'. If we use the criterion

$\mathrm{d}\tilde{p}/\mathrm{d}\tilde{x}_1 = 0$ as the criterion for $\tilde{x}_{1\,\text{exit}}$, then we have

$$1+\tilde{x}_{1\,\text{exit}}^2 = 2(\tilde{v}_{r1}+\tilde{v}_{r2})^{-1} \tag{13.1.49}$$

which for $\tilde{v}_{r1}=\tilde{v}_{r2}$ is consistent with (13.1.44). We can integrate (13.1.48) and use (13.1.49) to give

$$\tilde{p}(\tilde{x}_{1\,\text{exit}})-\tilde{p}(-\infty) = \int_{-\infty}^{\tilde{x}_{1\,\text{exit}}} \frac{\mathrm{d}\tilde{p}}{\mathrm{d}\tilde{x}_1}\,\mathrm{d}\tilde{x}_1 = 0 \tag{13.1.50}$$

which provides a unique solution for $\tilde{x}_{1\,\text{exit}}$, or equivalently for $(\tilde{v}_{r1}+\tilde{v}_{r2})$. The relevant transcendental equation for $\tilde{x}_{1\,\text{exit}}$ is

$$(1+3x^2)x+(3x^2-1)(1+x^2)(\tfrac{1}{2}\pi+\tan^{-1}x)=0 \tag{13.1.51}$$

whence

$$\tilde{x}_{1\,\text{exit}} = 0{\cdot}475 \tag{13.1.52}$$

The velocity field $\tilde{v}_1$ is given by

$$\frac{\partial\tilde{q}}{\partial\tilde{x}_3}=\frac{1}{\tilde{h}}\frac{\partial\tilde{q}}{\partial\hat{x}_3}=\frac{3}{4}\left[(\tilde{v}_{r1}+\tilde{v}_{r2})-\frac{2}{\tilde{h}}\right]\hat{x}_3^2 + \frac{1}{2}(\tilde{v}_{r1}-\tilde{v}_{r2})\hat{x}_3+\frac{1}{4}\left[\frac{6}{\tilde{h}}-(\tilde{v}_{r1}+\tilde{v}_{r2})\right] \tag{13.1.53}$$

(13.1.52) yields

$$\tilde{h}_{\text{exit}} = (1+\tilde{x}_{1\,\text{exit}}^2) = 1{\cdot}226 \tag{13.1.54}$$

For $\tilde{v}_{r1}=\tilde{v}_{r2}$, $\tilde{v}_1$ is symmetric about $\tilde{x}_3=0$. It first becomes zero when

$$(1+\tilde{x}_1^2) = 3(1+\tilde{x}_{1\,\text{exit}}^2)$$

i.e.

$$\tilde{x}_1 = -1{\cdot}65 \tag{13.1.55}$$

For all $\tilde{x}_1 < -1{\cdot}65$, there is a reversal of flow along $\tilde{x}_3=0$, and the positions of zero $\tilde{v}_1$ are given by

$$\hat{x}_3^2 = \frac{x_1^2-2{\cdot}678}{3\tilde{x}_1^2-0{\cdot}678} \tag{13.1.56}$$

which tends as $\tilde{x}_1\to\infty$ to $\hat{x}_3=\pm 3^{-\frac{1}{2}}$. From our earlier discussion of typical geometries for calenders we may expect to find circumstances in which flow reversal takes place within the lubrication region. We shall discuss this regime of flow reversal in connection with the rolling bank.

The pressure field $\tilde{p}$ is given by

$$\tilde{p}-\tilde{p}_a=3\int_{\tilde{x}_1}^{\tilde{x}_{1\,\text{exit}}}\left\{\frac{1}{1{\cdot}226(1+\tilde{x}_1^2)^2}-\frac{1}{(1+\tilde{x}_1^2)^3}\right\}d\tilde{x}_1$$

$$=\frac{3}{8}\left[\frac{(0{\cdot}332\tilde{x}_1^2-2{\cdot}130)\tilde{x}_1}{(1+\tilde{x}_1^2)^2}+0{\cdot}322(\tan^{-1}\tilde{x}_1-\tan^{-1}0{\cdot}475)\right.$$
$$\left.+(1{\cdot}678\times 0{\cdot}475)/1{\cdot}226\right] \tag{13.1.57}$$

This is plotted diagramatically in Fig. 13.4. Note that the profile is antisymmetric about $\tilde{p}(0)-\tilde{p}_a$, and that the integral under the curve gives the load on each of the rolls.

The energy equation can be written approximately as (cf. eqn (11.2.4) leading to (11.2.40))

$$\mathrm{Gz}_0\left(\tilde{h}^2\tilde{v}_1\frac{\partial\tilde{T}}{\partial\tilde{x}_1}-\tilde{h}\frac{d\tilde{h}}{d\tilde{x}_1}\tilde{v}_1\hat{x}_3\frac{\partial\tilde{T}}{\partial\hat{x}_3}+\tilde{h}\hat{v}_3\frac{\partial\tilde{T}}{\partial\hat{x}_3}\right)=\frac{\partial^2\tilde{T}}{\partial\hat{x}_3^2}+\mathrm{Br}\left(\frac{\partial\tilde{v}_1}{\partial\hat{x}_3}\right)^2 \tag{13.1.58}$$

where

$$\tilde{T}=(T-T_0)/T^*_{\text{oper}}=(T-T_0)/(T_1-T_0) \tag{13.1.59}$$

$$\mathrm{Gz}_0=v^*h_0^2/4X\kappa,\qquad \mathrm{Br}=\eta^*v^{*2}/\alpha(T_1-T_0) \tag{13.1.60}$$

and only the dominant terms for conduction and generation are included. It is customary, particularly in situations where $\tilde{h}\equiv 1$, to neglect all but the first convective term. A much more reliable and physically justifiable procedure is to regard $\tilde{T}$ as $\tilde{T}(\tilde{x}_1, \tilde{q})$. The approxi-

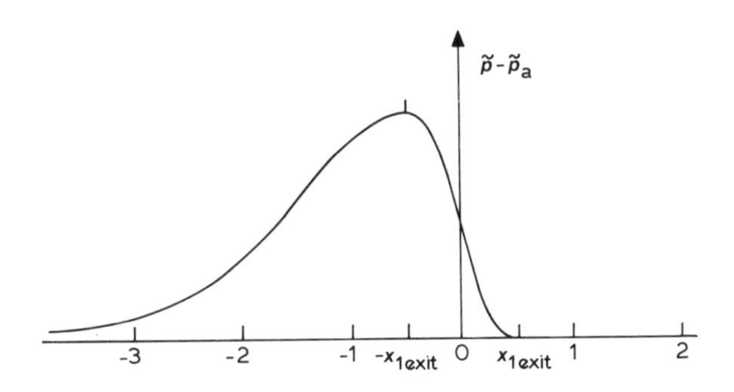

Fig. 13.4. Diagrammatic pressure profile for a Newtonian fluid in a calender nip when $\tilde{v}_{r1}=\tilde{v}_{r2}$ according to (13.1.57).

mate form of the energy equation becomes, provided $\partial\tilde{q}/\partial\tilde{x}_3 > 0$,

$$\mathrm{Gz}_0\,\tilde{h}\frac{\partial\tilde{T}}{\partial\tilde{x}_1} = \frac{\partial}{\partial\tilde{q}}\left(\frac{\partial\tilde{q}}{\partial\hat{x}_3}\frac{\partial\tilde{T}}{\partial\tilde{q}}\right) + \mathrm{Br}\left[\frac{\partial\tilde{q}}{\partial\hat{x}_3}\frac{\partial}{\partial\tilde{q}}\left(\frac{\partial\tilde{q}}{\partial\tilde{x}_3}\right)\right]^2 \tag{13.1.61}$$

where $\partial\tilde{q}/\partial\hat{x}_3$ is given by (13.1.53) and (13.1.47) as a function (implicit admittedly) of $\tilde{q}$ and $\tilde{x}_1$. This is perfectly well-behaved so long as $\partial\tilde{q}/\partial\hat{x}_3$ is not zero or negative. The boundary conditions for (13.1.61) are

$$\tilde{T}(\tilde{x}_1, \pm 1) = 0, \qquad \tilde{T}(\tilde{x}_{1\ \mathrm{entry}}, \tilde{q}) = \tilde{T}_{\mathrm{entry}}(\tilde{q}) \tag{13.1.62}$$

where $\tilde{T}_{\mathrm{entry}}(\tilde{q}) \equiv 0$ in the absence of more precise information from a solution within the rolling-bank region. From (13.1.55), it is clear that

$$0 > \tilde{x}_{1\ \mathrm{entry}} > -1{\cdot}65 \tag{13.1.63}$$

if (13.1.62) is to be relevant. The case when $\tilde{v}_1$ and $\partial\tilde{q}/\partial\tilde{x}$ are negative over part of the velocity profile will be considered in the next section.

It is fairly clear that the procedure adopted here for the constant-viscosity situation could be followed for any viscous fluid model, in particular the power-law approximation. A pressure profile similar to that given in Fig. 13.4 would be obtained, and an energy equation similar to (13.1.61) could be derived, with suitable alteration to the generation term (involving Br). Analytical details will not be given here (see Exercise 13.1.5 for the simplest extension to the Newtonian case), because important practical situations involve non-negligible values of B and Na, while the purely viscous assumption may be inaccurate.

EXERCISE 13.1.1. Verify result (13.1.22) and obtain estimates for $T^*_{\mathrm{adiab}}/T^*_{\mathrm{rheol}}$ in the rolling bank using the values given in (13.1.13) and $x_1 \approx 7X$.

EXERCISE 13.1.2. Verify eqn (13.1.32) showing that $\tilde{\eta} = \eta/\eta^*$ and interpreting the functional dependence on $\tilde{D}, \tilde{T}$ by reference to the fluid model (11.2.25) as an example.

EXERCISE 13.1.3. Verify the result (13.1.27) taking care to interpret all signs carefully in (13.1.26). Interpret the directions of the loads F_1 and F_2 in (13.1.39) and evaluate the lateral loads caused by the forces leading to the torques L_i in (13.1.36).

Show that for the Newtonian case with $W_1 = W_2 = W$, $d_1 = d_2 = d$, satisfying (13.1.49)

$$L_1 = L_2 \approx 1{\cdot}2\eta_0 W d^{\frac{5}{2}} h_0^{-\frac{1}{2}}$$

EXERCISE 13.1.4. Derive eqn (13.1.51) and verify (13.1.52) using (13.1.48)–(13.1.50). Check results (13.1.55), (13.1.56) and (13.1.57).

EXERCISE 13.1.5. Consider the case of flow of a temperature-independent power-law fluid given by eqn (3.3.5). Carry out the analysis equivalent to that leading to relations (13.1.47)–(13.1.51) for a Newtonian fluid ($\nu = 1$) for $d_1 = d_2 = d$, $W_1 = W_2 = W$. Show that the equivalent of (13.1.51) becomes

$$\int_{-\infty}^{x} \left[\frac{1}{(1+x^2)(1+y^2)^{2\nu}} - \frac{1}{(1+y^2)^{2\nu+1}}\right] dy = 0$$

Continue the analysis for the cases $\nu = \frac{1}{2}, \frac{1}{4}$.

13.2 ROLLING BANK MECHANICS

The solution given for the special case of a constant-viscosity fluid showed that reverse flow can arise within the nip, i.e. within the region where the lubrication approximation can apply. Indeed, if we let $\tilde{x}_1 \to -\infty$ in (13.1.50) with $\tilde{v}_{r1} = \tilde{v}_{r2}$,

$$\frac{\partial \tilde{q}}{\partial \hat{x}_3} \to \tfrac{1}{2}\tilde{v}_{r1}\tilde{x}_1^2(3\hat{x}_3^2 - 1) \tag{13.2.1}$$

and

$$\tilde{q} \to \tfrac{1}{2}\tilde{v}_{r1}\tilde{x}_1^2(\hat{x}_3^2 - 1)\hat{x}_3 \tag{13.2.2}$$

Note that this is zero for $\hat{x}_3 = -1$, 0 and 1, and so the actual flow through the nip, which is $0(1)$, is negligible compared with the $0(\tilde{x}_1^2)$ flows described by (13.2.2). When $\hat{x}_3 = -1/\sqrt{3}$, a maximum value is reached, given by

$$\tilde{q} \sim \tilde{v}_{r1}\tilde{x}_1^2/3\sqrt{3} \tag{13.2.3}$$

The similarity solution (13.2.1) is shown in Fig. 13.5(a) as a velocity profile and the solution (13.2.2) in terms of streamlines in Fig. 13.5(b).

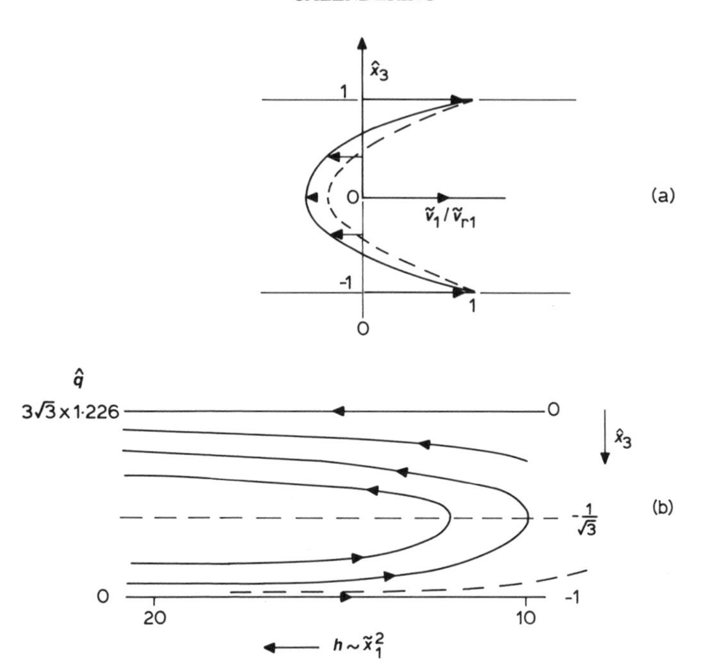

Fig. 13.5. (a) ——, asymptotic dimensionless velocity profile; ---, effect of 0(1) terms. (b) ——, asymptotic dimensionless streamline pattern; ---, limiting streamline when 0(1) terms are included.

In practice, the region of reversed streamlines will be bounded on the right when $\tilde{h}\rightarrow 1$ and on the left, where $\tilde{h}\gg 1$, by the physical extremity of the rolling bank. Furthermore, the 0(1) terms neglected in (13.2.1) will cause some of the flow in Fig. 13.5(b), near the line $\hat{x}_3=-1$, to pass through the nip. A diagrammatic view of the flow field, when all the incoming material enters on roll 1 is shown in Fig. 13.6. This includes two contra-rotating vortices VX_1 and VX_2, which in the perfect plane-flow situation described here represent trapped material. Half of the incoming flow remains effectively attached to the surface of roll 1 and passes directly through the nip. The other, upper, half of the incoming flow separates, at the inner stagnation point $\tilde{x}_{1\mathrm{st.i.}}$, from the roll surface region and passes back, next to the centre line, between the recirculating vortices, until it reaches the outer stagnation point $\tilde{x}_{1\mathrm{st.o.}}$ whence it passes over the outer surface of the rolling bank to attach itself to the surface of roll 2; it then remains next to the roll until it has passed through the nip. The flow near $\tilde{x}_{1\mathrm{st.i.}}$ can be well approximated

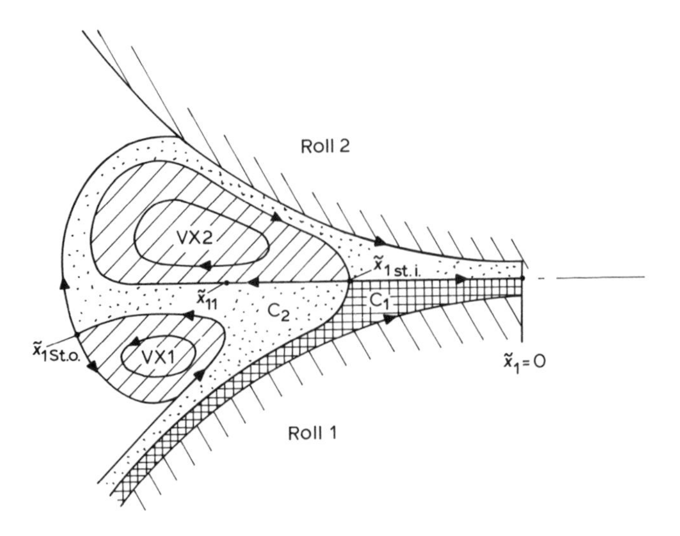

Fig. 13.6. Expanded (i.e. not to scale) diagrammatic view of rolling bank. Recirculating vortices, VX1 and VX2. Lower half of through flow, C1. Upper half of through flow, C2.

by the lubrication approximation, but that in the region near $\tilde{x}_{1\text{st.o.}}$ is essentially two-dimensional and will be governed, in the Newtonian case, by non-trivial solutions of $\nabla^4 \tilde{q} = 0$.

13.2.1 Thermal Effects: Flow into the Nip

A crude representation of thermal effects can be gained by noting that Na, or Br, as described by either (13.1.11) or (13.1.60) are independent of $\tilde{h}$, and so generation effects will scale uniformly with conduction effects through the whole region. However, convective effects (defined by Gz) increase with $\tilde{h}$, as can be seen from (13.1.60) directly by noting that the local Graetz number

$$\text{Gz}_{\tilde{x}_1} = \text{Gz}_0 \tilde{h}^2 \tag{13.2.4}$$

varies as $\tilde{h}^2$. If the rolling bank is very large, then a Graetz number describing recirculation flow in VX_2, for example, for that portion of the vortex lying beyond $\tilde{x}_1$, can be proposed in the form

$$\text{Gz}_1(\tilde{x}_1 \ll \tilde{x}_{1\text{st.i.}}) = \text{Gz}_0 \left[\int_{-\infty}^{\tilde{x}_1} \frac{d\tilde{x}_1}{\tilde{h}^2} \right]^{-1} \approx 3\,\text{Gz}_0\, |\tilde{x}_1^3| \tag{13.2.5}$$

This suggests that the rolling-bank flow can be truncated as far as thermal effects are concerned at some relatively large value of $|\tilde{x}_1|$, $|\tilde{x}_{1\infty}|<|x_{1\text{st.o.}}|$, and the temperature $\tilde{T}$ on any incoming ($\tilde{v}_1>0$) streamline put equal to that on the corresponding outgoing streamline ($\tilde{v}_1<0$); corresponding streamlines are those with the same value of $\tilde{q}$. This is the argument used in Subsection 11.4.2 for $\text{Gz}_\text{D}\geqslant 0(1)$. This procedure is clearly justified in that both the rate of generation of heat and the heat conducted to the walls in a slice of width $d\tilde{x}_1$ are proportional to $\tilde{h}^{-1}$. The total amount of heat generated (i.e. the mechanical work done) and the total heat lost to the wall in the rolling bank beyond $\tilde{x}_1$ will thus be $0(\tilde{x}_1^{-1})$ for large $\tilde{x}_1$.

The general argument will be similar, though not so strong, in the case of other viscous fluids. It will not be seriously affected by temperature variations of viscosity.

Lastly we note from (13.1.19) that the Deborah number varies as $\tilde{x}_1^{-3}$ and so the Newtonian approximation will be relevant for elastic fluids provided $|\tilde{x}_1|$ is large enough.

We can thus deduce that the most important mechanical consequences of the rolling bank can be understood, and calculated, in terms of a finite region of reversed flow near the nip. If X is large enough, the lubrication approximation will be applicable everywhere in this region, so the only difference from the nip-flow mechanics will arise in the treatment of the energy equation. The flow field in Fig. 13.6 is composed of four regions which we consider separately.

Region C1. In this region, $\partial\tilde{q}/\partial\tilde{x}_3>0$ everywhere and so equation (13.1.61) can be applied for all $\tilde{x}_1$. However, it is obvious that $T^*_\text{oper}\rightarrow 0$ is a very likely situation if T_0 is to be defined as the entering temperature of a layer that has been attached to a roll at temperature T_1 from $\tilde{x}_1\rightarrow -\infty$. It is therefore wiser to use T^*_rheol to define $\tilde{T}$ as in (11.4.44); Br in (13.1.61) then becomes Na (when the factor 2 in (13.1.11) is removed). If $\text{Na}\ll 1$, and the velocity field is independent of the temperature field, then the obvious equation to solve is that for $\hat{T}_\text{C1}=\tilde{T}/\text{Na}$, which will be $0(1)$ because it is based on T^*_gen. The boundary and initial conditions for this will be

$$\hat{T}_\text{C1}(\tilde{x}_1,-1)=\hat{T}_\text{C1}(\tilde{x}_{10},\tilde{q})=0; \qquad \hat{T}_\text{C1}(\tilde{x}_1,0)=\hat{T}_\text{C2}(\tilde{x}_1,0) \quad (13.2.6)$$

when $\hat{T}_\text{C2}$ is the interfacial value of $\tilde{T}/\text{Na}$ in region C2. Forward integration from some value of $\tilde{x}_{10}$, less than the value at which the incoming sheet (C1+C2) meets the recirculating vortex VX_1 will be possible. We discuss later the possibility that the temperature distribution at $\tilde{x}_{1\infty}$ is not uniform and is the result of some earlier nip flow.

Region C2. In this region, $\partial\tilde{q}/\partial\tilde{x}_3$ changes sign twice and so forward integration of eqn (13.1.61) is not immediately possible. The curved path shown in Fig. 13.6 is idealized in Fig. 13.7 into five separate zones. In zone 1, eqn (13.1.60) can be integrated forward from $\tilde{x}_{1\infty}$ to $\tilde{x}_{11}$ subject to boundary and initial conditions

$$\left.\begin{array}{ll}\hat{T}_{C2}(\tilde{x}_1,0)=\hat{T}_{C1}(\tilde{x}_1,0); & \hat{T}_{C2}(\tilde{x}_1,1)=\hat{T}_{VX1}(1)\\ \hat{T}_{C2}(\tilde{x}_{1\infty},\tilde{q})=0 & \end{array}\right\} \quad (13.2.7)$$

where the first of these is the same as the last of (13.2.6) and $\tilde{T}_{VX1}$ is a constant value to be determined by solution in region VX1. For consistency we should put

$$\tilde{x}_{10}=\tilde{x}_{1\infty} \quad (13.2.8)$$

In zone 2, which goes from $\tilde{x}_{11}$ to $\hat{x}_{1\mathrm{st.i.}}$, we have to overcome the difficulty that $\tilde{v}_1=0$ and reverses at some different value $\tilde{x}_{1r}$ for each

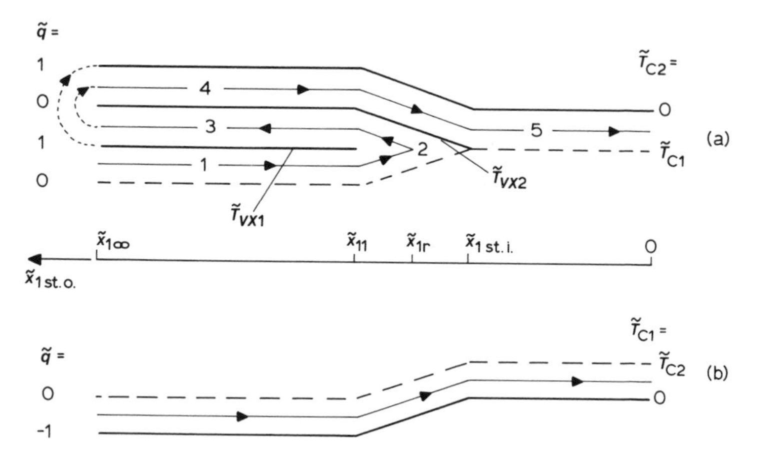

Fig. 13.7. Idealized representation of regions (a) C2 and (b) C1 in rolling bank for solution of the energy equation. Region C2 is split into 5 zones. Within zones 1, 3, 4, and 5, a parabolic partial-differential equation for temperature $\tilde{T}_{C2}$ can be employed in traditional fashion. In zone 2 a different method has to be used, neglecting conduction.

$0 \leqslant \tilde{q} \leqslant 1$. The simplest way to do this is to suppose that the flow is entirely adiabatic in which case each streamline can be considered separately. The relevant equation becomes, for the constant viscosity case,

$$\mathrm{Gz}_0 \tilde{v} \frac{\partial \hat{T}_{C2}}{\partial \tilde{x}_1} = \mathrm{Na}\left(\frac{\partial \tilde{v}_1}{\partial \hat{x}_3}\right)^2 \tag{13.2.9}$$

with

$$\tilde{v} = (\tilde{v}_1^2 + \tilde{v}_3^2)^{\frac{1}{2}} \operatorname{sn} \tilde{v}_1 \tag{13.2.10}$$

thus avoiding the $\tilde{v}_1 = 0$ difficulty.

This can be integrated from $\tilde{x}_{11}$ to $\tilde{x}_{1r}$ and back again to give the change in $\hat{T}_{C2}$,

$$[\hat{T}_{C2}(\tilde{x}_{11}, \tilde{q})]_1^3$$

in going from $\tilde{x}_{11}$ at the end of region 1 to $\tilde{x}_{11}$ at the beginning of region 3. Difficulties arise for $\tilde{q} = 0$ in that $\tilde{T}_{C2}$ should satisfy the boundary conditions

$$\left.\begin{aligned} \hat{T}_{C2}(\tilde{x}_1, 0) &= \hat{T}_{C1}(\tilde{x}_1, 0) \text{ for } \tilde{v}_1 > 0 \\ \hat{T}_{C2}(\tilde{x}_1, 1) &= \hat{T}_{VX2}(1) \text{ for } \tilde{v}_1 < 0 \end{aligned}\right\} \tag{13.2.11}$$

throughout $\tilde{x}_{11} < \tilde{x}_1 < \tilde{x}_{1\mathrm{st.i.}}$ where $\hat{T}_{VX2}$ is a constant. An alternative, which would be mathematically consistent even if physically unrealistic, would be to replace them by a no-heat-flux condition on the regions C1 and VX2 over the range $\tilde{x}_1$ involved. For C1 this would be simply

$$\frac{\partial \hat{T}_{C1}}{\partial \tilde{q}}(\tilde{x}_1, 0) = 0 \tag{13.2.12}$$

instead of the last of (13.2.6). For VX2 the relevant form is discussed below.

If the parabolic equation for $\hat{T}$ in zones 1, 3, 4 and 5 is solved by numerical methods—and finite-difference forms will be appropriate for the $(\tilde{x}_1, \tilde{q})$ rectangular regions—then it is sensible to look for a step-by-step routine that will link region 1 smoothly to region 3. Algorithms to avoid the difficulties arising at $\tilde{x}_{1r}$ are best chosen to suit the explicit or implicit marching routine used for all interior points.

In zone 3, the parabolic equation (13.1.61) can again be used but integrated in the $-\tilde{x}_1$-direction. The boundary conditions to be satisfied in the range $\tilde{x}_{1\infty} < \tilde{x}_1 < \tilde{x}_{11}$ are

$$\hat{T}_{C2}(\tilde{x}_1, 0) = \hat{T}_{VX2}(0); \qquad \hat{T}_{C2}(\tilde{x}_1, 1) = \hat{T}_{VX1}(1) \tag{13.2.13}$$

with the initial conditions at $\tilde{x}_{11}$ given by the solution to (13.2.9). This will yield the final conditions $\tilde{T}_{C2}(\tilde{x}_{1\infty}, \tilde{q})$ which become the initial conditions for zone 4. In zone 4, which runs from $\tilde{x}_{1\infty}$ to $\tilde{x}_{1\text{st.i.}}$, the boundary conditions for (13.1.60) are

$$\hat{T}_{C2}(\tilde{x}_1, 0) = \hat{T}_{VX2}(0); \qquad \hat{T}_{C2}(\tilde{x}_1, 1) = 0 \tag{13.2.14}$$

and in zone 5, from $\tilde{x}_{1\text{st.i.}}$ to 0 they are

$$\hat{T}_{C2}(\tilde{x}_1, 0) = \hat{T}_{C1}(\tilde{x}_1, 0); \qquad \hat{T}_{C2}(\tilde{x}_1, 1) = 0 \tag{13.2.15}$$

Region VX1. Within this closed region, it is reasonable to assume that $\tilde{T}_{VX1}$ is a function of $\tilde{q}$ only. By going back to eqn (13.1.58) and writing it in the approximate form corresponding to (13.1.61),

$$\mathrm{Gz}_0 \frac{(\tilde{v} \cdot \nabla)}{\tilde{v}} \hat{T}_{VX1} = \frac{1}{\tilde{h}^2 \tilde{v}} \frac{\partial^2 \hat{T}}{\partial \hat{x}_3^2} + \frac{1}{\tilde{h}^2 \tilde{v}} \left(\frac{\partial \tilde{v}_1}{\partial \hat{x}_3} \right)^2 \tag{13.2.16}$$

we can integrate around a closed streamline to eliminate the left-hand side as was done in Subsection 11.4.1, Gz_D small. The right-hand side then becomes the ordinary differential equation

$$\left[\oint \frac{1}{\tilde{h}^2} \frac{\partial \tilde{q}}{\partial \hat{x}_3} \, d\tilde{x}_1 \right] \frac{\partial^2 \tilde{T}_{VX1}}{\partial \tilde{q}^2} + \left[\oint \frac{1}{\tilde{h}^2 \tilde{v}} \frac{\partial^2 \tilde{q}}{\partial \hat{x}_3} \, d\tilde{x}_1 \right] \frac{\partial \tilde{T}_{VX1}}{\partial \tilde{q}} = \oint \frac{d\tilde{x}_1}{\tilde{h}^4 |\tilde{v}|} \left(\frac{\partial^2 \tilde{q}}{\partial \hat{x}_3^2} \right)^2 \tag{13.2.17}$$

with boundary conditions

$$\frac{\partial \tilde{T}_{VX1}}{\partial \tilde{q}} = 0 \text{ for } \tilde{q}_{\max} \tag{13.2.18}$$

$$\frac{\partial \tilde{T}_{VX1}}{\partial \tilde{q}} \oint \frac{d\tilde{x}_1}{\tilde{v}} = \oint \frac{1}{\tilde{v}} \frac{\partial \tilde{T}_{C2}}{\partial \tilde{q}} \, d\tilde{x}_1 \text{ for } \tilde{q} = 1 \tag{13.2.19}$$

The integrals in (13.2.17) and (13.2.19) run from $\tilde{x}_{1\infty}$ to $\tilde{x}_{11}$ and back again.

Region VX2. This is analogous to region VX1. Equation (13.2.17) can be taken over directly but with the maximum range in $\tilde{x}_1$ now extended to $[\tilde{x}_{1\infty}, \tilde{x}_{1\text{st.i.}}]$.

Boundary condition (13.2.18) becomes

$$\frac{\partial \tilde{T}_{VX2}}{\partial \tilde{q}} = 0 \text{ for } \tilde{q}_{min} \tag{13.2.20}$$

while (13.2.19) has to be replaced to suit the solution scheme adopted in region C2. If the adiabatic form (13.2.9) is used then it becomes

$$\frac{\partial \tilde{T}_{VX2}}{\partial \tilde{q}} \oint \frac{\partial \tilde{x}_1}{\tilde{v}} = \left\{ \int_{\tilde{x}_{11}}^{\tilde{x}_{1\infty}} + \int_{\tilde{x}_{1\infty}}^{\tilde{x}_{1st.i.}} \right\} \frac{1}{\tilde{v}} \frac{\partial \tilde{T}_{C2}}{\partial \tilde{q}} \, d\tilde{x}_1 \text{ for } \tilde{q} = 0 \tag{13.2.21}$$

Unfortunately, the solutions for the four regions are coupled because $\tilde{T}_{VX2}(0)$ and $\tilde{T}_{VX1}(1)$ which are used in boundary conditions (13.2.7), (13.2.13) and (13.2.14) are not given independently of the solution in region C2 but are implicitly connected with it through boundary conditions (13.2.19) and (13.2.21). In practice, it would be necessary to guess $\tilde{T}_{VX2}(0)$ and $\tilde{T}_{VX1}(1)$, solve separately in the regions VX1, VX2, C1+C2(1) ($\tilde{x}_1 < \tilde{x}_{11}$), C2(3)+C2(4) and C2($\tilde{x}_{11} < \tilde{x}_1 < x_{1st.i.}$), and then adjust the choices of $\tilde{T}_{VX2}(0)$ and $\tilde{T}_{VX1}(1)$ if (13.2.19) and (13.2.21) were not satisfied. It will be noted that the solution scheme given above incorporates that described in Subsection 13.1.1 in connection with the Newtonian case, eqn (13.1.61), in that the region C1+C2(5) ($x_1 > x_{1st.i.}$) is the nip flow.

For a numerical treatment of the flow of an Ellis fluid with temperature dependent coefficients, see Dobbels & Mewis (1977).

13.2.2 Transverse Flow Far from the Nip

In practice, it is obviously impossible to maintain fully two-dimensional (x_1, x_3) flow in the rolling bank. The thickness of the layer entering from the left in Fig. 13.2(a) will not be exactly equal to the thickness emerging on the roll on the right-hand side. Indeed, if it were, the take-off roll might as well have been placed before and not after passage through the nip. The rolling bank must therefore perform a redistribution role in the x_2-direction as indicated in Fig. 13.2(b).

A simplified treatment of this problem can be given for steady flow of a Newtonian (constant viscosity) fluid using the lubrication approximation. This is always possible if X/h_0 is large enough as can be seen from (13.1.4).

This means that we are now supposing there to be three distinct lubrication-flow regions within the flow field. The innermost, the nip region, determines the volume flow rate in the x_1-direction through the nip, i.e. q_1. The relevant pressure gradients are those shown in Fig. 13.4, which imply that $|x_1| \leqslant 0(X)$. The intermediate region, shown in Fig. 13.7, extends to much larger values of $|\tilde{x}_1|$, but is such that

$$|x_{1\infty}|h_0/X^2 \ll 1, \qquad |x_{1\infty}|/X \gg 1 \tag{13.2.22}$$

The outer zone which stretches to $\tilde{x}_{1\text{st.o.}}$ is such that

$$|x_{1\text{st.o.}}|h_0/X^2 \ll 1, \qquad \tilde{x}_{1\infty}/\tilde{x}_{1\text{st.o.}} \ll 1 \tag{13.2.23}$$

This is just about possible if $h_0/d = 10^{-4}$, in that some overlap is admissible.

The determining feature in this x_2 flow will be the variation with

$$\tilde{x}_2 = x_2/X \tag{13.2.24}$$

of the extent of the rolling bank

$$\tilde{x}_{1\text{st.o.}} = \tilde{X}_1(\tilde{x}_2) \tag{13.2.25}$$

as shown in Fig. 13.8. We define

$$\left.\begin{aligned} q_2(x_1, x_2) &= \int_{-\frac{1}{2}h}^{\frac{1}{2}h} v_2(x_1, x_2, x_3)\,\mathrm{d}x_3 \\ \text{and} \quad \tilde{q}_2 = 2q_2/v^* h_0, &\qquad \tilde{v}_2 = v_2/v^* \end{aligned}\right\} \tag{13.2.26}$$

with $\tilde{p}$ still given by (13.1.23).

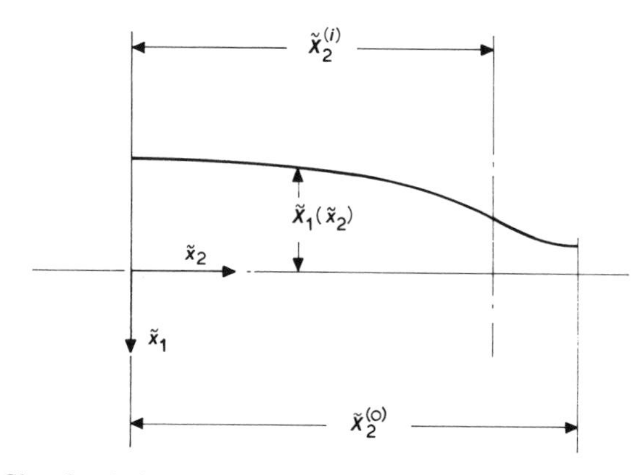

Fig. 13.8. Sketch of the extent of the rolling bank on the entry side of the calender nip. One half of the full width is shown. The entering sheet width is $2\tilde{X}_2^{(i)}$ and the calendered sheet width is $2\tilde{X}_2^{(o)}$.

The dynamical equation of motion for flow in the x_2-direction is given by (8.1.18) which in dimensionless form becomes

$$\frac{\partial^2 \tilde{v}_2}{\partial \hat{x}_3^2} = \tilde{h}^2 \frac{\partial \tilde{p}}{\partial \tilde{x}_2} \tag{13.2.27}$$

with boundary conditions $\tilde{v}_2(\pm 1) = 0$. It follows that

$$\tilde{q}_2 \sim -\frac{2}{3}\tilde{h}^3 \frac{\partial \tilde{p}}{\partial \tilde{x}_2} \tag{13.2.28}$$

However, from (13.1.48), by integration we get

$$\tilde{p}(\tilde{x}_1, \tilde{x}_2) \sim \int_{\tilde{X}_1}^{\tilde{x}_1} \frac{3}{2} \frac{(\tilde{v}_{r1} + \tilde{v}_{r2})}{\tilde{x}_1^4} \, \mathrm{d}\tilde{x}_1 \tag{13.2.29}$$

Thus

$$\frac{\partial \tilde{p}}{\partial \tilde{x}_2} \sim -\frac{3}{2} \frac{(\tilde{v}_{r1} + \tilde{v}_{r2})}{\tilde{X}_1^4} \frac{\mathrm{d}\tilde{X}_1}{\mathrm{d}\tilde{x}_2} \tag{13.2.30}$$

The total dimensionless volume-flowrate in the $\tilde{x}_2$-direction is given by

$$\begin{aligned} \tilde{Q}_2(\tilde{x}_2) &\approx \int_{\tilde{X}_1}^{0} \tilde{q}_2(\tilde{x}_1, \tilde{x}_2) \, \mathrm{d}\tilde{x}_1 = \int_{\tilde{X}_1}^{0} \frac{(v_{r1} + \tilde{v}_{r2})\tilde{x}_1^6}{\tilde{X}_1^4} \frac{\mathrm{d}\tilde{X}_1}{\mathrm{d}\tilde{x}_2} \, \mathrm{d}\tilde{x}_1 \\ &\approx -\frac{(\tilde{v}_{r1} + \tilde{v}_{r2})}{7} \tilde{X}_1^3 \frac{\mathrm{d}\tilde{X}_1}{\mathrm{d}\tilde{x}_2} = -\frac{(\tilde{v}_{r1} + \tilde{v}_{r2})}{28} \frac{\mathrm{d}\tilde{X}_1^4}{\mathrm{d}\tilde{x}_2} \end{aligned} \tag{13.2.31}$$

Because the flow is steady and the dimensionless flowrate/unit length in the $\tilde{x}_1$-direction out of the nip is, from (13.1.31),

$$\tilde{q}_1 = \tilde{q}(1) - \tilde{q}(-1) = 2$$

the total flowrate will be

$$\tilde{Q}_1 = 4\tilde{X}_2^{(o)} \tag{13.2.32}$$

where $\tilde{X}_2^{(o)}$ is the half-width of the calendered sheet. If the entering sheet is assumed to be of uniform thickness, then the entering flowrate/unit length will be

$$2\tilde{X}_2^{(o)}/\tilde{X}_2^{(i)} = 2\mathrm{O_c} \tag{13.2.33}$$

By simple mass conservation we deduce that

$$\frac{\partial \tilde{Q}_2}{\partial \tilde{x}_2} \sim \left\{ \begin{array}{lll} 2(\mathrm{O_c} - 1) & \text{for} & |\tilde{x}_2| < \tilde{X}_2^{(i)} \\ -2 & \text{for} & \tilde{X}_2^{(i)} \leqslant |\tilde{x}_2| < \tilde{X}_2^{(o)} \end{array} \right\} \tag{13.2.34}$$

whence

$$\tilde{Q}_2 \sim \begin{Bmatrix} 2(\mathrm{O_c}-1)\tilde{x}_2 & \text{for} \quad |\tilde{x}_2| < \tilde{X}_2^{(i)} \\ 2\mathrm{O_c}\tilde{X}_2^{(i)} - 2\tilde{x}_2 & \text{for} \quad \tilde{X}_2^{(i)} \leqslant |\tilde{x}_2| < \tilde{X}_2^{(o)} \end{Bmatrix} \tag{13.2.35}$$

Using (13.2.31) and integrating we get

$$\tilde{X}_1^4 \sim \begin{cases} \tilde{X}_{10}^4 - \dfrac{28}{(\tilde{v}_{r1}+\tilde{v}_{r2})}(\mathrm{O_c}-1)\tilde{x}_2^2 & \text{for} \quad |\tilde{x}_2| < \tilde{X}_2^{(i)} \\ \tilde{X}_{10}^4 + \dfrac{28}{(\tilde{v}_{r1}+\tilde{v}_{r2})}(\tilde{X}_2^{(o)}\tilde{X}_2^{(i)} - 2\tilde{X}_2^{(o)}\tilde{x}_2 + \tilde{x}_2^2) & \text{for} \quad \tilde{X}_2^{(i)} \leqslant |\tilde{x}_2| < \tilde{X}_2^{(o)} \end{cases} \tag{13.2.36}$$

One parameter, $\tilde{X}_{10}$, is not specified and indeed remains an arbitrary part of the solution, which could be regarded as a slowly-varying function of time, for example. Note also that from (13.1.49) and (13.1.54) $28/(\tilde{v}_{r1}+\tilde{v}_{r2}) \sim 17$.

13.2.3 Elastic Effects

From the argument based on relation (13.1.22) we must accept that an inelastic, purely viscous, theory for the rolling bank is unlikely to be quantitatively correct. The analysis given in Subsections 13.2.1 and 13.2.2 should therefore be regarded as illustrative, providing the basic qualitative features of the flow. As we have seen earlier, very few analytical solutions have been obtained for elastic fluid models. If the Deborah and Weissenberg numbers are both $\ll 1$, then perturbation expansions about the Newtonian solution can be carried out, just as Na or B expansions can be carried out for weakly-coupled thermal effects. For cases where De, Ws, Na or $\mathrm{B} \geqslant 0(1)$, numerical methods suggest themselves, though it must be emphasized that almost all of those successfully carried out still rely on a great deal of geometrical symmetry in the flow field. The rolling bank, as actually observed in many situations, is often such that the relevant length scales in the x_1-, x_2- and x_3-directions do not differ by orders of magnitude. In particular, the first of inequalities (13.2.23) often does not apply. This means that the full complexity of observed flow fields in rolling banks would be a daunting task to model computationally, even if a relatively simple Maxwell or Jeffreys fluid model were used.

Various authors have considered certain consequences of elastic forces (e.g. Paslay, 1955; Chong, 1968). The simplest effect is that of difference of normal stresses in a lubrication flow. For cases where the Weissenberg number

$$\mathrm{Ws} = \Lambda v^*/h_0 \tag{13.2.37}$$

(where Λ is suitably defined in terms of Ψ_1 and η from Section 4.2) is large, the difference of normal stresses $t_{11} - t_{33}$ and $t_{33} - t_{22}$ (in the coordinate system used here) will be large compared with t_{13}. Thus the relevant criterion for their neglect in the nip region becomes

$$\mathrm{Ws}(\mathrm{d}h/\mathrm{d}x_1) = \Lambda v^* X \ll 1 \tag{13.2.38}$$

This is apparently the same criterion as developed earlier leading to (13.1.20), though, also as explained earlier, the definition of Λ will depend upon the type of deformation involved. The contribution they make to the load in (13.1.39) will be of the order of De. It is possible to solve for a perturbed flow in the lubrication region on the basis of a viscometric flow model more easily than by carrying out the analyses for a memory-integral relation.

EXAMPLE 13.2.1. Estimate the contribution to the load on the rolls in a calender caused by the first normal stress difference.

To do this, it is necessary to go back to the original stress-equilibrium equation $\boldsymbol{\nabla}\boldsymbol{.}\mathbf{T} = 0$ as used in the lubrication equation, writing out all the relevant terms. The x_1 component becomes

$$\frac{\mathrm{d}}{\mathrm{d}x_3} t_{13}^{\mathrm{E}} = \frac{\mathrm{d}}{\mathrm{d}x_1}(p - t_{11}^{\mathrm{E}}) \text{ (instead of (8.1.7))}$$

where t_{13}^{E} is treated as a very slowly-varying function of x_1 and the right-hand side is, to first order, constant. For a viscometric flow (and this is what is relevant in the lubrication approximation, as demonstrated in Pearson (1967))

$$t_{13}^{\mathrm{E}} = \eta(D),\ t_{11}^{\mathrm{E}} - t_{33}^{\mathrm{E}} = \Psi_1(D)D^2$$

where

$$D = \mathrm{d}v_1/\mathrm{d}x_3, \text{ from Section 4.2}$$

so the requirement that $q_1 = \int_{-h/2}^{h/2} v_1 \, \mathrm{d}x_3 = \text{constant}$ yields D as a function of x_3 for any given x_1 and $p - t_{33}^{\mathrm{E}} - \Psi_1(D)D^2$ as a function of x_1.

The load now becomes in the symmetric case

$$v_{r1} = v_{r2},$$

$$F_1 = F_2 = \int_{-\infty}^{x_{1\,\mathrm{exit}}} (p - t_{33}^{\mathrm{E}} - p_{\mathrm{a}})\, \mathrm{d}x_1 \text{ (evaluated at } x_3 = \pm\tfrac{1}{2}h)$$

instead of (13.1.39), it being assumed that t_{11}^{E} and t_{33}^{E} are negligible at both $x_{1\,\mathrm{exit}}$ and $-\infty$; indeed, the boundary condition (13.1.45) implies that they are identically zero at $x_{1\,\mathrm{exit}}$, while the Newtonian limit will be relevant for $x_1 \to -\infty$ since $D \to 0$.

It follows that the contribution to the load of the difference of normal stresses is just

$$\int_{-\infty}^{x_1\,\mathrm{exit}} \Psi_1(D) D^2\, \mathrm{d}x_1$$

because the solution obtained previously from (13.1.32) for p will now apply to $p - t_{33}^{\mathrm{E}} - \Psi_1(D)D^2$.

From (13.1.23), (13.1.32) and (13.1.39) it follows that the main pressure contribution to F_1 is $0(X^2\eta^* v^*/h_0^2)$. That from the normal forces is $0(\mathrm{Ws}\, X\eta^* v^*/h_0)$ where

$$\mathrm{Ws} = \Psi_1(D)D^2/\eta(D)D$$

as defined in Exercise 13.2.2. The ratio of these two contributions is $X/h\mathrm{Ws}$ so unless $\mathrm{Ws} \gg 1$, the original lubrication approximation implies that the difference of normal stresses only leads to a small correction. It may, however, be verified that it is larger than the error for a Newtonian fluid implied by the lubrication approximation, which is of order h^2/X^2.

EXERCISE 13.2.1. Derive the results (13.2.34)–(13.2.36).

EXERCISE 13.2.2. Show that if the Weissenberg number is defined as

$$\mathrm{Ws} = \frac{t_{11} - t_{22}}{t_{12}} \text{ evaluated at } D = v^*/h_0$$

for lubrication flow in a channel of depth h_0 with characteristic velocity v^*, then the characteristic time in the definition (13.2.26) is given by

$$\Lambda = \Psi_1(D)D^2/\eta(D)$$

where η and Ψ_1 are defined in (4.2.2) and (4.2.3).

EXERCISE 13.2.3. Evaluate the contribution to the load for a third-order fluid (eqn (3.3.11)) when $|(\delta_1+\beta_1)v^*/h_0| \ll |\gamma_0|$, showing that it is linear in γ_0. The Newtonian solution given in Section 13.1 can be used once it has been shown that the viscosity of a third-order fluid is effectively Newtonian when $(\delta_1+\beta_1)v^{*2}/h_0^2 \ll \eta_0$.

EXERCISE 13.2.4. Carry out the same analysis for a general quasi-linear viscoelastic fluid, given by eqn (3.4.1). It is necessary to evaluate $\eta(D)$ and $\Psi_1(D)$ showing that the former is constant for all $m(s)$ leading to a finite stress.

13.3 FREE FLOW HEAT TRANSFER

When calendered material passes through any but the last nip in a calender train, it remains in contact with one of the rolls. If successive nips are diametrically opposed on the roll in question, then the time spent in contact will be

$$t_i \approx \pi / W_i \tag{13.3.1}$$

where it has been assumed that $x_{1\,\text{exit}} \ll d_i$. The temperature profile, i.e. $\tilde{T}(\hat{x}_3)$, at $\tilde{x}_{1\,\text{exit}}$, is assumed given by the solution to (13.1.61) or its equivalent for a non-Newtonian system. There is no flow of the material relative to the calender roll for $\tilde{x}_1 > \tilde{x}_{1\,\text{exit}}$, and so the temperature field will be given by a simple heat-conduction equation

$$\text{Gz}_i \frac{\partial \tilde{T}}{\partial \tilde{t}} = \frac{\partial^2 \tilde{T}}{\partial \hat{x}_3^2} \tag{13.3.2}$$

where

$$\tilde{t} = t/t_i \tag{13.3.3}$$

$$\text{Gz}_i = h_i^2/4\kappa t_i \tag{13.3.4}$$

h_i being the thickness of the calendered sheet, with boundary conditions

$$\tilde{T}(-1) = \tilde{T}_i, \qquad \frac{\partial \tilde{T}(1)}{\partial \hat{x}_3} = \text{Bi}_i(\tilde{T} - \tilde{T}_a) \tag{13.3.5}$$

$\tilde{T}_a$ being the ambient air temperature and Bi_i the relevant Biot number for heat transfer from the polymer surface to the ambient air.

The Graetz† number Gz_i can be written in many alternative ways; using $q_1 = \frac{1}{2}h_i d_i W_i$, we get

$$\mathrm{Gz}_i = q_1^2/\pi\kappa d_i^2 W_i \tag{13.3.6}$$

using $\tilde{q}_1 = q_1/\frac{1}{2}d_i W_i h_0$, $v^* = q_1/h_0$, we get

$$\mathrm{Gz}_i = v^* h_0^2 \tilde{q}_1/2\pi d_i \kappa \tag{13.3.7}$$

If we compare (13.3.7) with (13.1.60) we see that

$$\frac{\mathrm{Gz}_i}{\mathrm{Gz}_0} = \frac{2X}{\pi d_i}\tilde{q}_i \ll 1 \tag{13.3.8}$$

since $X \ll d_i$ and the other factors are all of order unity. Indeed, if we evaluate Gz_i for the values given in (13.1.13) we get

$$\mathrm{Gz}_i = 0{\cdot}4 \tag{13.3.9}$$

which implies that a fully-developed, $\tilde{T} = \tilde{T}(\hat{x}_3)$, solution is relevant. Thus any heating in the nip is dissipated by contact, and hence thermal exchange, with the roll to which the sheet adhered before it reaches the next nip. If the nip is the final one then it is necessary for the take-off roll to be a sufficient distance, $0(d_i)$, from the nip for thermal equilibrium to be attained.

† In the form (13.3.4), Gz_i is seen to be a Fourier number.

Chapter 14

Coating

This title covers a wide range of operations. A useful description of the more common continuous operations used to coat a moving uniform substrate is given in Middleman (1977, Chapter 8). The objective of achieving a uniform coating layer can be attained directly:

(a) by steady withdrawal of the web from a bath of the coating fluid, which leads to two-sided coating (Fig. 14.1), or
(b) by extrusion, from a slot die, of a curtain of melt which drops on to the moving web, to give a single-sided coating (Fig. 14.2);

or, indirectly in a two- or multi-stage process:

(c) by passing the web through the nip between two rollers, one of which has previously been coated with the coating fluid, so that part of the coating layer on the roll is transferred to the web; the coating on the roll is usually obtained by take-up from a bath of fluid into which the roll dips (or through a train of rolls, the last of which dips into the bath) (Fig. 14.3);
(d) by passing the web under tension over a rotating roll which has previously been coated with fluid (Fig. 14.4). This is sometimes termed kiss-coating.

In cases (c) and (d) the web and the coating roll can either be moving in the same direction, in which case the fluid transferred to the web passes through the nip, or in opposite directions in which case it does not. In both cases, the question arises as to how much of the fluid arriving on the coating roll is transferred to the web and how much remains on the roll. Also, the question arises as to whether the inlet side is flooded or not; if it is not, then the coating roll acts as a

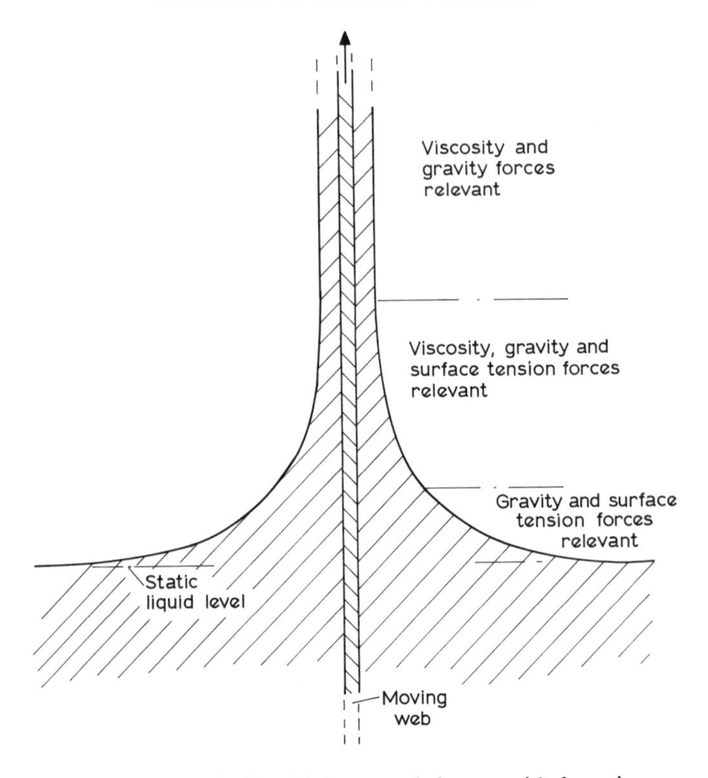

Fig. 14.1. Vertically withdrawn web for two-sided coating.

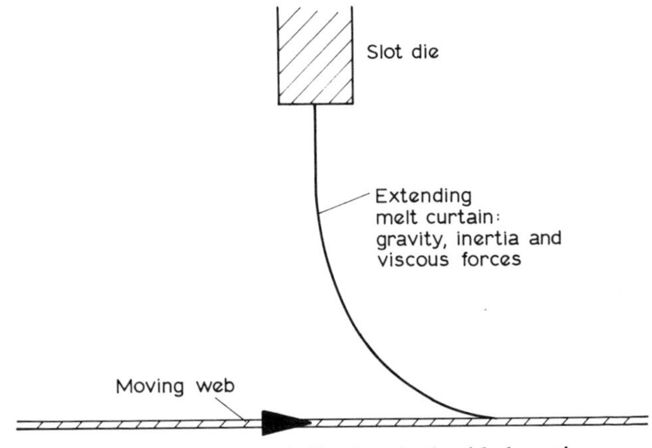

Fig. 14.2. Falling melt film for single-sided coating.

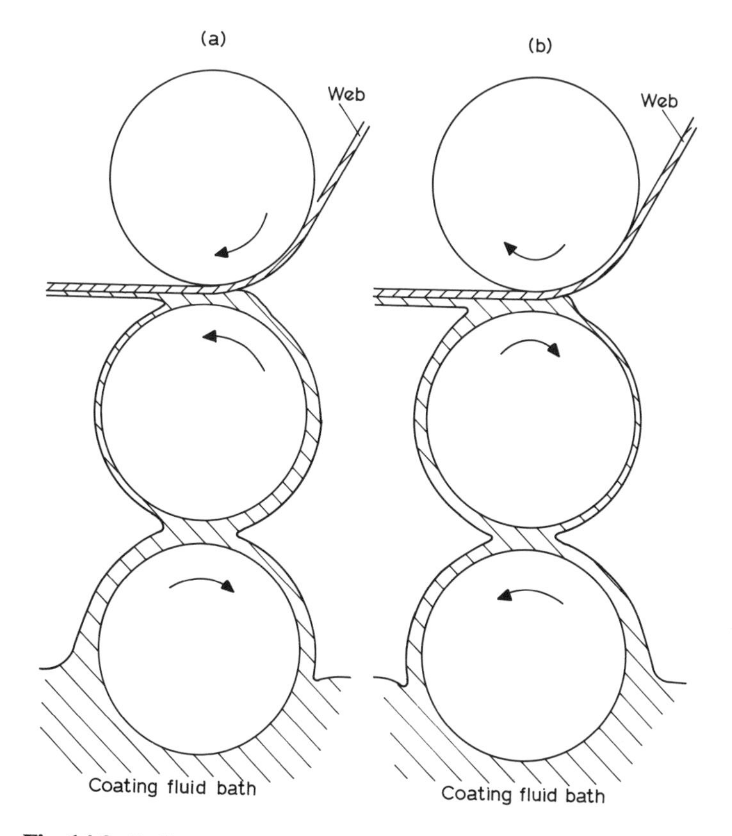

Fig. 14.3. Roller coating. (a) Co-directional. (b) Contra-directional.

separately metered applicator, as does the curtain-coating extruder in case (b). Otherwise, the coating process itself provides the metering.

In practice, various devices, including static doctor blades, can be used to achieve necessary metering. If the doctor blade is applied at the final stage of the web-coating operation, then strictly speaking the critical operation is no longer that of roll coating.

The mechanics of coating flows (including all roll-coating operations) are to a large extent covered by the two-dimensional lubrication approximation; the only exceptions arise in the direct coating operations where inertia, gravity and, to some extent, surface tension are as important as viscous forces. We concentrate here on the roll- and

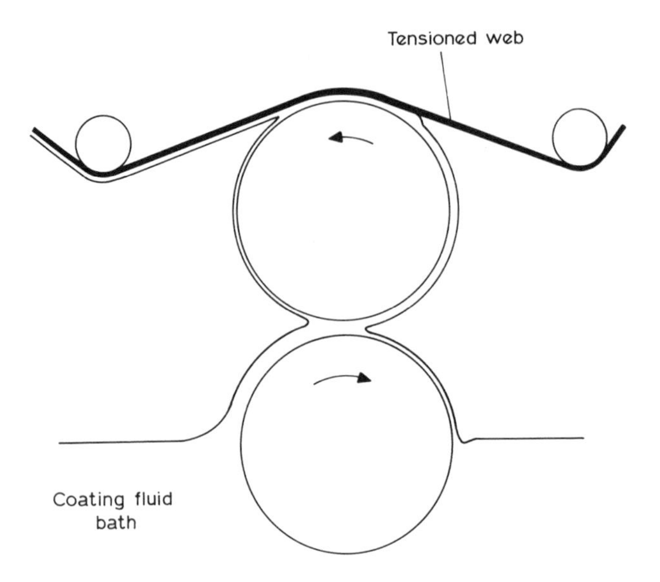

Fig. 14.4. Kiss coater.

kiss-coating operations; the decoupled (low B, Na) analyses of Subsection 13.1.1 can be used for the former and a similar one can be developed for the latter. Extrusion film coating is similar to film casting, a process which is covered in Chapter 16. (The detailed mechanics of flow in the region of attachment will not, however, be considered.)

Because the object of the process is to get adhesion to the web, the coated layer will not separate smoothly from either nip surface on the exit side, and so the exit boundary condition (13.1.44), used in calendering, will not be adequate. We discuss this matter in the next section.

The stability of the rolling operation was not discussed in the previous chapter on calendering. It is found that coating flows for relatively inviscid Newtonian fluids lead to fluid-mechanical instabilities at the exit. These are discussed separately in a later section, where three-dimensional perturbation flows are superposed on the basic plane flow solutions.

14.1 ENTRY AND EXIT FLOWS: EFFECTIVE END CONDITIONS FOR THE LUBRICATION APPROXIMATION

We have four cases to consider, as shown in Fig. 14.5.

(A) Here a dry web moves into the nip in the same direction as the coated roll surface. This is the situation considered in Fig. 13.6, and all the theory given in Section 13.2 applies in this, provided there is a region of recirculation. If the metering action of the coating roll is very good, it is possible that no reverse flow takes place and there are no closed streamlines. This is shown in Fig. 14.5(a). The effective entry point $\tilde{x}_{1\,\text{entry}}$ must be such that

$$|\tilde{x}_{1\,\text{entry}}| < |\tilde{x}_{1\text{st.i.}}| \tag{14.1.1}$$

where $\tilde{x}_{1\text{st.i.}}$ is as defined in Fig. 13.6. The entry boundary condition will then be

$$\tilde{p}(\tilde{x}_{1\,\text{entry}}) = \tilde{p}_a \tag{14.1.2}$$

instead of the first of (13.1.46). Under these circumstances $q_1 = v_{r1}h_{-\infty}$ will be determined by the incoming thickness $h_{-\infty}$ on the coated roll.

It will be noted that $x_{1\,\text{entry}}$ in the lubrication theory corresponds locally to the occurrence of a moving contact line C_A on the dry web. If any viscous effects in the ambient air can be neglected then the local fluid streamline can be expected to meet the web in a cusp. However, if viscous effects in the ambient air are included, then singularities arise if a solution consistent with traditional fluid mechanics and the no-slip boundary condition is sought. In any case no local two-dimensional solution to the entry flow-field taking account of free-surface effects has been obtained, so relation (14.1.2) is as accurate as any so far proposed.

(B) Here a coated web leaves the nip in the same direction as the wetted coating roller. The situation shown in Fig. 13.2 on the exit side no longer applies. The free surface that forms must have a stagnation point S_B where the streamline from upstream divides. If surface tension σ is non-negligible, i.e. if the capillary number

$$\text{Ca} = \eta v^*/\sigma \tag{14.1.3}$$

is not very large, then the influence of capillary forces has to be included in the pressure boundary condition to the lubrication equation. This has been treated by Pitts & Greiller (1961), and a general form for the exit boundary condition for the lubrication approximation

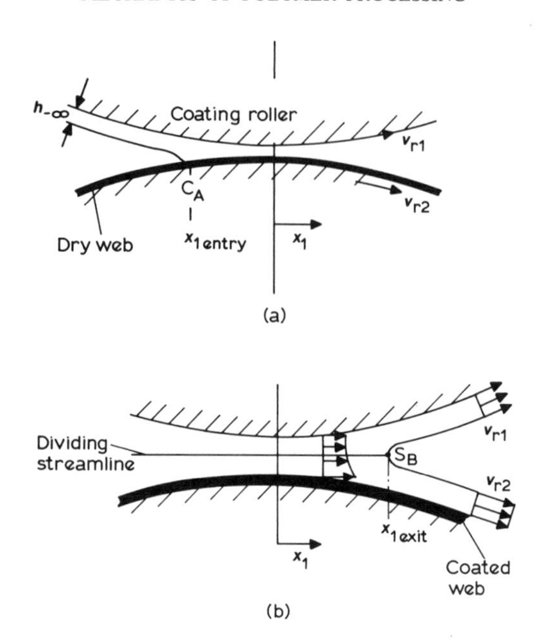

Fig. 14.5. Entry and exit flow geometries. (a) Entry: type A. (b) Exit: type B. A and B refer to co-directional motion of web and roll. (c) Entry: type C. (i) Overall view; (ii) expanded view of vortex region showing streamlines. (d) Exit: type D, showing streamlines when internal stagnation point is present. C and D refer to reverse roll coating.

can, in principle, be written as

$$\tilde{p}(\tilde{x}_{1\,\text{exit}}) = \tilde{p}_a - (\text{Ca}\,\tilde{X}\tilde{h})^{-1} f_B(\text{Ca}, \tilde{h}, \tilde{h}') \tag{14.1.4}$$

where $f_B = 0(1)$ and $\tilde{X} = X/h_0$. In practical cases $(\text{Ca}\,\tilde{X})^{-1}$ will be of order 10^{-1} or less.

A further boundary condition is provided by the requirement that S_B be a stagnation point.

For the Newtonian case, it may be shown (see Example 14.1.1) that this leads to

$$\tilde{h}^2[\tfrac{2}{3}(\tilde{v}_{r1} - \tilde{v}_{r2})^2 + (\tilde{v}_{r1} + \tilde{v}_{r2})^2] - 8\tilde{h}(\tilde{v}_{r1} + \tilde{v}_{r2}) + 12 = 0 \tag{14.1.5}$$

and so defines $\tilde{x}_{1\,\text{exit}}$, because $\tilde{v}_{r1}$ and $\tilde{v}_{r2}$ are known if $\tilde{q}_1 = 2$ is specified.

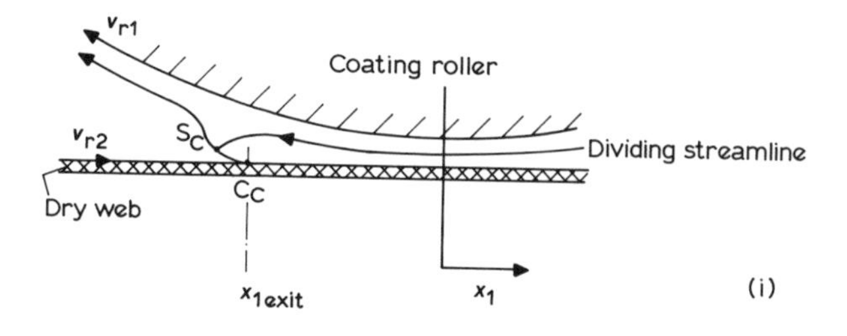

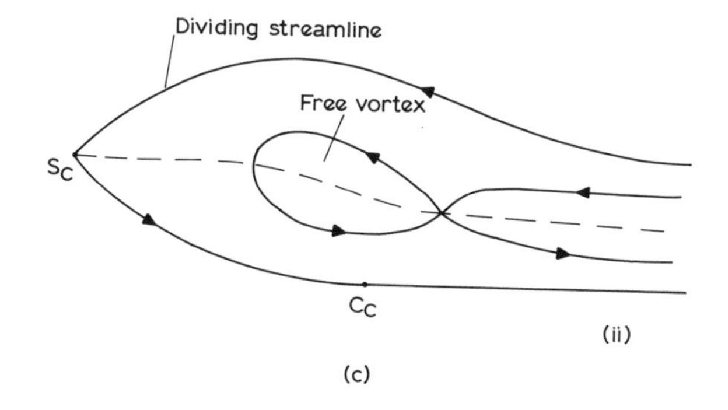

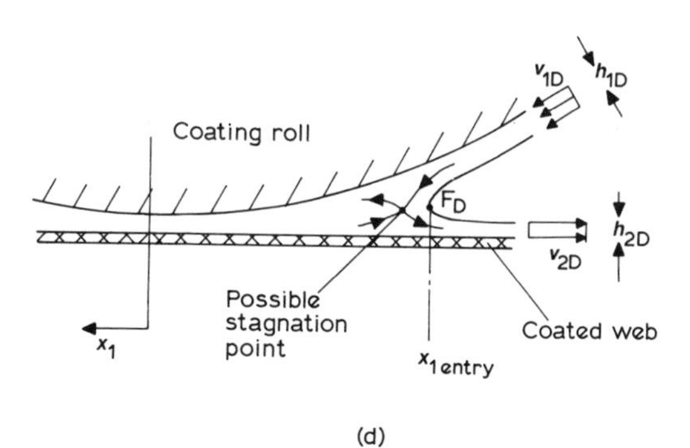

Fig. 14.5. *contd.*

The continuity of pressure requirement using (14.1.4) becomes

$$\tilde{p}(\tilde{x}_{1\ \mathrm{exit}}) - \tilde{p}(\tilde{x}_{1\ \mathrm{entry}}) = \int_{\tilde{x}_{1\ \mathrm{entry}}}^{\tilde{x}_{1\ \mathrm{exit}}} \frac{\mathrm{d}\tilde{p}}{\mathrm{d}\tilde{x}_1}\,\mathrm{d}\tilde{x}_1 = -(\mathrm{Ca}\,\tilde{X}\tilde{h}_{\mathrm{exit}})^{-1} f_{\mathrm{B}} \quad (14.1.6)$$

and so provides an implicit relation for $\tilde{x}_{1\ \mathrm{entry}}$. In this way the boundary conditions A and B are linked.

(C) Here the dry web moves into the nip as the coating roller moves away from it. The situation is shown in Fig. 14.5(c) where the free streamline has a stagnation point S_C and joins the web in a cusp at the moving contact line C_C. The local difficulties at C_C are the same as discussed in A above for C_A.

As in A we use the boundary condition

$$\tilde{p}(\tilde{x}_{1\ \mathrm{exit}}) = \tilde{p}_a \tag{14.1.7}$$

but we have no analogue of eqn (14.1.5) because the stagnation point S_C is now well out of the true lubrication region and is more of the nature of $\tilde{x}_{1\mathrm{st.o.}}$ in Fig. 13.6. In the lubrication sense, the streamline arriving at S_C cannot be identified with $\tilde{v}=0$ as it could in case B for S_B. Indeed, it is kinematically possible that a trapped vortex arises on the web entry side.

(D) Here a coated roll moves into a nip from which a coated web is emerging. The free streamline arrives with decreasing positive velocity ($\leqslant v_{1D}$), reaches an innermost position at F_D at which point its forward velocity (in the lubrication sense) is zero; its velocity then decreases further until it reaches the outgoing web speed of v_{2D}. As in case B the entry pressure boundary condition can be written

$$\tilde{p}(\tilde{x}_{1\ \mathrm{entry}}) = \tilde{p}_a - (\mathrm{Ca}\ \tilde{X}\tilde{h})^{-1} f_D(\mathrm{Ca}, \tilde{h}, \tilde{h}') \tag{14.1.8}$$

where $f_D = 0(1)$. The condition that F_D is a point of zero velocity can be written

$$\int_{-1}^{1} (\tilde{v}_1 + |\tilde{v}_1|)\, \mathrm{d}\hat{x}_3 = 2\tilde{q}_1/\tilde{h}_{\mathrm{entry}} \tag{14.1.9}$$

which prescribes $\tilde{x}_{1\ \mathrm{entry}}$ since $\tilde{q}_1$, the entering flowrate, is specified. The equivalent of (14.1.6) is

$$\int_{\tilde{x}_{1\mathrm{entry}}}^{\tilde{x}_{1\mathrm{exit}}} \frac{\mathrm{d}\tilde{p}}{\mathrm{d}\tilde{x}_1}\, \mathrm{d}\tilde{x}_1 = (\mathrm{Ca}\ \tilde{X}\tilde{h}_{\mathrm{entry}})^{-1} f_D \tag{14.1.10}$$

which provides a relatively simple relation for $\tilde{x}_{1\ \mathrm{exit}}$, and links the boundary conditions C and D.

It is kinematically possible that a stagnation point arises for $0 > \tilde{x}_1 > \tilde{x}_{1\ \mathrm{entry}}$, and that F_D lies well outside the region of applicability of the lubrication approximation. There is little evidence on this point.

It must be emphasized that the fine details of the flow patterns in the neighbourhood of $x_{1\ \mathrm{entry}}$ and $x_{1\ \mathrm{exit}}$ have not been examined here and indeed do not seem amenable to analytic solution in any generality. All that have been proposed above are apparently crude end boundary conditions to a lubrication approximation for the flow in the nip. It should be pointed out, however, that in any particular case f_B and f_D could in principle absorb all the fine details of the flow if it were known. Having obtained values for $x_{1\,\mathrm{entry}}$ and $x_{1\,\mathrm{exit}}$ in any particular case, it is always possible to discover *a posteriori* whether the lubrication approximation applies at those points or not. Fortunately the effect of the f_B and f_D terms in (14.1.6) and (14.1.8) decreases with $\tilde{h}$ and so is felt most strongly when $\tilde{h}\to 1$, i.e. when the lubrication approximation does apply.

There have been many other suggestions made in the literature for end boundary conditions to the lubrication approximation. Savage (1977) discusses the proposals of Swift (1932), Steiber (1933), Hopkins (1957), Coyne & Elrod (1970, 1971) and mentions the more profound work of Pitts & Greiller (1961), Bretherton (1961) and Taylor (1963). The subject cannot be said to be well understood.

More recently, the possibility of obtaining 'exact' solutions, particularly for the Newtonian fluid model, by numerical computation (Pearson & Richardson, 1983, Chapter 6) has led many workers to tackle free surface coating flows in fine detail, taking account of complex geometrical features of the coating equipment. In many cases inertia cannot be neglected and boundary layers have to be considered (Cerro and Scriven, 1980); this leads to a very different approach to that taken using lubrication theory and is taken to be outside the scope of this text, which is devoted almost exclusively to highly viscous flows.

EXAMPLE 14.1.1. The derivation of exit boundary conditions for roll coating using a Newtonian fluid.

From (13.1.63) we can write the velocity profile at any point in the lubrication flow as

$$\tilde{v}_1=\frac{3}{4}\left[(\tilde{v}_{r1}+\tilde{v}_{r2})-\frac{2}{\tilde{h}}\right]\hat{x}_3^2+\frac{1}{2}(\tilde{v}_{r1}-\tilde{v}_{r2})\hat{x}_3+\frac{1}{4}\left[\frac{6}{\tilde{h}}-(\tilde{v}_{r1}+\tilde{v}_{r2})\right]$$

satisfying

$$\tilde{v}_1(1)=\tilde{v}_{r1},\qquad \tilde{v}_1(-1)=\tilde{v}_{r2}$$

The least value for this, provided $\tilde{v}_{r1}$ and $\tilde{v}_{r2}$ are positive, which is true for this case, is given by

$$\frac{d\tilde{v}_1}{d\hat{x}_3}=\frac{3}{2}\left[(\tilde{v}_{r1}+\tilde{v}_{r2})-\frac{2}{\tilde{h}}\right]\hat{x}_3+\frac{1}{2}(\tilde{v}_{r1}-\tilde{v}_{r2})=0$$

or

$$\hat{x}_3 = -\frac{\tilde{h}(\tilde{v}_{r1} - \tilde{v}_{r2})}{3[(\tilde{v}_{r1} + \tilde{v}_{r2})\tilde{h} - 2]}$$

Clearly $\hat{x}_3$ will lie in the range $(-1, 1)$ only if $\tilde{h}$ is large enough.

A stagnation point in the lubrication-approximation flow field will arise when $\tilde{v}_1 = 0$ at $\hat{x}_3$ defined above, i.e. if

$$0 = \frac{1}{12}\frac{\tilde{h}(\tilde{v}_{r1} - \tilde{v}_{r2})^2}{[(\tilde{v}_{r1} + \tilde{v}_{r2})\tilde{h} - 2]} - \frac{1}{6}\frac{\tilde{h}(\tilde{v}_{r1} - \tilde{v}_{r2})^2}{[(\tilde{v}_{r1} + \tilde{v}_{r2})\tilde{h} - 2]} + \frac{1}{4}\left[\frac{6}{\tilde{h}} - (\tilde{v}_{r1} + \tilde{v}_{r2})\right]$$

which is the same as (14.1.5). This provides one relation between $\tilde{q}_1$ and $\tilde{h}_{\text{exit}}$, i.e. between $\tilde{q}_1$ and $\tilde{x}_{1\,\text{exit}}$, if it is assumed that flow up to S_B is governed by eqn (13.1.53). Clearly then (14.1.5) is immediately consistent with lubrication theory, in that it is based on a lubrication-theory velocity-profile.

However, (14.1.4) which represents a modification to (13.1.46) to take account of surface tension forces—and implicitly of departures from the lubrication approximation flow field in the neighbourhood of the stagnation point—is less easy to develop in a specific form. The matter is discussed in Pearson (1960), where the simplest approach is based on the balance of forces on either side of the meniscus region, yielding

$$[p(x_{1\,\text{exit}}) - p_a]h(x_{1\,\text{exit}}) = -2\sigma$$

Written out in the form (14.1.4) this yields $f_B = \frac{1}{2}$. Pitts and Greiller's (1961) solution is more complex and involves the velocity distribution near the stagnation point. An estimate of the contribution involved can be obtained by considering the velocity gradient $\partial v_1/\partial x_1$ and the associated normal-force difference $4\eta\partial v_1/\partial x$ (cf. pure shear flow: Section 4.3.2) which has to be balanced by a component of p_a. If the lubrication-flow-field solution is used, then

$$\eta\frac{\partial v_1}{\partial x_1} = \frac{\eta v^*}{X}\frac{\partial \tilde{v}_1}{\partial \tilde{x}_1} = \frac{\eta v^*}{X}\tilde{h}'\frac{d\tilde{v}_1}{d\tilde{h}} = 0\left(\frac{\eta v^*}{X}\right)$$

The dimensionless form that is relevant means that the contribution to $\tilde{p}$ is $0(X^{-2})$, and so it will influence f_B by a factor of order Ca/X; according to the lubrication approximation terms of order $\tilde{X}^{-2}$ are to be neglected, and so, as expected, only local variations in v_1 which are much larger, say a factor of $\tilde{X}$ larger, can be expected to influence the entire solution. These can be calculated only by solving, as stated in the text, a two-dimensional flow problem around the meniscus.

EXERCISE 14.1.1. Derive relation (14.1.9) on the basis of a lubrication flow of type (13.1.53) at $x_{1\ \mathrm{entry}}$, noting that

$$\int_{\hat{x}_3(\mathrm{F_D})}^{1} \tilde{v}_1\,\mathrm{d}\hat{x}_3 = \tilde{v}_{1\mathrm{D}}\tilde{h}_{1\mathrm{D}} = \tilde{q}_1$$

where $\tilde{h}_{1\mathrm{D}}$ is the prescribed thickness on the entering roll and

$$\int_{-1}^{1} \tilde{v}_1\,\mathrm{d}\hat{x}_3 = 2/\tilde{h}_{\mathrm{entry}}$$

Note that the thickness of the coating is given by

$$\tilde{h}_{2\mathrm{D}} = (2-\tilde{q}_1)/\tilde{v}_{2\mathrm{D}}$$

where $\tilde{v}_{2\mathrm{D}}<0$, $\tilde{q}_1>2$.

Express the result in dimensional terms and explain its implications; in particular, show how (14.1.9) refers to a non-flooded situation and how $v_{1\mathrm{D}}h_{1\mathrm{D}} = q_{\mathrm{entry}}$ determines q_1 and $v_{2\mathrm{D}}h_{2\mathrm{D}}$.

Other exercises can be found at the end of Chapter 8 of Middleman (1977), which will be found of great value to those wishing to study the process of coating in any depth. Recent analyses for Newtonian fluids are to be found in Benkreira *et al.* (1981) and Greener and Middleman (1981).

14.2 LUBRICATION APPROXIMATION FOR FLOW BETWEEN A FLEXIBLE TENSIONED WEB AND A ROLLER

The main difference between this situation shown in Fig. 14.6(a) and the one analysed in Section 13.1 and Subsection 13.2.1 is that the channel width $h(x_1)$ is not specified in advance. The curvature of the web has to be such that it balances the pressure in the fluid layer at any given position x_1. The web is held against the coating roll by the resolved component of the tensile forces, $2F \sin \phi$, that arises when the web turns through an angle 2ϕ (see Fig. 14.6(a)) on passing over the roll. F here is the tensile force/unit width in the web. In the absence of any fluid layer the web will be in contact with the roll over a circumferential distance ϕd, where d is the roll diameter. The

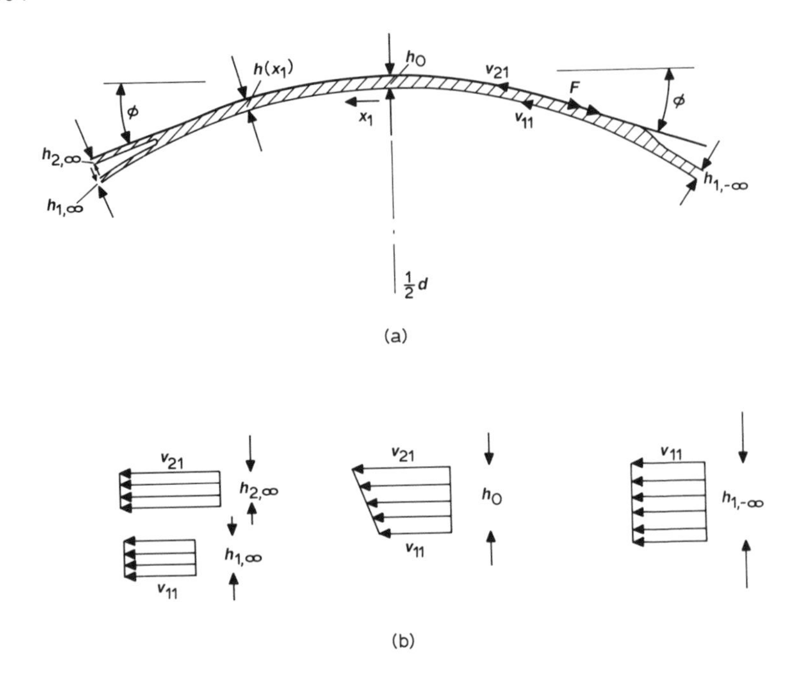

Fig. 14.6. Kiss-coating operation: co-directional. (a) Geometry of tensioned web passing over moving circular-cylindrical roll. (b) Velocity profiles far from entry and exit regions.

peripheral speed of the roll is v_{11} and of the web v_{21}. We suppose that

$$h \ll \phi d \tag{14.2.1}$$

and that

$$\mathrm{d}h/\mathrm{d}x_1 \ll 1 \tag{14.2.2}$$

where x_1 is a curvilinear coordinate measured along the web or roll. We consider a steady flow situation.

The curvature H of the web is given approximately by

$$H(x_1) = \frac{2}{d} - \frac{\mathrm{d}^2 h}{\mathrm{d}x_1^2} \tag{14.2.3}$$

when use is made of (14.2.2). The pressure within the fluid layer is such that

$$p(x_1) - p_\mathrm{a} = FH(x_1) \tag{14.2.4}$$

The flux q_1, which is constant in the steady state, will be related locally to p and h by equations of the form (8.1.22) or (8.1.23). In the case of relatively-low-viscosity coating fluid we can usually assume that $T_{S1} = T_{S2} =$ constant and that the fluid exhibits a power-law viscosity. For a Newtonian fluid the dimensionless forms (13.1.47), (13.1.48) would apply.

If ϕd is sufficiently large we can expect h to be essentially constant, at h_0 say, over most of the region

$$x_{1\,\text{entry}} < x_1 < x_{1\,\text{exit}} \tag{14.2.5}$$

This ensures that the pressure will be constant, p_0 say, over the region for which $h = h_0$.

From (14.2.4)

$$p_0 = 2F/d + p_a \tag{14.2.6}$$

while mass conservation requires that

$$q_1 = v_{11} h_{1,-\infty} \tag{14.2.7}$$

where $h_{1,-\infty}$ is the thickness of the fluid layer on the inlet side of the coating roll. Since $dp/dx = 0$ when $h = h_0$, we must have a linear velocity profile $v_1(x_1)$, as shown in Fig. 14.6(b), and so

$$q_1 = \tfrac{1}{2}(v_{11} + v_{21}) h_0 \tag{14.2.8}$$

or

$$h_0 = 2 v_{11} h_{1,-\infty} / (v_{11} + v_{21}) \tag{14.2.9}$$

The case shown in Fig. 14.6 corresponds to case A at inlet and case B at exit. Consequently p must fall to p_a at $x_{1\,\text{entry}}$ according to (14.1.2) and to the value given by the equivalent of (14.1.4) at $x_{1\,\text{exit}}$. A simple approximation is provided by

$$p = p_a - 2\sigma / \{h(x_{1\,\text{exit}}) - h_{1,\infty} - h_{2,\infty}\} \tag{14.2.10}$$

We expect

$$F/d \ll \sigma/h \tag{14.2.11}$$

There must therefore be a region in which dp/dx is significantly different from zero and for which d^2h/dx_1^2 will therefore be appreciable. From (14.2.3) and (14.2.4) we get an ordinary differential equation

$$\frac{d^2h}{dx_1^2} = \frac{2}{d} - \frac{p(x_1)}{F} + \frac{p_a}{F} \tag{14.2.12}$$

which on differentiation yields

$$\frac{dp}{dx_1} = -F\frac{d^3h}{dx_1^3} \qquad (14.2.13)$$

This can be substituted into (8.1.23) suitably expressed to yield a single equation for h.

If we use the Newtonian case as an example, and put $v_{11} = v_{21}$ for simplicity (it is in any case a very reasonable situation to consider), (13.1.48) in the form $q_1 = v_{11}h - (h^3/12\eta_0)(dp/dx_1)$ and (14.2.13) yield

$$\frac{d^3h}{dx_1^3} = \frac{12\eta_0 v_{11}}{F}\frac{(h^* - h)}{h^3} \qquad (14.2.14)$$

where h^* can obviously be identified with $h_0 = h_{1,-\infty}$ in the case where $p \to p_0$ for all x_1 far from the exit and entry regions.

The entry boundary condition to this equation is provided by

$$p(x_{1\ \text{entry}}) = p_a \qquad (14.2.15)$$

and the exit boundary conditions by (14.2.10) and (14.1.5) which become simply (see Exercise 14.2.2)

$$h(x_{1\ \text{exit}}) = 3h_0, \qquad p(x_{1\ \text{exit}}) = p_a - \sigma/h_0 \qquad (14.2.16)$$

Unlike the calendering case we are now seeking two separate solutions of (14.2.14) each of which asymptotes for $|x_1| \to \infty$ to the result (14.2.6), (14.2.9), with both h_0 and p_0 specified. The actual positions $x_{1\ \text{entry}}$ and $x_{1\ \text{exit}}$ are largely determined by the geometry of the roller system controlling the web, but this is not important if the entry and exit flows are decoupled. Equation (14.2.14) may be made dimensionless by writing

$$\hat{h} = (h - h_0)/h_0, \qquad \hat{x}_1 = x_1(12\eta_0 v_{11}/Fh_0^3)^{\frac{1}{3}} \qquad (14.2.17)$$

to give

$$\frac{d^3\hat{h}}{d\hat{x}_1^3} = -\frac{\hat{h}}{(1+\hat{h})^3} \qquad (14.2.18)$$

(14.2.10), (14.2.15) and (14.2.16) become

$$\hat{x}_1 = \hat{x}_{1\ \text{exit}}: \frac{d^2\hat{h}}{d\hat{x}_1^2} = 2\left(\frac{F}{12\eta_0 v_{11}}\right)^{\frac{2}{3}}\frac{h_0}{d}\left\{1 + \frac{\sigma d}{2Fh_0}\right\} \qquad (14.2.19)$$

$$\hat{x}_1 = \hat{x}_{1\ \text{entry}}: \frac{d^2\hat{h}}{d\hat{x}_1^2} = 2\left(\frac{F}{12\eta_0 v_{11}}\right)^{\frac{2}{3}}\frac{h_0}{d} \qquad (14.2.20)$$

Two length scales have been provided for x_1: a geometrical one, ϕd, used in (14.2.1) and a dynamical one, $h_0(F/12\eta_0 v_{11})^{\frac{1}{3}}$, used in (14.2.17). The asymptotic inner region with $p=p_0, h=h_0$ can be expected to arise when

$$d \gg h_0(F/12\eta_0 v_{11})^{\frac{1}{3}} \qquad (14.2.21)$$

assuming ϕ to be 0(1). Condition (14.2.2) implies that

$$F/12\eta_0 v_{11} \gg 1 \qquad (14.2.22)$$

Some insight into the exact solution of (14.1.18) subject to (14.2.19), (14.2.20) can be gained by considering $\hat{h} \ll 1$. The three solutions are then (see, for example, Bretherton, 1961)

$$\hat{h} \sim e^{-\hat{x}_1}, e^{\frac{1}{2}\hat{x}_1}\sin(\sqrt{3}\hat{x}_1/2), e^{\frac{1}{2}\hat{x}_1}\cos(\sqrt{3}\hat{x}_1/2) \qquad (14.2.23)$$

For $\hat{x}_1 \ll 0$, there is one dominant exponential solution, while for $\hat{x}_1 \gg 0$ there are two exponentially oscillatory solutions. By having the coefficients of the separate solutions small enough, $\hat{h}$ can be made arbitrarily small except for $|\hat{x}_1|$ sufficiently large. The problem is then split into two non-linear solutions based on a one-parameter asymptote $Ae^{-\hat{x}_1}$ for the entry region and a two-parameter asymptote $B\exp(\frac{1}{2}\hat{x}_1)\sin\left(\frac{\sqrt{3}}{2}\hat{x}_1+C\right)$ for the exit region. Thus one entry boundary condition and two exit boundary conditions provide the right constraints.

Full solutions to the non-linear equations can be obtained by numerical computation, but it must be remarked that they will only be as accurate as the boundary conditions.

EXERCISE 14.2.1. Show that (14.2.10) is obtained by supposing the meniscus at exit to be a hemisphere joined smoothly to the asymptotic 'planar' surfaces of the fluid coatings leaving the nip. Compare with the result given in Example 14.1.1 to show how inaccurate the arguments involved are likely to be.

EXERCISE 14.2.2. Consider the flow analysed in Section 14.2 (kiss coating) for a Newtonian fluid with $v_{11}=v_{21}$. Show that the scale velocity v^* defined by (13.1.30) becomes simply v_{11} in this case, and so the dimensionless wall velocities, corresponding to $\tilde{v}_{r1}$ and $\tilde{v}_{r2}$ in (14.1.5), are simply unity. Hence show that the dimensionless exit width $\tilde{h}(\tilde{x}_{1\,\mathrm{exit}})$ is given by

$$\tilde{h}^2-4\tilde{h}+3=0$$

and that the relevant root is given by $\tilde{h}=3$, thus leading to the first of (14.2.16).

Use symmetry to show that

$$h_{1,\infty}=h_{2,\infty}=\tfrac{1}{2}h_0$$

in (14.2.10), and hence obtain the second of (14.2.16).

14.3 STABILITY AND SENSITIVITY OF PLANE COATING FLOWS

All the analysis given so far for calendering and coating flows has assumed that the flow is steady two-dimensional in the (x_1, x_3) plane, except for the transverse x_2 flow described in Subsection 13.2.2. The lubrication approximation effectively removes any apparent dependence upon flow in the x_3-direction, though it was pointed out in Section 14.1 that the inlet and exit boundary conditions must in fact incorporate the consequences of local flow fields that involve fully two-dimensional effects.

The coupling between the entry flow and the exit flow, particularly in the case of coating systems when thermal effects are negligible, arises simply through the pressure field; in the case of kiss-coating, q_1 can be varied over a wide range, whereas in roll-coating it has to lie between relatively narrow limits for any given minimum nip gap.

The objective of the coating† operations is to provide a uniform layer on the web. It is therefore important to consider what will be the effect of small disturbances on the inlet flow; in particular, whether these will be damped out by the lubrication flow under the nip. On rather obvious and elementary grounds it is clear that upstream disturbances will have very little effect in the roll-coating case, particularly if a large rolling bank exists on the inlet side.‡ It is not immediately clear whether the same effect will be true in the kiss-coating mechanism described in Section 14.2, because of web deformability. A diagrammatic indication of what is involved in both cases is given in Fig. 14.7, and a sensitivity analysis is described briefly in Subsection 14.3.2 for the kiss-coating system.

Quite separate from the effect of inlet disturbances is the question of

† The same is true in essence of the calendering process.
‡ Indeed this is the attraction of the process.

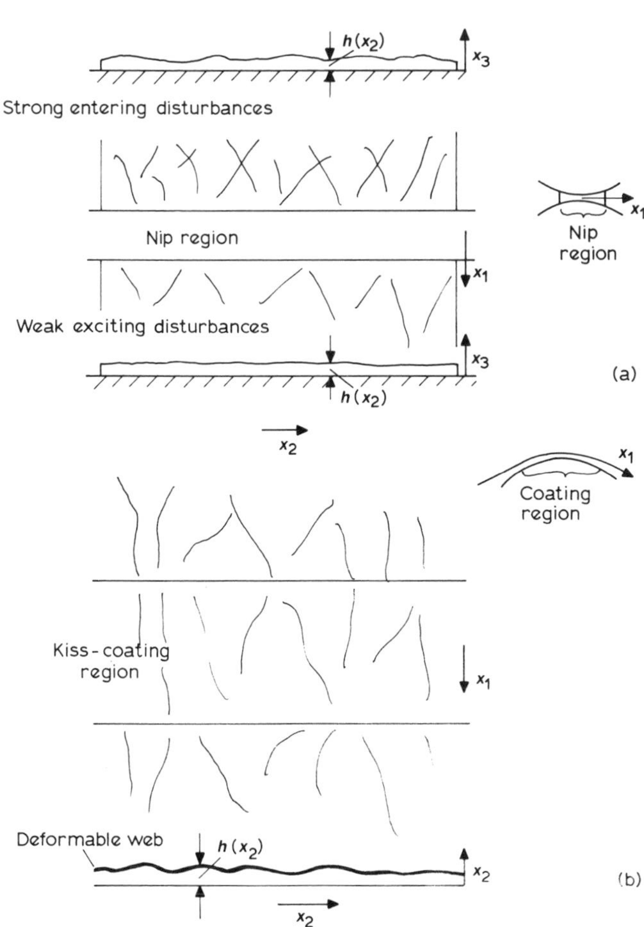

Fig. 14.7. Effect of disturbances. (a) Two-roll coating system showing the rapid decay of disturbances entering the nip-region. (b) Kiss-coating system showing the possibility of a disturbance $h(x_2) \neq h_0$ passing essentially without decay through the (longer) region of contact.

stability of the flows to small random disturbances on the exit side. There is considerable experimental evidence (Pitts & Greiller, 1961; Pearson, 1960) that roll-coating (and other allied) flows involving fluids that properly wet both roll and web surfaces are generally unstable to disturbances with x_2 dependence. It is fairly clear on physical grounds that in such cases surface tension is a stabilizing force, so the case

Ca$\rightarrow$0 might be expected to be the most unstable. The formal structure of a linearized stability analysis is given in the next section. This should not be taken as a definitive theory. It is intended merely to demonstrate the principles involved in obtaining such a theory.

14.3.1 Instability of the Roll Coating Operation

We first consider the undisturbed solution given earlier for calendering, using the Newtonian model, defined by (13.1.47) and (13.1.48). These dimensionless relations are based on v^* given by (13.1.30), i.e. q_1/h_0; for definiteness we suppose that boundary conditions (14.1.2), (14.1.4) and (14.1.5) are obeyed, that q_1 has been prescribed, and that $\tilde{v}_{r1} = \tilde{v}_{r2}$. (14.1.5) then yields (cf. Exercise 14.2.2)

$$\tilde{h}^{(0)}_{\text{exit}} = 3\tilde{v}_{r1}^{-1} \tag{14.3.1}$$

where the (0) represents the unperturbed solution, from which it follows that

$$\tilde{x}^{(0)}_{1\,\text{exit}} = [3\tilde{v}_{r1}^{-1} - 1]^{\frac{1}{2}} \tag{14.3.2}$$

If we suppose that $q_1^{(0)}$ corresponds to $\tilde{x}^{(0)}_{1\,\text{entry}} \rightarrow -\infty$, then we can use (14.1.6) and the relation (13.1.48) to give

$$3\int_{-\infty}^{(3\tilde{v}_{r1}^{-1}-1)^{\frac{1}{2}}} \left\{ \frac{\tilde{v}_{r1}}{(1+\tilde{x}_1^2)^2} - \frac{1}{(1+\tilde{x}_1^2)^3} \right\} d\tilde{x}_1 = -(\text{Ca}\,\tilde{X}\tilde{h}_{\text{exit}})^{-1} f_{\text{B}} \tag{14.3.3}$$

from which it follows that

$$\tilde{v}_{r1} = \tilde{v}_{r1}(\text{Ca}^{-1} = 0) + 0(\text{Ca}^{-1}) \tag{14.3.4}$$

Middleman (1977, p. 194) has shown that $\tilde{v}_{r1}^{-1} = 1{\cdot}3$.

We now consider small disturbances $\tilde{\mathbf{q}}^{(1)}(\tilde{x}_1, \tilde{x}_2, \tilde{t})$, $\tilde{p}^{(1)}(\tilde{x}_1, \tilde{x}_2, \tilde{t})$, $\tilde{x}^{(1)}_{1\,\text{exit}}(\tilde{x}_2, \tilde{t})$ where

$$\left.\begin{aligned} \tilde{\mathbf{q}} &= (1+\tilde{q}_1^{(1)}, \tilde{q}_2^{(1)}),\ |\tilde{q}^{(1)}| \ll 1 \\ \tilde{p} &= \tilde{p}^{(0)} + \tilde{p}^{(1)},\ |\tilde{p}^{(1)}| \ll 1 \\ \tilde{x}_{1\,\text{exit}} &= \tilde{x}^{(0)}_{1\,\text{exit}} + \tilde{x}^{(1)}_{1\,\text{exit}},\ |\tilde{x}^{(1)}_{1\,\text{exit}}| \ll 1 \end{aligned}\right\} \tag{14.3.5}$$

$$\tilde{t} = tv^*/X,\ \tilde{x}_2 = x_2/X \tag{14.3.6}$$

From continuity and the lubrication equations (see (8.1.28) and Example 8.1.2)

$$\frac{\partial \tilde{q}_1^{(1)}}{\partial \tilde{x}_1}+\frac{\partial \tilde{q}_2^{(1)}}{\partial \tilde{x}_2}=0 \tag{14.3.7}$$

$$\frac{\partial \tilde{q}_1^{(1)}}{\partial \tilde{x}_1}=-\frac{3\tilde{q}_1^{(1)}}{\tilde{h}^3},\qquad \frac{\partial \tilde{p}^{(1)}}{\partial \tilde{x}_2}=-\frac{3\tilde{q}_2^{(1)}}{\tilde{h}^3} \tag{14.3.8}$$

from which it follows that

$$\frac{\partial}{\partial \tilde{x}_1}\left(\tilde{h}^3\frac{\partial \tilde{p}^{(1)}}{\partial \tilde{x}_1}\right)+\tilde{h}^3\frac{\partial^2 \tilde{p}^{(1)}}{\partial \tilde{x}_2^2}=0 \tag{14.3.9}$$

The boundary conditions for the disturbance flow must now be considered. Since $\tilde{x}_{1\text{ entry}}$ has been taken to be the flooded condition we suppose

$$\tilde{\mathbf{q}}^{(1)},\tilde{p}^{(1)}\to 0 \quad \text{as} \quad x_1\to -\infty \tag{14.3.10}$$

The pressure boundary condition equivalent to (14.1.4) must now contain a term caused by the curvature of $\tilde{x}_{1\text{ exit}}$ in the $(\tilde{x}_1, \tilde{x}_2)$ plane. This yields (see Example 14.3.1)

$$\tilde{p}^{(1)}(\tilde{x}^{(0)}_{1\text{ exit}})=-\tilde{x}^{(1)}_{1\text{ exit}}\left\{\frac{\partial \tilde{p}^{(0)}}{\partial \tilde{x}_1}-(\text{Ca}\,\tilde{X}\tilde{h}^2)^{-1}\frac{\mathrm{d}\tilde{h}}{\mathrm{d}\tilde{x}_1}f_{\text{B}}\right\}_{\tilde{x}_1=\tilde{x}^{(0)}_{1\text{ exit}}}$$

$$-(\text{Ca}\,\tilde{X})^{-1}\frac{\partial^2\tilde{x}^{(1)}_{1\text{ exit}}}{\partial \tilde{x}_2^2}f^*_{\text{B}} \tag{14.3.11}$$

where f_{B} and f^*_{B} are assumed to be constant.

A second boundary condition at $\tilde{x}^{(0)}_{1\text{ exit}}$ can be obtained by modifying the argument used to obtain (14.1.5). The meniscus tip at $\hat{x}_3=0$ will now move according to the local velocity $\tilde{v}_1(\tilde{x}_{1\text{ exit}}, \tilde{x}_2, 0, \tilde{t})$ to yield $\partial\tilde{x}^{(1)}_{1\text{ exit}}/\partial t$; S_{B} will no longer be a stagnation point.

From (13.1.32) it follows, by integration, that

$$\frac{\partial \tilde{q}}{\partial \hat{x}_3}=\tilde{h}\tilde{v}_1=\tilde{h}^3\frac{\mathrm{d}\tilde{p}}{\mathrm{d}\tilde{x}_1}\hat{x}_3^2+C \tag{14.3.12}$$

where C is a constant given in this symmetrical case by

$$\tilde{h}\tilde{v}_{\text{r}1}=\tilde{h}^3\frac{\mathrm{d}\tilde{p}}{\mathrm{d}\tilde{x}_1}+C \tag{14.3.13}$$

for $\hat{x}_3=\pm 1$. If we now put $\hat{x}_3=0$ we obtain, according to the argument used above (see Exercise 14.3.2)

$$\frac{\partial \hat{x}_{1\text{ exit}}}{\partial \tilde{t}} = \left(-\tilde{h}^2 \frac{\partial \tilde{p}}{\partial \tilde{x}_1} + \tilde{v}_{r1}\right)_{\tilde{x}_1 = \tilde{x}_{1\text{exit}}} \tag{14.3.14}$$

On linearization this yields

$$\frac{\partial \tilde{x}^{(1)}_{1\text{ exit}}}{\partial \tilde{t}} = -\left[\tilde{x}^{(1)}_{1\text{ exit}} \frac{\mathrm{d}}{\mathrm{d}\tilde{x}_1}\left(\tilde{h}^2 \frac{\mathrm{d}\tilde{p}^{(0)}}{\mathrm{d}\tilde{x}_1}\right) + \tilde{h}^2 \frac{\partial \tilde{p}^{(1)}}{\partial \tilde{x}_1}\right]_{\tilde{x}_1 = \tilde{x}^{(0)}_{1\text{ exit}}} \tag{14.3.15}$$

It can be seen at once that eqn (14.3.9) together with boundary conditions (14.3.10), (14.3.11) and (14.13.15) form a linear system with an arbitrary scale. As is usual in such situations we take a Fourier decomposition in the $\tilde{x}_2$-direction and look for solutions of the form

$$x^{(1)}_{1\text{ exit}} = \mathcal{Re}\{\exp(\tilde{u}\tilde{t} + \mathrm{i}\tilde{n}\tilde{x}_2)\} \tag{14.3.16}$$

$$\tilde{p}^{(1)} = \mathcal{Re}\{\exp(\tilde{u}\tilde{t} + \mathrm{i}\tilde{n}\tilde{x}_2)\tilde{P}^{(1)}(\tilde{x}_1)\} \tag{14.3.17}$$

where $\tilde{n}$ is a dimensionless wave number and $\tilde{u}$ a dimensionless growth rate.

(14.3.9) becomes

$$\frac{\mathrm{d}^2\tilde{P}^{(1)}}{\mathrm{d}\tilde{x}_1^2} + \frac{6\tilde{x}_1}{1+\tilde{x}_1^2}\frac{\mathrm{d}\tilde{P}^{(1)}}{\mathrm{d}\tilde{x}_1} - \tilde{n}^2\tilde{P}^{(1)} = 0 \tag{14.3.18}$$

while (14.3.10), (14.3.11) and (14.3.15) become

$$\tilde{P}^{(1)} \to 0 \quad \text{as} \quad \tilde{x}_1 \to -\infty \tag{14.3.19}$$

and

$$\tilde{P}^{(1)} = \frac{3}{\tilde{h}^3} - \frac{3\tilde{v}_{r1}}{\tilde{h}^2} + (\mathrm{Ca}\,\tilde{X})^{-1}\left(\tilde{n}^2 f_{\mathrm{B}}^* + \frac{1}{\tilde{h}^2}\frac{\mathrm{d}\tilde{h}}{\mathrm{d}\tilde{x}_1} f_{\mathrm{B}}\right) \tag{14.3.20}$$

$$\frac{\mathrm{d}\tilde{P}^{(1)}}{\mathrm{d}\tilde{x}_1} = -\frac{1}{\tilde{h}^2}\left(\tilde{u} + \frac{3}{\tilde{h}^2}\frac{\mathrm{d}\tilde{h}}{\mathrm{d}\tilde{x}_1}\right) \tag{14.3.21}$$

when $\tilde{x}_1 = \tilde{x}^{(0)}_{1\text{ exit}}$.

Clearly for any given $\tilde{n}$, $\tilde{u}$ will be given by (14.3.21), while if $\tilde{u}$ is given, the system leads to eigenvalues $\tilde{n}$. The demarcation between stable ($\mathcal{Re}\,\tilde{u} < 0$) and unstable ($\mathcal{Re}\,\tilde{u} > 0$) situations is given by $\mathcal{Re}\,\tilde{u} = 0$. Clearly if $\tilde{n}$ is real, $\tilde{u}$ will be real and so $\tilde{u} = 0$ gives the neutral stability curve for $\tilde{n}$ as a function of $(\mathrm{Ca}\,\tilde{X})^{-1}$ assuming f_{B} and f_{B}^* are essentially constant.

This system has not been solved numerically partly because so much uncertainty is attached to the boundary conditions at $x_{1\,\text{exit}}$ in the perturbed case. It is not known whether the stability curve is very sensitive to the precise form of these boundary conditions.

An analytic solution was given by Pearson (1960) for the associated problem of the stability of flow under a wedge-shaped spreader, and an encouraging comparison was made with experiments based on the most unstable value of $\tilde{n}$, i.e. that value that maximized $\tilde{u}$.

Figure 14.8 gives some indication of the type of disturbances observed, with the complication that $\mathscr{Im}(\tilde{u})$ may not be zero in reality. However, most observations are made when the base flow has developed very strong secondary flows, under circumstances where the lubrication approximation is wholly inadequate to describe the flow pattern, particularly in the ribs that form beyond $x_{1\,\text{exit}}$.

The Newtonian analyses above could in principle be carried through for non-Newtonian fluid models; where the lubrication approximation can be properly applied, a purely viscous model would suffice, but significant elastic effects would be relevant near stagnation points on free surfaces or at cusps.

It is not at all clear why roller-coating flows are apparently more unstable than calendering flows; the effect of surface-tension forces is

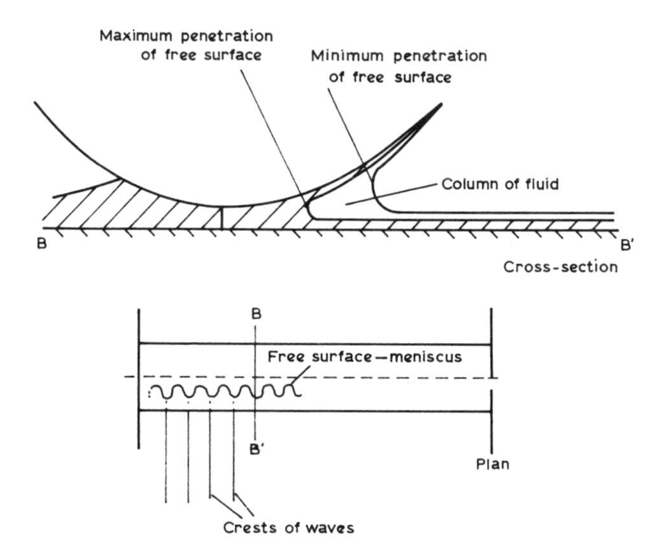

Fig. 14.8. Formation of free surface when ribbing occurs.

essentially negligible in the calendering case, so that cannot represent a stabilizing factor. If the linearized boundary conditions corresponding to (13.1.44) and (13.1.46) are developed, then we find that we get

$$\frac{\partial \tilde{p}^{(1)}}{\partial \tilde{x}_1} = \tilde{x}^{(1)}_{1\,\mathrm{exit}} \frac{3\tilde{v}_{r1}}{2} \frac{\mathrm{d}}{\mathrm{d}\tilde{x}_1}\left(\frac{1}{\tilde{h}^2}\right)_{\tilde{x}^{(0)}_{1\,\mathrm{exit}}} \tag{14.3.22}$$

$$\tilde{p}^{(1)} = -\tilde{x}^{(1)}_{1\,\mathrm{exit}}\left(\frac{\mathrm{d}\tilde{p}^{(0)}}{\mathrm{d}\tilde{x}_1}\right)_{\tilde{x}^{(0)}_{1\,\mathrm{exit}}} \tag{14.3.23}$$

without apparently any contribution from $\partial \tilde{x}^{(1)}_{1\,\mathrm{exit}}/\partial \tilde{t}$. This precludes any possibility of a full instability analysis, except to the extent that $\tilde{v}_{r1}$ (i.e. q_1) can be regarded as the parameter that can be varied to satisfy the neutral stability eigenvalue problem, when $\tilde{x}_{1\,\mathrm{exit}}$ is eliminated between (14.3.22) and (14.3.23).

14.3.2 Sensitivity of the Tensioned Web Coating Operation

Because the gap between web and roll can adjust itself to variations in the entering depth of coating fluid, we now consider the response of the system to a slightly perturbed film depth.

Unsteady Plane Disturbances

These are described by

$$h(x_1 < x_{1\,\mathrm{entry}}) = h_{1,-\infty} + \mathcal{Re}\,\{\mathcal{H}_0 \exp[ik(x_1 - v_{11}t)]\} \tag{14.3.24}$$

where $|\mathcal{H}_0| \ll h_{1,-\infty}$ and, for the case $v_{11} = v_{21}$, we can write $h_{1,-\infty} = h_0$.

We look for a time-varying perturbation to the mean flow situation given by the solution to (14.2.14). We therefore write

$$\left.\begin{aligned} h(x > x_{1\,\mathrm{entry}}) &= h^{(0)}(x_1) + \mathcal{Re}\,\{\mathcal{H}^{(1)}(x_1)\exp(-ikv_{11}t)\} \\ p(x > x_{1\,\mathrm{entry}}) &= p^{(0)}(x_1) + \mathcal{Re}\,\{\mathcal{P}^{(1)}(x_1)\exp(-ikv_{11}t)\} \\ q_1(x > x_{1\,\mathrm{entry}}) &= v_{11}h_0 + \mathcal{Re}\,\{\mathcal{Q}^{(1)}(x_1)\exp(-ikv_{11}t)\} \end{aligned}\right\} \tag{14.3.25}$$

$$x_{1\,\mathrm{entry}} = x^{(0)}_{1\,\mathrm{entry}} + \mathcal{Re}\{\mathrm{X}^{(1)}_{\mathrm{entry}} \exp(-ikv_{11}t)\} \tag{14.3.26}$$

and reconsider the relations satisfied by h, p and q_1, together with the boundary conditions at $x_{1\,\mathrm{entry}}$ and $x_{1\,\mathrm{exit}}$.

The pressure relation (14.2.12) is retained, and so we get

$$\mathcal{P}^{(1)} = -F\frac{\mathrm{d}^2\mathcal{H}^{(1)}}{\mathrm{d}x_1^2} \tag{14.3.27}$$

The flow relation similar to (13.1.48) is also unaltered; so from

$$\frac{\partial p}{\partial x_1}=\frac{12\eta_0}{h^3}\{v_{11}h-q_1(x_1,t)\}$$

we get

$$\frac{\mathrm{d}\mathscr{P}}{\mathrm{d}x_1}=-\frac{12\eta_0}{h^{(0)4}}\{h^{(0)}\mathscr{Q}^{(1)}+v_{11}(2h^{(0)}-3h_0)\mathscr{H}^{(1)}\} \tag{14.3.28}$$

From mass conservation, eqn (8.1.25), we get

$$\frac{\partial h}{\partial t}+\frac{\partial q_1}{\partial x_1}=0$$

or

$$-\mathrm{i}kv_{11}\mathscr{H}^{(1)}+\frac{\mathrm{d}\mathscr{Q}^{(1)}}{\mathrm{d}x_1}=0 \tag{14.3.29}$$

We may combine these into a single fourth-order differential equation for $\mathscr{H}^{(1)}$, namely

$$\frac{\mathrm{d}^4\mathscr{H}^{(1)}}{\mathrm{d}x_1^4}+\frac{3}{h^{(0)}}\frac{\mathrm{d}h^{(0)}}{\mathrm{d}x_1}\frac{\mathrm{d}^3\mathscr{H}^{(1)}}{\mathrm{d}x_1^3}+\frac{12\eta_0v_{11}(3h_0-2h^{(0)})}{Fh^{(0)4}}\frac{\mathrm{d}\mathscr{H}^{(1)}}{\mathrm{d}x_1}$$
$$-\frac{12\eta_0v_{11}}{Fh^{(0)5}}\left[3h_0\frac{\mathrm{d}h^{(0)}}{\mathrm{d}x_1}+\mathrm{i}kh^{(0)2}\right]\mathscr{H}^{(1)}=0 \tag{14.3.30}$$

$\mathscr{H}^{(1)}$ will be complex.

We retain the requirement that $p(x_{1\,\mathrm{entry}})=p_\mathrm{a}$ and so, linearizing, using (14.3.26) we obtain

$$\mathscr{P}^{(1)}(x^{(0)}_{1\,\mathrm{entry}})=-\frac{\mathrm{d}p^{(0)}}{\mathrm{d}x_1}(x^{(0)}_{1\,\mathrm{entry}})\chi^{(1)}_{\mathrm{entry}} \tag{14.3.31}$$

If we use the conservation-of-mass condition, it can be shown that

$$v_{11}\mathscr{H}_0\exp(\mathrm{i}kx^{(0)}_{1\,\mathrm{entry}})-\mathscr{Q}^{(1)}(x^{(0)}_{1\,\mathrm{entry}})$$
$$+\mathrm{i}kv_{11}\{h_0-h^{(0)}(x^{(0)}_{1\,\mathrm{entry}})\}\chi^{(1)}_{\mathrm{entry}}=0 \tag{14.3.32}$$

When $\chi^{(1)}_{\mathrm{entry}}$ is eliminated between (14.3.31) and (14.3.32) using (14.3.27) and (14.3.29) we retrieve a single boundary condition for $\mathscr{Q}^{(1)}$ in terms of $\mathscr{H}_0\exp(\mathrm{i}kx^{(0)}_{1\,\mathrm{entry}})$ which without loss of generality could be taken equal to unity. This is an integro-differential condition on $\mathscr{H}^{(1)}$.

Another boundary condition that arises for geometrical reasons is

$$\frac{\mathrm{d}\mathcal{H}^{(1)}}{\mathrm{d}x_1}(x^{(0)}_{1\,\mathrm{entry}})=0 \tag{14.3.33}$$

assuming that the tangent plane to the web away from the roll in a fixed (laboratory) coordinate system is effectively unaltered by the small change $\chi^{(1)}_{\mathrm{entry}}$.

Similarly at $x_{1\,\mathrm{exit}}$ we retain (14.2.10) which yields

$$\mathcal{P}^{(1)}(x^{(0)}_{1\,\mathrm{exit}})=-\left\{\frac{\mathrm{d}p^{(0)}}{\mathrm{d}x_1}-\frac{\sigma}{h_0^2}\frac{\mathrm{d}h^{(0)}}{\mathrm{d}x_1}\right\}_{x=x^{(0)}_{1\,\mathrm{exit}}}\chi^{(1)}_{\mathrm{exit}}-\frac{\sigma}{h_0^2}\mathcal{H}^{(1)}(x^{(0)}_{1\,\mathrm{exit}}) \tag{14.3.34}$$

when use is made of (14.2.16).

The motion of the tip of the meniscus requires that

$$\frac{\mathrm{d}\mathcal{P}^{(1)}}{\mathrm{d}x_1}=-\left(\frac{\mathrm{d}^2p^{(0)}}{\mathrm{d}x_1^2}+\frac{2}{3h_0}\frac{\mathrm{d}h^{(0)}}{\mathrm{d}x_1}\frac{\mathrm{d}p^{(0)}}{\mathrm{d}x_1}-\frac{4\eta_0ikv_{11}}{9h_0^2}\right)_{x_1=x^{(0)}_{1\,\mathrm{exit}}}\chi^{(1)}_{\mathrm{exit}}-\frac{2}{3}\frac{\mathcal{H}^{(1)}}{h_0}\frac{\mathrm{d}p^{(0)}}{\mathrm{d}x_1} \tag{14.3.35}$$

which is the equivalent of (14.3.15).

The analogue of (14.3.32) can be written

$$v_{11}\mathcal{H}_1\exp(ikx^{(0)}_{1\,\mathrm{exit}})-\mathcal{Q}^{(1)}(x^{(0)}_{1\,\mathrm{exit}})-ikv_{11}\{h_0-h^{(0)}(x^{(0)}_{1\,\mathrm{exit}})\}\chi^{(1)}_{\mathrm{exit}}=0 \tag{14.3.36}$$

where the thickness of fluid carried away by both roll and web is

$$h(x_1>x_{1\,\mathrm{exit}})=\tfrac{1}{2}[h_0-\mathcal{R}e\,\{\mathcal{H}_1\exp[ik(x_1-v_{11}t)]\}] \tag{14.3.37}$$

(14.3.35) and (14.3.34) can be combined by elimination of $\chi^{(1)}_{\mathrm{exit}}$, to give a boundary condition for $\mathcal{H}^{(1)}$ using (14.3.27).

A last boundary condition is provided by

$$\frac{\mathrm{d}\mathcal{H}^{(1)}}{\mathrm{d}x_1}(x^{(0)}_{1\,\mathrm{exit}})=0 \tag{14.3.38}$$

The fourth-order differential equation (14.3.30) is thus provided with four boundary conditions, (14.3.33), (14.3.38), one from (14.3.34) and (14.3.35)—all of which are homogeneous in $\mathcal{H}^{(1)}$—and one from (14.3.31) and (14.3.32), which introduces the entry perturbation $\mathcal{H}_0$.

The resultant perturbation $\mathcal{H}_1$ follows from (14.3.36) and the amplification factor

$$\mathrm{Am} = |\mathscr{H}_1/\mathscr{H}_0| \tag{14.3.39}$$

is the main result of the analysis.

It is relatively easy to investigate the damping of the disturbance wave (14.3.24) when the resultant form involving $\mathscr{H}^{(1)}$ passes through the essentially-constant undisturbed region with $h^{(0)} \approx h_0$, for then we can suppose that

$$\mathscr{Q}^{(1)} = \mathscr{Q}_0^{(1)} \exp \mathrm{i}Kx_1 \tag{14.3.40}$$

$$\mathscr{H}^{(1)} = \mathscr{H}_0^{(1)} \exp \mathrm{i}Kx_1 \tag{14.3.41}$$

where $K(k)$ is complex. From (14.3.29) we get

$$\mathrm{i}kv_{11}\mathscr{H}_0^{(1)} = \mathrm{i}K\mathscr{Q}_0^{(1)} \tag{14.3.42}$$

and from (14.3.27) and (14.3.28), by elimination of $\mathscr{P}^{(1)}$,

$$\mathrm{i}FK^3\mathscr{H}_0^{(1)} = \frac{12\eta_0}{h_0^3}(v_{11}\mathscr{H}_0^{(1)} - \mathscr{Q}_0^{(1)}) \tag{14.3.43}$$

By elimination of $\mathscr{Q}_0^{(1)}$ we can write

$$K = k\left(1 + \frac{\mathrm{i}Fh_0^3K^4}{12\eta_0 v_{11}k}\right) \tag{14.3.44}$$

and if $Fh_0^3k^3/12\eta_0 v_{11} \ll 1$ this can be approximated by

$$\mathrm{i}K = \mathrm{i}k - Fh_0^3k^4/12\eta_0 v_{11} \tag{14.3.45}$$

The damping effect over a length ϕd is then given by substitution in (14.3.41) and is

$$\mathrm{Am} \sim \exp(-Fh_0^3k^4\phi d/12\eta_0 v_{11}) \tag{14.3.46}$$

We already know that (14.2.21) must apply and so the argument of the exponential will be large or small in absolute value depending on whether

$$k^4d^4 > 12\eta_0 v_{11}d^3/Fh_0^3 \gg k^3d^3 \tag{14.3.47}$$

or not, ϕ being regarded as of order unity.

It can readily be seen from (14.3.46) that an unstable situation would arise if $\mathrm{Am} \to \infty$, but in view of the negative sign that is not to be expected in practice. The case of $Fh_0^3k^3/12\eta_0 v_{11} \gg 1$ is discussed in Exercise 14.3.7.

Steady Ribbed Disturbances
These are described by

$$h(x_1 < x_{1\,\text{entry}}) = h_0 + \mathscr{H}_0 \sin nx_2 \tag{14.3.48}$$

and as above we expect

$$h(x_1 > x_{1\,\text{exit}}) = \tfrac{1}{2}(h_0 + \mathscr{H}_1 \sin nx_2) \tag{14.3.49}$$

where $\mathscr{H}_0$, $\mathscr{H}_1 \ll h_0$, the amplification factor is still defined by (14.3.39), and n is a wave number.

Disturbances will arise in the flow between web and roll, which to the same order of linearization can be written

$$h = h^{(0)}(x_1) + \mathscr{H}^{(1)}(x_1)\sin nx_2 \tag{14.3.50}$$

$$q_1 = q_{10} + \mathscr{Q}_1^{(1)}(x_1)\sin nx_2; \qquad q_2 = \mathscr{Q}_2^{(1)}(x_1)\cos nx_2 \tag{14.3.51}$$

$$p = p^{(0)}(x_1) + \mathscr{P}^{(1)}(x_1)\sin nx_2 \tag{14.3.52}$$

It is important to note that if the web is assumed to be tensioned in the x_1-direction but not in the x_2-direction, then (14.2.12) is unaltered and (14.3.27) applies.

As before, we can use the mass conservation relation

$$\frac{\partial q_1}{\partial x_1} + \frac{\partial q_2}{\partial x_2} = 0$$

and the flow equation

$$q_1 = v_{11}h - \frac{h^3}{12\eta_0}\frac{\partial p}{\partial x_1}; \qquad q_2 = -\frac{h^3}{12\eta_0}\frac{\partial p}{\partial x_2}$$

to give a fourth-order ordinary differential equation for $\mathscr{H}^{(1)}$

$$\frac{d^4\mathscr{H}^{(1)}}{dx_1^4} + \frac{3}{h^{(0)}}\frac{dh^{(0)}}{dx_1}\frac{d^3\mathscr{H}^{(1)}}{dx_1^3} - n^2\frac{d^2\mathscr{H}^{(1)}}{dx_1^2} + \frac{12\eta_0 v_{11}(3h_0 - 2h^{(0)})}{Fh^{(0)4}}\frac{d\mathscr{H}^{(1)}}{dx_1} - \frac{36\eta_0 v_{11}}{Fh^{(0)5}}\frac{dh^{(0)}}{dx_1}\mathscr{H}^{(1)} = 0 \tag{14.3.53}$$

The boundary conditions for $\mathscr{H}^{(1)}$ will be, using the same arguments as for the unsteady case, where now

$$x_{1\,\text{entry,exit}} = x^{(0)}_{1\,\text{entry,exit}} + \chi^{(1)}_{\text{entry,exit}} \sin nx_2 \tag{14.3.54}$$

are

$$x_{1\,\mathrm{entry}}^{(0)}: \mathscr{Q}^{(1)} = v_{11}\mathscr{H}_0 \tag{14.3.55}$$

$$\mathscr{P}^{(1)} = -\frac{\mathrm{d}p^{(0)}}{\mathrm{d}x_1}\chi_{\mathrm{entry}}^{(1)} \tag{14.3.56}$$

$$\frac{\mathrm{d}\mathscr{H}^{(1)}}{\mathrm{d}x_1} = 0 \tag{14.3.57}$$

$$x_{1\,\mathrm{exit}}^{(0)}: \mathscr{Q}^{(1)} = v_{11}\mathscr{H}_1 \tag{14.3.58}$$

$$\mathscr{P}^{(1)} = -\left\{\frac{\mathrm{d}p^{(0)}}{\mathrm{d}x_1} - \frac{\sigma}{h_0^2}\frac{\mathrm{d}h^{(0)}}{\mathrm{d}x_1} - 2n^2\sigma\right\}\chi_{\mathrm{exit}}^{(1)} - \frac{\sigma}{h_0^2}\mathscr{H}^{(1)} \tag{14.3.59}$$

$$\frac{\mathrm{d}\mathscr{P}^{(1)}}{\mathrm{d}x_1} = -\left(\frac{\mathrm{d}^2p^{(0)}}{\mathrm{d}x_1^2} + \frac{2}{3h_0}\frac{\mathrm{d}h^{(0)}}{\mathrm{d}x_1}\frac{\mathrm{d}p^{(0)}}{\mathrm{d}x_1} + \frac{4\eta_0 n v_{11}}{9h_0^2}\right)\chi_{\mathrm{exit}}^{(1)} - \frac{2}{3}\frac{\mathscr{H}^{(1)}}{h_0}\frac{\mathrm{d}p^{(0)}}{\mathrm{d}x_1} \tag{14.3.60}$$

$$\frac{\mathrm{d}\mathscr{H}^{(1)}}{\mathrm{d}x_1} = 0 \tag{14.3.61}$$

These seven conditions suffice to specify $\chi_{\mathrm{entry}}^{(1)}$, $\chi_{\mathrm{exit}}^{(1)}$, $\mathscr{H}_1$ and to yield four boundary conditions for (14.3.53), in terms of $\mathscr{H}_0$.

The limiting case provided by $h \sim h_0$ yields

$$\frac{\mathrm{d}^4\mathscr{H}^{(1)}}{\mathrm{d}x_1^4} - n^2\frac{\mathrm{d}^2\mathscr{H}^{(1)}}{\mathrm{d}x_1^2} + \frac{12\eta_0 V_1}{Fh_0^3}\frac{\mathrm{d}\mathscr{H}^{(1)}}{\mathrm{d}x_1} = 0 \tag{14.3.62}$$

which has the important and physically obvious solution

$$\mathscr{H}^{(1)} = \text{constant} \tag{14.3.63}$$

There is thus no damping of steady transverse disturbances over the central region. A non-negligible amplification factor can therefore be expected. The situation would be altered if the web were tensioned in the x_2-direction.

It is now by no means clear that there cannot be values of $(nh_0, 12\eta_0 v_{11}/F, \mathrm{Ca})$ for which $\mathrm{Am}(nh_0, 12\eta_0 v_{11}/F, \mathrm{Ca})$ tends to infinity. This would correspond to instabilities of the type discussed in Subsection 14.3.1. This situation has not been examined.

EXAMPLE 14.3.1. Derivation of boundary condition (14.3.11).

We suppose the exit line to be given by (14.3.5) and substitute in

(14.1.4) to give

$$\tilde{p}(\tilde{x}^{(0)}_{1\,\text{exit}}+\tilde{x}^{(1)}_{1\,\text{exit}})\sim\tilde{p}_{\text{a}}-f_{\text{B}}/\{\text{Ca}\,\tilde{X}\tilde{h}(\tilde{x}^{(0)}_{1\,\text{exit}}+\tilde{x}^{(1)}_{1\,\text{exit}})\}$$

assuming that f_{B} is constant. On expansion of both sides we obtain

$$\tilde{p}^{(0)}(x^{(0)}_{1\,\text{exit}})+\tilde{x}^{(1)}_{1\,\text{exit}}\left(\frac{\mathrm{d}\tilde{p}^{(0)}}{\mathrm{d}\tilde{x}_1}\right)_{x^{(0)}_{1\,\text{exit}}}+\tilde{p}^{(1)}(\tilde{x}^{(0)}_{1\,\text{exit}})$$

$$\sim\tilde{p}_{\text{a}}-\frac{f_{\text{B}}}{\text{Ca}\,\tilde{X}}\left(\frac{1}{\tilde{h}(\tilde{x}^{(0)}_{1\,\text{exit}})}-\frac{\tilde{x}^{(1)}_{1\,\text{exit}}}{\tilde{h}^2(\tilde{x}^{(0)}_{1\,\text{exit}})}\left(\frac{\mathrm{d}\tilde{h}}{\mathrm{d}\tilde{x}_1}\right)_{x^{(0)}_{1\,\text{exit}}}\right)$$

The terms independent of $\tilde{x}^{(1)}_{1\,\text{exit}}$ or $\tilde{p}^{(1)}$ are separately balanced because they represent the boundary condition for the undisturbed flow. The terms involving $p^{(1)}$ and $x^{(1)}_{1\,\text{exit}}$ must either balance separately or else be balanced by any additional surface tension forces. To estimate the full effect of the latter we consider eqn (14.2.4) with F replaced by σ and consider any extra contribution to H arising from $\tilde{x}_{1\,\text{exit}}$ being a function of $\tilde{x}_2$. It is known that H is the sum of the curvatures of a surface in any two mutually perpendicular planes; the term involving Ca^{-1} in (14.1.4) represents the contribution of the curvature in the (x_1, x_3) plane. We must therefore add a contribution from curvature in the (x_1, x_2) plane, which by analogy with (14.2.3) can be written as $\mathrm{d}^2x^{(1)}_{1\,\text{exit}}/\mathrm{d}x_2^2$ at the tip of the meniscus, $x_3=0$. The dimensionless pressure difference associated with this

$$\Delta\tilde{p}=-(\text{Ca}\,\tilde{X})^{-1}f^*_{\text{B}}\frac{\mathrm{d}^2\tilde{x}^{(1)}_1}{\mathrm{d}\tilde{x}_2^2}$$

the f^*_{B} being of order 1, indeed = 1 following the above argument.

Boundary condition (14.3.11) follows directly.

EXERCISE 14.3.1. Verify relations (14.3.7)–(14.3.9).

EXERCISE 14.3.2. Derive boundary condition (14.3.15). Note that the rate of change of position of the exit line, $\partial x_{1\,\text{exit}}/\partial t$ is taken to be the instantaneous value of the velocity at the meniscus tip, $v_1(x_{1\,\text{exit}}, x_3=0)$.

EXERCISE 14.3.3. Derive relations (14.3.18) to (14.3.21) substituting for $\tilde{h}$ from (14.3.1). Use (13.1.48) to obtain $\mathrm{d}\tilde{p}^{(0)}/\mathrm{d}\tilde{x}_1$.

EXERCISE 14.3.4. Derive the analogues to (14.3.8) and (14.3.9) for a power-law fluid. Example 8.1.1 is a suitable starting point, and Subsection 10.2.1 is also helpful. Note that a perturbation about the steady flow is involved and so the situation is simpler than it would be if $\tilde{q}^{(1)}$ were 0(1). The difference from Subsection 10.2.1 is provided by the movement of the boundaries.

EXERCISE 14.3.5. Show that (14.3.32) is obtained by considering (i) the increase in material arriving at $x_{1\,\text{entry}}$, based on $v_{11}h$, (ii) the increase in material leaving $x_{1\,\text{entry}}$, based on $q_{1\,\text{entry}}$ and (iii) the increase in volume required to fill the gap between free surface just upstream of the entry point $x_{1\,\text{entry}}$ and the dry entering web.

The expansion technique used in Example 14.3.1 is required.

Show also that (14.3.33) follows when (14.2.1), (14.2.2) and (14.2.21) are invoked and it is assumed that the web passes through a fixed line at a distance of $0(d)$ from the roll.

EXERCISE 14.3.6. Derive boundary conditions (14.3.34) and (14.3.35).

EXERCISE 14.3.7. Examine the solution of (14.3.44) when $Fh_0^3k^3/12\eta_0 v_{11} \gg 1$ and show that the equivalent of (14.3.46) will be given by

$$\text{Am} \sim \exp\left\{-\left(\frac{12\eta_0 v_{11} k}{Fh_0^3}\right)^{\frac{1}{4}} \phi d \sin \pi/8\right\}$$

and so the criterion for damping will be whether

$$12\eta_0 v_{11} k d^4 / Fh_0^3 \gg 1$$

or not. This is to be compared with (14.3.47).

Interpret these conditions in terms of length scales and deduce when damping is to be expected.

EXERCISE 14.3.8. Derive eqn (14.3.53) and boundary conditions (14.3.55)–(14.3.61). Note particularly that a new argument has to be applied in order to derive (14.3.60), namely that there is to be no net motion of the interface normal to it, and so there is effectively a component of v_{11} in the x_2-direction yielding the third term in the parentheses in (14.3.60).

C—STRETCHING

Stretching here refers to those processes where polymer in the form of filaments or sheets is stretched or drawn continuously. The original filament or sheet can be obtained by melt extrusion through dies, as described in Chapter 10, or by calendering between rolls as described in Chapter 13. The extrudate is molten, and therefore has to be solidified (or cross-linked) during the process. For thermoplastics this means cooling and freezing; this can be done either

(a) by air cooling in the ambient environment, or
(b) by water cooling, usually by passage into a water bath but sometimes by contact with a water curtain, or
(c) by casting onto a metal roll.

The second has the disadvantage that the frozen fibre or sheet has to be subsequently dried; the last, that it can apply only to flat sheets; air cooling is therefore the most common.

For part of the process the fibre or sheet is unsupported externally, and so control of geometry and steadiness of operation require it to be in a state of tension. The tensioned fibre or sheet is mechanically self-supported. Being molten, the material will extend, and so the fibre or sheet will get thinner. In general, it is convenient to achieve significant reductions in thickness during the free-flow stage of the process; on the one hand, this is because direct extrusion of thin fibres or sheet would call for unacceptably large rates of shear in the die, leading to overlarge pressure drops and melt fracture; on the other, extensional free flow of cool material can lead to desirable molecular orientation and improved mechanical properties. Indeed, the latter requirement sometimes calls, in the case of linear polyolefins (HDPE,

PP), polyamides (Nylon) and polyesters (PET), for a further cold-drawing stage.

Sheets are extruded either from slot dies (Section 10.2), in which case the output remains essentially planar during the whole process and can be wound up directly, or from annular dies (Section 10.3) in which case the process is axisymmetric up to freezing and leads to a 'lay-flat' tube† which can be slit as required. Fibre spinning is discussed in the next chapter. The flat and axisymmetric geometries for film (thin sheet) production are treated in separate chapters thereafter. In both of these, the thin-sheet approximation presented in Section 8.2 is used. A fully satisfactory account of fibre and film production requires careful consideration of crystallization processes; these are particularly important in the former case. They are not considered in detail here, and readers are referred to other texts for a proper discussion of these phenomena.

† An exception to this arises in the case of rubber sheet extrusion when the tube is slit at or near the die and the two edges so formed are pulled apart to convert the tube into a flat sheet while molten.

Chapter 15

Fibre Spinning

15.1 THE PROCESS

The object of fibre spinning is to produce long thin tough polymeric filaments that are used largely by the textile industry, but are now becoming more important as reinforcement for composite plastics. In general, they are of constant cross-section, usually circular, and are made of a single homogeneous material. The process is a continuous high-speed one. It is carried out either with melt, if the material melts without decomposition at a suitable temperature, or with a solution if it does not.

Melt spinning consists of steady extrusion of hot melt through a series of small holes in a plate (spinnerets) into ambient air, the resulting extrudates being simultaneously extended and wound up on a rapidly rotating drum (godet). Freezing takes place between spinneret and godet. Large extension ratios, 0 (100), rapid cooling, 0 (1000 $K\,s^{-1}$) and high speeds, 0 (10 $m\,s^{-1}$) and more, are involved. It is used for polyamides, polyesters, polystyrene and polyolefins.

Solution dry-spinning involves a similar procedure carried out on a solution of the fibre-forming polymer, such as cellulose acetate or PVC, in a volatile solvent. Solidification is achieved by evaporation of the solvent into a gaseous environment during draw down.

Solution wet-spinning uses a non-volatile solvent and extrusion takes place into a bath containing non-solvents. A variety of physical and chemical processes lead to coagulation and solidification of the fibre, e.g. of cellulose.

Fibres that are formed by any of the above three methods are usually strengthened by a further controlled drawing process carried out at relatively high stress and low temperature in the solid phase. Extension ratios of from 2 to 8 are usually involved in this latter process, leading to considerable axial orientation of the fibre-forming material, and a corresponding increase in tensile modulus. The primary spinning and the secondary plastic-drawing process can be undertaken sequentially on the same spinning line, but they may be considered as separate operations in most cases. There will be a small but sometimes important built-in orientation, achieved just before freezing, in the primary spinning process.

This chapter will concentrate on the melt spinning process. The factors that are relevant are:

(a) extrusion through a short die, i.e. the spinneret hole, including any die swell and velocity-profile adjustment;
(b) rapid axisymmetric extension to large strains under conditions of
(c) rapid temperature changes and hence large variations in rheological behaviour. (For polymers going through a rubber–glass transition, the extensional viscosity may change by an order of magnitude); non-isothermal rheological effects may be significant;
(d) crystallization (in some cases) under conditions of high stress and rapid cooling;
(e) inertia, gravity and air-drag forces which will be relevant to the overall thread-line force balance.

The treatment given here will not be so exhaustive as the account given in the excellent text of Ziabicki (1976) from which it borrows freely; readers are also recommended to consult Petrie (1979) and Denn (1980). From a kinematical point of view, the main spinning flow is relatively simple, in that it can be treated as a one-dimensional flow, with velocity, pressure and temperature being functions of the axial coordinate only. From a rheological point of view, the flow is an unsteady uniaxial extensional flow. Overall the process is meant to be steady-state. Stability of the process has been extensively analysed and investigated experimentally. Causes for thread-line breakdown will be discussed.

The solid drawing step is less well understood. Its stability appears not to have been investigated.

Some attention has been paid to bi- or multi-component spinning, whereby two or more molten streams are combined at the spinneret, and to non-circular cross-sections.

15.2 SIMPLE MODEL EQUATIONS

The melt-spinning process is shown in its most elementary (single filament) form in Fig. 15.1. The part of most interest is the molten draw-down zone where rapid elongation and cooling take place. The holes are typically 0·1–1 mm in diameter, the molten draw-down zone is of order 0·1 m long (sometimes longer) and the entire thread-line greater than 1 m long. With draw-down ratios of several hundred, the diameter of the wound-up filament is of order 10 μm. Wind-up speeds vary from 1 m s^{-1} for thick monofilaments to as much as 100 m s^{-1} for fine fibres. Flight times in the molten region will therefore range typically from 10^{-3} to 10^{-1} s. Extension rates will be of order 100 s^{-1} (which is large by the standards of most laboratory measuring devices), and length-to-diameter ratios of order 1000.

The figures given above make it clear that in steady-state operation the (dimensionless) rate of change of filament radius with axial distance

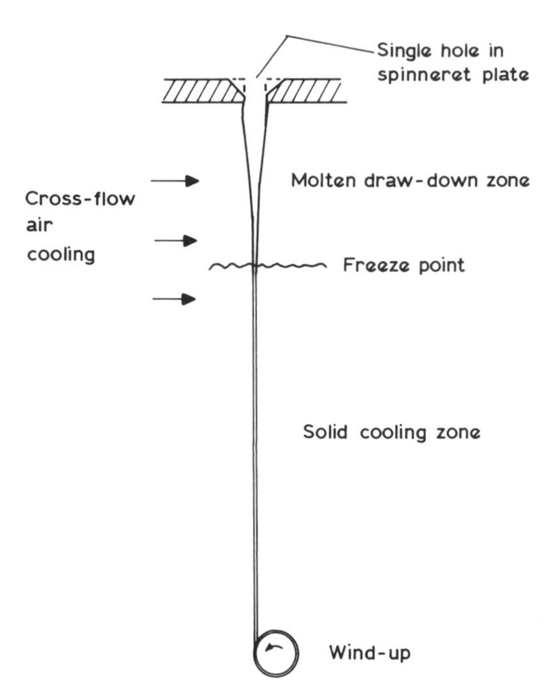

Fig. 15.1. Schematic diagram of melt-spinning process. A single spinning threadline is shown.

is very small and so radial velocity components will be very small compared with axial components. This will also be true even in most cases of unsteady operation. Figure 15.2(a) shows the region close to the spinneret where the axial velocity v_z changes from being strongly sheared in the r-direction to being essentially independent of r, as shown in Fig. 15.2(b). It may also be shown that the Graetz number

$$\mathrm{Gz} = Q/\pi\kappa l_f \tag{15.2.1}$$

where Q is the total flowrate and l_f is the length of the draw-down region, is of order one, which means that the flow will be fully-developed, in the sense that the bulk melt temperature will be governed by the local melt surface temperature and vice versa. In fact, the filament temperature will be largely governed by heat transfer to the ambient environment, which introduces fluid mechanical aspects in the outer flow field that will not be discussed here.

It is therefore reasonable to suppose that, except in a region of order 0·1 mm in length close to the spinneret, the velocity and temperature fields will be as shown in Fig. 15.2(b). The diameter, $d_f(z, t)$ of the filament, is such that

$$|\partial d_f/\partial z| \ll 1 \tag{15.2.2}$$

and the dependent variables defining the flow can be written, in terms of an (r, ϕ, z) coordinate system,

$$\mathbf{v} \approx \{0, 0, v_z(z, t)\}; \qquad T \approx T(z, t) \tag{15.2.3}$$

to first-order in (d_f/l_f), i.e. in $\partial d_f/\partial z$. The advantage of such a simple representation is that integration over the cross-section is trivially achieved for quantities independent of r by using the multiplication factor $\frac{1}{4}\pi d_f^2$.

The use of the very crude approximation (15.2.3) was originally suggested by observation, which provides conclusive evidence of (15.2.2). It is not a trivial matter to derive formally all the consequences of (15.2.2) on the full set of conservation and constitutive equations and associated free-boundary conditions, particularly if the relevant lowest-order approximations are to be provided from a perturbation expansion in (d_f/l_f). The matter was considered in Matovich & Pearson (1969) and the analysis repeated by various other authors (most recently by Schultz & Davis, 1982). The most elementary approach is to show that solutions based on (15.2.3) are self-consistent (see Example 15.2.1). For our purposes, it may be assumed that all

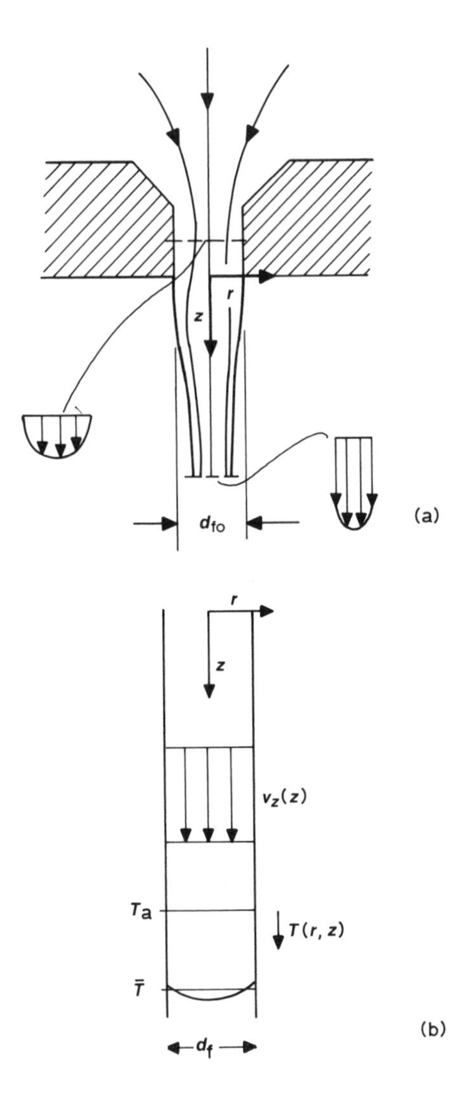

Fig. 15.2. Flow in threadline. (a) Streamlines passing through spinneret, showing change in profile achieved in a distance of order d_{f0}. (b) Flat axial velocity (v_z) and nearly flat temperature (T) profiles at positions far from the spinneret.

industrial processes of importance are very well modelled by a unidirectional flow field $v_z(z, t)$, with spinning carried out vertically downwards. Where it becomes relevant, the radial velocity component v_r can be obtained from $\partial v_z/\partial z$.

The integrated mass, momentum and energy equations can thus be written

$$\frac{\partial}{\partial t}(\rho_m d_f^2) \approx -\frac{\partial}{\partial z}(\rho_m v_z d_f^2) \tag{15.2.4}$$

$$\frac{\partial}{\partial t}(\rho_m d_f^2 v_z) + \frac{\partial}{\partial z}(\rho_m d_f^2 v_z^2) \approx \rho_m d_f^2 g + \frac{\partial}{\partial z}(d_f^2 t_{zz}) + 4 d_f t_{rz}^{(a-d)} + 4\frac{\partial}{\partial z}(\sigma_m d_f) \tag{15.2.5}$$

acceleration = gravity + continuum tension + air drag + surface tension

$$\frac{\partial}{\partial t}(\rho_m \Gamma_m d_f^2 T) + \frac{\partial}{\partial z}(\rho_m \Gamma_m d_f^2 v_z T) \approx 4 d_f h_a (T_a - T) + (t_{zz} - t_{rr})\frac{\partial v_z}{\partial z} d_f^2 \tag{15.2.6}$$

accumulation + convection = heat transfer + generation

where

$$t_{rr} \approx t_{\phi\phi} \approx -2\sigma_m / d_f \tag{15.2.7}$$

In the momentum equation (15.2.5), the effect of gravity (acting in the z-direction), surface tension and air drag are included; the surface tension σ_m may be supposed a known function of T; however, the axial stress at the interface $t_{rz}^{(a-d)}$ due to air drag must be specified as a function of z and t. Various correlations are given for the skin friction coefficient C_f in Table 3.7 of Ziabicki (1976) where

$$\tfrac{1}{2} C_f v_z^2 \rho_a = t_{rz}^{(a-d)} \tag{15.2.8}$$

Here ρ_a is the density of the ambient environment.

Ziabicki's Table 3.5 shows that surface-tension effects are small, although the contribution of air drag to the overall thread-line tension can be important. It is important to note that the air drag term in (15.2.5) can be of the same order of magnitude as the rheological force term $(d_f^2 t_{zz})_{,z}$ without weakening the assumption (15.2.3); eqn (15.2.5) remains self-consistent provided $t_{rz}^{(a-d)} \ll t_{zz}$, which will be true because of (15.2.2) (see Example 15.2.5 in this context).

In the energy equation (15.2.6), the conduction term is replaced by a surface heat-transfer term involving a heat-transfer coefficient h_a. Various correlations for the Nusselt number Nu are given in Table 3.3 of Ziabicki (1976) where

$$\mathrm{Nu} = h_a d_f / \alpha_m \tag{15.2.9}$$

The generation term is small for melt spinning.

Equation (15.2.7) is no more than a normal-stress boundary condition. Its effect can be included in eqn (15.2.5) by writing $t_{zz} - t_{rr}$ for t_{zz} in the continuum-tension term and 2 for 4 in the surface-tension term. This then makes the second term on the right-hand side directly dependent on the rheological equation of state, and has eliminated the isotropic component of stress, p, from the momentum equation.

The rate of deformation tensor is, to the order of approximation involved in (15.2.3), diagonal with

$$\dot{e}_1 = -2\dot{e}_2 = -2\dot{e}_3 = \partial v_z / \partial z \tag{15.2.10}$$

using the definitions of eqn (4.3.1) and identifying x_1 with z. Viewed from a fixed laboratory frame of reference, $\dot{e}_1$ is a function of z and t; for steady flow, $\dot{e}_1$ is a function of z only. However, viewed from a moving frame fixed in the fluid, $\dot{e}_1$ is a function of t even for steady flow. The deformation history of any particle while in the thread-line is that of unsteady locally-uniform elongation; it is a simple example of the general class described by (4.6.1). For times sufficiently far back, a more complex deformation history, associated with flow through the spinneret, would be relevant. If, however, the memory of the fluid is sufficiently short, then, for most of the thread-line flow, the effective history can be regarded as purely elongational. It may then be shown that $\mathbf{C}_t^t(s)$, in eqn (3.2.13) for a simple fluid, is given to the necessary order of approximation by the history of $\partial v_z / \partial z$ alone. The tensor function $\mathscr{H}^{\mathrm{E}}$ can then be replaced by a scalar functional $\tilde{\hbar}_e$ whereby

$$t_{zz}(z, t) - t_{rr}(z, t) = \mathop{\tilde{\hbar}_e}_{s=0}^{s=\infty} [c_{t_{zz}}^t(z, s)] \tag{15.2.11}$$

with

$$c_{t_{zz}}^t(z, s) = \left\{ \exp \int_0^s \frac{\partial v_z}{\partial z}(u, z)\, \mathrm{d}u \right\}^2 \tag{15.2.12}$$

where $\dfrac{\partial v_z}{\partial z} \equiv \left(\dfrac{\partial v_z}{\partial z}\right)^t = \dot{e}^t$ is a function of z, t and s. It is sometimes simpler to write, instead of (15.2.11) and (15.2.12)

$$t_{zz}(z, t) - t_{rr}(z, t) = \mathop{\mathscr{h}_{\mathrm{e}}}_{s=0}^{s=\infty} [\dot{e}^{t}(z, s)] \tag{15.2.13}$$

It was earlier pointed out that the non-isothermal nature of fibre spinning is important industrially and so the density and temperature history may be important. A suitably general rheological equation of state can be written formally as

$$t_{zz} - t_{rr} = \mathop{\mathscr{f}}_{s=0}^{s=\infty} [\dot{e}^{t}(s), T^{t}(s), \rho_{\mathrm{m}}^{t}(s); T(t), \rho_{\mathrm{m}}(t)] \tag{15.2.14}$$

If ρ_{m} is taken as constant, or as a simple known function of T only, then eqns (15.2.4)–(15.2.7) and (15.2.14) together with specified forms for C_{f} and Nu provide a determinate set of equations for the 'unknown' quantities d_{f}, v and T. The free-surface boundary conditions at $r = \frac{1}{2}d_{\mathrm{f}}$ are implicit in the system of equations. A particular flow field would then be defined by suitable inlet conditions, at $z = 0$ say, exit conditions, at $z = z_1$ say, and initial conditions, at $t = 0$ say, for an unsteady flow. It is at once clear that difficulties arise in the constitutive equation (15.2.14) if s is not effectively infinite for s corresponding to the point of entry ($z' = 0$) or the initial instant ($t' = 0$). A way round this problem is to suppose that the stress component $t_{zz} - t_{rr}$ is prescribed at entry or at the initial instant.

If however, crystallization is taking place with the same time scale as that relevant rheologically or dynamically, then a separate set of equations relating degree of crystallization (and hence ρ) to $\dot{e}$ and T has to be provided and solved simultaneously with (15.2.4)–(15.2.7) and (15.2.14). This will be of particular importance if solidification is achieved by crystallization rather than vitrification, and if a large extent of super-cooling arises in the thread-line. The matter is discussed at length in Ziabicki (1976, Chapter 2, Section B). Most of the literature on the mechanics of fibre spinning avoids the extra complexity of crystallization or a cooling-rate dependent glass transition temperature (T_{g}) by assuming a fixed solidification temperature, i.e. essentially glassy behaviour with a fixed T_{g}.

EXAMPLE 15.2.1. Derivation, justification and extension of the fibre flow approximation.

The crude velocity field (15.2.3i) can be analysed in terms of the local continuity equation, (2.2.7), written in the form

$$\frac{1}{r}\frac{\partial}{\partial r}(rv_r)+\frac{1}{r}\frac{\partial v_\theta}{\partial \theta}+\frac{\partial v_z}{\partial z}=0$$

If $v_z = v_z(z, t)$, $v_\theta \equiv 0$ and $v_r = 0$ at $r = 0$, the last two being consequences of axisymmetry, then direct integration yields

$$v_r = -\tfrac{1}{2}r\frac{\partial v_z}{\partial z}$$

This is the term neglected in (15.2.3i). Its maximum value (negative) arises when $r = \frac{1}{2}d_f$. If it is neglected, then (15.3.1), based on (15.2.4) with ρ_m constant shows that

$$v_z = 4d_f^{-2}Q\pi^{-1}$$

and so

$$\frac{\partial v_z}{\partial z} = -\frac{2}{d_f}\frac{\partial d_f}{\partial z}v_z$$

The maximum ratio of the term neglected in (15.2.3i) to that retained is

$$\frac{1}{4}\frac{d_f}{v_z}\frac{\partial v_z}{\partial z} = \frac{1}{2}\frac{\partial d_f}{\partial z}$$

which by definition, (15.2.2), is very small, the estimate given earlier being 0 (10^{-3}).

It is useful to consider the dimensionless variables

$$\hat{r} = 2r/d_f, \qquad \hat{z} = z/l_f$$

in terms of which, using the above relation for v_r,

$$\frac{\partial}{\partial z} \rightarrow \frac{1}{l_f}\frac{\partial}{\partial \hat{z}} - \frac{2v_r}{v_z d_f}\frac{\partial}{\partial \hat{r}} \qquad \frac{\partial}{\partial r} \rightarrow \frac{2}{d_f}\frac{\partial}{\partial \hat{r}}$$

in order to investigate the accuracy of (15.2.3ii). The local energy equation, neglecting generation and treating ρ as constant, can be written

$$\frac{\partial T}{\partial t}+v_r\frac{\partial T}{\partial r}+v_z\frac{\partial T}{\partial z}=\kappa\left\{\frac{1}{r}\frac{\partial}{\partial r}\left(r\frac{\partial T}{\partial t}\right)+\frac{\partial^2 T}{\partial r^2}\right\}$$

where $T=T(r, z, t)$. If we write $T=T^*(\hat{r}, \hat{z}, t)$, then this equation becomes

$$\frac{\partial T^*}{\partial t}+\frac{v_z}{l_f}\frac{\partial T^*}{\partial \hat{z}}=\frac{4\kappa}{d_f}\left\{\frac{1}{\hat{r}}\frac{\partial}{\partial \hat{r}}\left(\hat{r}\frac{\partial T^*}{\partial \hat{r}}\right)+\frac{d_f^2}{4l_f^2}\frac{\partial^2 T^*}{\partial \hat{z}^2}\right\}$$

For steady flow the first term drops out, while the last term is $0\ (d_f^2/l_f^2)$ of the previous one, and so negligible by comparison. In crude terms $\delta d_f/\delta z \sim d_f/l_f$. Writing the flowrate as

$$Q=\tfrac{1}{4}\pi d_f^2 v_z$$

we recover the elementary heat conduction equation

$$\mathrm{Gz}\frac{\partial T^*}{\partial \hat{z}}=\frac{1}{\hat{r}}\frac{\partial}{\partial \hat{r}}\left(\hat{r}\frac{\partial T^*}{\partial \hat{r}}\right)$$

where $\mathrm{Gz}=Q/\pi\kappa l_f$ is the Graetz number defined in (15.2.1). The unusual characteristic in this case is that the boundary condition is a mixed one

$$\alpha\left(\frac{\partial T^*}{\partial \hat{r}}\right)_{\hat{r}=1}=\tfrac{1}{2}h_a d_f\{T_a-T(\hat{r}=1)\}$$

or, in terms of the dimensionless temperature $\tilde{T}$ defined in (15.3.11) and the Nusselt number Nu defined in (15.2.9)

$$\frac{\partial \tilde{T}}{\partial \hat{r}}=\tfrac{1}{2}\,\mathrm{Nu}\,\tilde{T}$$

The initial condition is $\tilde{T}(\hat{z}=0)=1$.

Being a linear equation with linear boundary conditions, a formal analytical solution can be obtained in terms of eigenfunctions. However, the objective here is to obtain an order-of-magnitude estimate of

$$\frac{\tilde{T}(z,0)-\tilde{T}(z,1)}{\tilde{T}(z,1)}$$

which must be $\ll 1$ for (15.2.3ii) to be a reasonable approximation. From Table 3.4 of Ziabicki (1976), an experimental estimate for a multi-filament yarn gives $\mathrm{Nu}\approx\frac{1}{3}$. We know that $\partial\tilde{T}/\partial\hat{r}=0$ at $r=0$ and so taking

$$\tilde{T}(z,0)-\tilde{T}(z,1)=-\frac{1}{2}\left\{\frac{\partial \tilde{T}}{\partial \hat{r}}(z,0)+\frac{\partial \tilde{T}}{\partial \hat{r}}(z,1)\right\}$$

as a very crude approximation, we recover

$$\tilde{T}(z,0)-\tilde{T}(z,1)<0{\cdot}1\tilde{T}(z,1)$$

which is sufficient for many engineering purposes.

So far, we have only shown that the filament-spinning approximation is kinematically justified. It has yet to be shown how acceptable it is dynamically. The simplest view to take is that if $\partial d_f/\partial z \ll 1$, then segments of the moving fibre are very nearly segments of circular cylinders which experience unsteady, uniform elongation. In Subsection 4.3.1 a steady uniform elongation flow was described, and a uniform elongational ramp flow mentioned in Section 4.6, leading to the time-dependent extensional viscosity η_e^+. It can be shown directly that these satisfy the equations of motion exactly, provided inertia and gravity can be neglected. Similar arguments can be used here to show that the normally dominant terms in the equations of motion are satisfied exactly. Thus if a Newtonian fluid approximation is used, the Navier–Stokes equations yield, in the steady case for convenience,

$$\rho\left(v_r\frac{\partial v_r}{\partial r}+v_z\frac{\partial v_r}{\partial z}\right)=\eta_0\left[\frac{\partial}{\partial r}\left(\frac{1}{r}\frac{\partial}{\partial r}(rv_r)\right)+\frac{\partial^2 v_r}{\partial z^2}\right]-\frac{\partial p}{\partial r}$$

$$\rho\left(v_z\frac{\partial v_z}{\partial z}\right)=\rho g+\eta_0\frac{\partial^2 v_z}{\partial z^2}-\frac{\partial p}{\partial z}$$

where the fact that v_z is independent of r has been made use of. On substituting $v_r=-\frac{1}{2}rv_z'$ into the first equation and using the $(\tilde{r},\hat{z})$ coordinates it immediately becomes clear that

$$\frac{\partial p/\eta_0}{\partial \hat{r}}=0\left(\frac{d_f^2}{l_f^2}\frac{v_{z\,\max}}{l_f}(1+\text{Re})\right)$$

where Re is defined in (15.3.14), and that all the terms in the r momentum equation are much smaller than those in the z momentum equation, which yields

$$\frac{\partial p/\eta_0}{\partial \hat{z}}=0\left(\frac{d_f}{l_f}\frac{v_{z\,\max}}{l_f}(1+\text{Re})\right)$$

In a general non-dimensionalization, it would be predicted that

$$\left|\hat{\nabla}\left(\frac{p}{\eta_0}\right)\right| = 0\left(\frac{v_{z\,\max}}{l_f}(1+\mathrm{Re})\right)$$

i.e. an order-of-magnitude greater in l_f/d_f than arises from the velocity field chosen. The argument is completed by considering the boundary condition at $\hat{r}=1$, namely

$$\mathbf{T}\cdot\mathbf{n}=\mathbf{0}$$

The components of $\mathbf{n}$ are $[0(1), 0, 0(d_f/l_f)]$ and so the r and z components of the boundary condition yield

$$p-2\eta_0\frac{\partial v_r}{\partial r}=\eta_0\frac{\partial v_r}{\partial z}0\left(\frac{d_f}{l_f}\right)=0\left(\frac{d_f^2}{l_f^2}\eta_0\frac{v_{z\,\max}}{l_f}\right)$$

and

$$\left(p-2\eta_0\frac{\partial v_z}{\partial z}\right)0\left(\frac{d_f}{l_f}\right)=\eta_0\frac{\partial v_r}{\partial z}=0\left(\frac{d_f}{l_f}\right)$$

The first of these yields $p\sim-\eta_0(\partial v_z/\partial z)$. The second yields an equation relating $v_z(z)$ to $d_f(z)$ which, however, is not the same as that obtained in (15.2.5), and so provides a criterion for estimating the magnitude of $\partial v_z/\partial r$, a term neglected in calculating t^{E}_{rz}, which is therefore $0(d_f/l_f)$, and will be proportional to r so as to yield a term independent of r in the local z-momentum equation.

Formal asymptotic expansion of the local equations of motion and continuity together with boundary conditions shows that the problem is mathematically singular, because the highest order derivatives are an order-of-magnitude smaller in l_f/d_f than other terms present.

EXERCISE 15.2.1. Evaluate Gz for the values $d_{f0}=0{\cdot}25$ mm, $v_{z0}=0{\cdot}3\ \mathrm{m\,s^{-1}}$, $\kappa=10^{-7}\ \mathrm{m^2\,s^{-1}}$, $l_f=0{\cdot}5$ m.

EXERCISE 15.2.2. Show that only one convective term is obtained in the energy equation for T^* in Example 15.2.1 because the $(\hat{r}, \hat{z})$ coordinate system ensures that the surfaces $\hat{r}=$ constant are stream tubes.

EXERCISE 15.2.3. Show that if Nu is taken to be constant, separable solutions of the form

$$\tilde{T}_{(i)}=J_0(\beta^{(i)}\hat{r})\exp(-\hat{z}\beta^{(i)2}/\mathrm{Gz})$$

where

$$J_0(\beta^{(i)}) = -2\beta^{(i)} J_0'(\beta^{(i)})/\mathrm{Nu}$$

can be obtained for $\tilde{T}$ as defined in Example 15.2.1. Verify that for $\mathrm{Nu} = \frac{1}{3}$, the smallest positive value of $\beta^{(i)}$ is such that $\beta^{(i)} \sim \frac{1}{2}$ and that this gives

$$\{\tilde{T}_{(1)}(0, \hat{z}) - \tilde{T}_{(1)}(1, \hat{z})\}/T_{(1)}(1, \hat{z}) \sim 0{\cdot}1$$

as estimated in the Example.

Show also that this leads to $\tilde{T}(1, 1)/\tilde{T}(1, 0) \sim e^{-\frac{1}{4}\mathrm{Gz}}$. If $T_a = 30°\mathrm{C}$, $T_m = 120°\mathrm{C}$, $T_0 = 275°\mathrm{C}$, show that Gz based on l_f, the length of the molten part of the threadline, will be about 4, which is consistent with observations (cf. value obtained in Exercise 15.2.1 above). (See Ziabicki, 1976, p. 78 *et seq.* in this context.)

EXERCISE 15.2.4. Derive eqns (15.2.4)–(15.2.6) by considering a segment of the threadline between z and $z + dz$, where z and $z + dz$ are fixed, and using overall mass, momentum and energy balances. Neglect terms of order (d_f/l_f) smaller than those included, and note that eqn (15.2.7) supposes the ambient air to exert no stress across the filament surface. Some care needs to be taken in obtaining the term involving σ_m in eqn (15.2.5) because there are two contributions from σ_m: one directly as a surface force in the axial direction, and one indirectly through (15.2.7) as a contribution to pressure $p(z)$.

EXERCISE 15.2.5. Show that $|t_{rz}^{(a-d)}| \leqslant 0(d_f/l_f)\,|t_{zz}|$ in eqn (15.2.5) by integrating the relevant terms from $z = 0$ to $z = l_f$. Show that this leads to

$$\frac{\partial v_z}{\partial r} \leqslant 0\left(\frac{d_f}{l_f}\right)\frac{\partial v_z}{\partial z}$$

which is the same order-of-magnitude estimate as was obtained at the end of Example 15.2.1 for $\partial v_z/\partial r$ resulting from the axial component of the stress normal to the surface in the stress boundary condition at $r = \frac{1}{2}d_f$.

EXERCISE 15.2.6. Express $\dot{e}^t(z, s)$ in eqn (15.2.13) in terms of $\dot{e}(z, t)$. It will prove useful to use an equivalent of (15.3.4), so that

$$\left(\frac{\partial v_z}{\partial z}\right)^t \quad \text{can be written as} \quad \frac{\partial v_z}{\partial z}(z_s^t(z, s), t - s)$$

where

$$\frac{\partial z_s^t}{\partial s} = -v_z(z_s^t, t-s); \qquad z_s^t(z, 0) = z$$

Prove that (15.2.13) and (15.3.5) are consistent, but that the Lagrangean form (15.2.12) is necessary for unsteady flow with $\partial v_z^t/\partial z$ correctly interpreted following a fluid particle, i.e. that

$$\left(\frac{\partial v_z}{\partial z}\right)^t = \left(\frac{\partial v_z(z, t-s)}{\partial z}\right)$$

with $(t-s)$ held constant and $z = z_s^t$.

15.3 STEADY-STATE SOLUTIONS

Successful industrial fibre-spinning lines are by intention as steady as possible. Steady-state solutions have therefore been carefully investigated.

Equation (15.2.4) can be integrated directly to give

$$\rho_m v_z d_f^2 = 4\dot{M}/\pi \tag{15.3.1}$$

where $\dot{M}$ is the mass flowrate. This can be used to eliminate d_f. Equations (15.2.5) and (15.2.6) then become, when the heat generation term is neglected and Γ_m is taken to be constant,

$$\frac{dv_z}{dz} = \frac{g}{v_z} + \frac{d}{dz}\left(\frac{t_{zz} - t_{rr}}{\rho_m v_z}\right) + \frac{2\pi^{\frac{1}{2}} t_{rz}^{(a-d)}}{\dot{M}^{\frac{1}{2}} \rho_m^{\frac{1}{2}} v_z^{\frac{1}{2}}} + \frac{2\pi^{\frac{1}{2}}}{\dot{M}^{\frac{1}{2}} \rho_m^{\frac{1}{2}}} \frac{d}{dz}(\sigma_m v_z^{-\frac{1}{2}}) \tag{15.3.2}$$

and

$$\frac{dT}{dz} = \frac{2\pi^{\frac{1}{2}} h_a}{\dot{M}^{\frac{1}{2}} \Gamma_m} \frac{(T_a - T)}{\rho_m^{\frac{1}{2}} v_z^{\frac{1}{2}}} \tag{15.3.3}$$

The history of deformation can be obtained by recognizing that

$$\frac{\partial}{\partial s} \equiv -v_z \frac{\partial}{\partial z} \tag{15.3.4}$$

and so, for example,

$$c_{t_{zz}}^t(z, s) \to \{v_z(z_s)/v_z(z)\}^2 \tag{15.3.5}$$

where

$$\frac{\partial z_s}{\partial s} = -v_z(z_s) \tag{15.3.6}$$

This representation is similar to that discussed in Kase & Ikko (1981), and is a simpler form of that used in Exercise 15.2.6. Further progress requires consideration of end boundary conditions, and the use of dimensionless variables.

The exit boundary condition is provided by the wind-up at $z = z_1$, say. Given the low load on it, it can be taken to be a fixed velocity boundary condition

$$v_z(z_1) = v_{z1} \tag{15.3.7}$$

If the fibre is wound up as a melt, or freezing takes place instantaneously at a fixed position, such as the surface of a water bath, then the position of z_1 is unambiguously defined. If, however, as is usually the case in fibre spinning, v_z first reaches v_{z1} at the solidification point $z = l_f$, somewhere along the thread-line, the latter point cannot be specified *a priori* and is implicitly given by the solution of the governing equations.

The 'inlet' boundary condition to be applied at $z = 0$ depends upon the precise nature of the flow rearrangement that takes place at the outlet to the spinneret. Recently Fisher *et al.* (1980) (see also Pearson & Richardson, 1983, Chapter 7), have investigated the subject for a Newtonian jet and have concluded that, in the absence of better data, it is reasonable to suppose that the die swell that arises without thread-line tension is balanced by the draw-down effect of the axial tension in the fibre. This will not necessarily be the case with highly elastic melts, but in most industrial processes, elastic effects will be relatively unimportant at the spinneret, where the temperature is high and extension rates are relatively low. We can then write

$$d_f(0) = d_{f0}$$

the diameter of the spinneret hole, and so from (15.3.1)

$$v_z(0) = v_{z0} = 4\dot{M}/\pi\rho_m d_{f0}^2 \tag{15.3.8}$$

A first dimensionless group, the draw ratio, specific to this flow, is then defined as

$$\mathrm{Dr} = v_{z1}/v_{z0} \tag{15.3.9}$$

and is usually large. The last boundary condition will be provided by

$$T(0) = T_0 \tag{15.3.10}$$

The fact that $\mathrm{Dr} \gg 1$ makes choice of suitable length and velocity scales difficult. The simplest choice is d_{f0}, l_f and v_{z1}: this will be reasonable if $\partial v_z/\partial z$ is the same order of magnitude everywhere along the thread line. (Consideration of the exact solution, (15.3.20) below, for an isothermal and very viscous Newtonian fluid suggests that d_{f0}, $l_f/\ln \mathrm{Dr}$ and v_{z0} might be more suitable. However, experiments on isothermal highly elastic materials suggest that $\partial v_z/\partial z = \text{constant}$ is approached for high stretch rates.)

A stress scale $\eta_e v_{z1}/l_f$ can be obtained from the elongational viscosity η_e evaluated at the elongation rate $\dot{e} = v_{z1}/l_f$ and temperature T_0. A temperature scale is provided by $(T_0 - T_a)$. Thus with

$$\left.\begin{aligned} \tilde{v}_z = v_z/v_{z1}, \qquad \hat{z} = z/l_f, \qquad \tilde{T} = (T - T_a)/(T_0 - T_a) \\ \tilde{t}_{zz} - \tilde{t}_{rr} = (t_{zz} - t_{rr}) l_f / \eta_e(v_{z1}/l_f) v_{z1} \end{aligned}\right\} \tag{15.3.11}$$

eqns (15.3.2) and (15.3.3) become

$$\left.\begin{aligned} \frac{\mathrm{d}}{\mathrm{d}\hat{z}}\left(\frac{\tilde{t}_{zz} - \tilde{t}_{rr}}{\tilde{v}_z}\right) = \mathrm{Re}\frac{\mathrm{d}\tilde{v}_z}{\mathrm{d}\hat{z}} + \frac{\mathrm{Re}}{\mathrm{Fr}}\frac{1}{\tilde{v}_z} \\ + \frac{4\tilde{t}_{rz}^{(\mathrm{a-d})}\,\mathrm{Dr}^{\frac{1}{2}}\, l_f}{d_{f0}}\frac{1}{\tilde{v}_z^{\frac{1}{2}}} + \frac{2\,\mathrm{Ca}^{-1}\,\mathrm{Dr}^{\frac{1}{2}}\, l_f}{d_{f0}}\frac{\mathrm{d}}{\mathrm{d}\hat{z}}\left(\frac{1}{\tilde{v}_z^{\frac{1}{2}}}\right) \end{aligned}\right\} \tag{15.3.12}$$

$$\frac{\mathrm{d}\tilde{T}}{\mathrm{d}\hat{z}} + \frac{\mathrm{Nu}}{\mathrm{Dr}^{\frac{1}{2}}\,\mathrm{Gz}}\frac{\tilde{T}}{\tilde{v}_z^{\frac{1}{2}}} = 0 \tag{15.3.13}$$

where

$$\left.\begin{aligned} \mathrm{Re} = \rho_m v_{z1} l_f/\eta_e \\ \mathrm{Fr} = v_{z1}^2/g l_f \end{aligned}\right\} \tag{15.3.14}$$

$$\mathrm{Ca}^{-1} = \sigma_m/\eta_e v_{z1} \tag{15.3.15}$$

and Gz is given by (15.2.1).

The boundary conditions become

$$\tilde{v}_z(0) = \mathrm{Dr}^{-1}; \qquad \tilde{v}_z(1) = 1 \tag{15.3.16}$$

If, as is indeed the case for most laboratory situations, Re, Re/Fr, $\mathrm{Ca}^{-1}\, l_f/d_{f0}$ and $t_{rz}^{(\mathrm{a-d})} l_f/d_{f0}$ are small, then the momentum conservation equation (15.3.12) becomes essentially a constant-tension relation

$$(\tilde{t}_{zz} - \tilde{t}_{rr})/\tilde{v}_z = \tilde{F} \tag{15.3.17}$$

where

$$\tilde{F} = Fl_f/Q\eta_e \tag{15.3.18}$$

Commercial practice pushes v_{z1}, l_f and $1/d_{f0}$ to the largest values possible and so the left-hand side of (15.3.12) may not dominate, particularly close to the solidification point. The argument for dropping terms on the basis of the size of the various dimensionless groups, such as Re, Fr, Ca^{-1} and $t_{rz}^{(a-d)}l_f/\eta_e(\dot{e})d_{f0}$, usually assumes that the dimensionless variables involved are all 0(1). However, $\tilde{v}_z$ varies from Dr^{-1} to 1 and so it is difficult to know *ab initio* what the balance of forces will be in practice along the thread line.

Further progress depends upon being able to specify $(\tilde{t}_{zz} - \tilde{t}_{rr})$ as a function of $\tilde{v}_z$ and $\tilde{T}$. If $\tilde{T}$ enters significantly into the rheological equation of state, then (15.3.17), or (15.3.12), and (15.3.13) are coupled.

15.3.1 Newtonian and Other Purely Viscous Fluids

The solution for a fluid of large constant viscosity, in an isothermal environment, is very simple to derive. Equation (15.3.17) becomes, using $\eta_e = 3\eta_0$,

$$\frac{d\tilde{v}_z}{d\hat{z}} = \tilde{F}\tilde{v}_z \tag{15.3.19}$$

whence, using (15.3.8) and (15.3.9),

$$\tilde{v}_z = e^{\tilde{F}\hat{z}}/\mathrm{Dr} \tag{15.3.20}$$

From (15.3.7) it follows that $\mathrm{Dr} = e^{\tilde{F}}$ and hence

$$F = 3Q\eta_0 \ln \mathrm{Dr}/l_f \tag{15.3.21}$$

If l_f and Dr are given then F is predicted. (However, if Dr is not specified and F is fixed, the velocity scale used in (15.3.11) is inappropriate.)

The various Rivlin–Eriksen tensors defined by (3.3.13) can be calculated for the solution (15.3.20) as

$$\mathbf{A}_1 = \begin{pmatrix} -\dfrac{dv_z}{dz} & 0 & 0 \\ 0 & -\dfrac{dv_z}{dz} & 0 \\ 0 & 0 & 2\dfrac{dv_z}{dz} \end{pmatrix}, \quad \mathbf{A}_N = \begin{pmatrix} 0 & 0 & 0 \\ 0 & 0 & 0 \\ 0 & 0 & 2\times 3^{N-1}\left(\dfrac{dv_z}{dz}\right)^N \end{pmatrix} \tag{15.3.22}$$

using

$$\frac{D}{Dt} = v_z \frac{d}{dz}$$

A Deborah number† for the flow can best be defined by

$$\mathrm{De} = \Lambda\left(\frac{dv_z}{dz}\right)^* \tag{15.3.23}$$

This varies by a factor Dr along the thread line. Its maximum value is given by

$$\mathrm{De}_{\max} = \Lambda v_{z1} \ln \mathrm{Dr}/l_f \tag{15.3.24}$$

which is close to the value $\Lambda v_{z1}/l_f$ derived from the dimension scales used earlier. Assuming that the successive extra terms that are included in a Coleman–Noll *N*th-order-fluid (Subsection 3.3.3) expansion have coefficients that are proportional to Λ^N, it can be seen from (15.3.22) and (15.3.23) that all terms in the expansion will be of order 1 at the same value of De, or more realistically, that, for small values of De, the second-order approximation will first become valid near the wind-up.

† This is the simple definition of an elasticity number introduced in (7.2.8), and can be written in the form

$$\mathrm{De} = \Lambda[\mathrm{tr}(\mathbf{A}_1^2)/6]^{\frac{1}{2}}$$

If the idea behind the definition given in (8.2.11) is used instead, then we could write

$$\begin{aligned}\mathrm{De} &= \Lambda[\mathrm{tr}(\mathbf{A}_2^2)/6]^{\frac{1}{2}}/[\mathrm{tr}(\mathbf{A}_1^2)/6]^{\frac{1}{2}} \\ &= \sqrt{6}\Lambda(dv/dz)^*\end{aligned}$$

which is, apart from the factor $\sqrt{6}$, the same as the value based on (7.2.8).

However, it must be remembered that the initial condition at $z=0$ presupposes a change from tube flow to elongational flow in a short region and so $\mathrm{d}v_z/\mathrm{d}r$ will be very large in that region. Hence the elastic contribution should be large at the spinneret also.

These elastic effects will be discussed in later subsections.

If a non-isothermal situation is considered using a viscosity given by

$$\eta = \eta_{0\mathrm{a}} \exp \zeta_{\mathrm{m}}(T_{\mathrm{a}} - T) \tag{15.3.25}$$

and for convenience Nu is taken to be proportional to $v_z^{\frac{1}{2}}$ (i.e. to $\mathrm{Re}_z^{\frac{1}{2}}$ where Re_z is the local Reynolds number based on z, v_z and $(\eta/\rho)_{\mathrm{air}}$), (15.3.13) becomes

$$\frac{\mathrm{d}\tilde{T}}{\mathrm{d}\hat{z}} + \mathrm{St}\,\tilde{T} = 0 \tag{15.3.26}$$

where

$$\mathrm{St} = \mathrm{Nu}(\tilde{v}_z = 1)/\mathrm{Dr}^{\frac{1}{2}}\,\mathrm{Gz} \tag{15.3.27}$$

with solution

$$\tilde{T} = \mathrm{e}^{-\mathrm{St}\,\hat{z}} \tag{15.3.28}$$

The momentum equation (15.3.17) now becomes

$$\frac{1}{\tilde{v}_z}\frac{\mathrm{d}\tilde{v}_z}{\mathrm{d}\hat{z}} = \tilde{F} \exp(\mathrm{B}\mathrm{e}^{-\mathrm{St}\,\hat{z}}), \qquad \mathrm{B} = \zeta_{\mathrm{m}}(T_0 - T_{\mathrm{a}}) \tag{15.3.29}$$

with solution (Matovich, 1966)

$$\tilde{v}_z = \mathrm{Dr}^{f(\hat{z})}$$
$$f(\hat{z}) = -\left(\int_{\hat{z}}^{1} \mathrm{e}^{\mathrm{B}\exp(-\mathrm{St}\,x)}\,\mathrm{d}x \Big/ \int_0^1 \mathrm{e}^{\mathrm{B}\exp(-\mathrm{St}\,x)}\,\mathrm{d}x\right) \tag{15.3.30}$$

There is experimental evidence (Kase & Matsuo, 1965) that the exponential form for $\tilde{T}$, which is independent of draw-down rate, is observed in practice. For a relatively Newtonian material like PET or Nylon, the solution (15.3.30) is not unrealistic. It must be remembered that the material freezes when $T = T_{\mathrm{m}}$, i.e. when

$$\hat{z} = \frac{1}{\mathrm{St}} \ln\left(\frac{T_0 - T_{\mathrm{a}}}{T_{\mathrm{m}} - T_{\mathrm{a}}}\right) \tag{15.3.31}$$

This value gives the relevant value for l_f in the case of a fibre freezing on the thread line (see Exercise 15.3.4).

Similarly, solutions can be obtained for power-law fluids where

$$\eta_e = K_{ea} \exp[\zeta_m(T_a - T)] \left|\frac{dv_z}{dz}\right|^{\nu-1} \tag{15.3.32}$$

If η_e^* is based on $\dot{e} = v_{z1}/l_f$ and $T = (T_m + T_0)/2$ for definiteness, and the dimensionless variables (15.3.11) and the solution (15.3.28) are used, then (15.3.17) becomes

$$\frac{1}{\tilde{v}_z}\left(\frac{d\tilde{v}_z}{d\hat{z}}\right)^{\nu} = \tilde{F} \exp(\mathrm{B}e^{-\mathrm{St}\,\hat{z}}) \tag{15.3.33}$$

with solution

$$\tilde{v}_z = \left[\frac{\int_0^1 \exp\left(\frac{\mathrm{B}}{\nu} e^{-\mathrm{St}\,x}\right) dx}{\int_0^{\hat{z}} \exp\left(\frac{\mathrm{B}}{\nu} e^{-\mathrm{St}\,x}\right) dx + \mathrm{Dr}^{(1-\nu)/\nu} \int_{\hat{z}}^1 \exp\left(\frac{\mathrm{B}}{\nu} e^{-\mathrm{St}\,x}\right) dx}\right]^{v/(1-v)} \tag{15.3.34}$$

The power-law form (15.3.32) must be regarded as a quite distinct approximation from that relevant for simple shear when $\nu < 1$. There is, in fact, little evidence that such a form actually applies in fibre spinning; it can be verified that the constant-extension-rate 'asymptote' is approached as $\nu \to \infty$ in the isothermal case, and so $\nu > 1$ rather than $\nu < 1$ would be an appropriate crude approximation.

15.3.2 Viscoelastic Fluids

The most general rheological equation of state that we need consider is (15.2.14). The only result based on a fully general formulation is that given by Moore & Pearson (1975, Appendix 2) who show that the isothermal, incompressible form (15.2.13) leads, in an effectively infinitely-long fibre, to a 'universal' solution when rheological forces dominate. Although of mathematical interest, such a result is unlikely to be practically important.

The most realistic model that has been used is that of Matsui & Bogue, mentioned earlier in Subsection 5.2.3 (5.2.28), which is a non-isothermal quasilinear development of the rubber-like fluid. Recent work (Laun, 1978; Wagner, 1979*a*; Wagner *et al.*, 1979, see also Subsection 5.2.2) has shown the isothermal form to correspond reasonably well with the observed behaviour of real polymer melts like PE provided the kernel function, $\mathring{m}$ in (5.2.15), for example, is obtained from small-amplitude oscillation measurements and a non-linear strain-dependent factor, such as h in (5.2.15), is included. The basic strain measure used is the Finger tensor $\mathbf{C}_t^{t-1}$.

Petrie (1979, Section 6.1) discusses solutions for a range of differential ('rate') models (themselves exhaustively discussed in his Chapter 2), though few can be exhibited in analytical form except for the extremes of $\mathrm{De} \ll 1$ and $\mathrm{De} \gg 1$ (De is defined above in (15.3.24)). The case De small is worth considering in some detail because it introduces the important issue of inlet boundary conditions. Petrie, following Matovich & Pearson (1969), perturbs the Newtonian solution given above as (15.3.20). However, he chooses to regard v_{z0} and F, the in-line tension, as fixed rather than taking v_{z0} and v_{z1} to be fixed, as is more natural for a laboratory spinning operation. His constraint on the in-line tension leads in his case to a singular perturbation problem involving inner and outer expansions. A different approach will be adopted here, which yields a result more nearly related to that of Matovich and Pearson.

Low Deborah number expansions are conveniently based on a 'retarded motion' expansion, which in practice involves a sequence of Nth-order-fluid models (Subsection 3.3.3). Thus, for example, the 3rd-order model, eqn (3.3.11), can be written in dimensionless form as

$$\tilde{\mathbf{T}}^{\mathrm{E}} = (1+\tilde{\delta}_1 \operatorname{tr} \tilde{\mathbf{A}}_1^2)\tilde{\mathbf{A}}_1 + \tilde{\beta}_0\tilde{\mathbf{A}}_1^2 + \tilde{\gamma}_0\tilde{\mathbf{A}}_2 + \tilde{\beta}_1(\tilde{\mathbf{A}}_1 \cdot \tilde{\mathbf{A}}_2 + \tilde{\mathbf{A}}_2 \cdot \tilde{\mathbf{A}}_1) + \tilde{\gamma}_1\tilde{\mathbf{A}}_3 \tag{15.3.35}$$

where

$$\tilde{\mathbf{A}}_N = \left.\frac{\mathrm{d}^N \mathbf{C}_t(\tilde{t}')}{\mathrm{d}\tilde{t}'^N}\right|_{\tilde{t}'=\tilde{t}} = \mathbf{A}_N \Lambda^N = \mathrm{De}^N\, \hat{\mathbf{A}}_N$$

$$\tilde{t} = t/\Lambda, \qquad \mathrm{De} = \Lambda v_{z1}/l_{\mathrm{f}} \tag{15.3.36}$$

$$\tilde{\mathbf{T}}/\mathrm{De} = \hat{\mathbf{T}}^{\mathrm{E}} = \mathbf{T}^{\mathrm{E}} l_{\mathrm{f}}/\eta_0 v_{z1}, \qquad \tilde{\beta}_0 = \beta_0/\eta_0\Lambda, \qquad \tilde{\gamma}_0 = \gamma_0/\eta_0\Lambda$$

$$\tilde{\beta}_1 = \beta_1/\eta_0\Lambda, \qquad \tilde{\gamma}_1 = \gamma_1/\eta_0\Lambda^2, \qquad \tilde{\delta}_1 = \delta_1/\eta_0\Lambda^2$$

If the dimensionless coefficients $\tilde{\eta}_0$, $\tilde{\beta}_{N-2}$, $\tilde{\gamma}_{N-2}$, $\tilde{\delta}_{N-2}$, etc. are taken to be of order unity (which is consistent with supposing the fluid to be characterized by a single material time scale) then $\tilde{\mathbf{A}}_N$ will be $0(\mathrm{De}^N)$, and the difference between the 2nd-order approximation and the 3rd-order approximation will be 0(De) smaller than the smallest terms retained in the 2nd-order approximation. It is assumed here that $\left(\frac{l_f}{v_{z1}}\right)^N \mathbf{A}_N = \hat{\mathbf{A}}_N$ is of order unity, in that the time scale defining $\tilde{\mathbf{A}}_N$ is that of the motion, namely l_f/v_{z1}. It may be shown that all the 'differential models', such as the Oldroyd 8-constant model (3.3.22) or the spectrally-summed Maxwell model (3.3.32), can be written in the same retarded-motion form provided De is small enough (where Λ is taken to be the largest time that can be obtained from the constants in the constitutive equation).

In the steady fibre-spinning situation

$$\tilde{\mathbf{A}}_1 = \mathrm{De}\,\hat{\mathbf{A}}_1 = \mathrm{De}\begin{pmatrix} -\frac{d\tilde{v}_z}{d\hat{z}} & 0 & 0 \\ 0 & -\frac{d\tilde{v}_z}{d\hat{z}} & 0 \\ 0 & 0 & 2\frac{d\tilde{v}_z}{d\hat{z}} \end{pmatrix} \tag{15.3.37}$$

and from (3.3.13)

$$\tilde{\mathbf{A}}_2 = \mathrm{De}^2\,\hat{\mathbf{A}}_2 = \mathrm{De}^2\left(\tilde{v}_z \frac{d\hat{\mathbf{A}}_1}{d\hat{z}} + \hat{\mathbf{A}}_1^2\right) \tag{15.3.38}$$

or, more generally,

$$\tilde{\mathbf{A}}_N = \mathrm{De}^N\,\hat{\mathbf{A}}_N = \mathrm{De}^N\left(\tilde{v}_z \frac{d\hat{\mathbf{A}}_{N-1}}{d\hat{z}} + \hat{\mathbf{A}}_1 \cdot \hat{\mathbf{A}}_{N-1}\right) \tag{15.3.39}$$

(15.3.37), (15.3.38) and (15.3.39) can be substituted into (15.3.35) to yield

$$\begin{aligned}\tilde{\mathbf{T}}_N^E = {} & \mathrm{De}\,\hat{\mathbf{A}}_1 + \mathrm{De}(\tilde{\beta}_0\hat{\mathbf{A}}_1^2 + \tilde{\gamma}_0\hat{\mathbf{A}}_2) + \mathrm{De}^2\{(\tilde{\delta}_1\,\mathrm{tr}\,\hat{\mathbf{A}}_1^2)\hat{\mathbf{A}}_1 \\ & + \tilde{\beta}_1(\hat{\mathbf{A}}_1 \cdot \hat{\mathbf{A}}_2 + \hat{\mathbf{A}}_2 \cdot \hat{\mathbf{A}}_1) + \tilde{\gamma}_1\hat{\mathbf{A}}_3\}\end{aligned} \tag{15.3.40}$$

whence

$$\begin{aligned}\tilde{t}_{zz} - \tilde{t}_{rr} = {} & \mathrm{De}\left\{3\frac{d\tilde{v}_z}{d\hat{z}}\right\} + \mathrm{De}^2\left\{3(\tilde{\beta}_0 + \tilde{\gamma}_0)\left(\frac{d\tilde{v}_z}{d\hat{z}}\right)^2 \right. \\ & \left. + 3\tilde{\gamma}_0\tilde{v}_z\frac{d^2\tilde{v}_z}{d\hat{z}^2}\right\} + 0(\mathrm{De}^3)\end{aligned} \tag{15.3.41}$$

A perturbation expansion can be obtained by writing

$$\tilde{v}_z = \tilde{v}_z^{(1)} + \mathrm{De}\, \tilde{v}_z^{(2)} + \mathrm{De}^2\, \tilde{v}_z^{(3)} + \ldots \tag{15.3.42}$$

$$\tilde{F} = \mathrm{De}\, \tilde{F}^{(1)} + \mathrm{De}^2\, \tilde{F}^{(2)} + \mathrm{De}^3\, \tilde{F}^{(3)} + \ldots \tag{15.3.43}$$

substituting (15.3.41)–(15.3.43) into (15.3.17) and equating corresponding powers of De. The dynamical equation (15.3.17) now yields (15.3.19) as the lowest order approximation

$$\frac{\mathrm{d}\tilde{v}_z^{(1)}}{\mathrm{d}\hat{z}} = \tilde{F}^{(1)} \tilde{v}_z^{(1)} \tag{15.3.44}$$

with solution given by (15.3.20),

$$\tilde{v}_z^{(1)} = \mathrm{e}^{\ln \mathrm{Dr}\, \hat{z}}/\mathrm{Dr}, \qquad \tilde{F}^{(1)} = \ln \mathrm{Dr} \tag{15.3.45}$$

satisfying the boundary conditions (15.3.16), i.e.

$$\tilde{v}_z^{(1)}(0) = \mathrm{Dr}^{-1}, \qquad \tilde{v}_z^{(1)}(1) = 1 \tag{15.3.46}$$

The terms of order De^2 yield the inhomogeneous equation

$$\frac{\mathrm{d}\tilde{v}_z^{(2)}}{\mathrm{d}\hat{z}} - \tilde{v}_z^{(2)} \tilde{F}^{(1)} = \tilde{v}_z^{(1)} \tilde{F}^{(2)} + (\tilde{\beta}_0 + \tilde{\gamma}_0) \left(\frac{\mathrm{d}\tilde{v}_z^{(1)}}{\mathrm{d}\hat{z}} \right)^2 + \tilde{\gamma}_0 \tilde{v}_z^{(1)} \frac{\mathrm{d}^2 \tilde{v}_z^{(1)}}{\mathrm{d}\hat{z}^2} \tag{15.3.47}$$

with homogeneous boundary conditions

$$\tilde{v}_z^{(2)}(0) = \tilde{v}_z^{(2)}(1) = 0 \tag{15.3.48}$$

On substituting for $\tilde{F}^{(1)}$ and $\tilde{v}_z^{(1)}$, eqn (15.3.47) can be written

$$\frac{\mathrm{d}}{\mathrm{d}\hat{z}} (\tilde{v}_z^{(2)}\, \mathrm{e}^{-\ln \mathrm{Dr}\, \hat{z}}) = \frac{\tilde{F}^{(2)}}{\mathrm{Dr}} + \{\tilde{\beta}_0 + 2\tilde{\gamma}_0\} \left(\frac{\ln \mathrm{Dr}}{\mathrm{Dr}} \right)^2 \mathrm{e}^{\ln \mathrm{Dr}\, \hat{z}} \tag{15.3.49}$$

with solution

$$\tilde{v}_z^{(2)} = \{\tilde{\beta}_0 + 2\tilde{\gamma}_0\} \frac{\ln \mathrm{Dr}}{\mathrm{Dr}^2} \{ -[(\mathrm{Dr} - 1)\hat{z} + 1]\, \mathrm{e}^{\ln \mathrm{Dr}\, \hat{z}} + \mathrm{e}^{2 \ln \mathrm{Dr}\, \hat{z}} \} \tag{15.3.50}$$

$$\tilde{F}^{(2)} = \{\tilde{\beta}_0 + 2\tilde{\gamma}_0\} \frac{\ln \mathrm{Dr}(1 - \mathrm{Dr})}{\mathrm{Dr}} \tag{15.3.51}$$

satisfying (15.3.48).

The process can be continued for terms of order De^N, $N > 2$, in each case yielding corrections $v_z^{(N)}$, $\tilde{F}^{(N)}$ that satisfy the homogeneous boundary condition $\tilde{v}_z^{(N)}(0) = v_z^{(N)}(1) = 0$.

What Petrie's singular perturbation expansion shows is that rapid 'elastic' adjustments can be made, near the entry point, in particular on a length scale

$$L_e = v_{0z}\Lambda = v_{1z}\Lambda/\mathrm{Dr} = l_f\,\mathrm{De}/\mathrm{Dr} \tag{15.3.52}$$

Thus the introduction of viscoelastic effects compounds the difficulties inherent in analysing the matching of a simple shear flow near the walls of the spinneret to the elongational flow in the thread line proper. Many of the issues involved are discussed in Section 7.4.

The solution given by (15.3.42), (15.3.43) which appears to give at least asymptotically valid approximations for small enough De is uniquely given by the two boundary conditions (15.3.16). Yet the order of the full ordinary differential equation in $\tilde{v}_z$ is higher than second when (15.3.41) is substituted into (15.3.17) and so further boundary conditions on $\tilde{v}_z$ should be required to get a unique solution. This is the origin of Petrie's singular expansion, and is probably tantamount to saying that Dr in (15.3.16) is itself an unknown of the problem to be determined by the singular flow near the origin $z=0$.

The high Deborah number limit raises issues of the type discussed in Section 7.2 in connection with entry flow. For a Maxwell-type fluid, the result (cf. (7.2.37) *et seq.*)

$$\overset{\square}{\mathbf{T}}{}^{\mathrm{E}} = \mathbf{0} \tag{15.3.53}$$

was obtained for the leading term. This, however, was in the case where the kinematics was completely specified. In the thin-fibre case, such a limiting solution might allow of velocity discontinuities, corresponding to instantaneous extension of the filament. If a Jeffreys-type fluid (or the more general Oldroyd 8-constant model (3.3.22)) is employed, then instantaneous extension is not possible. However, the same problem arises for all viscoelastic models: the velocity boundary conditions (15.3.16) are not sufficient to prescribe the flow in the region $0 \leqslant \hat{z} \leqslant 1$.

Important features of the viscoelastic thread-line flow can be illustrated by using the model chosen by Denn *et al.* (1975)

$$\mathbf{T}^{\mathrm{E}} + \lambda \overset{\triangledown}{\mathbf{T}}{}^{\mathrm{E}} = \lambda G(\mathbf{A}_1 - \mu \mathbf{A}_1^2), \qquad \mu \geqslant 0 \tag{15.3.54}$$

When $\mu = 0$, the upper convected Maxwell model, identical as explained earlier to the Lodge rubber-like liquid with single relaxation time λ, is recovered. The viscosity η has been written as λG to illustrate the role played by the elastic modulus G when $\mathrm{De} \propto \lambda \to \infty$.

Using the dimensionless variables (15.3.11) and

$$\mathrm{De} = \lambda v_{z1}/l_f \tag{15.3.55}$$

$$\tilde{\mu} = \mu/\lambda \tag{15.3.56}$$

eqns (15.3.17) and (15.3.54) can be shown to yield

$$\tilde{v}_z - \frac{\tilde{v}_z'}{\tilde{F}} + \mathrm{De}\left\{\tilde{v}_z\tilde{v}_z' - \frac{\tilde{v}_z^2\tilde{v}_z''}{\tilde{v}_z'} + \frac{\tilde{\mu}}{\tilde{F}}\tilde{v}_z'^2\right\} - \mathrm{De}^2\left\{2\tilde{v}_z\tilde{v}_z'^2 - \frac{\tilde{\mu}}{\tilde{F}}(2\tilde{v}_z'^3 + \tilde{v}_z\tilde{v}_z'\tilde{v}_z'')\right\} = 0 \tag{15.3.57}$$

The equation is an ordinary second-order differential equation and so a solution satisfying the boundary conditions (15.3.46) can in principle be found for arbitrary $\tilde{\mu}$, De and $\tilde{F}$. $\tilde{\mu}$ and De are given by the operating conditions and the material parameters, but $\tilde{F}$ can be freely chosen at this stage.

One limit is obtained by supposing $\tilde{F}$ to remain $\ll$ De while De $\rightarrow \infty$. This yields

$$\tilde{v}_z\tilde{v}_z'^2 - \frac{\tilde{\mu}}{\tilde{F}}(\tilde{v}_z'^3 + \tfrac{1}{2}\tilde{v}_z\tilde{v}_z'\tilde{v}_z'') = 0 \tag{15.3.58}$$

or

$$\hat{\mu}\frac{\mathrm{d}}{\mathrm{d}\tilde{v}_z}\left(\tilde{v}_z^2\frac{\mathrm{d}\tilde{v}_z}{\mathrm{d}\hat{z}}\right) - \tilde{v}_z^2 = 0 \tag{15.3.59}$$

where

$$\hat{\mu} = \frac{1}{2}\frac{\tilde{\mu}}{\tilde{F}} \tag{15.3.60}$$

with solution

$$\tilde{v}_z^3 = \frac{1 - \mathrm{e}^{-1/\hat{\mu}}\,\mathrm{Dr}^{-3}}{1 - \mathrm{e}^{-1/\hat{\mu}}} - \frac{(1 - \mathrm{Dr}^{-3})}{1 - \mathrm{e}^{-1/\hat{\mu}}}\,\mathrm{e}^{-\hat{z}/\hat{\mu}} \tag{15.3.61}$$

For $\hat{\mu} \rightarrow 0$, this yields $\tilde{v}_z \approx 1$ almost everywhere with a rapid rise to $\tilde{v}_z = 1$ near $z = 0$. For $\hat{\mu} \rightarrow \infty$, the solution tends to

$$\tilde{v}_z = \mathrm{Dr}^{-3} + (1 - \mathrm{Dr}^{-3})\hat{z} \tag{15.3.62}$$

The Maxwell fluid limit $\tilde{\mu} \rightarrow 0$ is obviously compatible with $\hat{\mu} \rightarrow 0$ and so a velocity discontinuity at $\hat{z} = 0$ is not inexplicable.

An alternative limiting procedure, used by Denn and others (see Denn, 1977) is to let F, and hence $\tilde{F} \to \infty$. This limit was introduced in order to study the case of large Dr, and it was argued that $\mathrm{Dr} \to \infty$ would imply $F \to \infty$ on elementary physical grounds. Equation (15.3.57) then yields

$$\tilde{v}_z \tilde{v}_z' + \mathrm{De}\{\tilde{v}_z (\tilde{v}_z')^2 - \tilde{v}_z^2 \tilde{v}_z''\} - 2\mathrm{De}^2\, \tilde{v}_z (\tilde{v}_z')^3 = 0 \qquad (15.3.63)$$

which surprisingly has an exact but implicit solution (Pearson & Petrie, 1985), given by

$$\tilde{v}_z + \tfrac{1}{2}c \ln\left\{\frac{(\tilde{v}_z^2 - c)^3(1-c)^3}{(1-c)^3(\tilde{v}_z^3 - c^3)}\right\} - \sqrt{3}c\left\{\tan^{-1}\left(\frac{2\tilde{v}_z + c}{\sqrt{3}c}\right) - \tan^{-1}\left(\frac{2+c}{\sqrt{3}c}\right)\right\}$$
$$= 1 + (\hat{z} - 1)/\mathrm{De} \qquad (15.3.64)$$

where c is given by $\tilde{v}_z(0) = \mathrm{Dr}^{-1}$ such that $-\frac{1}{2} < c < 0$.

It may be shown that this general solution does not lead to real solutions satisfying (15.3.16) for all Dr. There is, as shown for the White–Metzner model in Fig. 15.10, a maximum value Dr(De) above which there are no solutions. This arises because eqn (15.3.63) is non-linear, and so solutions need not exist for arbitrary boundary conditions. It should be noted that eqn (15.3.63) is the one that is also relevant for an upper-convected Maxwell model, since all the terms in $\tilde{\mu}$ have been neglected. For those cases where Dr is small enough to permit a solution, the limiting linear form (15.3.62) is recovered, and this is consistent with the observations of Spearot & Metzner (1972) and other workers.

Chang & Lodge (1971) have suggested that the behaviour of an upper-convected Maxwell fluid at high draw-ratios can be understood in terms of a creep experiment. For large suddenly-imposed tensions a large instantaneous elongation occurs followed by an almost uniform rate of elongation (see Exercise 15.3.5). Petrie (1979, Chapter 6) has shown that solutions to (15.3.17) subject to fixed boundary conditions on velocity need not exist, and it may readily be shown that (15.3.17) does not allow of discontinuities in cross-section, and hence in axial velocity; hence any such discontinuity must correspond to a discontinuity in the governing equations. This is discussed at length in Pearson & Petrie (1985) where it is suggested that the adjustment region at the exit to the spinneret (which in the fibre-spinning approximation, $d_f/l_f \to 0$, is equivalent to a discontinuity at $\hat{z} = 0$) allows of an effective discontinuity in t_{rr} on emerging from the spinneret. If this argument is accepted then it implies that large values of Dr are achieved before the fibre-spinning approximations become relevant, and so Dr cannot be chosen independently of the upstream flow solution. In other words, it is not possible to treat De, $\tilde{F}$ and Dr as independent parameters in the fibre-spinning problem. This result seems to be true for both large and small De, provided $\text{De} \neq 0$.

If a Jeffreys-type model is employed then a discontinuity at $\hat{z}=0$ can no longer be expected; however, a higher-order equation even in the fibre-spinning approximation results, and additional boundary conditions are needed.

The asymptotic analyses and argument given above show how complex the interrelation between constitutive model, initial thread-line tension and filament dynamics can be. A possible way to tackle the initial-tension problem is to use an integral model of the type described in Sections 3.4 or 5.2, and to include the history of deformation prior to extrusion as part of the statement of the spinning situation. This is analogous to the treatment given in Section 7.4 for extrudate swell. Indeed if the short-capillary arguments—such as those leading to relations (7.4.19) and (7.4.20) for Es—are used then the deformation history is everywhere simple elongational and a unique self-consistent solution specifying $\tilde{F}$ should be obtained. This has not yet been undertaken, although Matsui & Bogue (1976) have used a crude simple-shear/elongation-flow discontinuity to provide a deformation history more relevant to a long capillary. Full details are not given by them.

Introduction of inertia prevents velocity discontinuities as Petrie (1979) shows.

More recent work has been based on numerical analyses (Pearson & Richardson, 1983, Chapter 6) that cover the exit region as well as the one-dimensional fibre-spinning region and involves all the problems inherent in analysis of extrudate swelling. Unfortunately, numerical instabilities have been encountered in most cases for Deborah numbers greater than a value of order unity, and so the issue of the high Deborah number limit remains unresolved.

EXERCISE 15.3.1. Show that the solution (15.3.20), (15.3.21) satisfies (15.3.17), (15.3.18) when

$$\tilde{t}_{rr} - \tilde{t}_{zr} = \partial\tilde{v}/\partial\hat{z}$$

Show that the latter equation is the correct one for a Newtonian fluid of constant viscosity $\eta_0 = \frac{1}{3}\eta_e$.

EXERCISE 15.3.2. Show that the solutions of (15.3.12) with (Re, Re/Fr, Ca^{-1}) taken to be (i) (Re, 0, 0), (ii) (0, Re/Fr, 0) and (iii) (0, 0, Ca^{-1}) successively, with $\eta_{\mathrm{e}}=$ constant, $t_{rz}^{(\mathrm{a-d})}=0$, and subject to (15.3.16) are

(i) $\tilde{v}_z(\hat{z})=C_1[(C_1\mathrm{Dr}+\mathrm{Re})\exp(-C_1\hat{z})-\mathrm{Re}]^{-1}$ where $(C_1+\mathrm{Re})=(C_1\mathrm{Dr}+\mathrm{Re})\exp(-C_1)$.

(ii) $\tilde{v}_z(\hat{z})=(\mathrm{Re}/2C_3^2\,\mathrm{Fr})\sinh^2[C_3(\hat{z}+Z_0)]$ where

$$Z_0=C_3^{-1}\sinh^{-1}\{C_3(2\mathrm{Fr}/\mathrm{Re}\,\mathrm{Dr})^{\frac{1}{2}}\}$$

and

$$C_3=\sinh^{-1}\{C_3(2\,\mathrm{Fr}/\mathrm{Re})^{\frac{1}{2}}\}-\sinh^{-1}\{C_3(2\,\mathrm{Fr}/\mathrm{Re}\,\mathrm{Dr})^{\frac{1}{2}}\}.$$

(iii) $\tilde{v}_z(\hat{z})=\mathrm{Dr}^{-1}[(1-\mathrm{Dr}\,l_{\mathrm{f}}/\mathrm{Ca}\,d_{\mathrm{f0}}C_4)\exp(C_4\hat{z})+\mathrm{Dr}\,l_{\mathrm{f}}/\mathrm{Ca}\,d_{\mathrm{f0}}C_4]^2$ where $\mathrm{Dr}^{\frac{1}{2}}=(1-\mathrm{Dr}\,l_{\mathrm{f}}/\mathrm{Ca}\,d_{\mathrm{f0}}C_4)\exp(C_4)+\mathrm{Dr}\,l_{\mathrm{f}}/\mathrm{Ca}\,d_{\mathrm{f0}}C_4$ respectively.

EXERCISE 15.3.3. Derive (15.3.22) from (3.3.13).

EXERCISE 15.3.4. Show that, from the solution (15.3.30),

$$l_{\mathrm{f}}=\frac{\mathrm{Dr}^{\frac{1}{2}}\,v_{z1}d_{\mathrm{f1}}^2}{\kappa\,\mathrm{Nu}(v_z=v_{z1},\,d_{\mathrm{f}}=d_{\mathrm{f1}})}\ln\left(\frac{T_0-T_{\mathrm{a}}}{T_{\mathrm{m}}-T_{\mathrm{a}}}\right)$$

from the requirement that $\hat{z}=1$ at the freeze point. Obtain the value of F. Carry out the same analysis for solution (15.3.34).

EXERCISE 15.3.5 (see Petrie, 1977). Show that, in the general fibre-spinning approximation, (15.3.54) with $\mu=0$ leads to the two constitutive equations

$$t_{zz}^{\mathrm{E}}+\lambda\left(\frac{\partial t_{zz}^{\mathrm{E}}}{\partial t}-v_z\frac{\partial t_{zz}^{\mathrm{E}}}{\partial z}-2\frac{\partial v_z}{\partial z}t_{zz}^{\mathrm{E}}\right)=2\lambda G\frac{\partial v_z}{\partial z}$$

$$t_{rr}^{\mathrm{E}}+\lambda\left(\frac{\partial t_{rr}^{\mathrm{E}}}{\partial t}-v_z\frac{\partial t_{rr}^{\mathrm{E}}}{\partial z}+\frac{\partial v_z}{\partial z}t_{rr}^{\mathrm{E}}\right)=-\lambda G\frac{\partial v_z}{\partial z}$$

Note that, in steady filament spinning, t_{rr}^{E}, t_{zz}^{E} are functions of z only and hence derive the form taken by (15.3.57) for $\mu=0$, using (15.3.17).

Now consider the case of unsteady extension of a uniform filament for which t_{rr}^{E}, t_{zz}^{E} are functions of t only and show that the constitutive equations become

$$t_{zz}^{E}+\lambda\left(\frac{dt_{zz}^{E}}{dt}+2\frac{d}{dt}\left(\ln\frac{A}{A_0}\right)t_{zz}^{E}\right)=-2\lambda G\frac{d}{dt}\left(\ln\frac{A}{A_0}\right)$$

$$t_{rr}^{E}+\lambda\left(\frac{dt_{rr}^{E}}{dt}-\frac{d}{dt}\left(\ln\frac{A}{A_0}\right)t_{rr}^{E}\right)=\lambda G\frac{d}{dt}\left(\ln\frac{A}{A_0}\right)$$

where $A(t)$ is the instantaneous cross-sectional area of the filament.

In a 'constant-force' experiment the axial force is

$$F=\begin{cases}F_0 & \text{for} \quad t\geqslant 0\\ 0 & \text{for} \quad t<0\end{cases}$$

Note that

$$t_{zz}^{E}-t_{rr}^{E}=F/A$$

Show that for all $t>0$ the dynamical equation becomes

$$\lambda^2\ddot{\gamma}=\lambda\dot{\gamma}\{1+\lambda\dot{\gamma}(1-3GA/F_0)-2\lambda^2\dot{\gamma}^2\}$$

where

$$\dot{\gamma}\equiv\frac{d\gamma}{dt}=\frac{d}{dt}\left(\ln\frac{A_0}{A}\right)$$

Show that, for long times, $A\to 0$, $\lambda\dot{\gamma}\to 1$.

Show that the constitutive equations can be written as

$$\frac{d}{dt}(A^2e^{t/\lambda}t_{zz}^{E})=-\lambda Ge^{t/\lambda}\frac{dA^2}{dt}$$

$$\frac{d}{dt}(A^{-1}e^{t/\lambda}t_{rr}^{E})=-\lambda Ge^{t/\lambda}\frac{d(A^{-1})}{dt}$$

and so, if $A=A_0$ for $t<0$,

$$t_{zz}^{E}-t_{rr}^{E}=Ge^{-t/\lambda}\left\{\frac{A_0^2}{A(t)^2}\left(1+\int_0^t e^{t'/\lambda}\frac{A(t')^2}{A_0^2}\frac{dt'}{\lambda}\right)-\frac{A(t)}{A_0}\left(1+\int_0^t e^{t'/\lambda}\frac{A_0}{A(t')}\frac{dt'}{\lambda}\right)\right\}$$

At $t=0$, by definition there is a discontinuity in F. Show that

$$\frac{F_0}{A_0}\left(\frac{A_0}{A(0)}\right)=G\left\{\left(\frac{A_0}{A(0)}\right)^2-\frac{A(0)}{A_0}\right\}$$

leading to a cubic equation

$$x^3 - \frac{F}{A_0 G} x^2 - 1 = 0$$

for the instantaneous elongational strain $A_0/A(0)$.

15.4 UNSTEADY SOLUTIONS. STABILITY

Most of the practical difficulties and weaknesses that arise in the fibre-spinning process come from unsteadiness. This can be of varying degrees of severity.

(a) Small fluctuations in filament thickness with a relatively large axial length scale; these may be due to variations in output rate or temperature at the spinneret, in the cooling process, or in the wind-up speed; no continuous process can be absolutely steady and so monitoring and control devices are normally used to sense departures from set conditions and to return them to the desired level. From the mathematical modelling point of view adopted in this text, such fluctuations can be studied by linearized sensitivity analysis: the effects of small variations in material parameters or operating conditions, such as η_0, ζ_m, v_{z0}, v_{z1}, Q, T_a, T_0, Nu on the final fibre diameter d_{f1} or its degree of orientation (or crystallinity) are obtained by linear perturbation methods.

(b) Large fluctuations in filament thickness with a correspondingly smaller axial length scale; these may be caused by small solid or highly-gelled particles in the extrudate, or by detachment of blobs of molten material from the extrudate swell region following accumulation on the exterior surface of the spinneret; they may also be caused by a phenomenon known as draw resonance, whereby small input variations can lead to large output variations (though with relatively large axial length scale). These effects can only be analysed fully using non-linear methods. Draw resonance will be discussed in some detail below; its onset is best described in terms of linearized stability analysis, which is very similar to the linearized sensitivity analysis mentioned under (a) above; successful non-linear theories have been based on the linearized stability analysis.

(c) Filament breakage. This is an extreme result of non-linear fluctuations and introduces the cohesive strength of the melt in question. As in the case of metals there can be both ductile and brittle fracture, and this will be briefly discussed below.

One well-known cause for filament instability and breakage in other contexts is capillary instability, discussed by Rayleigh and subsequently by many authors. It is unlikely to be relevant in most melt-spinning contexts and so will not be discussed further here.

Under certain circumstances, the flow through the spinneret can itself be unstable, and so give rise to relatively large fluctuations in v_{z0}, d_{f0} and Q. This matter has been treated in Section 9.1.

Further information can be obtained from Ziabicki (1976, 2.I, 2.IIIB and 3X), Petrie & Denn (1976), Pearson (1976*c*) and Denn & Pearson (1981).

15.4.1 Irregular Flow and Fracture

If the maximum cohesive strength of the material being spun is Σ_m, then 'brittle' fracture can be expected to occur whenever $t_{zz} - t_{rr} = \Sigma_m$, i.e. when

$$4F/\pi d_f^2 = \Sigma_m \tag{15.4.1}$$

However, a ductile form of fracture can arise (Petrie, 1975*a*) if a short segment l_0' of extrudate from the spinneret has an effective starting diameter, d_{f0}', sufficiently less than the steady-state value d_{f0}. Figure 15.3 illustrates diagrammatically the progress of such a short length (a

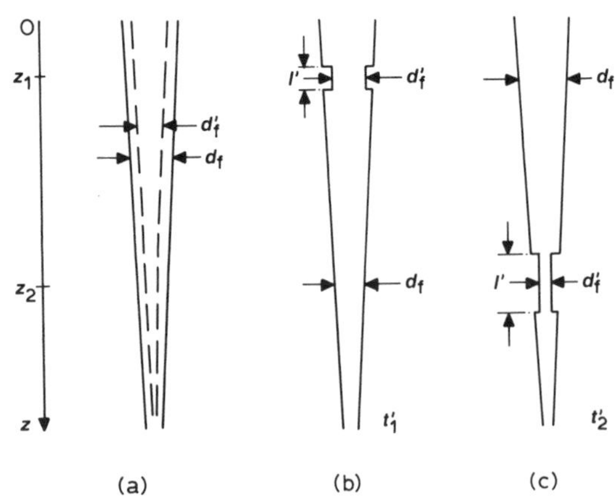

Fig. 15.3. Progress of weak region in filament spinning. (a) —— Profile of steady filament flow, d_f. --- Profile of unsteady weak spot, d_f'. (b), (c) Illustration of progress of weak spot of length $l'\{t'(z)\}$, $t_2'(z_2) > t_1'(z_1)$.

weak spot) as it passes down the thread-line. If the initial difference in diameter $d_{f0}-d'_{f0}$ is large enough, the weak spot parts before the wind-up is reached.

This can be shown analytically in the following fashion. We suppose that $l'_0 \ll l_f$ and so the tension in the thread-line is what it would be in the steady flow situation. If we take the isothermal Newtonian case for simplicity, then F is given by (15.3.21), and the flight time within the thread-line is given by (cf. eqn (15.3.6))

$$t_f = \int_0^{l_f} \frac{dz}{v_z(z)} \tag{15.4.2}$$

where

$$v_z = v_{z0}\,\mathrm{Dr}^{z/l_f} \tag{15.4.3}$$

Hence

$$t_f = \frac{l_f}{v_{z0}\ln \mathrm{Dr}}\left(1-\frac{1}{\mathrm{Dr}}\right) \tag{15.4.4}$$

If the initial thickness of the weak spot is given by $d'_{f0} < d_{f0}$, and we assume that F is constant, then

$$\frac{dv'_z}{dz} = \frac{1}{l'}\frac{dl'}{dt'} = \frac{F}{\eta_e}\frac{l'}{l'_0}\frac{4}{\pi d_{f0}^2} \tag{15.4.5}$$

is the equation governing the extension of the weak spot $(l'(t'))$. This may be integrated to give

$$\frac{1}{l'_0} - \frac{1}{l'} = \frac{4Ft'}{\pi\eta_e l'_0 d_{f0}^2} \tag{15.4.6}$$

Here t' is a Lagrangean time, as introduced by Kase (1981; see also Kase & Nakajima 1980; Petrie 1975*a*). Ductile fracture will arise if $l' \to \infty$ for $t' \leqslant t_f$, i.e. if

$$1 \leqslant \frac{4Fl_f}{v_{z0}\ln \mathrm{Dr}}\frac{(1-\mathrm{Dr}^{-1})}{\pi\eta_e d'^2_{f0}} = \frac{(1-\mathrm{Dr}^{-1})d_{f0}^2}{d'^2_{f0}} \tag{15.4.7}$$

or

$$d'_{f0} \leqslant d_{f0}(1-\mathrm{Dr}^{-1})^{\frac{1}{2}} \tag{15.4.8}$$

For large Dr this means that only a relatively small change in d_{f0} is needed to precipitate fracture.

In practice, breakage occurs because of (15.4.1), $l' \to \infty$ being incompatible with the requirement $l' \ll l_f$. Furthermore, $d_f(z)$ cannot be expected to show a discontinuity, but rather a slow variation with z at any fixed t. Nevertheless, the argument given above will apply to the thinnest portion.

15.4.2 Draw Resonance

The relevant unsteady flow equations are given in (15.2.4)–(15.2.6) above. The steady flow solutions are given in Section 15.3. Linearized perturbation equations are given in Appendix 1 for capillary flows, and in Section 14.3 for coating flows. Those relevant for fibre-spinning flow have been developed in a series of papers by Pearson & Matovich (1969), Kase (1974), Gelder (1971), Pearson & Shah (1972), Shah & Pearson (1972*a*, *b*, *c*), Fisher & Denn (1976), Ronca (1976*a*) and others.

The disturbance quantities a, b and c are defined by

$$\left.\begin{aligned} d_f(z,t) &= d_f^{(0)}(z)\{1+a(z,t)\} \\ v_z(z,t) &= v_z^{(0)}(z)\{1+b(z,t)\} \\ T(z,t) &= T^{(0)}(z)\{1+c(z,t)\} \end{aligned}\right\} \tag{15.4.9}$$

where $v_z^{(0)}$, $d_f^{(0)}$ and $T^{(0)}$ are the steady solutions in Section 15.3, and where

$$a, b, c \ll 1 \tag{15.4.10}$$

On substitution of (15.4.9) into (15.2.4)–(15.2.6) and on elimination of all terms of order (a^2, ab, b^2, ac, bc or c^2), a set of linear partial-differential equations for a, b and c are obtained. The solutions to these equations are determined by the initial and boundary conditions. For the sensitivity problem, a non-zero input boundary disturbance is imposed; for the stability problem, homogeneous (zero) boundary conditions are imposed, and so the problem becomes an eigenvalue one.

This is most simply illustrated for the isothermal Newtonian fluid (Pearson & Matovich, 1969). The steady solution (15.3.20) can be written

$$v_z^{(0)} = v_{z0} e^{\ln \mathrm{Dr}\, z/l_f} \tag{15.4.11}$$

(15.3.17), the continuity of tension condition, becomes

$$\frac{\partial v_z}{\partial z} = \frac{4F(t)}{3\pi\eta_0 d_f^2} \tag{15.4.12}$$

where $F(t)$ is now a function of time. The continuity of mass condition (15.2.4) becomes

$$\frac{\partial}{\partial t} d_{\mathrm{f}}^2 = \frac{\partial}{\partial z}(v_z d_{\mathrm{f}}^2) \tag{15.4.13}$$

Writing

$$\check{z} = \ln \mathrm{Dr}\, z/l_{\mathrm{f}}, \qquad \check{t} = v_{z0} \ln \mathrm{Dr}\, t/l_{\mathrm{f}} \tag{15.4.14}$$

and substituting (15.4.10) and (15.4.11) into (15.4.12) and (15.4.13) gives

$$\frac{\partial a}{\partial \check{t}} + \mathrm{e}^{\check{z}}\left(\frac{\partial a}{\partial \check{z}} + \frac{1}{2}\frac{\partial b}{\partial \check{z}}\right) = 0 \tag{15.4.15}$$

$$2a + b + \frac{\partial b}{\partial \check{z}} = f(\check{t}) \tag{15.4.16}$$

from which a single equation

$$\left(\frac{\partial}{\partial \check{t}} + \mathrm{e}^{\check{z}}\frac{\partial}{\partial \check{z}}\right)\frac{\partial a}{\partial \check{z}} = 0 \tag{15.4.17}$$

for a can be obtained. $f(\check{t})$ is an as yet undetermined constant.

The sensitivity problem corresponds in this simple case to a small input disturbance in velocity or diameter at the input, or in velocity at the output. By the usual method of Fourier decomposition, we can study separately the effect of small disturbances satisfying

$$\left.\begin{array}{lll} \text{(i)} & a(0, \check{t}) = a_0 \mathrm{e}^{\mathrm{i}\check{w}\check{t}}, & b(0, \check{t}) = b(\ln \mathrm{Dr}, \check{t}) = 0 \\ \text{(ii)} & b(0, \check{t}) = b_0 \mathrm{e}^{\mathrm{i}\check{w}\check{t}}, & b(\ln \mathrm{Dr}, \check{t}) = a(0, \check{t}) = 0 \\ \text{(iii)} & b(\ln \mathrm{Dr}, \check{t}) = b_1 \mathrm{e}^{\mathrm{i}\check{w}\check{t}}, & b(0, \check{t}) = a(0, \check{t}) = 0 \end{array}\right\} \tag{15.4.18}$$

where $\check{w}$ is a real positive frequency; a_0, b_0, b_1 can be complex, and it is assumed that the real part of the solution is taken. We can also study the effect of small variations in thread-line tension, through

$$\text{(iv)} \quad f(\check{t}) = f_0 \mathrm{e}^{\mathrm{i}\check{w}\check{t}}, \qquad a(0, \check{t}) = b(0, \check{t}) = 0 \tag{15.4.19}$$

Although all possible disturbances can be followed by suitable combinations of the boundary conditions (i)–(iv), this approach does not actually uncover the true mechanics of the thread-line near the spinneret, in the sense that a_0, b_0 and f_0 will be related in a specific way dependent upon the original source of the disturbance and the flow field in the readjustment region for which the fibre thread-line approximation (15.2.2), (15.2.3) does not apply. What we can expect to discover from the decomposition (i)–(iv) above is the region in $(a_0, b_0, b_1, f_0, \check{w})$ space for which the resulting thickness variations a_1 at the wind-up become large compared with the input disturbances $a_i = |a_0^2 + b_0^2 + b_1^2|^{\frac{1}{2}}$. We have an amplification factor

$$\mathrm{Am} = |a_1/a_i| \tag{15.4.20}$$

similar to that defined for coating flows in (14.3.39).

The phenomenon of draw resonance arises when a non-zero solution to (15.4.15), (15.4.16), i.e.

$$a(\ln \mathrm{Dr}, \check{t}) = a_1 e^{i\check{w}\check{t}}, \qquad a_1 \neq 0 \tag{15.4.21}$$

is possible for some real $\check{w}$ when there is no extraneous disturbance. An obviously draw-resonant situation based on the disturbances introduced in (15.4.18) would arise when

$$a_0 = b_0 = b_1 \equiv 0 \tag{15.4.22}$$

and indeed, almost all of the calculations carried out so far employ the homogeneous boundary conditions (15.4.22). It should be noted that these assume the readjustment (extrudate-swell or neck) region to be unaffected (to order a_1) by the non-zero disturbance $a(\check{z}, \check{t})$, $b(\check{z}, \check{t})$. Alternatives to (15.4.21) are hinted at in Pearson & Matovich (1969, Cases 1.C and 2.B) and in Ronca (1976), who introduces the concept of neck susceptivity, which is given by

$$\mathrm{Su} = \mathrm{d}b_0/\mathrm{d}f_0 \tag{15.4.23}$$

subject to

$$2a_0 + b_0 = 0 \tag{15.4.24}$$

Equation (15.4.24) implies that there is no accumulation of material in the neck, but a non-zero value of Su assumes that the extrudate swell varies with F, so

$$\mathrm{Su} \propto -\mathrm{d}\,\mathrm{Es}/\mathrm{d}F \tag{15.4.25}$$

Ronca takes the extreme values Su = 0 and Su = ∞ to illustrate possible effects.

Writing

$$a(\check{z}, \check{t}) = \tilde{a}(\check{z}) e^{i\check{w}\check{t}}, \qquad b(\check{z}, \check{t}) = \tilde{b}(\check{z}) e^{i\check{w}\check{t}} \tag{15.4.26}$$

where for the present $\check{w}$ is complex, substituting in (15.4.17), and using the first of (15.4.22) it follows by integration that

$$\tilde{a}(\check{z}) = A \int_0^{\check{z}} \exp(i\check{w}e^{-x})\, dx \qquad (15.4.27)$$

A being an arbitrary constant. From (15.4.14), using the second of (15.4.22), it then follows that

$$\tilde{b}(\check{z}) = -2A\left[\int_0^{\check{z}} \exp(i\check{w}e^{-x})\, dx + i\check{w}\int_0^{\check{z}} e^{-y}\int_0^{y} \exp(i\check{w}e^{-x})\, dx\, dy\right] \qquad (15.4.28)$$

Finally, using the third of (15.4.22) and rearranging, an implicit relation for $\check{w}$ in terms of Dr is obtained,

$$\int_{1/\mathrm{Dr}}^{1} e^{i\check{w}u}(1 - i\check{w}/\mathrm{Dr})u^{-1}\, du = e^{i\check{w}/\mathrm{Dr}} - e^{i\check{w}} \qquad (15.4.29)$$

Stability is ensured when $\mathscr{I}m(\check{w}) > 0$; instability arises when $\mathscr{I}m(\check{w}) < 0$. The neutral stability point is given by the least value of Dr for which $\mathscr{I}m(\check{w}) = 0$. It has been shown (Gelder, 1971) that this is given by

$$\mathrm{Dr}_{crit} = 20{\cdot}2 \qquad (15.4.30)$$

This prediction has been confirmed experimentally (see Denn, 1980) remarkably accurately and so suggests that for Newtonian fluids the assumption Su ~ 0, i.e. (15.4.22), is reliable. It may be shown (Denn, in Pearson & Richardson, 1983, Chapter 6; Pearson & Denn, 1981) that the limit Su $\to \infty$ is unconditionally stable.

The non-linear pair of equations (15.4.12), (15.4.13) subject to (15.4.22) can be solved, for Dr near Dr_{crit}, as shown in Fisher & Denn (1975). They note that, at any value of Dr, a set of eigenvalues $\check{w}_n$, corresponding to solutions of the linear perturbation equation (15.4.15), (15.4.16), (15.4.22), with progressively increasing values of $\mathscr{I}m(\check{w}_n)$, can be obtained from (15.4.29). Corresponding to these will be complex spatial functions $\tilde{a}_n(\check{z})$, $\tilde{b}_n(\check{z})$, and a general linear solution (real) will take the form

$$a = \sum_{n=1}^{\infty} \tilde{a}_n(\check{z})e^{i\check{w}\check{t}} + \tilde{a}_n^*(\check{z})e^{i\check{w}_n^*\check{t}} \qquad (15.4.31)$$

where (*) denotes a complex conjugate.

A suitable non-linear form is given by

$$a = \sum_{n=1}^{\infty} A_n(\check{t})\tilde{a}_n(\check{z}) + A_n^*(\check{t})\tilde{a}_n^*(\check{z}) \tag{15.4.32}$$

where the $A_n(\check{t})$ are complex functions of time to be obtained by direct numerical computation, which may be expected to tend to $\exp(i\check{w}_n\check{t})$ as

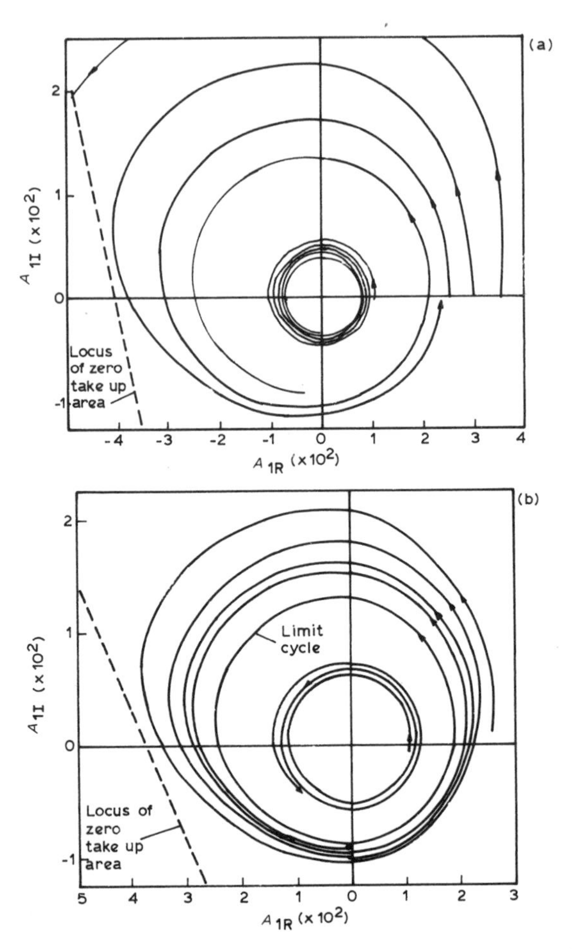

Fig. 15.4. Phase-plane trajectories for non-linear solution of Newtonian isothermal unsteady fibre spinning. A_{1R} and A_{1I} are the real and imaginary parts of the disturbance function A_1 defined by (15.4.31). Arrows indicate progress with time t. (a) Dr = 19·1 (b) Dr = 23·3. (From Fisher & Denn, 1975.)

$\mathrm{Dr} \to \mathrm{Dr}_{\mathrm{crit}}$. Because of the large value of $\mathscr{I}m(\check{w}_2)$, only the first term ($n = 1$) was retained by Fisher & Denn, who used a Galerkin approach to reduce the partial differential equations for a to two ordinary differential equations for A_1. Figure 15.4 (taken from Fisher & Denn) shows trajectories in the phase-plane (A_{1R}, A_{1I}) for different starting points. In (a), Dr is just below critical and so disturbances that are not so large that failure, basically of the type considered in Subsection 15.4.1, occurs at finite t will approach the steady state solution $A_1 \equiv 0$. In (b), Dr is just above critical, and it can be seen that disturbances approach a limit cycle with period close to that of the lowest eigenvalue $\check{w}_1(\mathrm{Dr}_{\mathrm{crit}})$. Figure 15.5 shows the actual values of normalized take-up force and cross-sectional area for Dr just above critical.

Figure 15.6 shows the ratio of maximum to minimum diameter for the limit cycle solution as a function of Dr, as given by Fisher & Denn. These compare well with results of Ishihara & Kase (1975).

The linearized perturbation scheme exemplified above for the simplest case can be applied to any of the steady-state solutions of eqns (15.3.1)–(15.3.3) by going back to the original equations (15.2.4)–(15.2.6). The neutral stability curves for several cases are given in

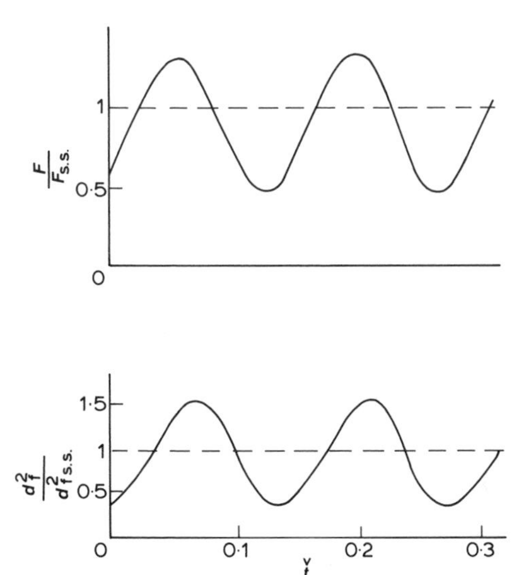

Fig. 15.5. Variation of tension F and cross-sectional area on a relative basis for the limit-cycle solution when Dr = 23·3 [Fig. 15.4(b)] (from Fisher & Denn, 1975).

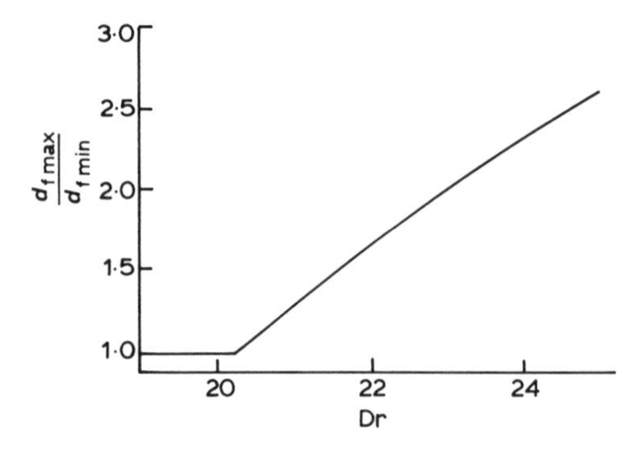

Fig. 15.6. Ratio of maximum to minimum diameter in the limit cycle solution as a function of Dr (from Fisher & Denn, 1975).

Pearson & Shah (1972), all using boundary conditions (15.4.22):

(a) Newtonian isothermal Re = 0(1), Re/We, Re/Fr, $\tilde{t}_{rz}^{(a-d)} \sim 0$ (includes inertial effects).
(b) Newtonian isothermal Re/Fr = 0(1), Re, Re/We, $\tilde{t}_{rz}^{(a-d)} \sim 0$ (includes gravity effects).
(c) Newtonian isothermal Re/We = 0(1), Re, Re/Fr, $\tilde{t}_{rz}^{(a-d)} \sim 0$ (includes surface tension effects since $\mathrm{We} = l_m v_{z1}^2 d_{f0}/\sigma_m \,\mathrm{Dr}^{\frac{1}{2}}$).

These are shown in Fig. 15.7 (with numerical improvements) for which it can be seen that inertia is strongly stabilizing, gravity more

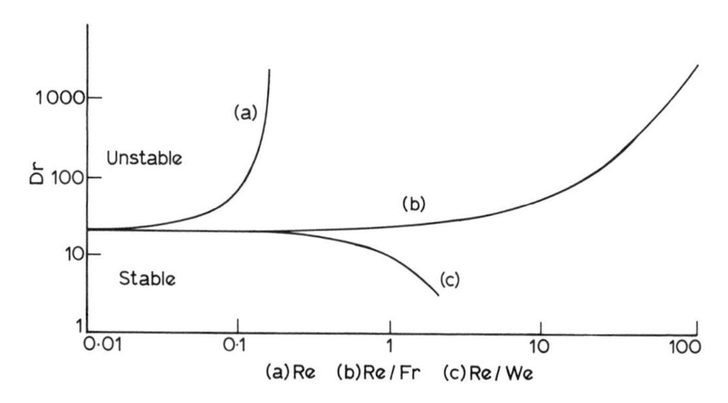

Fig. 15.7. Neutral stability curves for: (a) viscosity/inertia-dominated; (b) viscosity/gravity-dominated; (c) viscosity/surface tension-dominated fibre-spinning flows with no freezing. (From Shah & Pearson, 1972*b*.)

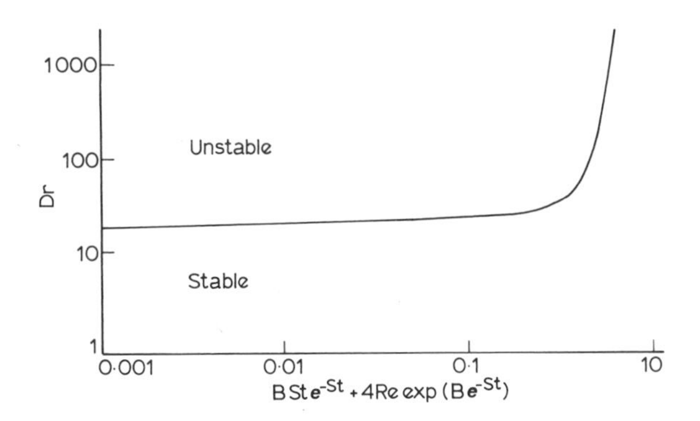

Fig. 15.8. Approximate neutral stability curve for temperature-dependent Newtonian fluid cooling without freezing in a fixed ambient environment (these results have since been improved by other workers). (From Shah & Pearson, 1972*b*.)

weakly so and surface tension destabilizing, the last-named result being happily consistent with the known phenomenon of Rayleigh instability of columns of fluid.

(d) Non-isothermal Newtonian without freezing (eqn (15.3.30)) provides the steady-state solution. Two separate dimensionless groups, B and St, are needed to describe the model; a composite parameter $\mathrm{B\,St\,e^{-St}}$ was found to lead to a useful grouping of neutrally-stable values of Dr about the line shown in Fig. 15.8.

An extension to include inertial effects led, less conclusively, to the alternative parameter $\mathrm{B\,St\,e^{-St}} + 4\,\mathrm{Re}\,\exp(\mathrm{Be^{-St}})$.

The stability of isothermal and non-isothermal power-law fluids without freezing is considered in Pearson & Shah (1974) (see steady-state solution (15.4.34)) from which it can be seen that $\nu > 1$ implies greater stability and $\nu < 1$ less stability to draw resonance, as shown in Fig. 15.9.

Fisher & Denn (1976), in modelling the laboratory experiments of Zeichner on a polystyrene melt, use a form of the White–Metzner model (3.3.35) where the relaxation time λ and hence the simple-shear viscosity is of power-law form. The Deborah number they use is

$$\mathrm{De}_0 = \frac{\mathrm{K}}{G}\left(\frac{v_{z0}}{l_{\mathrm{f}}}\right)^{\nu} \tag{15.4.33}$$

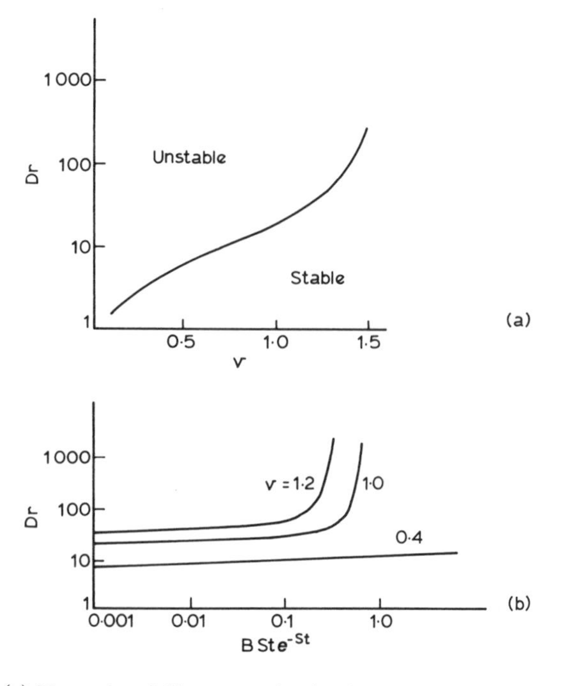

Fig. 15.9. (a) Neutral-stability curve for isothermal fibre-spinning of a power-law fluid as a function of the power-law index ν. (b) Approximate neutral-stability curves for non-isothermal fibre-spinning of power-law fluids cooling without freezing for various values of ν. (From Pearson & Shah, 1974.)

For the relatively small values of $\mathrm{Dr} = 0(10)$ they consider and with $\nu = \frac{1}{3}$, De_0 is not very different from De defined by (15.3.55), and is of order unity. Predictions compare relatively poorly with experiment. They also consider the stability of the steady solution to infinitesimal disturbances, and obtain the temporal eigenvalues corresponding to $\check{w}$ above. This leads to the stability diagram shown in Fig. 15.10. They have carried out non-linear dynamical calculations based on a form similar to (15.4.32) to yield limit cycles from which the ratio of maximum diameter to minimum diameter can be obtained. These were compared with low Deborah number flows involving PET, and excellent agreement was apparently obtained. It is difficult to comment on these published theories because relatively little detail is given.

A similar series of calculations applied to specific steady-state and

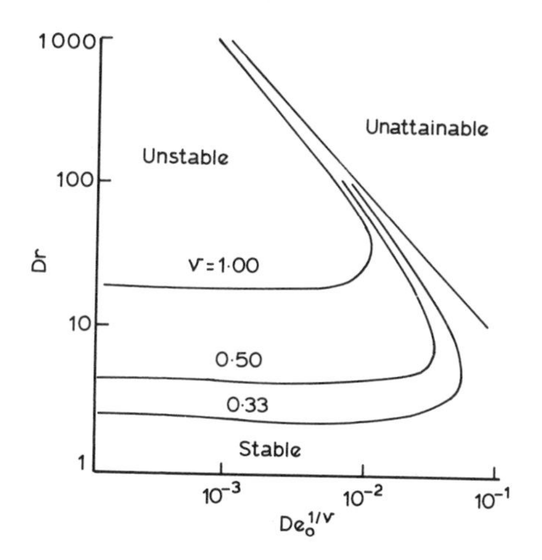

Fig. 15.10. Neutral-stability plots for isothermal fibre-spinning of White–Metzner fluid (from Fisher & Denn, 1975).

draw-resonance experiments have been reported by Kase and co-workers (Kase and Ikko, 1981; Kase, 1981; see also Denn & Pearson, 1981). As with the early calculations of Shah & Pearson, direct finite-difference integration procedures were used, rather than the eigen-function expansions employed by Denn and his co-workers. Very similar conclusions were drawn to those mentioned above. Kase's work is closely associated with practical fibre-spinning.

In practice, draw resonance does not always arise as it might be expected to according to the linear-stability calculation mentioned above. One reason for this is that the freezing phenomenon was not included. Pearson *et al.* (1976) used a very simple freezing model, implying infinite rates of crystallization or a sudden glass transition, to study the effect on stability of small variations of the freeze point; this implies that l_f will vary with time as $e^{i\tilde{\omega}\tilde{t}}$, because of the temperature perturbation $c(z, t)$ in (15.4.9). They concluded that fibre-spinning of a power-law fluid with no temperature dependence of viscosity would be unconditionally stable provided freezing took place at a fixed temperature on the thread-line and the frozen thread was completely rigid. Adjustments to the solution to include elasticity in the solid thread allowed of instability for very long thread-lines. In the limit of low elastic modulus, this corresponds to a constant tension wind-up boundary condition rather than a constant velocity. They note that their result is consistent with the theoretical predictions and experiments of Ishihara & Kase (1975) on the stability of freezing and non-freezing thread-lines.

Dynamical simulation of the unsteady flow of a Maxwell-model viscoelastic fluid is reported in Kase & Denn (1978).

EXERCISE 15.4.1. Carry out the analysis given in Subsection 15.4.1 for a power-law fluid as defined by (15.3.32) with $\zeta_m = 0$.

Show first that the undisturbed velocity becomes

$$\tilde{v}_z = \mathrm{Dr}^{-1}\{(\mathrm{Dr}^{1-1/\nu} - 1)\hat{z} + 1\}^{\nu/(\nu-1)}$$

hence that

$$t_f = \frac{l_f}{v_{z0}} \frac{(\nu-1)(1-\mathrm{Dr}^{-1/\nu})}{(\mathrm{Dr}^{1-1/\nu}-1)}$$

and finally, that the analogue of (15.4.8) is

$$d'_{f0} \leq d_{f0}(1-\mathrm{Dr}^{-1/\nu})^{\nu/2}\left[\frac{(\nu-1)}{(1-\mathrm{Dr}^{1/\nu-1})}\right]^{(\nu-1)/2}$$

EXERCISE 15.4.2. Express $f(t)$ introduced by (15.4.16) in terms of $F(t)$ given in (15.4.12) and its unperturbed value $F^{(0)}$ given by (15.3.21).

EXERCISE 15.4.3. Derive (15.4.29) from (15.4.28) and the third of (15.4.22).

EXERCISE 15.4.4. Show that the Deborah number De_0 defined in (15.4.33) would be a Weissenberg number, defined as $(t_{11} - t_{22})/t_{12}$, for a viscometric flow at shear rate v_{z0}/l_f. Show also that it can be written

$$\mathrm{De}_0 = \lambda(v_{z0}/l_f)v_{z0}/l_f$$

where λ is as defined in (3.3.35) and hence obtain its relationship to Dr defined in (15.3.55).

EXERCISE 15.4.5. Starting with the equations

$$\frac{\partial A}{\partial t} + \frac{\partial}{\partial z}(vA) = 0 \quad \text{and} \quad \frac{\partial v}{\partial z}A = \hat{F}$$

where $A \propto d_f^2$ and $\hat{F}$ is a suitable form of tension, show by elementary arguments that no draw resonance can take place at constant tension, i.e. $\hat{F} = \text{constant}$.

15.5 FIBRE DRAWING

This is a continuous cold drawing process (see Ziabicki (1976, Chapter 6) for a detailed account), leading to plastic deformation of the already spun, and partially oriented, fibre; elongation ratios of from 2 to 10, i.e. a Hencky strain of 1 or 2, are typical. The mechanical properties of the undrawn fibre can be studied in slow tensile experiments and usually fall into one of three categories, shown in Fig. 15.11, taken from Ziabicki (1976, Fig. 6.1). The conventional stress is based on the original cross-sectional area of the (undrawn) filament while the true stress takes account of the change in cross-sectional area that arises during drawing. The two most important classes of behaviour are exemplified by the conventional stress curves (b) and (c), the first of which is single-valued and unlikely to show any non-uniformity in a tensile test, whereas the second is multi-valued in strain for a limited range of stress, and leads to necking. This latter is illustrated in Fig. 15.12. Any given material can show both types of behaviour, depending on the temperature at which the tensile test is carried out. This is illustrated by Fig. 15.13, also taken from Ziabicki (1976, Fig. 6.2). Much of the early elongation is purely elastic, and

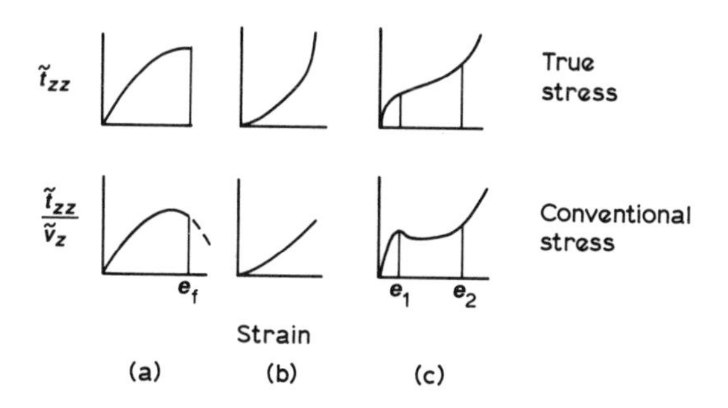

Fig. 15.11. Stress–strain curves for three different forms of elongational behaviour. (a) Stiff material showing plastic fracture at breaking elongation e_f. (b) Strain-hardening material. (c) Necking material showing turning point in true stress–strain curve. The elongation jumps from e_1 to e_2 across the neck (from Ziabicki, 1976).

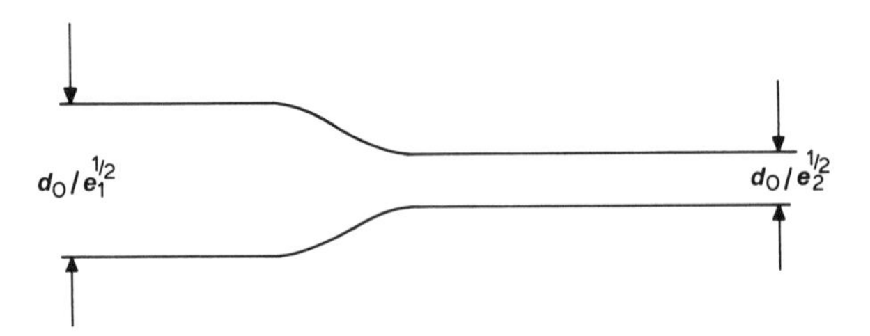

Fig. 15.12. Stable neck observed in slow tensile-test on fibre of initial diameter d_0. The change in diameter takes place in a length short compared with the total fibre length. This relates to case (c) in Fig. 15.11.

therefore recovered on release of stress in an isothermal experiment; elongation taking place during necking is essentially plastic, i.e. not recovered on release of stress in an isothermal experiment. However, the strong orientation imposed on the polymer chains during the necking process often means that additional recovery takes place on raising the temperature sufficiently close to the 'melt' point.

Commercial fibre-drawing lines are run very fast, at wind-up speeds of 50 m/s or more. Consequently viscous effects during drawing cannot be entirely neglected, while significant heating (of the order of 50 K) due to irreversible deformation takes place. Most successful drawing operations are based on localized regions of draw, often initiated by a slight rise in ambient temperature, e.g. by passing over a heated pin.

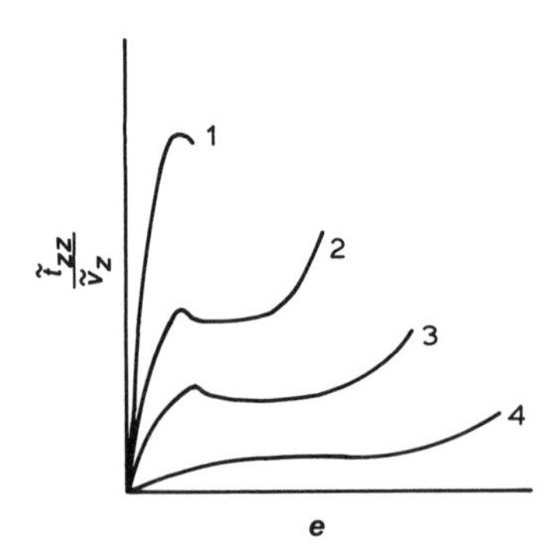

Fig. 15.13. Conventional stress/strain relation for polymeric fibre over a wide temperature range increasing from 1 to 4. Note that deformation changes from type (a) to type (c) and finally to type (b) (as defined in Fig. 15.11). Drawing takes place in the temperature range 2 to 3.

There is no satisfactory rheological theory for rapid cold-drawing and so there have been relatively few attempts to model the continuous process, of the type discussed in Sections 15.2–15.4 for the melt-spinning process: eqns (15.2.4)–(15.2.6), with boundary conditions (15.3.7) and (15.3.8), will still apply. Indeed, the entry boundary condition (15.3.8) can probably be applied with more confidence in the drawing process than in the melt-spinning one.

Although various forms of the Maxwell model have been applied to the drawing process, the observed rise of yield stress with elongation rate (Ziabicki, 1976, Fig. 6.5) suggests that a Kelvin–Voigt model is much more appropriate. This is one in which the true stress is the sum of an elastic stress based on the total elongation plus a viscous stress based on the instantaneous rate of extension. Plastic yield, or necking, can be accounted for by putting an upper limit G_{max} on the elastic stress component. The relevant elongational relation for a Kelvin–Voigt model, using the nomenclature of Section 15.3, in an isothermal situation (see Example 15.5.1)

$$\tilde{F}\tilde{v}_z = \frac{d\tilde{v}_z}{d\hat{z}} + \frac{\tilde{G}_{\text{max}}}{\text{Dr}} - H\left[\frac{\tilde{G}_{\text{max}}}{\text{Dr}} - \tilde{G}\left(\tilde{v}_z - \frac{1}{\text{Dr}}\right)\right] \tag{15.5.1}$$

where

$$G = \tilde{G}\eta_e v_{z\,1}/l_f \tag{15.5.2}$$

is the elastic modulus and

$$H[x] = \begin{cases} x & \text{if} \quad x > 0 \\ 0 & \text{if} \quad x \leqslant 0 \end{cases} \tag{15.5.3}$$

Figure 15.14 shows the rheological behaviour graphically for $d\tilde{v}/d\hat{z} = 0$. This can easily be solved to give

(i) for $\text{Dr}\,\tilde{v}_z - 1 \leqslant G_{\text{max}}/G = \hat{G}_{\text{max}}$

$$\tilde{v}_z = \frac{\tilde{F}\exp\{(\tilde{F}-\tilde{G})\hat{z}\} - \tilde{G}}{(\tilde{F}-\tilde{G})\text{Dr}} \tag{15.5.4}$$

(ii) for $\text{Dr}\,\tilde{v}_z - 1 > G_{\text{max}}/G$

$$\tilde{v}_z = \frac{\tilde{F}_0\exp\{\tilde{F}\hat{z}\} + \tilde{G}_{\text{max}}}{\tilde{F}\,\text{Dr}} \tag{15.5.5}$$

where $\tilde{F}_0$ is determined by the requirement that $\tilde{v}_z$ be continuous at

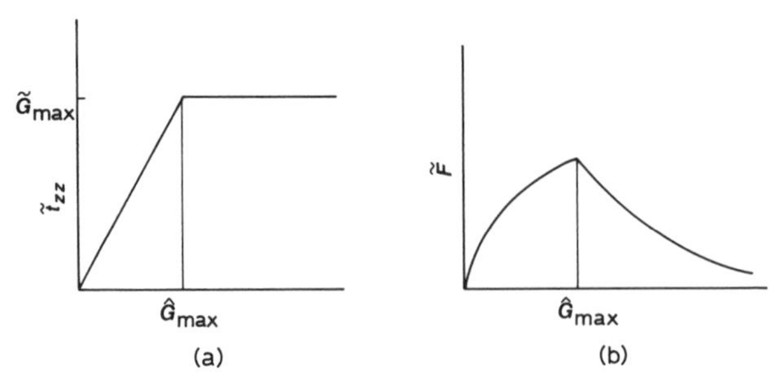

Fig. 15.14. Stress-elongation curves for Kelvin–Voigt model. (a) True stress, $\tilde{t}_{zz} = \tilde{F}\tilde{v}_z$. (b) Engineering stress, $\tilde{F} = \tilde{t}_{zz}/\tilde{v}_z$. The elongational strain $\hat{G} = \mathrm{Dr}\,\tilde{v}_z - 1$. (From Ziabicki, 1976.)

$\tilde{v}_z = \mathrm{Dr}^{-1}(1 + G_{max}/G)$, i.e.

$$\tilde{F}_0 = \tilde{F}\left[1 + \frac{\tilde{G}_{max}}{\tilde{G}} - \frac{\tilde{G}_{max}}{\tilde{F}}\right]^{\tilde{G}/(\tilde{G}-\tilde{F})} \tag{15.5.6}$$

$\tilde{F}$ is given by the requirement that $\tilde{v}_z(1) = 1$.

If $\tilde{G}$, $\tilde{G}_{max} \gg 1$, which is probably realistic, then the asymptotic solutions are

$$\text{(i)}\quad \tilde{v}_z \sim \left[1 - \left(1 - \frac{1}{\mathrm{Dr}}\right)\exp\{-\tilde{G}\hat{z}/\mathrm{Dr}\}\right] \text{ when}$$

$$\mathrm{Dr} \leqslant (1 + \hat{G}_{max}) \tag{15.5.7}$$

$$\text{(ii)}\quad \mathrm{Dr}\,\tilde{v}_z \sim \begin{cases} (1+\hat{G}_{max}) - \hat{G}_{max}\exp\{-\tilde{G}\hat{z}/(1+\hat{G}_{max})\} \\ \quad \text{for} \quad \tilde{v}_z \leqslant \mathrm{Dr}^{-1}(1+\hat{G}_{max}) \\ (1+\hat{G}_{max}) + [\mathrm{Dr} - (1+\hat{G}_{max})]\exp\{\hat{G}_{max}(\hat{z}-1)/(1+\hat{G}_{max})\} \\ \quad \text{for} \quad \tilde{v}_z > \mathrm{Dr}^{-1}(1+\hat{G}_{max}) \\ \quad \text{when } \mathrm{Dr} > (1+\hat{G}_{max}) \end{cases} \tag{15.5.8}$$

Case (i) represents a very rapid extension from $\tilde{v}_z = \mathrm{Dr}^{-1}$ to $\tilde{v}_z = 1$ near $\hat{z} = 0$, while case (ii) involves rapid extensions near both $\hat{z} = 0$ and $\hat{z} = 1$.

As expected the strain hardening case (i) (a form of behaviour (b) in Fig. 15.11) yields a near-discontinuity at $\hat{z} = 0$ that corresponds to the Maxwell-model result (15.3.61) with $\hat{\mu} \to 0$, when $\mathrm{De} \to \infty$. Both are extreme limits of the Jeffreys or White–Metzner models.

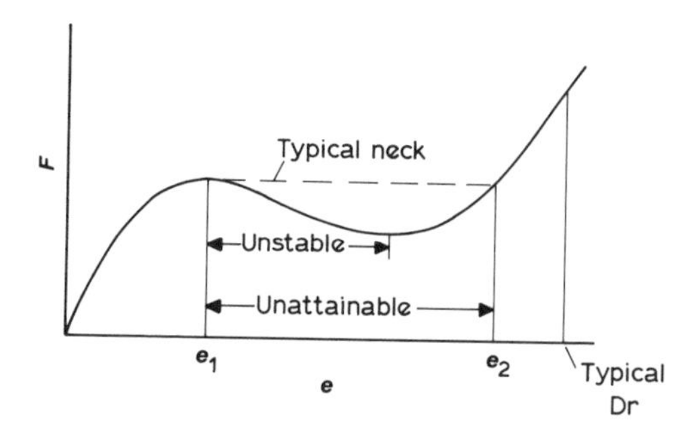

Fig. 15.15. Tension vs elongation for real drawing operation, taking account of adiabatic heat release. The region between e_1 and e_2 is unattainable in steady operation.

The plastic yielding case (ii) (an extreme form of behaviour (c) in Fig. 15.1) yields more complex behaviour as might be expected. It is not at all clear that the steady solution (15.5.8) is stable to small disturbances; indeed, it might be expected that it would be catastrophically unstable because the conventional (elastic) stress in (15.5.1) tends to zero as the strain tends to infinity. Real materials tend to show strain hardening (in terms of the conventional as well as the true stress) and so thread-line breakage would not necessarily occur, even if unsteady flow might arise. Irreversible dissipation of energy as heat tends to reduce the isothermal strain hardening effect, and in the limit of zero heat transfer an adiabatic rheological stress–strain curve becomes relevant (see Ziabicki, 1976, Fig. 6.10). It is therefore likely that the real fibre-drawing processes can be modelled rheologically by a conventional stress–strain curve of the type shown in Fig. 15.15; for slow drawing rates, when viscosity can be largely neglected, the portion in which dF/de is negative is essentially unstable, for reasons similar to those given in Section 9.1 (cf. Fig. 9.3); it may therefore be assumed that actual steady stable operations involve a draw ratio outside that region, and one above $\mathrm{Dr} = e_2$ if a substantial stable neck is to be formed. However, this matter has not been fully investigated. A recent paper by Bernstein & Zapas (1981) shows that BKZ models may be suitable for describing necking. Some interesting effects involving non-uniform extension are described in Adrianova *et al.* (1971).

EXAMPLE 15.5.1. Fibre-drawing induced by a sudden rise in temperature. The analysis given in Section 15.5 above is based on a given entry velocity $\tilde{v}_{z0} = \mathrm{Dr}^{-1}$, but also on a given initial strain of zero. This can be justified as a reasonable approximation either:

(i) by supposing that elongation can be restrained physically until $\hat{z} = 0$, say, by a pair of rollers gripping the solid fibre;
(ii) by assuming that a sudden sharp rise in temperature at $\hat{z} = 0$ causes a sudden decrease in both Young's modulus and tensile strength of the fibre.

On simple continuity grounds, the engineering strain at a subsequent stage of a steady drawing process will be given by

$$(\tilde{v}_z/\tilde{v}_{z0}) - 1 = \mathrm{Dr}\,\tilde{v}_z - 1$$

since $\tilde{v}_{z0} = \mathrm{Dr}^{-1}$ from (15.3.16).

The rheological model used in (15.5.1) supposes that in equilibrium, the elongational stress $t_{zz} = F(\tilde{v}_z/v_{z0})$ will be linearly proportional to the engineering strain, with Young's modulus G, up to a maximum stress G_{max}, which is equivalent to a maximum strain $\hat{G}_{max}$. Thereafter the material yields. There are two components to the stress during deformation: one from a purely viscous mechanism with elongational viscosity η_e and the other from the purely elasto–plastic behaviour already described.

Equations (15.5.4) and (15.5.5) represent direct integrals of the dynamical equation (15.5.1), where (15.5.4) satisfies the inlet boundary condition on velocity. The outlet boundary condition on (15.5.5) implies that

$$\tilde{F}\,\mathrm{Dr} = \tilde{F}_0 \mathrm{e}^{\tilde{F}} + \tilde{G}_{max}$$

which taken with (15.5.6) leads to a complex implicit relation for $\tilde{F}$ in terms of Dr, $\tilde{G}$ and $\tilde{G}_{max}$.

If $\tilde{G}$, and $\tilde{G}_{max} \gg 1$, which means that viscous forces are weak though not negligible, which is the circumstance likely to lead to necking, then it can be assumed that the true stress in the fibre can never exceed G_{max} by more than a very small amount, and that will arise when plastic flow takes place, i.e. when $\mathrm{Dr} > (1 + \hat{G}_{max})$. The asymptotic result (15.5.8) can be derived on the basis that $\tilde{F} \sim \tilde{F}_{max} = \hat{G}_{max}/(1 + \hat{G}_{max})$ and that the exponential terms in (15.5.4) and (15.5.5) are only significant near $\hat{z} = 0$ and $\hat{z} = 1$ respectively. The result (15.5.7) follows even more simply.

Chapter 16

Film Stretching and Casting

The production of flat thin sheets of thermoplastics (and to a much lesser extent, of rubbers) is a major activity in the polymer industry. Flexibility, toughness and often transparency are their important characteristics. The sheets produced vary in modulus from readily extensible elastomers to almost inextensible biaxially-drawn semi-crystalline thermoplastics. Typically thicknesses range from 20 μm to 2 mm, with lateral dimensions varying from 10^{-1} m for plastic bags to 10 m for wound rolls used in building operations. The ratio of width (or length) to thickness is therefore of order 10^3. The flat sheet approximation derived in Section 8.2 will therefore apply, even during the early stages of the process when the ratios are somewhat less.

16.1 FILM CASTING

This section considers the extrusion and extension of a flat sheet of molten polymer which freezes either by falling onto a chilled metal roll or by passing into a water bath. In both cases, the freezing line occurs at a fixed position below the slot die, where a fixed velocity is imposed. Figure 16.1 shows the general arrangements and Fig. 16.2 the plan view when flow is effectively vertical, with a (x_1, x_2)-coordinate system in the plane of the sheet. The initial width of the sheet is w_0 at the die-lips; the length of the region of extension is l. As in the case of fibre spinning, there will be a short adjustment region near the die-lips where a confined simple-shear flow involving a large velocity gradient in the x_3-direction changes to a largely extensional-flow field with no velocity gradient in the x_3-direction. This will be a region of extrudate swell of the type discussed in Section 7.4 for capillary flow. The

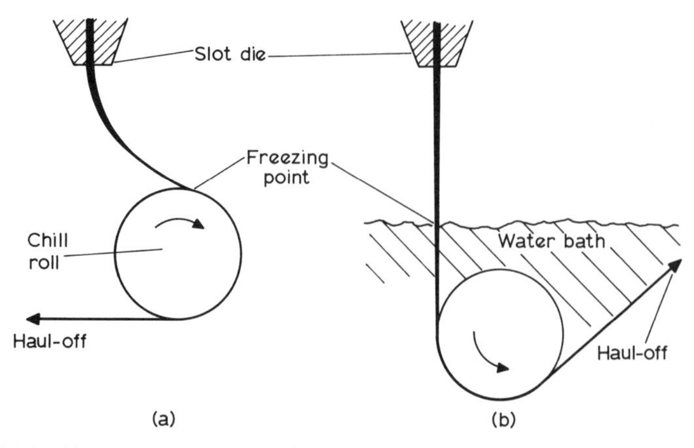

Fig. 16.1. (a) Chill-roll casting (sectional view) showing some gravitational bending of the extending sheet when the chill roll is not vertically below the die-lips. (b) Water-bath casting (sectional view) showing the sheet drawn vertically downwards.

approach of Tanner or Pearson and Trottnow can be used for the plane flow situation involved in slot extrusion. For our present purposes, we suppose that the sheet thickness h_0 and the exit velocity v_{10} are specified at the die-lips ($x_1 = 0$). The wind-up velocity v_{11} at $x_1 = l$ is also specified. We again write

$$\mathrm{Dr} = v_{11}/v_{10} \tag{16.1.1}$$

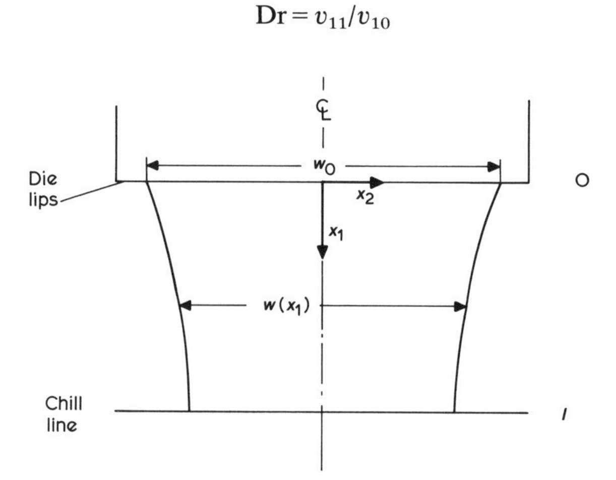

Fig. 16.2. Plan view showing the die-lips at $x_1 = 0$ and the chill line at $x_1 = l$; $x_2 = 0$ is the centre-line and $w(x_1)$ is the width of the sheet.

In most cases of interest, $w_0 > l \gg h_0$, and Dr will lie between 2 and 20. There is no need to achieve significant cooling of the melt between die-lips and water bath or chill roll, so heat transfer is not a dominant feature of the process, as it was in melt spinning of thin fibres.

A Stanton number can be defined as in (15.3.27), but this requires the Nusselt number to be known for the air-side heat-transfer process. A Graetz number for heat transfer within the sheet can be defined as

$$\mathrm{Gz} = qh^*/\kappa l \tag{16.1.2}$$

where q is the volume flux/unit width, and h^* is a characteristic sheet thickness. If the draw ratio Dr is large, then Gz will be relatively ill-defined. This distinguishes the sheet extension case from the fibre-drawing case; in the latter, as (15.2.1) shows, the Graetz number is well-defined. For $q = 10^{-3}\,\mathrm{m^2\,s^{-1}}$, $h^* = 10^{-3}\,\mathrm{m}$, $l = 1\,\mathrm{m}$ and $\kappa = 10^{-7}\,\mathrm{m^2\,s^{-1}}$, using (16.1.2), $\mathrm{Gz} = 10$, which is substantially higher than typical values obtained ($\mathrm{Gr} \leqslant 1$) in the fibre-spinning case and suggests that the temperature distribution within the extending sheet cannot be taken to be independent of x_3. However, the effect of such a temperature variation on an extensional sheet flow is much less important than in a corresponding confined shear flow; for simple calculation purposes a mean temperature

$$\bar{T}(x_1, x_2) = \int_{-\frac{1}{2}h}^{\frac{1}{2}h} T(x_1, x_2, x_3)\,\mathrm{d}x/h \tag{16.1.3}$$

is probably adequate for use in any rheological equation of state such as (8.2.21) defining the material; this equation of state provides the surface stress tensor $\mathbf{T}$ given by (8.2.6) in terms of a history of deformation provided by a surface velocity distribution (v_1, v_2), both being functions of (x_1, x_2) only. $\bar{T}$ can itself be estimated by using an appropriate Nusselt number for heat transfer to the ambient environment.

The feature that distinguishes our film casting flow from a plane flow in the (x_1, x_2) plane is the change in width $w(x_1)$ that occurs on drawing down, and the consequential variation in thickness h with lateral position x_2 at the freeze line. The objective of the industrial process is to achieve a constant thickness, h_1, at the wind-up; the starting point is usually a uniform flow at the die-lips providing a starting thickness h_0 independent of x_2. A simple mass balance yields

$$\bar{h}_1 = \frac{\rho_s w_0 h_0}{\rho_m w_1 \,\mathrm{Dr}} \tag{16.1.4}$$

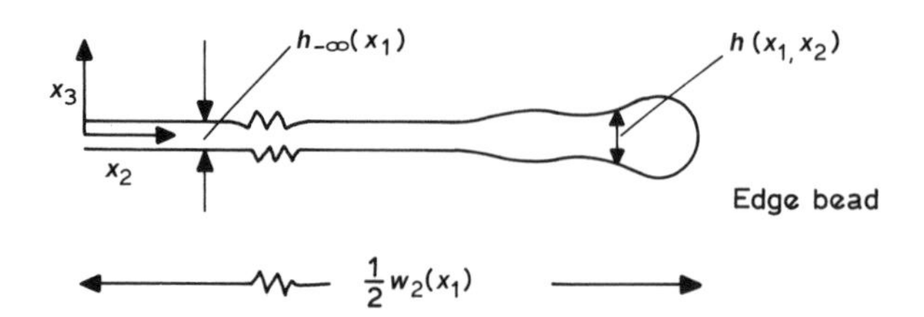

Fig. 16.3. Cross-section of an extending sheet showing thickening of the edge (edge beading) as observed in practice.

account being taken of the density change on freezing, neck-in from w_0 to w_1 and the x_1-velocity ratio Dr. It is observed in practice that the sheet forms an edge-beading as shown in Fig. 16.3, and that this apparently plays a role in reducing neck-in and promoting uniformity in h as a function of x_2; indeed, in some production lines, such a beading is deliberately created at the die-lips by opening out the die-gap at the ends of the slot-die; the thick edge on the frozen sheet is then trimmed off, and reprocessed.

The elastic nature of the flow can be estimated in terms of a Deborah number as defined in (15.3.24) for fibre-spinning.

16.1.1 Model Equations and Boundary Conditions: General Formulation

As in Section 8.2 we consider a surface velocity field

$$\mathbf{v} = \{v_1(x_1, x_2, t), v_2(x_1, x_2, t)\} \tag{16.1.5}$$

associated sheet thickness $h(x_1, x_2, t)$, and a tension tensor $\mathbf{T}(x_1, x_2, t)$ $(T_{ij} = ht_{ij}; i, j = 1, 2)$.

The mass conservation equation is

$$\frac{\partial}{\partial t}(\rho_m h) + \frac{\partial}{\partial x_1}(\rho_m h v_1) + \frac{\partial}{\partial x_2}(\rho_m h v_2) = 0 \tag{16.1.6}$$

The momentum conservation equation becomes, in scalar form

$$\frac{\partial}{\partial t}(\rho_m h v_1) + \frac{\partial}{\partial x_1}(\rho_m h v_1^2) + \frac{\partial}{\partial x_2}(\rho_m h v_1 v_2) = \rho_m g_1 h + \frac{\partial T_{11}}{\partial x_1} + \frac{\partial T_{12}}{\partial x_2} \tag{16.1.7}$$

$$\frac{\partial}{\partial t}(\rho_m h v_2) + \frac{\partial}{\partial x_1}(\rho_m h v_1 v_2) + \frac{\partial}{\partial x_2}(\rho_m h v_2^2) = \frac{\partial T_{12}}{\partial x_1} + \frac{\partial T_{22}}{\partial x_2}$$

where g_1 is the component of gravity in the x_1-direction.† Surface tension and air drag are neglected because they will be very small in practice. If inertia can also be neglected and the density is constant, then eqns (8.12.19) and (8.2.13) are recovered. The energy-conservation equation, using the averaged temperature defined by (16.1.3), becomes

$$\frac{\partial}{\partial t}(\rho_m \Gamma_h h\bar{T}) + \frac{\partial}{\partial x_1}(\rho_m \Gamma_m h\bar{T}v_1) + \frac{\partial}{\partial x_2}(\rho_m \Gamma_m h\bar{T}v_2) = 2h_a(T_a - \bar{T}) + T_{11}\frac{\partial v_1}{\partial x_1} + T_{12}\left(\frac{\partial v_1}{\partial x_2} + \frac{\partial v_2}{\partial x_1}\right) + T_{22}\frac{\partial v_2}{\partial x_2} \quad (16.1.8)$$

If generation is small, then the last three terms can be neglected, and a linear equation for $\bar{T}$ is obtained, based on a single-sided heat-transfer coefficient h_a.

The boundary conditions for the flow are:

(i) inlet, $x_1 = 0$: $v_1 = v_{10}(x_2, t)$, $v_2 = 0$

$$\bar{T} = \bar{T}_0(x_2, t),\ h = h_0(x_2, t) \quad (16.1.9)$$

(ii) exit, $x_1 = l$: $v_1 = v_{11}(t)$, $v_2 = 0$ (16.1.10)

(iii) edges, at $x_2 = \pm\frac{1}{2}w(x_1, t)$ if symmetry is preserved:

$$\mathbf{T}\cdot\hat{\mathbf{n}} = \mathbf{0}, \quad \text{where} \quad \hat{\mathbf{n}} = (-\sin\theta, \cos\theta),\ \tan\theta = \partial w/\partial x_1 \quad (16.1.11)$$

There is the additional kinematic condition for the position of the edge

$$\frac{\partial w}{\partial t} + v_1\frac{\partial w}{\partial x_1} = v_2 \quad (16.1.12)$$

The vector relation (16.1.11) requires that there be no force across the edge regarded as a linear boundary in the (x_1, x_2)-plane. To the extent that the sheet has finite thickness h, this is physically equivalent to an edge-surface perpendicular to the (x_1, x_2)-plane, as shown in Fig.

† This formulation presents no difficulties if x_1 is vertical, in which case $g_1 \equiv g$. If, however, the falling film is drawn away from the vertical, then it will not in general be plane; the present set of equations will, however, still be relevant when $v_2 \approx 0$, for x_1 can then be interpreted as a coordinate measured along the falling film in a vertical plane and g_1 will be the component of gravity in that direction; l will not be prescribable *a priori*.

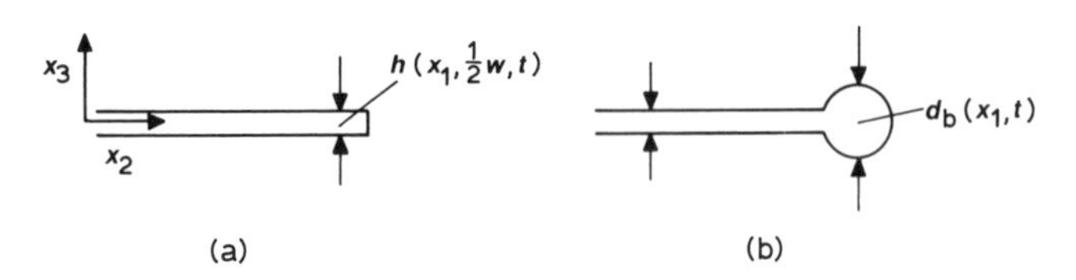

Fig. 16.4. Idealizations of the sheet edge. (a) Sharp edge—no beading. (b) Filamentous edge—circular beading.

16.4(a). Note that

$$\mathbf{T} \cdot \mathbf{k} \equiv 0 \tag{16.1.13}$$

by definition, where $\mathbf{k}$ is the unit vector in the x_3-direction, so condition (16.1.11) merely continues the zero-surface-force condition around the edge.

The set of equations and boundary conditions (16.1.6)–(16.1.13) have to be complemented by a rheological equation of state for $\mathbf{T}$ in terms of $\mathbf{v}$ and $\bar{T}$, e.g. (8.2.21) with $\bar{T}$ as an additional variable. Because $\mathbf{T}$ cannot in general be diagonalized by the same rotation of coordinates for all points in the (x_1, x_2)-plane, we cannot regard the flow field $\mathbf{v}$ as a purely extensional flow and so use simplified relations like (8.2.7), (8.2.8).

The relations (16.1.11) and (16.1.13) are inconsistent in the sense that the fundamental inequality

$$|\nabla h| \ll 1 \tag{16.1.14}$$

on which the thin sheet approximation Section 8.2 is based, fails at the edges. Certainly sharp corners at $x_2 = \pm\frac{1}{2}w$, $x_3 = \pm\frac{1}{2}h$ are incompatible with the existence of any non-zero surface tension. We are therefore led to try to model the observed edge beading. Figure 16.4(b) shows an idealization (itself unfortunately not devoid of its own corners), whereby a circular beading of diameter $d_b(x_1, t)$ is attached (in practice more smoothly than shown in the Figure) to the sheet that has slowly-varying thickness h up to the beading. The velocity parallel to the edge beading is taken to be continuous between bead and sheet, though the velocity perpendicular to the bead need not be continuous, the difference being accounted for by a growth in the mass flux down, or the local thickness of, the beading.

$$\mathbf{v}_b = \{v_{b1}(x_1, t), v_{b2}(x_1, t)\} \tag{16.1.15}$$

for the velocity of the axis of the bead and F_b for the total axial force in the beading, then (16.1.11) is replaced by

$$F_b \frac{\partial^2 w}{\partial x_1^2} \cos^3 \theta = \hat{\mathbf{n}} \cdot \mathbf{T} \cdot \hat{\mathbf{n}} = T_{11} \sin^2 \theta$$

$$- 2T_{12} \sin \theta \cos \theta + T_{22} \cos^2 \theta \quad (16.1.16)$$

and

$$\frac{\partial F_b}{\partial x_1} \cos \theta = (T_{22} - T_{11}) \sin \theta \cos \theta + T_{12}(\cos^2 \theta - \sin^2 \theta) \quad (16.1.17)$$

and (16.1.12) by

$$\frac{\partial w}{\partial t} + v_{b1} \frac{\partial w}{\partial x_1} = v_{b2} \quad (16.1.18)$$

The continuity of mass for the beading, corresponding to (15.2.4) for example, becomes

$$\frac{\partial}{\partial t}(\rho_m d_b^2) + \cos \theta \frac{\partial}{\partial x_1}(\rho_m d_b^2 v) = \frac{4h}{\pi}(\mathbf{v} - \mathbf{v}_b) \cdot \hat{\mathbf{n}} \quad (16.1.19)$$

where continuity of axial velocity requires that

$$v = v_{b1} \cos \theta + v_{b2} \sin \theta = v_1 \cos \theta + v_2 \sin \theta \quad (16.1.20)$$

v_1 and v_2 being evaluated at $x_2 = \frac{1}{2}w$. F_b will be given by a rheological equation for the bead similar to (15.2.14), where

$$F_b = \tfrac{1}{4}\pi d_b^2 t_{zz} \quad (16.1.21)$$

z being an axial coordinate. However, this formulation seems to be indeterminate at this stage, because only two equations, (16.1.19) and (16.1.20), have so far been added to define three new variables, d_b, v_{b1}, and v_{b2}. It appears that a further constraining equation such as $\mathbf{v} \equiv \mathbf{v}_b$ or $F_b = \text{constant}$ would be required to make the system determinate; $\mathbf{v}_b$ will obey the same boundary conditions as $\mathbf{v}$ at $x_1 = 0$ and l. $d_b(0)$ will be specified.

16.1.2 Simple Solutions

Steady flow is considered in this subsection.

If $l \gg w_0$, then the thin sheet can be treated as a thin slice cut out of a filament and its dynamics and kinematics will be equivalent. For negligible surface tension, (15.2.7) requires that $t_{rr} = t_{\phi\phi} = 0$ for a filament, which is equivalent to $t_{22}=t_{33}=0$ for a sheet, satisfying (16.1.13). The drawn-down ratio Dr will be achieved equally by a necking-in ratio $w_1/w_0=\mathrm{Dr}^{-\frac{1}{2}}$ and a thickness reduction $h_1/h_0=\mathrm{Dr}^{-\frac{1}{2}}$. This is not what is desired in practice, so $l<w_0$ in general.

If $l \ll w_0$, we may expect that the central portion of the sheet, given by $|w-x_2| \gg l$ will experience plane flow, with $v_2=0$. The dynamical situation is then very similar to that described for a filament. The correspondence will be similar to that between Subsection 4.3.1 (elongational flow) and Subsection 4.3.2 (pure shear flow). The thin-sheet approximation (cf. (15.2.2)), is

$$\frac{\mathrm{d}h}{\mathrm{d}x_1} \ll 1 \tag{16.1.22}$$

the conservation equations (16.1.6)–(16.1.8) become (cf. (15.3.1)–(15.3.3)),

$$\rho_{\mathrm{m}} v_1 h = \dot{m} \tag{16.1.23}$$

$$\frac{\mathrm{d}v_1}{\mathrm{d}x_1} = \frac{g}{v_1} + \frac{\mathrm{d}}{\mathrm{d}x_1}\left(\frac{t_{11}}{\rho_{\mathrm{m}} v_1}\right) \tag{16.1.24}$$

$$\frac{\mathrm{d}\bar{T}}{\mathrm{d}x_1} = \frac{2h_{\mathrm{a}}}{\dot{m}\Gamma_{\mathrm{m}}}(T_{\mathrm{a}}-\bar{T}) \tag{16.1.25}$$

where air-drag and surface tension have been neglected in (16.1.24) and generation is neglected in (16.1.25).

The relevant rheological equation of state will, however, involve two functionals, instead of the one given in (15.2.14), which we can write formally as

$$t_{11}-t_{jj} = \mathop{\mathcal{F}}_{\substack{1j\ \mathrm{PS}\\ s=0}}^{s=\infty} [\dot{e}^t(s), \bar{T}^t(s), \rho_{\mathrm{m}}^t(s); \bar{T}(t), \rho_{\mathrm{m}}(t)], \qquad j=2,3 \tag{16.1.26}$$

where

$$\dot{e} = \mathrm{d}v_1/\mathrm{d}x_1 \tag{16.1.27}$$

Of these only $t_{11}-t_{33}$ enters into the conservation equations used above; $t_{11}-t_{22}$ becomes relevant when considering the flow field near the edges. We note that $t_{33}=0$ from (16.1.13).

The arguments and solutions developed in Subsections 15.3.1 and 15.3.2 can be carried over directly, or adapted very simply, to the plane-flow situation. For example, the solution for an isothermal

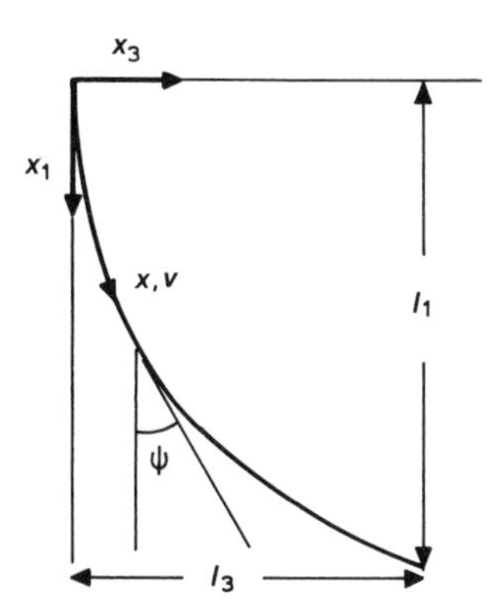

Fig. 16.5. Coordinate system for plane flow in a sheet drawn away from the vertical (x_1-axis). x is measured along the sheet from the die-lips at $(x_1, x_3) = (0, 0)$ and wind-up takes place at (l_1, l_3). v varies from v_0 at $(0, 0)$ to v_1 at (l_1, l_3).

Newtonian viscous fluid is unaltered as far as the velocity is concerned; if inertia is negligible, then

$$v_1 = v_{10}\,\mathrm{Dr}^{x_1/l}; \qquad h = h_0\,\mathrm{Dr}^{-x_1/l} \tag{16.1.28}$$

The stress t_{11} will differ in that, from eqn (4.3.10),

$$\eta_{\mathrm{PS}} = \tfrac{4}{3}\eta_{\mathrm{e}}, \quad \text{and the stress} \quad t_{22} = \tfrac{1}{2}t_{11}$$

The equations developed in Subsection 16.1.1 above need no further discussion if x_1 is vertical. However, for the plane-flow situation, when the extending sheet is drawn away from the vertical (case (a) of Fig. 16.1), a (curvilinear) coordinate system as shown in Fig. 16.5 is appropriate. Equations (16.1.23)–(16.1.24) become, using the Newtonian isothermal approximation,

$$\rho_{\mathrm{m}} v h = \dot{m} \tag{16.1.29}$$

$$\frac{\mathrm{d}v}{\mathrm{d}x} = \frac{g\cos\psi}{v} + \frac{\mathrm{d}}{\mathrm{d}x}\left(\frac{\eta_{\mathrm{PS}}}{\rho_{\mathrm{m}} v}\frac{\mathrm{d}v}{\mathrm{d}x}\right) \tag{16.1.30}$$

Balance of forces perpendicular to the sheet yields

$$\frac{\eta_{\mathrm{PS}}}{\rho_{\mathrm{m}}}\frac{\mathrm{d}v}{\mathrm{d}x}\left(\frac{\mathrm{d}\psi}{\mathrm{d}x}\right) = g\sin\psi + v^2\frac{\mathrm{d}\psi}{\mathrm{d}x} \tag{16.1.31}$$

The boundary conditions on v are not known directly as functions of ψ, and so solution is non-trivial.

No solutions have so far been presented for the edge flow either when $w \gg l$ or when $w \not\gg l$; even if a linear (Newtonian) rheology is assumed, the fact that $w(x_1)$ is a dependent variable makes the problem formulated in Subsection 16.1.1 basically non-linear. The problems associated with the edge-bead, in particular the non-uniqueness of the conditions (16.1.16), (16.1.19), compound this difficulty.

16.1.3 Stability

The similarity of eqns (16.1.23)–(16.1.25) and associated boundary conditions to eqns (15.3.1)–(15.3.3) and their boundary conditions means that the draw-resonance phenomenon in fibre spinning will have an analogue in wide-sheet extension. Indeed, the isothermal Newtonian result (15.4.30) can be taken over directly, the numerical factor being the same, for the extension of a very wide sheet. Other modes of instability are possible, since a non-zero disturbance v_2 is allowable, provided v_1 and h have an x_2-dependence of the form $\exp(inx_2)$ in the infinitesimal disturbance case (cf. the stability of coating flows, Subsection 14.3.2, eqn (14.3.50) in particular). The neutral stability curve in (Dr, n)-space has been examined by Yeow (1972, 1974), in which he shows that $\mathrm{Dr}_{\mathrm{crit}}$ occurs at $n=0$, and is therefore given by (15.4.30). More recent work (Yeow & Aird, 1981, unpublished) shows that the most unstable mode for a power-law fluid with $\nu > 1{\cdot}2$ yields $n > 0$. However, the critical draw ratio rises with ν, being 40 at $\nu = 1{\cdot}2$ and 91 for $\nu = 1{\cdot}5$.

EXERCISE 16.1.1. Derive eqns (16.1.16)–(16.1.20).

EXERCISE 16.1.2. Show that for $l \gg w_0$, the relation (16.1.13) is satisfied only to within the accuracy of the fibre-spinning approximation.

EXERCISE 16.1.3. Derive eqns (16.1.29)–(16.1.31). Show that, if inertia is neglected, then the horizontal component of the tension, $T \sin \psi$, is constant. It can be argued on elementary physical grounds, but also follows from the equations

$$\frac{\mathrm{d}\psi}{\mathrm{d}x} = \rho g \sin \psi / v, \qquad \frac{\mathrm{d}T}{\mathrm{d}x} = -\rho g \cos \psi / v$$

16.2 COLD DRAWING OF FILM

Cast thin sheet is not significantly oriented, particularly if the draw ratio after extrusion is relatively low and the Deborah number of the process small. Significant biaxial orientation and increase in modulus can be achieved by simultaneous (or sequential) stretching of the film in two perpendicular directions below its crystalline melt point—the process is used industrially for PET and PP. One device used is known as a stenter or tenter (a term borrowed from the textile industry), which grips the edges of the sheet as it is fed, from rollers, and carries it forward at increasing speed† and increasing width until the stretched sheet is drawn off at higher speed by a second set of rollers, the edges being then released by the stenter clamps.

Figure 16.6 shows the geometrical arrangement. The imposed movement of the clamps fixes the width $w_s(x_1)$ and the forward movement $v_{s1}(x_1, \frac{1}{2}w_s)$ at the edges. This is chosen so that it matches the roll

† In many cases, the forward draw, i.e. increasing speed, is undertaken separately before the sideways draw, i.e. increasing width, the latter being carried out at constant forward speed.

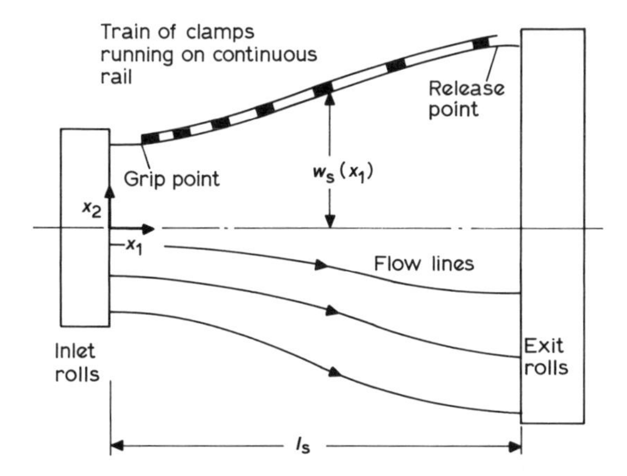

Fig. 16.6. Plan view of stenter. The inlet and exit rolls and the stenter clamps are driven. The clamps increase in velocity with x_1, to match the inlet and exit roll speeds; the width w_s is fixed by the geometry of the clamp railway.

speeds v_{s10} and v_{s11}. If it is assumed that v_{s1} is independent of x_2 and

$$v_{s2} = \frac{v_{s1}}{w_s}\frac{dw_s}{dx_1}x_2 \tag{16.2.1}$$

then the thickness h is also a function of x_1 only given by

$$\frac{h}{h_0} = \frac{w_{s0}}{w_s}\frac{v_{s10}}{v_{s1}} \tag{16.2.2}$$

The object of the process is certainly to achieve h_1 constant at the exit. This will only be even approximately true if $l_s \gg w_s$, where l_s is the length of the stretching zone or, more accurately stated, if

$$0 < \frac{dw_s}{dx_1} \ll 1 \tag{16.2.3}$$

The flow field defined by (16.2.1), (16.2.2) subject to (16.2.3) is a general form of the extensional flow (4.3.1) where $\dot{\boldsymbol{\gamma}}$ is a function of time. It is better in the case of cold drawing to consider it in terms of the extension ratios

$$\gamma_1 = v_{s1}/v_{s10}, \qquad \gamma_2 = w_s/w_{s0}, \qquad \gamma_3 = h/h_0 \tag{16.2.4}$$

For low-Deborah-number† flows, a purely elastic or elasto-plastic model, in which $\mathbf{T}^E$ is a function of $\boldsymbol{\gamma}$ only, provides the appropriate rheology. For higher-Deborah-number flows, a Kelvin–Voigt type of model, as used in simplified form in Section 15.5, is probably more relevant. One such is given in Funt (1975); it suffers formally from being appropriate only in the irrotational flow situation defined by (4.3.1), and was chosen to match transient experimental observations obtained on small rectangular samples stretched uniformly. The tension tensor can be taken as diagonal in the (x_1, x_2, x_3)-coordinate system of Fig. 16.6, and in our present notation, the Funt model can be written

$$\frac{T_i}{h} = G(\bar{T})\left\{E_i + \lambda(\bar{T})\frac{dE_i}{dt}\right\}, \qquad i = 1, 2 \tag{16.2.5}$$

where

$$E_1 = \gamma_1 - \gamma_1^{-3}\gamma_2^{-2}, \qquad E_2 = \gamma_2 - \gamma_2^{-3}\gamma_1^{-2} \tag{16.2.6}$$

† The material time scale λ used in defining a Deborah number in this case is a retardation time, not a relaxation time as is usual.

It must be noted that if γ and hence $\mathbf{E}$ are taken to be functions of x_1 only, then

$$\frac{dE_i}{dt} = v_{s1}\frac{dE_i}{dx_1} \tag{16.2.7}$$

for steady flow, and $\mathbf{T}$ is also a function of x_1 only. The stress-equilibrium equation

$$\boldsymbol{\nabla}\,.\,\mathbf{T} = \mathbf{0} \tag{16.2.8}$$

which is relevant in this case because G is large compared with stresses caused by inertia, clearly cannot be satisfied except approximately on the basis that $\|d\mathbf{T}/dx_1\| \ll \|\mathbf{T}\|/w$. No attempt has been made to seek dynamically-non-trivial solutions satisfying (16.2.8), for these require that $\mathbf{v}_s$ be a function of x_1 and x_2, and hence the tensor $\mathbf{T}$ would lose its assumed diagonal form. The rheological equation (16.2.5) would then be inadequate, and would have to be replaced by a suitable rotationally invariant form. (This issue is discussed briefly in Pearson & Gutteridge, 1978.)

In a physical sense, the difference in total in-line tension at the inlet and exit rolls, namely $w_{s0}T_{10} - w_{s1}T_{11}$ has to be balanced by shear forces/unit length (T_{12}) generated by the stenter clamps at the edges of the sheet.

Under certain circumstances, e.g. with $w_s = w_{s0} = w_{s1}$ and with v_{s1} not prescribed at $x_2 = \pm\frac{1}{2}w_s$, a necking phenomenon similar to that described for a fibre in Section 15.5 can arise. This can be visualized by supposing Fig. 15.12 to represent the (x_1, x_3) cross-section of the stretching sheet. More complicated possibilities have not been investigated.

Chapter 17

Film Blowing

17.1 THE PROCESS

Film blowing is a method of producing thin sheets of thermoplastics rather more rapidly and economically than is possible with the casting process. The melt film-blowing process is based on simultaneous stretching and inflation of a moving tube of polymer melt extruded from an annular die of a type discussed in Section 10.3. The blowing 'bubble' of molten polymer is cooled while this is occurring and crystallizes or vitrifies at a 'frost-line' beyond which no significant deformation takes place. The inflated circular-cylindrical tube so formed is further cooled and then passed through a train of guide plates or rollers which flatten it sufficiently for it to go through a pair of driven rubber nip-rolls without crinkling. It is then wound onto cylindrical cores and sold or used as 'lay-flat' tubing; alternatively, the edges can be trimmed off after passing through the nip-rolls and two rolls of flat sheet can be wound from the lay-flat tube. The process is shown diagrammatically in Fig. 17.1. It is usually carried out vertically upward, sometimes vertically downward, and occasionally horizontally.

It is usual to aim at a uniform sheet-thickness and therefore at axisymmetry and steadiness of the molten portion of the bubble. Because of the change of symmetry in going through the guide rolls (axisymmetry to planar symmetry) a relatively long stretched length of frozen but still-cooling tube is maintained between frost-line and nip-rolls.

Typical parameters for the process are:

h_0, die gap thickness	1–2 mm
d_0, diameter at die-lips	50–500 mm

v_{10}, mean velocity at die-lips	10–50 mm s^{-1}
Bu, blow ratio $= d_f/d_0$	1·5–3
Dr, draw ratio $= v_{1f}/v_{10}$	5–15
l_f, axial distance to frost-line	0·250–5 m

It follows that the final film thickness

$$h_f = h_0/\text{Bu Dr} \tag{17.1.1}$$

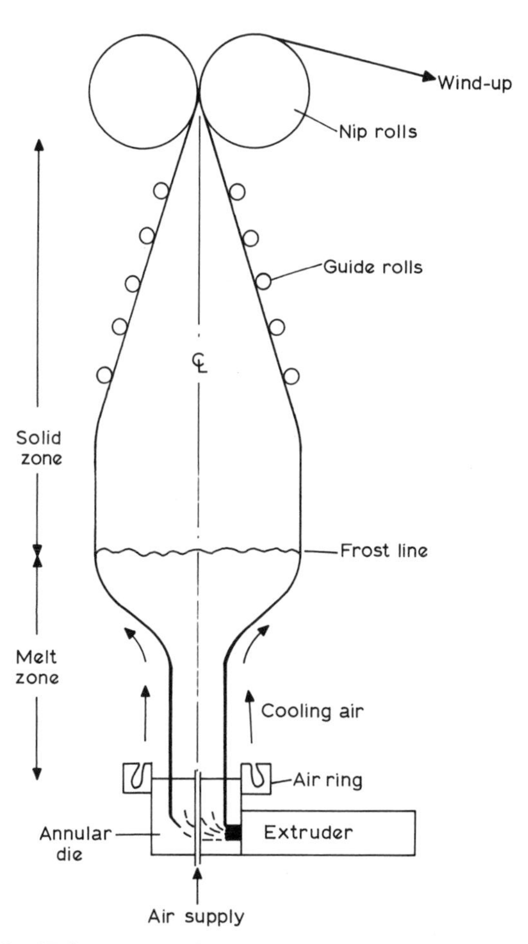

Fig. 17.1. General view of melt film-blowing process.

will be of order 50 μm, and so the film thickness h is everywhere small compared with the gross dimensions of the bubble: the flat-sheet approximation derived in Section 8.2 will therefore apply.

The flight time t_f in the molten region will be of the order of 10 s, during which the temperature of the melt will cool by about 75 K. A further cooling by about 100 K is necessary before winding up; these figures depend upon the nature of the polymer. The rate of cooling is therefore of the order of $10\ \mathrm{K\,s^{-1}}$, which is far less rapid than in the case of fibre spinning, and the Fourier number (the notation Gz is retained for consistency with arguments in other chapters)

$$\mathrm{Gz} = h_0^2/\kappa t_f \tag{17.1.2}$$

is of order unity, so that the film can be supposed of essentially uniform temperature through its thickness.

In practice, cooling is usually achieved by blowing an annular jet of air onto the film from an external axisymmetric air-ring; this can be reinforced by an internal air-cooling system. In the case of some very thick extruded tube, as for the production of large bags, cooling is achieved by a water ring or spray; this, though, has obvious disadvantages because the material has to be dried before it is wound up.

The bubble is in principle unsupported between die-lips and guide system, radial stretching (i.e. blow-up) being achieved by means of a small internal overpressure ($p_0 \approx 50\ \mathrm{N\,m^{-2}}$) and axial stretching by means of an axial tension at the frost line ($F_f \approx 10$ N). In practice, surprisingly large deviations from axisymmetry arise without leading to similarly large variations in film thickness. In some cases, internal formers are used over which the blowing bubble is effectively drawn. External constraints, often in the form of irises, are now commonly used.

The melt film-blowing process is a very important one commercially, a substantial fraction of polyolefin production (LDPE, HDPE and PP) being converted thereby into wrapping film. The quality of the output is usually measured by the uniformity of film thickness (variations of $> \pm 5\%$ with a length scale varying from 10 mm to 10 m are still common) and the optical clarity, which depends upon the degree and type of crystallinity achieved (in crystallizable polymers). Uniformity of film thickness requires steady uniform output from the screw extruder driving the flow, a well-designed die held at a steady uniform temperature with uniform die gap, even air cooling and low friction at the guide rolls or plates.

Although considerable attention has been given to screw and die design, much less is understood about the air cooling process, which is often relatively uncontrolled.

A related process, which involves cold drawing by bubble inflation and extension, is used to produce thin (*c.* 25 μm) films of biaxially oriented materials (namely PP and some PET). In this process, illustrated in Fig. 17.2, the extruded tube is first chilled to below the freezing temperature to provide a solid circular pipe, and is then heated by radiant elements to just below the crystallization temperature, at which point it expands and draws very rapidly under the relatively high stress to which it is subjected by the haul-off rolls. The draw- and blow-ratios are chosen to be more nearly equal than in the melt film-blowing process, and are each typically about 8.

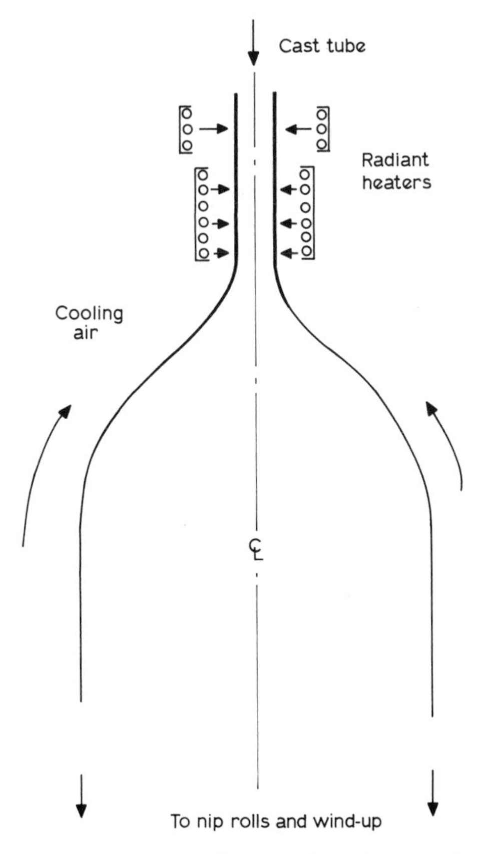

Fig. 17.2. General view of biaxial-draw film-blowing process.

The rheological behaviour of the materials being blown (hot or cold) can, in some measure, be inferred from the degree of shrinkage that arises when the lay-flat tube is annealed in an oven at a temperature just below the softening point. It is found that a remarkably high proportion (50% even) of the deformation can be recovered in some cases: the material has, in some sense, deformed elastically to that extent; however, it is by no means clear what this implies for the stress/strain (or rate of strain) relationship during deformation. The latter is unsteady in body coordinates and more complex than in the case of fibre spinning; there are two principal axes of extension along which the rates of extension vary independently with time (blow and draw do not take place with similar histories). These axes remain fixed in the body only provided axisymmetry is maintained. One of the difficulties experienced in analysing the blown film process (even in its ideal axisymmetric form) is that the functional relationships (8.2.7) and (8.2.8), or their non-isothermal analogues like (16.1.26), defining the material rheology, have not been evaluated experimentally at all widely. Such formal numerical solutions as have been obtained, and in some cases compared with experiment, have therefore been based on unsubstantiated rheological equations, and on measured rather than calculated temperature distributions (Pearson & Richardson, 1983, Chapter 8). The relaxation-time spectrum for the polyolefins used in the melt-blowing process is relatively wide and spans the characteristic flight-time (10 s) and extension flow time (1–5 s). A mean value for relaxation time is such that the Deborah number for the flow is of order 1 and, for some purposes, may be regarded as large compared to one.

The question of sensitivity to small disturbances and of overall stability has been little examined. In practice, any one blowing unit only works satisfactorily within relatively narrow limits of Bu and Dr but the reasons for this are little understood.

Further details on the process can be obtained from Schenkel (1966).

17.2 MODEL EQUATIONS AND BOUNDARY CONDITIONS FOR MELT FILM BLOWING

General formulation of mechanical equations for the blowing film viewed as a thin sheet must be founded on a careful description of the geometry of a moving surface. Although this subject, differential geometry, is a well-established branch of mathematics, it is, nevertheless, a complicated one, giving rise to related co- and contra-variant tensors in two and three dimensions, and is avoided by most workers in engineering fluid mechanics. It is therefore usual, and indeed much simpler, to avoid general formulations wherever possible by the use of special coordinate systems which exploit to the full any symmetries of the flows in question. This is analogous to the use of a limited range of orthogonal curvilinear coordinates instead of general coordinates in continuum-mechanical theories. Thus, for axisymmetric flow, orthogonal surface coordinates, corresponding to the principal directions of surface force $\mathsf{T} = h\mathbf{T}$ can be used. Once non-axisymmetric disturbances are considered, however, a more elaborate formulation becomes necessary. Those interested in the subject will find an elegant and brief account in Aris (1962, Chapters 9 and 10): the hidden complexities of the beguilingly simple relations (8.2.13) and (8.2.15) that were stated earlier are there made apparent.

The simplest representation for our purposes uses the (r, ϕ, z) circular cylindrical coordinate system shown in Fig. 17.3. The z axis is in the 'machine' direction, along the axis of the annular die, which in a symmetrical system passes through the centre of the nip between the haul-off rolls. The position of the inner surface of the melt film bubble

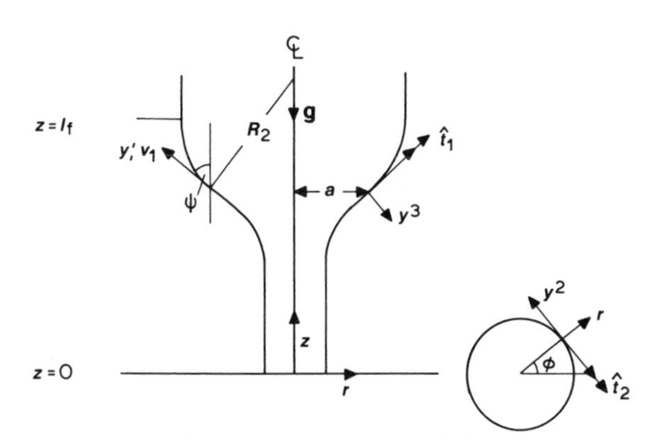

Fig. 17.3. Coordinate systems: (r, ϕ, z) based on centre-line axis of process; (y^1, y^2, y^3) based on local tangent plane to blowing sheet. R_2 is always such that one centre of curvature lies on the axis. $\hat{t}_1$ and $\hat{t}_2$ are principal tensions in the film. ψ is the angle between the y^1 and z-axes. l_f is the axial length of the blowing region. **g** indicates the direction of gravity.

is then given by

$$r = a(\phi, z, t) \tag{17.2.1}$$

If the flow is steady and axisymmetric, then a becomes a function of z only. A surface-coordinate system (y^1, y^2) is provided by the local values of (z, ϕ), the coordinate net being orthogonal when a is independent of ϕ. For the latter case, we use the physical-component representation

$$\mathbf{v}(z, t) = (v_1, v_2, v_3) \tag{17.2.2}$$

for the velocity in the (y^1, y^2, y^3)-directions, y^3 being in a direction normal to the surface; $v_2 = 0$ except for the case of a rotating die, and $v_3 = 0$ for steady flow.

The metric tensor for the (y^1, y^2) system is given by

$$\begin{pmatrix} g_{11} & 0 \\ 0 & g_{22} \end{pmatrix} = \begin{pmatrix} 1 + (da/dz)^2 & 0 \\ 0 & a^2 \end{pmatrix} \tag{17.2.3}$$

where

$$ds^2 = g_{11}(dy^1)^2 + g_{22}(dy^2)^2 \tag{17.2.4}$$

is the square of the length of a differential line element (dy^1, dy^2) along the surface. g_{33} is arbitrary at this stage, but can conveniently be taken as h^2 so that $y^3 = 0, 1$ are the inner and outer surfaces of the sheet.

It may readily be verified that the Reynolds number of the flow defined by the parameters quoted in Section 17.1 using a typical viscosity η_0 of 10^4 N s m^{-2} is very small, 0 $(<10^{-2})$, and so inertia effects may be neglected. Similarly, surface-tension forces, which are of the order of the inverse of the capillary number, Ca^{-1}, times as large as viscous forces can be shown to be negligible. The contribution of gravity forces relative to viscous forces will be given by the ratio

$$\mathrm{Re}/\mathrm{Fr} = \rho g l^{*2}/\eta^* v^* \tag{17.2.5}$$

which may be calculated to be large† or small relative to unity, depending on the choices made for l^*, v^* and η^*. This suggests that gravity forces be included in the analysis and that their actual significance be judged by comparison of the solutions obtained when they

† In practice, gravity forces cannot dominate viscous forces, or the bubble would collapse.

are included and neglected, respectively. This has the effect of including body-force terms ($\mathbf{F}_t$ and F_n) in the mechanical balance equations (8.2.12) and (8.2.15) which can be formally altered to read

$$\nabla_{surf} \cdot \mathbf{T} + \mathbf{F}_t = 0 \tag{17.2.6}$$

and

$$\frac{\hat{t}_1}{R_1} + \frac{\hat{t}_2}{R_2} = \Delta p + F_n \tag{17.2.7}$$

where $\mathbf{F}_t$ is the tangential or surface force vector and F_n is the normal component. Δp is the difference in air pressure between the inside and outside of the bubble.

The local mechanics for the process is given by the set of equations (8.2.15), (8.2.19), (8.2.20) and (8.2.21), slightly modified. However, for our purposes here, it is worth deriving independently and much more simply an integrated set of equations relevant for axisymmetric steady bubble-flow. The mass-conservation equation becomes simply

$$2\pi\rho v_1 ah = \dot{M} = \rho\dot{V} \tag{17.2.8}$$

where $\dot{M}$ is the constant mass flowrate. If the machine direction is vertically upward, an axial force balance in the z-direction yields

$$\left.\begin{aligned} &2\pi a\hat{t}_1 \cos\psi - \pi a^2 \Delta p = F \\ &\text{where} \\ &\frac{dF}{dz} = 2\pi\rho gah \sec\psi \end{aligned}\right\} \tag{17.2.9}$$

$F(l_f)$ is the net axial force on the nip-rolls and guide system, and ψ is the angle between the y^1 and z axes, given by

$$\tan\psi = da/dz \tag{17.2.10}$$

(17.2.9) takes the place of (17.2.6). The local form, (17.2.7), is retained, and writing

$$R_1 = -(\cos\psi \, d\psi/dz)^{-1}, \qquad R_2 = a \sec\psi \tag{17.2.11}$$

it becomes

$$\Delta p + \rho gh \sin\psi = \hat{t}_2 \cos\psi \, a^{-1} - \hat{t}_1 \cos\psi \, d\psi/dz \tag{17.2.12}$$

The energy-conservation equations, neglecting generation, can be based on the mean temperature $\bar{T}(z)$. Using heat transfer coefficients

$h_a^{(i)}$ and $h_a^{(o)}$ for the inner and outer surfaces of the blowing film, and ambient inner and outer temperatures $T_a^{(i)}$ and $T_a^{(o)}$, the equation for $\bar{T}$ becomes

$$\dot{M}\Gamma_m \,\mathrm{d}\bar{T}/\mathrm{d}z = \{h_a^{(i)}(T_a^{(i)} - \bar{T}) + h_a^{(o)}(T_a^{(o)} - \bar{T})\} \times 2\pi a \sec\psi \tag{17.2.13}$$

Although (17.2.13) is formally very simple, a choice of the T_a's and h_a's is a major difficulty; it is, however, not a problem specific to polymer melt mechanics, but is part of the much wider problem of radiant and convective heat transfer as met in Chapters 15 and 16 above.

The unknown principal stresses

$$t_{11} = \hat{t}_1/h, \qquad t_{22} = \hat{t}_2/h \tag{17.2.14}$$

will be given by the rheological equation of state in terms of the histories of the principal strains

$$e_1 = v_1/v_1(0), \qquad e_2 = a/a_0, \qquad e_3 = h/h_0 \tag{17.2.15}$$

and/or the rates of strain (physical components)

$$\dot{e}_1 = \cos\psi \frac{\mathrm{d}v_1}{\mathrm{d}z}, \qquad \dot{e}_2 = \frac{v_1 \cos\psi}{a}\frac{\mathrm{d}a}{\mathrm{d}z}, \qquad \dot{e}_3 = \frac{v_1 \cos\psi}{h}\frac{\mathrm{d}h}{\mathrm{d}z} \tag{17.2.16}$$

The e_i are the elements of the (diagonal) Finger tensor. As expected, from (17.2.8), it follows that

$$\rho e_1 e_2 e_3 = \text{constant} \tag{17.2.17}$$

The $\dot{e}_i$ are the $\partial_i v_i'$ of (8.2.3). This representation will fail for that part of the deformation history preceding extrusion.

This failure is the same problem as discussed in connection with fibre spinning in Subsection 15.3.2 and has not been adequately resolved for flow in the neighbourhood of the extrusion-swell region; Petrie (1979, Subsection 6.2.1) has, however, shown that for many rheological equations of state, additional boundary conditions on the stress at the origin ($z = 0$) of extensional flow suffice to make solutions unique. Similar arguments can be advanced for the film-blowing case, as discussed in Petrie (1973, 1974). See, in this context, the calculations of Winter & Fischer (1981).

The boundary conditions on the flow field need careful discussion.

We suppose the inlet to the bubble flow to be at, or very near, the die exit $z=0$, where

$$a=\tfrac{1}{2}d_0, \qquad h=h_0, \qquad \bar{T}=T_0 \quad \text{and} \quad v_1=v_{10} \qquad (17.2.18)$$

Neither ψ nor $\mathbf{T}$ will be prescribed. It will be supposed that shear deformation ceases at the frost-line, which is the exit to the bubble flow; taking into account any shrinkage due to cooling between frost-line and haul-off roll, which will, in practice, be of order 0·1 or less, the in-line velocity can be specified at the frost-line, i.e.

$$v_1(l_f)=v_{1f}=\mathrm{Dr}\, v_{10} \qquad (17.2.19)$$

while

$$\mathrm{d}a/\mathrm{d}z=\tan\psi(l_f)=0 \qquad (17.2.20)$$

(This boundary condition, which is the steady form of $\mathrm{D}a/\mathrm{D}t=0$, appears less than obvious to some workers, who find it difficult to reject a flow in which $\mathrm{d}a/\mathrm{d}z$ is discontinuous at $z=l_f$.)

In the absence of substantial supercooling, which is much less likely for film blowing than in the case of fibre spinning, the temperature at the frost-line is known, i.e.

$$\bar{T}(l_f)=T_m \qquad (17.2.21)$$

The commercial process also requires

$$a(l_f)=\mathrm{Bu}\, a_0 \qquad (17.2.22)$$

to be specified.

Inlet and exit requirements on the bubble flow thus provide 8 boundary conditions, viz. (17.2.18)–(17.2.22). These are to be applied to the five equations (17.2.8)–(17.2.10), (17.2.12) and (17.2.13); the system is made determinate by allowing suitable choice of the parameters $\dot{M}$, l_f, F_f and Δp.

If Δp is specified, then Bu will be determined by the solution; if F_f is specified, then Dr will be determined by the solution; if shock cooling is applied at a fixed l_f, then $\bar{T}$ has a discontinuity at $z=l_f$; $\dot{M}$ is effectively given by the product of h_0, a_0 and v_{10}.

EXERCISE 17.2.1. Show that if $l^*=l_f$ and $v^*=\tfrac{1}{2}(v_{10}+v_{1f})$, then the range of values for Re/Fr given by (17.2.15) using $\eta^*=10^4\,\mathrm{N\,s\,m^{-2}}$ and values for l_f, v_{10} and v_{1f} from Section 17.1, includes numbers both large and small compared with unity.

17.3 STEADY AXISYMMETRIC SOLUTIONS

It is quite clear that the system of equations and boundary conditions developed in Section 17.2 is highly non-linear. No purely analytical progress can be made; all solutions so far obtained have been purely numerical, i.e. obtained by the use of digital computation (see Pearson & Richardson, 1983, Chapter 8). Uncertainty in the heat-transfer and rheological specification of the system has meant that these solutions have largely qualitative significance.

17.3.1 Newtonian Isothermal

This was the first situation investigated and, in many ways, the most carefully analysed. The rheological relations become (Pearson & Petrie, 1970*b*, *c*), for ρ constant,

$$\hat{t}_1 = 2\eta_0(\dot{e}_1 - \dot{e}_3)h, \qquad \hat{t}_2 = 2\eta_0(\dot{e}_2 - \dot{e}_3)h \tag{17.3.1}$$

or, using (17.2.8) and (17.2.16)

$$\left.\begin{aligned} \hat{t}_1 &= -2\eta_0 v_1 \cos\psi\left(2\frac{\mathrm{d}h}{\mathrm{d}z} + \frac{h}{a}\frac{\mathrm{d}a}{\mathrm{d}z}\right) \\ \hat{t}_2 &= 2\eta_0 v_1 \cos\psi\left(\frac{h}{a}\frac{\mathrm{d}a}{\mathrm{d}z} - \frac{\mathrm{d}h}{\mathrm{d}z}\right) \end{aligned}\right\} \tag{17.3.2}$$

If dimensionless variables

$$\tilde{F} = Fa_0/\eta_0\dot{V}, \qquad \tilde{a} = a/a_0, \qquad \tilde{h} = h/h_0, \qquad \tilde{z} = z/a_0, \qquad \tilde{l}_\mathrm{f} = l_\mathrm{f}/a_0 \tag{17.3.3}$$

are used, and we write

$$\mathrm{Pd} = \pi a_0^3 \Delta p/\eta_0\dot{V}, \qquad \mathrm{Wt} = 2\rho_0 g h_0/\Delta p \tag{17.3.4}$$

as dimensionless parameters defining the effects of internal air pressure and gravity, then v_1 can be eliminated using (17.2.8) and (17.3.2) can be substituted into (17.2.9) and (17.2.12) to give

$$2\tilde{a}^2(\tilde{F} + \tilde{a}^2\mathrm{Pd})\tilde{a}_{\tilde{z}\tilde{z}} = 6\tilde{a}_{\tilde{z}} + \tilde{a}\sec^2\psi\{\tilde{F} + \tilde{a}^2\,\mathrm{Pd}[1 - 2(2 + \mathrm{Wt}\,\tilde{h}\sin\psi)]\} \tag{17.3.5}$$

$$\tilde{F}_{\tilde{z}} = \mathrm{Pd}\,\mathrm{Wt}\,\tilde{a}\tilde{h}\sec\psi \tag{17.3.6}$$

$$4\tilde{h}_{\tilde{z}}/\tilde{h} = -2\tilde{a}_{\tilde{z}}/\tilde{a} - \sec^2\psi(\tilde{F} + \tilde{a}^2\,\mathrm{Pd}) \tag{17.3.7}$$

$$\tilde{a}_{\tilde{z}} = \tan\psi \tag{17.3.8}$$

subject to boundary conditions

$$\left.\begin{array}{r}\tilde{a}(0)=\tilde{h}(0)=1; \\ \tilde{a}(\tilde{l}_f)\tilde{h}(\tilde{l}_f)=\mathrm{Dr}^{-1},\ \psi(\tilde{l}_f)=0\end{array}\right\} \tag{17.3.9}$$

In this formulation, $\tilde{l}_f$, Pd and Wt are supposed known, but Bu is unspecified. $\tilde{F}_f$ is determined by Dr. For computational purposes, it is simpler to drop the third of (17.3.9) and suppose $\tilde{F}_0$ known also. It should be noted that (17.3.9) represents two-point boundary conditions: step-by-step integration can be carried out either from $\tilde{z}=0$, i.e. by guessing ψ_0 or from $\tilde{z}=l_f$ in which case Bu and Dr have to be guessed. For the case Wt = 0, which is exhaustively examined in Pearson & Petrie (1970*b*), only Bu need be guessed, because eqn (17.3.5) then becomes independent of $\tilde{h}$, and so Dr is not involved in its solution; eqn (17.3.7) is then integrated from $\tilde{z}=0$ by simple quadrature, knowing $\tilde{a}$; $\tilde{F}$ is constant. Solution for prescribed values of Bu and Dr for given h_0, a_0, η_0, $\dot{V}$ requires a search through various values for Pd (the product Pd Wt being fixed) and $\tilde{F}_0$ to satisfy the boundary conditions (17.3.9) together with

$$\tilde{a}(\tilde{l}_f)=\mathrm{Bu} \tag{17.3.10}$$

The weakness of this viscous approach is that polyolefins near the crystallization point are clearly elastic, particularly for extensions of order 3 or 4 in the engineering or linear measure (of order 1 or 2 in the logarithmic or Hencky measure). As explained in Section 17.1, the Deborah number for melt film-blowing is of order unity or larger in many cases, and so low-Deborah-number expansions about the Newtonian solution (relevant for slightly elastico-viscous flows) may fail to be realistic approximations.

17.3.2 Elastic Isothermal

This can be viewed as the situation approached in the high-Deborah-number limit for Maxwell-type fluids; if a purely elastic solution can be found, it might perhaps be used as the first term in a De^{-1} expansion (or whatever more complex expansion parameter going to zero with De^{-1} proves relevant).

There are, of course, an infinity of possible elastic equations of state. Petrie (1975*b*) used the simplest linear form

$$\mathbf{T}^{\mathrm{E}}=G\mathbf{G}_0(t-0)=G\{\mathbf{C}_t^{t-1}(t-0)-\mathbf{1}\}=G\mathbf{H}_t^t(t-0) \tag{17.3.11}$$

where $\mathbf{C}_t^t$ is the Finger strain and $\mathbf{H}_t^t$ is defined in (3.4.7). For steady axisymmetric incompressible flow, the tensors are diagonal with

$$\mathbf{C}_t^{t-1} = \begin{bmatrix} a_*^2 h_*^2/a^2h^2 & 0 & 0 \\ 0 & a^2/a_*^2 & 0 \\ 0 & 0 & h^2/h_*^2 \end{bmatrix} \tag{17.3.12}$$

where a_*, h_*, $t-0$ have been written instead of a_0, h_0 and t to indicate that the film is pre-strained at the beginning of the bubble flow ($t=0$, $a=a_0$, $h=h_0$). This means that $\mathbf{T}^E$ is not zero at $t=0$; indeed, it would be very unrealistic to suppose that it could be so. This raises the same problem as was discussed at length in Subsection 15.3.2 for fibre spinning and referred to in Section 17.2 above. It also means that a^* and h^* have to be determined by the complete solution of the problem; clearly they are related to Δp and F. From (17.3.11) and (17.3.12), it follows that

$$\hat{t}_1 = Gh\left(\frac{h_*^2 a_*^2}{h^2 a^2} - \frac{h^2}{h_*^2}\right), \qquad \hat{t}_2 = Gh\left(\frac{a^2}{a_*^2} - \frac{h^2}{h_*^2}\right) \tag{17.3.13}$$

G is the elastic modulus, and may quite reasonably be taken to be independent of $\bar{T}$. Substitution of (17.3.13) into (17.2.9) and (17.2.12), using now

$$\tilde{F} = F/2\pi G a_0 h_0, \qquad \mathrm{Pd} = \tfrac{1}{2} a_0 \Delta p / h_0 G \tag{17.3.14}$$

with $\tilde{a}$, $\tilde{h}$, $\tilde{z}$ and Wt defined as in (17.3.3), (17.3.4), gives

$$\left(\frac{\tilde{h}_*^2 \tilde{a}_*^2}{\tilde{h}^2 \tilde{a}^2} - \frac{\tilde{h}^2}{\tilde{h}_*^2}\right) = \frac{\sec\psi}{\tilde{a}\tilde{h}} (\tilde{F} + \tilde{a}^2\,\mathrm{Pd}) \tag{17.3.15}$$

$$\frac{\mathrm{d}\tilde{F}}{\mathrm{d}\tilde{z}} = \mathrm{Pd}\,\mathrm{Wt}\,\tilde{a}\tilde{h}\sec\psi \tag{17.3.16}$$

and

$$\tilde{a}(\tilde{F} + \tilde{a}^2\,\mathrm{Pd})\frac{\mathrm{d}\psi}{\mathrm{d}\tilde{z}} = \mathrm{Pd}(2 + \mathrm{Wt}\,\tilde{h}\sin\psi) + \tilde{a}\tilde{h}\cos\psi\left(\frac{\tilde{a}^2}{\tilde{a}_*^2} - \frac{\tilde{h}^2}{\tilde{h}_*^2}\right) \tag{17.3.17}$$

where

$$\frac{\mathrm{d}\tilde{a}}{\mathrm{d}\tilde{z}} = \tan\psi$$

By comparison with the Newtonian case, we see that the differential order of the system has been reduced by one, but that two additional unknown parameters have been introduced, namely a^* and h^*. The problem is now no longer uniquely posed if we wish to satisfy the boundary conditions (17.3.9) and (17.3.10) and allow a^*, h^*, $\tilde{F}_f$ and Pd to be freely chosen. Petrie (1976*b*) argues that the choice of a^* has little effect on the shape, $\tilde{a}$, of the solutions obtained, though it affects the stress levels. An alternative approach is to argue that the nature of the flow field within the die and in the annular die swell region is likely to provide a constraint on a^*, h^*.

13.3.3 Elastico-viscous Materials

The equations relevant for an Oldroyd 8-constant model, eqn (3.3.22), were developed by Petrie (1973). The only components relevant in this axisymmetric analysis are the principal (diagonal) components t_k^E, which obey the rheological equation of state

$$t_k^E+\lambda_1 v_z \cos\psi \frac{dt_k^E}{dz}-2\mu_1\dot{e}_k t_k^E+\mu_0\left(\sum_i t_i^E\right)\dot{e}_k+\nu_1\left(\sum_i t_i^E\dot{e}_i\right)$$

$$=2\eta_0\left\{\dot{e}_k+\lambda_2 v_z\cos\psi\frac{d\dot{e}_k}{dz}-2\mu_2\dot{e}_k^2+\nu_2\left(\sum_i e_i^2\right)\right\} \quad (17.3.18)$$

with no summation on k. The relevant surface 'tensions' $\hat{t}_1$ and $\hat{t}_2$ are given by

$$\hat{t}_1=h(t_1^E-t_3^E), \qquad \hat{t}_2=h(t_2^E-t_3^E) \quad (17.3.19)$$

The set of equations (17.2.8)–(17.2.12), (17.2.16), (17.3.18) and (17.3.19) describe the system, which can be reduced to seven first-order non-linear ordinary differential equations, as compared with the fourth-order system (17.3.5)–(17.3.8) for a Newtonian fluid. Three additional boundary conditions are therefore required. By extension of the arguments used for fibre spinning and in Subsection 17.3.2 above, these can be equated to the three initial values t_k^E $(z=0)$, which will themselves be determined by the flow. Petrie considered the special case of the Maxwell model (with $\lambda_1=\mu_1$, $\mu_0=\nu_1=\lambda_2=\mu_2=\nu_2=0$) which reduces the order of the system by one. In fact, he took $\mathrm{Wt}\equiv 0$, so that he then had a fifth-order system, and, in principle, allowed λ_1 and η_0 to be functions of the rate-of-strain invariant $\sum_i \dot{e}_i^2$. He suggests that his numerically-integrated solutions were relatively insensitive to initial conditions on the stress, and apparently used the particular constraints

$$\sum_i t_i^{\mathrm{E}} = 0, \qquad t_2^{\mathrm{E}} = 0 \quad \text{at} \quad z = 0 \tag{17.3.20}$$

which were sufficient to make his system determinate.

The relevant Deborah number is given by

$$\mathrm{De} = \lambda_1 v_{10}/a_0 \tag{17.3.21}$$

which, for his example, was as small as 0·1, but this was based on $\lambda_1 \sim 0{\cdot}3$ s. In view of the fact that a large amount of the strain is found to be recoverable on annealing, a value of De as large as unity would probably be more realistic for much thermoplastic film blowing.

An attempt to relate the rheology of the material being stretched to its linear viscoelastic spectrum using a strain dependent factor has been made by Wagner (1976*a*, 1978, 1979*b*), Wagner & Laun (1978), Wagner & Stephenson (1979), Laun (1978), who then applied his model to the film blowing process (Wagner, 1976*b*).

17.3.4 Non-isothermal Solutions and Comparison with Experiment

The first comparisons for LDPE were presented by Petrie (1975*b*), who took account of changes in density and viscosity with temperature in what was otherwise the Newtonian model of Subsection 17.3.1 above. The temperature profile was taken from experiment, the viscosity variation was exponential as in (15.3.25) and the density variation linear as in (9.1.6). He used the experimental results of Ast (1973, 1974), Farber (1973) and Farber & Dealy (1974). To check on the effect of elasticity he also obtained a solution based on the elastic model (17.3.11) to compare with Ast's results. The experimental observations lay between the purely elastic and purely viscous predictions, a result which Petrie regarded as satisfactory. The shape of the solutions is not sensitive to the absolute magnitude of the elastic modulus or the viscosity η_0 provided Bu and Dr are prescribed; however, the resulting values of F and Δp can, in principle, be compared with observations. Petrie found the value predicted for Δp to correspond far better in the elastic than the viscous case, emphasizing the likely importance of elastic strain.

Wagner (1976*b*) carried out very similar comparisons to those of Petrie, over a wider range of conditions.

EXERCISE 17.3.1. Derive (17.3.18) from (3.3.23).

EXERCISE 17.3.2. Show that a possible film-blowing solution for a Newtonian fluid when $\mathrm{Wt} \sim 0$ is given by $a = a_0$. Derive the relevant value for Δp and the solution for $h(z)$, $v(z)$ for given Dr. Note that such a solution can be obtained for arbitrary heat-transfer conditions in (17.2.13) and express the solution in terms of an arbitrary distribution $\eta_{PS}(z)$ for the relevant pure-shear viscosity.

17.4 UNSTEADY AND ASYMMETRIC FLOWS

As in the case of fibre spinning (Section 15.4) or film casting (Subsection 16.1.4) it is possible that the steady solutions predicted in Section 17.3 will be unstable. In particular, we can consider that an axisymmetric equivalent of draw-resonance (Subsection 15.4.2) might arise, on the grounds that the disturbance with greatest symmetry proved the most unstable in Yeow's (1974) investigation into sheet extension of Newtonian fluids. Yeow (1976) has examined this mode of instability for a Newtonian isothermal-model flow, using a constant value of Δp, i.e. Pd, and linear perturbation methods. (In practice, blown-film production units operate as often as not with a fixed mass of internal air rather than with a constant internal overpressure; this would lead, in general, to a coupling between an unsteady Δp and unsteady bubble kinematics.) Analysis of this situation involves consideration of an unsteady metric for the surface coordinates used. The effect of small departures from axisymmetry can also be examined by the method of linear perturbations, to take account specifically of small variations in output and temperature at the die lips. The complications involved in a proper description of the space coordinate metric then become much greater than in the unsteady axisymmetric case, and the two will be discussed separately.

17.4.1 Linearized Axisymmetric Disturbances

The flow field is described by the dimensionless variables†

$$\left.\begin{aligned} \tilde{a}(\tilde{z}, \tilde{t}) &= \bar{\tilde{a}}(\tilde{z})\{1 + A(\tilde{z})\exp(\mathrm{i}\tilde{w}\tilde{t})\} \\ \tilde{h}(\tilde{z}, \tilde{t}) &= \bar{\tilde{h}}(\tilde{z})\{1 + H(\tilde{z})\exp(\mathrm{i}\tilde{w}\tilde{t})\} \\ \tilde{v}_1(\tilde{z}, \tilde{t}) &= \bar{\tilde{v}}_1(\tilde{z})\{1 + V_1(\tilde{z})\exp(\mathrm{i}\tilde{w}\tilde{t})\} \end{aligned}\right\} \qquad (17.4.1)$$

† Strictly, the terms $A(\tilde{z})\exp(\mathrm{i}\tilde{w}\tilde{t})$, etc., should be written $\mathcal{R}e\,\{A(\tilde{z})\exp(\mathrm{i}\tilde{w}\tilde{t})\}$, etc., with $A(\tilde{z})$ regarded as complex.

where

$$\tilde{v}_1 = v_1/v_{10}, \qquad \tilde{t} = tv_{10}/a_0 \tag{17.4.2}$$

are dimensionless variables, $\bar{\tilde{a}}$, $\bar{\tilde{h}}$ and $\bar{\tilde{v}}_1$ represent the steady solution of Subsection 17.3.1, $\tilde{w}$ is a dimensionless frequency and $|A|$, $|H|$ and $|V_1| \ll 1$. We choose as surface coordinates, as before, $(y^1, y^2) \equiv (\tilde{z}_1, \phi)$. From (17.2.3) *et seq.*, we obtain, using (17.4.1), the linearized form of the diagonal metric tensor as

$$\mathbf{g} = \begin{bmatrix} 1 + \left(\dfrac{d\bar{\tilde{a}}}{d\tilde{z}}\right)^2 + 2\dfrac{d\bar{\tilde{a}}}{d\tilde{z}}\dfrac{d(\bar{\tilde{a}}A)}{d\tilde{z}} & 0 & 0 \\ 0 & \bar{\tilde{a}}^2 + 2\bar{\tilde{a}}^2 A & 0 \\ 0 & 0 & \bar{\tilde{h}}^2 + 2\bar{\tilde{h}}^2 H \end{bmatrix} \tag{17.4.3}$$

where the $\exp(i\tilde{w}\tilde{t})$ has been left out for convenience and A and H must be regarded as complex. The (physical) components of the dimensionless rate-of-strain tensor (see, for example, Aris, 1962, p. 228) become

$$\left.\begin{aligned} \tilde{e}_1 &= \frac{1}{g_{11}^{\frac{1}{2}}}\frac{\partial g_{11}^{\frac{1}{2}}}{\partial \tilde{t}} + \frac{1}{g_{11}^{\frac{1}{2}}}\frac{\partial \tilde{v}_1}{\partial \tilde{z}} \\ \tilde{e}_2 &= \frac{1}{g_{22}^{\frac{1}{2}}}\frac{\partial g_{22}^{\frac{1}{2}}}{\partial \tilde{t}} + \frac{1}{g_{11}^{\frac{1}{2}}}\frac{\partial}{\partial \tilde{z}}(\ln g_{22}^{\frac{1}{2}})\tilde{v}_1 \\ \tilde{e}_3 &= \frac{1}{g_{33}^{\frac{1}{2}}}\frac{\partial g_{33}^{\frac{1}{2}}}{\partial \tilde{t}} + \frac{1}{g_{11}^{\frac{1}{2}}}\frac{\partial}{\partial \tilde{z}}(\ln g_{33}^{\frac{1}{2}})\tilde{v}_1 \end{aligned}\right\} \tag{17.4.4}$$

which, on substitution of (17.4.3) and linearization, yield

$$\left.\begin{aligned} (\tilde{e}_1 - \bar{\tilde{e}}_1)\exp(-i\tilde{w}\tilde{t}) &= \cos\psi\frac{d(\bar{\tilde{v}}_1 V_1)}{d\tilde{z}} \\ &\qquad - (\bar{\tilde{e}}_1 - i\tilde{w})\sin\psi\cos\psi\frac{d(\bar{\tilde{a}}A)}{d\tilde{z}} \\ (\tilde{e}_2 - \bar{\tilde{e}}_2)\exp(-i\tilde{w}\tilde{t}) & \\ = \cos\psi\frac{\bar{\tilde{v}}_1}{\bar{\tilde{a}}}\frac{d(\bar{\tilde{a}}A)}{d\tilde{z}} &+ \bar{\tilde{e}}_2\left(V_1 - A - \cos\psi\sin\psi\frac{d(\bar{\tilde{a}}A)}{d\tilde{z}}\right) + i\tilde{w}A \\ (\tilde{e}_3 - \bar{\tilde{e}}_3)\exp(-i\tilde{w}\tilde{t}) & \\ = \cos\psi\frac{\bar{\tilde{v}}_1}{\bar{\tilde{h}}}\frac{d(\bar{\tilde{h}}H)}{d\tilde{z}} &+ \bar{\tilde{e}}_3\left(V_1 - h - \cos\psi\sin\psi\frac{d(\bar{\tilde{a}}A)}{d\tilde{z}}\right) + i\tilde{w}H \end{aligned}\right\} \tag{17.4.5}$$

where the $\bar{\bar{e}}_i$ are given by (17.2.16). The dimensionless stresses will be given by a dimensionless form of (17.3.1). With ρ constant, the mass-conservation equation (17.2.8) becomes

$$\tilde{e}_1+\tilde{e}_2+\tilde{e}_3=0 \tag{17.4.6}$$

The dimensionless radii-of-curvature become

$$\tilde{R}_1=\frac{R_1}{a_0}=-\frac{\sec^2\psi}{(\mathrm{d}^2\bar{\bar{a}}/\mathrm{d}\tilde{z}^2)}$$

$$\times\left[1+3\sin\psi\cos\psi\frac{\mathrm{d}(\bar{\bar{a}}A)}{\mathrm{d}\tilde{z}}-\frac{\mathrm{d}^2(\bar{\bar{a}}A)}{\mathrm{d}\tilde{z}^2}\bigg/\frac{\mathrm{d}^2\bar{\bar{a}}}{\mathrm{d}\tilde{z}^2}\right]$$

$$\tilde{R}_2=\frac{R_2}{a_0}=\bar{\bar{a}}\sec\psi\left[1+A+\sin\psi\cos\psi\frac{\mathrm{d}(\bar{\bar{a}}A)}{\mathrm{d}\tilde{z}}\right] \tag{17.4.7}$$

Using (17.4.5), (17.4.7) and a dimensionless form of (17.3.1) in the dynamical equation (17.2.7) with $F_n\equiv 0$ yields a linear relation between A, H and V_1, involving the known functions $\bar{\bar{a}}$, $\bar{\bar{h}}$ and $\bar{\bar{v}}_1$. Similarly, (17.4.5) and (17.3.1) substituted into the second dynamical equation (17.2.9) with $\bar{F}$ constant, yields a second linear relation between A, H and V_1. The continuity condition (17.4.6) provides the third. These prove to form a fourth-order system of linear differential equations, with $\tilde{F}=\bar{\bar{F}}\{1+f\exp(\mathrm{i}\tilde{w}\tilde{t})\}$. The relevant boundary conditions for these three equations, in the stability calculations, are

$$A(0)=H(0)=V_1(0)=0 \tag{17.4.8}$$

$$V_1(\bar{l}_f)=0 \tag{17.4.9}$$

and the linearized form of $\mathrm{D}\tilde{a}/\mathrm{D}\tilde{t}=0$ at $\tilde{l}_f$, which becomes

$$\bar{\bar{v}}_1\frac{\mathrm{d}(\bar{\bar{a}}A)}{\mathrm{d}\tilde{z}}+\mathrm{i}\tilde{w}\bar{\bar{a}}A=0 \quad \text{at} \quad \tilde{z}=l_f \tag{17.4.10}$$

The complete range of steady solutions possible are given by three parameters, Bu, Dr and $\tilde{l}_f$, or alternatively, by $\tilde{F}$, Pd and $\tilde{l}_f$. Yeow (1976) examined the neutral-stability surface, i.e. that for which $\mathcal{I}m\ \tilde{w}=0$, to a very limited extent. However, for realistic values of Bu and $\tilde{l}_f$, the least value of Dr involved appeared to be unrealistically large, i.e. $>10^3$.

A sensitivity analysis could be carried out taking at least one of $A(0)$, $H(0)$ and $V_1(0)$ to be non-zero, thus covering axisymmetric pulsations with a suitable real value for $\tilde{w}$.

17.4.2 Linearized Non-axisymmetric Disturbances

A more complex flow field is now considered, namely

$$\begin{aligned}\tilde{a}(\tilde{z},\tilde{t},\phi)&=\bar{\tilde{a}}(\tilde{z})\{1+A(\tilde{z})\exp(i\tilde{w}\tilde{t}+im\phi)\}\\ \tilde{h}(\tilde{z},\tilde{t},\phi)&=\bar{\tilde{h}}(\tilde{z})\{1+H(\tilde{z})\exp(i\tilde{w}\tilde{t}+im\phi)\}\\ \tilde{v}_1(\tilde{z},\tilde{t},\phi)&=\bar{\tilde{v}}_1(\tilde{z})\{1+V_1(\tilde{z})\exp(i\tilde{w}\tilde{t}+im\phi)\}\\ \tilde{v}_2(\tilde{z},\tilde{t},\phi)&=\bar{\tilde{v}}_1(\tilde{z})\{V_2(\tilde{z})\exp(i\tilde{w}\tilde{t}+im\phi)\}\end{aligned} \qquad (17.4.11)$$

there now being a velocity disturbance $\tilde{v}_2$ in the ϕ-direction with m integral.

The metric of the (y^1, y^2) coordinate net, as defined in Section 17.2 is no longer diagonal, being given by

$$\begin{bmatrix} 1+\left(\dfrac{\partial\tilde{a}}{\partial\tilde{z}}\right)^2 & \dfrac{\partial\tilde{a}}{\partial\tilde{z}}\dfrac{\partial\tilde{a}}{\partial\phi} \\ \dfrac{\partial\tilde{a}}{\partial\tilde{z}}\dfrac{\partial\tilde{a}}{\partial\phi} & \tilde{a}^2+\left(\dfrac{\partial\tilde{a}}{\partial\phi}\right)^2 \end{bmatrix} \qquad (17.4.12)$$

The (y^1, y^2) directions are no longer the lines of curvature, and therefore no longer the most suitable for describing the dynamical equations, which in the absence of inertial terms are those of shell theory (Novozhilov, 1964). A relevant orthogonal net for surface coordinates is one whose metric is diagonal and whose second fundamental tensor form is also diagonal (see, for example, Sokolnikoff, 1964). It will be noted that the off-diagonal terms in (17.4.12) are $0(|A|)$ and so small. Hence the orthogonal net lying along the lines of curvature will be inclined to the (y^1, y^2) net by only a small angle, of order $|A|$. The transformation between any such net, defining a coordinate system (u^1, u^2) and the (y^1, y^2) system, can be written in the form

$$\frac{\partial y^\alpha}{\partial u^\beta}=\begin{bmatrix}1+\text{I} & \text{II}\\ \text{III} & 1+\text{IV}\end{bmatrix} \qquad (17.4.13)$$

where I–IV are terms of order $|A|$. The details of the relations obeyed by I–IV need not be given here (they can be found in Yeow, 1972). At any given point on the bubble surface, given by $\tilde{a}$ in (17.4.11), the (u^1, u^2, u^3) metric can be chosen, when in linearized form, to be the same as (17.4.3). The components of the rate-of-strain tensor can then be expressed in the u^i coordinate system in terms of the variables in (17.4.11), where V_1 and V_2 refer to the undeformed y^i coordinate system. The important difference is that $\tilde{e}_{12}$ is no longer zero, but of order $|V_1|$. The dimensionless radius of curvature $\tilde{R}_1$ is unaltered from (17.4.7) but $\tilde{R}_2$ has an additional term Am^2 in the bracket on the right-hand side. The relation (17.2.7) is unchanged, but the linearized dynamical equation will be affected by changes in $\tilde{t}_{22}$ and $\tilde{R}_2$. Two more dynamical equations are required, which come from the surface vector equation (17.2.6). These can be written (Novozhilov, 1964)

$$\left.\begin{aligned}\frac{\partial}{\partial u^1}(\tilde{t}_{11}g_{22}^{\frac{1}{2}})+\frac{1}{g_{11}^{\frac{1}{2}}}\frac{\partial}{\partial u^2}(t_{12}g_{11})-\frac{\partial g_{22}^{\frac{1}{2}}}{\partial u^1}\tilde{t}_{22}=0\\ \frac{1}{g_{22}^{\frac{1}{2}}}\frac{\partial}{\partial u^1}(\tilde{t}_{12}g_{22})+\frac{\partial}{\partial u^2}(\tilde{t}_{22}g_{11}^{\frac{1}{2}})-\frac{\partial g_{11}^{\frac{1}{2}}}{\partial u^2}\tilde{t}_{11}=0\end{aligned}\right\} \tag{17.4.14}$$

The extra boundary conditions required are

$$V_2(0)=V_2(\tilde{l}_f)=0 \tag{17.4.15}$$

The system of equations is now seventh-order. As in the axisymmetric case, both stability and sensitivity analyses can be carried out. Very limited investigations were undertaken by Yeow (1972) from which it might tentatively be concluded that $m=1$ modes are more unstable than $m=0$ modes.

EXERCISE 17.4.1. Show that there is, in general, no possible film-stretching solution with $a=a_0$. Argue from the dynamical relations (17.5.8) and (17.5.9).

17.5 THE COLD-DRAW FILM-BLOWING PROCESS

An alternative to immediate inflation of the tube extruded from the die is provided by the cold-draw system in which the tube is first chilled and then reheated to a temperature just below the melting point when rapid biaxial extension takes place. This process, shown in Fig. 17.2, leads to a highly-oriented material. The rheological behaviour of the material will be wholly different from that relevant for melt film-blowing of the same material (PP for example), and the stress levels will be much higher. The subject has been considered by Pearson & Gutteridge (1978), who based their work on a specific industrial process, and used a rheological model proposed by Funt (1975). For our purposes here, we can describe it by

$$e_i - \frac{1}{e_i^3 e_j^2} = \int_{-\infty}^{t} \frac{\hat{t}_i(t')/h(t')}{\lambda\{T(t')\}G\{T(t')\}} \exp\left(-\int_{t'}^{t} \frac{dt''}{\lambda\{T(t'')\}}\right) dt' \quad (17.5.1)$$

where

$$(i, j) = (1, 2) \quad \text{or} \quad (2, 1)\dagger$$

it being assumed that, for all times for which λ is not effectively infinite (T being sufficiently low), the deformation history will be irrotational extension along the axes 1 and 2. The particular forms taken for $G(T)$ and $\lambda(T)$ were derived empirically as

$$G(T) = G_0\left(1 - \frac{T}{T_m}\right), \lambda(T) = \lambda_m + \lambda_0\left(1 - \frac{T}{T_m}\right), T < T_m \quad (17.5.2)$$

The engineering strains e_1 and e_2 refer to an undeformed state as in Subsection 17.3.2, i.e. (17.3.12) where

$$e_1 = (C_{11}^{-1})^{\frac{1}{2}}, \qquad e_2 = (C_{22}^{-1})^{\frac{1}{2}} \quad \text{with} \quad e_1 e_2 e_3 = 1$$

in the case of nearly-incompressible materials.

The stress-equilibrium equations (17.2.9) and (17.2.12) will still apply. The temperature field will in this case be supposed known, i.e. $\bar{T}(z)$ will be a given parametric function. For convenience $\bar{T}(z)$ was taken to be fixed, $\bar{T}_*$, for all $z < 0$, and so an elastically-stretched (undeforming) tube reaches the plane $z = 0$, at which point it is supposed to suffer a sudden discontinuity in temperature that precipitates stretching. The initial conditions will therefore be similar to those discussed in Subsection 17.3.2. In this case, we can suppose the dimensions and forward velocity of the unstretched tube to be known, i.e. a_*, h_*, v_{1*} to be given.

From (17.5.1) it is clear that the effective strains are

$$e_{1\,\text{eff}} = \tilde{v}_1 - \tilde{v}_1^{-3}\tilde{a}^{-2}, \qquad e_{2\,\text{eff}} = \tilde{a} - \tilde{a}^{-3}\tilde{v}_1^{-2} \quad (17.5.3)$$

where $\tilde{v}_1 = v_1/v_{1*}$, $\tilde{a} = a/a_*$ are the dimensionless engineering strains. Using the new strain measure $e_{1\,\text{eff}}$ and $e_{2\,\text{eff}}$, the integral relation (17.5.1) can be converted into the linear Kelvin–Voigt model, essen-

† Compare eqn (5.2.28) which relates to an elastico-viscous fluid. Equation (17.5.1) is the same model as (16.2.5) and relates to a visco-elastic solid.

tially (16.2.5) again, defined by

$$t_i/G = e_{i\ \mathrm{eff}} + \lambda \dot{e}_{i\ \mathrm{eff}} \tag{17.5.4}$$

where G and λ are functions of temperature. Kinematically, it is clear that

$$\mathrm{d}z = v_1 \cos\psi\, \mathrm{d}t \tag{17.5.5}$$

and so (17.5.4) becomes

$$t_i/G = e_{i\ \mathrm{eff}} + \lambda v_1 \cos\psi\, \mathrm{d}e_{i\ \mathrm{eff}}/\mathrm{d}z \tag{17.5.6}$$

It is convenient to work in dimensionless variables

$$\tilde{z} = z/a_*, \qquad \tilde{F} = F/\pi a_*^2 \Delta p, \qquad \tilde{G} = Gh_*/a_*\Delta p, \qquad \tilde{\lambda} = \lambda v_{1*}/a_* \tag{17.5.7}$$

The stress equations (17.2.9) and (17.2.12) now become

$$\frac{\mathrm{d}\tilde{v}_1}{\mathrm{d}\tilde{z}} = \frac{\sec^2\psi(\tilde{F}+\tilde{a}^2) - (1-\tilde{v}_1^{-4}\tilde{a}^{-2})\sec\psi\,\tilde{G} - 2\tilde{\lambda}\tan\psi\,\tilde{G}\tilde{a}^{-3}\tilde{v}_1^{-3}}{\tilde{\lambda}\tilde{G}(1+3\tilde{v}_1^{-4}\tilde{a}^{-2})} \tag{17.5.8}$$

$$\begin{aligned}\frac{\mathrm{d}\psi}{\mathrm{d}\tilde{z}} = (-2\tilde{a} + (1-\tilde{a}^{-4}\tilde{v}_1^{-2})\tilde{v}_1^{-1}\tilde{G}\cos\psi + \tfrac{1}{2}\tilde{\lambda}\tilde{G}\tilde{a}^{-1}\sin 2\psi(1+3\tilde{a}^{-4}\tilde{v}_1^{-2})\\ + 2\tilde{\lambda}\tilde{a}^{-3}\tilde{v}_1^{-3}\cos^2\psi\tilde{G}\,\mathrm{d}\tilde{v}_1/\mathrm{d}z)/(\tilde{F}+\tilde{a}^2)\end{aligned} \tag{17.5.9}$$

with (17.2.10) providing a third equation to specify the three unknown variables $\tilde{v}_1$, ψ and $\tilde{a}$. The boundary conditions will be

$$\tilde{z} = 0-: \tilde{G}(\bar{T}) = \frac{\tilde{a}_0^5\tilde{v}_{10}^3}{(\tilde{a}_0^4\tilde{v}_{10}^2 - 1)} = \frac{\tilde{F}+\tilde{a}_0^2}{2(1-\tilde{a}_0^{-2}\tilde{v}_{10}^{-4})} \tag{17.5.10}$$

which follow from (17.5.3), (17.5.4) with $\dot{e}_{1\ \mathrm{eff}} = 0$, (17.2.9) with F constant and $\psi = 0$, and (17.2.12);

$$\tilde{z} = \tilde{z}_\mathrm{f}: \psi = 0, \tilde{v}_1 = \mathrm{Dr}, \tilde{a} = \mathrm{Bu} \tag{17.5.11}$$

If F and Δp are prescribed, Dr and Bu follow from the solution and vice versa, assuming a solution exists.

At $z = 0$, $\tilde{a}$ and $\tilde{v}_1$ will be continuous (unless $\lambda \equiv 0$); an elementary force balance shows that ψ will also be continuous. However, there will be discontinuities in $\mathrm{d}\tilde{v}_1/\mathrm{d}\tilde{z}$ and $\mathrm{d}\psi/\mathrm{d}\tilde{z}$, which can be deduced from eqns (17.5.8) and (17.5.9).

Solutions were obtained by numerical computation for both the viscoelastic ($\lambda \neq 0$) and the elastic ($\lambda = 0$) cases. The Deborah number, as defined in (17.3.21), is simply $\tilde{\lambda}$, and in this case, was of order unity. The solutions obtained, as shown in Fig. 17.4, were totally different as regards $\tilde{v}_1(\tilde{z})$, the visco-elastic one being much closer to observations. The role of $\tilde{\lambda}$, which is related to λ_2 in the Jeffreys fluid when $\lambda_1 \rightarrow \infty$ with η_0/λ_1 fixed, is obviously an important one.

EXERCISE 17.5.1. Derive (17.5.8)–(17.5.11). Carry out the force balance to show that ψ is continuous at $z = 0$: it suffices to take a small element of length $\mathrm{d}z$ from $\phi = 0$ to $\phi = 2\pi$ that includes the line $z = 0$; if ψ is discontinuous then a singularity in hoop stress at $z = 0$ ensues.

17.6 SIMILARITY AND SCALE-UP

From the earlier sections, it is clear that theoretical predictions for bubble behaviour in blown-film processes are difficult to make. In practice, the necessary value for Bu is obtained (by trial and error) by the operator altering the amount of air contained within the bubble; Dr can be set directly by adjusting the speed of the haul-off rolls once the necessary output through the die has been established (or vice versa). Relatively little precise control of the cooling system is achieved. Although there are indications that cooling and stretch rates near the crystallization point for crystallizable polymers can affect the crystallinity and morphology of the frozen film, no well-documented investigations of this aspect of the process have been published. (Some results on orientation and stress at the frost-line are given in Choi *et al.* (1980).) It is therefore not surprising that most blown-film processes have been developed on-line.

It is worth considering how experiments carried out on very small-scale plant, particularly with new polymers, could be used to develop larger-scale processes. The question is one of scale-up, which is best investigated by means of dimensionless equations and groups. It will be assumed that geometrical similarity is sought, i.e. that Dr, Bu and $\tilde{l}_f$ will be held fixed in the scale-up. This still allows a_0, h_0 and v_{10} to be varied independently provided Wt and De are held constant (or remain negligible). Pd must be constant for geometrical similarity, so Wt constant requires that, for the case of an Oldroyd 8-constant isothermal model, for example, $a_0^2\rho_0 g/\eta_0 v_{10}$ remain constant in scale-up. If

(17.3.21) is used as the definition of De then $\lambda_1 v_{10}/a_0$ has also to be held constant in scale-up. These two requirements are incompatible unless one of Wt and De can be neglected, or unless $\eta_0/\rho_0 g\lambda_1$ scales as a_0. Obviously $\rho_0 g$ will vary little, so either two different materials, but with similar rheological properties in the sense that they can both be characterized by η_0 and λ_1, are used, or different temperatures are used to alter the ratio η_0/λ_1. For a simple Maxwell model, this ratio is

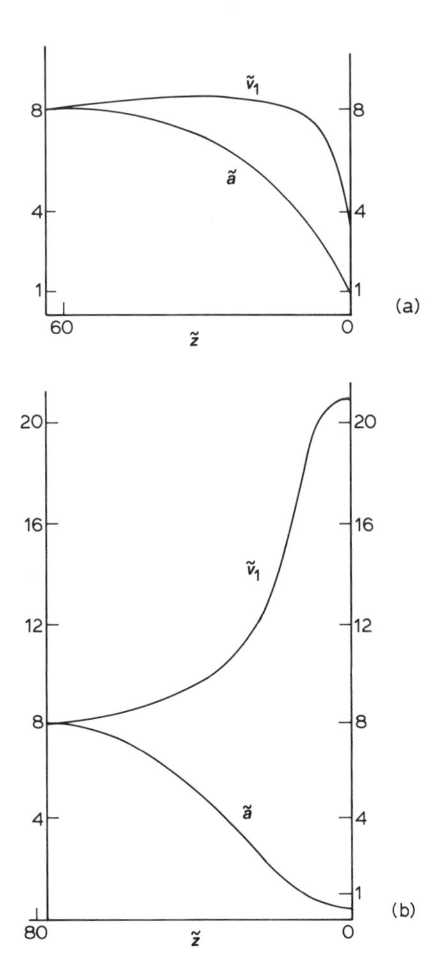

Fig. 17.4. Typical solutions from Pearson & Gutteridge (1978) for biaxial film-blowing process. (a) Viscoelastic model. (b) Elastic model.

G, the elastic modulus, which is generally regarded as varying relatively slowly with temperature.

One useful opportunity for using scale-up laws will be when $\mathrm{Wt} \ll 1$. The temperature field should be similar, but this would be relatively easy to achieve if v_{10} scaled as a_0 (thus ensuring constancy of the Deborah number), and h_0 were unaltered. It would merely require that the heat-transfer coefficients $h_a^{(i)}$ and $h_a^{(o)}$ of (17.2.13) be functions of $\tilde{z}$, with $T_a^{(i)}$ and $T_a^{(o)}$ held constant.

When a purely viscous approximation becomes relevant, i.e. when $\mathrm{De} \ll 1$, then more flexibility is possible. For a fixed temperature field, v_{10} will scale as a_0^2 to keep Wt constant. If, on the other hand, v_{10} is to scale as a_0, for example, then η_0 must increase with a_0: this can be achieved by reducing the temperature at the die-lips for increasing a_0 and adjusting the heat-transfer coefficients accordingly so as to achieve the same relative variation in η_0 with $\tilde{z}$. In practice, geometrical similarity is seldom achieved when units of different lay-flat width are operated.

EXERCISE 17.6.1. Investigate the requirements for geometrically similar scale-up in the case of cold-draw film-blowing as described in Section 17.5. Show that it is essentially the same as for an Oldroyd fluid model.

Part IV

CYCLIC PROCESSES

Chapter 18

Mixing

18.1 INTRODUCTION

Mixing is a process, essential to polymer processing, that is inseparable from the melting, heating, pumping and forming operations discussed so far. The processor's objective is usually to obtain as intimate and uniform a mixture of the various components of any compound as possible, so any mixing that is achieved adventitiously is entirely beneficial. However, in many circumstances, one or more specific mixing stages have to be introduced into the full process in order to obtain a sufficiently homogeneous material product.

It is not an entirely straightforward matter to describe mixing, a somewhat imprecise term used to cover a variety of physical changes. It is not strictly a part of mechanics as understood here, and so readers who wish to master the details should study one or more of the texts, chapters or articles devoted to the subject in the literature.†

The range of physical effects that are relevant can best be understood by considering various examples.

(1) The mixing of highly pigmented (master batch) granules of a given polymer with unpigmented (virgin) granules of the same polymer. This usually involves a tumbling operation, which ideally leads to a random distribution of pigmented particles in the mixture of solid particles.

A random distribution is one in which the probability of finding a pigmented particle at any location (occupied by some particle) in the

† See, for example, Tadmor & Gogos (1979, Chapters 7 and 11), Middleman (1977, Chapter 12), McKelvey (1962, Chapter 12), Frados (1976, Chapter 29), Funt (1976).

(random) mixture is equal to the proportion of pigmented particles in the mixture as a whole. In other words, no part of the whole can be expected to be systematically different (i.e. distinguishable) from any other. It is important to note that a random distribution is not the same as an ordered distribution; the latter is obtained, for example, when pigmented particles are inserted in a very definite and geometrically uniform fashion.

(2) The mixing of molten polymer formed by the melting of the mixture of solid particles (virgin and master batch) described in (1) above. This is usually carried out in a screw extruder and is termed laminar or simple mixing. Visual observation shows that the kinematics of screw extrusion converts the isolated, near spherical, blobs of strong colour in the compacted feed into long thin streaks or sheets of coloured melt interleaved with sheets of uncoloured melt. As the thickness of the sheets decreases, so the eye ceases to notice the distinction, until an apparently homogeneous, i.e. uniformly coloured, extrudate is formed.

(3) The blending of feed material, whether in granule, powder or molten form, whose molecular weight or molecular weight distribution varies from batch to batch. The object here is to even out differences that occur on a large length or timescale. A sudden discontinuity in the properties of the material supplied at the feed hopper of a screw extruder, for example, will be smeared out if the times taken by individual particles of polymer passing through the extruder differ (as indeed they will) over a relatively wide range. In formal terms, we say that the residence-time distribution will be broad, and the process involved is a form of distributive mixing.

(4) The dispersion of aggregates of carbon-black particles throughout a rubber matrix. Apart from any mixing of the aggregates as such with the rubber, a stress-induced break-up of the aggregates is involved. This is usually termed dispersive mixing. A similar result is often desired in the blending of polymers which are not miscible in a thermodynamical sense.

(5) The homogenization of temperature variations that are produced within a polymer melt by the mixing and pumping processes described in earlier chapters. This is usually achieved by a gentle local blending action involving static or moving mixing devices.

The above examples illustrate the following basic processes involved in mixing.

(A) Convective redistribution of essentially distinct or large elements of material. This is obviously true in the mixing of solid granules where each granule maintains its identity and relative motion of the granules is the only effect possible. It is also the dominant factor in large-scale blending that depends upon broad residence-time distributions: for example, if a deliberate and significant amount of back-mixing is designed into a twin-screw extruder (see Section 12.2), this can be regarded as contributing to convective mixing.

(B) Shear-induced extension of material surfaces or interfaces. This is caused by local velocity gradients and is the main mechanism whereby the granular appearance of the feed involved in Example (2) above is slowly converted into almost imperceptible striations.

The special case of uniform simple shearing motion is almost invariably chosen for diagrammatic illustration by authors discussing laminar mixing; in practice, such uniform steady flow fields are not sufficiently representative of processing flows (as far as mixing is concerned) to be considered as wholly typical. An approach based on a turbulent velocity field (Danckwerts, 1952; Batchelor, 1959) or general deformation field (Ottino *et al.*, 1979) provides equally useful quantitative estimates of surface stretching in many processing flows, particularly where rotating mixing head or epicyclic gear trains are employed.

Both convective and shear mixing are kinematical processes: only a knowledge of the velocity fields is necessary to predict their effects.

(C) Stress-induced dispersion of distinct material particles. This is a process whereby internal interfaces are ruptured and thus new ones created at or above a critical shear-stress level. This involves the dynamics of the mixing flow and obviously refers to situations where the purely passive surface stretching involved in laminar mixing is restrained by internal forces associated with interfaces.

Thus, in the intensive blending of dissimilar molten polymers, the separate domains of the dispersed phase resist deformation because of the equivalent of surface tension, while in the dispersion of carbon-black particles the same role is played by short range forces between primary particles within the aggregates.

(D) Diffusion of mass or heat. This is a molecular process that has already been explicitly introduced in our continuum formulation for the energy equation as Fourier's law (eqn (2.2.10)) and makes a significant contribution to thermal homogenization on a small length scale (see Example (5) above). Mass transfer by diffusion is only significant for relatively small molecules, such as plasticizers or other additives, or for segments of a polymer chain, as in heat sealing or the healing of weld lines. It can be said to have local significance at best. The combined effects of (B) and (D) are discussed in Batchelor (1959) and Ottino *et al.* (1979).

Criteria for mixedness can only be described in terms of relevant length scales. There are several of these to consider.

Ultimate particle size. For a pure fusible polymer this is the size of the polymer molecule (a few tens of nm) and usually below any length of interest in the process or the product. For a loaded resin or rubber compound, it is the fundamental particle size of the filler (carbon-black, glass, wood or mineral; powder or fibre) once the latter has been dispersed (of the order of μm) and may well be appreciable both visually and mechanically. For several pure polymers like uPVC and PTFE that are rarely fully melted during processing, the relevant particle size appears to be determined by the polymerization process itself, and internal grain boundaries can be relevant mechanically, particularly as regards fracture.

Sample size. Mixedness can be assessed in terms of any quantifiable characteristic of a finite portion of the material in question. For example, for a partly opaque polymer film, this can be the amount of light transmitted through any specified surface area; for a resin loaded with small glass beads, it can be the number of such beads in a given volume; for a polymer blend, it can be the proportion of one of the components in a similar given volume. In each case the sample size is important. Variations in the chosen property that show up for small samples may be imperceptible for larger samples. The length scale at which sample variations become irrelevant can crudely be termed the scale of mixing.

Population size. Equally important can be the amount of material from which the (small) samples discussed above are taken. Thus the total amount of polymer film subjected to light transmission tests (over portions of its surface area) may be that required to make a single plastic bag; or the total amount of glass loaded resin used for sampling the glass content may be a set of twenty-four nylon gear wheels. The mixing process may be arranged to be very efficient within this amount of material (typically a discrete fabricated article) but yet allow of small unacceptable variations between separate articles. The sum total of all the articles produced in one year, say, can be regarded as one population from which samples (individual articles) are taken, while each separate article can be regarded as a different population (with a very much smaller length scale) from which small portions of the article are taken as samples. Having decided upon the relevant population and sample sizes, and the quantifiable characteristics to be used, the intensity of mixing can be assessed in terms of the distribution about the mean of that characteristic: as mixing improves, so the standard deviation will decrease.

Only the processor and end user can decide on the sample and population scales that will be relevant in their particular application and on the intensity of mixing that must be achieved.

From the above concepts, it is reasonable to suppose that the efficacy of a mixer will depend upon the following factors.

(1) Moments of the residence-time distribution, in particular the first two. This will give the mean residence time and the spread of residence times, the second of which is a measure of back-mixing, itself one of the best forms of convective or distributive mixing. A broad distribution leads to good long-term mixing.

(2) Mean total shear: over the mean residence time. This gives a good indication of the total amount of laminar mixing sustained by each element of material on passage through the mixer. However, a broad residence-time distribution is likely to lead to a broad total-shear distribution, which is a bad characteristic for a mixer, in that the amount of local mixing sustained will vary widely from element to element; to be adequate for all elements of material, the mixer has to be wastefully over-effective for most.

(3) Maximum difference of the principal stresses sustained on passage through the mixer. The degree of dispersion, i.e. the ultimate size reduction of aggregates or coherent blobs, achieved will depend upon this maximum principal-stress difference.† Note that once broken up, further mixing of the separated parts of an aggregate can be achieved by the process of laminar mixing. As in (2) above, a wide spread in the distribution of maximum stresses achieved will be a bad characteristic for a dispersive mixer, though an excessive amount of high stress mixing will be unnecessarily wasteful. In cases where the mechanics of dispersion has been analysed in detail, the importance of this single scalar dynamical quantity has been confirmed.

† It may be shown that this maximum principal-stress difference is indeed a measure of the maximum shear stress, which is equally relevant for simple shearing, extensional and more complex flows.

(4) Proximity to heat-transfer surfaces. Where mixing is largely a matter of thermal homogeneity and where temperature control is important—as it is in most polymer processes—the ultimate requirement is for each material element to pass sufficiently close for a sufficiently long time to a metal surface held at a uniform temperature. This requirement, seldom stated, probably lies behind the design and efficiency of most of the static and rotary mixing devices mounted downstream of screw extruders, including breaker plates and gauzes.

This last requirement is unlike the other mixing requirements in that it is much less a matter of statistics and is capable of fairly precise quantitative assessment in terms of Brinkman, Nahme and Graetz numbers. Furthermore, flow patterns that are good for requirement (4) may contribute little to requirements (1) and (2) (distributive and laminar mixing) and vice versa. Figure 18.1 shows how the contacting of the dividing streamline in a T-bend with the wall surface contributes much more to thermal mixing than to laminar mixing.

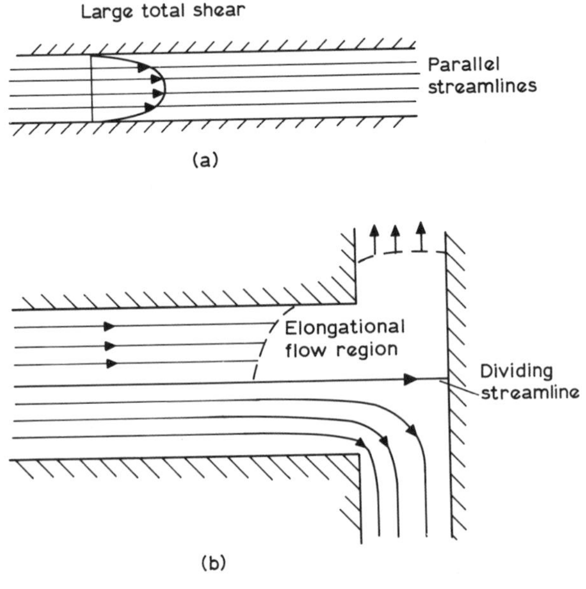

Fig. 18.1. (a) Steady simple shear in a long pipe: good laminar mixing: poor thermal homogenization. (b) 'Elongational' flow at a T-bend: little contribution to laminar mixing; important contribution to thermal homogenization.

The question of how close is the relation between total laminar shear and quality of mixing is a complex one. The viscometric flow shown in Fig. 18.1(a) is an admirable flow field for reducing the striation thickness when the striations start perpendicular to the streamlines. However, annular layers initially oriented parallel to the pipe axis are not mixed at all by flow through the pipe. Thus flows with constant stretch history (see Section 4.1) are not always ideal for mixing.

Different types of mixer will now be discussed separately.

18.2 ROLL MILLING

The 2-roll mill is still used as a mixing device in the rubber industry. It provides the third of the important examples of rolling mentioned at the beginning of Section B. It is shown in Fig. 18.2.

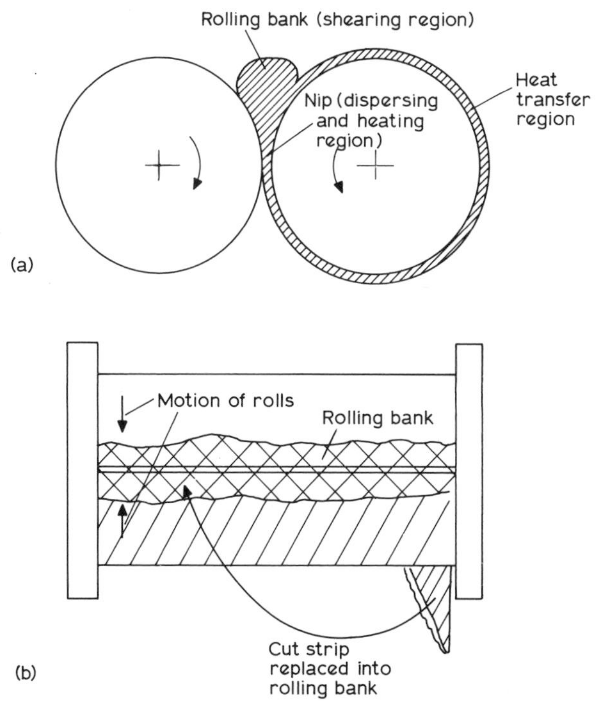

Fig. 18.2. Rolling mill. (a) Cross-sectional view of roll mills showing large rolling bank. (b) Plan view from above showing 'convective' effect of manual intervention.

The kinematics and mechanics of flow through the nip between two rotating cylinders has already been discussed exhaustively in Chapters 13 (Calendering) and 14 (Coating) for a homogeneous material. The main differences between those processes and milling is that the latter is usually concerned with a less homogeneous and more elastic material than the first two and that it develops with time, being more nearly a batch process than a continuous one.

It is possible to start with gum rubber, carbon-black, oil and various additives in separate measured quantities and to mix them on a roll-mill. The mechanics and kinematics of that non-uniform process is much too complex to discuss fully, though there are certain characteristic regimes of operation worth describing (see Tokita & White, 1966; White & Tokita, 1968) in qualitative terms. The more usual application is the incorporation of small quantities of curing agent into an already compounded but unvulcanized rubber. All of the factors mentioned earlier can be illustrated for this simple example. The solid powder has to be distributed evenly in a rough way; it then has to be intimately mixed and dispersed; meanwhile the temperature of the material has to be maintained relatively low to prevent premature curing.

In practice, convective or distributive mixing is often achieved by manual intervention: the operator cuts a strip of rubber emerging on the downstream side of the nip, removing it from the roll to which it had adhered and feeds it back into the rolling bank at a different axial position; this is shown diagrammatically in Fig. 18.2(b). The effect of this, as discussed in Subsection 13.2.2, is to cause slow lateral flow, with relatively low shear rates, within the rolling bank towards the depleted regions. As the greater portion of the compound, at any moment, lies within the rolling bank, this slow motion leads to a gradual equalization of the proportion of curing agent within the material as a function of the axial coordinate (x_2 in Fig. 13.2(b)), and thus corrects for any unevenness in the way the powder was initially sprinked onto the rolling bank of rubber. Note that an important factor is the relocation, without significant shearing, of a discrete portion of material (essentially physical mechanism (1) in Section 8.1).

Laminar, or shear, mixing is achieved within both the rolling bank and the nip and is due to shearing in the (x_1, x_3) plane, using the coordinate system of Fig. 13.1. Assuming the lubrication-approximation Newtonian solution for the velocity field given by

(13.1.53), we obtain, using (13.1.23), the local rate of shear

$$\frac{\partial v_1}{\partial x_3} = \frac{\partial^2 q}{\partial x_3^2} = \frac{2q_1}{h_0^2 \tilde{h}^2} \frac{\partial^2 \tilde{q}}{\partial \hat{x}_3^2}$$

$$= \frac{2q_1}{h_0^2} \frac{1}{(1+\tilde{x}_1^2)} \left[\left\{ \tfrac{3}{2}(\tilde{v}_{r1} + \tilde{v}_{r2}) - \frac{3}{(1+\tilde{x}_1^2)} \right\} \hat{x}_3 + \tfrac{1}{2}(\tilde{v}_{r2} - \tilde{v}_{r1}) \right] \quad (18.2.1)$$

The total rate of shearing/unit axial length within the region $x_{1\,\text{entry}} < x_1 < x_{1\,\text{exit}}$ can be written (but see Example 18.2.1 for a more detailed discussion)

$$\frac{q_1 X}{h_0} \int_{\tilde{x}_{1\,\text{entry}}}^{\tilde{x}_{1\,\text{exit}}} d\tilde{x}_1 \int_{-1}^{1} \left| \left\{ \tfrac{3}{2}(\tilde{v}_{r1} + \tilde{v}_{r2}) - \frac{3}{(1+\tilde{x}_1^2)} \right\} \hat{x}_3 + \tfrac{1}{2}(\tilde{v}_{r2} - \tilde{v}_{r1}) \right| d\hat{x}_3 = \frac{q_1 X}{h_0} \int_{\tilde{x}_{1\,\text{entry}}}^{\tilde{x}_{1\,\text{exit}}} d\tilde{x}_1 \left| \tfrac{3}{2}(\tilde{v}_{r1} + \tilde{v}_{r2}) - \frac{3}{(1+\tilde{x}_1^2)} + \frac{(\tilde{v}_{r2} - \tilde{v}_{r1})^2 (1+\tilde{x}_1^2)}{6[(\tilde{v}_{r1} + \tilde{v}_{r2})(1+\tilde{x}_1^2) - 2]} \right| \quad (18.2.2)$$

Clearly if $\tilde{x}_{1\,\text{entry}} \to -\infty$, this integral diverges, which suggests that the main contribution to laminar shear comes from the rolling bank, and so the nip-gap setting is to some extent irrelevant to the efficiency of the device as a laminar mixer provided only that

$$|x_{1\,\text{entry}}| \gg X \gg h_0 \quad (18.2.3)$$

On the other hand, the maximum shear stress is achieved when the shear rate (18.2.1) is a miximum. It may be shown that the latter is given by

$$\max\left(\frac{\partial v_1}{\partial x_3}\right) = \frac{q_1}{h_0^2} \max[\{3 - \tfrac{1}{2}(\tilde{v}_{r1} + \tilde{v}_{r2})(3-f)\}, \tfrac{1}{48}(\tilde{v}_{r1} + \tilde{v}_{r2})^2 (3+f)^2] \quad (18.2.4)$$

where f, the frictioning factor, defined by (13.1.41), is taken to be positive and is rarely larger than 1/4. The minimum value, for $\frac{1}{2}(\tilde{v}_{r1} + \tilde{v}_{r2})$ is about 4/5 and the maximum is 1, so the maximum shear stress will be of the form $a\eta_0 v^*/h_0$ with a fairly close to 1. This maximum value lies within the nip.

The total rate of generation of heat/unit axial length is given by

$$\dot{\mathscr{G}}=\frac{2\eta_0 q_1^2 X}{h_0^3}\int_{\tilde{x}_{1\,\text{entry}}}^{\tilde{x}_{1\,\text{exit}}} \mathrm{d}\tilde{x}_1 \int_{-1}^{1} \frac{1}{(1+\tilde{x}_1^2)}\left[\left\{\tfrac{3}{2}(\tilde{v}_{r1}+\tilde{v}_{r2})-\frac{3}{1+\tilde{x}_1^2}\right\}\hat{x}_3 + \tfrac{1}{2}(\tilde{v}_{r2}-\tilde{v}_{r1})^2\right]\mathrm{d}\hat{x}_3 = \frac{2\eta_0 q_1^2 X}{h_0^3}\int_{\tilde{x}_{1\,\text{entry}}}^{\tilde{x}_{1\,\text{exit}}} \frac{1}{(1+\tilde{x}_1^2)} \times\left[6\left\{\frac{\tilde{v}_{r1}+\tilde{v}_{r2}}{2}-\frac{1}{(1+\tilde{x}_1^2)}\right\}^2+\tfrac{1}{2}(\tilde{v}_{r2}-\tilde{v}_{r1})^2\right]\mathrm{d}\tilde{x}_1 \quad (18.2.5)$$

This integral does not diverge in the same way that the integral in (18.2.2) for the total shear diverged as $x_{1\,\text{entry}}\to-\infty$; from (13.1.54) for the case $\tilde{v}_{r1}=\tilde{v}_{r2}=(1+\tilde{x}^2_{1\,\text{exit}})=(1{\cdot}226)^{-1}$, we get

$$\dot{\mathscr{G}}\approx\tfrac{1}{8}\eta_0 W_1^2 d_1^{\frac{5}{2}} h_0^{-\frac{1}{2}} \quad (18.2.6)$$

The main contribution to the generation of heat comes from within the nip region, i.e. the region for which $|\tilde{x}_1|\leqslant 1$.

The full rate of heat transfer to the calender rolls is a more complex calculation. However, a good approximation is obtained by supposing the loss from the rolling bank and nip regions to be insignificant compared with the loss taking place while the material remains in contact with one of the two rolls (usually the hotter one, roll 1 say) in its 'calendered' thickness. This is the situation discussed in Section 13.3, and requires that

$$d\gg X \quad (18.2.7)$$

which is a weaker requirement than (18.2.3). The simplest calculation that can be carried out is to suppose that the material leaves the nip with a mean temperature $\bar{\bar{T}}_{\text{exit}}$ and re-enters the rolling bank with a lower mean temperature $\bar{\bar{T}}_{\text{entry}}$. The relevant Graetz number Gz_1 is defined by (13.3.4).

If $\text{Gz}_1\ll 1$, then it is easily shown that

$$\bar{\bar{T}}_{\text{entry}}\sim\tilde{T}_1 \quad (18.2.8)$$

and

$$\dot{\mathscr{H}}\sim\tfrac{1}{2}W_1 d_1 h\rho\Gamma(\bar{T}_{\text{exit}}-T_1) \quad (18.2.9)$$

If $\text{Gz}_1\gg 1$, then the penetration depth of the cooling is much smaller than h_1 and $\dot{\mathscr{H}}$ can be approximated by (see Example 18.2.2)

$$\dot{\mathscr{H}}\sim 2^{\frac{1}{2}} d_1\rho\Gamma(\kappa W_1)^{\frac{1}{2}}(\bar{T}_{\text{exit}}-T_1) = \tfrac{1}{2}\pi^{-\frac{1}{2}} W_1 d_1 h_1 \rho\Gamma\,\text{Gz}_1^{-\frac{1}{2}}(\bar{T}_{\text{exit}}-T_1) \quad (18.2.10)$$

The relation

$$\dot{\mathcal{G}} = \dot{\mathcal{H}} \tag{18.2.11}$$

gives $\bar{T}_{exit}$ directly. For $Gz_1 \gg 1$, we use the further relation

$$\tfrac{1}{2}W_1 d_1 h_1 \rho \Gamma(\bar{T}_{exit} - \bar{T}_{entry}) = \dot{\mathcal{G}} = \dot{\mathcal{H}} \tag{18.2.12}$$

to give

$$\bar{T}_{entry} = \left\{1 - \left(\frac{4\pi}{Gz_1}\right)^{\frac{1}{2}}\right\}\bar{T}_{exit} \tag{18.2.13}$$

In order to evaluate the efficiency of the device we must know the volume/unit axial length A_{mat} of the material held on the mill and the time of mixing t_{mix}. To the order of approximation used here

$$A_{mat} = h_0 X \tilde{x}_{1\,entry}(1 + \tfrac{1}{3}\tilde{x}_{1\,entry}^2) \approx \tfrac{1}{3}h_0 X \tilde{x}_{1\,entry}^3 \tag{18.2.14}$$

The mean total laminar shear sustained by any element is then given by

$$Mx = \frac{t_{mix}}{A_{mat}}\{\text{expression of type (18.2.2)}\} \tag{18.2.15}$$

while the mean number of passes through the nip

$$Np = q_1 t_{mix}/A_{mat} \tag{18.2.16}$$

The mean output of the mill will be

$$\bar{\dot{V}} = LA_{mat}/t_{mix} \tag{18.2.17}$$

where L is the length of the rolls, while the work done/unit volume will be

$$\bar{W}_{mix} = L\dot{\mathcal{G}}/\bar{\dot{V}} \tag{18.2.18}$$

For any particular system, we can expect the requirements to be a lower limit on Mx, Np and the maximum shear stress separately. The temperature constraint can either be a simple upper bound on $\bar{T}_{exit}$, or some combination of $\bar{T}_{exit}$, $\bar{T}_{entry}$ and t_{mix}. This will be given by the curing characteristics of the mixture. It is advantageous to maximize $\bar{\dot{V}}/L$ and to minimize $\bar{W}_{mix}$. At our disposal are W_1, h_0 and $\tilde{x}_{1\,entry}$ (i.e. A_{mat}) according to the usual operating adjustments provided.

This immediately suggests that T_1 should be as low as possible, and the lower limit for this is usually set by the requirement that the material should adhere to the rolls and be drawn into the nip.

In practice, optimum operating conditions are usually obtained empirically. If they are known to be on the edge (in terms of the mixing constraints) of the range of acceptable operating conditions, then it is possible to scale up from one piece of equipment to a larger one (or vice versa) according to whichever criteria still have to be first satisfied for an optimum solution. To do this, it has to be noted that

$$\text{Maximum shear stress} \propto \eta_0 W_1 d_1 h_0^{-1}$$

$$\left.\begin{aligned} \mathrm{Mx} &\propto W_1 t_{\mathrm{mix}} d_1 h_0^{-1} \tilde{x}_{1\,\mathrm{entry}}^{-2} \\ \mathrm{Np} &\propto W_1 t_{\mathrm{mix}} d_1^{\frac{1}{2}} h_0^{-\frac{1}{2}} \tilde{x}_{1\,\mathrm{entry}}^{-3} \\ \bar{T}_{\mathrm{exit}} - T_1 &\propto \begin{cases} \text{(i)} \;\; W_1 d_1^{\frac{3}{2}} h_0^{-\frac{3}{2}} (\eta_0/\rho\Gamma) \\ \text{(ii)} \;\; W_1^{\frac{3}{2}} d_1^{\frac{3}{2}} h_0^{-\frac{1}{2}} (\eta_0/\rho\Gamma\kappa^{\frac{1}{2}}) \end{cases} \end{aligned}\right\} \tag{18.2.19}$$

where

(i) $$W_1 h_0^2 \kappa^{-1} \ll 1$$

(ii) $$W_1 h_0^2 \kappa^{-1} \gg 1$$

and that

$$\left.\begin{aligned} \bar{\dot{V}}/L &\propto d_1^{\frac{1}{2}} h_0^{3/2} \tilde{x}_{1\,\mathrm{entry}}^{3} t_{\mathrm{mix}}^{-1} \\ \bar{W}_{\mathrm{mix}} &\propto \eta_0 W_1^2 d_1^2 h_0^{-2} t_{\mathrm{mix}} \tilde{x}_{1\,\mathrm{entry}}^{-3} \end{aligned}\right\} \tag{18.2.20}$$

It should be emphasized that all the results given so far in this section are for a temperature-independent Newtonian fluid and should be regarded as no more than illustrative. Real elastomers have very highly shear-dependent viscosities in the viscometric flows characteristic of the lubrication approximation, and are not insensitive to the temperature differences relevant in milling. Much of the analysis can be repeated, sometimes analytically, and certainly numerically, for power-law temperature-dependent fluids. The effect of elasticity is discussed below.

The fact that a low value of $\bar{T}$ is desirable both from the point of view of preventing early cure and for increasing η for dispersive mixing means that long relaxation times will be relevant in roll milling. Tokita & White (1966) have provided a useful distinction between various regimes of behaviour; their classification is given in Fig. 18.3 and some stimulating theoretical arguments are given there and in White & Tokita (1968). A related, but necessarily much briefer, discussion can be given here in terms of the three characteristic kinematic time scales of the roll-mill flow. These are as follows.

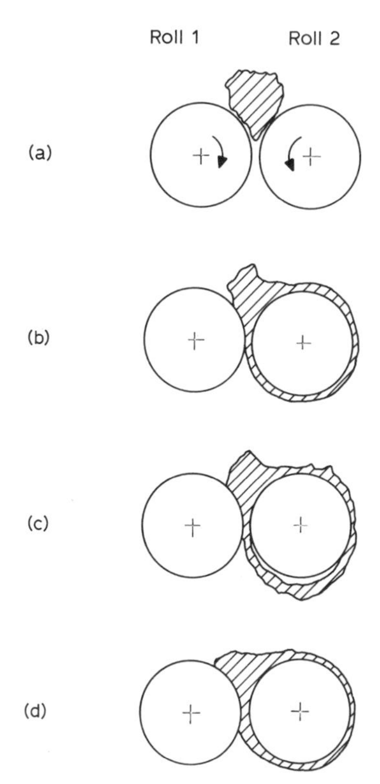

Fig. 18.3. Crude diagrammatic cross-sectional representation of rubber on mill as $T_1 > T_2$ increases (or equivalently W_1 decreases) from (a) to (d); (a) slip at boundary between stiff (very high relaxation time) rubber and cold metal roll; (b) highly elastic, tight band of rubber around slightly warmer roll; (c) loose band of coherent but melt-fractured rubber passing through still warmer roll; (d) relatively smooth band of viscoelastic rubber adhering to warm roll. (From Tokita & White, 1966.)

(1) The time taken to pass from nip exit back into the rolling bank. This gives

$$t_{\text{rot}} = W^{-1} \tag{18.2.21}$$

and if the relevant relaxation time (for the stretched rubber band in regime (b)) is λ_b, then a relaxation number

$$\text{Rx} = \lambda_b W \tag{18.2.22}$$

can be defined.

(2) The time taken to travel a length X which is the significant length scale (in the lubrication approximation) over which 0(1) changes occur; in particular if a flow field of the converging (elongational) type described in Section 7.2 arises in the entry region to the nip, then

$$t_{\text{elong}} = X/Wd \tag{18.2.23}$$

and a Deborah number, for elongation, can be defined as

$$\text{De} = \lambda_e/t_{\text{elong}} = \lambda_e W/(d/h_0)^{\frac{1}{2}} \tag{18.2.24}$$

(3) The minimum time taken for unit shear in the nip, given by

$$t_{\text{shear}} = h_0/Wd \tag{18.2.25}$$

which leads to a Weissenberg number

$$\text{Ws} = \lambda_s/t_{\text{shear}} = \lambda_s Wd/h_0 \tag{18.2.26}$$

for the simple-shear flow in the nip.

From our earlier discussions in Chapter 5 of the rheology of polymer melts, we can deduce that

$$\lambda_b \geqslant \lambda_e > \lambda_s \tag{18.2.27}$$

with λ_e a slowly-decreasing function of t_{elong}^{-1} and λ_s a more rapidly-decreasing function of t_{shear}^{-1}. Thus we cannot assert without further analysis that

$$\text{Ws} \gg \text{De} \gg \text{Rx} \tag{18.2.28}$$

though we may suspect this to be the case. What we can more confidently assume, as do Tokita and White, is that all three dimensionless groups will decrease as T_1 increases, i.e. as we go from regimes (a) to (d).

Regime (b) requires that $\text{Rx} > 1$ or else the band would not stay taut. The fact that De is probably $\gg 1$ ensures that the material emerging from the nip is elongated in the flow direction. Indeed the argument suggests that right through the nip, there will be a largely elongational core flow lubricated by a highly sheared flow close to the rolls. The fact that $\text{Ws} \gg 1$ ensures that the simple-shear viscosity will be greatly reduced from its low-shear-rate value in the lubricating zones and so the necessary flat velocity profile is achieved. This is shown diagrammatically in Fig. 18.4. No satisfactory dynamical treatment of this flow regime has yet been given. It is related to the very high-pressure/high-shear-rate lubrication flows relevant in elastohydrodynamic lubrication flows where the role of the relaxation time has been investigated.

Regime (c) is clearly related to the phenomenon of melt fracture described in Section 9.1. Earlier investigations into pressure flow in pipes or channels have suggested that unstable flow (from an elastico-viscous point of view) arises when Ws or De exceeds a value that is greater than unity though not $\gg 1$; this would yield a criterion for instability in going from regime (d) to regime (c). Regime (d) can then be thought of as a largely viscous flow, with $\mathrm{De} \leqslant 1$. No dynamical treatment of this flow breakdown has been given. The sag shown in regime (c) is consistent with $\mathrm{Rx} \ll 1$.

EXAMPLE 18.2.1. Estimation of the shear imposed on fluid elements passing through the nip of a two-roll mixer.

To do this, we start by examining (18.2.1) carefully. For the special, symmetric, case $\tilde{v}_{r1} = \tilde{v}_{r2} < 1$, we see that the sign of the shear rate is given by the sign of $\{\tilde{v}_{r1}(1+\tilde{x}_1^2) - 1\}\hat{x}_3$. Thus for any given $\tilde{x}_1$, it changes sign at $\hat{x}_3 = 0$; also it changes sign at $\hat{x}_1 = -\tilde{x}_{1\,\mathrm{exit}}$ using the result (13.1.49). It is obvious that provided the flow is everywhere symmetric about $\hat{x}_3 = 0$, then shear histories are equal and opposite on either side of $\hat{x}_3 = 0$. It is on the basis of that rather trivial observation that the modulus signs have been used in eqn (18.2.2); otherwise, for the symmetric case, the total integral would have been zero, an obviously

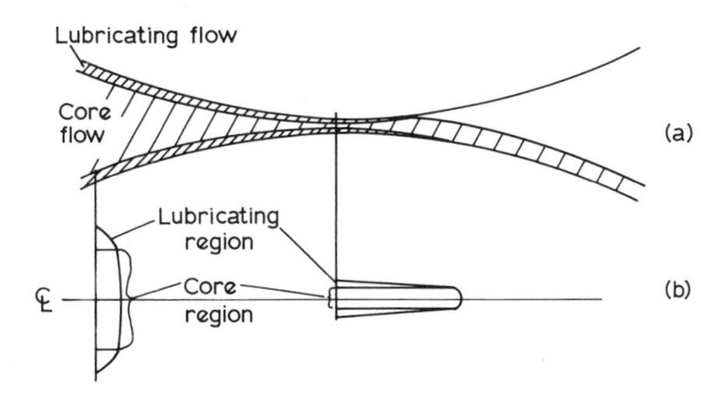

Fig. 18.4. Simple-elongational-core and highly-sheared-outer-flow model for high De flow through roll mills. (a) Cross-section. (b) Velocity profiles at two stations. (From Funt, 1976.)

unphysical result if total rate of shearing is to be used as a measure of mixing for the flow as a whole.

A more subtle approach must be used in connection with the reversal of sign at $-\tilde{x}_{1\,\text{exit}}$. To understand the issue involved, the behaviour of individual fluid elements (or more strictly, sheets) as they pass through the nip must be examined. To do this we consider the history of shear of a typical element (denoted by $\tilde{q}$) as it moves from $\tilde{x}_1 \to \infty$ through $-\tilde{x}_{1\,\text{exit}}$ to $\tilde{x}_{1\,\text{exit}}$: this requires us to evaluate

$$\int_{x_{1\,\text{entry}}}^{x_{1\,\text{exit}}} \frac{dx_1}{v(q)}\left(\frac{\partial v_1}{\partial x_3}\right)(q) = \frac{X}{h_0}\int_{\tilde{x}_{1\,\text{entry}}}^{\tilde{x}_{1\,\text{exit}}} \frac{d\tilde{x}_1}{1+\tilde{x}_1^2}\frac{1}{\tilde{v}}\frac{\partial \tilde{v}_1}{\partial \hat{x}_3}$$

for constant $\tilde{q}$ provided $v(q)>0$. In principle, the cubic equation (13.1.47) provides the relevant $\hat{x}_3(\tilde{q}, \tilde{x}_1)$, and indeed, for $\tilde{v}_{r1} = \tilde{v}_{r2}$, an algebraic form

$$\hat{x}_3 = \frac{\tilde{q}}{(\tilde{h}\tilde{v}_{r1}-1)}\left\{\left[1+\left(1-\frac{1-\frac{1}{3}\tilde{v}_{r1}\tilde{h}}{(1-\tilde{v}_{r1}\tilde{h})\tilde{q}^2}\right)^{\frac{1}{2}}\right]^{\frac{1}{3}} + \left[1-\left(1-\frac{1-\frac{1}{3}\tilde{v}_{r1}\tilde{h}}{(1-\tilde{v}_{r1}\tilde{h})\tilde{q}^2}\right)^{\frac{1}{2}}\right]^{\frac{1}{3}}\right\}$$

can be given. This can be substituted into the integral above to yield the required result. This will not yield the same result as (18.2.2) which has effectively replaced $\dfrac{\partial v_1}{\partial \hat{x}_3}$ by $\left|\dfrac{\partial v_1}{\partial \hat{x}_3}\right|$.

Another important point is that the simple integral given above does not allow for recirculation as is implied if $x_{1\,\text{entry}} < x_{1\text{st.i.}}$ (see Fig. 13.6 for definition of $x_{1\text{st.i.}}$).

However, on the basis that $\tilde{x}_{1\,\text{entry}} = 0(1)$, it can be seen that passage through the nip yields a total shear of order $(X/h_0) \gg 1$. The result given by (18.2.2) can be interpreted as meaning that for large rolling banks a volume of $0(\tilde{x}_1^3)$ is being subjected to a total mean rate of shear of $0(\tilde{x}_1)$; thus individual elements are being subjected to a mean rate of shear of $0(\tilde{x}_1^{-2})$. Each of them will spend on average an $0(\tilde{x}_1^{-2})$ proportion of their time in the nip, and so purely on order-of-magnitude grounds, the rolling bank flow will contribute the same order-of-magnitude shear as the nip flow.

EXAMPLE 18.2.2. Estimate of the rate of heat transfer to the roll in a two-roll mixer when $Gz_1 \gg 1$.

We shall suppose that the outer surface of the melt layer of thickness h_1 attached to roll 1 does not lose heat to its surroundings and that the layer has a uniform temperature, $\bar{T}_{\text{exit}}$, as it emerges from the nip. We suppose also that it is in contact with the roll over a distance πd_1 or, equivalently, for a time $2\pi/W_1$ before it joins the rolling bank. Finally, because $\mathrm{Gz}_1 \gg 1$ we suppose that heat can only diffuse effectively to the roll surface, held at temperature T_1, from regions that are relatively close to the roll, i.e. over distances small compared with h_1. It follows that the rate of heat transfer is independent of h_1 and is given by the simplest of unsteady heat-conduction problems, namely that for a half space $x>0$, initially at a uniform temperature $\bar{T}_{\text{exit}}$, whose surface $x=0$ is suddenly (at $t=0$) held at temperature T_1.

The governing equation can be written

$$\frac{\partial T}{\partial t} = \kappa \frac{\partial^2 T}{\partial x^2}$$

$$T(0, x) = \bar{T}_{\text{exit}}, \qquad T(t, 0) = T_1$$

and the total heat flow/unit area in time $t_1 = 2\pi/W_1$ is given by

$$\frac{\dot{\mathcal{H}} t_1}{\pi d_1} = \int_0^{t_1} \alpha \left(\frac{\mathrm{d}T}{\mathrm{d}x}\right)_{x=0} \mathrm{d}t$$

The solution to the equation and given boundary conditions for T is

$$T = \bar{T}_{\text{exit}} - (\bar{T}_{\text{exit}} - T_1)\operatorname{erfc}(x/2\sqrt{\kappa t})$$

whence

$$\left(\frac{\mathrm{d}T}{\mathrm{d}x}\right)_{x=0} = \frac{(\bar{T}_{\text{exit}} - T_1)}{(\pi \kappa t)^{\frac{1}{2}}}$$

and so

$$\dot{\mathcal{H}} = \frac{2\pi^{\frac{1}{2}} d_1}{t_1^{\frac{1}{2}}} \kappa^{\frac{1}{2}} \rho \Gamma (\bar{T}_{\text{exit}} - T_1)$$

which leads to (18.2.10).

EXERCISE 18.2.1. Carry out the integration with respect to $\mathrm{d}\hat{x}_3$ in eqn (18.2.2).

EXERCISE 18.2.2. Follow through the same argument for $\tilde{v}_{r1} > \tilde{v}_{r2} > 0$, as given in Example 18.2.1.

EXERCISE 18.2.3. Confirm that, if material makes on average many passes through the nip during its time on a two-roll mill, uniformity of mixing is assisted by unsteadiness in the rolling bank flow.

EXERCISE 18.2.4. Derive the result (18.2.4) (where the two local maxima lie at different values of x_1) and show that the range given for $\frac{1}{2}(\tilde{v}_{r1}+\tilde{v}_{r2})$ of $\sim[\frac{4}{5}, 1]$ follows from (13.1.49), (13.1.50) and (13.1.54).

EXERCISE 18.2.5. Derive the results (18.2.5) and (18.2.6) providing an accurate value for the coefficient approximated by 1/8 in (18.2.6).

EXERCISE 18.2.6. Carry through the analysis leading to results (18.2.1)–(18.2.20) for a power-law fluid when $\tilde{v}_{r1}=\tilde{v}_{r2}$. Start from Example 8.1.1 and Exercise 13.1.5.

Show, in particular, that the integral in (18.2.2) still diverges and that $\dot{\mathscr{G}}$ is still finite. Show also that the maximum shear stress can still be written as proportional to $\eta v^*/h_0$ when η is suitably defined.

Consider how the argument would be affected if $\tilde{v}_{r1}\neq\tilde{v}_{r2}$ without attempting detailed calculations.

18.3 INTERNAL MIXERS

These are batch machines that were developed primarily for the rubber industry to masticate raw rubber (reduce its high molecular weight fraction) and to disperse carbon-black uniformly within the rubber. The typical internal mixer consists of a closed chamber within which rotate a pair of intermeshing rotors. Figures 18.5 and 18.6 (taken from Funt, 1976) show the basic arrangement of the rotors in the chamber and examples of the rotors themselves for well-known versions. A lucid account of their mode of action is given in both Funt (1976, Chapter 5) and Tadmor & Gogos (1979, Section 11.9), with the former providing the best analysis of the changes that take place within the material over the mixing cycle. Both authors take the view that a complete dynamical analysis of the process is not feasible. It is nevertheless possible to interpret the mode of action of the mixer in terms of the various concepts discussed above, and to compare it directly with the roll mill analysed in the previous section.

We turn first—as has been done throughout this text—to the

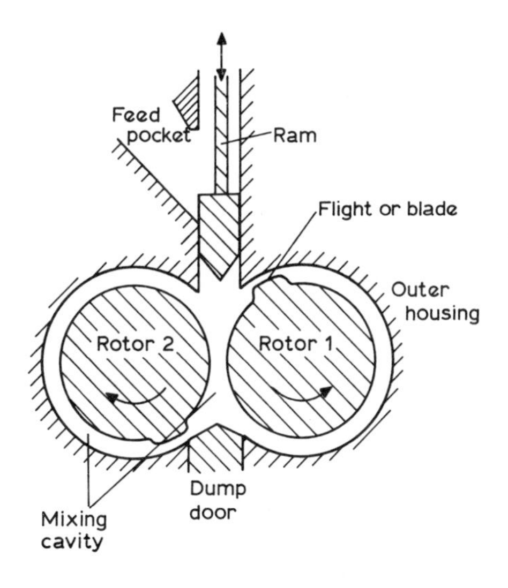

Fig. 18.5. Schematic cross-sectional view of an internal mixer.

geometry of the system. An idealized system is shown in Fig. 18.7. The relevant radial length scales are:

d_{cav}—the diameter of each of the cylindrical cavities;

h_1—the depth of the main 'channel' between the rotor and the cavity wall;

h_2—the clearance between the flight on each rotor and the cavity wall. For convenience we suppose that this is also the clearance between each flight and the opposing rotor.

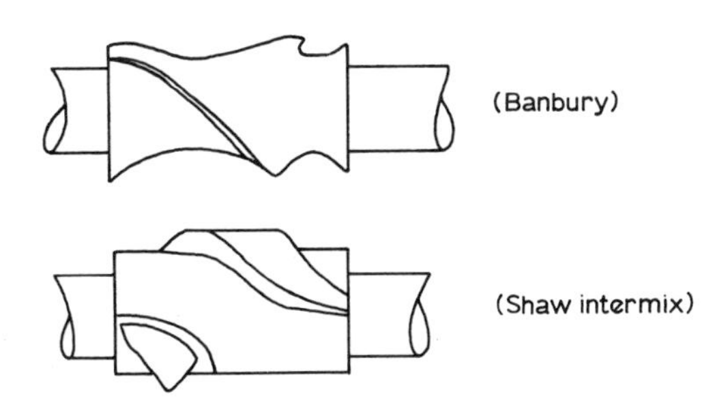

Fig. 18.6. Typical rotor shapes for internal mixers.

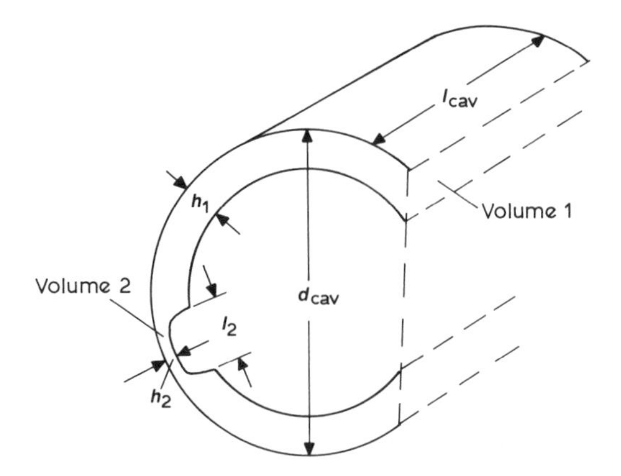

Fig. 18.7. Idealized dimensions for internal mixer.

A fair approximation is provided by the inequalities

$$h_2 \ll h_1 \ll d_{cav} \tag{18.3.1}$$

These can be compared with the inequalities

$$h_0 \ll X(<x_{1\ entry}) \ll d_1 \tag{18.3.2}$$

that arose in roll mills. The length of the rotors, l_{cav}, will be of the same order as d_{cav}, while the length of the clearance gap, l_2, will be of order h_1.

Intensive (dispersive) mixing will take place in the flight clearances, each of volume

$$V_2 = l_{cav} h_2 l_2$$

Laminar mixing will take place mainly in the channels between rotor and cavity wall, each of volume

$$V_1 = \pi l_{cav} h_1 d_{cav}$$

Distributive mixing, in an axial direction, will take place partly in the channels, because the flights are not purely axial, and partly in any free volume V_0 left above the intermeshing region, where

$$V_0 \leqslant 0(l_{cav} d_{cav}^2)$$

depending on the volume actually loaded in the batch.

Heat will be generated mainly in volume V_1, being transferred to both the walls of the cavity and the rotors, particularly if they are heated or cooled by water circulating in channels provided close to the rubber/metal surfaces. One of the characteristics of high-speed, short-cycle operation is that overall thermal equilibrium is not attained over most of the cycle: the bulk temperature of the mixture increases monotonically in general at the earliest stages of the cycle, the metal is hotter than the material fed in and so generation $\dot{\mathcal{G}}$ and heat transfer $\dot{\mathcal{H}}$ both tend to increase the temperature, while just before dumping the rubber compound is at its hottest and tends to transfer heat to the metal of the machine. $\dot{\mathcal{G}}$ and $\dot{\mathcal{H}}$ then move into balance.

Changes in the rheological properties of the mixture also occur during the mixing cycle. Dispersion of the carbon-black in the rubber tends to increase the viscosity and relaxation time, while adhesion at the metal surfaces clearly changes when slippery additives are incorporated into the gum rubber. Reduction in molecular weight of the rubber by mastication will reduce viscosity and relaxation time.

In general therefore, the torque on the motors—indicative of energy dissipation—will tend to rise initially as the carbon-black is incorporated and then fall as the temperature rises and the rubber is masticated. In modern equipment the charge is usually force-fed by the ram into the nip between the rotors, though in older unpressurized machines the charge would often sit above the nip without being drawn in until its temperature had risen sufficiently by conduction from the metal surfaces to adhere to them. Under the latter circumstances, a sequence of events rather similar to those described in Fig. 18.3 (a)–(d), could be expected to occur.

The possible operating variables for a typical internal mixer are batch size (varied little in practice), heat-exchange-fluid temperature, rotor speed and mixing time. There is no equivalent of varying the nip gap h_0 in a roll mill. However, the various length scales can be adjusted in the design problem.

Rules governing scale-up can be obtained once the relevant criteria for mixing have been decided upon.

18.4 CONTINUOUS MIXERS: EXTRUDERS

Both single- and twin-'screw' extruders are used for mixing, with the latter being the more common. The twin-screw extruder analysed in Chapter 12 was of particularly simple geometry, in that the helical channels were assumed to be of constant cross-section. An obvious

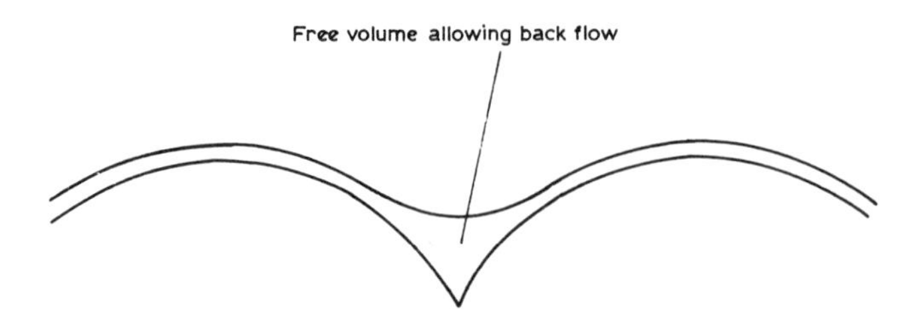

Fig. 18.8. Axially continuous free volume in continuous mixer.

characteristic of many of the screw extruders used for mixing (or compounding as it is often called) is the remarkable variety and variation of channel geometry, involving interruptions to the screw flight, reversals of helicity, barriers in the main channel, grooves in the barrel and outsize clearances.

We may analyse the twin-screw extruder of Chapter 12 in terms of the mixing function listed as A–D in Section 18.1. Convective redistribution, interpreted in terms of a broad residence-time distribution, is obviously encouraged by back mixing. This was exemplified in Fig. 12.12 for the back flow associated with the tetrahedron gap and with any additional axially continuous free volume in the intermeshing region. This is shown in Fig. 18.8. Laminar mixing is achieved in all the channels: whether it suffices in any given case is a question to be answered by either experiment or calculation. Stress-induced dispersive mixing probably takes place most effectively in the calender gaps—it was earlier suggested in Subsection 12.2.2 that these gaps were responsible for melting also and that proper design should ensure a sufficient number of passages through such a gap for the typical material element. Lastly, thermal mixing and control is achieved by having each element of material pass through relatively narrow channels, such that the flow Graetz number is sufficiently small to ensure good heat transfer to the bounding metal surfaces.

Many of the geometrical complications adopted in commercial continuous screw mixers can be interpreted as special devices aimed at avoiding the possibility that a proportion of the material held in the centre of the C-shaped volumes (see Fig. 12.4) will pass virtually unmixed from the feed pocket to the nose end of the screw in a helically symmetric system. All of those mentioned above will reduce the pumping efficiency of the screw and so will increase both the mean residence time and the total mean shear. Interruptions to the flight will promote distributive mixing in the channel cross-section, and will effectively exchange material between different positions along the channel length (separated by roughly πd_s). Barriers in the main channel will provide regions of intensive (dispersive) mixing through which every element of material must pass, and will simultaneously improve thermal homogeneity (see for example Martin 1972).

Full dynamical and thermal analysis is even less practicable for specialized mixing screws than in the case of internal mixers, though the principles applied in Chapters 8, 9, 10, 11, 12 and 13 and in Section 18.2 above can be used with profit in special cases.

Chapter 19

Injection Moulding

19.1 THE PROCESS

Injection moulding is the most complex of the processes yet discussed. It is essentially cyclic and in many ways the relevant flows are more unsteady than in any of the situations considered earlier. The usual sequence of operations involved is:

(A) Preplasticization of the polymer. For hard thermoplastics the feed will usually be granular; for elastomers, it can be either a long, effectively continuous strip (or rope) fed from a roll, or cut pieces fed from a hopper; for hard thermosets, it will usually be a powder, though in this last case there may not be any preplasticizing as such.

In most modern equipment, a single-screw extruder is used for preplasticizing. The plasticized polymer (or compound) is accumulated in a reservoir. This may be the front end of the screw barrel itself—in which case the screw is driven back by the pressure in the reservoir—or a separate chamber with a piston which is driven back as the reservoir fills.

(B) Injection of the preplasticized polymer. This is a rapid operation whereby the molten polymer in the reservoir is rammed through a series of channels into the relevant cavities of the closed mould. For thermoplastics, the mould is colder than the injected polymer melt; for cross-linking elastomers or thermosetting resins, the mould is hotter than the injected compound. In most cases the preplasticizing screw is also the ram—it thus reciprocates as well as rotates and is fitted at the nose end with a non-return (ring) valve.

(C) Cooling or curing of the injected material. Once the mould is filled, the material is held under pressure within the mould cavity until it has frozen (or crystallized) in the case of thermoplastics, or cured (cross-linked) in the case of cross-linked elastomers or thermosets.

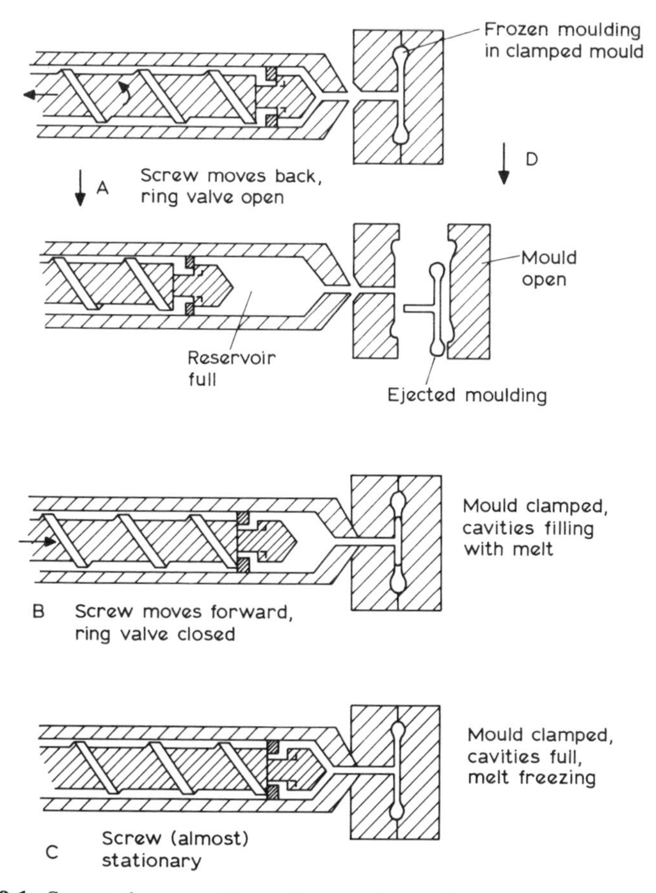

Fig. 19.1. Stages in operation of a reciprocating screw pre-plasticating injection-moulding machine.

(D) Ejection of the solidified material. This is a purely mechanical operation involving opening of the mould and extraction, or ejection, of the frozen or cross-linked part.

The operations A–D are shown diagrammatically in Fig. 19.1. From our point of view, the screw preplasticization step will not be basically different from that discussed in Chapter 11, although the dwell periods and the reciprocating action will lead to differences in detail, particularly important when estimating the temperature of the material in the reservoir prior to injection. These are discussed in Section 19.2.

The injection process involves higher pressure drops and flowrates than those discussed in the similar process of continuous die extrusion (Chapter 10) and so heat generation will be more significant. Indeed it plays a vital and desirable role, whether the material is thermoplastic or cross-linking: for the former it ensures that premature crystallization or vitrification, leading to only partially full mould cavities, does not arise; for the latter, it raises the temperature sufficiently for chemical cross-linking to take place relatively rapidly once the mould is filled.

One of the main objectives in die design is to achieve a desirable shape of extrudate: the flow pattern within the die channels and particularly near the die tips is all important. This will be less true for the flow into and within the mould: in injection moulding the prime objective is simply to fill the mould (although questions of part stability once it has been ejected mean that flow patterns are not unimportant). In general it can be said that temperature calculations are more important for moulding flows than for die flows and streamline calculation less so. This will be reflected in our treatment of injection moulding flows in Section 19.3. The dimensional requirements on the moulded part are provided by the dimensions of the mould cavity.

The cooling or cross-linking process is primarily one of heat conduction and heat release (or absorption) due to physical or chemical change. To first order the material is static. However, the density changes taking place during freezing or cross-linking mean that slow, highly viscoelastic, small-amplitude flows take place after the mould is filled. A mould-holding pressure is maintained to ensure that it remains filled and that the material does not shrink away from the mould cavity surfaces. These factors are considered in Section 19.4.

Because the injection moulding cycle involves a great number of mechanically or hydraulically driven motions of the metal parts of the injection moulding machine, more attention is often paid to the mechanics of the machine than to the mechanics of the material flow. The intention in this text is to discuss the latter rather than the former. Similarly because large temperature differences are often deliberately imposed between different parts of the metal surrounding the polymer, and because heating and cooling of the metal mould is often itself cyclic, more attention is sometimes paid to heat transfer within the metal than within the polymer. We shall be concerned here with the latter rather than the former. It should be realised however, that a full understanding of the performance of any given injection-moulding machine with any particular set of moulds cannot be achieved without simultaneous consideration of the mechanics of both machine and polymer flow, and of the overall heat balance.

There is, in practice, a greater demand for predictive calculations for new mould system designs than for new die designs. This is because continuous extrusion processes tend to have much longer runs in any given configuration than analogous moulding processes; one of the chief advantages of moulding is that new, precisely-dimensioned parts can be obtained, once a mould has been made, simply by exchanging the moulds fitted to the platens of a given injection machine. However, because moulds are themselves expensive to make and the cost of manufacture of moulded parts is strongly influenced by cycle time, it is important to be sure in advance that the mould can be adequately filled and the part chilled or cured within a specified time. We shall therefore describe in Section 19.5 Computer-Aided Design methods for injection moulding systems: this section will include a discussion of simple methods for network simulation.

19.2 PLASTICIZING

The most important method is screw preplasticizing. In some cases, single- or twin-screw extruders are used to fill a separate reservoir chamber through a non-return valve; except for the intermittency of their operation, their behavior can be analysed as in Chapters 11 and 12. In most cases, however, modern injection moulding machines employ a reciprocating single-screw extruder that contains its own reservoir at the nose end of the screw; the barrel ends in a nozzle that is chosen to fit the (sprue) entry to the mould channel system.

Analysis of the flow mechanics for this intermittent single-screw extrusion process involves some changes from that given in Chapter 11. The relevant differences will be outlined here.

19.2.1 Geometry

The geometry will be essentially the same as that shown in Fig. 11.2. The L/D ratio will usually and the compression ratio will often be lower, largely because the same quality of extrudate is not required as in extrusion. For some cold-fed rubber machines, there may be no compression ratio at all.

It proves convenient to choose a further frame of reference that is

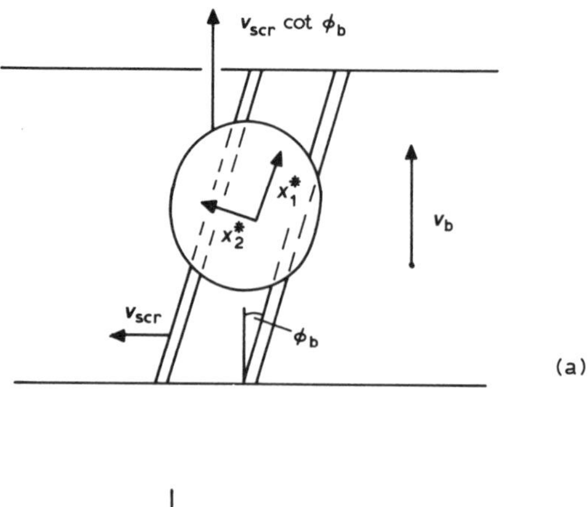

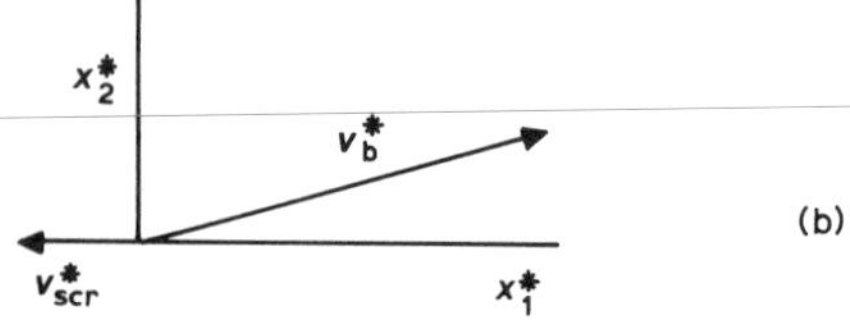

Fig. 19.2. (a) Unrolled geometry for single-screw in barrel, showing motions, relative to (x_1, x_2) coordinate system of Fig. 11.3(b), of barrel (v_b), screw (v_{scr}) and of new O frame of reference, with coordinate system (x_1^*, x_2^*). (b) Motion of barrel (v_b^*) and screw (v_{scr}^*) relative to new coordinate system x^*.

axially fixed but which rotates with the screw. This retains a fixed axial coordinate position for the feed pocket, but means that the position of the nose of the screw and the depth of the screw at any axial position vary with time as the screw retracts. This is best explained by Fig. 19.2, the analogue of Fig. 11.3, which considers the unrolled approximation. The object of this particular coordinate system is to simplify the calculations for quasi-steady operation when v_{scr} is constant and where screw flow analysis is carried out for x_1^* increasing from its (fixed value) at the feed pocket. The relevant relative barrel and screw velocities are given by

$$\mathbf{v}_b^* = (v_b \cos\phi_b - v_{scr}\cot\phi_b\cos\phi_b,\ v_b\sin\phi_b - v_{scr}\cos\phi_b,\ 0)$$
$$\mathbf{v}_{scr}^* = (-v_{scr}\operatorname{cosec}\phi_b,\ 0,\ 0) \tag{19.2.1}$$

It is assumed that v_b rapidly reaches its set value once the screw starts rotating. The motion of the screw pocket is along the same line as in (11.1.2) but with different period.

One kinematical result can be mentioned here, relating the output of the screw to its backwards velocity. Using the nomenclature of Chapter 11,

$$Q_{T1} = \tfrac{1}{4}\pi d_b^2 v_{scr} \tag{19.2.2}$$

If the moving $\mathbf{x}^*$ system is used then

$$Q_{T1} = Q_{T1}^* + v_{scr} \operatorname{cosec} \phi_b w_c h_s \tag{19.2.3}$$

where

$$w_c \operatorname{cosec} \phi_b = \cot \phi_b w_{ca} = \pi d_b w_c/(w_c + w_f)$$

A reasonable value for Q_{T1} is $\frac{1}{4} v_b w_c h_s$ and so, from (19.2.1),

$$\frac{v_{sc}}{v_b} \approx \frac{w_c h_s}{d_b^2} \approx \frac{h_s}{3d_b} \ll 1 \tag{19.2.4}$$

as expected, and Q_{T1}^* will not differ by much from Q_{T1} when $h_s/d_b \ll 1$. The effect of screw retraction will be minor.

19.2.2 Feed Zone

The frictional analysis can be carried out in the $\mathbf{x}^*$ frame and coordinate system with $\mathbf{v}_b^*$ and χ_b^* taking the place of $\mathbf{v}_b$ and χ_b in Subsection 11.2.1 (Fig. 11.6), and v_s^* being the relevant unknown velocity of the solid block. The volumetric delivery of the screw will be

$$\dot{V} = \dot{V}^* + v_{scr} \operatorname{cosec} \phi_b w_c h_s \tag{19.2.5}$$

The calculations for $\dot{V}^*$ would be as in Subsection 11.2.1. There is no effect of time variations in v_{scr}.

The viscous lubricating-layer analysis carried out in Subsection 11.2.2 leads to a steady similarity solution outlined in Subsection 11.2.3. This can be simply modified to deal with the case of v_{scr} constant but non-zero (see Example 19.2.1). The differences will arise only through the boundary condition (11.2.51i) which becomes $\tilde{v}_1^s(0) = -\tilde{v}_0^*$, with v_{W1}^* in (11.2.34) and $\tilde{v}_0^*$ suitably defined. To the extent that the analysis of Subsection 11.2.3 has neglected the consequences of the apparent movement of the feed pocket—described by the paragraph ending with (11.1.2)—slow changes with time of v_{scr} will not lead to additional inaccuracies, particularly if (19.2.4) holds. The extent to which such a solution, which implies a point $\tilde{x}_1^* = 0$ at which $\tilde{h} \to 0$, will ever be relevant in a preplasticizing screw that usually has a dwell period will depend upon whether the barrel is below the melting or softening point of the polymer.

When the screw is not rotating, as will happen during part of the injection-moulding cycle, heat will be conducted from the barrel through the lubricating layers to the solid bed. Such melting as occurs will increase the depth of the lubricating layers. This can be calculated by a one-dimensional heat-transfer analysis with a moving melting interface (the classical Stefan problem). For our purposes here, we need only an estimate of the depth of the molten layer if the dwell time (inactive interval) is t_d.

For $\mathrm{Sf} \gg 1$, where Sf is as defined in (11.2.44), a linear temperature profile within the melted layer can be used as a reasonable approximation, and so

$$\frac{\mathrm{d}h}{\mathrm{d}t} \approx \frac{\kappa_m}{\mathrm{Sf}\, h} \qquad (19.2.6)$$

with a solution

$$h_d = (2\kappa_m t_d/\mathrm{Sf})^{\frac{1}{2}} \qquad (19.2.7)$$

for the layer thickness at the end of the dwell time. This is the thickness that the steady lubricating layer of Subsection 11.2.3 will have achieved in a length of order $v_W t_d$, using relation (11.2.37). This length can be compared with the distance the lubricating layer is effectively convected during the active (screw rotating) part of the cycle; this length will be of order $\frac{1}{2}v_W t_a$ where t_a is the active time interval. Thus unless $t_a \gg t_d$, which is not to be expected in general, the melting effect of conduction will always ensure that very thin viscous layers governed by steady-similarity solutions will not arise. In any case, our mechanical analysis of the pre-melting zone is inadequate, because of the unquantified effect of melt percolation from the lubricating layer into the voids of the solid bed: experiment has shown, as discussed earlier, that this significantly delays formation of a melt pool.

19.2.3 Plasticating Zone

It is sufficient for our purposes to consider the 6-zone model, for this will probably cover most cases. We have already seen that the steady-state model is not relevant for all melt layers shallower than the value given by (19.2.7). Taking as representative values

$$t_d = 10\ \mathrm{s}, \qquad \kappa_m = 10^{-7}\ \mathrm{m^2\,s^{-1}}, \qquad \mathrm{Sf} = 10 \qquad (19.2.8)$$

$$h_d \approx 0{\cdot}5\ \mathrm{mm} \qquad (19.2.9)$$

This is clearly large compared with representative values of h_F or h_C and is as large at h_D. For screws less than 60 mm in diameter, h_D will be a significant fraction of $\frac{1}{2}h_s$ over much of the screw. Hence straightforward heat conduction will have a significant effect on the melting process.

The analysis of Subsection 11.3.1 can be modified to take account of the unsteady nature of the melting process, by treating w_A, h_C, h_D and h_E and by implication p, $\mathbf{v}$ and T as functions of time. The overall mass balance (11.3.1) would have to include the effect of density changes and becomes

$$\frac{\partial \dot{M}^*_T}{\partial x^*_1} = -\frac{\partial m^*_T}{\partial t} \tag{19.2.10}$$

where

$$m^*_T = \int \rho \, d\mathscr{A} \tag{19.2.11}$$

(cf. definition (11.4.5)). Similarly (11.3.5) becomes

$$\frac{\partial \dot{M}^*_{A1}}{\partial x^*_1} + \frac{\partial m^*_A}{\partial t} = -\sum \dot{m}^*_{AK} \tag{19.2.12}$$

Relation (11.3.7) now involves $\mathbf{v}^*_b$ and $\mathbf{v}^*_{scr}$, but otherwise the equations for regions A and B are unaltered. Since the flow in region F is fully developed, a locally steady-state solution will apply with only minor changes to (11.3.8)–(11.3.16) involving $\mathbf{v}^*_b$ and $\mathbf{v}^*_{scr}$.

The equations for region C that are to be changed are (11.3.18iv), (11.3.20), (11.3.22) and (11.3.24), which become

$$\rho_m v^*_{Crel\,3}(x^*_{A4}, h_s - h^*_C) = \rho_s v^*_{A3} + (\rho_s - \rho_m)\, \partial h^*_C/\partial t \tag{19.2.13}$$

$$\left(\alpha \frac{\partial T^*}{\partial x^*_{A3}}\right)\Bigg|^{h_s - h^*_C +}_{h_s - h^*_C -} = \rho_s \xi_s \left(v^*_{A3} + \frac{\partial h^*_C}{\partial t}\right) \tag{19.2.14}$$

$$\rho_m \Gamma_m \left(v^*_{Crel\,4} \frac{\partial T^*}{\partial x^*_{A4}} + v^*_{Crel\,3} \frac{\partial T^*}{\partial x^*_{As}} + \frac{\partial T^*}{\partial t}\right) = \alpha \frac{\partial^2 T^*}{\partial x^{*2}_{A3}} + \eta \left(\frac{\partial v^*_{Crel4}}{\partial x^*_{A3}}\right)^2 \tag{19.2.15}$$

$$\frac{\partial \dot{m}^*_{C4}}{\partial x^*_{A4}} = \rho_s v^*_{A3} + (\rho_s - \rho_m) \frac{\partial h^*_C}{\partial t} = \rho_m v^*_{Crel\,3}$$

where x^*_{A4} has been suitably defined. Because $\partial h^*_C/\partial t$ now arises, initial conditions $h^*_C(x^*_{A4}, 0)$ and T^* have to be prescribed which are obtained from solution of the heat conduction equation that is obtained when $\mathbf{v}^*$, $\mathbf{v}^*_b$ and $\mathbf{v}^*_{scr}$ are put identically equal to zero. This becomes

$$\frac{\partial T^*}{\partial t} = \kappa_m \frac{\partial^2 T^*}{\partial x^{*2}_{A3}} \tag{19.2.16}$$

with boundary conditions

$$T^*(h_s) = T_b, \qquad T^*(h_s - h^*_C) = T_m \tag{19.2.17}$$

and with

$$\frac{\partial h^*_C}{\partial t} = \xi_s \rho_m \left(\alpha \frac{\partial T^*}{\partial x^*_{A3}} \right) \Bigg|^{h_s - h^*_C +}_{h_s - h^*_C -} \tag{19.2.18}$$

giving the time development of h^*_C. The initial conditions for (19.2.16) and (19.2.18) are themselves given by the solution of (19.2.15) and (19.2.14) at $t = t_a$. Clearly, the values obtained for h^* and T^* after a complete cycle of length $t_a + t_d$ will not necessarily be the same as at the beginning; in practice we expect them to asymptote to specific values.

Interestingly, this illustrates one possible reason why it takes many cycles before any injection moulding operation settles down to a strictly periodic sequence; this latter is the equivalent of the asymptotic steady operating condition of continuous extrusion.

The equations for regions D and E have to be amended in the same way as those for region C.

It is not clear whether accurate computational solution of the full unsteady equations for a specific situation would be worth while, in view of the complexity of the system of equations; it is however worth pointing out that the extra independent variable t does not make an order of magnitude difference to the calculations involved, since stepping forward in time is to some extent equivalent to a single iteration in the completely steady form of the non-linear equations.

19.2.4 Metering Zone

This zone will be the most important in determining the temperature of the melt fed into the reservoir, and, if the compression ratio is large enough, in determining the flowrate. Amendments can be made to all the analyses described in Section 11.4 to cover the intermittency of the operation, though it is not necessary to do so exhaustively and in detail here.

An effective Graetz number Gz_i was introduced in Section 13.3, eqn (13.3.4), and can be similarly introduced here, in the form

$$Gz_d = h_s^2/4\kappa t_d \tag{19.2.19}$$

If $Gz_d \ll 1$, then thermal equilibrium is reached during the dwell time and $T^* = T_b = T_s$ everywhere at the beginning of the active phase. If $Gz_d \gg 1$, then thermal conduction is almost negligible and so the dwell period will have little effect on the velocity and temperature profiles during the active phase. It is, however, likely that Gz_d will be 0(1) in many cases of interest.

$A \gg 1$, $Gz_{d.c.}$ small

The relevant equation for T^* becomes

$$\rho_m \Gamma_m \left(v_2^* \frac{\partial}{\partial x_2^*} + \frac{\partial}{\partial t} \right) T^* = \frac{\partial}{\partial x_3^*} \left(\alpha_m \frac{\partial T^*}{\partial x_3^*} \right) + \eta(D^*, T^*) D^{*2} \tag{19.2.20}$$

with no change to the boundary conditions on T^* but, for the present, arbitrary initial conditions. The momentum equations remain unaltered but with boundary conditions reflecting the changes to $\mathbf{v}_b^*$ and $\mathbf{v}_{scr}^*$. For $Gz_{c.c.} \gg 1$, an additional term

$$\rho_m \Gamma_m \left\{ \left(\frac{1}{|v_2|} \right)_{x_3^*+} + \left(\frac{1}{|v_2|} \right)_{x_3^*-} \right\} \frac{\partial T^*}{\partial t}$$

has to be added to the left-hand side of (11.4.14), with similar changes to (11.4.16) and so on up to (11.4.42). The problem has not been worked through.

For $Gz_{c.c.} \ll 1$, which is the situation most thoroughly investigated so far, we use (19.2.20) without the term involving $v_2^* \, \partial/\partial x_2^*$ which leads to

$$\frac{\partial^2 \tilde{T}^*}{\partial \tilde{x}_3^{*2}} = -\text{Na}^* \, e^{-\tilde{T}^*} \left\{ \left(\frac{\partial \tilde{v}_1^*}{\partial \tilde{x}_3^*} \right)^2 + \left(\frac{\partial \tilde{v}_2^*}{\partial \tilde{x}_3^*} \right)^2 \right\}^{\frac{1}{2}(\nu+1)} + \frac{\partial \tilde{T}^*}{\partial \tilde{t}} \tag{19.2.21}$$

instead of (11.4.49) where $\tilde{t} = t\kappa_m/h_s^2$. This leads to a computational problem simpler than the one tackled by Yates (1968) and governed by eqn (11.4.58). The heat conduction equation and boundary conditions relevant for the dwell time are obtained by putting $\tilde{\mathbf{v}}^* \equiv 0$, so solutions relevant over a full cycle (or many successive cycles) can readily be obtained.

$\underline{A \gg 1, \mathrm{Gz}_{\mathrm{d.c.}} = 0(1)}$

This is probably the most relevant situation, and one that would be worth tackling along the lines described for $\mathrm{Gz}_\mathrm{D} = 0(1)$ in Subsection 11.4.1. However, a real difficulty arises in the $\mathbf{x}^*$ coordinate system because $\tilde{v}_1^*$ is necessarily <0 near the screw. The relevant equation for T^* becomes in this case

$$\mathrm{Gz}^*_{\mathrm{d.c.}}\ \tilde{v}_1^* \frac{\partial \tilde{T}^*}{\partial \tilde{x}_1^*} + \mathrm{Gz}_\mathrm{a}^* \frac{\partial \tilde{T}^*}{\partial \tilde{t}^*} = \frac{\partial^2 \tilde{T}^*}{\partial \tilde{x}_3^{*2}} + \mathrm{Na}^*\, \mathrm{e}^{-\tilde{T}^*} \left\{ \left(\frac{\partial \tilde{v}_1^*}{\partial \tilde{x}_3^*} \right)^2 + \left(\frac{\partial \tilde{v}_2^*}{\partial \tilde{x}_3^*} \right)^2 \right\}^{\frac{1}{2}(\nu+1)} \tag{19.2.22}$$

where

$$\mathrm{Gz}_\mathrm{a}^* = h_\mathrm{s}^2/\kappa_\mathrm{m} t_\mathrm{a} \quad \text{and} \quad \tilde{t}^* = t/t_\mathrm{a} \tag{19.2.23}$$

There is no possible alternative choice of direction for the convective term that does not show a change in sign of the corresponding velocity $\mathbf{v}^*$, because of the true flow back implied by the retraction velocity $\mathbf{v}_\mathrm{scr}$ of the screw. However, if the length of the metering zone can be considered constant, i.e. independent of screw retraction, then the original $\mathbf{x}$ coordinate system fixed in the screw, but now moving axially, rather than the $\mathbf{x}^*$ system moving relative to the screw but fixed axially, can be used and $\tilde{v}_z\ \partial/\partial \tilde{z}$ takes the place of $\tilde{v}_1^*\ \partial/\partial \tilde{x}_1^*$ in (19.2.22).

Figure 19.3 illustrates schematically the ranges for which the various equations (11.4.49), (19.2.22) and (19.2.21) will be relevant during successive cycles following start up. Use of the forms (i) and (iii) will represent approximations; the size of the regions corresponding to (i) and (iii) will depend upon the values of Gz_d, Gz_a and $\mathrm{Gz}_{\mathrm{d.c.}}$. Note that the mean temperature of the preplasticizing screw output to the reservoir is to be calculated at $\tilde{x}_1 = \mathrm{Gz}_{\mathrm{d.c.}}$. The example shown in the figure suggests that only the solution corresponding to (19.2.21) is required. However, for much smaller values of $\mathrm{Gz}_{\mathrm{d.c.}}$ and Gz_a, the solution of (11.4.49) could be reasonably accurate over much of the active cycle, and would therefore dominate. It is assumed that as $t/(t_\mathrm{a} + t_\mathrm{d})$ becomes large, an asymptotically-regular periodic situation will ensue.

19.2.5 Overall Calculations for Output to Reservoir

All the calculation techniques devised for the various zones of an extruder have supposed the flowrate Q_T to be given; but, as was explained in Section 11.5, a full analysis must predict Q_T so as to

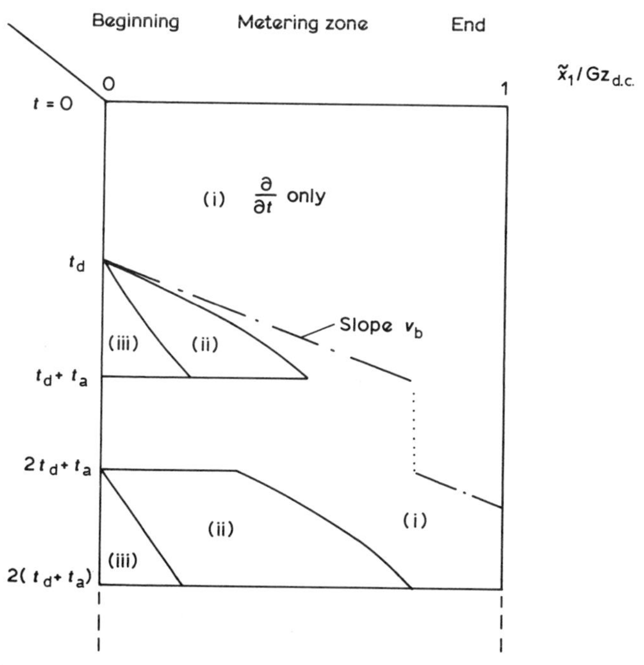

Fig. 19.3. Regions of importance for: (i) pure conduction: unsteady calculation using eqn (19.2.21); (ii) combined conduction and convection: calculation requiring full eqn (19.2.22); (iii) pure convection: steady flow calculation using eqn (11.4.49).

satisfy the boundary conditions on pressure p, or more relevantly in this case the overall force balance on the screw. In the former case, no axial motion of the screw was possible and so the requisite 'operating point' was obtained by balancing the pressure rise along the screw against the pressure drop through the die. In the present case the dominant components of the axial force balance will be the reservoir pressure p_{res} acting over the entire cross-section of the screw, the viscous drag (shear stress in the melt) at the flight tips caused by the axial movement of the screw and any resistance to motion, F_{mach} imposed mechanically or hydraulically at the drive-end of the screw. The last named will be determined by the design of the machine in question and not by polymer melt mechanics. An estimate of the total viscous drag is given by (see Example 19.2.2)

$$F_{drag} = 2(\eta v_{scr}/h_f)\pi d_b l_m w_f/(w_c + w_f) \qquad (19.2.24)$$

where η is evaluated for $T=T_b$, $D=(v_{scr}^2+v_b^2)^{\frac{1}{2}}/h_f$ and l_m is the 'wetted' length of the screw. If we take the volumetric delivery of the screw to be half that of the drag flow delivery then the pressure drop over a (notional) metering section of length l_m and depth h_s will be (see Exercise 19.2.5)

$$p \approx 8\eta v_b l_m/h_s^2 \tag{19.2.25}$$

where η is now evaluated at $D=v_b/h_s$.

If we take this to be the major component of the reservoir pressure, then the force exerted by the melt in the reservoir on the screw will be

$$F_{res} \approx 2\pi\eta d_b^2 v_b l_m/h_s^2 \tag{19.2.26}$$

The balance equation for the screw will be

$$F_{mach}+F_{drag}=F_{res} \tag{19.2.27}$$

Using the relationship (19.2.4) derived earlier for v_{scr}, which is consistent with the volumetric delivery implied by (19.2.25), we obtain the ratio

$$\frac{F_{drag}}{F_{res}} \approx \frac{\eta_{flight}}{\eta_{channel}} \frac{h_s^3}{30 d_b^2 h_f} \tag{19.2.28}$$

where $w_f/(w_f+w_c)\approx 8$. For the figures given in Table 11.1, $h_s^3 \approx d_b^2 h_f$, while for all $\nu \geqslant 1$, we expect $\eta_{flight} \leqslant \eta_{channel}$. Hence we can expect

$$F_{drag} \ll F_{res} \tag{19.2.29}$$

EXAMPLE 19.2.1. Similarity solution for developing lubricating layer close to feed hopper during quasi-static operation of preplasticizing screw.

We consider the situation described by (19.2.1) and use the similarity variables developed in Subsection 11.2.3. The velocity v_{W1} in (11.2.34) remains the velocity of the solid bed relative to the $\mathbf{x}^*$ coordinate system, called v_{W1}^* in the text above, and so $\tilde{v}_0^* = v_{scr}^*/v_{W1}^*$.

The only change to the set of equations and boundary conditions (11.2.47)–(11.2.54) is that

$$\tilde{v}_1^s(0) = -\tilde{v}_0^*$$

If $\nu=1$, $\mathrm{B}=0$ and (11.2.54) applies, then it can readily be shown that

$$\tilde{v}_1^s(\hat{x}_3) = -\tilde{v}_0^* + (1+\tilde{v}_0^*)\hat{x}_3$$

$$\tilde{v}_3^s(\hat{x}_3) = \tfrac{1}{2}(1+\tilde{v}_0^*)\hat{x}_3^3$$

and so $\tilde{T}^s$ satisfies the equation

$$\left[\frac{\mathrm{Sf}}{A}\frac{\mathrm{d}}{\mathrm{d}\hat{x}_3}+\left\{\tfrac{1}{2}(1+\tilde{v}_0^*)\left(\hat{x}_3-\frac{\tilde{v}_0^*}{1+\tilde{v}_0^*}\right)^2-\tfrac{1}{2}\frac{\tilde{v}_0^{*2}}{(1+v_0^*)}\right\}\right]\frac{\mathrm{d}\tilde{T}^s}{\mathrm{d}\hat{x}_3}+\mathrm{Sf}\frac{\mathrm{Gn}}{A}(1+\tilde{v}_0^*)^2=0$$

with

$$\tilde{T}^s(0)=1, \qquad \tilde{T}^s(1)=0, \qquad \frac{\mathrm{d}\tilde{T}^s}{\mathrm{d}\hat{x}_3}(1)=\frac{\rho_\mathrm{m}A}{2\rho_\mathrm{s}}(\tilde{v}_0^*-1)$$

This is an eigenvalue problem yielding

$$A(\mathrm{Sf}, \mathrm{Gn}, \rho_\mathrm{m}/\rho_\mathrm{s}, \tilde{v}_0^*)$$

EXAMPLE 19.2.2. Estimation of F_{drag} in (19.2.24).

Most of the dissipation of energy through viscous effects arises because of the screw rotation and leads to a torque as calculated in Example 11.5.1. The viscous forces leading to a net axial force on the screw will arise partly from pressure flow in the screw channels and partly from the relative axial motion of the screw in the barrel. The former can be no larger than given by $\pi d_\mathrm{b} h_\mathrm{s} p_{\mathrm{res}}$, which is small compared with $F_{\mathrm{res}}=\frac{1}{4}\pi d_\mathrm{b}^2 p_{\mathrm{res}}$, because $h_\mathrm{s} \ll d_\mathrm{b}$; F_{res} has been defined in this way in getting from (19.2.25) to (19.2.26). Thus, if F_{drag} is to be significant compared with F_{res}, e.g. in (19.2.27), then F_{drag} must be dominated by the component due to the relative motion v_{scr}. A crude approximation is thus given by drag flow in both the main channel and the flight clearance. The relative contribution will be in the ratio

$$\frac{\eta_{\mathrm{channel}} w_\mathrm{c} h_\mathrm{f}}{\eta_{\mathrm{flight}} w_\mathrm{f} h_\mathrm{s}}$$

The geometry of screws usually ensures that $w_\mathrm{c} h_\mathrm{f} < h_\mathrm{s} w_\mathrm{f}$ but not by a very large factor (4, say, using Table 11.1). Thus if η is significantly shear-rate dependent, the main channel contribution may be as large as or even larger than the flight clearance contribution. Taking the relevant shear rates to be $v_\mathrm{b}/h_\mathrm{s}$ and $v_\mathrm{b}/h_\mathrm{f}$, and $\nu=\frac{1}{2}$, then

$$\frac{\eta_{\mathrm{channel}}}{\eta_{\mathrm{flight}}}=\left(\frac{h_\mathrm{s}}{h_\mathrm{f}}\right)^{\frac{1}{2}}\sim 6$$

using the values in Table 11.1, and the channel contribution will be 50% larger than the flight clearance contribution.

The shear stress at the barrel wall, in the flight clearance regions, in the z-direction, is given approximately by

$$\eta(v_{\mathrm{rel}}/h_{\mathrm{f}})v_{\mathrm{scr}}/h_{\mathrm{f}}$$

where $v_{\mathrm{rel}}=(v_{\mathrm{b}}^2+v_{\mathrm{scr}}^2)^{\frac{1}{2}}$.

If this is integrated over the area of the barrel wall opposite the screw flight and doubled then (19.2.24) is obtained. In any case the result (19.2.29) will not be affected.

EXERCISE 19.2.1. Verify that the coordinate system described in Subsection 19.2.1 and shown in Fig. 19.2 is consistent with the velocities (19.2.1).

To do this it is necessary to recognize that the steady motion $v_{\mathrm{scr}}\cot\phi_{\mathrm{b}}$ of coordinate system $\mathbf{x}^*$ relative to coordinate system $\mathbf{x}$, which seeks to account for the motion of the feed pocket relative to the screw may be small relative to the periodic 'saw-tooth' motion $v_{\mathrm{b}}\cos\phi_{\mathrm{b}}$ of feed pocket relative to the screw neglected in Chapter 11, because the net effect of the latter is assumed to average out in the feed zone.

EXERCISE 19.2.2. Obtain an eigenvalue integral equation for A.

EXERCISE 19.2.3. Show how x_5 can be redefined to take account of v_{scr} in the analysis for region F, and hence derive the relevant forms for $\dot{M}_{\mathrm{F1}}^*$ and $\dot{m}_{\mathrm{F2}}^*$ (cf. eqns (11.3.13) and (11.3.14)).

EXERCISE 19.2.4. Extend the arguments leading to Fig. 19.3 to include the feed and compression zones. Note that Fig. 19.3 has been cut off before a truly asymptotic, though periodic, regime has been established as $t\to\infty$.

EXERCISE 19.2.5. Show that (19.2.25) follows with 10 instead of 8 if a Newtonian temperature-independent fluid is considered and $A\gg 1$. Use the fact that the drag flow, taking $v_{\mathrm{scr}}\ll v_{\mathrm{b}}$, can be written as $q=\frac{1}{2}v_{\mathrm{b}}h_{\mathrm{s}}$, and so the pressure flow has to be $-\frac{1}{4}v_{\mathrm{b}}h_{\mathrm{s}}$. Show that this requires a parabolic pressure-flow field with $v_{1\max}=-(3/8)v_{\mathrm{b}}$ and hence a wall shear stress of $-(3/2)\eta v_{\mathrm{b}}/h_{\mathrm{s}}$. The pressure gradient $\partial p/\partial x_1$ can then be written $3\eta v_{\mathrm{b}}/h_{\mathrm{s}}^2$. Note that l_{m} is an axial length and so show that the relevant pressure drop is, for a square screw with $(w_{\mathrm{ca}}+w_{\mathrm{fa}})=d_{\mathrm{b}}$,

$$\Delta p=3\sqrt{1+\pi^2}\,\eta v_{\mathrm{b}}l_{\mathrm{m}}/h_{\mathrm{s}}^2$$

which is the required result.

The factor 8 is used in (19.2.25) because any degree of shear-rate dependence of viscosity will probably reduce the relevant factor, while the effect of v_{scr} is to reduce the drag flow in the proportion $2v_{scr}/v_b$, and so less pressure flow is required.

This implies that unless F_{mach} provides the dominant constraint to retraction of the screw, the reservoir pressure will be too small to affect the delivery of the screw, and $p_{res} \sim 0$.

It may be concluded that provided $Br_{d.c.}$, $Gz_{d.c.}$, Gz_a and Gz_d are sufficiently small (<1 is adequate), the polymer delivered to the reservoir will be effectively at the barrel temperature T_b. This is the ideal circumstance and leads to a uniform material temperature in the reservoir at the injection stage. Significant variation in material temperature within the reservoir may arise in the case of large machines cycled too rapidly at high screw speed.

19.3 INJECTION

Mechanical analysis is primarily concerned with the time taken to fill various parts of the runner-mould system, and the temperature developed in the polymer during this process. For most machines, the injection part of the cycle takes place at constant input rate† (with very rapid rise to that rate) until a given maximum pressure is developed in the reservoir (against the ram) and then at constant reservoir pressure.‡ The portion of the runner-mould system that is filled with polymer increases with time and so an unsteady representation of the flow process must be considered. It may be expected that the flowrate will decrease during the constant-pressure phase.

The geometry of the system of cavities can be reasonably represented in most cases by a series of circular tubes (of uniform or slowly varying cross-section), and disc-like or planar narrow channels of slowly varying depth, connected to one another by bends or deliberate constrictions (as in the case of the gates to the mould cavities proper).

† More strictly it is the movement of the ram or screw that is controlled; because of compressibility and back flow, this is not the same as the flowrate into the mould.

‡ The extent to which this objective is achieved in practice depends upon the design and quality of the hydraulic equipment used to drive the ram or screw. The use of micro-computers has now opened up the possibility of imposing any time-varying pressure profile or alternatively any screw displacement profile.

19.3.1 Tube Flow

Uniform tube flow has already been discussed in Section 7.1, together with entry flow in Section 7.2. The role of Graetz and generation numbers was introduced in the former and of Deborah number in the latter. The applicability of lubrication theory to elastic fluid flow in tubes of slowly varying section was shown in Section 8.1 to depend upon the Deborah number. We shall suppose for simplicity here that all axisymmetric flow channels can be modelled either as being of very slowly varying cross-section, and thus satisfying (8.1.6) or as involving sharp changes as shown in Fig. 7.1(a) with $\theta_e = 0(1)$.

We consider now circular-cylindrical tube flow: we take over eqns (7.1.12), (7.1.16), (7.1.25)–(7.1.32), (7.1.47) together with (7.1.18)–(7.1.20), (7.1.43) and define

$$\hat{v}_r = 2(\mathrm{L/D})\tilde{v}_r, \qquad \hat{t} = \tilde{t}/2(\mathrm{L/D}) \tag{19.3.1}$$

so as to ensure that all important terms in the equations will be 0(1). We define

$$\mathrm{B} = |T_\mathrm{W} - T_0|/T^*, \qquad T^* = T^*_{\mathrm{rheol}} \tag{19.3.2}$$

as in (8.1.43), and so obtain

$$\frac{1}{\tilde{r}}\frac{\partial}{\partial \tilde{r}}(\tilde{\rho}\tilde{r}\hat{v}_r) + \frac{\partial}{\partial \hat{z}}(\tilde{\rho}\tilde{v}_z) = -\frac{\partial \tilde{\rho}}{\partial \hat{t}} \tag{19.3.3}$$

$$\frac{\partial \hat{p}}{\partial \tilde{r}} \approx 0 \tag{19.3.4}$$

$$\frac{1}{\tilde{r}}\frac{\partial}{\partial \tilde{r}}(\tilde{r}\tilde{t}^{\mathrm{E}}_{\mathrm{rz}}) - \frac{\partial \hat{p}}{\partial \hat{z}} \approx 0 \tag{19.3.5}$$

$$\begin{aligned} \mathrm{Gz}\,\tilde{\rho}\Gamma_\mathrm{p}\left(\frac{\partial \tilde{T}}{\partial \hat{t}} + \hat{v}_r\frac{\partial \tilde{T}}{\partial \tilde{r}} + \tilde{v}_z\frac{\partial \tilde{T}}{\partial \hat{z}}\right) &= \frac{1}{\tilde{r}}\frac{\partial}{\partial \tilde{r}}\left(\tilde{\alpha}\tilde{r}\frac{\partial \tilde{T}}{\partial \tilde{r}}\right) \\ &+ \mathrm{Na}\left[-\hat{p}\left(\frac{1}{\tilde{r}}\frac{\partial}{\partial \tilde{r}}(\tilde{r}\hat{v}_r) + \frac{\partial \tilde{v}_z}{\partial \hat{z}}\right) + \tilde{t}_{rz}\frac{\partial \tilde{v}_z}{\partial \tilde{r}} \right. \\ &\left. - \left(\frac{\partial \ln \tilde{\rho}}{\partial \ln \tilde{T}}\right)_{\tilde{\mathrm{p}}}\left(\frac{\partial \hat{p}}{\partial \hat{t}} + \hat{v}_r\frac{\partial \hat{p}}{\partial \tilde{r}} + \tilde{v}_z\frac{\partial \hat{p}}{\partial \hat{z}}\right)\right] \end{aligned} \tag{19.3.6}$$

where the wall boundary conditions are

$$\tilde{v}_r = \tilde{v}_z = 0, \qquad \tilde{T} = \mathrm{B} \quad \text{at} \quad \tilde{r} = 1 \tag{19.3.7}$$

For the present we assume that B is above the freezing or vitrification point for the polymer. The contrary situation will be discussed later under channel flow. Equation (19.3.6) is a more complete form of (7.1.72).

The single relevant non-diagonal stress component $\tilde{t}_{rz}$ can be assumed to be given by a viscous relation

$$\tilde{t}_{rz} = \tilde{\eta}(\hat{p}, \tilde{T}, \partial\tilde{v}_z/\partial\tilde{r})\partial\tilde{v}_z/\partial\tilde{r}$$

where

$$\tilde{\eta} = \eta(p, T, \partial v_z/\partial r)/\eta(0, T^*, D^*) \tag{19.3.8}$$

The initial conditions are not immediately obvious. Where material had been at rest at ambient pressure and temperature $\tilde{T}^{(0)}(\tilde{r}, \tilde{z})$ we would expect

$$\tilde{\rho} = \rho_e(0, \tilde{T}^{(0)}) \quad \text{at} \quad \hat{t} = 0 \tag{19.3.9}$$

according to (2.2.2). There is always a slightly ill-defined period during which the ram builds up to its set forward speed and the material in the reservoir is compressed. During this period, it is not easy to specify either the pressure drop $\partial\hat{p}/\partial\hat{z}$ or the mass flow rate

$$\dot{M} = 2\pi \int_0^{\frac{1}{2}d_c} r v_z \rho \, \mathrm{d}r \tag{19.3.10}$$

There is a further difficulty introduced by the fact that the position at which $\hat{p} = \hat{p}_a$ ($=0$ say) moves forward during injection, so that the effective initial point in time increases with distance from the nozzle. Put differently L/D varies with time as the tube fills from $\hat{z} = 0$ to $\hat{z} = 1$.

The entry temperature condition on $\tilde{T}(\tilde{r}, 0, \hat{t})$ will depend upon the nature of the flow passage leading to $\hat{z} = 0$ and cannot be prescribed in any generality.

Lastly there is the possibility that the equilibrium thermodynamic relation (2.2.2) will not be adequate under conditions of relatively rapid changes of pressure and temperature. We are helped here by the fact that the pressure scale is L/D times the viscous stress scale and so questions of whether p itself is well-defined do not arise. The differential relation (4.2.42) would then be relevant. The pressure variation of viscosity may also be important.

Assuming that the equilibrium thermodynamic density $\rho_e(p, T)$ is representative of the relations between pressure, temperature and density variations, and that η can be regarded as a known function

$\eta(p, T)$, we can define three further temperature scales

$$T^*_{\text{expan}} = \rho_e \Big/ \left|\frac{\partial \rho_e}{\partial T}\right| \tag{19.3.11}$$

$$T^*_{\text{comp}} = 1 \Big/ \frac{\partial \rho}{\partial p} \Gamma_p \tag{19.3.12}$$

$$T^*_{\text{visc}} = \eta / \rho_e \Gamma_p (\partial \eta / \partial p) \tag{19.3.13}$$

where

$$T^*_{\text{rheol}} = \eta / |\partial \eta / \partial T| \text{ as in (7.1.18)}$$

The dimensionless system of equations and boundary conditions (19.3.3)–(19.3.9) now involves 6 dimensionless groups

$$\text{Gz} = T^*_{\text{gen}} / T^*_{\text{adiab}}, \qquad \text{Br} = T^*_{\text{gen}} / T^*_{\text{op}}$$

$$\text{Na} = T^*_{\text{gen}} / T^*_{\text{rheol}}, \qquad \text{Vp} = T^*_{\text{adiab}} / T^*_{\text{visc}} = T^*_{\text{gen}} / T^*_{\text{visc}} \, \text{Gz} \tag{19.3.14}$$

$$\text{Ex} = T^*_{\text{gen}} / T^*_{\text{expan}}, \qquad \text{Cm} = T^*_{\text{adiab}} / T^*_{\text{comp}} = T^*_{\text{gen}} / T^*_{\text{comp}} \, \text{Gz}$$

where T^*_{gen} is defined in (7.1.19), T^*_{op} in (7.1.21) and T^*_{adiab} in (7.1.40). Br is defined in (7.1.22) also. Note that Co, as defined in (6.1.29), can be equivalent to Ex defined in (19.3.14) above. Of these Vp will be of order 10^{-1} and Cm of order 10^{-2} except where pressure-induced phase changes lead to large changes in viscosity or density. It is therefore usual to neglect compression, though with certain materials Vp is 0(1) and care must be exercised in overall calculations. The magnitudes of the other 4 groups (Gz, Na, Br and Ex) are very dependent upon the value taken for T^*_{gen} which is itself largely determined by the value taken for η in (7.1.19), i.e. η^* in (7.1.32). Provided Na or Br are $\leqslant 0(1)$, then the viscosity variations implied by the expected rise in temperature will not affect the order of magnitude of the four remaining dimensionless groups. However, if $\text{Na} \gg 1$, then the viscosity reduction caused by the crudely calculated temperature rise T^*_{gen} will be of order $e^{-\text{Na}}$. In practice the temperature rise will be lower than this. Indeed we can arrive at a corrected value for Na by requiring that

$$\text{Na}_{\text{corr}} = \text{Na}_{\text{uncorr}} \, e^{-\text{Na}_{\text{corr}}} \tag{19.3.15}$$

which yields, for $\text{Na}_{\text{uncorr}} \gg 1$, $\text{Na}_{\text{corr}} \sim \ln \text{Na}_{\text{uncorr}}$. In most cases, even $\ln \text{Na}$ will be $\geqslant 3$. T^*_{gen} can then be replaced by $T^*_{\text{gen}} / \text{Na}_{\text{uncorr}}$. The relevant value of Ex will become Ex/Na, which will be of order 10^{-1} at most. However, it is worth remembering that the operating temperature difference represented by B in (19.3.7) is not altered by the above argument. It is also worth remembering that the viscosity is very shear-rate dependent, particularly for highly viscous polymers, and so Na will not be as sensitive to flow rate as would a Newtonian fluid.

Solution of the dimensionless equations (19.3.3)–(19.3.9) can be carried out by direct numerical computation. Alternatively the dimensional forms can be used directly. The advantage of purely numerical solution is that boundary and inlet conditions can be made time dependent, while the temperature distribution (as a function of x_3) at the advancing front can be varied in accordance with whatever local solution for flow and heat transfer is assumed to hold. Comparison with experimental observations can be achieved by suitable choice of the constitutive functions η, ρ, Γ_p and α. The disadvantage of purely numerical computation is that writing and testing a generally applicable program is a major task, particularly if full flexibility in specification of initial, boundary and entry conditions and of constitutive functions is to be achieved. If many of the relevant physical effects are negligible, then the program can be unnecessarily unwieldy, while if Na is very large then sharp shear layers with very large temperature gradients arise, which the grid or network chosen for the calculation may not resolve. There is therefore an advantage in seeking limited, representative, analytical solutions to investigate the physical effects that are likely to arise, as a guide to later numerical computation or as a means of rapid calculation in special cases. The following have been obtained and could be useful in overall analyses as discussed in Section 19.5. We seek P, $\dot{V}$ (or $\dot{M}$), $\bar{T}$ relations in each case.

I. Na, B, Ex, Cm, Vp $\ll 1$

Here the velocity field is decoupled from the temperature field, and so is independent of Gz. The solutions given for a capillary rheometer in Section 7.1 will apply. For the power law fluid (3.3.5),† it may readily be shown (e.g. Skelland, 1967, p. 110, or Example 7.1.1) that

$$\dot{V} \equiv Q = \frac{\pi d_c^{3+1/\nu}}{(3+1/\nu)} \left(\frac{P}{2^{2+3\nu} K l_c} \right)^{1/\nu} \tag{19.3.16}$$

with the more general viscous result given by (7.1.60).

† Throughout this chapter as elsewhere in this text, great care has to be taken in dealing with the power-law fluid model which is often written with a factor $\dot{\gamma}^{\nu}$. If $\dot{\gamma} > 0$ no difficulties arise, but if $\dot{\gamma} < 0$, then the relevant form $|\dot{\gamma}|^{\nu-1}\dot{\gamma}$ is cumbersome to carry through all manipulations. It should be understood in what follows that $|\dot{\gamma}|^{\nu}$ is sometimes written instead of $-|\dot{\gamma}|^{\nu-1}\dot{\gamma}$.

The temperature field is obtained from the energy equation (19.3.6) by substituting the velocity field $\hat{v}_r = 0$, $\tilde{v}_z$ given by (7.1.57). It is now much more sensible to define $T^* = T^*_{op}$ and so we write

$$\hat{T} = \tilde{T}/\mathrm{B} \tag{19.3.17}$$

to give the linear partial differential equation for $\hat{T}$

$$\mathrm{Gz}\left(\frac{\partial \hat{T}}{\partial \hat{t}} + \tilde{v}_z \frac{\partial \hat{T}}{\partial \hat{z}}\right) = \frac{1}{\tilde{r}}\frac{\partial}{\partial \tilde{r}}\left(\tilde{r}\frac{\partial \hat{T}}{\partial \tilde{r}}\right) + \mathrm{Br}\,\tilde{\eta}\left(\frac{\partial \tilde{v}_z}{\partial \tilde{r}}\right)^2 \tag{19.3.18}$$

with boundary condition

$$\hat{T} = \hat{T}_\mathrm{W}(\hat{z}, \hat{t}) \tag{19.3.19}$$

where $\hat{T}_\mathrm{W} = 0(1)$.

$\mathrm{Gz} \ll 1$. The fully-developed situation is achieved for times long enough to exhaust the tube. The governing equation (19.3.17) reduces to the ordinary differential equation

$$\frac{1}{\tilde{r}}\frac{\mathrm{d}}{\mathrm{d}\tilde{r}}\left(r\frac{\mathrm{d}\hat{T}}{\mathrm{d}\tilde{r}}\right) = -\mathrm{Br}\,\tilde{\eta}\left(\frac{\mathrm{d}\tilde{v}_z}{\mathrm{d}\tilde{r}}\right)^2 \tag{19.3.20}$$

with a solution for $\hat{T}_\infty - \hat{T}_\mathrm{W}$ that is independent of $\hat{z}$ and $\hat{t}$, provided $\mathrm{d}\hat{T}_\mathrm{W}/\mathrm{d}\hat{t} \not\gg 1$, and is linearly dependent on Br.

Integration yields

$$\hat{T}_\infty(\tilde{r}) - \hat{T}_\mathrm{W} = \mathrm{Br}\left[-\ln \tilde{r}\int_0^{\tilde{r}} \tilde{r}\tilde{\eta}\left(\frac{\mathrm{d}\tilde{v}_z}{\mathrm{d}\tilde{r}}\right)^2 \mathrm{d}\tilde{r} - \int_{\tilde{r}}^1 \ln \tilde{r}\,\tilde{r}\tilde{\eta}\left(\frac{\mathrm{d}\tilde{v}_z}{\mathrm{d}\tilde{r}}\right)^2 \mathrm{d}\tilde{r}\right] \tag{19.3.21}$$

For the power law fluid (3.3.5) it follows that

$$\hat{T}_\infty(\tilde{r}) - \hat{T}_\mathrm{W} = \mathrm{Br}(1 - \tilde{r}^{3+1/\nu})\,|\hat{T}_\mathrm{W} - \hat{T}_0|\,\{2(3+1/\nu)\} \tag{19.3.22}$$

where

$$\mathrm{Br} = \frac{16KQ^2(4Q/\pi d_c^3)^{\nu-1}}{\pi^2 d_c^4 \kappa\,|T_\mathrm{W} - T_0|}$$

$\mathrm{Gz} = 0(1)$. This is the developing situation where all three terms in the equation are important. If $\mathrm{Br} \gg 1$, then a dimensionless temperature based on T^*_{gen} automatically makes the dimensionless multiplier of the last term in (19.3.18) unity, with the wall boundary temperature equal to Br^{-1}, i.e. close to zero.

Writing

$$\hat{T} = \hat{T}_\infty(\tilde{r}) + \hat{T}_D(\tilde{r}, \hat{z}, \hat{t}) \qquad (19.3.23)$$

where $\hat{T}_\infty$ is the solution of (19.3.20) gives

$$\mathrm{Gz}\left(\frac{\partial \hat{T}_D}{\partial \hat{t}} + \tilde{v}_z \frac{\partial \hat{T}_D}{\partial \hat{z}}\right) = \frac{1}{\tilde{r}} \frac{\partial}{\partial \tilde{r}}\left(\tilde{r} \frac{\partial \hat{T}_D}{\partial \tilde{r}}\right) \qquad (19.3.24)$$

with

$$\hat{T}_D(1, \hat{z}, \hat{t}) = 0 \qquad (19.3.25)$$

$$\hat{T}_D(\tilde{r}, \hat{z}, 0) = -\hat{T}_\infty \qquad (19.3.26)$$

$$\hat{T}_D(\tilde{r}, 0, \hat{t}) = -\hat{T}_\infty - \hat{T}_W \qquad (19.3.27)$$

We may not, in general, decompose $\hat{T}_D$ into a time-dependent and an axial-position-dependent part. The initial condition (19.3.26) supposes that the flow is initiated after a long dwell time.

As in Subsection 19.2.4 (see Fig. 19.3), there will be regions in $(\hat{z}, \hat{t})$ space for which only the $\partial/\partial\hat{t}$ term, and only the $\partial/\partial\hat{z}$ term, will be relevant on the left-hand side of (19.3.24). These correspond, respectively, to the early stages after instantaneous start-up in a long tube and to the final stage after all temporal thermal transients have decayed away. The solution to the former is the well-known solution to the heat conduction equation expressed in the form

$$\hat{T}_T(\tilde{r}, \hat{t}) = \sum_i \hat{T}_{Ti} J_0(\tilde{r}\tilde{\alpha}_i) e^{-\hat{t}\tilde{\alpha}_i^2/\mathrm{Gz}} \qquad (19.3.28)$$

where $J_0(z)$ is the zero-order Bessel function and $\tilde{\alpha}_i$ are its zeros in ascending order (Abramowitz & Stegun, 1964, Section 9).

The $\hat{T}_{Ti}$ satisfy the requirement

$$\sum_i \hat{T}_{Ti} J_0(\tilde{r}\tilde{\alpha}_i) = -\hat{T}_\infty \qquad (19.3.29)$$

which leads to

$$\hat{T}_{Ti} = \frac{-2\int_0^1 \hat{T}_\infty(\tilde{r})\tilde{r} J_0(\tilde{r}\tilde{\alpha}_i)\, d\tilde{r}}{|J_1(\tilde{\alpha}_i)|^2} \qquad (19.3.30)$$

If $\tilde{v}_z(\tilde{r})$ is similarly expressed in the form

$$\tilde{v}_z(\tilde{r}) = \sum_i \tilde{v}_{zi} J_0(\tilde{r}\tilde{\alpha}_i)$$

then

$$\frac{\bar{T}}{T^*_{\text{gen}}} = \frac{\pi}{\dot{V}} \sum_i \{\tilde{v}_{zi} \hat{T}_{\text{T}i} / [J_1(\tilde{\alpha}_i)]^2\} e^{-\hat{t}\tilde{\alpha}_i^2/\text{Gz}} + \tilde{T}_\infty$$

The solution to the latter can also be obtained in terms of eigenfunctions as

$$\hat{T}_{\text{D}}(\tilde{r}, \hat{z}) = \sum_i \hat{T}_{\text{D}i}(\tilde{r}) e^{-\hat{z}/\text{Gz}\hat{z}_i} \tag{19.3.31}$$

with the eigenvalues chosen to satisfy

$$\frac{1}{\tilde{r}} \frac{\text{d}}{\text{d}\tilde{r}} \left(\tilde{r} \frac{\text{d}\hat{T}_{\text{D}i}}{\text{d}\tilde{r}} \right) + \frac{1}{\hat{z}_i} \tilde{v}_z \hat{T}_{\text{D}i} = 0 \tag{19.3.32}$$

and the homogeneous boundary conditions

$$\frac{\text{d}\hat{T}_{\text{D}i}}{\text{d}\tilde{r}} = 0 \quad \text{at} \quad \tilde{r} = 0, \qquad \hat{T}_{\text{D}i} = 0 \quad \text{at} \quad \tilde{r} = 1 \tag{19.3.33}$$

The values $\hat{T}_{\text{D}i}(0)$ are provided by the requirement

$$\sum \hat{T}_{\text{D}i}(\tilde{r}) = -\hat{T}_\infty - \hat{T}_{\text{W}} \tag{19.3.34}$$

This representation fails for small values of $\hat{z}$ if $\hat{T}_{\text{W}} \neq 0$. A discontinuity in temperature arises at $\hat{z} = 0$, $\tilde{r} = 1$ which leads to a thermal boundary layer.

$\text{Gz} \gg 1$. This is the case mentioned above for $\hat{z} \ll 1$. (If $\hat{z}$ is used instead of l in the definition of Gz, a local value of Gz is obtained for each axial position in the tube.) For steady flow a Lévêque-style solution is now relevant. For most of the flow field, i.e. except in a thin boundary layer, an adiabatic approximation is suitable, and so we write, neglecting the conduction term,

$$\tilde{v}_z \frac{\partial \check{T}}{\partial \hat{z}} = \tilde{\eta} \left(\frac{\text{d}\tilde{v}_z}{\text{d}\tilde{r}} \right)^2 \tag{19.3.35}$$

where

$$\check{T} = T / T^*_{\text{adiab}} \tag{19.3.36}$$

This can be integrated to give

$$\check{T} = \tilde{\eta} \left(\frac{\text{d}\tilde{v}_z}{\text{d}\tilde{r}} \right)^2 \hat{z} / \tilde{v}_z \tag{19.3.37}$$

This solution must fail as $\tilde{r}\rightarrow 1$, when $\tilde{v}_z\rightarrow 0$. There will be a boundary layer near $\tilde{r}=1$ in which conduction will be important. The relevant local variable is given by

$$\hat{r}=\mathrm{Gz}^{\frac{1}{3}}(1-\tilde{r}) \tag{19.3.38}$$

The velocity $\tilde{v}_z$ is suitably represented by

$$\tilde{v}_z \sim -\left(\frac{\mathrm{d}\tilde{v}_z}{\mathrm{d}\tilde{r}}\right)_{\mathrm{W}}\frac{\hat{r}}{\mathrm{Gz}^{\frac{1}{3}}}=\left|\frac{\mathrm{d}\tilde{v}_z}{\mathrm{d}\tilde{r}}\right|_{\mathrm{W}}\frac{\hat{r}}{\mathrm{Gz}^{\frac{1}{3}}} \tag{19.3.39}$$

and so eqn (19.3.18) becomes

$$\left|\frac{\mathrm{d}\tilde{v}_z}{\mathrm{d}\tilde{r}}\right|_{\mathrm{W}}\hat{r}\frac{\partial\hat{T}}{\partial\hat{z}}=\frac{\partial^2\hat{T}}{\partial\hat{r}^2}+\frac{\mathrm{Br}}{\mathrm{Gz}^{\frac{2}{3}}}\tilde{\eta}\left(\frac{\mathrm{d}\tilde{v}_z}{\mathrm{d}\tilde{r}}\right)^2 \tag{19.3.40}$$

where the last term is effectively a constant for $\hat{r}=0(1)$.

Similarly solutions based on the similarity variable

$$\tilde{y}=\hat{r}/\tilde{z}^{\frac{1}{3}} \tag{19.3.41}$$

can be used to satisfy the boundary condition

$$\hat{T}(\tilde{y}=0)=\hat{T}_{\mathrm{W}} \tag{19.3.42}$$

and to match the adiabatic solution (19.3.37) as $\tilde{y}\rightarrow\infty$. We therefore write

$$\hat{T}=\hat{T}_{\mathrm{B}}(\tilde{y})+\frac{\mathrm{Br}}{\mathrm{Gz}^{\frac{2}{3}}}\tilde{\eta}\left|\frac{\mathrm{d}\tilde{v}_z}{\mathrm{d}\tilde{r}}\right|_{\mathrm{W}}\hat{z}^{\frac{2}{3}}\hat{T}_{\mathrm{A}}(\tilde{y}) \tag{19.3.43}$$

where

$$\begin{aligned}&\hat{T}_{\mathrm{B}}(0)=T_{\mathrm{W}}; \qquad \hat{T}_{\mathrm{B}}\rightarrow 0 \quad \text{as} \quad \tilde{y}\rightarrow\infty\\&\hat{T}_{\mathrm{A}}(0)=0; \qquad \hat{T}_{\mathrm{A}}\rightarrow\tilde{y}^{-1} \quad \text{as} \quad \tilde{y}\rightarrow\infty\end{aligned} \tag{19.3.44}$$

The last of these is chosen to match (19.3.37).

From (19.3.40), (19.3.41) and (19.3.43), we can obtain the governing ordinary differential equations for $\hat{T}_{\mathrm{A}}$ and $\hat{T}_{\mathrm{B}}$. These are

$$\frac{\mathrm{d}^2\hat{T}_{\mathrm{B}}}{\mathrm{d}\tilde{y}^2}+\frac{1}{3}\left|\frac{\mathrm{d}\tilde{v}_z}{\mathrm{d}\tilde{r}}\right|_{\mathrm{W}}\tilde{y}^2\frac{\mathrm{d}\hat{T}_{\mathrm{B}}}{\mathrm{d}\tilde{y}}=0 \tag{19.3.45}$$

with solution

$$\hat{T}_{\mathrm{B}}=\hat{T}_{\mathrm{W}}\left\{1-\frac{\displaystyle\int_0^{\tilde{y}}\exp\left\{-\frac{1}{9}\left|\frac{\mathrm{d}\tilde{v}_z}{\mathrm{d}\tilde{r}}\right|_{\mathrm{W}}\tilde{y}^3\right\}\mathrm{d}\tilde{y}}{\displaystyle\int_0^{\infty}\exp\left\{-\frac{1}{9}\left|\frac{\mathrm{d}\tilde{v}_z}{\mathrm{d}\tilde{r}}\right|_{\mathrm{W}}\tilde{y}^3\right\}\mathrm{d}\tilde{y}}\right\}=\hat{T}_{\mathrm{W}}\left\{1-\frac{\displaystyle\int_0^{\frac{1}{9}|\mathrm{d}\tilde{v}_z/\mathrm{d}\tilde{r}|\tilde{y}^3}\mathrm{e}^{-x}x^{-\frac{2}{3}}\,\mathrm{d}x}{\displaystyle\int_0^{\infty}\mathrm{e}^{-x}x^{-\frac{2}{3}}\,\mathrm{d}x}\right\}$$

$$=\hat{T}_{\mathrm{W}}\left\{1-P\left(\frac{1}{3},\frac{1}{9}\left|\frac{\mathrm{d}\tilde{v}_z}{\mathrm{d}\tilde{r}}\right|_{\mathrm{W}}\tilde{y}^3\right)\right\} \quad (19.3.46)$$

where $P(a, x)$ is the incomplete Gamma function (Abramowitz & Stegun, 1964, Subsection 6.5.1) and

$$\frac{\mathrm{d}^2\hat{T}_{\mathrm{A}}}{\mathrm{d}\tilde{y}^2}+\frac{1}{3}\left|\frac{\mathrm{d}\tilde{v}_z}{\mathrm{d}\tilde{r}}\right|_{\mathrm{W}}\tilde{y}^2\frac{\mathrm{d}\hat{T}_{\mathrm{A}}}{\mathrm{d}\tilde{y}}-\frac{2}{3}\left|\frac{\mathrm{d}\tilde{v}_z}{\mathrm{d}\tilde{r}}\right|_{\mathrm{W}}\tilde{y}\hat{T}_{\mathrm{A}}=\left|\frac{\mathrm{d}\tilde{v}_z}{\mathrm{d}\tilde{r}}\right|_{\mathrm{W}} \quad (19.3.47)$$

By using the transformation

$$x=\frac{1}{9}\left|\frac{\mathrm{d}\tilde{v}_z}{\mathrm{d}\tilde{r}}\right|_{\mathrm{W}}\tilde{y}^3 \quad (19.3.48)$$

this can be shown to lead to an inhomogeneous confluent-hypergeometric function equation (see Section 13, Abramowitz & Stegun)

$$x\frac{\mathrm{d}^2\hat{T}_{\mathrm{A}}}{\mathrm{d}x^2}+\left(\tfrac{2}{3}+x\right)\frac{\mathrm{d}\hat{T}_{\mathrm{A}}}{\mathrm{d}x}-\tfrac{2}{3}\hat{T}_{\mathrm{A}}=\left[\frac{1}{9x}\left|\frac{\mathrm{d}\tilde{v}_z}{\mathrm{d}\tilde{r}}\right|_{\mathrm{W}}\right]^{\frac{1}{3}} \quad (19.3.49)$$

with a purely formal solution, satisfying (19.3.44),

$$\begin{aligned}\hat{T}_{\mathrm{A}}=M\left(-\frac{2}{3},\frac{2}{3},-\frac{1}{9}\left|\frac{\mathrm{d}\tilde{v}_z}{\mathrm{d}\tilde{r}}\right|_{\mathrm{W}}\tilde{y}^3\right)\int_{-\frac{1}{9}|\mathrm{d}\tilde{v}_z/\mathrm{d}\tilde{r}|_{\mathrm{W}}\tilde{y}^3}^{0}\left[-\frac{1}{9x}\left|\frac{\mathrm{d}\tilde{v}_z}{\mathrm{d}\tilde{r}}\right|_{\mathrm{W}}\right]^{\frac{1}{3}}\\ \times\frac{U(-\frac{2}{3},\frac{2}{3},x)}{Wr(-\frac{2}{3},\frac{2}{3},x)}\mathrm{d}x-U\left(-\frac{2}{3},\frac{2}{3},-\frac{1}{9}\left|\frac{\mathrm{d}\tilde{v}_z}{\mathrm{d}\tilde{r}}\right|_{\mathrm{W}}\tilde{y}^3\right)\int_{-\frac{1}{9}|\mathrm{d}\tilde{v}_z/\mathrm{d}\tilde{r}|\tilde{y}^3}^{0}\\ \times\frac{\left[-\frac{1}{9x}\left|\frac{\mathrm{d}\tilde{v}_z}{\mathrm{d}\tilde{r}}\right|_{\mathrm{W}}\right]^{\frac{1}{3}}M(-\frac{2}{3},\frac{2}{3},x)}{Wr(-\frac{2}{3},\frac{2}{3},x)}\mathrm{d}x\end{aligned} \quad (19.3.50)$$

where

$$Wr=U\frac{\mathrm{d}M}{\mathrm{d}x}-M\frac{\mathrm{d}U}{\mathrm{d}x} \quad (19.3.51)$$

The value for $\Delta\bar{T}$ differs asymptotically from $P/\rho\Gamma$ by a term of order $\mathrm{Gz}^{-\frac{2}{3}}$. This solution is given in Richardson (1979).

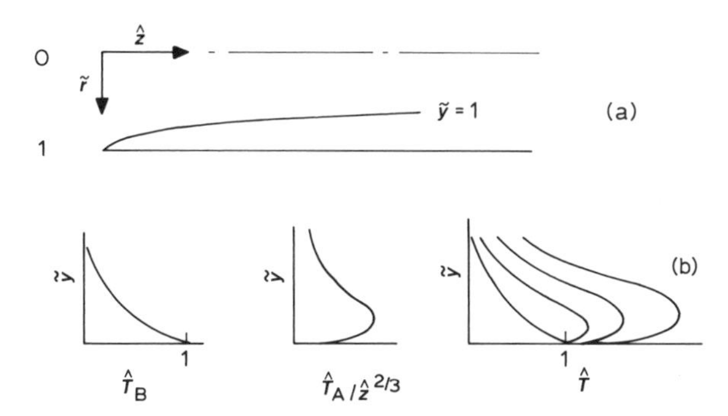

Fig. 19.4. The boundary-layer structure of the temperature field. (a) Growth of boundary layer, $\tilde{y}=1$, in $(\tilde{r}, \hat{z})$ plane. (b) The dimensionless temperature functions $\hat{T}_A/\hat{z}^{\frac{2}{3}}$, $\hat{T}_B$ and $\hat{T}$ of eqn (19.3.43) in crude form.

Figure 19.4 shows the boundary layers in diagrammatic form.

A similar but simpler development can be given for the transient start-up problem, corresponding to $\hat{t} \ll \mathrm{Gz}$. It involves well-known solutions to the heat conduction equation (see Exercise 19.3.4).

II. Na, B<1; Ex, Cm, Vp≪1

Here the velocity field is to be regarded as weakly coupled to the temperature field through the wall temperature and generation. A perturbation expansion for $\mathbf{v}$ and T

$$\mathbf{v}=\mathbf{v}^{(0)}+\mathbf{v}^{(1)}(\mathrm{Na}, \mathrm{B}; \mathrm{Gz}, \mathrm{Br})+\mathbf{v}^{(2)}(\mathrm{Na}^2, \mathrm{BNa}, \mathrm{B}^2; \mathrm{Gz}, \mathrm{Br})+\ldots$$
$$T=T^{(0)}(\mathrm{Br}, \mathrm{Gz})+T^{(1)}(\mathrm{Na}, \mathrm{B}; \mathrm{Gz}, \mathrm{Br})+\ldots \tag{19.3.52}$$

can be postulated, where the $\mathbf{v}^{(K)}$ and $T^{(K)}$ are assumed to be linear functions of $\mathrm{Na}^L\mathrm{B}^{K-L}, L=0,\ldots,K$, and otherwise of order unity. These can be substituted into the eqns (19.3.3)–(19.3.8) with $\tilde{\rho}, \tilde{\Gamma}_p, \tilde{\alpha}$ treated as constant, $\tilde{\eta}$ independent of $\hat{p}$, and the terms involving Na $\hat{p}$ and $\mathrm{Na}(\partial \ln \tilde{\rho}/\partial \ln \tilde{T})_{\tilde{p}}$ in (19.3.6) neglected. The solutions $T^{(0)}$, $\mathbf{v}^{(0)}$ are those discussed in I above. The equations for $\mathbf{v}^{(K)}$ and $T^{(K)}$ become linear inhomogeneous partial-differential equations with homogeneous boundary conditions.

The relations between P, $\dot{V}$ and $\bar{T}$ can be written typically as

$$X=X^{(0)}(Y; \mathrm{Gz}, \mathrm{Br})+X^{(1)}(\mathrm{Na}, \mathrm{B}; Y, \mathrm{Gz}, \mathrm{Br})+\ldots$$

where Y stands for either P or $\dot{V}$ and X for $\bar{T}$ and the other of P or $\dot{V}$.

III. $\mathrm{Na} \not< 1$, $\mathrm{Gz} \to 0$; Ex, Cm, $\mathrm{Vp} \ll 1$
This is the generally fully-developed case. It may be treated as an unsteady problem. Most interest has, however, centred on the steady case $\mathrm{Na} > 1$, particularly when $\mathrm{Na} \gg 1$, with

$$\eta = K_{\mathrm{m}} \mathrm{e}^{-\zeta_{\mathrm{m}}(T - T_{\mathrm{W}})} \left| \frac{\mathrm{d}\tilde{v}_z}{\mathrm{d}\tilde{r}} \right|^{\nu - 1} \tag{19.3.53}$$

Martin (1967) has given a partly analytical solution, demonstrating the interesting feature that $\dot{V}$ is not a single-valued function of P, or more strictly of the pressure gradient $\mathrm{d}p/\mathrm{d}z$, for fixed d_{b}.

If we define

$$\mathrm{Na} = \zeta_{\mathrm{m}} K_{\mathrm{m}} (\tfrac{1}{2} d_{\mathrm{c}})^{1-\nu} v^{*1+\nu} / \alpha \tag{19.3.54}$$

$$\tilde{p}_{\tilde{z}} = (\tfrac{1}{2} d_{\mathrm{b}})^{1+\nu} (\mathrm{d}p/\mathrm{d}z) / K_{\mathrm{m}} v^{*\nu} \tag{19.3.55}$$

and

$$\overset{x}{T} = \tilde{T}/\mathrm{Na} = \zeta_{\mathrm{m}}(T - T_{\mathrm{W}})/\mathrm{Na} \tag{19.3.56}$$

the stress equilibrium and energy equations become†

$$\frac{1}{\tilde{r}} \frac{\mathrm{d}}{\mathrm{d}\tilde{r}} \left\{ \mathrm{e}^{-\mathrm{Na}\overset{x}{T}} \tilde{r} \left(\frac{\mathrm{d}\tilde{v}_z}{\mathrm{d}\tilde{r}} \right)^{\nu} \right\} = -\tilde{p}_{\tilde{z}} \tag{19.3.57}$$

$$\frac{1}{\tilde{r}} \frac{\mathrm{d}}{\mathrm{d}\tilde{r}} \left(\tilde{r} \frac{\mathrm{d}\overset{x}{T}}{\mathrm{d}\tilde{r}} \right) + \mathrm{e}^{-\mathrm{Na}\overset{x}{T}} \left(\frac{\mathrm{d}\tilde{v}_z}{\mathrm{d}\tilde{r}} \right)^{1+\nu} = 0 \tag{19.3.58}$$

with boundary conditions

$$\tilde{v}_z(1) = \overset{x}{T}(1) = 0 \tag{19.3.59}$$

To get a relationship between $\dot{V}$ and $\mathrm{d}p/\mathrm{d}z$, it is convenient to take

$$v^* = \dot{V}/\pi(\tfrac{1}{2} d_{\mathrm{c}})^2 \tag{19.3.60}$$

$\tilde{v}_z$ can be eliminated between (19.3.57) and (19.3.58) to give the equation

$$\frac{\mathrm{d}^2 u}{\mathrm{d}v^2} = -C\mathrm{e}^u \tag{19.3.61}$$

† For convenience $\tilde{p}_{\tilde{z}}$ will be taken as positive, and so $(\mathrm{d}\tilde{v}_z/\mathrm{d}\tilde{r})$ will be positive also.

where

$$\left.\begin{aligned} v &= \ln \tilde{r} \\ u &= (3+1/\nu)v + \mathrm{Na}\,\overset{x}{T}/\nu \\ C &= (\tfrac{1}{2}\tilde{p}_{\tilde{z}})^{1+1/\nu}\,\mathrm{Na}/\nu \end{aligned}\right\} \tag{19.3.62}$$

This gives a temperature profile

$$\tilde{T} = \nu \ln\{2(3+1/\nu)^2\gamma(C\tilde{r}^{3+1/\nu}+\gamma)^{-2}\} \tag{19.3.63}$$

where

$$\gamma^2 + 2\gamma\{C-(3+1/\nu)^2\} + C^2 = 0 \tag{19.3.64}$$

This bounds the value of C in terms of ν for which γ is real $[C \leqslant \frac{1}{2}(3+1/\nu)^2]$. $\tilde{v}_z$ and hence $\dot{V}$ follow by successive integration of (19.3.57). Equations (19.3.60), (19.3.54) and (19.3.55) then provide a relation between $\tilde{p}_{\tilde{z}}$, $\dot{V}$ and ν. (Note that Na has disappeared as an independent parameter.) The dimensionless curves provided by Martin (1967) are reproduced schematically in Fig. 19.5 as

$$\frac{2\zeta_m}{\pi}\left(\frac{K_m 2^{1+3\nu}}{\alpha\nu^\nu d_c^{1+3\nu}}\right)^{1/(1+\nu)} \dot{V} \quad \text{vs} \quad \frac{1}{2}\left(\frac{\nu^\nu d_c^{1+3\nu}}{\alpha^\nu K_m 2^{1+3\nu}}\right)^{1/(1+\nu)} \frac{\mathrm{d}p}{\mathrm{d}z}$$

for various values of ν.

Martin observes that 'the velocity profiles corresponding to the upper branches of the curves in (Fig. 19.5) all contain points of inflection strongly suggesting that such flows, if physically realizable, would be unstable'. He also shows that if the logarithms of $\dot{V}$ and $\mathrm{d}p/\mathrm{d}z$ are plotted (see his Fig. 8) for various values of K_m, then straight lines are effectively attained over most of the lower branches of the curves, suggesting that moderate values of Na will not lead to incorrect values of ν in capillary rheometry.

The paradox of the double-valuedness of $\dot{V}$ in terms of $\mathrm{d}p/\mathrm{d}z$ is often more apparent than real. In any real flow it is P, i.e. the integral of $\mathrm{d}p/\mathrm{d}z$ over the entire length of the tube, that is prescribed. Even for tubes long enough for Gz to be small, when Na is large enough to lead to solutions on the upper branch of the curve, the major contribution to P may still come from the pressure gradient near $\hat{z}=0$. Nevertheless for long enough tubes double-valuedness will arise.

Values for $\bar{T}$ have not been presented for $\mathrm{Na} \neq 0$.

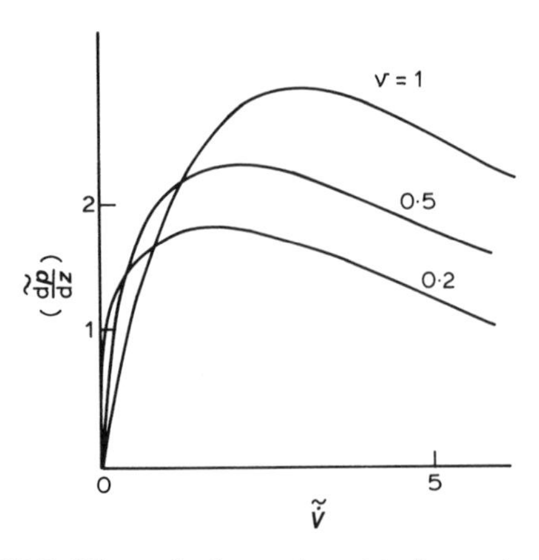

Fig. 19.5. Dimensionless volumetric flow rate

$$\tilde{\dot{V}} = \frac{2\zeta_m}{\pi}\left(\frac{K_m 2^{1+3\nu}}{\alpha\nu^{\nu} d_c^{1+3\nu}}\right)^{1/1+\nu} \dot{V}$$

and associated dimensionless pressure gradient

$$\widetilde{\left(\frac{dp}{dz}\right)} = \frac{1}{2}\left(\frac{\nu^{\nu} d_c^{1+3\nu}}{\alpha^{\nu} K_m 2^{1+3\nu}}\right)^{1/1+\nu} \frac{dp}{dz}$$

(From Martin, 1967, Fig. 1.)

IV. $Na \gg Gz \gg 1$; Ex, Cm, $Vp \ll 1$

This section covers the early stage of development of the flow discussed in III, whether it be start-up from rest or the inlet region of a steady flow. The case $Gz \gg 1$ has already been considered in I above, where even with $Na = 0$ a relatively complex thermal boundary layer had to be described asymptotically in terms of a similarity variable $\tilde{y}$. In this case, the interaction between the temperature field and the flow field will make their dependence on B non-linear unless $B \ll 1$. The problem for $B = 0$, $\nu = 1$ is discussed at length in Ockendon (1979), who extended the similarity solution given by Pearson (1977). Their solutions were described for a channel flow, though they apply equally well in an asymptotic sense to a pipe flow.

The uniform time-dependent case for

$$Na^{-1} \ll \tilde{t} \ll 1 \tag{19.3.65}$$

and no shear-rate dependence of viscosity ($\nu = 1$) is a solution of

(19.3.5) and (19.3.6) in the restricted form

$$\mathrm{Pe}\,\frac{\partial \overset{x}{T}}{\partial \tilde{t}} = \frac{1}{\tilde{r}}\frac{\partial}{\partial \tilde{r}}\left(\tilde{r}\frac{\partial \overset{x}{T}}{\partial \tilde{r}}\right) + \tilde{t}_{\mathrm{rz}}\frac{\partial \tilde{v}_{\mathrm{z}}}{\partial \tilde{r}}$$

where

$$\tilde{t}_{\mathrm{rz}} = \mathrm{e}^{-\mathrm{Na}\overset{x}{T}}\frac{\partial \tilde{v}_{\mathrm{z}}}{\partial \tilde{r}}$$

with

$$\overset{x}{T}(1) = \tilde{v}_{\mathrm{z}}(1) = 0$$

Using the similarity variable

$$\hat{y} = \mathrm{Pe}^{\frac{1}{2}}(1-\tilde{r})/\tilde{t}^{\frac{1}{2}} \tag{19.3.66}$$

and writing

$$\left.\begin{aligned} \overset{x}{T} = \overset{x}{T}(\hat{y}), \qquad \tilde{v}_{\mathrm{z}} = \tilde{v}_{\mathrm{z}}(\hat{y}) \\ \frac{\partial \hat{p}}{\partial \hat{z}} = f(\tilde{t}, \mathrm{Na}) \end{aligned}\right\} \tag{19.3.67}$$

gives the single equation, when $\hat{y} = 0(1)$,

$$\frac{\mathrm{d}^2 \overset{x}{T}}{\mathrm{d}\hat{y}^2} + \tfrac{1}{2}\hat{y}\frac{\mathrm{d}\overset{x}{T}}{\mathrm{d}\hat{y}} + \tfrac{1}{4}\mathrm{e}^{\mathrm{Na}\overset{x}{T}}f^2\,\tilde{t} = 0 \tag{19.3.68}$$

from which it follows that

$$f = -2F(\mathrm{Na})\,\tilde{t}^{-\frac{1}{2}} \tag{19.3.69}$$

(Note that $\hat{y} \to \infty$ when $(1-\tilde{r}) = 0(1)$.) The solution obtained depends upon the *a priori* assumption, later to be shown to be consistent, that $\overset{x}{T} \to 0$, $\tilde{v}_{\mathrm{z}} \to$ constant, 1, when $\hat{y} \to \infty$. This requires that $\dot{V} \approx \pi v^*(\frac{1}{2}d_{\mathrm{c}})^2$, as defined in (19.3.60), i.e. that the region over which $\tilde{v}_{\mathrm{z}}$ is substantially different from unity is a negligible proportion of the cross-sectional area of the tube. It also assumes that the heat generated adiabatically in the core region ($\tilde{v}_{\mathrm{z}} = 1$) of the flow is negligible compared with the heat generated in the boundary region. From (19.3.5), noting that $(\partial \tilde{v}_{\mathrm{z}}/\partial \tilde{r})$ is only substantially different from zero in the boundary layer, it follows by integration that

$$\int_0^\infty \frac{\mathrm{d}\tilde{v}_{\mathrm{z}}}{\mathrm{d}\tilde{y}}\,\mathrm{d}\tilde{y} = 1 = F(\mathrm{Na})\int_0^\infty \mathrm{e}^{\mathrm{Na}\overset{x}{T}}\,\mathrm{d}\hat{y}$$

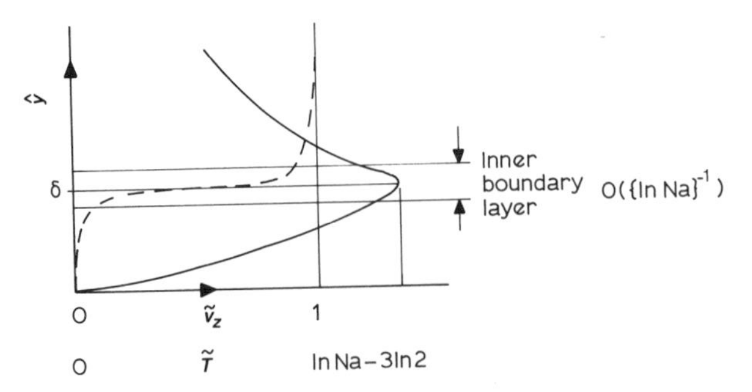

Fig. 19.6. Dimensionless temperature and velocity profiles for unsteady uniform flow for Na ≫ Gz ≫ 1. ——, $\tilde{T}(\hat{y})$; – – –, $\tilde{v}_z(\hat{y})$.

whence

$$\frac{d^2\overset{x}{T}}{d\hat{y}^2}+\tfrac{1}{2}\hat{y}\frac{d\overset{x}{T}}{d\hat{y}}+e^{Na\overset{x}{T}}\Big/\left(\int_0^\infty e^{Na\overset{x}{T}}\,d\hat{y}\right)^2=0 \qquad (19.3.70)$$

Solution of (19.3.70) requires an appreciation of the double-layer structure of the boundary layer, as shown schematically in Fig. 19.6.

The maximum temperature $Na\,\overset{x}{T}=\tilde{T}=\ln Na-3\ln 2$ is attained at a point $\hat{y}=\delta\approx 0{\cdot}955$. The factor $e^{Na\overset{x}{T}}$ in the generation term varies from 1 at $\hat{y}=0$ and as $\hat{y}\to\infty$ to $\frac{1}{8}$ Na at $\hat{y}=\delta$; the only significant generation therefore takes place in a region very close to the position of maximum temperature; similarly because the viscosity varies as $e^{-Na\overset{x}{T}}$, virtually all of the shear takes place in the same region. Thus the velocity goes from zero (asymptotically) for $\hat{y}<\delta-0(1/\ln Na)$ to unity for $\hat{y}>\delta+0(1/\ln Na)$ as is shown in Fig. 19.6. Within the boundary layer of width $0(1/\ln Na)$, the dominant terms in (19.3.70) are the first and the last. For $\hat{y}$ outside the boundary leyer, the dominant terms are the first two. It proves possible to match the inner and outer asymptotic solutions, giving

$$\left.\begin{array}{ll}\tilde{T}\sim\ln Na\,\mathrm{erfc}(\tfrac{1}{2}\hat{y}), \quad \tilde{v}_z\sim 1; & \hat{y}>0{\cdot}955\\ \tilde{T}\sim\ln\{\tfrac{1}{2}Na\,e^{Y}/(1+e^{Y})^2\}, & \tilde{v}_z=e^{Y}/(1+e^{Y});\\ \quad Y=(\hat{y}-\delta)\,e^{-\frac{1}{4}\delta^2}\pi^{-\frac{1}{2}}\ln Na=0(1) & \\ \tilde{T}\sim\ln Na\,\mathrm{erf}(\tfrac{1}{2}\hat{y}), \quad \tilde{v}_z\sim 0; & \hat{y}<0{\cdot}955\end{array}\right\} \qquad (19.3.71)$$

We can show that the outer boundary condition $\tilde{T}\to 0$ is true only to $0(1)$, and that the asymptotic solution (19.3.71) is bounded below for some small $\tilde{t}>0$ in the sense that the solution is ill defined as $\tilde{t}\to 0$.

When $\tilde{t}$ becomes 0(1) the cross-sectional area for which $\tilde{v}_z \neq 0$ is significantly reduced: the curvature terms in eqns (19.3.5) and (19.3.6) become relevant, while the outer boundary condition on $\tilde{v}_z$ increases above unity. As $\tilde{t} \to \infty$, so the solution $\overset{x}{T}_\infty$ of III, defined by (19.3.56) and (19.3.63), must be approached. the details have not been worked out.

The steady developing case for $\mathrm{Na}^{-1} \ll \mathrm{Gz}^{-1} \ll 1$ can be treated similarly. Equation (19.3.6) is used in the form

$$\mathrm{Gz}\left(\hat{v}_r \frac{\partial \overset{x}{T}}{\partial \tilde{r}} + \tilde{v}_z \frac{\partial \overset{x}{T}}{\partial \hat{z}}\right) = \frac{1}{\tilde{r}} \frac{\partial}{\partial \tilde{r}}\left(\tilde{r} \frac{\partial \overset{x}{T}}{\partial \tilde{r}}\right) + \tilde{t}_{rz} \frac{\partial \tilde{v}_z}{\partial \tilde{r}}$$

and the similarity variable

$$\check{y} = (1 - \tilde{r})\check{z}^{-\frac{1}{2}} \quad \text{where} \quad \check{z} = \hat{z}/\mathrm{Gz} \tag{19.3.72}$$

is used to provide the appropriate boundary layer. Because both convective terms contribute to the solution, the two independent variables are taken to be $\overset{x}{T}(\check{y})$ and a form of dimensionless stream function $q(\check{y})$ where

$$\tilde{v}_z = -\frac{1}{\check{z}^{\frac{1}{2}}} \frac{\mathrm{d}\check{q}}{\mathrm{d}\check{y}}, \qquad \hat{v}_r = \frac{1}{2\check{z}^{\frac{1}{2}}}\left(\check{y} \frac{\mathrm{d}\check{q}}{\mathrm{d}\check{y}} - \check{q}\right) \tag{19.3.73}$$

On substitution into (19.3.5) and (19.3.6), as before, the two equations

$$\frac{\mathrm{d}^2 \check{q}}{\mathrm{d}\check{y}^2} = -K\, \mathrm{e}^{\mathrm{Na}\overset{x}{T}} \tag{19.3.74}$$

$$\frac{\mathrm{d}^2 \overset{x}{T}}{\mathrm{d}\check{y}^2} + \tfrac{1}{2}\check{q} \frac{\mathrm{d}\overset{x}{T}}{\mathrm{d}\check{y}} + K^2\, \mathrm{e}^{\mathrm{Na}\overset{x}{T}} = 0 \tag{19.3.75}$$

with boundary conditions

$$\left.\begin{aligned} &\overset{x}{T}(0) = \check{q}(0) = (\mathrm{d}\check{q}/\mathrm{d}\check{y})(0) = 0 \\ &\overset{x}{T} \to 0, \qquad \frac{\mathrm{d}\check{q}}{\mathrm{d}\check{y}} \to -1 \quad \text{as} \quad \check{y} \to \infty \end{aligned}\right\} \tag{19.3.76}$$

K is given by

$$\int_0^\infty \frac{\mathrm{d}^2 \check{q}}{\mathrm{d}\check{y}^2}\, \mathrm{d}\check{y} = 1 = K \int_0^\infty \mathrm{e}^{\mathrm{Na}\overset{x}{T}}\, \mathrm{d}\check{y}$$

whence

$$\frac{\mathrm{d}^2 \overset{x}{T}}{\mathrm{d}\check{y}^2} + \tfrac{1}{2}\check{q} \frac{\mathrm{d}\overset{x}{T}}{\mathrm{d}\check{y}} = \mathrm{e}^{\mathrm{Na}\overset{x}{T}} \Big/ \left(\int_0^\infty \mathrm{e}^{\mathrm{Na}\overset{x}{T}}\, \mathrm{d}\check{y}\right)^2 \tag{19.3.77}$$

Equation (19.3.77) is the same as (19.3.70) except that $\hat{y}$ is replaced by $\check{q}$. Solution of (19.3.77) leads to the same double-layer structure as for (19.3.70), with a maximum temperature, still given by $\tilde{T}=\ln\frac{1}{8}\,\text{Na}$, at $\check{y}\simeq\pi^{\frac{1}{2}}$. Matching of inner and outer expansions gives

$$\tilde{T}\sim\ln\text{Na}\,\text{erfc}(\tfrac{1}{2}\check{y}-\tfrac{1}{2}\pi^{\frac{1}{2}}),\qquad \tilde{v}_z\sim 1;\qquad \check{y}>\pi^{\frac{1}{2}} \tag{19.3.78}$$

$$\tilde{T}\sim\ln\text{Na}\,\check{y}/\pi^{\frac{1}{2}},\qquad \tilde{v}_z\sim 0;\qquad \check{y}<\pi^{\frac{1}{2}} \tag{19.3.79}$$

with (19.3.71) unaltered except that

$$Y=(\check{y}-\pi^{\frac{1}{2}})\pi^{-\frac{1}{2}}\ln\text{Na} \tag{19.3.80}$$

The local pressure gradient can be integrated (see Example 19.3.1) to give

$$P=-\int_0^l \frac{\mathrm{d}p}{\mathrm{d}z}\,\mathrm{d}z=\frac{8}{\pi^{\frac{3}{2}}}\,\text{Gz}^{\frac{1}{2}}\,K_{\text{m}}\,\frac{l_{\text{c}}\dot{V}}{(\frac{1}{2}d_{\text{c}})^4}\,\frac{\ln\text{Na}}{\text{Na}} \tag{19.3.81}$$

where K_{m} is now the Newtonian viscosity at $\tilde{T}=0$.

However this integral includes a $z^{-\frac{1}{2}}$ singularity in $\mathrm{d}\hat{p}/\mathrm{d}z$ as $z\to 0$, which is unreal. It may be better written as

$$P=8\sqrt{2}\rho\Gamma\ln\text{Na}/\zeta_{\text{m}}\pi^{\frac{1}{2}}\,\text{Gz}^{\frac{1}{2}}$$

Ockendon (1979) has tackled the difficult problem of matching the similarity solution (19.3.78)–(19.3.80) to the asymptotically isothermal $T=0$ flow at $z=0$ without fully resolving it. Figure 19.7 gives the pressure gradient $\partial\hat{p}/\partial\hat{z}$ and layer thickness $2\text{Na}\,\varepsilon/\ln\text{Na}$ (where ε is the position of maximum temperature in terms of units of $\tilde{y}=(1-\tilde{r})$) as a function

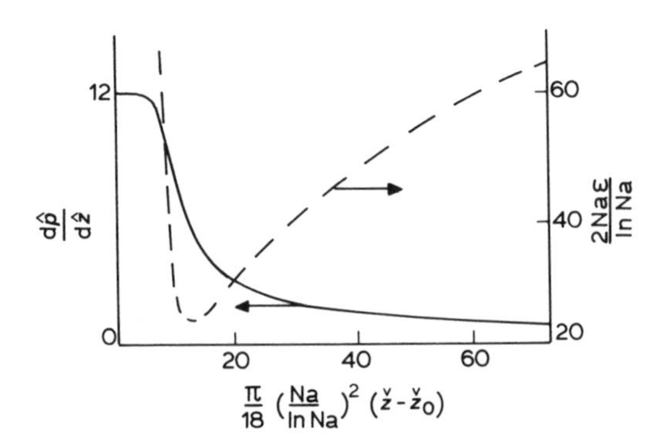

Fig. 19.7. Dimensionless pressure gradient (——) and layer thickness (– – –) in terms of distance downstream. (From Ockendon, 1979.)

of downstream distance

$$\frac{\pi}{18}\left(\frac{\text{Na}}{\ln \text{Na}}\right)^2(\check{z}-\check{z}_0)$$

where $\check{z}_0$ is an unknown origin of $0(\text{Na}^{-\frac{3}{2}})$. The most interesting feature of this solution is the initial reduction in size of the boundary layer to a minimum value of about $12 \ln \text{Na}/\text{Na}$ before it grows as $(\check{z}-\check{z}_0)^{\frac{1}{2}}$. Although $\check{z}_0$ is only $0(\text{Na}^{-\frac{3}{2}})$, the contribution of $\mathrm{d}\hat{p}/\mathrm{d}\hat{z}$ in the region where it is close to unity is obviously very considerable and different from the similarity solution; this makes application of the asymptotic result (19.3.81) liable to considerable error in any practical situation. The asymptotic result also has the potentially unstable characteristic that P decreases as Q increases.

V. $\text{Br} \ll 1$, $|\text{B}| \not\ll 1$, $\text{Gz} \gg 1$; Ex, Cm, $\text{Vp} \ll 1$

This corresponds to a situation where there is a rheologically significant difference between the entry and wall temperatures, where generation is relatively unimportant and where the flow is in the early stages of development: it has been analysed by Ockendon & Ockendon (1977) for B both positive and negative, when $\nu = 1$. However, their solution is described for a uniform channel and covers the case $\text{Gz} = 0(1)$ also, so discussion of it will be deferred until the next subsection, particularly since it is of most practical importance when $\text{B} < 0$ and thermoplastics are being modelled.

VI. $\text{Vp} = 0(1)$; $\text{Gz} < 1$

In many materials used in injection moulding

$$\eta/(\partial\eta/\partial p)_{\text{T}} = 0(500 \text{ bar}) \tag{19.3.82}$$

and so substantial changes in viscosity can be expected, in going from reservoir to melt front, caused by pressure changes alone. The viscosity relation

$$\eta = K_{\text{m}} \mathrm{e}^{-\zeta_{\text{m}}(T-T_{\text{W}})} \mathrm{e}^{\tilde{\omega}_{\text{m}} p}\left(\left|\frac{\mathrm{d}v_z}{\mathrm{d}r}\right|\right)^{\nu-1} \tag{19.3.83}$$

will therefore be used now instead of (19.3.53). If $\zeta_{\text{m}}(T-T_{\text{W}})$ and $\tilde{\omega}_{\text{m}} p$ are taken to be zero, then the pressure drop P_0 in a capillary of length l_{c} and diameter d_{c} for a flow rate $\dot{V}$, on the basis of uniaxial fully-developed flow (i.e. $\text{Gz} \ll 1$), is known to be

$$P_0 = \frac{2K_{\text{m}}\dot{V}^{\nu} l_{\text{c}}}{(\frac{1}{2}d_{\text{c}})^{3\nu+1}}\left(\frac{3\nu+1}{\pi\nu}\right)^{\nu} = P_0' l_{\text{c}} \tag{19.3.84}$$

(see for example Denn, 1981, or Example 7.1.1 earlier). This is the expected unique power-law relation between P_0 and $\dot{V}$. If now ζ_m is still taken to be zero but the $\tilde{\omega}_m$ contribution to (19.3.83) is included, the pressure drop P for the same flow rate $\dot{V}$ is easily seen to follow from

$$\frac{dp}{dz} = -P_0' e^{\tilde{\omega} p} \tag{19.3.85}$$

integrated from $z=0$ to $z=l_c$, which yields (assuming $p_{out}=0$)

$$P = -\frac{1}{\tilde{\omega}_m} \ln(1 - P_0 \tilde{\omega}_m) \tag{19.3.86}$$

Clearly for given $\tilde{\omega}_m$, P_0 cannot be greater than $\tilde{\omega}_m^{-1}$ and a singularity (choking) arises for $P_0 = \tilde{\omega}_m^{-1}$. Alternatively put, an exponential pressure dependence with no temperature dependence leads to an actual output $\dot{V}$, which is related to the $\tilde{\omega}_m = 0$ output $\dot{V}_0$, according to

$$\left(\frac{\dot{V}}{\dot{V}_0}\right) = \frac{1 - e^{-\tilde{\omega}_m P}}{\tilde{\omega}_m P} = \frac{P_0}{P} \tag{19.3.87}$$

The term $\zeta_m(T - T_W)$ can be made arbitrarily small by making $d_c \to 0$, so the choking phenomenon is a real one.

If, however, as will be the case in many injection-moulding situations, the isothermal approximation $T \sim T_W$ is replaced by an adiabatic one (Gz $\gg$ 1) then T is independent of T_W and depends upon T_0, the reservoir temperature, and the pressure loss only. For a drop in pressure of Δp, there will be a mean temperature rise of $\Delta p / \rho_m \Gamma_{vm}$. If T_W in (19.3.83) is taken to be T_0, and T to be $\bar{T}$, then the effective viscosity can be written as

$$\eta = K_m \, e^{-\zeta_m P / \rho_m \Gamma_{vm}} \, e^{(\tilde{\omega}_m + \zeta_m / \rho_m \Gamma_{vm}) p} \left| \frac{dv_z}{dr} \right|^{\nu - 1} \tag{19.3.88}$$

(When $p = P$ at the inlet to the capillary, $\bar{T} = T_0$ and there is no contribution from ζ_m. When $p = 0$ at the outlet, $\bar{T} = P / \rho_m \Gamma_{vm}$ and there is no contribution from $\tilde{\omega}_m$.) The important point is that the unknown reservoir pressure is now implicitly involved in the outlet viscosity.

The ratio

$$\rho_m \Gamma_{vm} \tilde{\omega}_m / \zeta_m = T^*_{rheol} / T^*_{visc} = \mathrm{Gz\ Vp/Na} \tag{19.3.89}$$

is 0(1) for most polymers. The ratio

$$\zeta_m P / \rho_m \Gamma_{vm} \approx \mathrm{Na/Gz} \tag{19.3.90}$$

The distinction between Γ_v and Γ_p is not taken to be significant. Equation (19.3.85) can now be replaced by

$$\frac{dp}{dz} = -P_0' \, e^{-P/\Pi_m^*} e^{p/\Pi_m} \tag{19.3.91}$$

where P_0' can be regarded as specifying the flow rate $\dot{V}$ (or vice versa) and

$$\Pi_m^* = \rho_m \Gamma_{vm}/\zeta_m, \qquad \Pi_m^{-1} = \tilde{\omega}_m + \zeta_m/\rho_m \Gamma_{vm} \tag{19.3.92}$$

On integration,

$$P/\Pi_m = -\ln[1 - P_0 e^{-P/\Pi_m^*}/\Pi_m] \tag{19.3.93}$$

which has no singularity, and has the approximate solution

$$P \simeq \Pi_m^* \ln(P_0/\Pi_m) \tag{19.3.94}$$

when $P_0 \gg \Pi_m$. It should be emphasized that the result (19.3.94), through (19.3.88), depends upon the simultaneous and therefore inconsistent approximations $\mathrm{Gz} \ll 1$, and $\mathrm{Gz} \gg 1$, and so must be regarded instead as having significance at most when $\mathrm{Gz} = 0(1)$.

Useful results for freezing flow are presented in Sampson & Gibson (1981).

19.3.2 Channel Flow

Channel flow can be treated in just the same way as tube flow above, with only minor differences in the equations and dimensionless groups involved. It has been introduced in Subsection 10.2.1 (slot dies) in the decoupled situation, following the arguments given in Section 8.1 on the lubrication approximation. The essential feature of our treatment here is that we consider only plane flow in a uniform channel of depth h and length l, and so write

$$\mathbf{v} = \{v_1(x_1, x_3, t), 0, v_3(x_1, x_3, t)\}, \qquad T = T(x_1, x_3, t) \tag{19.3.95}$$

with

$$T(0, x_3, t) = T_0, \qquad T(x_1, \pm\tfrac{1}{2}h, t) = T_W \tag{19.3.96}$$

as entry and boundary conditions on the temperature. Equation (19.3.96) is an idealized approximation. An initial condition on $T(x_1, x_3, 0)$ may or may not be relevant for unsteady flows. The quantities of interest will be P, $\bar{T}$ and $\dot{m}$ ($=\dot{M}$/width in the x_2

direction) or its volumetric equivalent q. We note once again that the flow field may initially be empty of fluid and so the full problem is a moving-front problem. This will be particularly true when the tube walls are cooled below the melting temperature, i.e. $T_W < T_m$. We shall concentrate on the freezing problem, i.e. thermoplastics, in this subsection. Some of this account parallels the work of Janeschitz–Kreigl (1979).

The equations describing the flow can be written, using (7.1.24) for the energy equation, as

$$\frac{\partial(\rho_m v_1)}{\partial x_1}+\frac{\partial(\rho_m v_3)}{\partial x_3}=-\frac{\partial \rho_m}{\partial t} \tag{19.3.97}$$

$$\frac{\partial t_{13}}{\partial x_3}\approx\frac{\partial p}{\partial x_1} \tag{19.3.98}$$

$$\rho_m\Gamma_{pm}\left(v_3\frac{\partial T}{\partial x_3}+v_1\frac{\partial T}{\partial x_1}+\frac{\partial T}{\partial t}\right)\approx\frac{\partial}{\partial x_3}\left(\alpha_m\frac{\partial T}{\partial x_3}\right)+t_{13}\frac{\partial v_1}{\partial x_3}$$
$$-p\left(\frac{\partial v_3}{\partial x_3}+\frac{\partial v_1}{\partial x_1}\right)-\left(\frac{\partial \ln \rho}{\partial \ln T}\right)_p\left(\frac{\partial p}{\partial T}+v_1\frac{\partial p}{\partial x_1}+v_3\frac{\partial p}{\partial x_3}\right) \tag{19.3.99}$$

The entry condition will be (19.3.96) but the boundary condition will involve a solidified layer if $T_W < T_m$. We suppose that the freezing interface is at $x_3 = \pm x_{3m}$. It follows that

$$T(\pm x_{3m}) = T_m, \qquad \mathbf{v}(\pm x_{3m}) = \mathbf{0} \tag{19.3.100}$$

provided the rate of crystallization (or vitrification) is rapid enough, and the rate of change of x_{3m} is slow enough.

Within the solidified layer, $\mathbf{v}\equiv 0$ and a heat condition equation will be relevant, i.e.

$$\frac{\partial}{\partial x_3}\left(\alpha_s\frac{\partial T}{\partial x_3}\right)=\rho_s\Gamma_{ps}\frac{\partial T}{\partial t}+\left(\frac{\partial \ln \rho}{\partial \ln T}\right)_p\frac{\partial p}{\partial t} \tag{19.3.101}$$

satisfying (19.3.96ii) and (19.3.100i). A further interfacial boundary condition on the temperature, similar to (11.2.21), can be written

$$-\left(\alpha_m\frac{\partial T}{\partial x_3}\right)_{x_3=\pm x_{3m-}}+\left(\alpha_s\frac{\partial T}{\partial x_3}\right)_{x_3=\pm x_{3m+}}\equiv\left[\alpha\frac{\partial T}{\partial x_3}\right]_{\pm x_{3m}}$$
$$=\pm\rho_s\xi_s\frac{dx_{3m}}{dt} \tag{19.3.102}$$

Note that dx_{3m}/dt is included in (19.3.89) but neglected in (19.3.100ii) because α_m and α_s are relatively small (see also Exercise 19.3.9). Using dimensionless variables

$$\left.\begin{aligned}
&\tilde{x}_3 = x_3/\tfrac{1}{2}h, \qquad \tilde{x}_1 = x_1/\tfrac{1}{2}h, \qquad \hat{x}_1 = x_1/l\\
&\tilde{v}_1 = v_1/v^*, \qquad \hat{v}_3 = v_3 l/\tfrac{1}{2}hv^*, \qquad \tilde{q} = q/q_0\\
&\tilde{t} = tv^*/\tfrac{1}{2}h, \qquad \hat{t} = tv^*/l, \qquad \hat{T} = (T - T_m)/(T_0 - T_m)\\
&\tilde{t}_{13} = \tilde{\eta}(\hat{p}, \hat{T}, \partial\tilde{v}_1/\partial\tilde{x}_3)\, \partial\tilde{v}_1/\partial\tilde{x}_3\\
&\tilde{\alpha}_{s,m} = \alpha_{s,m}/\alpha_m(T_m), \qquad \tilde{\rho}_{s,m} = \rho_{s,m}/\rho_m(T_m)\\
&\tilde{\Gamma}_{ps,m} = \Gamma_{ps,m}/\Gamma_{pm}(T_m), \qquad \tilde{\eta} = \frac{\eta(p, T, \partial v_1/\partial x_3)}{\eta(0, T_m, D^*)}\\
&\tilde{p} = p/\eta(0, T_m, D^*)D^*, \qquad \hat{p} = \tilde{p}\tfrac{1}{2}h/l
\end{aligned}\right\} \tag{19.3.103}$$

where

$$v^* = q_0/h, \qquad D^* = q_0/\tfrac{1}{2}h^2 \tag{19.3.104}$$

$$\left.\begin{aligned}
&\mathrm{Na} = \eta(T_m, D^*)v^{*2}\zeta_m/\alpha_m\\
&\mathrm{Br} = \eta(T_m, D^*)v^{*2}/(T_0 - T_m)\alpha_m\\
&\mathrm{B_s} = (T_m - T_W)\zeta_m, \qquad \mathrm{B_m} = (T_0 - T_m)\zeta_m\\
&\mathrm{Sf} = \rho_s\xi_s/\rho_m\Gamma_{pm}(T_0 - T_m), \qquad \mathrm{Gz} = \frac{v^*h^2}{4\kappa l}
\end{aligned}\right\} \tag{19.3.105}$$

(19.3.96i) and (19.3.97)–(19.3.102) become

$$\frac{\partial(\tilde{\rho}_m\tilde{v}_1)}{\partial\hat{x}_1} + \frac{\partial(\tilde{\rho}_m\hat{v}_3)}{\partial\tilde{x}_3} = -\frac{\partial\tilde{\rho}_m}{\partial\hat{t}} \tag{19.3.106}$$

$$\frac{\partial\tilde{t}_{13}}{\partial\tilde{x}_3} = \frac{\partial\hat{p}}{\partial\hat{x}_1} \tag{19.3.107}$$

$$\begin{aligned}
\mathrm{Gz}\,\tilde{\rho}_m\tilde{\Gamma}_{pm}\left(\tilde{v}_1\frac{\partial\hat{T}}{\partial\hat{x}_1} + \hat{v}_3\frac{\partial\hat{T}}{\partial\tilde{x}_3} + \frac{\partial\hat{T}}{\partial\hat{t}}\right) &= \frac{\partial}{\partial\tilde{x}_3}\left(\alpha_m\frac{\partial\hat{T}}{\partial\tilde{x}_3}\right)\\
&\quad + \mathrm{Br}\left[\tilde{t}_{13}\frac{\partial\tilde{v}_1}{\partial\tilde{x}_3} - \hat{p}\left(\frac{\partial\hat{v}_3}{\partial\tilde{x}_3} + \frac{\partial\tilde{v}_1}{\partial\hat{x}_1}\right)\right.\\
&\quad \left. - \left(\frac{\partial\ln\rho}{\partial\ln T}\right)_p\left(\frac{\partial\hat{p}}{\partial\hat{t}} + \tilde{v}_1\frac{\partial\hat{p}}{\partial\hat{x}_1} + \hat{v}_3\frac{\partial\hat{p}}{\partial\tilde{x}_3}\right)\right]
\end{aligned} \tag{19.3.108}$$

$$\tilde{T}(\pm\tilde{x}_{3m}) = 0, \qquad \mathbf{v}(\pm\tilde{x}_{3m}) = 0 \tag{19.3.109}$$

$$\tilde{T}(\hat{x}_1=0)=1 \tag{19.3.110}$$

$$\frac{\partial}{\partial \tilde{x}_3}\left(\tilde{\alpha}_s \frac{\partial \hat{T}}{\partial \tilde{x}_3}\right)=\mathrm{Gz}\,\tilde{\rho}_s \tilde{\Gamma}_{ps}\frac{\partial \hat{T}}{\partial \hat{t}}+\mathrm{Br}\left(\frac{\partial \ln \rho}{\partial \ln T}\right)_p \frac{\partial \hat{p}}{\partial \hat{t}} \tag{19.3.111}$$

$$\hat{T}(\pm 1)=-\mathrm{B_s}/\mathrm{B_m} \tag{19.3.112}$$

$$\left[\tilde{\alpha}\frac{\partial \hat{T}}{\partial \tilde{x}_3}\right]_{\pm \tilde{x}_{3m}}=\pm \mathrm{Sf}\,\mathrm{Gz}\frac{\mathrm{d}\tilde{x}_{3m}}{\mathrm{d}\hat{t}} \tag{19.3.113}$$

If the melt front is at $\hat{x}_1=\hat{x}_{1f}<1$, then

$$\frac{\mathrm{d}\tilde{x}_{1f}}{\mathrm{d}\hat{t}}=\tilde{q} \tag{19.3.114}$$

The 'initial' condition will depend upon the time $\hat{t}_f$ at which any position $\hat{x}_1$ of the channel is filled. Assuming that q is constant, so $\tilde{q}\equiv 1$, and that the flow is supposed fully mixed on reaching the melt point, it follows from (19.3.114) that

$$\hat{T}(\hat{x}_1, \tilde{x}_3, \hat{t}=\hat{x}_1)=\hat{\hat{T}}(\hat{x}_1) \tag{19.3.115}$$

For $\hat{t}<\hat{x}_1$, the channel is not filled at $\hat{x}_1$. For $\hat{t}>\hat{x}_1$, eqns (19.3.106)–(19.3.113) apply. Clearly $\hat{\hat{T}}(0)=1$ from (19.3.110). This aspect is considered in greater detail in Subsection 19.4.1.

The constitutive relation (19.3.8) is only altered by $\partial\tilde{v}_z/\partial\tilde{r}$ becoming $\partial\tilde{v}_1/\partial\tilde{x}_3$. If the power law relation is used,

$$\left.\begin{aligned}\tilde{\eta}&=\mathrm{e}^{-\mathrm{B_m}\hat{T}}|\partial\tilde{v}_1/\partial\tilde{x}_3|^{\nu-1}\\ \eta(0, T_\mathrm{m}, D^*)&=K_\mathrm{m}(v^*/\tfrac{1}{2}h)^{\nu-1}\end{aligned}\right\} \tag{19.3.116}$$

The set of equations (19.3.106)–(19.3.116) involves the dimensionless groups Gz, Br, $\mathrm{B_m}$, $\mathrm{B_s}$ and Sf directly and Ex, Vp and Cm implicitly. The latter three may be regarded as small, as before, but ρ_s/ρ_m, Γ_{ps}/Γ_{pm} and α_s/α_m are kept as 0(1) constants. Solution by direct numerical computation is again possible. Simplified analytical and semi-analytic solutions may be derived.

I. Na, $\mathrm{B_m}$, Ex, Cm, Vp $\ll 1$

Because $\tilde{x}_{3m}$ is an unknown of the problem, the set of equations and boundary conditions (19.3.106)–(19.3.113) is necessarily coupled and in principle must be solved simultaneously. For the steady situation $\tilde{x}_{3m}$ is a function of $\hat{x}_1$ only; this will apply sufficiently far back from the melt front provided $\tilde{q}\equiv 1$, constant. Close to the melt front $\hat{x}_1=\hat{t}$, $\tilde{x}_{3m}$

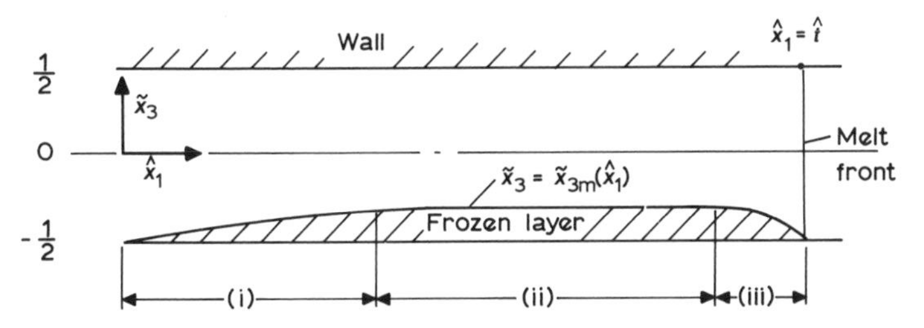

Fig. 19.8. Instantaneous cross-section of frozen layer in channel during injection moulding. (i) Steady inlet region: $|d\tilde{x}_{3m}/d\hat{x}_1| < 0$; (ii) steady uniform region: $d\tilde{x}_{3m}/d\hat{x}_1 \sim 0$; (iii) unsteady front region.

will be a function of time since $\tilde{x}_{3m}(\hat{x}_1 = \hat{t}) = 0$. The situation is illustrated diagrammatically in Fig. 19.8, where three zones are distinguished. The inlet zone (i), in which the layer thickness $(\frac{1}{2} - \tilde{x}_{3m})$ increases, necessarily arises because conduction to the wall must cool the melt as it flows past near the wall provided $\tilde{x}_{3m}$ is close enough to $\frac{1}{2}$. The steady uniform region (ii) arises if Gz is small enough for the flow to become fully-developed, in which case the heat conducted to the wall, through the frozen layer, is exactly balanced by the heat generated within the flowing fluid. The unsteady adjustment region (iii) can be treated independently of the others and is likely to be short.

$\mathrm{Gz} \ll 1$. Provided $\mathrm{Br} \neq 0$ (which in practice is never the case anyway although it is sometimes convenient to make this approximation), this will have as an asymptotic limit the fully-developed profile determined by

$$\frac{d}{d\tilde{x}_3}\left[\eta\left(\frac{d\tilde{v}_1}{d\tilde{x}_3}\right)\frac{d\tilde{v}_1}{d\tilde{x}_3}\right] = \hat{p}_{,\hat{x}_1}, \quad \text{constant} \tag{19.3.117}$$

$$\frac{d}{d\tilde{x}_3}\left(\tilde{\alpha}_m \frac{d\hat{T}}{d\tilde{x}_3}\right) + \mathrm{Br}\,\tilde{\eta}\left(\frac{d\tilde{v}_1}{d\tilde{x}_3}\right)^2 = 0 \tag{19.3.118}$$

with

$$\left.\begin{aligned} \hat{T}(\tilde{x}_{3m}) &= \tilde{v}_1(\tilde{x}_{3m}) = 0 \\ \frac{d\hat{T}}{d\tilde{x}_3}(0) &= \frac{d\tilde{v}_1}{d\tilde{x}_3}(0) = 0 \end{aligned}\right\} \tag{19.3.119}$$

and (see Exercise 19.3.10)

$$\left(\tilde{\alpha}_m \frac{d\hat{T}}{d\tilde{x}_3}\right)_{\tilde{x}_{3m}} = -\frac{\mathrm{B_s}}{\mathrm{B_m}}\frac{\tilde{\alpha}_s}{(1 - \tilde{x}_{3m})} \tag{19.3.120}$$

bearing in mind that

$$\tilde{q} = 2\int_0^{\tilde{x}_{3m}} \tilde{v}_1 \, d\tilde{x}_3 = 1 \tag{19.3.121}$$

Because of symmetry, only positive values of $\tilde{x}_3$ are considered. If the power-law relation (19.3.116) with $B_m \equiv 0$ is used and $\tilde{\alpha}_m \equiv 1$ then (19.3.117) and (19.3.118) yield

$$\frac{d^2\hat{T}}{d\tilde{x}_3^2} + \mathrm{Br}(\tilde{x}_3\,|\hat{p}_{,\hat{x}_1}|)^{1+1/\nu} = 0 \tag{19.3.122}$$

where

$$\frac{d\tilde{v}_1}{d\tilde{x}_3} = \mathrm{sgn}\,\hat{p}_{,\hat{x}_1}(|\hat{p}_{,\hat{x}_1}|\tilde{x}_3)^{1/\nu} \tag{19.3.123}$$

Equation (19.3.123) can be integrated to give, using (19.3.119),

$$\tilde{v}_1 = \frac{\nu}{\nu+1}\,|\hat{p}_{,\hat{x}_1}|^{1/\nu}(\tilde{x}_{3m}^{1+1/\nu} - \tilde{x}_3^{1+1/\nu}) \tag{19.3.124}$$

where $\mathrm{sgn}\,\hat{p}_{,\hat{x}_1}$ is taken to be -1. Further integration using (19.3.121) gives

$$|\hat{p}_{,\hat{x}_1}| = \left(\frac{2\nu+1}{2\nu}\right)^{\nu} \tilde{x}_{3m}^{-(2\nu+1)} \tag{19.3.125}$$

Equation (19.3.122) can then be integrated to give

$$\hat{T} = \mathrm{Br}\,\frac{\nu}{2(3\nu+1)}\left(\frac{2\nu+1}{2\nu}\right)^{\nu} \tilde{x}_{3m}^{-2\nu}\left\{1 - \left(\frac{\tilde{x}_3}{\tilde{x}_{3m}}\right)^{3+1/\nu}\right\} \tag{19.3.126}$$

and (19.3.120) yields

$$\frac{\tilde{x}_{3m}^{2\nu+1}}{1-\tilde{x}_{3m}} = \left(\frac{\nu}{2\nu+1}\right)\mathrm{Br}\left(\frac{B_m}{B_s\tilde{\alpha}_s}\right) = Z \tag{19.3.127}$$

If Z is very small,

$$\tilde{x}_{3m} \sim Z^{1/(2\nu+1)} \tag{19.3.128}$$

If Z is very large

$$\tilde{x}_{3m} \sim 1 - 1/Z \tag{19.3.129}$$

i.e. the thickness of the solid layer is $1/Z$.

$\text{Gz} = 0(1)$. This is the developing situation, and it is of interest to consider the case where Br is vanishingly small, because this has some application to situations where premature freezing can occur, i.e. to cases where $\tilde{x}_{3m} = 0$ for a finite value of $\tilde{x}_1$. In practice, such flows are not steady, and will be determined at their later stages by overall pressure drop rather than by flowrate. Nevertheless, the steady solution is worth considering because it can be investigated analytically to some extent.

Although it is not strictly correct to neglect terms in $\hat{v}_3$, by supposing $d\tilde{x}_{3m}/d\hat{x}_1$ to be sufficiently small, it is possible to carry out an analysis based on a stream function (see Example 19.3.2 for details) that is equivalent to neglecting the terms in $\hat{v}_3$ and using a variable $\hat{x}_3 = \tilde{x}_3/\tilde{x}_{3m}$ instead of $\tilde{x}_3$. With these changes, eqns (19.3.123)–(19.3.125) will still apply, though (19.3.122) has to be replaced by

$$\frac{\partial^2 T}{\partial \tilde{x}_3^2} = \text{Gz}\left(\frac{2\nu+1}{2\nu+2}\right)\frac{1}{\tilde{x}_{3m}}\left\{1-\left(\frac{\tilde{x}_3}{\tilde{x}_{3m}}\right)^{1+1/\nu}\right\}\frac{\partial \hat{T}}{\partial \hat{x}_1} \qquad (19.3.130)$$

while (19.3.102) can be written, as in (19.3.120) with $\tilde{\alpha}_m = 1$,

$$\left(\frac{\partial \hat{T}}{\partial \tilde{x}_3}\right)_{\tilde{x}_{3m}} = -\frac{\text{B}_s\tilde{\alpha}_s}{\text{B}_m(1-\tilde{x}_{3m})} \qquad (19.3.131)$$

It is natural to look for a similarity solution

$$\hat{T} = \hat{X}_1(\hat{x}_1)\hat{X}_3(\tilde{x}_3/\tilde{x}_{3m}), \qquad \tilde{x}_{3m} = \tilde{x}_{3m}(\hat{x}_1) \qquad (19.3.132)$$

so that, from (19.3.130), putting $\hat{x}_3 = \tilde{x}_3/\tilde{x}_{3m}$,

$$\frac{\hat{X}_3''}{\hat{X}_3\{1-\hat{x}_3^{1+1/\nu}\}} = \frac{\text{Gz}(2\nu+1)}{(2\nu+2)}\frac{\hat{X}_1'}{\hat{X}_1}\tilde{x}_{3m} = -C_1,\ C_1 \quad \text{a constant} \qquad (19.3.133)$$

and from (19.3.131)

$$\hat{X}_3'(1) = -\frac{\text{B}_s\tilde{\alpha}_s}{\text{B}_m}\frac{\tilde{x}_{3m}}{(1-\tilde{x}_{3m})\hat{X}_1} \qquad (19.3.134)$$

Both C_1 and $\hat{X}_3'(1)$ are as yet undetermined while

$$\hat{X}_3(1) = 0 \qquad (19.3.135)$$

Substituting for $\hat{X}_1$ from (19.3.134) into (19.3.133) gives

$$\frac{\tilde{x}_{3m}'}{(1-\tilde{x}_{3m})} = -\frac{(2\nu+2)}{(2\nu+1)}\frac{C_1}{\text{Gz}} = -C_2 \qquad (19.3.136)$$

with integral

$$\tilde{x}_{3m} = 1 - \exp C_2(\hat{x}_1 - \hat{x}_{10}) \tag{19.3.137}$$

(19.3.133) can also be used to solve for $\hat{X}_3$. Using (19.3.135) and $\hat{X}_3'(0) = 0$, a least value for $|\hat{X}_3(1)|$ can be derived.

Actually an infinity of values for $\hat{X}_3'(1)$ can be expected to exist that will satisfy the equations and boundary conditions. We would expect the smoothest profile, with no points of inflection, to be the most likely asymptotically-valid similarity solution. This point would have to be resolved computationally (see Lee & Zerkle, 1969).

Note that $\tilde{x}_{3m}$ tends to zero as $\hat{x}_1 \to -\infty$, but also $= 0$ for a finite value of $\hat{x}_1 = \hat{x}_{10}$, which in practice will be >1. The temperature varies with $\hat{x}_1$ proportionally to

$$\hat{X}_1 = -\frac{\mathrm{B_s}\tilde{\alpha}_\mathrm{s}}{\mathrm{B_m}\hat{X}_3'(1)}\frac{\tilde{x}_3}{(1-\tilde{x}_{3m})} = -\frac{\mathrm{B_s}\tilde{\alpha}_\mathrm{s}}{\mathrm{B_m}\alpha_\mathrm{m}\hat{X}_3'(1)}[\exp\{-C_2(\hat{x}_1 - \hat{x}_{10})\} - 1] \tag{19.3.138}$$

Clearly this solution will not necessarily match smoothly to the entry condition $\hat{X}_1(0) = \text{constant}$. Near $\hat{x}_1 = 0$, a high Gz solution, discussed below, will be relevant. The possibility of selecting joint values of C_2 and $\hat{x}_{10}$ to best match the latter solution has not been investigated. When $\mathrm{Br} \neq 0$ or $\tilde{x}_{3m} \not\ll 1$, a similarity solution of the above type is not possible, and so full numerical solution (of the equation obtained at the end of Example 19.3.2) would be relevant. The phenomenon of $\tilde{x}_{3m} \to 0$ could no longer be expected with $\mathrm{Br} \neq 0$.

$\mathrm{Gz} \gg 1$. Because conduction will play little part except in the neighbourhood of the wall, we expect $1 - \tilde{x}_{3m} \ll 1$. The core velocity distribution will be largely unchanged and so we consider the 'slow' development of the frozen layer within the thermal boundary layer. We therefore consider a modification to the analysis given in Subsection 19.3.1 (I, $\mathrm{Gz} \gg 1$). An adiabatic solution will still hold in the core to a good approximation.

For the power law fluid this yields

$$\check{T} - \check{T}(0) = \left|\frac{\partial \tilde{v}_1}{\partial \tilde{x}_3}\right|^{\nu+1} \frac{\hat{x}_1}{\tilde{v}_1} = \frac{(\nu+1)\tilde{x}_3^{1+1/\nu}\hat{x}_1}{\nu(1-\tilde{x}_3^{1+1/\nu})} \tag{19.3.139}$$

where

$$\check{T} = \left(\frac{2\nu}{2\nu+1}\right)^{\nu} \frac{\mathrm{Gz}}{\mathrm{Br}} \hat{T} \tag{19.3.140}$$

and so $\hat{T}$ will be negligible for $\mathrm{Br/Gz} \ll 1$ unless $(1 - \tilde{x}_3^{1+1/\nu}) = 0\ (\mathrm{Br/Gz})$.

In the thermal boundary layer we use the similarity boundary-layer variable

$$\tilde{y} = \mathrm{Gz}^{\frac{1}{3}}(\tilde{x}_{3m} - \tilde{x}_3)/\hat{x}_1^{\frac{1}{3}} \qquad (19.3.141)$$

and, following (19.3.43), the form

$$\hat{T} - 1 = \hat{T}_{\mathrm{B}}(\tilde{y}) + \frac{\mathrm{Br}}{\mathrm{Gz}^{\frac{2}{3}}}\,\tilde{\eta}_{\mathrm{W}} \left|\frac{\partial v_1}{\partial x_3}\right|_{\mathrm{W}} \hat{x}_1^{\frac{2}{3}}\hat{T}_{\mathrm{A}}(\tilde{y}) \qquad (19.3.142)$$

where

$$\begin{aligned} &\hat{T}_{\mathrm{B}}(0) = -1; \qquad \hat{T}_{\mathrm{B}} \to 0 \quad \text{as} \quad \tilde{y} \to \infty \\ &\hat{T}_{\mathrm{A}}(0) = 0; \qquad \hat{T}_{\mathrm{A}} \to \tilde{y}^{-1} \quad \text{as} \quad \tilde{y} \to \infty \end{aligned} \qquad (19.3.143)$$

The solutions (19.3.46) and (19.3.50) are recovered. From them, $(\partial\hat{T}/\partial\tilde{x}_3)_{\tilde{x}_{3m}}$ can be calculated, and by use of (19.3.131), $\tilde{x}_{3m}$ follows. Thus

$$1 - \tilde{x}_{3m} = \frac{\mathrm{B_s}\tilde{\alpha}_{\mathrm{s}}\hat{x}^{\frac{1}{3}}}{\mathrm{B_m}\mathrm{Gz}^{\frac{1}{3}}}\left[\frac{9}{2(1+\frac{1}{2}\nu)}\right]^{\frac{1}{3}}\left[\int_0^\infty \mathrm{e}^{-x^3}\,\mathrm{d}x\right]\left\{1 - 0\left(\frac{\mathrm{Br}}{\mathrm{Gz}^{\frac{2}{3}}}\,\hat{x}_1^{\frac{2}{3}}\right)\right\} \qquad (19.3.144)$$

where the $\hat{x}_1^{\frac{1}{3}}$ term comes from $\hat{T}_{\mathrm{B}}$ and the $0(\mathrm{Br}\,\hat{x}_1/\mathrm{Gz}^{\frac{2}{3}})$ contribution comes from $\hat{T}_{\mathrm{A}}$ (see also Richardson, 1983).

Looking back to (19.3.124) and (19.3.125) we see that the effect of a growing frozen layer will involve a change in $(\partial\tilde{v}_1/\partial\tilde{x}_3)$ of order $(\hat{x}_1/\mathrm{Gz})^{\frac{1}{3}}$, which is to be compared with the $0(\mathrm{Br}\,\tilde{x}_1/\mathrm{Gz})$ term due to heat generation. If $\mathrm{Br} < \mathrm{Gz}^{\frac{2}{3}}$, then an expansion in terms of $(\tilde{x}_1/\mathrm{Gz})^{\frac{1}{3}}$ is the way to include the narrowing effect of the frozen layer on the channel flow; in this way, a matching with the solution (19.3.137, 19.3.138) can be attempted.

II. Na, B < 1; Ex, Cm, Vp ≪ 1

The arguments given for tube flow apply in this case also.

III. Na ≮ 1, Gz → 0; Ex, Cm, Vp ≪ 1

This is the general fully-developed case. The equations are those for I, Gz ≪ 1 above, (19.3.117)–(19.3.121), where $\tilde{\eta}$ is to be regarded as $\tilde{\eta}(\hat{T}, \mathrm{d}\tilde{v}_1/\mathrm{d}\tilde{x}_3)$. If the power law relation (19.3.116) is used then eqns (19.3.122) and (19.3.123) become

$$\frac{\mathrm{d}^2\overset{x}{T}}{\mathrm{d}\tilde{x}_3^2} + \mathrm{e}^{-\mathrm{Na}\overset{x}{T}}\left(\left|\frac{\mathrm{d}\tilde{v}_1}{\mathrm{d}\tilde{x}_3}\right|^{\nu+1}\right) = 0 \qquad (19.3.145)$$

where

$$\left|\frac{d\tilde{v}_1}{d\tilde{x}_3}\right| = e^{\text{Na}\overset{x}{T}}\,|\hat{p}_{,\hat{x}_1}\tilde{x}_3| \tag{19.3.146}$$

and $\overset{x}{T}$ is as defined in (19.3.56).

Unfortunately no general solution of these equations has been obtained, so the equivalent of Fig. 19.5 would have to be obtained numerically. This would be only part of the problem in this case, because the interface position $\tilde{x}_{3m} \leqslant 1$ will vary along the relevant solution curve. Only for sufficient high q, i.e. Br, will $\tilde{x}_{3m} = 1$.

IV. $\text{Na} \gg \text{Gz} \gg 1$; Ex, Cm, $\text{Vp} \ll 1$

The steady developing case is virtually the same as that discussed in Section 19.3 I–IV above. The solutions of Ockendon (1979) and Pearson (1977) can be applied directly (they were derived for channel flow) provided the wall is assumed to be at $\tilde{x}_{3m}$, (assuming $|1-\tilde{x}_{3m}| \ll 1$), the latter being given by (19.3.120).

The uniform unsteady case can also be treated, the solution having possible relevance to the melt-front situation. Writing

$$\hat{y} = (\tilde{x}_{3m} - \tilde{x}_3)/\tilde{t}^{\frac{1}{2}}, \qquad \tilde{x}_{3m} = X_m \tilde{t}^{\frac{1}{2}} \tag{19.3.147}$$

$$\overset{x}{T}_{m,s}(0) = 0, \qquad \overset{x}{T}_s(-X_m) = -\frac{B_s}{B_m \text{Br}}, \qquad \overset{x}{T}_m(\infty) = \frac{1}{\text{Br}} \tag{19.3.148}$$

and putting

$$\tilde{\rho}_s = \tilde{\rho}_m = 1 \tag{19.3.149}$$

for simplicity we can use the similarity forms given by (19.3.67); eqns (19.3.68)–(19.3.71) can be carried over directly for $\overset{x}{T}_m$, the melt temperature field, provided $\text{Br}^{-1} \ll \ln \text{Na}/\text{Na}$ and $X_m \ll 1$; the equation for $\overset{x}{T}_s$ is simply

$$\tilde{\alpha}_s \frac{d^2\overset{x}{T}_s}{d\hat{y}^2} + \tfrac{1}{2}\hat{y}\frac{d\overset{x}{T}_s}{d\hat{y}} = 0 \tag{19.3.150}$$

with solution, satisfying (19.3.148),

$$\overset{x}{T}_s = -\frac{B_s}{B_m \text{Br}} \frac{\text{erf}(-\frac{1}{2}\hat{y}\tilde{\alpha}_s^{-\frac{1}{2}})}{\text{erf}(\frac{1}{2}X_m\tilde{\alpha}_s^{-\frac{1}{2}})} \tag{19.3.151}$$

Equation (19.3.113) then yields

$$\frac{1}{2}\frac{\mathrm{Sf\,Pe}\sqrt{\pi}}{\mathrm{Br}}X_m = \frac{\mathrm{B_s}\tilde{\alpha}_s^{\frac{1}{2}}}{\mathrm{B_m\,Br\,erf}(\frac{1}{2}X_m\tilde{\alpha}_s^{-\frac{1}{2}})} - \frac{\ln \mathrm{Na}}{\mathrm{Na}} \qquad (19.3.152)$$

If we approximate $\mathrm{erf}(\frac{1}{2}X_m\tilde{\alpha}_s^{-\frac{1}{2}})$ by $1/\sqrt{\pi}\,X_m\tilde{\alpha}_s^{-\frac{1}{2}}$ then (19.3.152) can be solved for X_m

$$X_m = \frac{-\ln \mathrm{Na} + \sqrt{(\ln \mathrm{Na})^2 + 2\pi\,\mathrm{Sf\,Pe\,B_m B_s}\tilde{\alpha}_s}}{\sqrt{\pi}\,\mathrm{Sf\,Pe\,B_m}} \qquad (19.3.153)$$

The above applies if Na is large. If Na is small, then a much simpler solution, determined by conduction alone is relevant, namely

$$X_m = \frac{1}{\sqrt{\pi}\,\mathrm{Sf\,Pe}}(\sqrt{1 + 2\pi \mathrm{B_s}\tilde{\alpha}_s\,\mathrm{Sf\,Pe/B_m}} - 1) \qquad (19.3.154)$$

V. $\mathrm{Br} \ll 1$, $\mathrm{B_m} \gg 1$; Ex, Cm, $\mathrm{Vp} \ll 1$

For a cooled wall we are interested in the case $\mathrm{B_s} > 0$. Ockendon & Ockendon (1977) have provided a useful solution for this case when there is formally no freezing, i.e. $\mathrm{B_s} < 0$ or $\mathrm{Sf} = 0$; the latter corresponds fairly closely to some glassy polymers, e.g. polystyrene.

First they show that the boundary-layer analysis given in 19.3.2 I, $\mathrm{Gz} \gg 1$ (the Lévêque solution for $\mathrm{B} \ll 1$) can be extended to cover arbitrary B. By writing

$$\tilde{y} = \mathrm{Gz}^{\frac{1}{3}}(1 - \tilde{x}_3)/\hat{x}_1^{\frac{1}{3}} \qquad (19.3.155)$$

using a stream function $\tilde{q} = \hat{x}_1^{\frac{2}{3}}\hat{q}(\tilde{y})$ such that

$$\partial\tilde{q}/\partial\hat{x}_1 = -\hat{v}_3; \qquad \partial\tilde{q}/\partial\tilde{x}_3 = \tilde{v}_1 \qquad (19.3.156)$$

and allowing $\tilde{\eta} = \tilde{\eta}(\hat{T})$, with η^* chosen to make the entry pressure gradient unity, they obtain the boundary-layer equations

$$\tilde{\eta}\{\hat{T}(\tilde{y})\}\frac{\mathrm{d}^2\hat{q}}{\mathrm{d}\tilde{y}^2} = 1; \qquad \frac{\mathrm{d}^2\hat{T}}{\mathrm{d}\tilde{y}^2} + \tfrac{2}{3}\,\hat{q}\,\frac{\mathrm{d}\hat{T}}{\mathrm{d}\tilde{y}} = 0 \qquad (19.3.157)$$

subject to boundary conditions

$$\hat{T}(0) = \hat{q}(0) = \hat{q}'(0) = 0 \qquad (19.3.158)$$

where

$$\hat{T} \to 1; \qquad \hat{q} \to \tilde{y}^2 \quad \text{as} \quad \tilde{y} \to \infty \qquad (19.3.159)$$

In principle this can be solved numerically for arbitrary $\tilde{\eta}$.

They then consider the case when $|\mathrm{B}| \gg 1$, $\mathrm{B}<0$ and show that the entry-zone thermal boundary layer is $0(|\mathrm{B}|)$ thick,† but that the velocity is essentially zero except in an outer portion $c_0 = 0(1)$ thick, in terms of $\tilde{y}$; in terms of $\tilde{x}_3$ the layers are $0(|\mathrm{B}|\,\hat{x}_1^{\frac{1}{3}}/\mathrm{Gz}^{\frac{1}{3}})$ and $0(\hat{x}_1^{\frac{1}{3}}/\mathrm{Gz}^{\frac{1}{3}})$ thick respectively, which requires that

$$\hat{x}_1 \ll |\mathrm{B}|^{-3}\,\mathrm{Gz} \tag{19.3.160}$$

i.e. $\mathrm{Gz} \gg |\mathrm{B}|^3$.

The velocity at the edge of the boundary layer matches the essentially undisturbed parabolic velocity profile relevant for $\hat{T} \sim 1$.

Next they show that the notion of a detached layer in which the temperature $\hat{T}$ changes significantly from 1 and the velocity alters from zero on the wall side to a parabolic profile on the inner constant temperature side can be extended to the region in which

$$\hat{x}_1 = 0(\mathrm{Gz}/|\mathrm{B}|^3) \tag{19.3.161}$$

The layer is $0(|\mathrm{B}|^{-1})$ thick at a distance $0(1)$ from the wall given by

$$\tilde{x}_3 = a(\check{x}_1) > 0 \tag{19.3.162}$$

where

$$\check{x}_1 = |\mathrm{B}|^3\,\hat{x}_1/\mathrm{Gz} \tag{19.3.163}$$

Writing

$$\check{y} = |\mathrm{B}|\,(\tilde{x}_3 - a), \qquad \check{Y} = \check{y}/c_0(1-a) \tag{19.3.164}$$

a solution in the form

$$q = |\mathrm{B}|\,\tilde{q} = q(\check{x}_1)Q(\check{Y}), \qquad \hat{T} = \hat{T}(\check{Y}) \tag{19.3.165}$$

can be obtained, where

$$q = c_0^2(1-a)^2 a^{-2}; \qquad \mathrm{d}q/\mathrm{d}\check{x}_1 = 2/3c_0(1-a) \tag{19.3.166}$$

This can be solved with the entry condition $a \to 1$ and

$$\check{x}_1 \to 0$$

to give

$$\check{x}_1/3c_0^3 = -\ln a + \tfrac{1}{2}a^{-2} - 2a^{-1} + \tfrac{3}{2} \tag{19.3.167}$$

† This may appear confusing, in that $|\mathrm{B}| \gg 1$, but the units are in a stretched variable, so the boundary layer is apparently thick.

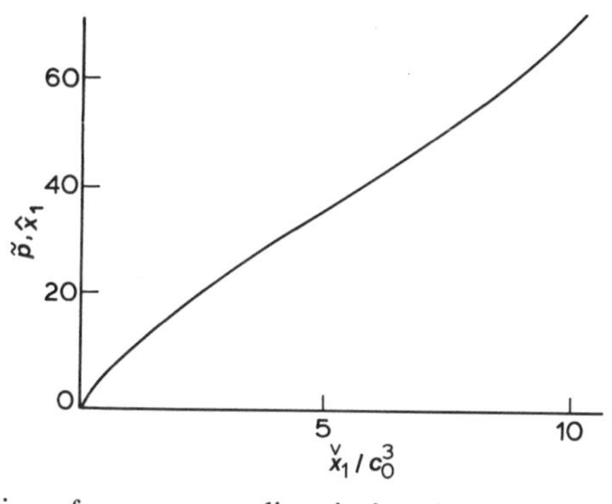

Fig. 19.9. Variation of pressure gradient in free-layer region when $B \ll 1$ (from Ockendon & Ockendon, 1977).

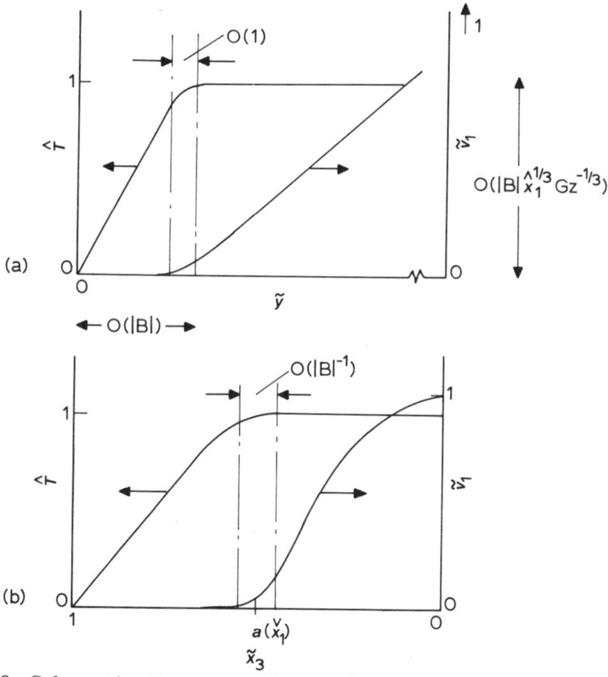

Fig. 19.10. Schematic diagrams of boundary layers for $B \ll -1$. (a) $Gz \gg |B|$. Here the thin thermal boundary layer, almost stagnant, takes the temperature from 0 to its core value. The velocity adjusts from near zero to an almost constant shear region in a still thinner layer separating the core from the thermal boundary layer. The layers grow as $\hat{x}_1^{\frac{1}{3}}$. (b) $Gz = 0(|B|^3)$. Here the stagnant thermal 'boundary' layer occupies a thickness $1-a$ that is a substantial fraction of the channel half-width. The free layer giving the velocity adjustment to the parabolic core profile is $0(|B|^{-1})$ thick. The dependence of a upon $\check{x}_1$ is given by eqn (19.3.167).

when

$$\check{x}_1 \to \infty, \qquad a \sim \left(\frac{2}{3}\frac{\check{x}_1}{c_0^3}\right)^{-\frac{1}{2}}$$

$$\check{x}_1 \to 0, \qquad a \sim 1 - \check{x}_1^{\frac{1}{3}}/c_0$$

The pressure gradient $\tilde{p}_{,\check{x}_1} = -a^{-3}$, and is plotted in Fig. 19.9 as a function of $\check{x}_1$ (taken from Fig. 3 of Ockendon & Ockendon). The free layer structure is illustrated in Fig. 19.10. (This differs from Fig. 4 in Ockendon & Ockendon which appears to be incorrect. Readers are warned that Section 3 of that reference contains numerous typographical errors which, together with a very condensed style and imprecise notation, make it difficult to follow.)

The last stage in their analysis concerns the stage when $a \to 0$ and a hot jet develops near $\tilde{x}_3 = 0$ in which conduction cannot be neglected, and the centre-line temperature decreases from 1 towards 0. This would lead to very large pressure drops or, in practice as the melt front reached the relevant values of $\check{x}_1$, to very much slower rates of flow, i.e. to a rapid reduction in Gz and hence freezing off. The details of this have not been worked out.

19.3.3 Disc Flow

We consider here axisymmetric flow in the channel of uniform depth between two flat parallel discs fed from their centre point. Figure 19.11 illustrates the situation. The important feature is that the flow field necessarily depends upon both the r and z coordinates, so an asymptotic fully-developed flow field is not a feature of the idealized system. We can expect the lubrication approximation to hold even here if

$$h \ll R_i < R_m < R_0 \qquad (19.3.168)$$

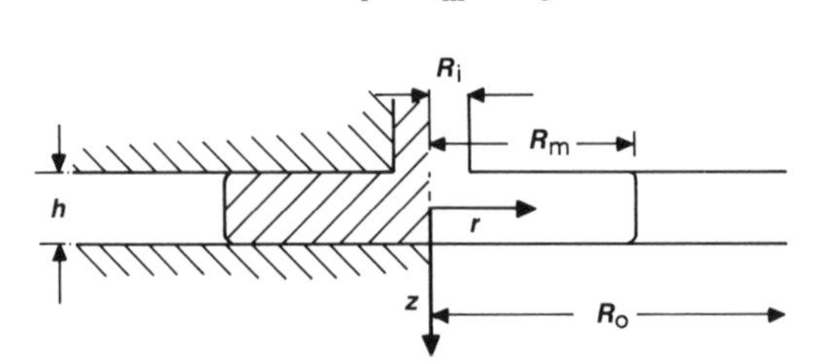

Fig. 19.11. Elementary geometry for disc flow when $R_0 > R_m \gg h$.

and over most of the flow field if $h \ll R_0$. In many practical situations R_i will not be much larger than h though for most of the filling process $R_m \gg h$.

The velocity distribution will be given by

$$\mathbf{v} = (v_r, v_\phi, v_z) = \{v_r(r, z, t), 0, v_z(r, z, t)\} \tag{19.3.169}$$

with $v_z \ll v_r$ and the temperature by $T(r, z, t)$ with boundary and initial conditions idealized as in Subsection 19.3.2 to

$$T(r, \pm\tfrac{1}{2}h, t) = T_W, \qquad T(R_i, z, t) = T_0 \tag{19.3.170}$$

The mass conservation relation becomes

$$2\pi r \int_{-\frac{1}{2}h}^{\frac{1}{2}h} \rho v_r \, \mathrm{d}z = \dot{M} \tag{19.3.171}$$

or, if ρ is constant,

$$\bar{v} = \frac{1}{h} \int_{-\frac{1}{2}h}^{\frac{1}{2}h} v_r \, \mathrm{d}z = \dot{V}/2\pi r h \tag{19.3.172}$$

This shows that $\bar{v}$, which in tube or channel flow was used as v^*, is now inversely proportional to the radius r. Clearly if $R_0 \gg R_i$, there will be no satisfactory characteristic velocity, and so definition of Gz and Na is not unique.

A form for Gz can be obtained by noting its inverse dependence on length of the flow path and so writing

$$\mathrm{Gz}_{\mathrm{disc}}^{-1} = \frac{2\pi\kappa}{\dot{V}h} \int_{R_i}^{R_0} r \, \mathrm{d}r = \frac{\kappa\pi}{\dot{V}} \frac{(R_0^2 - R_i^2)}{h}$$

or

$$\mathrm{Gz} \sim \frac{\dot{V}h}{\kappa\pi R_0^2} \tag{19.3.173}$$

which differs only by a factor of 2 from its (local) value at R_0. The relevant point is that Gz_r decreases as r^{-2} not as r^{-1}. Similarly the local value Na_r will decrease as r^{-2}.

In the lubrication approximation the pressure field $p(r, z, t)$ will enter only as $\partial p/\partial r$, balancing the shear stress variation

$$\frac{\partial}{\partial z}\left(\eta \frac{\partial v_r}{\partial z}\right)$$

both being regarded as functions of r only. The analysis will differ from that given in Subsection 19.3.2 (or Subsection 19.3.1) above for the various regimes I–VI only in the sense that $\bar{v}_r$, which takes the place of $\bar{v}_1$ (or $\bar{v}_z$), will vary (as r^{-1}) with r. This introduces the same sort of complications (though more extreme) that arise when the channel depth h_c varies slowly with x_1, or the tube diameter d_c with z. Similar dimensionless equations can be obtained by using h and R_0 as length scales, $\dot{V}/2\pi R_0 h$ as the velocity scale, and $\dot{V}/2\pi R_0 h^2$ as the shear-rate scale (thus yielding a pressure scale).

By using the variables

$$x = \tfrac{1}{2}r^2, \qquad u = rv_r \tag{19.3.174}$$

a pair of equations more suited to the disc problem are obtained, i.e.

$$\frac{\partial p}{\partial x} = \frac{1}{2x}\frac{\partial}{\partial z}\left(\eta\frac{\partial u}{\partial z}\right) \tag{19.3.175}$$

and

$$u\frac{\partial T}{\partial x} = \kappa\frac{\partial^2 T}{\partial z^2} + \frac{\eta}{2\rho\Gamma x}\left(\frac{\partial u}{\partial z}\right)^2 \tag{19.3.176}$$

where

$$\int_{-h/2}^{h/2} u \, \mathrm{d}z = \bar{u}h = \dot{V}/2\pi \tag{19.3.177}$$

and

$$\tfrac{1}{2}R_\mathrm{i}^2 \leqslant x \leqslant \tfrac{1}{2}R_0^2$$

Thus if a temperature-independent power-law approximation is used for η and generation can be neglected, u is independent of x, and the same eigenfunctions can be used for T as in the plane channel case. The details are discussed in Stevenson (1976).

The limits of applicability of the lubrication approximation may be more severe than (19.3.168) suggests. This can best be seen by considering the full r-equation of motion, but neglecting inertia and gravity terms, i.e.

$$\frac{\partial(p - t_{rr}^\mathrm{E})}{\partial r} = \frac{t_{rr}^\mathrm{E} - t_{\phi\phi}^\mathrm{E}}{r} + \frac{\partial t_{rz}^\mathrm{E}}{\partial z} \tag{19.3.178}$$

In low-Na plane-channel flow, the components t_{rr}^E and $t_{\phi\phi}^\mathrm{E}$ (or rather their equivalents t^E_{11}, t^E_{33}) are zero, but in disc flow there will be non-zero components due to

$$d_{rr} = \partial v_r/\partial r \quad \text{and} \quad d_{\phi\phi} = v_r/r$$

as well as the major t^E_{rz} component associated with

$$d_{rz} = \tfrac{1}{2}\,\partial v_r/\partial z$$

For a Newtonian fluid, with η constant, the t^E_{rr} and $t^E_{\phi\phi}$ terms will be $0(h/r)$ smaller than the t^E_{rz} term, and can be neglected because of (19.3.168), except perhaps in a small neighbourhood of $r = R_i$ when $R_i = 0(h)$.

For a highly elastomeric material, where η_{PS} may be orders of magnitude larger than η_{shear}, then

$$t^E_{\phi\phi} - t^E_{rr} = 4\eta_{PS} v_r/r, \qquad (19.3.179)$$

evaluated on the centre plane $z = 0$ where v_r and $-\partial v_r/\partial r$ are maximum, may be of the same order as, or larger than,

$$t^E_{rz} = \eta_{shear}\,\partial v_r/\partial z \qquad (19.3.180)$$

evaluated at the wall, using the velocity field given by the lubrication approximation, even though (19.3.168) holds. This will be true whenever

$$\left(\frac{\eta_{PS} h}{\eta_{shear} r}\right) \geqslant 0(1) \qquad (19.3.181)$$

η_{PS} referring to $D^* = v^*/r$ and η_{shear} to $D^* = v^*/h$. The inequality (19.3.181) will hold for sufficiently small r or sufficiently large v^*.

A complete solution of the problem in these circumstances is not feasible though it is interesting to speculate on the nature of the flow when the inequality sign in (19.3.181) is replaced by $\gg$. This is the case when the elastic hoop stress caused by the $d_{\phi\phi}$ stretching dominates over the more usual viscous shear stress caused by d_{rz}. Simplified viscous and elastic limits corresponding to $\ll$ and $\gg$ in (19.3.181) are shown in Fig. 19.12. A more realistic, intermediate, situation is illustrated in Fig. 19.13. Here the core flow, which carries almost all the volume flux $\dot{V}$ in a plane flow of type (b), Fig. 19.12, is sandwiched between two thin lubricating layers of type (a), Fig. 19.12. The model assumes that the fluid in the lubricated layers is in a different physical form from that in the core region, and that the velocity gradient changes very sharply in the neighbourhood of $z = \frac{1}{2}h_{el}$ (h_{el} will be a

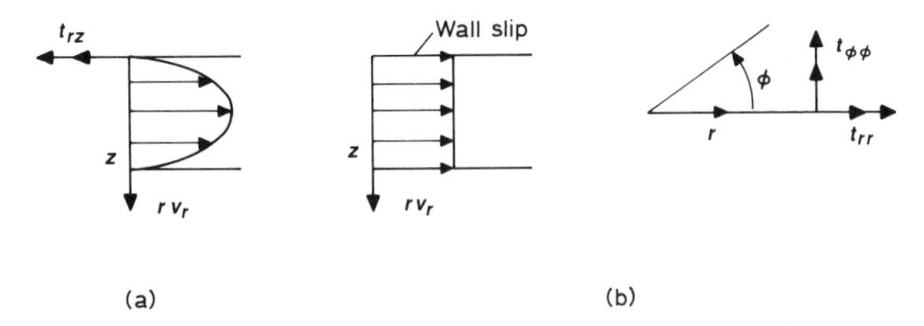

Fig. 19.12. Asymptotic limits for disc flow. (a) Lubrication flow: t_{rz} dominates: rv_i is a function of z only. (b) Plane elastic flow: $t_{\phi\phi} - t_{rr}$ dominates; $rv_r = \dot{V}/2\pi h$, constant. Slip at the walls is effectively assumed.

slowly-varying function of r). The idea, similar to that introduced in Section 7.2, is described in detail in Hull (1981*a*) and more briefly for the case of conical flow in Hull *et al.* (1981).

A simple illustrative calculation can be undertaken assuming that there is an actual discontinuity, and that the fluids in the two regions behave as Newtonian fluids of very different, but constant, viscosities. Indeed, if $\eta_{el} \gg \eta_{sh}$ then $h_{sh} \ll h$, and the shear in the lubricating layers will be effectively constant. Furthermore the core velocity will be effectively $\dot{V}/2\pi rh = q/r$, say. It follows that

$$-\frac{\partial}{\partial r}(p - t_{rr}^{E}) \sim 2\eta_{sh} q/h_{sh}^2 \quad \text{in lubricating region}$$
$$\sim 4\eta_{el} q/r^2 \quad \text{in core region} \qquad (19.3.182)$$

The relevant lubricating depth h_{sh} is thus given by

$$h_{sh}^2 \sim r^2(\eta_{sh}/4\eta_{el}) \qquad (19.3.183)$$

Note that the volume flux in the lubricating layer changes and so this crudest model is tacitly assuming that material moves from the core to the lubricating layer, changing its rheological properties as it does so. Note also that the stress is not continuous at the interface, both t_{rz} and

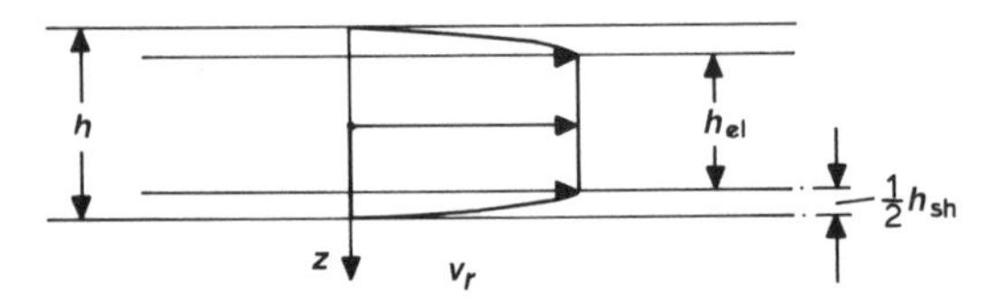

Fig. 19.13. Lubricated core flow for mainly elastic response (disc flow).

$(p - t_{zz}^{E})$ showing jumps, jumps that are, however, small compared with p. A more careful analysis is given in Example 19.3.3.

The conclusion to be drawn from the result (19.3.183) is that the dynamics of flow close to a pin gate, i.e. $R_i = 0(h)$, will not be described by the lubrication approximation, and that R_0 may have to be very much larger than h before the overall pressure drop can be adequately calculated on the basis of the lubrication approximation.

19.3.4 Stability

The developing flow fields discussed in Subsections 19.3.1–19.3.3, particularly those where Na or B are large, show large temperature gradients with consequential viscosity gradients. An extreme case is the one illustrated in Fig. 19.6 where almost all the shear takes place in a narrow layer close to the tube (or plane channel) walls. The velocity profile exhibits a point of inflection $\partial^2 \tilde{v}_z / \partial \hat{y}^2 = 0$. This is often regarded as a potential source of instability, the classic case being one of inertial instability in plane Poiseuille flow, the criterion for instability having been enuciated by Rayleigh and later Tollmien (Lin, 1955, p. 53); the nature of the most unstable disturbances is then assumed to be of the type discussed in Appendix 1, eqn (A.1.5). In the highly coupled situation of interest here, a basic associated temperature perturbation

$$T_1(r, \phi, z, t) = \mathscr{Re}[C(r)\{k(z - ct) + in\phi\}]$$

would have to be included in the analysis. However, even this would not provide a fully rigorous approach, because the unperturbed velocity and temperature fields are themselves 'slowly' varying functions of z (in the spacially developing case) or t (in the temporally developing case). The situation is very similar to that arising in boundary-layer stability theory (see, for example, Stuart 1963, pp. 564–5) which has been the subject of much detailed work in recent years.

No application to highly-viscous problems of the type considered here has been made. The essence of these linear and non-linear approaches is that they are local, the characteristic length scale being d or h, and not l. A much cruder approach has been taken by Shah & Pearson (1974; see Mhaskar *et el.*, 1977, for its application to the mould-filling problem). This was mentioned briefly in Subsection 10.3.5 above and can be outlined at this point because the basic mechanism is likely to be relevant to mould filling even though the detailed calculations are unlikely to be quantitatively applicable.

The treatment differs from traditional linearized-stability analyses in that a global approach is used, in which the overall boundary conditions to the flow are relevant (in the x_1, x_2 plane for the channel flow of Subsection 19.3.2 or the r, ϕ plane for the disc flow of Subsection 19.3.3). The dynamics is expressed in terms of the local flux vector

$$\mathbf{q}(x_1, x_2) = \int_{-\frac{1}{2}h}^{\frac{1}{2}h} v_{1,2}(x_1, x_2, x_3)\, \mathrm{d}x_3 \qquad (19.3.184)$$

and the heat transfer in terms of a Nusselt number operating on the mixing-cup mean temperature $\bar{T}$.

Assuming constant density the integrated equations are taken to be

$$\boldsymbol{\nabla} \cdot \mathbf{q} = 0 \qquad (19.3.185)$$

$$\mathbf{q}\,|\mathbf{q}|^{\nu-1} = -\mathcal{K}_{\mathrm{W}}\, \mathrm{e}^{\zeta_{\mathrm{W}}(\bar{T}-T_{\mathrm{W}})} \boldsymbol{\nabla} p \qquad (19.3.186)$$

$$\frac{\partial \bar{T}}{\partial t} + (\mathbf{q} \cdot \boldsymbol{\nabla})\bar{T} = \mathcal{H}(T_{\mathrm{W}} - \bar{T}) + \mathcal{G}_{\mathrm{W}}\,|\mathbf{q}|^{\nu+1}\, \mathrm{e}^{-\zeta_{\mathrm{W}}(\bar{T}-T_{\mathrm{W}})} \qquad (19.3.187)$$

where $\boldsymbol{\nabla}$ operates in the plane of the channel or disc, and $\mathcal{K}_{\mathrm{W}}$, $\mathcal{H}$ and $\mathcal{G}_{\mathrm{W}}$ are constants dependent on the material parameters K_{m}, $\rho\Gamma$, and α and on the channel depth h. The boundary conditions for the case of channel flow in a rectangular mould of length l and width $\mathrm{G}_4 l$, fed over the entire channel width at the end $x_1 = 0$, can be taken to be

$$\left.\begin{aligned} p(0, x_2) = P, \qquad \bar{T}(0, x_2) = T_0, \qquad p(l, x_2) = 0 \\ q_2(x_1, 0) = q_2(x_1, \mathrm{G}_4 l) = 0 \end{aligned}\right\} \qquad (19.3.188)$$

The relations may be made dimensionless as in (19.3.103) by using $q^* = \dot{V}/\mathrm{G}_4 l$, $p^* = \mathcal{K}_{\mathrm{W}}/q^{*\nu} l$, $l^* = l$ and $T^* = \zeta_{\mathrm{W}}^{-1}$. The steady-state solution will be given by

$$\tilde{\mathbf{q}}^{(0)} = (1, 0); \qquad \tilde{\bar{T}}^{(0)}(\hat{x}_1); \qquad \hat{p}^{(0)}(\hat{x}_1) \qquad (19.3.189)$$

and the linearized disturbance terms by

$$\begin{aligned} \tilde{\mathbf{q}}^{(1)} &= \mathcal{R}e\left[\left\{\hat{n}\tilde{Q}^{(1)}(\hat{x}_1), -\mathrm{i}\frac{\mathrm{d}\tilde{Q}^{(1)}}{\mathrm{d}\hat{x}_1}\right\} \exp(-\mathrm{i}\hat{n}\hat{x}_2 + \hat{c}\hat{t})\right] \\ \tilde{\bar{T}}^{(1)} &= \mathcal{R}e[\tilde{T}^{(1)}(\hat{x}_1)\exp(-\mathrm{i}\hat{n}\hat{x}_2 + \hat{c}\hat{t})] \qquad (19.3.190) \\ \hat{p}^{(1)} &= \mathcal{R}e[\tilde{P}^{(1)}(\hat{x}_1)\exp(-\mathrm{i}\hat{n}\hat{x}_2 + \hat{c}\hat{t})] \end{aligned}$$

where

$$|\tilde{Q}^{(1)}, \tilde{T}^{(1)}, \tilde{P}^{(1)}| \ll 1$$

and $\hat{n}$, $\hat{c}$ are real. The form for $\tilde{\mathbf{q}}^{(1)}$ automatically satisfies (19.3.185).

On substituting into (19.3.186)–(19.3.188) the following relations are obtained.

$$\left.\begin{aligned} \mathrm{Gz}\frac{\mathrm{d}\bar{\bar{T}}^{(0)}}{\mathrm{d}\hat{x}_1} &= -\bar{\bar{T}}^{(0)} + \mathrm{Gn}\,\mathrm{e}^{-\bar{\bar{T}}^{(0)}}; \qquad \bar{\bar{T}}^{(0)}(0) = \mathrm{B} \\ \hat{p}^{(0)} = \int_{\hat{x}_1}^{1} \mathrm{e}^{-\bar{\bar{T}}^{(0)}}\,\mathrm{d}\hat{x}_1; \qquad \hat{P}^{(0)} &= \hat{p}^{(0)}(0); \qquad 0 = \hat{p}^{(0)}(1) \end{aligned}\right\} \quad (19.3.191)$$

and

$$\mathrm{Gz}\frac{\mathrm{d}^3\tilde{Q}^{(1)}}{\mathrm{d}\hat{x}_1^3} + (1+\bar{\bar{T}}^{(0)} + \hat{c}\,\mathrm{Gz})\frac{\mathrm{d}^2\tilde{Q}^{(1)}}{\mathrm{d}\hat{x}_1^2} + \Big\{\nu\hat{n}^2\,\mathrm{Gz} + \mathrm{Gz}\frac{\mathrm{d}^2\bar{\bar{T}}^{(0)}}{\mathrm{d}\hat{x}_1^2}$$
$$+\hat{c}\,\mathrm{Gz}\frac{\mathrm{d}\bar{\bar{T}}^{(0)}}{\mathrm{d}\hat{x}_1} + \left(1+\bar{\bar{T}}^{(0)} + \mathrm{Gz}\frac{\mathrm{d}\bar{\bar{T}}^{(0)}}{\mathrm{d}\hat{x}_1}\right)\frac{\mathrm{d}\bar{\bar{T}}^{(0)}}{\mathrm{d}\hat{x}_1}\Big\}\frac{\mathrm{d}\tilde{Q}^{(1)}}{\mathrm{d}\hat{x}_1}$$
$$+\{\nu(1+\hat{c}\,\mathrm{Gz}) - \bar{\bar{T}}^{(0)}\}\hat{n}^2\tilde{Q}^{(1)} = 0$$

with

$$\frac{\mathrm{d}\tilde{Q}^{(1)}}{\mathrm{d}\hat{x}_1} = 0, \qquad \frac{\mathrm{d}^2\tilde{Q}^{(1)}}{\mathrm{d}\hat{x}_1^2} = \nu\hat{n}^2\tilde{Q}^{(1)} \quad \text{at} \quad \hat{x}_1 = 0 \qquad (19.3.192)$$

$$\frac{\mathrm{d}\tilde{Q}^{(1)}}{\mathrm{d}\hat{x}_1} = 0 \quad \text{at} \quad \hat{x}_1 = 1 \quad (\text{and } \hat{n}\ \mathrm{G}_4 = n\pi,\ n \text{ integral})$$

where†

$$\mathrm{Gz} = q^*/\mathcal{H}l, \qquad \mathrm{Gn} = \mathcal{G}_{\mathrm{W}}q^{*\nu}\zeta_{\mathrm{W}}/\mathcal{H}, \qquad \mathrm{B} = (T_0 - T_{\mathrm{W}})\zeta_{\mathrm{W}} \qquad (19.3.193)$$

We can study the behaviour of $\hat{P}^{(0)}$ as a function of v^*, i.e. of Gz with $\mathrm{Gn}\,\mathrm{Gz}^{-\nu}$ held constant. Typical shapes are shown in Fig. 19.14 (taken from Shah & Pearson, 1974, III, Fig. 2). The important result is that for B sufficiently large, there is a range of Gz for which $\mathrm{d}\hat{P}^{(0)}/\mathrm{d}\,\mathrm{Gz}$ is negative. This, as was discussed in Section 9.1, can be expected to be unstable. Figure 19.15 (taken from Shah & Pearson, 1974, II, Fig. 2), plots the range of Gz as a function of B for which this is the case for various values of ν, with $\mathrm{Gn} = 0$. We can also study the neutral-stability curves for the disturbance field $\tilde{Q}^{(1)}$, i.e. those joint values of B, Gz, $\hat{n}$ that satisfy (19.3.192) when $\hat{c} = 0$, for various ν. The least values of B are plotted in Fig. 19.16 (taken from Shah & Pearson, 1974, II, Fig. 3)

† Note that Gn is defined differently here from the usage in Shah & Pearson (1974).

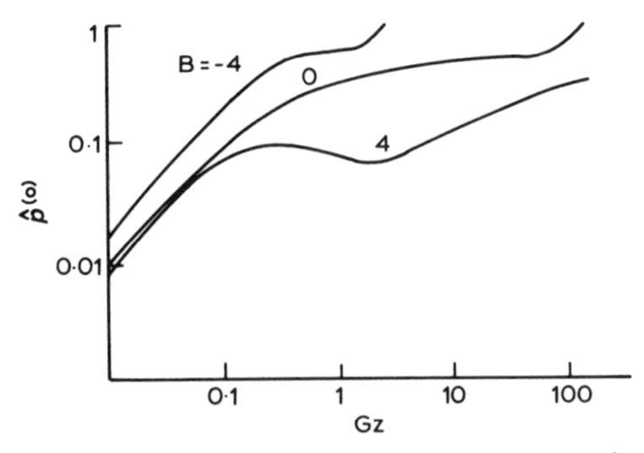

Fig. 19.14. Pressure-drop vs flowrate curves showing $d\hat{P}^{(0)}/d\,Gz<0$ for a range of Gz near 1 when $B=4$. $Gn\,Gz^{-\nu}=10$.

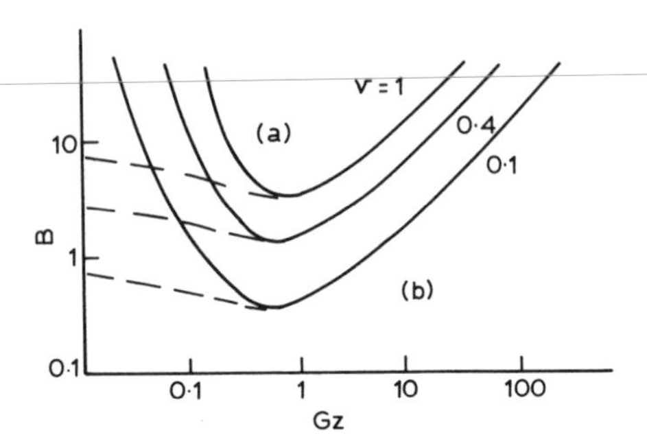

Fig. 19.15. Separation of (B, Gz) plane into regions where: (a) $d\hat{P}^{(0)}/d\,Gz<0$ (areas above ——); (b) $d\hat{P}^{(0)}/d\,Gz>0$ (areas below ——); – – –, lower values of Gz for $\hat{P}^{(0)}$ at which $d\hat{P}^{(0)}/d\,Gz=0$. Various power-law fluids.

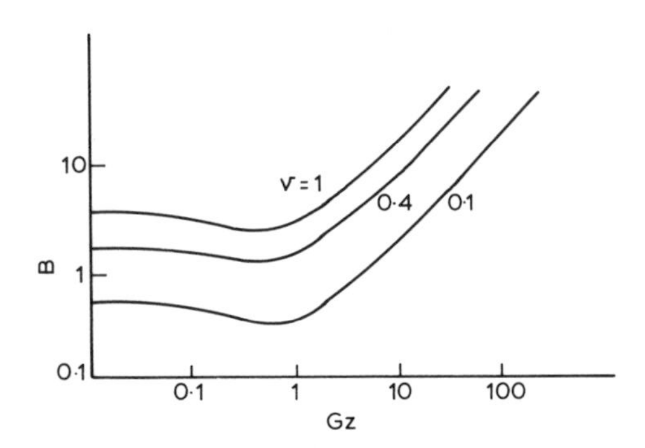

Fig. 19.16. Neutral stability curves for power-law fluids for various values of ν.

for $\mathrm{Gn}=0$. Above the neutral stability curves $\hat{c}>0$ for some $\hat{n}$ while below them $\hat{c}<0$ for all $\hat{n}$. Comparison of the two sets of curves in Figs. 19.15 and 19.16 shows most satisfactory correspondence, as would be expected on physical grounds. Corresponding curves can be obtained for other values of $\mathrm{Gn}\,\mathrm{Gz}^{-\nu}$. These results have related to a full channel. Corresponding results for a channel that is filling, i.e. one where

$$l=q^*t$$

for which the last boundary condition in (19.3.192) is replaced by

$$\hat{n}^2\tilde{Q}^{(1)}+\frac{\mathrm{d}^2\tilde{Q}^{(1)}}{\mathrm{d}\hat{x}_1^2}-\frac{\mathrm{d}\bar{\bar{T}}^{(0)}}{\mathrm{d}\hat{x}_1}\frac{\mathrm{d}\tilde{Q}^{(1)}}{\mathrm{d}\hat{x}_1}=0 \quad \text{at} \quad \hat{x}_1=1 \qquad (19.3.194)$$

are given in Mhaskar, Pearson & Shah (1977).

EXAMPLE 19.3.1. Estimate the total pressure drop in case IV of Section 19.3.

We note first that

$$\frac{\partial\hat{p}}{\partial\hat{z}}=2\mathrm{e}^{-\mathrm{Na}\check{T}}\frac{\mathrm{d}\tilde{v}_z}{\mathrm{d}\tilde{r}}=\frac{2K(\mathrm{Na})}{\check{z}^{\frac{1}{2}}}$$

from (19.3.75) by analogy with (19.3.68): $\check{z}^{\frac{1}{2}}$ in (19.3.72) takes the role of $\check{t}^{\frac{1}{2}}$ in (19.3.66). Clearly $K(\mathrm{Na})$ must be estimated. We use the results (19.3.71) subject to (19.3.80) to yield

$$K=1\Big/\int_0^\infty \mathrm{e}^{\mathrm{Na}\check{T}}\,\mathrm{d}\check{y}=\frac{\mathrm{d}Y}{\mathrm{d}\check{y}}\Big/\int_{-\ln\mathrm{Na}}^\infty \mathrm{e}^{\mathrm{Na}\check{T}}\,\mathrm{d}Y$$

$$\sim\frac{\ln\mathrm{Na}}{\pi^{\frac{1}{2}}}\Big/\int_{-\infty}^\infty \tfrac{1}{2}\mathrm{Na}\frac{\mathrm{e}^Y}{(1+\mathrm{e}^Y)^2}\,\mathrm{d}Y\sim\frac{2\ln\mathrm{Na}}{\pi^{\frac{1}{2}}\mathrm{Na}}$$

We now write, using the non-dimensionalizing forms (7.1.42), and on substitution for K in $\mathrm{d}\hat{p}/\mathrm{d}\hat{z}$ above,

$$\frac{\mathrm{d}p}{\mathrm{d}z}=\frac{\eta_0 v^*}{(\frac{1}{2}d_c)^2}\frac{\mathrm{d}\hat{p}}{\mathrm{d}\hat{z}}=\frac{4\eta_0 v^*}{(\frac{1}{2}d_c)^2}\frac{\ln\mathrm{Na}(\mathrm{Gz})^{\frac{1}{2}}l_c^{\frac{1}{2}}}{\mathrm{Na}\,\pi^{\frac{1}{2}}z^{\frac{1}{2}}}$$

If we integrate from $z=0$ to $z=l_c$ we obtain

$$P=-\int_0^{l_c}\frac{\mathrm{d}p}{\mathrm{d}z}=\frac{8\eta_0 v^* l_c(\mathrm{Gz})^{\frac{1}{2}}}{(\frac{1}{2}d_c)^2}\frac{\ln\mathrm{Na}}{\mathrm{Na}}$$

Using the result (19.3.60) the relation (19.3.81) is obtained. Note that the apparently linear dependence upon $\dot{V}$ is of only formal significance because $\mathrm{Gz} \propto \dot{V}$ and $\mathrm{Na} \propto \dot{V}^2$. Hence in this approximation

$$P \propto \dot{V}^{-\frac{1}{2}} \ln \dot{V}$$

For large $\dot{V}$ this is a decreasing function of $\dot{V}$, thus demonstrating the same type of behaviour as shown in Fig. 19.5.

The result of Ockendon (1979) where the singularity at $z=0$ is replaced by an inlet distance of order $\mathrm{Na}^{-\frac{3}{2}}$ yields a pressure drop which is marginally smaller than calculated above, namely $P \propto \dot{V}^{-\frac{1}{2}}$, so the prediction that P will decrease with $\dot{V}$ still holds.

EXAMPLE 19.3.2. Derivation of equations describing the developing but steady situation when $\tilde{x}_{3m}$ is a (slowly-varying) function of $\hat{x}_1$.

It is convenient to use a stream function $\hat{q}$ $(\hat{x}_1, \hat{x}_3)$, where $\hat{x}_3$ is defined just after (19.3.132), such that

$$\tilde{v}_1 = -\frac{1}{\tilde{x}_{3m}} \frac{\partial \hat{q}}{\partial \hat{x}_3}, \qquad \hat{v}_3 = \frac{\partial \hat{q}}{\partial \hat{x}_1} - \frac{\tilde{x}'_{3m}}{\tilde{x}_{3m}} \hat{x}_3 \frac{\partial \hat{q}}{\partial \hat{x}_3}$$

This can be shown to satisfy the continuity equation (19.3.106) with $\tilde{\rho}_{\mathrm{m}} = 1$.

From the usual lubrication argument, which only requires that $\tilde{x}_{3m} \leqslant 0(1)$, eqn (19.3.107) can still be used, with solutions (19.3.123)–(19.3.125), which yields

$$\hat{q} = -\left[\frac{2\nu+1}{2\nu+2} \hat{x}_3 - \frac{\nu}{2\nu+2} \hat{x}_3^{2+1/\nu}\right]$$

The energy equation (19.3.108) then becomes

$$\frac{\partial^2 \hat{T}}{\partial \hat{x}_3^2} = \mathrm{Gz}\left[\left(\frac{2\nu+1}{2\nu+2}\right)\tilde{x}_{3m}(1-\hat{x}_3^{1+1/\nu})\frac{\partial \hat{T}}{\partial \hat{x}_1} + \tilde{x}'_{3m}\hat{x}_3\left\{\left(\frac{2\nu+1}{\nu+1}\right) - \left(\frac{3\nu+1}{2\nu+2}\right)\hat{x}_3^{1+1/\nu}\right\}\frac{\partial \hat{T}}{\partial \hat{x}_3}\right] - \mathrm{Br}\, \tilde{x}_{3m}^{-2\nu}\left(\frac{2\nu+1}{2\nu}\right)^{1+\nu} \hat{x}_3^{1+1/\nu}$$

If $\mathrm{Br} \ll 1$ and $\tilde{x}'_{3m} \ll \tilde{x}_{3m}$, then eqn (19.3.130) is effectively recovered. Otherwise numerical solution of the above partial differential equation for $\hat{T}$ subject to the boundary conditions (19.3.131)

$$\hat{T}(\hat{x}_1, 1) = 0 \quad \text{and} \quad \hat{T}(0, \hat{x}_3) = \hat{T}_0(\hat{x}_3)$$

would be necessary.

EXAMPLE 19.3.3. The two-fluid model for injection into a shallow disc.

Figure 19.13 shows the flow field considered where $h_{sh} \ll h_{el} \sim h$. An axisymmetric lubrication flow is assumed in the sense that the flow field is dominated by the velocity $v_r(r, t)$. The relevant exact dynamical equation is

$$\frac{\partial p}{\partial r} = \frac{1}{r}\frac{\partial}{\partial r}(rt^{E}_{rr}) + \frac{\partial}{\partial z}t^{E}_{rz} - \frac{t^{E}_{\phi\phi}}{r}$$

(being the r component of $\nabla . \mathbf{T} = 0$). This can be rewritten directly in the form (19.3.178).

A crude two-fluid model can be proposed whereby

(i) in the region $\frac{1}{2}h > |z| > \frac{1}{2}h_{el}$, i.e. a region of thickness $\frac{1}{2}h_{sh}$, the velocity is given by

$$v_r = v_{\text{interf}}(h - 2|z|)/h_{sh}$$

giving a uniform shear rate of $-2v_{\text{interf}}/h_{sh}$.

(ii) in the region $0 < |z| < \frac{1}{2}h_{el}$, the velocity is given by

$$v_r = v_{\text{interf}}$$

From continuity

$$\dot{V} = 2\pi r v_{\text{interf}}(h_{el} + \tfrac{1}{2}h_{sh})$$

whence

$$v_{\text{interf}} = \frac{\dot{V}}{\pi r(2h - h_{sh})} \sim \frac{\dot{V}}{2\pi r h}$$

The rate of deformation tensors in the two regions become, taking $z > 0$ for definiteness,

$$\text{(i)} \quad \mathbf{D}^{(i)} \sim \frac{\dot{V}}{2\pi h r^2}\begin{bmatrix} -(h-2z)/h_{sh} & 0 & -r/h_{sh} \\ 0 & (h-2z)/h_s & 0 \\ -r/h_{sh} & 0 & 0 \end{bmatrix}$$

$$\text{(ii)} \quad \mathbf{D}^{(ii)} \sim \frac{\dot{V}}{2\pi h r^2}\begin{bmatrix} -1 & 0 & 0 \\ 0 & 1 & 0 \\ 0 & 0 & 0 \end{bmatrix}$$

The region (i) is seen to be an almost viscometric (Section 4.2) flow field, while (ii) is seen to be an almost pure shear field (Subsection 4.3.2). A Deborah number for the flow, as discussed in Section 8.1 and defined in (8.1.6), can be written in this case as

$$\mathrm{De} \sim \Lambda \dot{V}/2\pi h r^2$$

and is seen to be related to d_{rr} or $d_{\phi\phi}$, and applies in both regions (i) and (ii). The relevant Weissenberg number, as defined in (8.1.7), for the 'lubricating' region (i), can be written

$$\mathrm{Ws} \sim \Lambda \dot{V}/2\pi h h_{\mathrm{sh}} r$$

Clearly in region (i) where Λ is common for De and Ws, Ws/De $\sim r/h_{\mathrm{sh}} \gg 1$, and so the viscometric approximation is entirely acceptable. In region (ii) Λ may, as argued earlier, be much larger than in region (i) and so the Deborah number contribution may not be negligible. These effects of elasticity must be accounted for in the constitutive relations used in the two regions (i) and (ii).

In region (i), a power-law viscosity is relevant making

$$t_{rz}^{\mathrm{E}} = -K_{\mathrm{m}}(2\,|d_{rz}|)^{\nu} = -K_{\mathrm{m}}(\dot{V}/\pi h h_{\mathrm{sh}} r)^{\nu}$$

with

$$t_{rr}^{\mathrm{E}} - t_{zz}^{\mathrm{E}} = 4\Psi_1(2d_{rz})d_{rz}^2 \quad \text{as in (4.2.3)}$$

$$t_{rr}^{\mathrm{E}} - t_{\phi\phi}^{\mathrm{E}} = 4\Psi_2(2d_{rz})d_{rz}^2 \quad \text{as in (4.2.4)}$$

In region (ii), a Maxwell-type constitutive relation becomes relevant, which gives a history-dependent pure shear viscosity of the type described by (4.3.9) prescribing

$$t_{\phi\phi}^{\mathrm{E}} - t_{rr}^{\mathrm{E}} = \eta_{\mathrm{PS}}(d_{\phi\phi})d_{\phi\phi} \sim \eta_{\mathrm{PS}}\dot{V}/2\pi h r^2$$

with a similar form for $t_{rr}^{\mathrm{E}} - t_{zz}^{\mathrm{E}}$.

The dynamics of the flow is determined principally by region (ii) in which the term

$$\left|\frac{\partial t_{rz}^{\mathrm{E}}}{\partial z}\right| \sim 0 \ll \left|\frac{t_{rr}^{\mathrm{E}} - t_{\phi\phi}^{\mathrm{E}}}{r}\right| \sim \eta_{\mathrm{PS}}\dot{V}/2\pi h r^3$$

Hence from (19.3.178)

$$p - t_{rr}^{\mathrm{E}} \sim \frac{\eta_{\mathrm{PS}}\dot{V}}{4\pi h}\left[\frac{1}{r^2} - \frac{1}{R_0^2}\right]$$

At the interface between (i) and (ii), continuity of $p - t_{zz}^{\mathrm{E}}$ and t_{rz}^{E} should strictly speaking apply.

From the solution obtained in region (ii)

$$p-t_{zz}^{E}=p-t_{rr}^{E}+(t_{rr}^{E}-t_{zz}^{E})$$

is a known function of r, while $t_{rz}^{E}\sim 0$.

It is therefore a question of whether a flow of the type chosen for (i) is consistent with the given interfacial conditions on $p-t_{zz}^{E}$ and t_{rz}^{E}.

The relevant value of $p-t_{rr}^{E}$ in (i) will be given by $(p-t_{zz}^{E})+(t_{zz}^{E}-t_{rr}^{E})$. The first of these, $p-t_{zz}^{E}$, will everywhere approximate to its value at the interface—as can be verified by considering with z-momentum equation—while $t_{zz}^{E}-t_{rr}^{E}$ is known in terms of d_{rz} in (i). Thus

$$\frac{\partial(p-t_{rr}^{E})}{\partial r}\sim\frac{\eta_{PS}\dot{V}}{2\pi hr^{3}}+\frac{4\partial}{\partial r}[(\Psi_2-\Psi_1)d_{rz}^{2}]$$

The term

$$\frac{t_{rr}^{E}-t_{\phi\phi}^{E}}{r}\sim\frac{4(\Psi_1-\Psi_2)}{r}d_{rz}^{2}$$

These must be balanced by the term $\partial t_{rz}^{E}/\partial z$. On purely dimensional grounds we suppose that this last will be of order (t_{rz}^{E}/h_{sh}), for at the walls, we suppose t_{rz}^{E} to be given by the power law from $-K_m(\dot{V}/\pi hh_{sh}r)^{\nu}$ and at a distance h_{sh} from the wall, it is necessarily zero. For convenience we write

$$\mathrm{Ws}=|(t_{rr}^{E}-t_{\phi\phi}^{E})/t_{rz}^{E}|$$

The terms in (19.3.178) have orders of magnitude

$$\frac{\eta_{PS}\dot{V}}{hr^{3}},\ \frac{K_m}{h_{sh}}\left(\frac{\dot{V}}{hh_{sh}r}\right)^{\nu}\quad\text{and}\quad\frac{K_m}{r}\left(\frac{\dot{V}}{hh_{sh}r}\right)^{\nu}\mathrm{Ws}$$

Their relative magnitudes are therefore

$$1,\ \frac{K_m}{\eta_{PS}}\left(\frac{\dot{V}}{hh_{sh}r}\right)^{\nu-1}\left(\frac{r}{h_{sh}}\right)^{2}\quad\text{and}\quad\frac{K_m}{\eta_{PS}}\left(\frac{\dot{V}}{hr^{2}}\right)^{\nu-1}\frac{r}{h_{sh}}\mathrm{Ws}$$

We expect in most cases that $1<\mathrm{Ws}<r/h_{sh}$ and so h_{sh} will be given by

$$\left(\frac{h_{sh}}{h}\right)^{\nu+1}=0\left\{\frac{K_m}{\eta_{PS}}\left(\frac{\dot{V}}{h^{2}r}\right)^{\nu-1}\left(\frac{r}{h}\right)^{2}\right\}=0\left\{\frac{\eta_{sh}}{\eta_{PS}}\left(\frac{r}{h}\right)^{2}\right\}$$

For this to be consistent with the requirement that $h_{sh}\ll h$, we require that

$$\frac{\eta_{sh}}{\eta_{PS}} = 0\left\{\left(\frac{h}{r}\right)^2\left(\frac{h_{sh}}{h}\right)^{\nu+1}\right\}$$

For h/r, h_{sh}/h each about 10^{-1} and $\nu=\frac{1}{3}$, η_{sh}/η_{PS} has to be about $\frac{1}{2}10^{-3}$. Surprisingly this can just about be achieved over operating shear rates with thermoplastics in the rubbery state. Note that this allows Ws to be of order 10^2 without altering the argument. If however h/r is of order 1, then the above argument will almost certainly apply.

EXERCISE 19.3.1. Show that

(i) $\hat{p} = p/\Gamma^* \rho^* T^*_{\text{adiab}}$
(ii) $\tilde{\eta} = \tilde{\eta}(\hat{p}=0)(1+\text{Vp}\,\hat{p})$
(iii) $\tilde{\rho} = (1+\text{Cm}\,\hat{p}+\tilde{T}\,\text{Ex/Na})$

and hence show how the various dimensionless groups given in (19.3.14) arise in eqns (19.3.3) and (19.3.6).

EXERCISE 19.3.2. Verify eqn (19.3.22) using the results of Example 7.1.1 and Exercise 7.1.1.

EXERCISE 19.3.3. (For those interested in mathematical analysis using special functions.)

Consider the cases $\nu=1$, $\frac{1}{2}$ and $\frac{1}{3}$ in (19.3.22) to give $\hat{T}_\infty(\tilde{r})$ in polynomial form.

Evaluate the integral in (19.3.30) in terms of integrals of $J_0(z)$.

Using the polynomial form for $\tilde{v}_z$ given by (7.1.69) evaluate the $\tilde{v}_{z1}$ as defined below (19.3.30).

Finally, assess the probability of expressing $\hat{T}_{\text{D}i}(\tilde{r})$ as expansions in terms of $J_0(\tilde{r}\tilde{\alpha}_i)$.

EXERCISE 19.3.4. Show that, for $\hat{t} \ll \text{Gz}$, eqn (19.3.18) is dominated by

$$\frac{\partial \check{T}}{\partial \hat{t}} = \frac{\partial^2 \check{T}}{\partial \check{r}^2} + \tilde{\eta}\left(\frac{\partial \tilde{v}}{\partial \tilde{r}}\right)^2$$

where $\check{r} = \text{Gz}^{\frac{1}{2}}(1-\tilde{r})$, with obvious boundary conditions $\check{T}(0,\hat{t})=0$, $\check{T}(\check{r},0)=0$.

Use a similarity variable

$$\check{y} = \check{r}/2\hat{t}^{\frac{1}{2}}$$

to write the solution in the form

$$\check{T}=\tilde{\eta}\left(\frac{\partial\tilde{v}}{\partial\tilde{r}}\right)^2\hat{t}T_{\text{trans}}(\check{y})$$

where

$$\frac{\mathrm{d}^2T_{\text{trans}}}{\mathrm{d}\check{y}^2}+2\check{y}\frac{\mathrm{d}T_{\text{trans}}}{\mathrm{d}\check{y}}+4=0$$

and $T_{\text{trans}}(0)=0$, $T_{\text{trans}}\to 1$ as $\check{y}\to\infty$.

Solve the equation subject to the given boundary conditions.

EXERCISE 19.3.5. Show that (19.3.68) follows from (19.3.5) and (19.3.6) for the special case considered in (19.3.66), (19.3.67) where $\tilde{r}$ is put equal to unity in the expression for $\tilde{t}_{rz}$. Verify that (19.3.70) holds.

EXERCISE 19.3.6. Verify that the asymptotic forms (19.3.71) satisfy eqn (19.3.70), paying particular attention to the order of magnitude of the last term on the left-hand side. Establish the order of magnitude of the terms neglected in the asymptotic forms (19.3.71).

EXERCISE 19.3.7. Show that (19.3.73) satisfies the continuity equation (19.3.3) when $\tilde{\rho}=1$.

EXERCISE 19.3.8. Derive (19.3.93) from (19.3.91) and hence (19.3.94). Explain why the adiabatic argument introduced for (19.3.88) is inconsistent with the fully-developed result (19.3.84).

EXERCISE 19.3.9. Show that the second of the boundary conditions (19.3.100) requires that

$$\frac{\partial x_{3m}}{\partial t}\ll v^*$$

the exact boundary condition being

$$v_3(x_{3m})=-\frac{(\rho_{\text{s}}-\rho_{\text{m}})}{\rho_{\text{m}}}\frac{\partial x_{3m}}{\partial t},\qquad v_1(x_{3m})=0$$

by analogy with (11.2.19) for example.

EXERCISE 19.3.10. Show that (19.3.120) makes use of the result that the relevant steady solution of (19.3.111), (19.3.109), (19.3.112) when the terms involving Br are neglected, is a linear temperature profile

$$\tilde{T} = -\frac{\mathrm{B_s}}{\mathrm{B_m}} \frac{(\tilde{x}_3 - \tilde{x}_{3m})}{(1 - \tilde{x}_{3m})}$$

Show that $\mathrm{B_s/B_m}$ need not be zero even if $\mathrm{B_m} = 0$ as is assumed for (19.3.122), and that in that case a suitable redefinition of T^*_{op} can lead to $\mathrm{B_s/B_m} = -1$.

EXERCISE 19.3.11. Show that if, in Subsection 19.3.2 IV, $X_m \not\ll 1$, then the relevant equation of $\overset{x}{T}_{\mathrm{m}}$ in (19.3.68) is

$$\frac{\mathrm{d}^2\overset{x}{T}}{\mathrm{d}\hat{y}^2} + \tfrac{1}{2}(\hat{y} - X_m)\frac{\mathrm{d}\overset{x}{T}}{\mathrm{d}\hat{y}} + \tfrac{1}{4}\mathrm{e}^{\mathrm{Na}\overset{x}{T}} f^2 \tilde{t} = 0$$

so that the solutions (19.3.71) are replaced by

$$\tilde{T} \sim \ln \mathrm{Na}\, \mathrm{erfc}\{\tfrac{1}{2}(\hat{y} - X_m)\} \qquad \hat{y} > \delta$$
$$\tilde{T} \sim \ln\{\tfrac{1}{2}\,\mathrm{Na}\, \mathrm{e}^{Y}/(1 + \mathrm{e}^{Y})^2\} \qquad Y = (\hat{y} - \delta)\mathrm{e}^{-\frac{1}{4}(\delta - X_m)^2}\pi^{-\frac{1}{2}} \ln \mathrm{Na}$$
$$\tilde{T} \sim \ln \mathrm{Na}\, |\mathrm{erf}\{\tfrac{1}{2}(\hat{y} - X_m)\} - \mathrm{erf}(-\tfrac{1}{2}X_m)| \qquad \hat{y} < \delta$$

where

$$2\,\mathrm{erf}\{\tfrac{1}{2}(\delta - X_m)\} + \pi^{-\frac{1}{2}}\,\mathrm{erf}(-X_m) = 1$$

Verify that (19.3.150) and (19.3.151) apply and hence deduce that (19.3.152) is replaced by

$$\frac{1}{2}\frac{\mathrm{Sf\,Pe}\,\pi^{\frac{1}{2}}}{\mathrm{Br}} X_m = \frac{\mathrm{B_s}\tilde{\alpha}_{\mathrm{s}}^{\frac{1}{2}}}{\mathrm{B_m\,Br\,erf}(\tfrac{1}{2}X_m\tilde{\alpha}_{\mathrm{s}}^{-\frac{1}{2}})} - \frac{\ln \mathrm{Na}}{\mathrm{Na}}\mathrm{e}^{-\frac{1}{4}X_m^2}$$

EXERCISE 19.3.12. Derive (19.3.154) by using (19.3.180), (19.3.68) with $f \equiv 0$ and (19.3.113).

EXERCISE 19.3.13. Derive (19.3.191) and (19.3.192) from (19.3.185)–(19.3.188) using (19.3.189), (19.3.190), showing that (19.3.185) is satisfied.

19.4 COOLING AND CURING

Once the injection process is complete, the material in the mould has to be solidified. For a cross-linking elastomer, this means holding it at an elevated temperature, relative to the material in the reservoir (typically 200°C and 100°C respectively), until the necessary chemical reactions have taken place. In a glassy or semi-crystalline polymer, a cooling process is involved, whereby the recently injected material (typically at 200–250°C) is cooled by conduction to the walls (held typically at 30°C) until it is sufficiently rigid to be ejected without change of shape.

Various factors, important in determining the quality of the final moulded article, are relevant. Most noteworthy among these is the change of volume consequent upon change in temperature, degree of cure and, above all, degree of crystallization. In the absence of any make-up flow of molten polymer during the cooling phase—driven by the mould-holding pressure—the moulded article will shrink away from the mould walls and varying shot size will be likely.

Rates of cooling and flow fields immediately prior to crystallization will affect the morphology of any semi-crystalline product, and can lead to orientation effects. Non-uniform temperature histories can lead to internal stresses and strains.

In cross-linking materials non-uniform temperature histories can lead to non-uniform cure, while elastic strain can be cured into the moulded article.

All of these effects will be examined in the following subsections.

19.4.1 Curing

The effects of the chemical cross-linking agents added to rubber compounds can be most simply described in terms of a scorch time, t_{scorch}, followed by a first-order cross-linking reaction determined by a reaction rate constant $k_{\text{c-1}}$. This is illustrated in Fig. 19.17, where long-term effects are also shown. In practice more than one first-order reaction may characterize the curing process and so the curve (b) may be formed of two, say, almost straight lines, the second relating to a slower cross-linking reaction. The quantities t_{scorch} and $k_{\text{c-1}}$ will both be functions of temperature, the simplest representation being given by

$$k_{\text{c-1}} = k_{\text{c-10}} e^{-E_{\text{c-1}}/kT} \tag{19.4.1}$$

with $E_{\text{c-1}}$ an activation energy and T the absolute temperature. Even

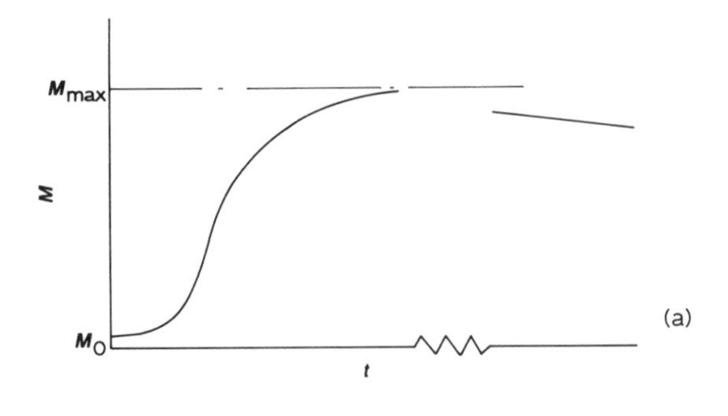

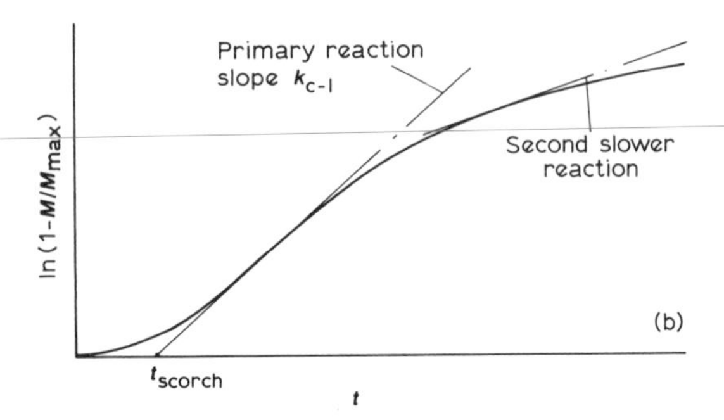

Fig. 19.17. Typical experimental relations for modulus (M) as a function of time (t) during curing at a fixed temperature. It is usually assumed that the modulus is proportional to cross-link density. (a) The initial modulus M_0 is associated largely with entanglements; the maximum modulus M_{max} is assumed to relate to complete exhaustion of the cross-linking agent; the slow long-term reduction in modulus is associated with thermal degradation of the rubber. (b) A linear portion of the curve of $\log(1-M/M_{max})$ can be extrapolated to give t_{scorch} at the point where it meets the time axis. The slope of this linear portion yields the chemical rate constant k_{c-1}.

in cases where the curve for the modulus M vs t cannot be adequately represented by a single first-order reaction, time–temperature superposition of the type implied by (19.4.1), is often possible.

The object of good moulding practice is to arrange for t_{scorch} to be reached just as the mould is filled, and for the moulding to be ejected when M reaches 80%, say, of M_{max}. To be more specific, this situation should be true for each material element in the moulding. It is therefore valuable to be able to predict the state of cure as a function of time and position before, during and after the injection process.

We therefore write

$$\frac{ds}{dt} = k_{scorch}(T) \tag{19.4.2}$$

where s, the degree of scorch, is zero in the material fed to the machine and is unity when $t = t_{scorch}$. This effectively defines t_{scorch} as k_{scorch} is known. For simplicity, we may assume that $k_{scorch} \sim 0$ for all $T < T_{scorch}$ and increases exponentially with T for all $T > T_{scorch}$. Clearly if the reservoir temperature, $T_{res} < T_{scorch}$, and the mould temperature $T_{mould} > T_{scorch}$, then the induction phase starts during injection. Once t_{scorch} is reached, the relevant kinetic equation becomes

$$\frac{dF}{dt} = k_{c-1}(T)(1-F) \tag{19.4.3}$$

where

$$F = (M - M_0)/(M_{max} - M_0) \tag{19.4.4}$$

and can be loosely termed the degree of cross-linking.

For the curing phase, which may be relatively long compared with the injection phase, we suppose that the material is motionless and that the temperature is given by a balance between heat transfer and chemical energy release (or absorption). For channel or disc-like moulds the relevant equation will be

$$\rho\Gamma \frac{\partial T}{\partial t} = \alpha \frac{\partial^2 T}{\partial x_3^2} + R_{c-1} \frac{dF}{dt} \tag{19.4.5}$$

If the mould fills at time $t = t_{fill}$, then the coupled equations (19.4.2) or (19.4.3) and (19.4.5) will have initial conditions either

$$s(\mathbf{x}, t_{fill}) = s_{fill}(\mathbf{x})$$

or

$$F(\mathbf{x}, t_{fill}) = F_{fill}(\mathbf{x}) \tag{19.4.6}$$

and

$$T(\mathbf{x}, t_{fill}) = T_{fill}(\mathbf{x})$$

and the boundary condition on (19.4.5) will be

$$T(x_3 = \pm\tfrac{1}{2}h, t) = T_W(t) \tag{19.4.7}$$

If $s_{\text{fill}} < 1$, i.e. (19.4.6i) is relevant, then we use the ancillary relation

$$F(t_{\text{scorch}}) = 0 \tag{19.4.8}$$

when switching from eqn (19.4.2) to eqn (19.4.3). Solution of these equations is relatively standard, and is most easily carried out by numerical computation.

The problem of prescribing s_{fill} (or F_{fill}) remains, and depends on the temperature history of each particle that finds itself at (x_1, x_3) at time t_{fill}. To simplify discussion of this matter, it will be supposed that s is prescribed at the entry to the mould, say s_0 (constant) and that the mould cavity is a channel of constant length l and depth h, i.e. the situation covered in Subsection 19.3.3. It will readily be appreciated that the main complication in deriving the temperature history relates to the, so far unresolved, problem of flow in the neighbourhood of the melt front. The latter moves with the mean velocity $v^* = q_0/h$. For Na, B≪1, the velocity profile $v_1(x_3)$ is independent of t and x_1. For

$$0 < |x_3| < x_{3\text{I}}, \quad \text{where} \quad v_1(x_{3\text{I}}) = v^* \quad \text{defines} \quad x_{3\text{I}}$$

material will be moving forward faster than the melt front while for

$$x_{3\text{I}} < |x_3| < \tfrac{1}{2}h$$

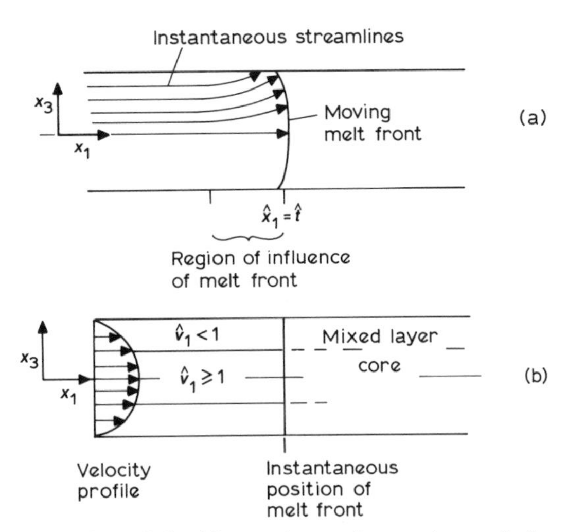

Fig. 19.18. Illustration of the kinematics at the moving melt front. (a) Curvature of streamlines near the front. (b) Distinction between core and mixed layers in region behind melt front.

the reverse will be true. At the melt front material in the former region will be fed into the latter region. This is shown in Fig. 19.18. In (a) the curvature of streamlines is shown whereby redistribution of material takes place. The region of influence of this curvature will have a length of order h, which is small compared with l. The instantaneous position of the melt front defines a core region for which

$$\hat{v}_1 = v_1 h/q \geqslant 1 \tag{19.4.9}$$

and a mixing region for which

$$\hat{v}_1 < 1 \tag{19.4.10}$$

As the interface passes through the position $\hat{x}_1$, the material moving into the core region is assumed to possess the properties, i.e. temperature and degree of scorch (or cross-linking), of the material arriving along the streamline at that particular value of $\tilde{x}_3$, whereas the material moving into the mixed layer is assumed to possess the mean of the properties for all the material moving into the mixed layer, i.e. independent of $\tilde{x}_3$ for all $|\tilde{x}_3| > \tilde{x}_3(\hat{v}_1 = 1)$. Once the melt front has passed by, all the material moves forward with the appropriate velocity $v_1(x_3)$.

At the moment the mould is filled there are thus two regions in $(\hat{x}_1, \tilde{x}_3)$ space, as illustrated in Fig. 19.19. In the larger, wedge-shaped, core region (a) is material that has not suffered mixing at the melt front; in the smaller, complementary, region (b) is material that has (by assumption) been fully mixed as the melt front passed by. Noting that the mould fills at time $\hat{t} = 1$, using the definition for $\hat{t}$ given in (19.3.103), the interface between the two regions lies at

$$\hat{v}_1(\tilde{x}_3) = \tilde{x}_{3\text{I}} \tag{19.4.11}$$

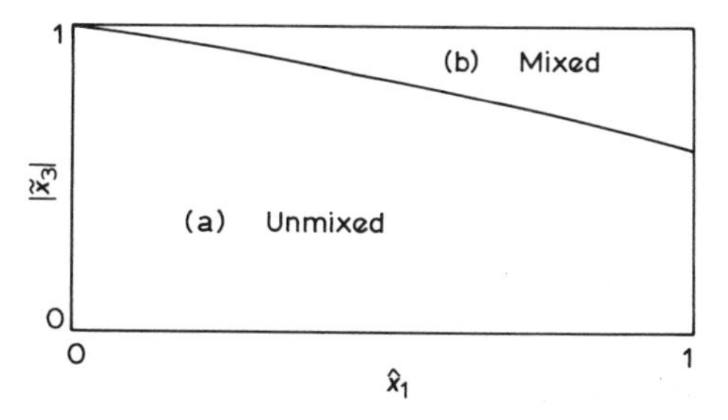

Fig. 19.19. Regions of material in mould flow that have (a) and have not (b) suffered mixing at the melt front at the moment t_{fill} ($\hat{t} = 1$).

For all material in region (a) therefore, typically at $(\hat{x}_{1P}, \tilde{x}_{3P})$ say, we have only to solve the equation

$$\left.\begin{aligned}\frac{ds}{d\hat{t}} &= k_{\text{scorch}}\{T[\hat{x}_1 = \hat{x}_{1P} + (\hat{t}-1)\hat{v}_1(\tilde{x}_{3P})\tilde{x}_{3P}]\} \\ s(1-\hat{x}_1/\hat{v}_1) &= s_0\end{aligned}\right\} \qquad (19.4.12)$$

to get $s(\hat{t}=1)$. If s reaches 1 for $\hat{t}=\hat{t}_{\text{scorch}}<1$, then eqn (19.4.3) has to be solved for $1>\hat{t}\geq\hat{t}_{\text{scorch}}$ with T defined as in (19.4.12) and

$$F(\hat{t}_{\text{scorch}}) = 0 \qquad (19.4.13)$$

For all material in region (b), typically at $(\hat{x}_{1Q}, \tilde{x}_{3Q})$ say, we can deduce that it was mixed in the melt front when the latter passed the station

$$\hat{x}_{1F}(\hat{x}_{1Q}, \tilde{x}_{3Q}) = \frac{\hat{x}_{1Q} - \hat{v}_1(\tilde{x}_{3Q})}{1-\hat{v}_1(\tilde{x}_{3Q})} \qquad (19.4.14)$$

From mass and property conservation arguments according to the mixing model described above the value of any intrinsic property in the mixed layer $(1>\tilde{x}_3\geq\tilde{x}_{3I})$ can be calculated. Thus, if $c(\tilde{x}_3)$ is the relevant value for material arriving at the instantaneous position of the interface, the bulk-flow mean-value of this property is given by

$$\bar{c} = \int_0^1 c(\tilde{x}_3)\hat{v}_1(\tilde{x}_3)\,d\tilde{x}_3 \qquad (19.4.15)$$

and the mixed value c_{mix} of this property in the mixed layer is given by

$$c_{\text{mix}} = \int_0^{\tilde{x}_{3I}} \{\hat{v}_1(\tilde{x}_3)-1\}c(\tilde{x}_3)\,d\tilde{x}_3 \Big/ \int_0^{\tilde{x}_{3I}} \{\hat{v}_1(\tilde{x}_3)-1\}\,d\tilde{x}_3 \qquad (19.4.16)$$

all evaluated at the position $\hat{x}_1$ at time $\hat{x}_1$.

Note that c could stand for s, F or T (assuming that ρ and Γ were constant) and that the model is not a unique one (see Exercise 19.4.2); it is merely a relatively simple unambiguous one that does not run counter to any reasonable kinematic or physical interpretations that have been proposed.

The values of T, s or F in material meeting the melt front at any particular values of $\hat{x}_1$ and $\tilde{x}_3$ can be calculated at a step-by-step process for successively larger values of $\hat{x}_1$, i.e. $\hat{t}$. For all $|\tilde{x}_3|<\tilde{x}_{3I}$, no mixing will be involved. For all $|\tilde{x}_3|>\tilde{x}_{3I}$, c_{mix} is given by the velocity weighted mean of the solution obtained for $|\tilde{x}_3|<\tilde{x}_{3I}$. This, therefore, provides the values of s or F at $\hat{x}_{1F}$ given by (19.4.14). Forward integration for all $\hat{t} > \hat{x}_{1F}$ can then be carried out using the analogues of eqns (19.4.12) or (19.3.3).

19.4.2 Crystallization and Shrinkage

It will be supposed here for simplicity that crystallization takes place in a definitive manner at a sharp interface between molten and solid material. Questions of annealing, of molecular-weight segregation, of supercooling, or of crystalline material suspended within the melt, are outside the scope of this analysis. It will also be supposed that the primary change in density takes place on crystallization, and that a significant latent heat is involved. The situation is therefore entirely consistent with that supposed relevant in Section 11.3 and in Sections 19.2 and 19.3 above. Modifications to the theory given below to take account of gradual changes in density can, if necessary, be made later.

The convective term in the energy equation, $|\rho \mathbf{v} \cdot \nabla U| \ll \rho |\partial U/\partial t|$, will be far from dominant in the cooling phase, i.e. $\mathrm{Gz} \ll 1$, while the generation term will be negligible, $\mathrm{Br} \ll 1$. Hence the equations governing the temperature can to a reasonable approximation be written

$$\left.\begin{aligned} \frac{\partial \hat{T}}{\partial \check{t}} &= \frac{\partial^2 \hat{T}}{\partial \tilde{x}_3^2} \qquad 0 < |\tilde{x}_3| \leqslant \tilde{x}_{3m} \\ \frac{\partial \hat{T}}{\partial \check{t}} &= \frac{\kappa_s}{\kappa_m} \frac{\partial^2 \hat{T}}{\partial \tilde{x}_3^2} \qquad \tilde{x}_{3m} < |\tilde{x}_3| < 1 \end{aligned}\right\} \tag{19.4.17}$$

where

$$\check{t} = (4\kappa_m/h^2)t \tag{19.4.18}$$

$$\left.\begin{aligned} \hat{T}(\pm\tilde{x}_{3m}) = 0, \qquad \hat{T}(\pm 1) = -\mathrm{B_s}/\mathrm{B_m} \\ \left[\tilde{\alpha}\frac{\partial \hat{T}}{\partial \tilde{x}_3}\right]_{\pm\tilde{x}_{3m}} = \pm \mathrm{Sf}\frac{\mathrm{d}\tilde{x}_{3m}}{\mathrm{d}\check{t}} \end{aligned}\right\} \tag{19.4.19}$$

and $\hat{T}(t=0)$ is given. Sf is defined in (19.3.105), while eqns (19.4.17)–(19.4.19) follow from those given in Subsection 19.3.2 earlier.

These equations neglect the slow changes in density involved. However, the values for $\hat{T}(\hat{x}_1, \tilde{x}_3, t)$ that are calculated according to (19.4.17)–(19.4.19) will imply a density field $\rho(\hat{x}_1, \tilde{x}_3, t)$ which alters with time, assuming that an equation of state such as (2.2.2) can be used. A value for pressure has to be provided.

The continuity equation (19.3.106) can then be used in integrated form to estimate the slow influx of material from the reservoir into the mould, which again is supposed to be of constant length l and depth h. If we use the approximation suggested above, that $\tilde{\rho}$ has two values (1 in the melt and $\tilde{\rho}_s > 1$ in the solid), then we can write

$$\int_{-1}^{1} \frac{\partial \tilde{\rho}}{\partial \hat{t}} \, d\tilde{x}_3 = -2(\tilde{\rho}_s - 1) \frac{\partial \tilde{x}_{3m}}{\partial \hat{t}} \tag{19.4.20}$$

and so

$$\frac{\partial}{\partial \hat{x}_1} \int_{-\tilde{x}_{3m}}^{\tilde{x}_{3m}} \tilde{v}_1 \, d\tilde{x}_3 = -2(\tilde{\rho}_s - 1) \frac{\partial \tilde{x}_{3m}}{\partial \hat{t}} \tag{19.4.21}$$

Note that $\partial \tilde{x}_{3m}/\partial \hat{t}$ is given by one of the internal boundary conditions (19.4.19) and will be of order

$$\mathrm{Sf}^{-1} \, \tilde{t}/\hat{t} = 1/\mathrm{Gz}\,\mathrm{Sf} \tag{19.4.22}$$

It can be seen that

$$v_1 = 0\left(\frac{4\kappa_m l \rho_m \Gamma_m T^*_{op}(\tilde{\rho}_s - 1)}{h^2 \rho_s \xi_s}\right) \tag{19.4.23}$$

which as expected does not involve v^* otherwise defined. Indeed the right-hand side of (19.4.23) can be used as a scale velocity v^*_{shr} for the pressure-holding phase. If $\tilde{x}_{3m}$ becomes $\ll 1$, then clearly an extra factor of $\tilde{x}_3^{-1}$ appears in the order of magnitude of v_1.

The value of $\int_{-\tilde{x}_{3m}}^{\tilde{x}_{3m}} \tilde{v}_1 \, d\tilde{x}_3$ is itself related through (19.3.107) to $\partial \hat{p}/\partial \hat{x}_1$ where $\tilde{t}_{13}$ is given by (19.3.103) and the scale for $\hat{p}$ is given by the viscous stress $\eta(D^*_{shr}) \, D^*_{shr}$ where $D^*_{shr} = 2v^*_{shr}/h$. By order of magnitude arguments, this gives a total pressure drop, in the shrinkage phase, across the mould of order

$$P^*_{shr} = 4\eta_{shr} v^*_{shr} l/h^2 \tag{19.4.24}$$

where h may have to be replaced by $h(1 - \tilde{x}_{3m})$ at the later stages. Clearly if $P^*_{shr} \ll P_{m-h}$, the mould holding pressure, then the shrinkage due to cooling is easily made up by additional inflow. However, if $P_{m-h} \leqslant 0(P^*_{shr})$ then it is likely that molten material will not flow into the mould fast enough to prevent true shrinkage at the $\hat{x}_1 = 1$ end of the mould. Indeed inflow will only occur up to that value of $\hat{x}_1$ which leads to an actual pressure drop according to (19.3.107), (19.4.21), (19.4.17)–(19.4.19) of P_{m-h}.

Usually $\frac{1}{2}P_{max} < P_{m-h} < P_{max}$, where P_{max} is the maximum pressure drop achieved during injection. Therefore initially, $P_{\mathrm{m-h}} \gg P^*_{\mathrm{shr}}$ and it is only when $\tilde{x}_{3m}$ (and $\hat{\hat{T}}$) have fallen due to significant crystallization that $P_{\mathrm{m-h}}$ falls below P^*_{shr}. Note that P^*_{shr} is being thought of as time dependent, with T^*_{op} being replaced by $\hat{\hat{T}}$.

The most likely place to get $\tilde{x}_{3m} \to 0$, i.e. a physical elimination of flow, is at the gate, where h is very small anyway. If this is very rapid, then the mould holding pressure has little effect on material shrinkage in the mould during the cooling phase. Voids must form somewhere in the mould cavity; the most likely result is that these would arise at the metal/polymer interface, and that the article ejected would be dimensionally smaller than the mould. This is not a desirable situation, particularly because it can lead to variable size and weight in the moulded article: if inserts are involved, poor adhesion or non-adhesion of polymer to inserts can occur, again a most undesirable result. The practice of deliberately heating the gate region has been recommended in certain cases to prevent premature freezing off.

The general argument followed above in going from (19.3.106) to (19.4.21) can be amended to cover the case of a glassy polymer being cooled or a cross-linking polymer being cured. Here $\tilde{x}_{3m}$ need not formally be included though the dependence of η upon T or the degree of cure will now become a dominant factor. In a glassy polymer, we would write

$$\int_{-1}^{1} \frac{\mathrm{d}\tilde{p}}{\mathrm{d}\hat{t}}\, d\tilde{x}_3 = \frac{\partial \hat{p}}{\partial \hat{t}} \int_{-1}^{1} \frac{p^*}{\rho}\left(\frac{\partial \rho}{\partial p}\right)_{\mathrm{T}} \mathrm{d}\tilde{x}_3 + \int_{-1}^{1} \frac{T^*}{\rho}\left(\frac{\partial \rho}{\partial T}\right)_{\mathrm{p}} \frac{\mathrm{d}\hat{T}}{\mathrm{d}\hat{t}}\, \mathrm{d}\tilde{x}_3 \tag{19.4.25}$$

Of these the second term on the right-hand side would probably dominate and an order of magnitude is immediately obtained from (19.3.106), (19.4.17) and (19.4.18) for

$$v^*_{\mathrm{shr}} = \frac{T^*_{\mathrm{op}}}{T^*_{\mathrm{expan}}} \frac{4\kappa^* l}{h^2} \tag{19.4.26}$$

using the definition given in (19.3.11).

If an approximate numerical solution is obtained, based on (19.4.20) or the temperature term in (19.4.25) then the relevant solution for $\hat{p}$ can be fed back into the analysis to obtain a better estimate for $\partial\tilde{\rho}/\partial\hat{t}$. However, it is doubtful whether such accuracy would be worth seeking in most cases. In this context see Chung & Ryan (1981) who give further relevant references.

It is useful to obtain an estimate of the total shear strain arising during the cooling phase. On the basis of (19.4.20), the order of magnitude for the additional shear strain will be

$$\gamma_{sh}^{*} = 0(l(\tilde{\rho}_s - 1)/h) \tag{19.4.27}$$

where it is assumed that most of the material has crystallized by the time make-up flow ceases. Although $(\tilde{\rho}_s - 1)$ may be of order 0·1, a large factor l/h will ensure that large shear strain is applied during the mould holding phase. This result justifies the use of a viscosity in the analysis (in defining $\tilde{t}_{13}$) rather than using an elastico-viscous model.

For an elastomer that changes little in volume during the curing phase, the reverse is more often true; the measured behaviour of $\hat{p}$ is found to follow a stress relaxation pattern with time, and the function $\eta^{-}(t)$ defined by (4.6.19) would be appropriate. A non-isothermal theory of stress relaxation which compared well with experiment for HDPE is given by Dietz & Bogue (1978).

Lastly, it is worth remarking that t_c, the actual time of cooling in the mould, will be chosen so that

$$\left.\begin{array}{ll} 4\kappa^{*}t_c/h^2 = 0(1 + B_s/B_m) & \text{if} \quad Sf \leqslant 0(1) \\ \text{or} \quad 4\kappa^{*}t_c/h^2 = 0(SfB_m/B_s) & \text{if} \quad Sf \gg 1 \end{array}\right\} \tag{19.4.28}$$

to ensure that most of the article ejected will be frozen.

19.4.3 Orientation and Frozen-in Strain

Microscopic examination of flat plastic moulded articles has shown (Dietz *et al.*, 1978) a pattern of orientation shown diagrammatically in Fig. 19.20. The outer (skin) layer displays an orientation apparently determined by the state of the material laid down from the melt front as the latter passes by the cold mould wall. The next layer displays an orientation consistent with the elastic shear strain that is induced in a Poiseuille-type flow (sheared unidirectionally) of a rubbery material. The central core layer is relatively weakly oriented and corresponds to solidification from an undisturbed melt.

Other effects have also been observed that are more characteristic of thermal stresses developed in a material solidifying at rest. Figure 19.21 illustrates how final cooling of a flat sheet that solidifies while significant symmetric temperature gradients exist in the material leads to compressive stresses on the outer skin and tensile stresses in the core. Any asymmetry in these patterns can lead to buckling of thin moulded articles, whether before or during annealing.

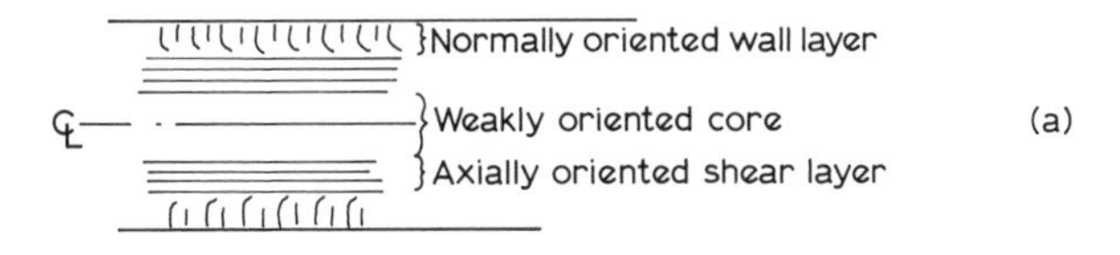

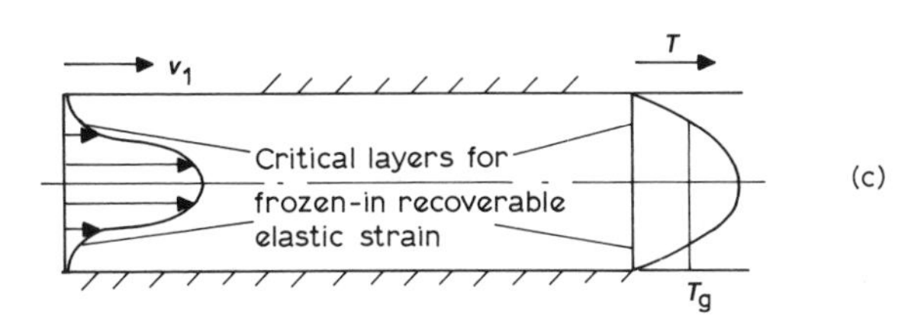

Fig. 19.20. Orientation in injection moulding. (a) Typical orientation in flat injection-moulded plaque. (b) Path line and deformation of fluid element reaching the side wall near the melt front, perhaps explaining the wall-layer orientation in (a) above. (c) Velocity and temperature profiles in a glassy polymer well behind the melt front; the critical layer is one for which $T \sim T_g$. (For a semi-crystalline polymer, the critical layer would correspond to the effective crystallization temperature.) The orientation in the critical layer correlates with that observed in the shear layer in (a) above.

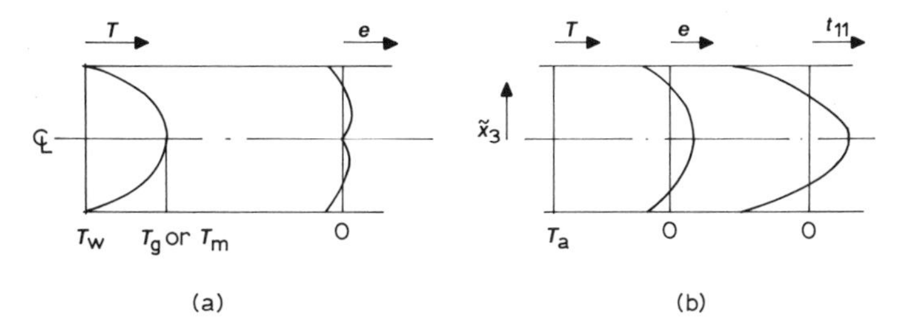

Fig. 19.21. Thermal stress development on cooling. The mean stress, $\frac{1}{2}\int_{-1}^{1} t_{11}\, d\tilde{x}_3$, is zero. (a) At solidification, a small axial thermal strain and stress will be present. (b) After cooling to the uniform ambient temperature, a larger axial thermal strain and stress develop.

The orientation, or stress anisotropy, of a transparent moulded article is often investigated by measuring its birefringence. It appears that birefringence can be caused either by the large elastic strains (arising through conformational molecular orientation at low modulus) or the small glassy strains (arising by bond stretching or bending at high modulus), so it is not always clear what the relationship between birefringence, actual stress and frozen-in elastic strain is.

The analysis of Dietz & White (1978; see also White & Dietz, 1979) is based on (1) the rheo-optical law

$$\mathbf{n}-\tfrac{1}{3}\,\mathrm{tr}\,\mathbf{n}\mathbf{1}=C(\mathbf{T}-\tfrac{1}{3}\,\mathrm{tr}\,\mathbf{T}\mathbf{1}) \qquad (19.4.29)$$

where $\mathbf{n}$ is the refractive-index tensor and C is the stress-optical coefficient; (2) the supposition that the value of $\mathbf{n}$ present in the material at the moment the temperature drops below a specified value, T_m say, is frozen into the material and characterizes the optical behaviour thereafter; (3) the stress relaxation relation

$$\mathbf{T}'=\mathbf{T}(0)\exp\left\{-\frac{1}{\lambda_0}\int_0^t \frac{\mathrm{d}t'}{a_T\{T(t')\}}\right\} \qquad (19.4.30)$$

which follows from a simplified form of (5.2.28) and which applies after time $t=0$, i.e. the moment the mould is filled. $\mathbf{T}(0)$ is determined by the flow field in the molten central region of the mould cavity, as was analysed in Subsection 19.3.2 above.

For material that freezes during the injection phase, a power-law relation for both t_{13} and $t_{11}-t_{33}$ is assumed. Observed values of birefringence are said to compare very satisfactorily with predictions based on these assumptions.

A theory of residual stresses in polymers is given in Williams (1981).

EXAMPLE 19.4.1. Development of thermal stress and strain.

The simplest case will be considered whereby a flat plaque cooling in a mould finally solidifies in an unstrained condition with a symmetric temperature distribution

$$\tilde{T}=\tilde{T}_0(\tilde{x}_3) \quad \text{where} \quad \tilde{T}_0(0)=0, \qquad \tilde{T}_0(\pm 1)=-1$$

At a much later stage, the temperature is everywhere equal, $\tilde{T}=-1$ without loss of generality, an internal strain field $e_1(\tilde{x}_3)\neq 0$ having developed.

If both the thermal expansion coefficient β and the biaxial elastic modulus E are constant, and it is supposed that the initial length l_0 has changed to $l_\infty < l_0$, we can require that

$$\bar{e}_1 = 1 - l_\infty / l_0 = -\beta(1 + \tilde{T}_0) + Et_{11}$$

where $t_{33} = 0$. The stress equilibrium requirement is that

$$\int_{-1}^{1} t_{11}\,\mathrm{d}\tilde{x}_3 = 0$$

whence

$$\bar{e}_1 = -\beta\left(1 + \int_0^1 \tilde{T}_0(\tilde{x}_3)\,\mathrm{d}\tilde{x}_3\right)$$

The local stress and strain then follow.

EXERCISE 19.4.1. Derive relations (19.4.11) and (19.4.14) by elementary kinematical arguments. Note that any material entering the mould for $|\tilde{x}_3| < \tilde{x}_{3\mathrm{I}}$ is not mixed, and so remains at the same $\tilde{x}_3$ position, while any material that is mixed at the melt front and is left at $|\tilde{x}_3| > \tilde{x}_{3\mathrm{I}}$ just after the melt front passes also moves forward at a constant $\hat{v}_1(\tilde{x}_3)$, whereas the melt front itself moves at unit speed.

EXERCISE 19.4.2. Derive the relation for c in region (b) that follows if no mixing is assumed to arise, i.e. if no streamlines are allowed to cross.

Show in particular that as the melt front passes any given position the position of a particle changes discontinuously from $y^- = 1 - x_3^-$ to $y^+ = 1 - x_3^+$ where

$$y^+ = \tfrac{3}{2}(y^-)^2 - \tfrac{1}{2}(y^-)^3$$

Show how this can be used in connection with (19.4.14) and verify that a step-by-step forward integration for T can thus be carried out.

EXERCISE 19.4.3. Obtain the result (19.4.26) by order-of-magnitude arguments.

EXERCISE 19.4.4. Obtain the order-of-magnitude estimates (19.4.28) on the basis of (19.4.17)–(19.4.19).

19.5 OVERALL CALCULATIONS. COMPUTER-AIDED DESIGN

The previous three subsections have dealt with separate parts of the injection moulding operation in an idealized fashion, stressing the unsteady nature of the process. A satisfactory description of the whole process requires that idealizations be suitably chosen and that interactions between the separate flow and heat-transfer processes be suitably described and accounted for. The most important features to be dealt with are:

(1) Representation of actual flow passages by an interconnected set, a tree, of idealized units. Nozzle, sprue and runners present less difficulty than mould cavities, which often possess intricate and asymmetric shapes and, because of inserts or weight-saving design, may be multiply-connected. (The possibility of producing complex shapes is one of the advantages of the moulding process.)

(2) Following the position of the melt front along each of the branches of the tree during the filling operation. At each node (junction point between units) which involves more than one exit unit, the flow splits into two or more streams. The movement of the melt front in each branch will be determined by the flowrates in them, and the flowrates will be determined by the requirement that the pressure be single-valued at each node. This introduces implicit relationships between the separate unit solutions which, except in very trivial situations, lead to iterative numerical solution schemes.

The nodal pressure requirement has to be modified if a significant pressure drop is associated with a node, for then it must be decided whether the drop arises on the inlet side, in which case it will be the same for each stream, or on the outlet side, in which case it may be different for each stream. Furthermore, nodes are by definition regions where the lubrication approximation does not apply, and so 'pressure losses' will involve deviatoric components of the stress tensor that will probably be of the same order as the total node 'pressure loss'. Small changes in geometry could then have proportionately large effects on stress distributions and hence flowrates into exits. This matter is of particular importance in gate design, which determines how, when and at what temperature material flows into the mould cavities. At present, an appreciation of these local effects is part of the mould designer's art, and has not been subjected to careful analysis or experiment, except for idealized studies on 'jetting'.

(3) Choice of temperature distribution for flow at entry to each unit. In the various models considered in Section 19.3, a fully-mixed approximation, i.e. uniform temperature profile, was most commonly applied. This would be consistent with highly 'turbulent' or, more realistically, strong secondary flow within the relevant node preceding entry. This would be less realistic for very smooth transition from unit to unit (i.e. one-to-one correspondence between streamlines in successive units) where an alternative adiabatic approximation, corresponding to continuity of temperature along streamlines, could be adopted. In the former approximation, only a knowledge of $\bar{T}$ is carried forward through a node, whereas in the latter, the full functional dependence of T upon the stream function q, say, at exit from one unit would be required to prescribe the (same) dependence of T upon q, or its analogue, on entry to a succeeding unit.

The difference between the two approximations is of great significance in connection with freezing off at gates. Smooth laminar flow through a gate with a temperature profile characteristic of a previous channel flow leads to a thicker frozen layer than that predicted for constant inlet temperature, which latter allows formally of a temperature discontinuity at the flow boundary. In simple engineering terms, it is a question of ensuring that the hot core fluid contacts the walls of the mould gates, melting and sweeping away any frozen material. Such an effect would be weaker during the (almost) passive phase of cooling (Section 19.4), but might still be significant.

(4) Choice of model regime for each unit. To the extent that flowrates and temperatures are implicitly determined by the overall solution, the relevant dimensionless groups, Gz, Na, Br, are also implicitly determined at any stage of filling. In general this will not lead to difficulty, because the basic regime involved for any unit, i.e. $\mathrm{Gz} \ll 1$, $0(1)$ or $\gg 1$ and $\mathrm{Na} \geqslant 0(1)$ or $\ll 1$, will be insensitive to small changes in the values involved. However, there will obviously be circumstances where any formal method of selection will cause a change in regime to occur during an iterative calculation; instability in the iterative process can then arise if the predictions of calculations for two overlapping regimes do not correspond satisfactorily.

This is illustrated in Fig. 19.22 where application of 2 asymptotic results, valid for $l \gg 1$ and $\ll 1$, in the two regions $l > 1$ and $l < 1$ leads to a discontinuity in the prediction for P as a function of l, and an apparent fall in P for an increase in l in going from A to B, i.e. an unstable situation. The exact relation is shown to be clearly monotonic,

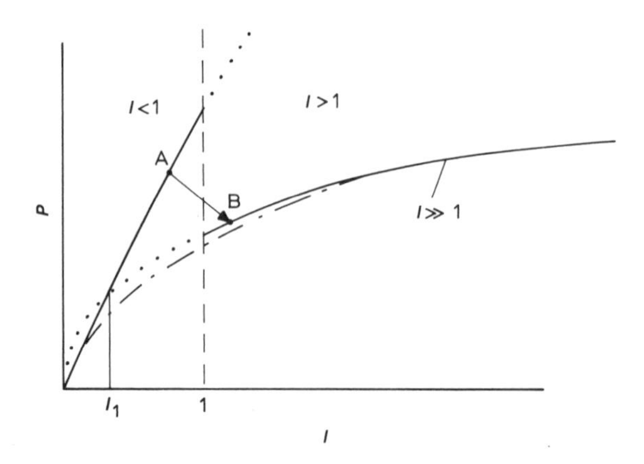

Fig. 19.22. Hypothetical pressure drop (P)/channel length (l) relation to illustrate possible difficulty in iterative calculations. —·—, exact relation; —···, asymptotic relations for $l \gg 1$ and $l \ll 1$.

as would be expected in any real situation. One simple way out of the difficulty would clearly be to apply the asymptotic results in the complementary regions $l < l_1$ and $> l_1$ instead. An alternative, equally simple method is to use the same regime approximation for each unit throughout a calculation, even though the final result might show that the original choice was inappropriate. Similarly, a choice may have to be made between a model dominated by initial values of variables, one that assumes a quasi-steady state and one that is both unsteady and non-uniform. This has been discussed in connection with the frozen-layer thickness in channel flow, Subsection 19.3.2, and analogously in Section 19.2 where Fig. 19.3 illustrates the three relevant regimes for pre-plasticizing screw extrusion.

(5) Choice of wall temperatures. A constant uniform mould temperature for the whole flow system can be shown to be inaccurate in practice, partly because the thermal conductivity of the mould metal is not infinitely large, but more importantly because the metal mould does not have infinite thermal capacity. Heat transfer between the mould and its outer environment, whether by radiation, electrical heating or the use of heat transfer fluids flowing through internal passages in the mould, is therefore a significant factor in determining both temporal and spatial variations in wall temperatures. In some cases, such variations are deliberately engineered.

These features are little different from those that arise in all the other idealizations of polymer-processing unit operations discussed in earlier chapters. They are brought into sharper focus here because of the essentially unsteady nature of injection moulding, and of the need to minimize cycle times.

Although computer-aided design programs were first evolved to predict extruder performance, it is clear that they are now most in demand as aids to mould designers (Pearson & Richardson 1983, Chapter 5). A brief description of the general procedure involved in computer simulation of rubber injection moulding is given in Richardson *et al.* (1980) while similar approaches are described in Byam *et al.* (1979) and in parts of Wang *et al.* (1975–81).

Such a simulator must accept, store and use:

(a) A simplified description of the flow passages and mould cavities.
(b) Operating conditions, e.g. reservoir and wall temperatures, initial flowrate, maximum pressure drop, time of onset and value of mould-holding pressure, cooling time, cycle time, i.e. sufficient information about any externally imposed conditions.
(c) Mathematical relationships for rheological, heat-transfer, phase change (freezing) or chemical (cross-linking) behaviour of the polymer melt.
(d) Mathematical models, and associated computational routines, for polymer melt flow in various units (and nodes), providing relationships between pressure, flowrate and melt temperature, based on the information provided by (a), (b) and (c). These routines will embody the solutions relevant for the various regimes described in Subsections 19.3.2–19.3.4, for each of the basic types of unit; or the more elaborate numerical solutions for more complex geometries (Hieber & Shen, 1980; Ryan & Chung, 1980; see also Kreuger & Tadmor, 1980).
(e) Mathematical models for freezing or cross-linking during the passive, mould-holding phase, according to the regimes described in Section 19.4, which use information from (a), (b) and (c), with initial temperatures provided by (d).

The computer program providing the simulation operates iteratively over a set of pre-determined time intervals tracing the flow of polymer into the mould channel system. At any given time the instantaneous melt front position and the velocity, pressure and temperature fields in those units that are not empty are calculated on the basis of their values at the previous time (i.e. one time interval earlier) and the model relations (d). If the melt front reaches an extremity of the tree, then that branch of the mould is filled; the program then switches for that branch to a computational routine based on the model relations (e). All branches may or may not fill before injection ceases; if they do not, an unsatisfactory moulding is predicted; if they do fill, but the moulding is not sufficiently frozen or cross-linked before ejection, then an unsatisfactory moulding is still predicted. If a fully frozen or cross-linked moulding is produced, a feasible, but not necessarily optimal, design has been produced.

It is not realistic to attempt automatic, i.e. computer-controlled, improvements to any chosen design. There are too many options available. It is therefore left to the designer to adjust geometrical parameters in (a), operating parameters in (b), and occasionally material parameters in (c), on the basis of evidence provided by earlier calculations and by experience, before proceeding to a further simulation. Indeed, in order to avoid overelaborate calculation routines in early simulations, it proves convenient to override the programmed choice of unit flow model—as described in (d) and (4) above—and use the simplest crude isothermal model. A strongly interactive computer simulation has therefore proved most useful in practice.

Chapter 20

Thermoforming and Blow Moulding

These two cyclic processes can conveniently be considered together. In both, a preformed sheet or tube of molten (or softened) plastic is pushed or inflated against the inside of a cold mould. In thermoforming the preform is usually a flat sheet clamped around its edges which is heated by radiant energy just before deformation; in blow moulding the preform is usually a nearly cylindrical tube (parison) that has just been extruded vertically downwards and is clamped top and bottom by the two halves of the mould. In thermoforming, deformation can be achieved by some combination of vacuum on the mould side of the sheet, excess air pressure on the side opposite to the mould, and movement of a plunger that pushes the warmed sheet towards the mould. This is shown diagrammatically in Fig. 20.1. A typical product is a tray-like container. In blow moulding, deformation is usually achieved by rapid air inflation of the closed volume formed within the clamped parison; the air supply is almost always provided through a spigot about which the neck of the parison is clamped. This is shown diagrammatically in Fig. 20.2. A typical product is a plastic bottle with a separately-moulded screw top. In both of these melt processes, deformation is rapid and Deborah numbers are high. The strains are largely extensional and of the same order as in film blowing, i.e. 0(10) or less. Elastic or viscoelastic behaviour is to be expected of the material during inflation.

Gravity forces will rarely be significant during inflation, though they may be important during extrusion of the parison in blow moulding. The material temperature can be considered constant during the forming process.

The cheapest and therefore commonest processes use sheets and parisons of uniform thickness. This leads to mouldings that are of

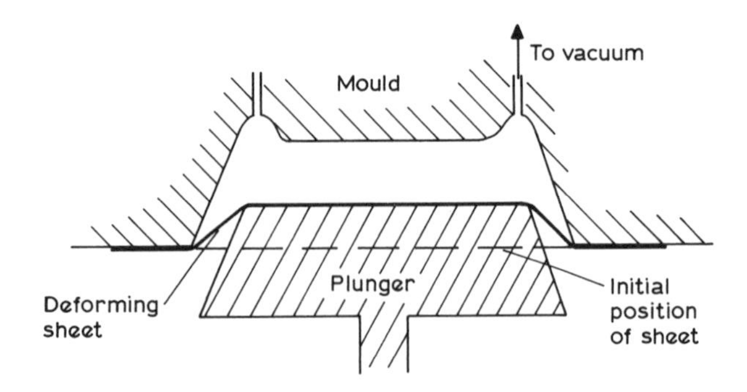

Fig. 20.1. Simple sketch of thermoforming operation.

variable (and necessarily reduced) thickness; because the resulting thickness variations are not optimal for mechanical strength, the resulting containers usually have to be heavier than would be necessary if optimal thickness variations could be designed into the processes. Some advantage can be gained during parison formation by having variable thickness around the cylinder cross-section, achieved by careful die-design, and variable thickness along the parison, achieved by overall output control during extrusion using a movable plunger or throttle within the die.

Recently, a stretch-blow moulding process has been developed for materials (like PET and PP) that can be cold-drawn. The preform is usually injection-moulded and so can be cooled close to the crystallization point before stretching in an axial direction using a plunger located at the base of the preform inner mould body, followed by radial inflation. This process demands very careful control of material temperature in the same way as do the cold-drawing process and the cold-draw bubble process described in Sections 16.2 and 17.5.

A full mechanical analysis of the unsteady inflation process of a viscoelastic material into an asymmetric mould would present the greatest analytic problems yet met in this text, even if the thin-film approximation were used throughout. The steps required in such an analysis would include:

(a) specification of the (inner) mould surface geometry;
(b) specification of the initial and subsequent instantaneous material (surface) geometry, with associated time-dependent metric of the surface coordinate system employed;

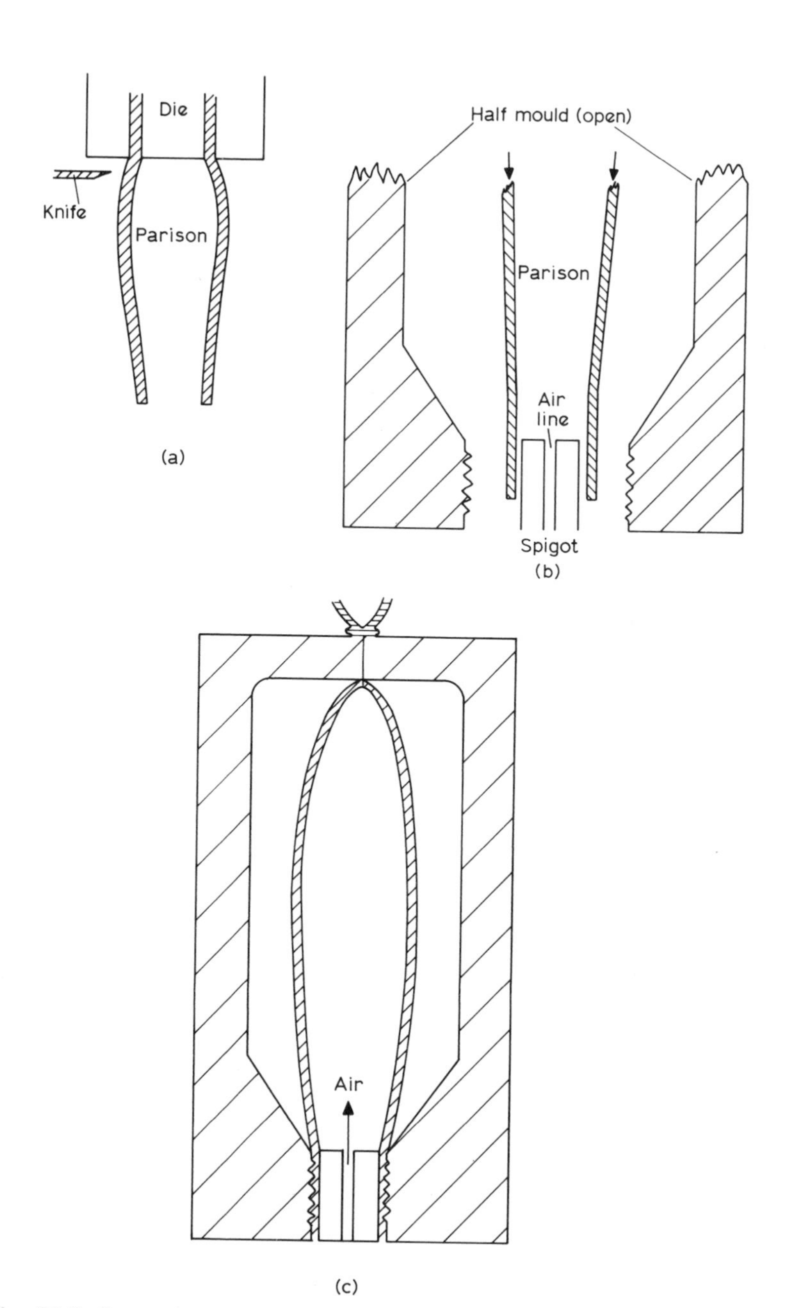

Fig. 20.2. Stages in blow moulding. (a) Extrusion of parison. (b) End of extruded parison moving over spigot between opened mould (halves). (c) Mould closed over parison and spigot with air inflating parison through channel in spigot.

(c) derivation of the relationship between the instantaneous strain and rate-of-strain fields for the material surface and the time-dependent metric in (b) (this includes the continuity relation);
(d) specification of the rheological equation of state for the surface flow, which involves three independent strain components in the surface;
(e) expression of the dynamical equations of motion in terms of the coordinates and metric employed in (b) and (c), and of the boundary conditions (constraints) on the flow field;
(f) step-by-step solution of the resulting sets of equations, for successive time intervals, starting from some initial configuration.

At any step in the calculation, the surface geometry is known. The dynamical equations and boundary conditions from (e) will yield the surface stress distribution. Knowing this, substitution in (d) yields an 'inverse' problem for the determination of the instantaneous strain-rates, which then, using (c), yields a further inverse problem for the determination of the time derivatives of the metric. Finally, the 'updated' metric yields from (b) the geometry (position) of the material surface for the next step in the calculation. The full calculation procedure, in general, will be considerably more difficult than, for example, the linearized analysis given in Subsection 17.4.2.

Such a general approach would necessarily involve the use of general tensor analysis, as described, for example, in Aris (1962, Chapter 10), which is beyond the scope of this account. Attention will be concentrated here on simple symmetric geometries that will allow the principles of mechanical analyses to be made clear.

20.1 THERMOFORMING

The simplest case to analyse is that of plane strain inflation of a sheet. Figure 20.3 defines the coordinate systems and variables involved when a long sheet of uniform thickness is inflated into a symmetric wedge-shaped trough.

The first part of the process involves the expansion of the strip through a continuous set of cylindrical shapes with cross-sections in the form of circular arcs, where the thickness of the sheet is a function of time only. The material is in contact with the mould only at the point E_1 where it is clamped. When the angle subtended at its axis by the

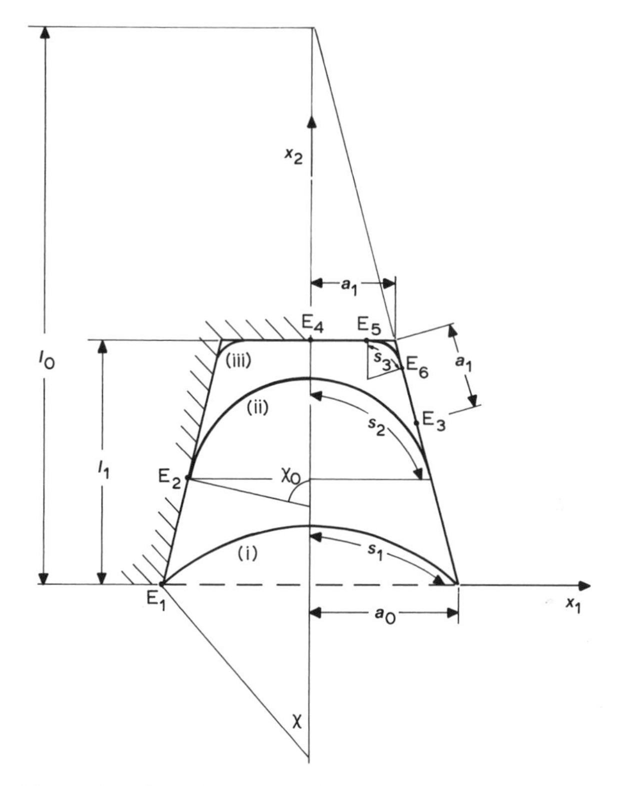

Fig. 20.3. Expansion of a strip of sheet held along lines through E_1 by inflation from below ($x_2 < 0$) into a mould in the form of a truncated wedge. – – –, initial position of strip. Typical positions at three successive stages are shown as (i), (ii) and (iii). (x_1, x_2) are fixed rectangular coordinates in the plane of strain. s is a surface coordinate in the (x_1, x_2)-plane such that $s = 0$ coincides with $x_1 = 0$. s_1, s_2 and s_3 are the half-lengths of 'free' material during the three phases of moulding.

circular arc, χ, reaches the value χ_0, the inflated strip becomes tangential at E_1 to the mould surface and further inflation causes the point of contact E_2 to move into the mould cavity. This second stage in the process still involves a circular-arc cross-sectional shape of uniform thickness for the part of the strip not in contact with the mould; different solutions will be obtained depending upon the extent to which the material lying along the mould surface between E_1 and E_2 can slip relative to the surface. The third stage begins when the inflating strip reaches E_4 on the plane of symmetry, i.e. when E_2 reaches E_3;

thereafter, the material is in contact with the mould along both E_1E_6 and E_4E_5. The free portion is still of uniform thickness.

The most interesting result of an analysis of this flow field is that the solution, as far as the thickness of the resulting moulding is concerned, is largely determined by kinematics and not by dynamics. This is wholly the case for either perfect slip or perfect adhesion at the mould walls.

The case for perfect slip is easily dealt with. At any stage there will be uniform tension in the s-direction; this is consistent with uniform stress and uniform thickness. The engineering strain e_1 in the s-direction will be simply

$$\text{(i)} \qquad \frac{s_1}{a_0} = \frac{\chi}{\sin\chi} \qquad 0 \leqslant \chi < \chi_0 \tag{20.1.1}$$

$$\text{(ii)} \qquad \frac{s_2 + E_1E_2}{a_0} = \frac{\chi_0}{\sin\chi_0} + \frac{x_2(1-\chi_0\cot\chi_0)}{a_0\sin\chi_0},$$

$$0 < x_2 \leqslant l_1 - a_1\sin\chi_0 \tag{20.1.2}$$

where x_2 refers to the position of E_2.

$$\text{(iii)} \qquad \frac{s_3 + E_1E_6 + E_4E_5}{a_0} = 1 + \frac{l_1}{a_0}\left(-\cot\frac{\chi_0}{2} + \frac{\chi_0}{2}\operatorname{cosec}^2\frac{\chi_0}{2}\right)$$

$$+\frac{x_2}{a_0}\operatorname{cosec}\frac{\chi_0}{2}\left(\sec\frac{\chi_0}{2} - \frac{\chi_0}{2}\operatorname{cosec}\frac{\chi_0}{2}\right),$$

$$l_1 - a_1\sin\chi_0 < x_2 < l_1 \tag{20.1.3}$$

where x_2 refers to the position of E_6.

At completion of the process the formed material is of uniform thickness h_1 where

$$h_1 = h_0a_0/(a_1 + l_1\operatorname{cosec}\chi_0) \tag{20.1.4}$$

The rate at which the moulding is formed will, of course, depend upon the rheology of the material and the inflation pressure.

The case of perfect adhesion leads to a variation in thickness over the mould surface. Corresponding to any instantaneous attachment point E_1, E_2 or E_6, there will be a characteristic strain $e_1(x_2)$, and a corresponding thickness

$$h(x_2) = h_0/e_1(x_2) \tag{20.1.5}$$

From (20.1.1), which still applies, we obtain

(i)
$$e_1(0) = \chi_0/\sin \chi_0 \tag{20.1.6}$$

In region (ii) we have

$$\frac{d}{dx_2}\left(\frac{s_2}{e_1}\right) = -\frac{1}{e_1 \sin \chi_0} \tag{20.1.7}$$

Writing $s_2 = (a_0 - x_2 \cot \chi_0)\chi_0 \operatorname{cosec} \chi_0$ and using (20.1.6), (20.1.7) can be solved to give

(ii)
$$e_1 = \frac{\chi_0}{\sin \chi_0}\left(1 - \frac{x_2}{a_0 \tan \chi_0}\right)^{1-\tan \chi_0/\chi_0} \tag{20.1.8}$$

At $x_2 = a_0 \tan \chi_0 - a_1(\tan \chi_0 + \sin \chi_0)$

$$e_1 = \frac{\chi_0}{\sin \chi_0}\left\{\frac{a_1}{a_0}(1 + \cos \chi_0)\right\}^{1-\tan \chi_0/\chi_0} \tag{20.1.9}$$

In region (iii) the analogue of (20.1.7) becomes

$$\frac{d}{dx_2}\left(\frac{s_3}{e_1}\right) = -\frac{2}{e_1 \sin \chi_0} \tag{20.1.10}$$

In this case we write

$$s_3 = \tfrac{1}{2}\chi_0 \operatorname{cosec}^2 \tfrac{1}{2}\chi_0(l_1 - x_2)$$

and use (20.1.9) to yield

$$e_1 = \frac{\chi_0}{\sin \chi_0}\left\{\frac{a_1}{a_0}(1 + \cos \chi_0)\right\}^{1-\tan \chi_0/\chi_0} \times \left\{\left(\frac{a_0}{a_1} - 1\right)\sec \chi_0 - \frac{x_2}{a_1}\operatorname{cosec} \chi_0\right\}^{1-\tan \frac{1}{2}\chi_0/\frac{1}{2}\chi_0} \tag{20.1.11}$$

The thinnest portion of the formed trough will be in the corner when E_5 meets E_6 when

$$h_{min} = 0 \tag{20.1.12}$$

This is obviously unacceptable physically and in practice will be avoided because a finite curvature at the corner will be imposed. Again, the rate at which the mould is filled will depend upon the rheology of the material.

If there is friction at the interface between the mould wall and the sheet expanded against it, then the rheology of the material will be

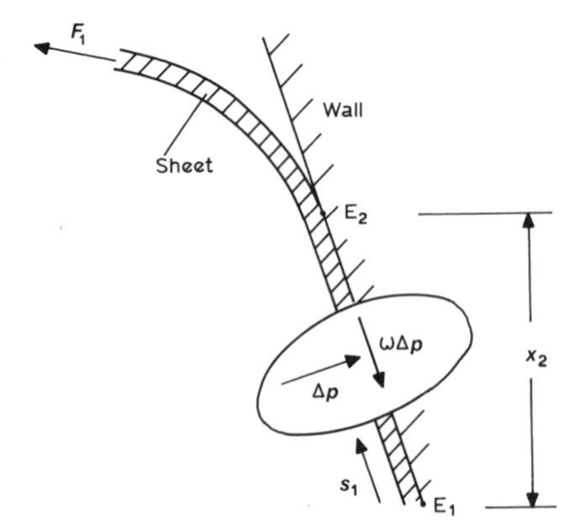

Fig. 20.4. Diagrammatic illustration of the frictional force acting between the mould (wall) and the sheet when the material between E_1 and E_2 is being elongated in the s_1 direction.

more strongly coupled to the kinematical problem. At any instant, the force F_1 along the sheet in the s_1 direction will be governed by the relations (see Fig. 20.4)

$$\left.\begin{aligned} \frac{dF_1}{ds_1} &= \omega\Delta p \qquad 0 \leqslant s_1 < x_2 \operatorname{cosec} \chi_0 \\ \frac{dF_1}{ds_1} &= 0 \qquad x_2 \operatorname{cosec} \chi_0 \leqslant s_1 \end{aligned}\right\} \tag{20.1.13}$$

where Δp is the inflation pressure and ω is the (not necessarily constant or uniform) coefficient of friction. This assumes that the force in the strip is extending the sheet in contact with the wall, and that the friction coefficient is that of dynamic friction. There may easily be a region

$$0 \leqslant s_1 < x_2^* \operatorname{cosec} \chi_0 < x_2 \operatorname{cosec} \chi_0$$

for which dF_1/ds_1 is given by the rheology of stress relaxation (i.e. no extension) and is less than the force provided by limiting friction. In either case the setting up of dynamical equations becomes much more complicated than the simple setting-up of the kinematic relations given above for $\omega = 0$ or $\rightarrow \infty$. Using order of magnitude arguments, it can be seen that $F_1 = 0(a_0\Delta p)$ and so if $\omega = 0(1)$, then the length scale over which friction might be relevant is $0(a_0)$.

EXAMPLE 20.1.1. Axisymmetric inflation of a disc-shaped sheet.

The simplest case (the analogue of the wedge-like trough) is a truncated cone (the shape of a drinking beaker, many of which are moulded rapidly in just such a fashion). The analysis of even the first stage of the process is much more elaborate than that given for the plane-strain case above, and necessarily involves the rheology of the sample. One of the earliest papers on the subject (Williams, 1970) used a purely elastic model of considerable simplicity, which will be adopted here. The development given here differs a little from his. Figure 20.5 defines the coordinates and variables to be used. $\hat{t}_1$ and $\hat{t}_2$ are the principal tensions, as used in (17.2.9) and (17.2.12) for the analysis of film blowing. Indeed the same mechanical equations hold. For stage (i), $F \equiv 0$ in (17.2.9) and so

$$2\hat{t}_1 \cos \psi - a\Delta p = 0 \tag{20.1.14}$$

while neglecting the gravitational terms in (17.2.12) yields

$$\Delta p = \hat{t}_2 \cos \psi a^{-1} + \hat{t}_1 \cos \psi \, \mathrm{d}\psi/\mathrm{d}z \tag{20.1.15}$$

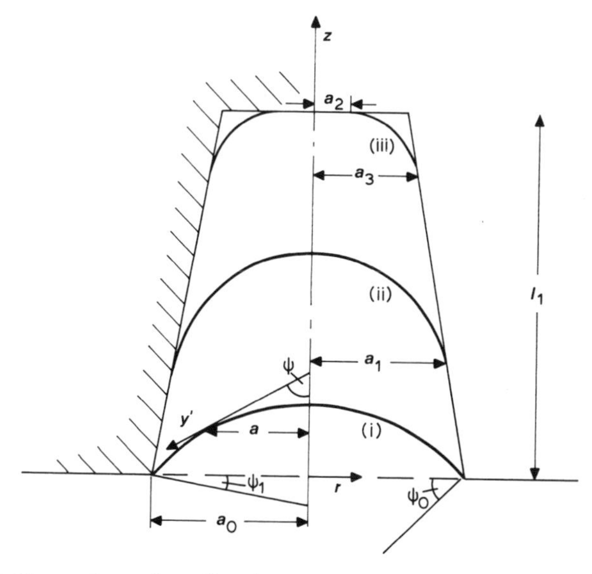

Fig. 20.5. Expansion of a disc-shaped sheet held at $z=0$, $r=a_0$ into a beaker-shaped mould, in the form of a truncated cone. – – –, initial position of disc; ——, subsequent positions of disc, at successive stages (i), (ii) and (iii). (z, r) are cylindrical polar coordinates, $2\psi_1$ is the cone angle.

Williams' (1970) elastic model yields the rheological relations

$$\hat{t}_1 = F_0 e_1/e_2, \qquad \hat{t}_2 = F_0 e_2/e_1 \qquad (20.1.16)$$

where e_1 and e_2 are the engineering strains in the y^1 and $y^2(\phi)$ directions, assuming that h_0 is independent of initial radial position r. However, the actual values e_1 and e_2 disappear when the product

$$\hat{t}_1 \hat{t}_2 = F_0^2 \qquad (20.1.17)$$

is considered. Using (20.1.14) and (20.1.15) to give $\hat{t}_1$ and $\hat{t}_2$ in terms of Δp and a, noting that

$$\frac{\mathrm{d}\psi}{\mathrm{d}z} = \frac{\mathrm{d}\psi}{\mathrm{d}a}\frac{\mathrm{d}a}{\mathrm{d}z} = -\tan\psi \frac{\mathrm{d}\psi}{\mathrm{d}a} \qquad (20.1.18)$$

yields a single relation for ψ in terms of a:

$$\left(\frac{2F_0}{\Delta p}\right)^2 2a = \frac{\mathrm{d}}{\mathrm{d}a}(a^4 \sec^2 \psi) \qquad (20.1.19)$$

Using as a necessary boundary condition

$$\psi(a=0) = \tfrac{1}{2}\pi \qquad (20.1.20)$$

and requiring $\hat{t}_1$ to be bounded at $a=0$ gives an integral

$$\cos\psi = a\Delta p/2F_0 \equiv a/f \qquad (20.1.21)$$

which represents a spherical cap. It should be noted at this point that this is an essentially fortuitous result coming from the rheology chosen. Although the stress field

$$\hat{t}_1 = \hat{t}_2 = F_0 \qquad (20.1.22)$$

is uniform and surprisingly independent of Δp, and thus of the radius of curvature, the actual strains are not constant. Around the line of attachment $e_1 = e_2 = 1$ and there is no deformation, but a jump in stress from the unstressed, undeformed state. Elsewhere

$$\frac{\mathrm{d}a}{\mathrm{d}r}\operatorname{cosec}\psi = e_1 = e_2 = \frac{a}{r} \qquad (20.1.23)$$

where r is the initial radius of the material at a. The boundary condition is

$$a(a_0) = a_0$$

Integration of (20.1.23) using (20.1.21) gives

$$r\{f+(f^2-a^2)^{\frac{1}{2}}\}=f+(f^2-a_0^2)^{\frac{1}{2}}a \tag{20.1.24}$$

whence

$$e_1=\frac{1+(1-\tilde{r}^2\cos^2\psi_0)^{\frac{1}{2}}}{1+\sin\psi_0} \tag{20.1.25}$$

where

$$\tilde{r}=r/a_0, \qquad \psi_0=\psi(a_0)\leqslant\psi_1 \tag{20.1.26}$$

For stage (ii) which begins when $\psi_0=\psi_1$, $F\equiv 0$ still, and the only change from the earlier results arises because Δp increases as contact arises for $z>0$, i.e. when $a=a_1<a_0$, according to

$$f_1=a_1\sec\psi_1 \tag{20.1.27}$$

where $a_1=a_0-z\tan\psi_1$. Simple continuity shows that, if there is no slip,

$$\frac{\mathrm{d}r_1}{\mathrm{d}a_1}=\frac{r_1}{a_1\sin\psi_1}, \quad \text{where} \quad r_1=r(a_1) \tag{20.1.28}$$

which may be integrated to yield

$$\frac{a_1}{a_0}=\left(\frac{r_1}{a_0}\right)^{\sin\psi_1} \tag{20.1.29}$$

The thickness of the beaker sides will then be

$$\frac{h}{h_0}=\left(\frac{r_1}{a_1}\right)^2=\left(\frac{a_1}{a_0}\right)^{2(\operatorname{cosec}\psi_1-1)} \tag{20.1.30}$$

until stage (iii) is reached at

$$\left.\begin{aligned} z&=l_1(1+\sin\psi_1)-a_0\cos\psi_1 \\ a_{3\,\max}=a_{1\,\min}&=a_0(1+\sin\psi_1)-l_1(1+\sin\psi_1)\tan\psi_1 \end{aligned}\right\} \tag{20.1.31}$$

At that stage, the value of e_1 for $r_2=a_2=0$ is given by

$$\left(\frac{a_1}{r_1}\right)\frac{2}{1+\sin\psi_0}$$

which can be derived directly from an analogue of (20.1.25).

During stage (iii), relation (20.1.14) becomes

$$2\hat{t}_1 \cos \psi = \Delta p(a - a_2^2/a) \tag{20.1.32}$$

while (20.1.15) is retained. On integration as before, using as boundary conditions

$$\psi(a_2) = \tfrac{1}{2}\pi, \qquad \hat{t}_1(a_2) \quad \text{bounded} \tag{20.1.33}$$

a solution

$$f_3^2(a^2 - a_2^2) + f_2 = (a^2 - a_2^2)^2 \sec^2 \psi \tag{20.1.34}$$

is obtained. Here a_2 is the radius of the base covered by the expanded sheet, but of course f_2 and f_3 are as yet unspecified. $\psi(a_3) = \psi_1$ provides one relation between them, i.e.

$$\cos^2 \psi_1 = \frac{(a_3^2 - a_2^2)^2}{f_3^2(a_3^2 - a_2^2) + f_2^4} \tag{20.1.35}$$

However, a_3 and a_2 are not obviously related in the way that E_5 and E_6 were in the previous example (see Fig. 20.3).

During stage (iii) e_1 and e_2 are no longer equal, as can be seen from (20.1.32) which leads to $\hat{t}_1 \neq \hat{t}_2$ when substituted into (20.1.15), using (20.1.34). Further progress can only be made if r can be found as a function of a for $a_2 \leqslant a \leqslant a_3$ by integration. There is no analytic form for this, and so numerical treatment would be required. Continuity imposes as boundary conditions

$$\left.\begin{aligned} \mathrm{d}r(a_2) &= \mathrm{d}a_2/e_1(a_2) \\ \mathrm{d}r(a_3) &= \mathrm{d}a_3/e_1(a_3)\sin \chi_1 \end{aligned}\right\} \tag{20.1.36}$$

Simultaneous satisfaction of (20.1.36) effectively provides the relevant additional equation for f_2 and f_3.

EXERCISE 20.1.1. Show that if, in Example 20.1.1, $\psi_1 = 0$, i.e. the beaker is cylindrical, then the thickness during stage (ii) varies as e^{-2z/a_0} instead of as given by (20.1.30).

20.2 BLOW MOULDING

The parison shown in Fig. 20.2 was deliberately drawn to be different from a perfectly uniform circular-cylindrical tube, though in practice many parisons approximate to that simple geometry. Asymmetries can arise for three reasons:

(a) Deliberate choice of a non-axisymmetric die giving variations in parison thickness at the die-lips for any fixed throughput and temperature.
(b) Deliberate variations in output with time, leading to variations in parison thickness and radius (when account is taken of extrudate swell) near the die-lips. This is equivalent to variations of thickness and radius with distance from the die-lips at any given stage in the extrusion of the parison (see Fig. 20.2(a)).
(c) Variations in thickness and radius of the parison due to draw-down caused by the weight of the part of the parison already extruded.

These aspects will be examined in the next subsection.

When the mould closes over the parison it is brought together along the mould separation line at the base, yielding a very complex asymmetry there, and is clamped about the spigot to form the top of the container neck. This latter is a form of compression moulding, which will not be further discussed. In many cases, the mould cross-section is not circular, and so inflation of the parison will seldom lead to simultaneous arrival of the extending tube at all points of the mould inner surface. Nor is the moulding exactly cylindrical: the cross-section will vary along its length. In general, therefore, a full description of the blow-moulding inflation process would involve material surfaces with both curvatures varying continuously over the surface. As explained earlier, this is beyond the scope of our analytical procedures. The most that we can hope to do is to consider idealized symmetrical systems that will provide upper and lower bounds to the thicknesses to be expected in the blown articles, and to provide estimates of the strains and strain rates achieved just before chilling at the mould surface, these providing indications of the molecular orientation (or frozen-in strain) in the blown containers. An alternative approach is provided in Dutta & Ryan (1982).

20.2.1 Parison Extrusion

The simplest process involves continuous extrusion of the parison using a single-screw extruder and a cross-head annular die of the type described in Section 10.3 and analysed as a slot die in Section 10.2. The extruder will be run at constant speed and so the output is varied

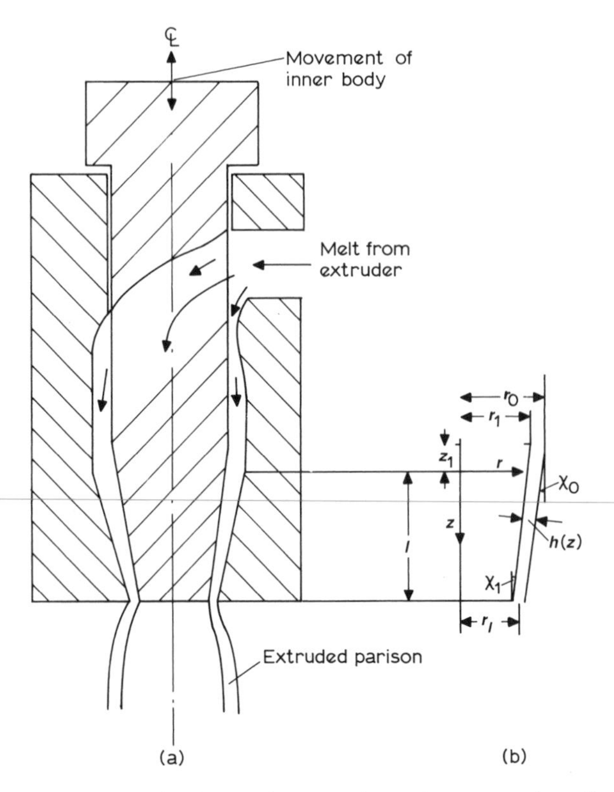

Fig. 20.6. Diagram of simplest axisymmetric parison extrusion die allowing adjustable die-lip geometry. (a) Overall sectional view through axis. (b) Coordinate system and parameters for die channel geometry.

by altering the die pressure; this is achieved by varying the geometry of the flow channels, often by increasing or decreasing the gap at the die-lips by the movement of the inner die body. This is shown in Fig. 20.6 for the simplest possible axisymmetric arrangement. A cylindrical polar coordinate system (r, z) is shown based on the axis of symmetry of the inner movable mandrel so that $z=0$ coincides with the beginning of the conical convergence of half angle χ_0 in the outer die body. The outer wall of the variable narrow channel leading up to the die-lips is given by

$$r = r_0 - \tan \chi_0 z \qquad 0 \leqslant z \leqslant l \qquad (20.2.1)$$

The almost matching truncated cone on the end of the movable mandrel has a slightly smaller half-angle χ_1, and the position of the inner wall of the narrow channel is given by

$$r = r_1 - \tan \chi_1(z + z_1) \tag{20.2.2}$$

over a range of z that includes $[0, l]$ provided z_1 is within the designed range. The channel gap can be written

$$h \simeq [(r_0 - r_1) - z(\tan \chi_0 - \tan \chi_1) + \tan \chi_1 z_1]\cos \tfrac{1}{2}(\chi_0 + \chi_1) \tag{20.2.3}$$

which decreases as z increases, but also decreases as z_1 decreases.

The relationship between pressure drop and flowrate can be obtained by the methods described in Section 10.2. Even for a power-law fluid an analytic solution in terms of K, ν, r_0, r_1, l, χ_0, χ_1 and z, is not possible (see Example 20.2.1).

The radius r_p and thickness h_p of the extruded parison will be determined primarily by

$$\begin{aligned} r_l &= \tfrac{1}{2}\{(r_1 + r_0) - \tan \chi_1 z_1 - (\tan \chi_0 + \tan \chi_1)l\} \\ h_l &= \{(r_0 - r_1) + \tan \chi_1 z_1 - (\tan \chi_0 - \tan \chi_1)l\} \end{aligned} \tag{20.2.4}$$

and

$$q_l = Q/2\pi r_l$$

where Q is the total volume flowrate; however r_p and h_p will not be the same as r_l and h_l because of extrudate swell. This phenomenon is discussed in Section 7.4 for a tubular (rod-like) extrudate. The annular situation was considered in Pearson & Trottnow (1978) who defined 2 extrudate swell parameters

$$B_r = r_p/r_l, \qquad B_h = h_p/h_l \tag{20.2.5}$$

Here r_p and h_p are the relaxed parison radius and thickness respectively. In that case, $\chi_1 = \chi_0 = 0$ and $h(z)$ was constant with $h \ll r_0$. Their predictions were that B_r and B_h would depend significantly on q_l and l for highly elastic fluids (being significantly greater than unity) but it is clear that they would also depend in the present case on $\chi_0 + \chi_1$, $\chi_0 - \chi_1$ and z_1. Such work as has been done experimentally confirms this, but no satisfactory theories have been advanced.

An important aspect of parison extrusion is that it is a cyclic operation. The die-flow and extrudate-swell analysis discussed briefly above relates to steady flow. However, variation of z_1 with time, and periodic severing of the parison (at intervals t_s say), means that time scales not considered in steady extrusion become relevant. In particular, the time t_r for re-establishment of steady flow after a step change in extrusion conditions may be of the order of, or much larger than, t_s. One reason for this will probably be that the timescale for establishment of thermal equilibrium in the die channel flow, $t_i = h^{*2}/4\kappa$, will be at least of order t_s. Flow rate calculations for Q based on equilibrium observations may, therefore, be substantially in error. Similarly steady-state correlations for B_r and B_h may prove to be inapplicable, particularly when the relaxation time λ of the melt is of order t_s.

Lastly, the shape of the parison when it is finally gripped by the two halves of the mould as they close, may be affected by extension under its own weight. Clearly the bottom tip will be unstressed, while the top end will only just have been extruded and will not have had the time to suffer viscous extension. Analysis of the situation is best carried out in a Lagrangean frame of reference (relating to material particles) rather than in an Eulerian frame of reference (relating to fixed points in space). If we take $t=0$ to be the time at which the parison is severed at the die lips, then t_s can be regarded as the time at which the parison is caught by the mould halves (and resevered). We refer to individual elements of the parison in terms of the time s at which they left the die-lips. The total extensional force on the s cross-section is clearly fixed and equal to the weight extruded between $t=0$ and $t=s$. The relevant rheology of the material is, therefore, that measured by a creep test in tension.† It is important to note that an unpressured hollow tube behaves like a rod in such an experiment: the radius $r(t, s)$ will always scale linearly with the thickness $h(t, s)$ and as the inverse square-root of the local length $l(t, s)$. Thus

$$e_z(t, s) = \frac{l(t, s)}{l(s, s)} = e_h^{-2} = \left[\frac{h(s, s)}{h(t, s)}\right]^2 = e_r^{-2} = \left[\frac{r(s, s)}{r(t, s)}\right]^2 \quad (20.2.6)$$

where e_z, e_h and e_r are the effective principal strains for an incompressible material. The creep test yields

$$e_z = t_{zz}(s) J_e(t, s; t_{zz}(s)) \quad (20.2.7)$$

where J_e is the extensional creep compliance (cf. J in (5.1.16)). J_e can be calculated for a specific rheological model. Relations (20.2.6) and (20.2.7) can only be strictly applied to a parison of variable thickness

† Only true if $e_z - 1 \leqslant 1$; otherwise there is a difference between the constant force and the constant stress experiment.

$h(z, t)$ and radius $r(z, t)$ when

$$h \ll r, \qquad \partial r/\partial z \ll 1 \tag{20.2.8}$$

(note that here an Eulerian frame has been used). Suppose that

$$v_0(s) = v_z(s, s), \qquad h_0(s) = h(s, s), \qquad r_0(s) = r(s, s) \tag{20.2.9}$$

are determined arbitrarily by the extrusion process. Then the extensional stress at the die-lips

$$t_{zz}(s, s) = \rho g \int_0^s v_0(u) r_0(u) h_0(u)\, \mathrm{d}u / r_0(s) h_0(s) \tag{20.2.10}$$

from which $e_z(t, s)$, $h(t, s)$ and $r(t, s)$ can be obtained using (20.2.6) and (20.2.7). The relationship between this Lagrangean (s) frame and the more usual Eulerian (z) frame is obtained explicitly by the transformation

$$z(t, s) = \int_s^t v_0(u) e_z(t, u)\, \mathrm{d}u \tag{20.2.11}$$

The inverse problem consists in specifying t_s, $h(z, t_s)$, $r(z, t_s)$ and requiring $v_0(s)$, $h_0(s)$ and $r_0(s)$ to be chosen to yield the required result. This is what is intended and attempted by the use of a variable-geometry die mentioned in the earlier part of this section. In practice, die design and mandrel movement are determined largely by trial and error. Any analytic progress would depend upon numerical techniques. Parison extrusion can also be achieved by ram ejection from a reservoir, as in injection moulding. In this case, the rate of output

$$Q \propto v_0 h_0 r_0, \quad \text{or equivalently } q_l,$$

can be varied independently of h_l and r_l in (20.2.4). In such cases, t_s is drastically reduced and so gravity effects may be made negligible.

If the blow-moulded article is not intended to have an axis of symmetry, e.g. its cross-section may be chosen to be almost square, then a final uniform wall thickness will not be consistent with a uniform parison wall thickness. The need then arises to achieve a variable $h_0(\phi)$ at the die-lip exit in the still circular-cylindrical ($r_0(\phi) =$ const) parison. The important point is that to avoid puckering of the parison, the mean extrudate velocity $v_0(\phi)$ shall not vary with ϕ, even though $h_0(\phi)$ does. A way of achieving this is illustrated very simply in Fig. 20.7, which is analogous to the unrolled (or slot-die) geometry

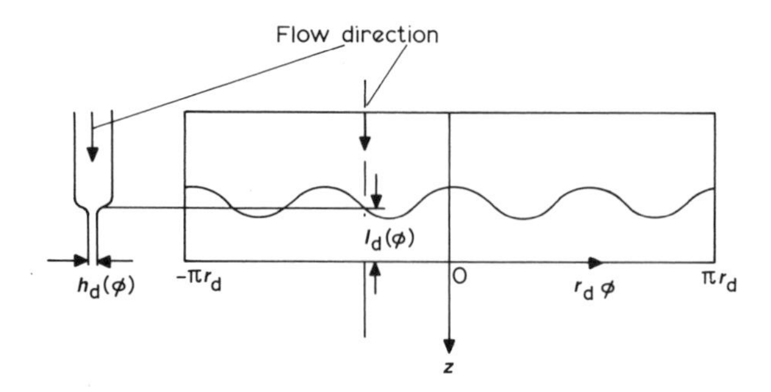

Fig. 20.7. Diagram of variable land length $l_d(\phi)$ and land depth $h_d(\phi)$ in parison die.

illustrated in Fig. 10.4. The main pressure drop is assumed to arise in the narrow part of the die channel, whose depth h_d and length l_d are functions of ϕ only. If we suppose that

$$h_d \ll l_d \ll \pi r_d \tag{20.2.12}$$

where r_d is the radius of the die-lips, then flow can be assumed to be in the z-direction only for all $z > -l(\phi)$. Assuming a power-law model as in (10.2.18), and writing

$$\Delta\tilde{p}_d = \frac{\Delta p_d(\phi)}{\Delta p_d(0)}, \quad \tilde{q}_d = \frac{q_d(\phi)}{\tilde{q}_d(0)}, \quad \tilde{h}_d = \frac{h_d(\phi)}{h_d(0)}, \quad \tilde{l}_d = \frac{l_d(\phi)}{l_d(0)} \tag{20.2.13}$$

where Δp_d is the pressure drop between $z = -l(\phi)$ and $z = 0$, and q_d is the flowrate/unit width, it is easily seen that

$$\Delta\tilde{p}_d = \tilde{q}^{\nu}\tilde{l}/\tilde{h}^{2\nu+1} \tag{20.2.14}$$

If we write

$$\tilde{u}_d = \tilde{q}_d/\tilde{h}_d \tag{20.2.15}$$

and require both $\tilde{u}_d$ and $\Delta\tilde{p}$ to be independent of ϕ, then

$$\tilde{l} = \tilde{h}^{\nu+1} \tag{20.2.16}$$

The design criterion (20.2.16) is modified if the die-swell parameters B_h and B_r, defined in (20.2.5), vary with $(\tilde{q}_d, \tilde{h}_d)$, for then it is $\tilde{u}_d/B_hB_r$ which is to be held constant and so

$$\tilde{l} = \tilde{h}^{\nu+1}/B_hB_r \tag{20.2.16*}$$

The meaning of B_r, when B_r is a function of ϕ, needs a little explanation. For $h_d \ll r_d$, it refers to lateral swell (in the ϕ direction) irrespective of the actual value of r_d. The parison could still be circular with an apparent $\bar{B}_r$ equal to the mean of B_r around the circumference ($-\pi < \phi \leq \pi$). In such circumstances, ϕ_d would correspond to a slightly different ϕ_p after extrusion swelling, but such refinements need not be pursued in detail here. For other aspects of parison control, see Prichatt *et al.* (1975).

20.2.2 Parison Blowing

The simplest situation to be considered here is that of cylindrical, but non-axisymmetric, blowing. This would arise in the central portion of a long uniform parison, i.e. where $h(t_s, z, \phi)$,† as defined in Subsection 20.2.1, depends upon ϕ alone, and can be written as $h_0(\phi)$, with $r(t_s, z, \phi)$ a constant, r_0 say. The subsequent stages of inflation would be represented by $h(t, \phi)$ and $r(t, \phi)$, $t > t_s$. A simple example is shown in Fig. 20.8. As in parison extension or thermoforming, a Lagrangean frame of reference is convenient. The initial ($t = t_s$) position of a material element will be represented by ϕ_0. As subsequent times, it is convenient to regard h, ϕ, and the circumferential stress as functions of t and ϕ_0.

In the example shown in Fig. 20.8, there is a first phase during which the parison remains circular and $r_0 < r = r(t) < r_1$; the circumferential tension $\hat{t}_2$ in the expanding parison is independent of ϕ and is given by

$$\hat{t}_2 = h(t, \phi_0)t_{\phi\phi} = \Delta p(t)r(t) \tag{20.2.17}$$

where $\Delta p(t)$ is the inflation overpressure. h will be related to $t_{\phi\phi}$ by the rheological equation of state for plane flow in an extending sheet. Continuity requires that

$$r_0 \int_0^{\phi_0} \frac{h_0(\chi)}{h(t, \chi)} \mathrm{d}\chi = r(t)\phi(t, \phi_0) \tag{20.2.18}$$

with

$$\phi\left(t, \frac{\pi}{4}\right) = \frac{\pi}{4} \quad \text{all} \quad t \tag{20.2.19}$$

† This presupposes an Eulerian (t, z, ϕ) frame of reference and coordinate system instead of the Lagrangean (t, s, ϕ) frame of reference and coordinate system, the connection being given by (20.2.11).

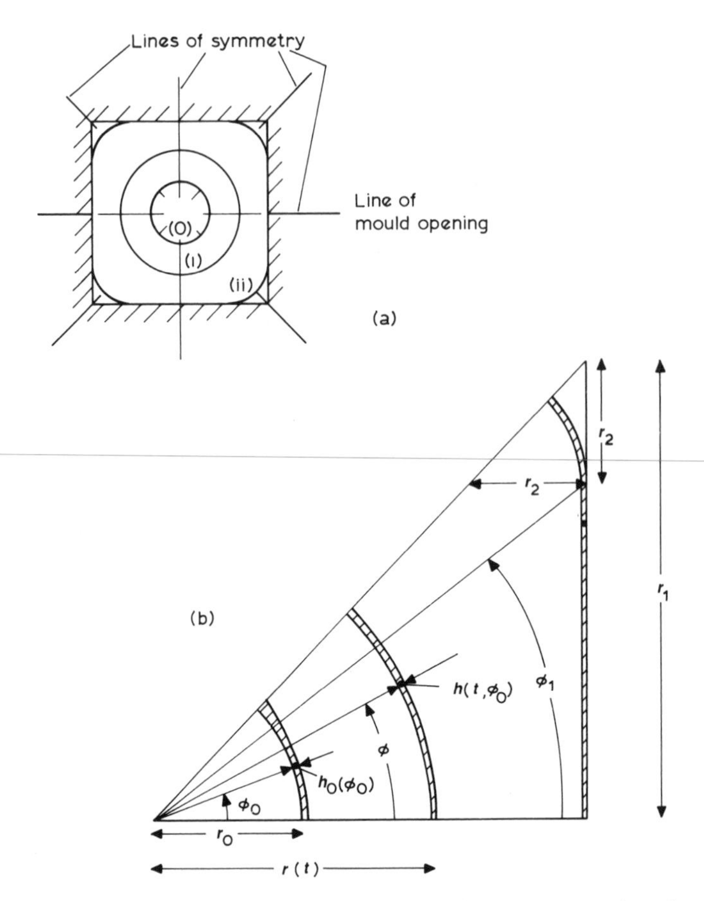

Fig. 20.8. Diagram of blow moulding operation into square-sectioned mould. (a) Entire cross-section for cylindrical region of plane inflation, showing successive portions of parison: (O) initial circular shape, $t = 0$; (i) expanding circular shape in first phase for $0 < t \leq t_1$; (ii) expanding set of circular arcs in second phase for $t > t_1$. (b) Detailed geometry of one octant. r_1, half-width of mould; r_0, initial radius; $r(t)$, radius during first phase; $r_2(t)$, radius during second phase. The angles ϕ_0, ϕ refer to initial and first phase positions of ●, an element of the parison having initial thickness $h_0(\phi_0)$ and subsequent thickness $h(t, \phi_0)$. The angle $\phi_1(t)$ refers to the instantaneous limiting position of contact between the parison and the mould wall.

Equations (20.2.17)–(20.2.27), together with the rheological equations of state and suitable initial conditions on the strain or stress, provide a means for calculating $r(t)$, $h(t, \phi_0)$ and $\phi(t, \phi_0)$.

If a Newtonian fluid model is used, then

$$t_{\phi\phi}e_\phi = 4\eta_0 \frac{\partial e_\phi}{\partial t} \tag{20.2.20}$$

where

$$e_\phi = h_0(\phi_0)/h(t, \phi_0) \tag{20.2.21}$$

From (20.2.21) it follows that

$$\frac{\Delta p(t) r_0}{\eta_0 \pi} \int_0^{\frac{1}{4}\pi} \frac{e_\phi(t, \chi)\, \mathrm{d}\chi}{h_0(\phi_0)} = \frac{1}{e_\phi^2} \frac{\partial e_\phi}{\partial t} \tag{20.2.22}$$

If we use, as a scaled dimensionless time, $\tilde{t}$ given by

$$\mathrm{d}\tilde{t} = \frac{\Delta p(t) r_0}{\pi \eta_0 h_0(0)} \int_0^{\pi/4} e\phi(t, \chi)\, \mathrm{d}\chi\, \mathrm{d}t, \qquad \tilde{t}(0) = 0 \tag{20.2.23}$$

and write

$$\tilde{h}_0(\phi) = h_0(\phi)/h_0(0) \tag{20.2.24}$$

then (20.2.20)–(20.2.22) yield

$$\frac{\partial}{\partial \tilde{t}}\left(\frac{1}{e_\phi}\right) = -[\tilde{h}_0(\phi)]^{-1} \tag{20.2.25}$$

where e_ϕ is now treated as a function of ϕ_0 and $\tilde{t}$, with solution, if $e_\phi(0, \phi_0) = 1$,

$$e_\phi = [1 - \tilde{h}_0^{-1}(\phi_0)\tilde{t}]^{-1} \tag{20.2.26}$$

Hence, when $r(t) = r_1$, i.e. at t_1 say, $h(t_1)$ is known as a function either of ϕ_0 or ϕ.

A second phase then takes place, whereby the parison is in contact with the mould surface for all $\phi < \phi_1 < \pi/4$ (in the octant $0 \leqslant \phi \leqslant \pi/4$). It will be assumed for simplicity that the parison adheres to the mould on touching it. Because of symmetry, the portion of the parison that is still inflating will have a cross-section that is an arc of a circle with radius

$$r_2 = r_1(1 - \tan \phi_1) \tag{20.2.27}$$

The point of contact ($r = r_1 \sec \phi_1$, $\phi = \phi_1$) will coincide with some specific (t, ϕ_0) such that there is a unique relation between (t, ϕ_0) and ϕ_1. Continuity requires that

$$dr_2 = -r_1 \sec^2 \phi_1 \, d\phi_1 = -r_0 e_\phi(t, \phi_0) \, d\phi_0 \qquad (20.2.28)$$

Equation (20.2.17) still holds provided $r(t)$ is interpreted as r_2 for $t > t_1$. Equation (20.2.18) is replaced by

$$\tfrac{1}{4}\pi r_2 = r_0 \int_{\phi_0(\phi_1)}^{\frac{1}{4}\pi} e_\phi(\tilde{t}, \phi_0) \, d\phi_0 \qquad (20.2.29)$$

where, in the second phase, (20.3.23) is replaced by

$$d\tilde{t} = \frac{\Delta p(t) r_2}{\pi \eta_0 h_0(0)} \, dt \qquad (20.2.30)$$

Equation (20.2.26) is thus retained. Hence the entire process can be followed, even if only by numerical solution of the implicit equations involved.

EXAMPLE 20.2.1. Use the results of Subsection 10.2.1 to calculate the pressure drop in the converging part of the channel $0 < z < l$ in Fig. 20.6 (Subsection 20.2.1). The radius and channel width are given by (20.2.2) and (20.2.3). Lubrication theory can be employed. The relevant relation is (10.2.18) where account is taken of the fact that $\mathbf{q}$ in (10.2.18) is inversely proportional to radius.

If Q is the total flowrate then

$$q(z) = q_l r_l / r = q_l r_l / \{r_1 - \tan \chi_1 (z + z_1)\}$$

and

$$\frac{\partial p}{\partial z} = -\frac{2K[2(2 + 1/\nu)]^\nu q_l^\nu r_l^\nu}{[(r_0 - r_1) - z(\tan \chi_0 - \tan \chi_1) + \tan \chi_1 z_1]^{2\nu+1} [r_1 - \tan \chi_1 (z + z_1)]^\nu}$$

This can be written as

$$\frac{\partial p}{\partial z} = -\frac{AQ^\nu}{(B - z)^{2\nu+1} (C - z)^\nu}$$

where A, B and C are parameters involving χ_0, χ_1, z_1, r_1 and r_0. Integrating formally

$$\Delta P = AQ^\nu \int_0^l \frac{dz}{(B - z)^{2\nu+1} (C - z)^\nu}$$

which can be expressed in terms of tabulated functions.

EXAMPLE 20.2.2. Calculation of J_e as defined in (20.2.7) for a Jeffreys fluid model. This is treated in Petrie (1979, Section 5.1).

We use the Jeffreys model given by (3.3.25) and (3.3.24) with $b=c=0$. We note at once that the creep (or constant force) elongational flow implied in Subsection 20.2.1 is irrotational and uniform (locally), so $\mathbf{W}=0$ and $\mathring{\mathbf{J}} \equiv \frac{d\mathbf{J}}{dt} \equiv \dot{\mathbf{J}}$ in (3.3.23).

We can thus write the effective rheological equation of state as

$$\mathbf{T}^E + \lambda_1(\dot{\mathbf{T}}^E - 2a\mathbf{D}\,.\,\mathbf{T}^E) = 2\lambda_1 G\{\mathbf{D} + \lambda_2(\dot{\mathbf{D}} - 2a\mathbf{D}^2)\}$$

where $\mathbf{D}$ is diagonal, whose (r, ϕ, z) components are given by

$$\left(-\frac{1}{2}\frac{\dot{e}_z}{e_z}, -\frac{1}{2}\frac{\dot{e}_z}{e_z}, \frac{\dot{e}_z}{e_z}\right)$$

using (20.2.6) to give e_r in terms of e_z. For the creep situation $\dot{\mathbf{T}}=0$ for $t>s$ and $\mathbf{T}$ has only one non-zero component, $t_{zz}=t_{zz}^E - t_{rr}^E$, t_{rr} being zero.

On substitution into the rheological equation of state, a second-order differential equation for $\dot{\gamma}=\dot{e}_z/e_z$ is obtained:

$$(1 - a\lambda_1\dot{\gamma} - 2a^2\lambda_1^2\dot{\gamma}^2 - \lambda_1\ddot{\gamma}/\dot{\gamma})t_{zz} = 3\lambda_1 G\{\dot{\gamma} + \lambda_2(\ddot{\gamma} - a\dot{\gamma}^2)$$
$$+ \lambda_1\lambda_2(\dddot{\gamma} - \ddot{\gamma}^2/\dot{\gamma} - 2a^2\dot{\gamma}^3 - a\ddot{\gamma}\dot{\gamma})\}$$

For constant stress, $t_{zz}=t_{zz}(s)$ is constant; for constant force

$$t_{zz}(t>s) = t_{zz}(s)e_z(t); \qquad e_z(s) = 1$$

Two initial conditions must be prescribed at $t=s$ to get a unique solution. Assuming a discontinuity in t_{zz} at $t=s$ such that $t_{zz}(t<s)=0$, the rheological equation of state yields a jump in $\dot{\gamma}$, namely

$$\dot{\gamma}(s) - \dot{\gamma}(s-0) = \dot{\gamma}(s) = t_{zz}(s)/3\lambda_2 G$$

which provides one initial condition. At the next level of approximation, i.e. for very small but not negligible extensions $e_z - 1$, a linear form of the rheological equation of state will apply, i.e.

$$\lambda_1\dot{t}_{zz}(s+) + t_{zz}(s+) = 3\lambda_1 G\{\dot{\gamma}(s+) + \lambda_2\ddot{\gamma}(s+)\}$$

or, using the first initial boundary condition with $\dot{t}_{zz}=0$,

$$\ddot{\gamma}(s) = \frac{t_{zz}(s)}{3\lambda_2^2\lambda_1}(\lambda_2 - \lambda_1)$$

in the limit as $s+$ coincides with s. This is the second required condition.

There is no general analytical solution to the second-order equation for $\dot{\gamma}$ but a numerical solution can be obtained, from which e_z can be obtained by simple integration.

J_e follows from the definition (20.2.7).

EXAMPLE 20.2.3. Parison extension for a Newtonian fluid model

A simple problem that can be solved is provided by $v_0(s)$, $h_0(s)$, $r_0(s)$ all being constant, and J_e given by a Newtonian viscous model of viscosity η_0. Equation (20.2.10) then yields, for the extensional stress at time t in material that was extruded at time s,

$$t_{zz}(t, s) = \rho g v_0 s e_z(t, s)$$

while the equation of motion yielding e_z is

$$t_{zz}(t, s)e_z = 3\eta_0\, \partial e_z/\partial t, \qquad e_z(s, s) = 1$$

This has the solution for

$$e_z = \left[1 - \frac{\rho g v_0}{3\eta_0}(st - s^2)\right]^{-1} \quad 0 < s < t \; = \left[\frac{h_0}{h(t, s)}\right]^2 = \left[\frac{r_0}{r(t, s)}\right]^2$$

with

$$z(s, t) = \tfrac{1}{2} v_0 t^{*2}(t^{*2} - t)^{-\frac{1}{2}}\left\{\tan^{-1}\frac{(t-2s)}{(t^{*2} - t^2)^{\frac{1}{2}}} + \tan^{-1}\frac{t}{(t^{*2} - t^2)^{\frac{1}{2}}}\right\}$$

where

$$t^* = (12\eta_0/\rho g v_0)^{\frac{1}{2}}$$

The dimensionless group determining the amount of draw-down is $(\rho g v_0 t_s^2/\eta_0)$ which is related to the group Re/Fr defined by (15.3.14), (15.3.15) in Chapter 15 on fibre spinning; t_s has to be interpreted as the flight time $(\sim l/v_{z1})$ in the latter.

A related problem involving a BKZ fluid model is treated in Wineman (1979).

EXERCISE 20.2.1. Show that the analytic solution is possible for $\nu = 1, \frac{1}{2}, \frac{1}{3}$ in Example 20.2.1.

EXERCISE 20.2.2. Show that an (unpressurized) parison that does not buckle circumferentially behaves like a rod provided $(dr_p/dz)_t$,

$(\partial h_p/\partial z)_t \ll 1$. To prove this, consider both inner and outer boundary conditions on t_{rr}. Compare the result with that given in Exercise 17.3.2.

Hence verify the results (20.2.6) where l is the arbitrary length of a specimen in a creep test associated with a short length extruded at time s in parison formation.

EXERCISE 20.2.3. Derive eqns (20.2.10) and (20.2.11).

EXERCISE 20.2.4. Consider extrusion of a parison of variable thickness as implied by the die shown in Fig. 20.7 and described by (20.2.16*) with B_r a function of ϕ_d, where ϕ_d defines the position of a streamline at the die-lips and ϕ_p the corresponding position of the same streamline after die swell.

Show these will be related, for a die that is symmetric about $\phi_d = 0$ (and thus equivalently about $\phi = \pi$), by

$$\phi_p = \pi \int_0^{\phi_d} B_r(\phi)\, d\phi \Big/ \int_0^{\pi} B_r(\phi)\, d\phi$$

EXERCISE 20.2.5. Derive relations (20.2.18), (20.2.22) and (20.2.26), noting that the results do not involve the rheology of the material provided only that it is homogeneous at any given time.

Appendix 1

Procedure for Obtaining Linearized Stability Equations for Capillary Flow

We consider the fully-developed solution (7.1.57) which we write

$$\mathbf{v}_0 = \{0, 0, v_{0z}(r)\}, \qquad v_{0z}(\tfrac{1}{2}d) = 0 \tag{A.1.1}$$

which satisfies (7.1.56), and gives rise to the unperturbed stress field

$$\mathbf{T}_0^{\mathrm{E}} = \begin{bmatrix} \sigma_{0r} & 0 & \tau_0 \\ 0 & \sigma_{0\phi} & 0 \\ \tau_0 & 0 & \sigma_{0z} \end{bmatrix} \tag{A.1.2}$$

where

$$\begin{aligned} \sigma_{0z} - \sigma_{0r} &= \Psi_1(\partial_r v_{0z})(\partial_r v_{0z})^2 \\ \sigma_{0r} - \sigma_{0\phi} &= \Psi_2(\partial_r v_{0z})(\partial_r v_{0z})^2 \\ \tau_0 &= \eta(\partial_r v_{0z})\partial_r v_{0z} \end{aligned} \tag{A.1.3}$$

and

$$\partial_r v_{0z} = \dot{\gamma}_0(\tfrac{1}{2}Pr/l) \tag{A.1.4}$$

with $\dot{\gamma}_0$ the inverse function defined in (7.1.56).

We now consider a small perturbation $\varepsilon\mathbf{v}_1$, where $\varepsilon \ll 1$, of the form

$$\mathbf{v}_1(r, \phi, z, t) = \mathscr{Re}\left[\left\{\mathrm{i}k\left(\frac{A}{r}\right) + im\left(\frac{B}{r}\right), -\frac{\mathrm{d}B}{\mathrm{d}r}, -\frac{1}{r}\frac{\mathrm{d}A}{\mathrm{d}r}\right\} \times \exp\{\mathrm{i}k(z - ct) + im\phi\}\right] \tag{A.1.5}$$

where $A(r)$, $B(r)$, $B'(r) \equiv \mathrm{d}B/\mathrm{d}r$, $A'(r) = \mathrm{d}A/\mathrm{d}r$ are complex, satisfying the boundary conditions

$$kA(\tfrac{1}{2}d) + mB(\tfrac{1}{2}d) = A'(\tfrac{1}{2}d) = B'(\tfrac{1}{2}d) = 0 \tag{A.1.6}$$

with $k(A/r) + m(B/r)$, B' and A'/r bounded at $r = 0$.

k (real) and m (integral) are wave numbers, c a complex wave speed.

The corresponding perturbation stress tensor $\varepsilon\mathbf{T}_1^{\mathrm{E}}$ will be a linear function or functional of $\mathbf{v}_1$ if we work to terms of order ε and neglect higher-order terms in ε. We can illustrate the procedure in two cases, dropping the $\mathcal{R}e\{\ \}$ for convenience.

A.1.1 JEFFREYS FLUID MODEL

Using the upper convected derivative

$$\mathbf{T}^{\mathrm{E}}+\lambda_1\overset{\triangledown}{\mathbf{T}}{}^{\mathrm{E}}=2\eta_0(\mathbf{D}+\lambda_2\overset{\triangledown}{\mathbf{D}}) \tag{A.1.7}$$

where

$$\overset{\triangledown}{\mathbf{J}}=\frac{\partial\mathbf{J}}{\partial t}+\mathbf{v}\,.\,\nabla\mathbf{J}+(\nabla\mathbf{v})^{\mathrm{T}}\,.\,\mathbf{J}+\mathbf{J}\,.\,(\nabla\mathbf{v}) \tag{A.1.8}$$

On substituting for $\mathbf{J}=\mathbf{J}_1+\varepsilon\mathbf{J}_1$ and linearizing, we obtain

$$\begin{aligned}
&\mathbf{T}_1^{\mathrm{E}}+\lambda_1\left(\frac{\partial\mathbf{T}_1^{\mathrm{E}}}{\partial t}+\mathbf{v}_0\,.\,\nabla\mathbf{T}_1^{\mathrm{E}}+(\nabla\mathbf{v}_0)^T\,.\,\mathbf{T}_1^{\mathrm{E}}+\mathbf{T}_1^{\mathrm{E}}\,.\,\nabla\mathbf{v}_0\right)\\
&\quad=-\lambda_1\{\mathbf{v}_1\,.\,\nabla\mathbf{T}_0^{\mathrm{E}}+(\nabla\mathbf{v}_1)^{\mathrm{T}}\,.\,\mathbf{T}_0^{\mathrm{E}}+\mathbf{T}_0^{\mathrm{E}}\,.\,(\nabla\mathbf{v}_1)\}\\
&\quad+2\eta_0\left\{\mathbf{D}_1+\lambda_2\left[\frac{\partial\mathbf{D}_1}{\partial t}+\mathbf{v}_0\,.\,\nabla\mathbf{D}_1+(\nabla\mathbf{v}_0)^T\,.\,\mathbf{D}_1+\mathbf{D}_1\,.\,\nabla\mathbf{v}_0\right.\right.\\
&\quad\left.\left.+\mathbf{v}_1\,.\,\nabla\mathbf{D}_0+(\nabla\mathbf{v}_1)^T\,.\,\mathbf{D}_0+\mathbf{D}_0\,.\,\nabla\mathbf{v}_1\right]\right\}
\end{aligned} \tag{A.1.9}$$

We can write, from (A.1.5)

$$\nabla\mathbf{v}_1=\exp\{ik(x-ct)+im\,\phi\}\begin{bmatrix}
ik\left(\dfrac{A}{r}\right)+im\left(\dfrac{B}{r}\right) & -B'' & -\left(\dfrac{A'}{r}\right)\\[2ex]
-\dfrac{km}{r}\left(\dfrac{A}{r}\right)-\dfrac{m^2}{r}\left(\dfrac{B}{r}\right)+\dfrac{B'}{r} & -im\dfrac{B'}{r}+\dfrac{ikA}{r^2}+im\dfrac{B}{r^2} & -im\dfrac{A'}{r^2}\\[2ex]
-k^2\left(\dfrac{A}{r}\right)-km\left(\dfrac{B}{r}\right) & -ikB' & -ik\dfrac{A'}{r}
\end{bmatrix} \tag{A.1.10}$$

and suppose that

$$\mathbf{T}_1^{\mathrm{E}} = \exp\{ik(z-ct)+im\phi\}\mathbf{P}_1^{\mathrm{E}}(r) \quad (\mathrm{A.1.11})$$

From this it follows that

$$\frac{\partial \mathbf{T}_1^{\mathrm{E}}}{\partial t} = -ikc\,\exp\{ik(z-ct)+im\phi\}\mathbf{P}_1^{\mathrm{E}}(r) \quad (\mathrm{A.1.12})$$

$$\mathbf{v}_0 \cdot \nabla \mathbf{T}_1^{\mathrm{E}} = ikv_{0z}(r)\exp\{ik(z-ct)+im\phi\}\mathbf{P}_1^{\mathrm{E}}(r) \quad (\mathrm{A.1.13})$$

Also

$$(\nabla \mathbf{v}_0)^T \cdot \mathbf{T}_1^{\mathrm{E}} + \mathbf{T}_1^{\mathrm{E}} \cdot (\nabla \mathbf{v}_0) = \dot{\gamma}_0(r) \times \begin{bmatrix} 0 & 0 & p_{1rr}^{\mathrm{E}}(r) \\ 0 & 0 & p_{1r\phi}^{\mathrm{E}}(r) \\ p_{1rr}^{\mathrm{E}}(r) & p_{1r\phi}^{\mathrm{E}}(r) & 2p_{1zz}^{\mathrm{E}}(r) \end{bmatrix} \exp\{ik(z-ct)+im\phi\} \quad (\mathrm{A.1.14})$$

$$(\mathbf{v}_1 \cdot \nabla)\mathbf{T}_0^{\mathrm{E}} = \exp\{ik(z-ct)+im\phi\}\left(\frac{ikA}{r}+\frac{imB}{r}\right)\frac{\partial}{\partial r}(\mathbf{T}_0^{\mathrm{E}}) \quad (\mathrm{A.1.15})$$

Similar results follow for the terms involving $\mathbf{D}_0$ and $\mathbf{D}_1$ so that $\exp\{k(z-ct)+im\phi\}$ becomes a factor of the entire equation (A.1.9) and can be dropped.

(A.1.9) then becomes a relatively simple set of six linear inhomogeneous equations for $p_{1rr}^{\mathrm{E}} \ldots p_{1zz}^{\mathrm{E}}$, with their coefficients being linear functions of

$$\mathrm{L}_1 = 1 + ik\{v_{0z}(r) - c\}\lambda_1 \quad \text{and} \quad \partial \mathrm{L}_1/\partial r \quad (\mathrm{A.1.16})$$

and the inhomogeneous terms being linear in A, A', A'', B, B', B'' with coefficients functions of L_1, $\partial \mathrm{L}_1/\partial r$ and $\partial^2 \mathrm{L}_1/\partial r^2$. They can be readily solved to give p_{1rr}^{E}, $p_{1r\phi}^{\mathrm{E}}$, p_{1rz}^{E}, $p_{1\phi\phi}^{\mathrm{E}}$, $p_{1\phi z}^{\mathrm{E}}$, p_{1zz}^{E}, as homogeneous linear functions of A, A', A'', B, B', B''. These can then be substituted into

$$\nabla \cdot \mathbf{T}_1 = \nabla \cdot \mathbf{T}_1^{\mathrm{E}} - \nabla p_1 = 0 \quad (\mathrm{A.1.17})$$

where p_1 can be written $\exp\{ik(z-ct)+im\phi\}\hat{p}_1(r)$. On eliminating $\hat{p}_1$, one third- and one fourth-order linear homogeneous (coupled) ordinary differential equation in A and B with coefficients functions of L_1 and its first four derivatives are obtained. These have to satisfy the boundary conditions (A.1.6) and the boundedness conditions at $r=0$.

The homogeneous nature of the equations for A and B leads to an eigenvalue problem for the parameters k, c when values for m and P/l are given. It is known that for small P/l, when eqn (A.1.7) effectively tends to that for a Newtonian fluid, there are no eigenvalues that have a positive or zero value for $\mathscr{I}m\, c$. Thus all small disturbances must be damped and the flow is stable. We seek the smallest positive value of P/l that will give a positive or zero value for $\mathscr{I}m\, c$. This gives the marginally stable flowrate. We may expect larger values of P/l (at least in the neighbourhood of P_{crit}/l) to give flow unstable to small disturbances, and so to lead to the sort of flow field described as melt flow instability.

A.1.2 KAYE–BKZ FLUID MODEL

$$\mathbf{T}^{\mathrm{E}} = \int_0^{\infty} \{\Phi_1(s, I_{C_t^t}, \mathrm{II}_{C_t^t})C_t^t + \Phi_{-1}(s, \mathrm{I}_{C_t^t}, \mathrm{II}_{C_t^t})C_t^{t-1}\,\mathrm{d}s \qquad (\mathrm{A.1.18})$$

From the flow field $\mathbf{v}_1$ given in (A.1.5), we can calculate the displacement history of individual fluid elements. Thus we write

$$\left.\begin{aligned} r'(\mathbf{r}_0, s) &= r_0 + \varepsilon\check{r}(s) \\ \phi'(\mathbf{r}_0, s) &= \phi_0 + \varepsilon\check{\phi}(s) \\ z'(\mathbf{r}_0, s) &= z_0 - v_{0z}s + \varepsilon\check{z}(s) \end{aligned}\right\} \qquad (\mathrm{A.1.19})$$

where

$$\left.\begin{aligned} \frac{\mathbf{d}\check{\mathbf{r}}}{\mathbf{ds}} &= \mathbf{v}_1(r', \phi', z') + \check{\mathbf{r}}\,.\,\nabla\mathbf{v}_0 \\ &= \exp\{ik(c - v_0)s\}\mathbf{v}_1(\mathbf{r}_0, 0) + \check{r}\dot{\gamma}_0\boldsymbol{\delta}_z \end{aligned}\right\} \qquad (\mathrm{A.1.20})$$

$\boldsymbol{\delta}_z$ being a unit vector in the $\mathbf{z}$-direction.

Clearly the factor $\exp\{ik(z_0 - ct_0) + cm\phi_0\}$ will arise in $\check{\mathbf{r}}$, and so may be dropped from the analysis temporarily.

We find that

$$\begin{aligned} \check{r}(s) &= \frac{(\mathrm{e}^{ik(c-v_0)s} - 1)\{kA(r_0) + mB(r_0)\}}{k(c - v_0)r_0} \\ \check{\phi}(s) &= \frac{\mathrm{i}(\mathrm{e}^{ik(c-v_0)s} - 1)B'(r_0)}{k(c - v_0)r_0} \qquad (\mathrm{A.1.21}) \\ \check{z}(s) &= \frac{\mathrm{i}(\mathrm{e}^{ik(c-v_0)s} - 1)A'(r_0)}{k(c - v_0)r_0} + \frac{\dot{\gamma}_0(kA_0 + mB_0)}{k(c - v_0)r_0}\left(\frac{\mathrm{e}^{ik(c-v_0)s} - 1}{\mathrm{i}k(c - v_0)} - s\right) \end{aligned}$$

We may now calculate the tensors $\mathbf{C}_t^t(s)$ and $\mathbf{C}_t^{t-1}(s)$ according to eqns

(B.6.7)–(B.6.12) and (B.5.7)–(B.5.12) of BA&H, using eqn (A.1.19) and (A.1.21) taking care to linearize in ε. The details need not be given in full, but a typical term becomes

$$C_{rr}^{-1}=(\dot{\gamma}_0 s)^2+2\varepsilon\left(\frac{\partial \check{r}}{\partial r_0}-\dot{\gamma}_0 s\frac{\partial \check{z}}{\partial r_0}\right)=C_{0rr}^{-1}+\varepsilon C_{1rr}^{-1} \qquad \text{(A.1.22)}$$

As in A.1.1, we find that all terms in $\mathbf{C}_t^t$ are linear in A, B and their derivatives. Similarly we find that

$$I_C=I_0+\varepsilon I_1, \qquad II_C=II_0+\varepsilon II_1 \qquad \text{(A.1.23)}$$

with I_1 and II_2 linear in A, B and their derivatives. Thus $\mathbf{T}_1^E$ can be expressed linearly in terms of A and B with coefficients given in terms of certain integrals over s. The rest of the argument goes exactly as from (A.1.17) in A.1.1 above.

It should be noted that neither the flowrate nor the mean pressure gradient is affected to order ε. The same would not be true if terms of order ε^2 and higher were included. At fixed t, the factor $\exp\{i(kz+m\phi)\}$ represents a helical disturbance with pitch equal to $\pi m\, d_c/k$ (and m starts), either right- or left-handed (in that the sign of m can be changed without affecting the solution). At fixed z, the factor $\exp\{i(-kct+m\phi)\}$ represents a rotation about the axis of the disturbance. Combining the two interpretations, we see that the full $(\exp\{ik(z-ct)+im\phi\}$ factor represents a corkscrew rotation. At fixed ϕ, the disturbance appears as a traveling wave.

A careful study of photographs leading to the diagrams given in Fig. 9.4 shows that both a single such disturbance and a superposition of right- and left-handed screw motions ($m=\pm|m|$) arise in practice.

Appendix 2

Perturbation Expansions for Thermal Effects

A.2.1 FULL LUBRICATION APPROXIMATION

We start from the full lubrication equations (8.1.17) and (8.1.21) for unidirectional pressure flow, between two stationary parallel plates at $x_3 = \pm\frac{1}{2}h$, of a temperature dependent Newtonian fluid whose viscosity is given by

$$\eta = \eta_w \exp[-\zeta(T - T_w)] \tag{A.2.1}$$

where $T(\pm\frac{1}{2}h) = T_w$ is the constant wall temperature.

Using an obvious non-dimensionalization:

$$\begin{aligned} &\tilde{x}_3 = x_3 h, \qquad \tilde{v}_1 = v_1 h/q, \qquad \tilde{p} = ph^2/\eta_w q \\ &\tilde{T} = \zeta(T - T_w) \end{aligned} \tag{A.2.2}$$

where

$$q = \int_{-\frac{1}{2}h}^{\frac{1}{2}h} v_1(x_3)\,\mathrm{d}x_3 \tag{A.2.3}$$

the relevant equations (8.1.77), (8.1.21) and (A.2.1) yield

$$\frac{\mathrm{d}}{\mathrm{d}\tilde{x}_3}\left((\exp - \tilde{T})\frac{\mathrm{d}\tilde{v}_1}{\mathrm{d}\tilde{x}_3}\right) = \frac{\mathrm{d}\tilde{p}}{\mathrm{d}\tilde{x}_1} \tag{A.2.4}$$

$$\frac{\mathrm{d}^2\tilde{T}}{\mathrm{d}\tilde{x}_3^2} = -\mathrm{Na}\exp(-\tilde{T})\left(\frac{\mathrm{d}\tilde{v}_1}{\mathrm{d}\tilde{x}_3}\right)^2 \tag{A.2.5}$$

where the Nahme number, cf. (8.1.39),

$$\mathrm{Na} = \eta_w \zeta q^2/\alpha h^2 \tag{A.2.6}$$

The boundary conditions become

$$\tilde{v}_1(\pm\tfrac{1}{2}) = \tilde{T}(\pm\tfrac{1}{2}) = 0 \tag{A.2.7}$$

Although (A.2.4), (A.2.5) have an analytic solution we consider the situation when

$$\mathrm{Na} \ll 1$$

and seek a perturbation solution about the case Na = 0, by using the expansions

$$\left.\begin{aligned} \tilde{v}_1 &= \tilde{v}_1^{(0)} + \mathrm{Na}\,\tilde{v}_1^{(1)} + \mathrm{Na}^2\,\tilde{v}_1^{(2)} + \dots \\ \tilde{T} &= \tilde{T}^{(0)} + \mathrm{Na}\,\tilde{T}^{(1)} + \mathrm{Na}^2\,\tilde{T}^{(2)} + \dots \\ \frac{\mathrm{d}\tilde{p}}{\mathrm{d}\tilde{x}_1} &= \tilde{p}_{,1}^{(0)} + \mathrm{Na}\,\tilde{p}_{,1}^{(1)} + \mathrm{Na}^2\,\tilde{p}_{,1}^{(2)} + \dots \end{aligned}\right\} \tag{A.2.8}$$

where we are regarding q as fixed. (An alternative non-dimensionalization would have been used if $\dfrac{\mathrm{d}\tilde{p}}{\mathrm{d}\tilde{x}_1}$ had been regarded as given, and $\tilde{q} = \dfrac{q}{v^* h}$ would have been perturbed instead of $\dfrac{\mathrm{d}\tilde{p}}{\mathrm{d}\tilde{x}_1}$.)

We note first that

$$\exp(-\tilde{T}) = \exp(-\tilde{T}^{(0)})\{1 + \mathrm{Na}\,\tilde{T}^{(1)} + \mathrm{Na}^2\,(\tilde{T}^{(2)} + \tfrac{1}{2}\tilde{T}^{(1)2}) + \dots\} \tag{A.2.9}$$

Substituting (A.2.8) and (A.2.9) into (A.2.3)–(A.2.5) yields

$$1 - \int_{-\frac{1}{2}}^{\frac{1}{2}} \tilde{v}_1^{(0)}(\tilde{x}_3)\,\mathrm{d}\tilde{x}_3 = \mathrm{Na} \int_{-\frac{1}{2}}^{\frac{1}{2}} \tilde{v}_1^{(1)}\,\mathrm{d}\tilde{x}_3 + 0(\mathrm{Na}^2) \tag{A.2.10}$$

$$\begin{aligned} \frac{\mathrm{d}}{\mathrm{d}\tilde{x}_3}\left(\exp(-\tilde{T}^{(0)})\frac{\mathrm{d}\tilde{v}_1^{(0)}}{\mathrm{d}\tilde{x}_3}\right) - \tilde{p}_{,1}^{(0)} &= \mathrm{Na}\left\{\frac{\mathrm{d}}{\mathrm{d}\tilde{x}_3}\left(\exp(-\tilde{T}^{(0)})\frac{\mathrm{d}\tilde{v}_1^{(1)}}{\mathrm{d}\tilde{x}_3}\right)\right. \\ &\quad \left. + \frac{\mathrm{d}}{\mathrm{d}\tilde{x}_3}\left(\tilde{T}^{(1)}\exp(-\tilde{T}^{(0)})\frac{\mathrm{d}\tilde{v}_1^{(0)}}{\mathrm{d}\tilde{x}_3}\right) - \tilde{p}_{,1}^{(1)}\right\} + 0(\mathrm{Na}^2) \end{aligned} \tag{A.2.11}$$

$$\frac{\mathrm{d}^2\tilde{T}^{(0)}}{\mathrm{d}\tilde{x}_3^2} = -\mathrm{Na}\left\{\frac{\mathrm{d}^2\tilde{T}^{(1)}}{\mathrm{d}\tilde{x}_3^2} + \exp(-\tilde{T}^{(0)})\left(\frac{\mathrm{d}^2\tilde{v}_1^{(0)}}{\mathrm{d}\tilde{x}_3}\right)^2\right\} + 0(\mathrm{Na}^2) \tag{A.2.12}$$

The boundary conditions (A.2.7) become simply

$$\left.\begin{aligned} \tilde{v}_1^{(0)} + \mathrm{Na}\,\tilde{v}_1^{(1)} = 0(\mathrm{Na}^2), \qquad \tilde{T}^{(0)} + \mathrm{Na}\,\tilde{T}^{(1)} = 0(\mathrm{Na}^2) \\ \text{at } \tilde{x}_3 = \pm\tfrac{1}{2} \end{aligned}\right. \tag{A.2.13}$$

From the 0(1) term in (A.2.12), together with the boundary conditions (A.2.13) for $\tilde{T}^{(0)}$, we see that

$$\tilde{T}^{(0)} = 0 \tag{A.2.14}$$

From the 0(1) term in (A.2.11) using (A.2.14) we obtain

$$\frac{d^2\tilde{v}_1^{(1)}}{d\tilde{x}_3^2} = \tilde{p}_{,1}^{(0)} \tag{A.2.15}$$

Together with the 0(1) term in (A.2.10) and the boundary conditions (A.2.13) for $\tilde{v}_1^{(0)}$, we deduce that

$$\tilde{v}_1^{(0)} = \tfrac{3}{2}(1-4\tilde{x}_3^2), \qquad \tilde{p}_{,1}^{(0)} = -12 \tag{A.2.16}$$

The zero-order solution (A.2.14), (A.2.16) may now be substituted back into the 0(Na) term in (A.2.12) to yield the inhomogeneous equation

$$\frac{d^2\tilde{T}^{(1)}}{d\tilde{x}_3^2} = -144\tilde{x}_3^2 \tag{A.2.17}$$

for $\tilde{T}^{(1)}$. This may be integrated using the 0(Na) term in the temperature boundary conditions (A.2.13) to yield

$$\tilde{T}^{(1)} = \tfrac{3}{4} - 12\tilde{x}_3^4 \tag{A.2.18}$$

(A.2.14), (A.2.16) and (A.2.18) may now be substituted into the 0(Na) term in (A.2.11) to give

$$\frac{d^2\tilde{v}_1^{(1)}}{d\tilde{x}_3^2} = \tilde{p}_{,1}^{(1)} + 9\text{–}720\tilde{x}_3^4 \tag{A.2.19}$$

which is subject to $\tilde{v}_1^{(1)}(\pm\tfrac{1}{2}) = 0$ from (A.2.13) and

$$\int_{-\frac{1}{2}}^{\frac{1}{2}} \tilde{v}_1^{(1)}\, d\tilde{x}_3 = 0$$

from (A.2.10). The solution is

$$\tilde{v}_1^{(1)} = -\tfrac{1}{8}\tilde{p}_{,1}^{(1)}(1\text{–}4\tilde{x}_3^2) + (\tfrac{9}{2}\tilde{x}_3^2 - 24\tilde{x}_3^6 - \tfrac{3}{4}); \qquad \tilde{p}_{,1}^{(1)} = -\tfrac{18}{7} \tag{A.2.20}$$

This completes the solution to first order in Na. It is clear that the procedure may be continued for higher order terms, in Na^2 etc. It will be noted that successively higher powers of $\tilde{x}_3^2$ are involved, leading in general to steeper gradients near $\tilde{x}_3 = \pm\tfrac{1}{2}$.

A.2.2 DEVELOPING FLOW

In this case we allow the temperature and velocity fields to depend upon

$$\hat{x}_1 = x_1 \kappa / qh \qquad \text{(A.2.21)}$$

as well as $\tilde{x}_3$. Equation (A.2.5) is then modified to include an extra term representing convection to yield

$$\tilde{v}_1 \frac{\partial \tilde{T}}{\partial \hat{x}_1} - \frac{\partial^2 \tilde{T}}{\partial \tilde{x}_3^2} = \text{Na} \exp(-\tilde{T}) \left(\frac{\partial \tilde{v}_1}{\partial \tilde{x}_3}\right)^2 \qquad \text{(A.2.22)}$$

Equations (A.2.3) and (A.2.4) are retained although formally the derivatives in (A.2.4) become partial derivatives and not total derivatives. The boundary conditions (A.2.7) have to be supplemented by an initial condition

$$\tilde{T}(0, \tilde{x}_3) = \tilde{T}_0(\tilde{x}_3) \qquad \text{(A.2.23)}$$

The same basic scheme of solution is used as in A.2.1 above except that $\tilde{T}^{(0)}$, $\tilde{v}_1^{(0)}$ are now given by the coupled set of equations

$$\frac{\partial}{\partial \tilde{x}_3}\left(\exp(-\tilde{T}^{(0)}) \frac{\partial \tilde{v}_1^{(0)}}{\partial \tilde{x}_3}\right) = \tilde{p}_{,1}^{(0)} \qquad \text{(A.2.24)}$$

$$\tilde{v}_1^{(0)} \frac{\partial \tilde{T}^{(0)}}{\partial \hat{x}_1} - \frac{\partial^2 \tilde{T}^{(0)}}{\partial \tilde{x}_3^2} = 0 \qquad \text{(A.2.25)}$$

$$1 - \int_{-\frac{1}{2}}^{\frac{1}{2}} \tilde{v}_1^{(0)} \, d\tilde{x}_3 = 0 \qquad \text{(A.2.26)}$$

This is essentially the problem treated in Section 19.3 when $\text{Br} \to 0$, $\xi_s \to \infty$ and $\text{Gz} = 0(1)$.

The simplest case arises when $\tilde{T}_0 = 0$, for then (A.2.14) and (A.2.16) are recovered.

The first difference arises in the equation for $\tilde{T}^{(1)}$ which becomes

$$\tfrac{3}{2}(1 - 4\tilde{x}_3^2) \frac{\partial \tilde{T}^{(1)}}{\partial \hat{x}_1} - \frac{\partial^2 \tilde{T}^{(1)}}{\partial \tilde{x}_3^2} = 144 \tilde{x}_3^2 \qquad \text{(A.2.27)}$$

with zero boundary conditions and arbitrary initial conditions.† Numerical solution is necessary, but clearly practicable, starting from $\hat{x}_1 = 0$.

The equation for $\tilde{v}_1^{(1)}$ involves the same terms from (A.2.11), although that involving $\tilde{T}^{(1)}$ is no longer that given in (A.2.19) but is a computed function of $\hat{x}_1$ instead.

† If it is assumed that B can be written as Na $\tilde{T}^*$, then a reasonable initial condition is given by

$$\tilde{T}^{(1)}(0, \tilde{x}_3) = \tilde{T}^*$$

Subsequent terms in the expansion involve the same differential operators for the unknown functions $\tilde{v}_1^{(k)}$ and $\tilde{T}_1^{(k)}$ and so lead to no further difficulties.

EXERCISE A.2.1. Continue the solution of (A.2.10)–(A.2.13) to terms of order Na^2.

Appendix 3

Extruder Equations in Helical Coordinates

The non-approximate coordinate system that proves most convenient to use for deep channels is one in which (see Zamodits, 1964)

(i) $$y^3 = r \tag{A.3.1}$$

so that circular cylinders coaxial with the barrel are (y^1, y^2) surfaces

(ii) $$y^1 = 2\pi z/(w_{ca} + w_{fa}) = z/p \tag{A.3.2}$$

where w_{ca} and w_{fa} are defined in Fig. 11.2 and whose sum is the pitch of the screw;

(iii) $$y^2 = \phi - y^1 + 2N\pi \tag{A.3.3}$$

where ϕ is the angular coordinate for a given cylindrical polar coordinate system. N is an integer chosen so that $0 < y^2 < 2\pi$. Lines of constant y^2 are such that $\phi - y^1 = \text{constant}$ with N increasing by one each time that ϕ jumps back from 2π to 0, and so are helices.

The boundaries of the screw channel become

$$\begin{aligned} &\text{barrel:} && y^3 = \tfrac{1}{2}d_b \\ &\text{screw root:} && y^3 = \tfrac{1}{2}d_b - h_s; \quad 0 \leqslant y^2 \leqslant w_{ca}/p \\ &\text{screw tip:} && y^3 = \tfrac{1}{2}d_b - h_f; \quad w_{ca}/p \leqslant y^2 \leqslant 2\pi \\ &\text{screw walls:} && \tfrac{1}{2}d_b - h_s < y^3 < \tfrac{1}{2}d_b - h_f; \quad y_2 = 0,\ w_{ca}/p \end{aligned} \tag{A.3.4}$$

The centre of the feed pocket can be taken as $y^1 = 0$ and the nose of the screw as $y^1 = l_s \operatorname{cosec} \phi_b$.

The depth of the screw channel is taken to be a slowly varying function of y^1 only, so

$$\frac{\partial h_s}{\partial y^2}=0; \qquad \frac{\partial h_s}{\partial y^1} \ll h_s \tag{A.3.5}$$

In the fully developed approximation that is relevant for deep channels, in which the side walls $y^2=$constant play a significant part, all physical quantities, such as v, D and T must be regarded as functions of y^2 and y^3.

This coordinate system is non-orthogonal and so the equations of motion, the energy equation and the constitutive relation have to be expressed in general tensor form. Either of the texts, Aris (1962) or BA & H (Appendixes A and B), can be consulted for full details. Here only the essential relationships are given.

The covariant components of the metric tensor are

$$\left.\begin{array}{lll} g_{11}=r^2+p^2, & g_{22}=r^2, & g_{33}=1; \\ g_{12}=g_{21}=r; & g_{13}=g_{31}=g_{32}=g_{23}=0 \end{array}\right\} \tag{A.3.6}$$

The Christoffel symbols are

$$\Gamma^2_{31}=\Gamma^2_{32}=r^{-1}; \qquad \Gamma^3_{11}=\Gamma^3_{12}=\Gamma^3_{22}=-r; \quad \text{all others}=0 \tag{A.3.7}$$

The simple continuity condition (2.2.7) becomes

$$v^k_{;k}=\frac{\partial v^k}{\partial x^k}+\Gamma^k_{ik}v^i=0 \tag{A.3.8}$$

where the covariant derivative is employed. For fully developed flow $v^k=v^k(y^2, y^3)$ this yields

$$\frac{\partial v^2}{\partial y^2}+\frac{\partial v^3}{\partial y^3}+\frac{v^3}{y^3}=0 \tag{A.3.9}$$

which is the equation for the secondary flow (v^2, v^3).

The components of the rate of deformation tensor can be written, again for fully developed flow, as

$$\begin{gathered} D^{11}=\frac{1}{p^2}\frac{\partial v^1}{\partial y^2}, \qquad D^{12}=D^{21}=\frac{1}{2}\left(\frac{r^2+p^2}{r^2p^2}\frac{\partial v^1}{\partial y^2}-\frac{1}{p^2}\frac{\partial v^2}{\partial y^2}\right) \\ D^{22}=\frac{r^2+p^2}{r^2p^2}\frac{\partial v^2}{\partial y^2}+\frac{v^3}{r^3}, \qquad D^{23}=D^{32}=\frac{1}{2}\left(\frac{r^2+p^2}{r^2p^2}\frac{\partial v^3}{\partial y^2}+\frac{\partial v^2}{\partial y^3}\right) \\ D^{13}=D^{31}=\frac{1}{2}\left(\frac{\partial v^1}{\partial y^3}-\frac{1}{p^2}\frac{\partial v^3}{\partial y^3}\right), \qquad D^{33}=\frac{\partial v^3}{\partial y^3} \end{gathered} \tag{A.3.10}$$

where r has been written for y^3 for convenience.

The second invariant of $\mathbf{D}$, which is needed for the power-law constitutive equation is obtained by the double contraction

$$D^2 = D^i_k D^k_i \equiv D^{il} D^{jk} g_{lk} g_{ji}$$

All the necessary components are given in (A.3.6) and (A.3.10).

The stress equilibrium equation $\nabla . \mathbf{T} = 0$ becomes, when the power-law equation (3.3.5) is used,

$$T^{ik}_{;k} \equiv (-pg^{ik} + \eta(D)D^{ik})_{;k} = 0 \qquad \text{(A.3.11)}$$

where

$$s_{;k} = \frac{\partial s}{\partial y^k}, \qquad A^{ik}_{;k} = \frac{\partial A^{ik}}{\partial y^k} + \Gamma^i_{lk} A^{lk} + \Gamma^k_{lk} A^{il} \qquad \text{(A.3.12)}$$

and again all the necessary Christoffel symbols are given by (A.3.7).

The energy equation (2.2.9) subject to (2.2.10) and $dU = \rho\Gamma \, d\Gamma \, dT$ can be written

$$\rho\Gamma v^k \frac{\partial T}{\partial y^k} = \alpha g^{ik} \left\{ \frac{\partial^2 T}{\partial x^i \, \partial x^k} + \Gamma^i_{kj} \frac{\partial T}{\partial x^k} \right\} + \eta(D) g_{ij} g_{kl} D^{ik} D^{jl} \qquad \text{(A.3.13)}$$

where account is taken of (A.3.8).

The flows within the extruder channel are given by

$$q_2 = \int_{\frac{1}{2}d_b - h_s}^{\frac{1}{2}d_b} y^3 v^2 \, dy^3 / \sin \phi_b \qquad \text{(A.3.14)}$$

where q_2 is the 'leakage' flux across the surfaces $y^2 = \text{const.}$ per unit length in the y^3 direction, and

$$Q_1 = \int_0^{w_{ca}/p} \int_{\frac{1}{2}d_b - h_s}^{\frac{1}{2}d_b} \left(y^3 v^1 + \frac{(y^3)^3}{(y^3)^2 + p^2} v^2 \right) dy^3 \, dy^2 \qquad \text{(A.3.15)}$$

where Q_1 is the output across surfaces $y^1 = \text{const.}$

EXERCISE A.3.1. Write out in full the relations (A.3.10), (A.3.11) and (A.3.13) for the special case in which all dependent variables are functions of y^3 only.

Appendix 4

Heat Transfer within an Extruder Screw: Order of Magnitude Analysis

A.4.1 TRANSFER BETWEEN SCREW AND BARREL IN STATIONARY SITUATION ASSUMING SCREW AND BARREL TEMPERATURES TO BE CONSTANTS, T_s and T_b RESPECTIVELY

The total heat flux will be

$$\dot{q}_{SB} \approx \pi\alpha_m(T_s - T_b)\left(\frac{w_{fa}}{h_f} + \frac{w_{ca}}{h_s}\right)l_s \qquad (A.4.1)$$

where h_f, h_s, w_{fa}, w_{ca}, d_b, l_s have the meanings indicated in Figs. 11.1–11.3, and α_m is the thermal conductivity of the melt.

For typical values given in Table 11.1, the term representing transfer across the flight clearance (w_{fa}/h_f) will be about 4 times as large as that representing transfer across the main channel (w_{ca}/h_s).

A.4.2 TRANSFER ALONG THE SCREW ASSUMING A TEMPERATURE T_0 LEVEL WITH THE FEED POCKET AND A TEMPERATURE T_s HALF WAY ALONG THE SCREW

The heat flux will be

$$\dot{q}_S \approx \tfrac{1}{2}\alpha_M\pi(d_b - 2h_s)^2(T_s - T_0)/l_s \qquad (A.4.2)$$

where α_M is the thermal conductivity of the screw.

If we assume that $\dot{q}_{SB} = \dot{q}_S$ then

$$\frac{T_s - T_b}{T_s - T_0} \approx \frac{\alpha_M}{10\alpha_m}\frac{h_s}{w_{ca}l_s^2}(d_b - 2h_s)^2 \approx \frac{\alpha_M}{10\alpha_m}\frac{h_s d_b}{l_s^2} \qquad (A.4.3)$$

Even if we assume that

$$\alpha_M/\alpha_m = 10^3 \tag{A.4.4}$$

which is large enough to cover most metal/melt combinations, typical values given in Table 11.1 lead to

$$(T_s - T_b)/(T_s - T_0) \approx 10^{-2} \tag{A.4.5}$$

From this order of magnitude result we deduce that conduction of heat across the melt is much more effective than conduction along the screw.

It also explains why the outer barrel is split into several separate zones for heating and cooling for otherwise large temperature gradients along the barrel could ensue.

A.4.3 TRANSFER UP SCREW FLIGHT COMPARED WITH TRANSFER ACROSS FLIGHT CLEARANCE

For equal heat transfer rates

$$\frac{T_s - T_f}{T_f - T_b} = \frac{\alpha_m h_s}{\alpha_M h_f} \approx \frac{1}{10} \tag{A.4.6}$$

using $\alpha_M/\alpha_m \approx 400$ instead of (A.4.4) and typical values from Table 11.1.

Hence it is reasonable to assume that the surface temperature of the screw is constant at any given axial position, even though it varies along the screw.

The consequence of the order of magnitude arguments in A.4.1–A.4.3 is that T_s will be locally determined by a zero net balance of heat transferred into the screw from the melt. If heat generated within the melt flows into the screw within the main channel, it will have to flow out into the melt at the flight tip, or vice versa.

A.4.4 HEAT GENERATION WITHIN THE MELT DUE TO THE RELATIVE MOTION OF BARREL AND SCREW

To obtain order of magnitude estimates, we can suppose that the rate of heat generation is that given by drag flow and we can calculate the latter in terms of the work done by the shear stress at the barrel wall, assuming the screw to be stationary.

For the flight clearance region, this gives

$$\dot{W}_F \propto \pi K_m \left(\frac{v_b}{h_f}\right)^{\nu} v_b w_{fa} l_s \tag{A.4.7}$$

and for the main channel region

$$\dot{W}_c \propto \pi K_m \left(\frac{v_b}{h_c}\right)^{\nu} v_b w_{ca} l_s \tag{A.4.8}$$

assuming uniform temperature.

The ratio of these

$$\frac{\dot{W}_c}{\dot{W}_F} = \left(\frac{h_f}{h_c}\right)^{\nu} \frac{w_{ca}}{w_f} \sim 3 \tag{A.4.9}$$

for the typical values used earlier, with $\nu = \frac{1}{3}$. Thus although we expect very high local rates of heat generation over the flight clearance, the bulk of the total heat generated will still be within the main channel.

A.4.5 ORDER OF MAGNITUDE ESTIMATE FOR SCREW TEMPERATURE

The calculations in Example A.4.1 below indicate that an upper limit for the screw temperature will be given by

$$T_s = T_b + \tfrac{1}{2} K_m v_b^{\nu+1} h_s^{1-\nu} / \alpha_m$$

which for many situations becomes large, of order 100 K. This is known to be unrealistically high. The arguments used in Exercise A.4.1 suggest that a more realistic upper bound will be given by

$$T_s = T_b + \tfrac{1}{6} K_m v_b^{\nu+1} h_s^{1-\nu} / \alpha_m$$

while for large extruders a true value may be closer to

$$T_s = T_b + \tfrac{1}{2} K_m v_b^{\nu+1} h_f^{1-\nu} / \alpha_m$$

EXAMPLE A.4.1. Calculate the fully developed temperature distribution in drag flow between 2 parallel plates at $x_3 = 0$, $x_3 = h_s$ with relative speed v_b held at temperatures T_b and T_s. Deduce the value for T_s that leads to no heat transfer at $x_3 = 0$. Using usual nomenclature, the velocity field will be

$$v(x_3) = v_b x_3 / h_s$$

and the fully developed temperature field will be given by

$$\alpha_m \frac{d^2T}{dx_3^2} = -K_m\left(\frac{v_b}{h_s}\right)^{\nu+1}; \qquad T(0) = T_s,\ T(h_s) = T_b$$

Integrating twice yields

$$T(x_3) = T_s + (T_b - T_s)x_3/h_s - \frac{K_m}{2\alpha_m}\left(\frac{v_b}{h_s}\right)^{\nu+1} x_3(x_3 - h_s)$$

For no heat transfer at $x_3 = 0$

$$\frac{(T_b - T_s)}{h_s} + \frac{K_m}{2\alpha_m}\left(\frac{v_b}{h_s}\right)^{\nu+1} h_s = 0$$

or

$$T_s - T_b = \tfrac{1}{2}K_m v_b^{\nu+1} h_s^{1-\nu}/\alpha_m$$

Note that this increases with h_s unless $\nu = 1$. For $\nu = \frac{1}{3}$ and $h_f/h_s = 40$, this means a ratio of over 10 for the temperature differences $T_s - T_b$ and $T_f - T_b$.

EXERCISE A.4.1. Show, using (A.4.8), the results of Example A.4.1 and interpreting (A.4.6) to mean that $T_f = T_s$, that

$$T_s - T_b \sim \tfrac{1}{6}K_m v_b^{\nu+1} h_s^{1-\nu}/\alpha_m$$

in a typical extruder. Note that this neglects any convective heat transfer within the screw channel, as discussed in Subsection 11.4.1 with $\mathrm{Gz}_c \gg 1$.

Appendix 5

Maillefer Screw Analysis

The basic geometry of the Maillefer screw is shown in Fig. A.5.1. The feed channel (1) has a decreasing width w_{c1} over the 'compression' region, in which for convenience the depth h_{s1} can be taken as constant. It is connected over a barrier (a second relieved flight of rather longer pitch) to a delivery channel (2) of increasing width w_{c2}, in which similarly for convenience the depth h_{s2} can be taken as constant. The barrier is intended to separate the unmelted solid feed from the molten product, the design being such that solid particles cannot pass over the barrier, while melt formed at or near the barrel surface will be conveyed, largely by drag flow, from channel 1 to channel 2. This screw involves 3 more geometrical parameters than the single screw discussed in Chapter 11, namely h_r the barrier clearance, h_{s2} the delivery channel depth in the transition section and $dw_{c2}/dx_1 = -dw_{c1}/dx_1 = \text{constant}$, a slope which is related to the difference in the helical angle of the two flights, F and R.

Analysis of flow in such a screw is complicated by the fact that any position x_1 within the transition section (from $w_{c2} = 0$ to $w_{c1} = 0$) one of six possible regions may be relevant. The feed channel may either be completely full of solid (case A) as in the feed zone of a single-screw extruder (11.2) or partly full of solid and partly full of melt (case B) as in the melting zone (11.3) or finally completely full of melt (case C) as in the metering zone of a single-screw extruder (11.4). The delivery channel may either be full of melt (case 1) as in the metering zone (11.4 again) or only partially full (case 2) as in vented screws (11.8). The three feed zone possibilities are shown in Figs. A.5.2 and A.5.3.

Clearly which of these six possibilities arise and in what order will depend upon the geometry of the screw, the operating conditions and the material properties. If solid is to be retained in the feed channel

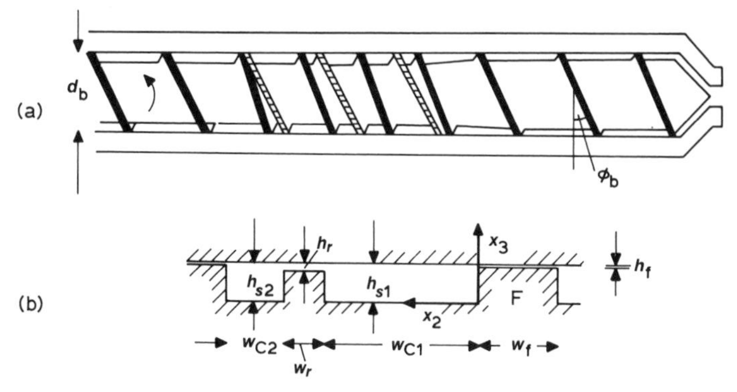

Fig. A.5.1. Diagrammatic view of Maillefer screw. (a) ▬, close fitting flight; ⊏⊐⊐, relieved flight. The number of turns in each of the zones (feed, plasticating and metering) would in practice be far larger. The curved arrow indicates direction of rotation. (b) Cross-section of channels showing nomenclature. (A rectangular cross-section is used for simplicity.)

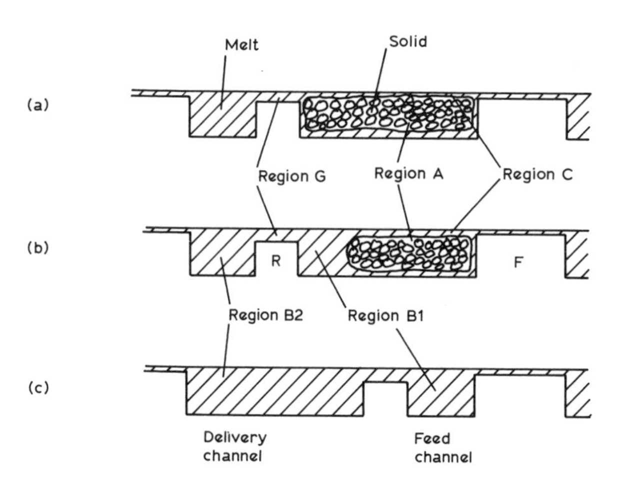

Fig. A.5.2. Three possible cases for the flow regime in the feed channel. For convenience, case 1 is shown for the delivery channel.

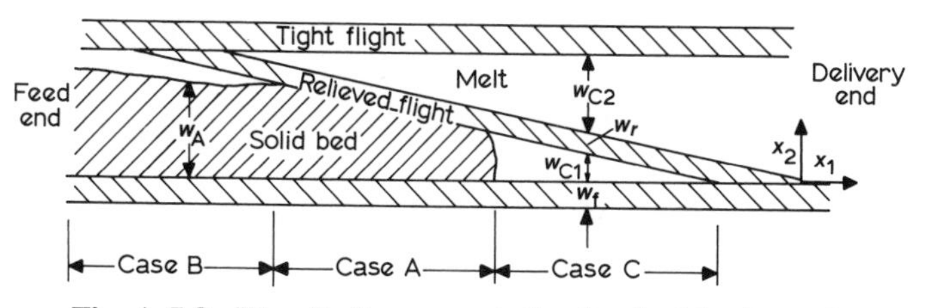

Fig. A.5.3. 'Unrolled' representation for double channel.

and melting is to be complete by the point at which $w_{c1}=0$, then the last part of the transition section must be describable by case C. Unless venting is required then high pumping efficiency suggests that case 1 should apply over much of the transition zone.

Outside the transition zone, the feed and delivery channels behave as they would in a single screw extruder. What is needed therefore is a simple mathematical model to describe the transition section. For simplicity the most elementary model possible is described here. The connection with the results of Chapter 11 will be obvious.

The following approximations will therefore be made:

(1) The channels are shallow and of uniform depth: $h_s \ll d_b$ and so the 'unrolling' implied by Figs A.5.1(b), A.5.2 and A.5.3 is justified.
(2) The channels are wide: $h_s \ll w_{c1}$, w_{c2} almost everywhere.
(3) The channels are long: w_{c1}, $w_{c2} \ll l_c$. Thus the lubrication approximation can be used everywhere.
(4) Rectangular channel cross-section with

$$w_{c1}=w_c(1-x_1/l_c); \qquad w_{c2}=w_c x_1/l_c \tag{A.5.1}$$

(5) Negligible flight clearance: $h_f \ll h_s$, h_r, thus ensuring no backflow.
(6) Steady flow.
(7) Negligible effect of lubricating layers (regions D and E of 11.3.1) at the screw surface in terms of the melting mechanics.
(8) Constant (temperature and pressure independent) material properties, including Newtonian melt viscosity η_0.

The variables of interest are the downstream mass flowrates $\dot{M}_{A1}$, $\dot{M}_{B11}$ and $\dot{M}_{B21}$, in the regions A (solid), B1 and B2 (melt) respectively (we neglect those in regions C and G by comparison to those in regions A, B1 and B2); the lateral mass flowrates/unit length from zone C into zone B1 and from zone G into zone B2, $\dot{m}_{C2}$ and $\dot{m}_{G2}$ respectively; the pressures p_1 and p_2 at either end of the relieved flight clearance; the width w_A of the solid bed and F_{AC1}, the downstream shear force/unit length exerted on the solid bed by the sheared melt in region C. Relations between these variables will be derived using mass and force balances and the approximations applied to regions A and C in 11.3.1. Thus

$$\dot{M}_{B11}+\dot{M}_{B21}+\dot{M}_{A1}=\dot{M}_T(\text{constant}) \tag{A.5.2}$$

$$\frac{d}{dx_1}\dot{M}_{B11}=\dot{m}_{C2}-\dot{m}_{G2} \tag{A.5.3}$$

$$\frac{d}{dx_1} M_{B21} = \dot{m}_{G2} \tag{A.5.4}$$

$$\dot{M}_{A1} = \rho_s v_{A1} w_A h_s \tag{A.5.5}$$

$$F_{AC1} = w_A h_s \frac{dp_1}{dx_1} \tag{A.5.6}$$

$$\dot{M}_{B11} = \tfrac{1}{2}\rho_m (w_{c1} - w_A) h_s \left(v_{b1} - \frac{h_s^2}{6\eta_0} \frac{dp_1}{dx_1} \right) \tag{A.5.7}$$

$$\dot{M}_{B21} = \tfrac{1}{2}\rho_m w_{c2} h_s \left(v_{b1} - \frac{h_s^2}{6\eta_0} \frac{dp_2}{dx_1} \right) \tag{A.5.8}$$

(provided the channel is full)

$$\dot{m}_{G2} = \tfrac{1}{2}\rho_m h_r \left\{ v_{b2} + \frac{h_2^2}{6\eta_0} \frac{(p_1 - p_2)}{w_r} \right\} \tag{A.5.9}$$

(provided the channel is full)

Equations (A.5.8) and (A.5.9) obviously apply in cases 1 and B or C respectively.

Using the result of Exercise 11.3.3 for large Sf (which for our present purposes may be taken as an additional simplification) gives

$$\dot{m}_{C2} = \{\rho_m \Gamma_m v_{b2} \alpha_m (T_b - T_m) w_A / \xi_s^*\}^{\frac{1}{2}} \tag{A.5.10}$$

and

$$F_{AC1} = \eta_0 (v_{b1} - v_{A1}) \{\rho_m v_{b1} w_A \xi^* / \alpha_m (T_b - T_m)\}^{\frac{1}{2}} \tag{A.5.11}$$

where η_0, v_{b2}, v_{b1}, ξ_s^*, ρ_m, α_m, T_b and T_m are as defined earlier.

The initial conditions are taken to be

$$x_1 = 0: \quad \dot{M}_{B11} = \dot{M}_{B21} = 0, \quad p_1 = 0 \tag{A.5.12}$$

$$x_1 = l_c: \quad \dot{M}_{B11} = \dot{M}_{A1} = 0 \tag{A.5.13}$$

As earlier we seek a solution in the form of $p_2(l_c)$ as a function of $\dot{M}_T$, with T_b and v_b as parameters.

The set of equations (A.5.2)–(A.5.12) apply only in the case B1. For case C1, $w_A \equiv 0$ and so (A.5.5), (A.5.6), (A.5.10) and (A.5.11) become identically zero, and therefore irrelevant. For case A1, which arises when $w_A = w_{C1}$, we replace (A.5.9) by

$$\dot{m}_{G2} = \dot{m}_{C2} \tag{A.5.9a}$$

The precise circumstances in which this will be relevant are not completely obvious: however if $p_1 - p_2$ would otherwise increase $\dot{m}_{G2}$ beyond $\dot{m}_{C2}$ and $\dot{m}_{C2}/v_{b2} < h_r$, it seems sensible to assume that (A.5.9a) will hold. For Fig. A.5.3, case A is shown as following between case B and case C; however this is not based on observation or calculation and should only be regarded as speculative. If case 2 applies then $dp_2/dx_1 = 0$, and (A.5.8) can be regarded as defining a $w_{c2} < w_c x_1 / l_c$.

Solution of these equations is a little more difficult than for a single channel single screw extruder because of the downstream boundary conditions (A.5.13). $p_2(0)$ is used as a variable parameter which has to be chosen so as to satisfy the boundary condition at $x_1 = l_c$. Even for the very simplified model given above, numerical solution proves necessary.

Appendix 6

Computed Results for Deep Melt-Filled Channels in Single-screw Extrusion (see Subsection 11.4.2)

Martin (1969) has given results obtained by finite-difference computation, for power-law temperature-dependent viscous fluids in fully-developed flow ($\mathrm{Gz}_{\mathrm{d.c.}} \ll 1$) for

$$A = 1, 2, 5; \qquad \nu = 1, 0{\cdot}6, 0{\cdot}3; \qquad \mathrm{Na} = 0, 4; \qquad \mathrm{Pe}_{\mathrm{c.c.}} = 0\text{–}5000; \qquad \tilde{p}_{,\tilde{x}_1} = 0, 1, 3$$

as defined in (11.4.1), (3.3.5), (11.4.46), (11.4.65) and (11.4.45) respectively. ϕ_b is everywhere 20° except in Fig. A.6.1.

A selection of these are shown in the accompanying figures, which are taken from his thesis, and which cannot be found in the readily available literature.

The stream function Ψ is such that

$$\tilde{v}_2 = \partial\Psi/\partial\tilde{x}_3, \qquad \tilde{v}_3 = -\partial\Psi/\partial\tilde{x}_2,$$

while

$$\hat{T} = \tilde{T}/\mathrm{Na} = (T - T_0)/T^*_{\mathrm{gen}}$$
$$\tilde{Q} = \tilde{Q}_{\mathrm{T1}}, \qquad \bar{\tilde{q}}_1 = \tilde{Q}_{\mathrm{T1}}/A$$

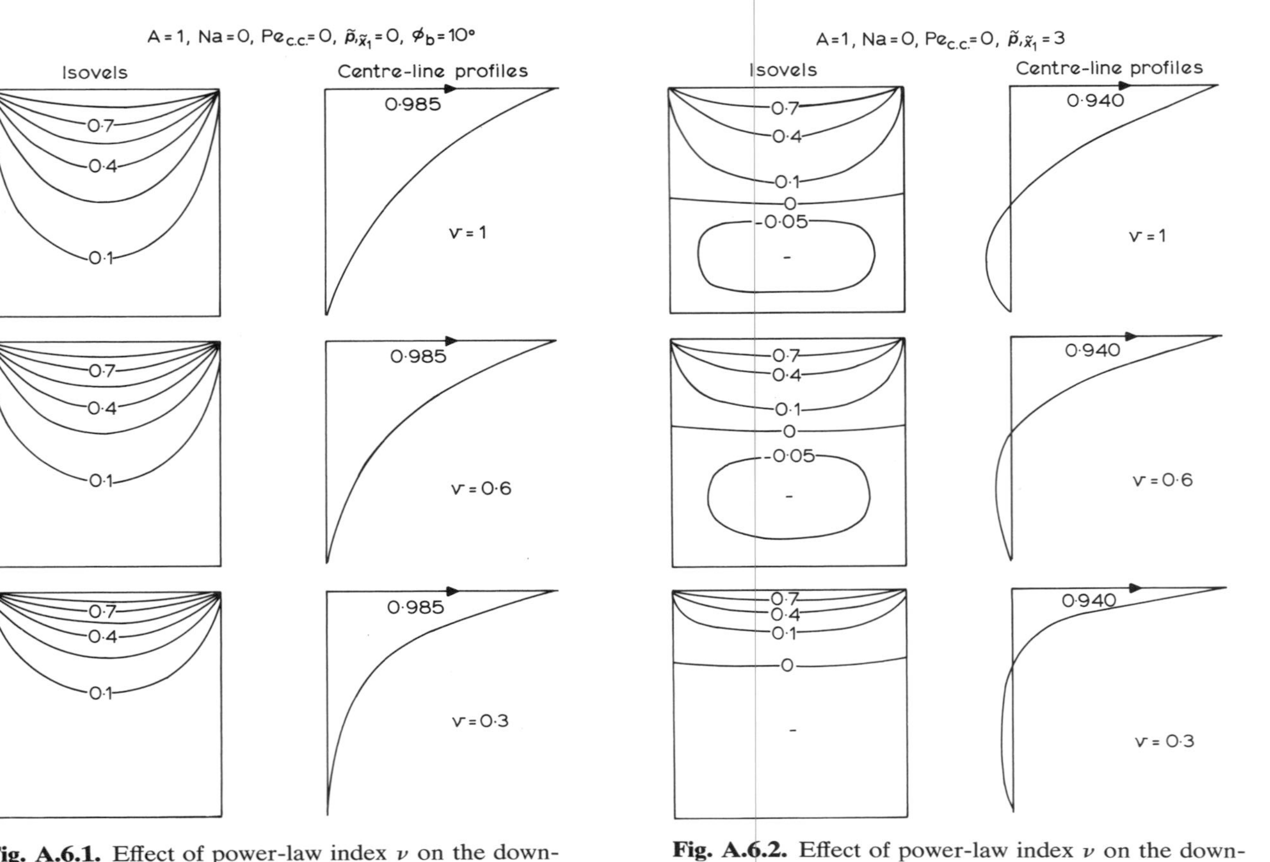

Fig. A.6.1. Effect of power-law index ν on the down-channel velocity $\tilde{v}_1$ for isothermal pure drag flow.

Fig. A.6.2. Effect of power-law index ν on the down-channel velocity $\tilde{v}_1$ for isothermal flow with pressure gradient.

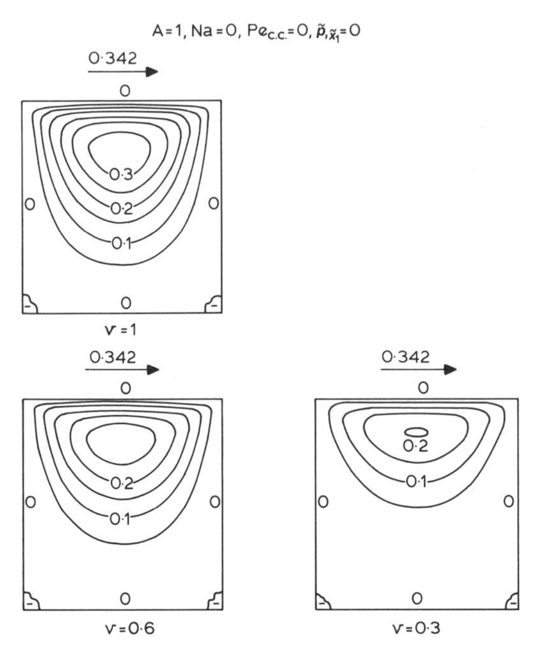

Fig. A.6.3. Behaviour of the transverse flow stream-function Ψ with changing power-law index, ν.

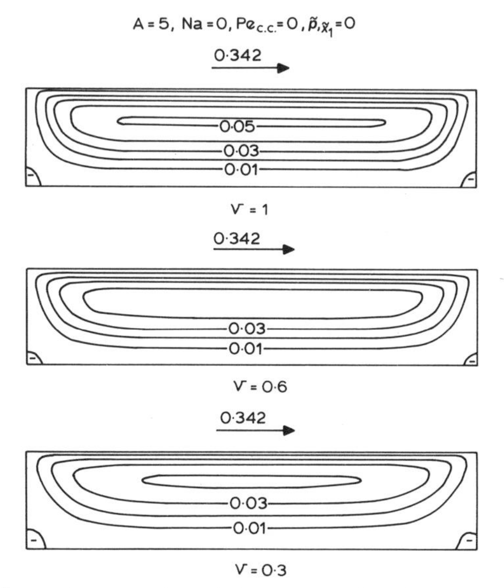

Fig. A.6.4. Effect of power-law index ν on the transverse flow stream-function, Ψ.

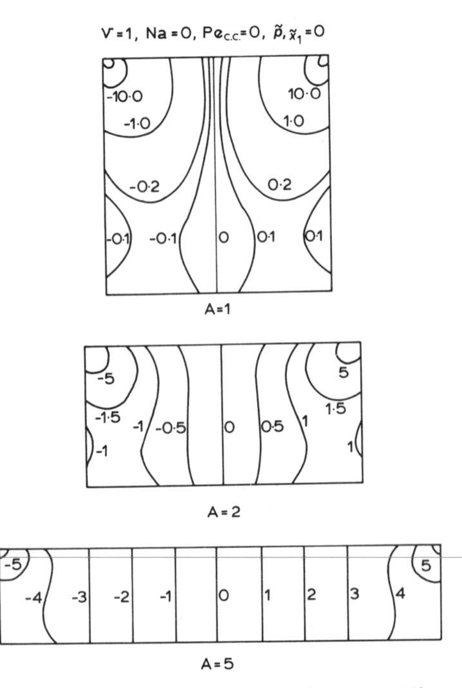

Fig. A.6.5. Dependence of the pressure field $\tilde{p}(\tilde{x}_2, \tilde{x}_3) - \tilde{p}(\frac{1}{2}A, 0)$ on the aspect ratio, A.

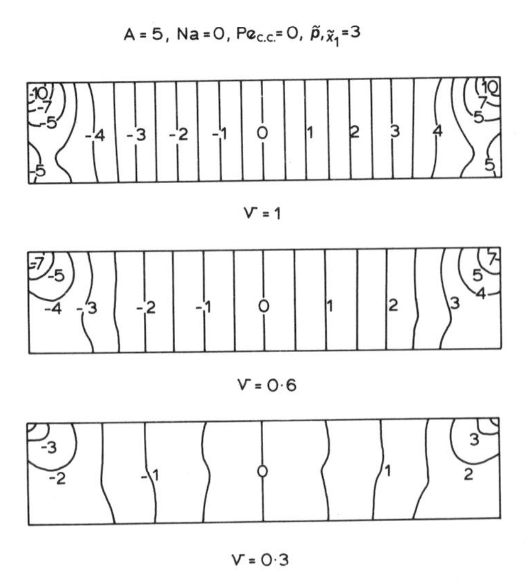

Fig. A.6.6. Changes in the pressure field $\tilde{p}(\tilde{x}_2, \tilde{x}_3) - \tilde{p}(\frac{1}{2}A, 0)$ with power-law index, ν.

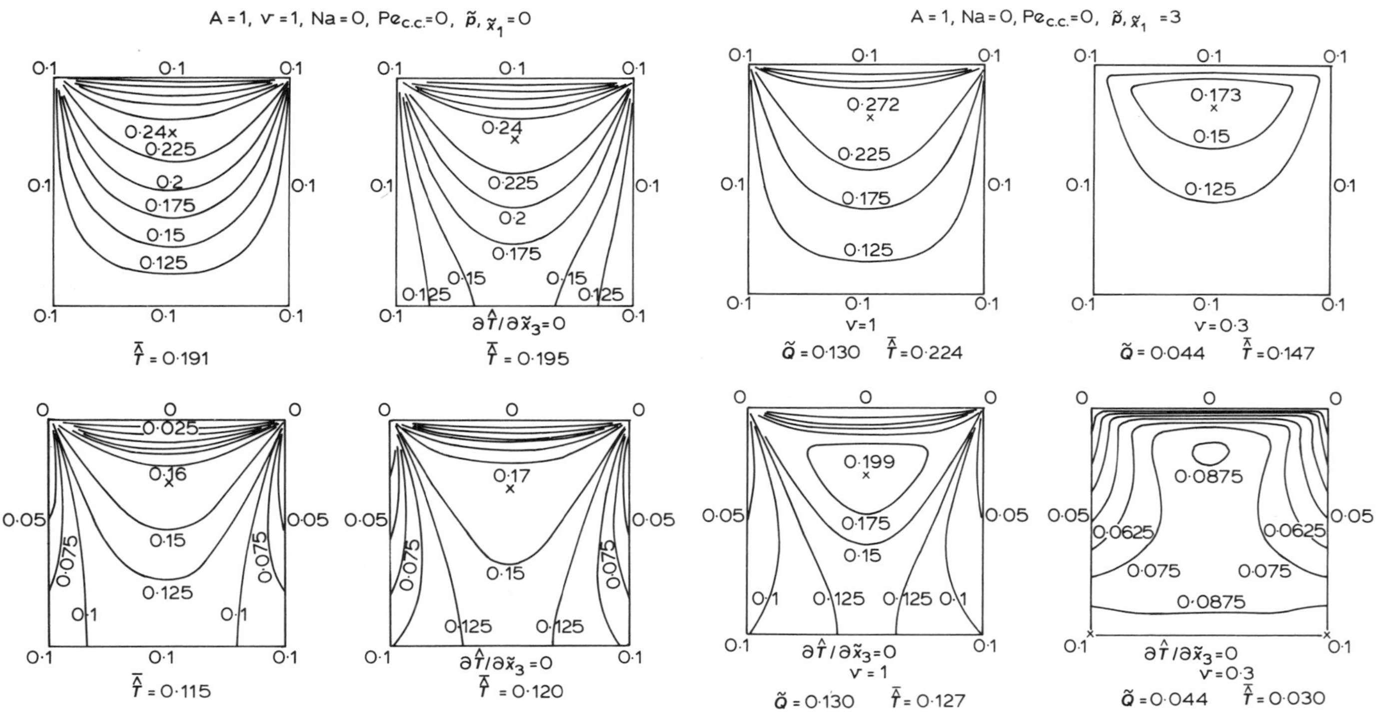

Fig. A.6.7. The effect of different boundary conditions on the temperature field $\hat{T} = \tilde{T}/\mathrm{Na}$, where $\tilde{T}$ is defined by eqn (11.4.44).

Fig. A.6.8. Temperature contours $\hat{T}$ for flows with a large back-pressure gradient.

Fig. A.6.9. The effect of different boundary conditions on the temperature field $\hat{T}$ with convection present.

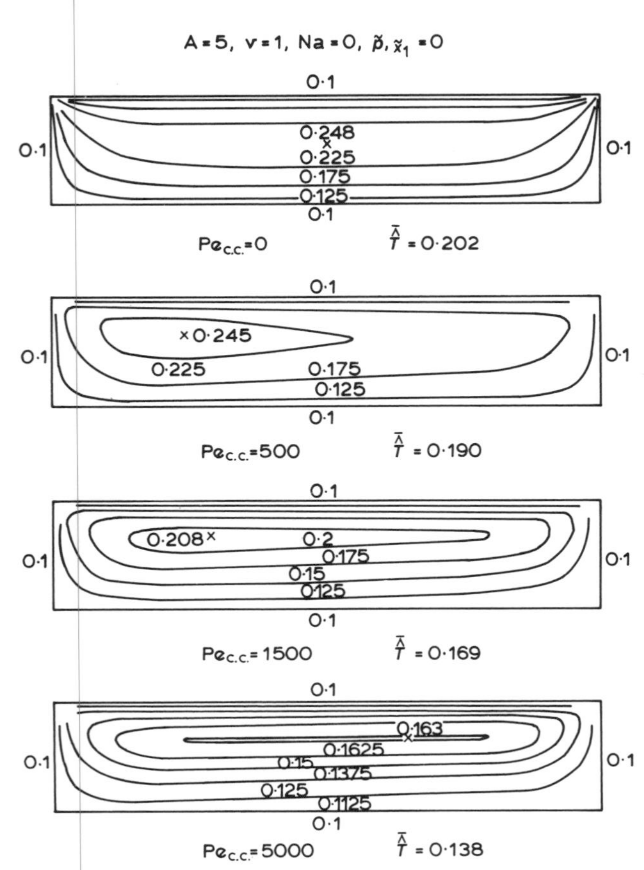

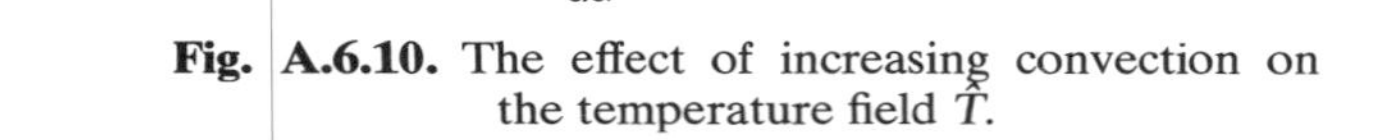

Fig. A.6.10. The effect of increasing convection on the temperature field $\bar{\hat{T}}$.

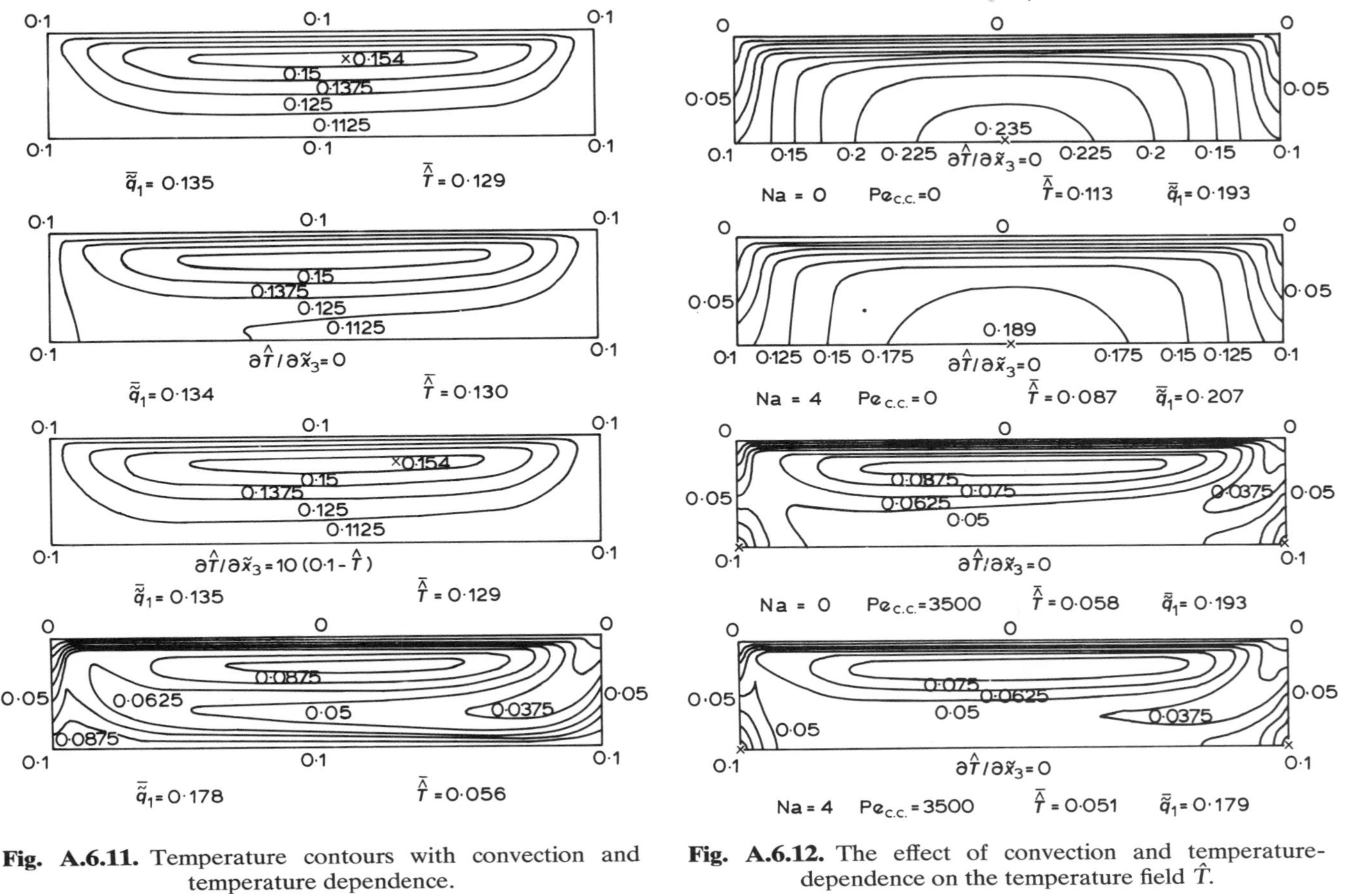

Fig. A.6.11. Temperature contours with convection and temperature dependence.

Fig. A.6.12. The effect of convection and temperature-dependence on the temperature field $\hat{T}$.

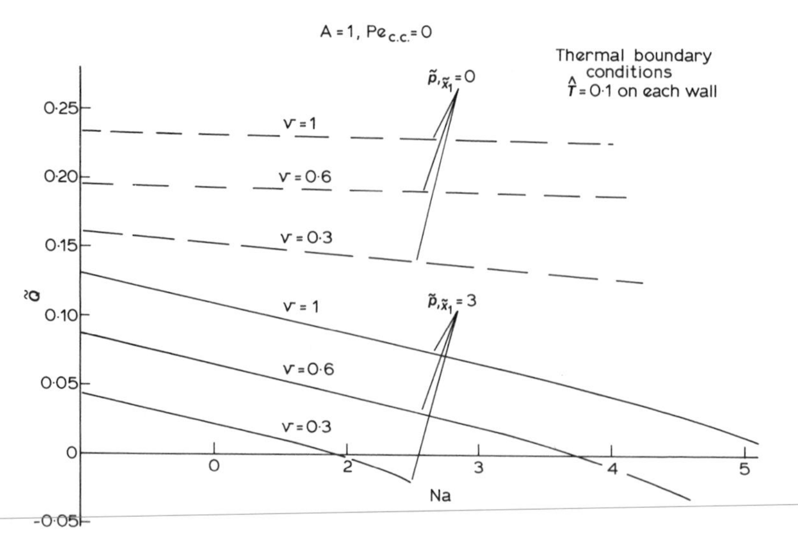

Fig. A.6.13. Flowrate dependence on the interaction between Na and $p_{,\tilde{x}_1}$ when convection is absent.

References

ABRAMOWITZ, M. & STEGUN, I. A. (1964). *Handbook of Mathematical Functions*. National Bureau of Standards, Washington.

ACIERNO, D., LAMANTIA, F. P., MARRUCCI, G. & TITOMANLIO, G. (1976). *J. non-Newt. Fluid Mech.*, **1,** 125.

ADRIANOVA, G. P., KECHETYAN, A. S. & KARGIN, V. A. (1971). Self-oscillation mechanism of necking on extension of polymers. *J. Polym. Sci.*, **A9,** 1919.

ARIS, R. (1962). *Vectors, Tensors and the Basic Equations of Fluid Mechanics*, Prentice Hall, London.

AST, W. (1973). The cooling process in the manufacture of blown film from low-density polyethylene (in German). *Kunststoffe*, **63,** 427.

AST, W. (1974). A contribution to the problems of air cooling in the blown film installation (in German). *ibid.*, **64,** 146.

ASTARITA, G. (1967). Two dimensionless groups relevant in the analysis of steady flows of viscoelastic materials. *Ind. Engng. Chem. Fund.*, **6,** 257.

ASTARITA, G. (1975). *An Introduction to Non-linear Continuum Mechanics*, Societa Editrice di Chimica, Milano.

ASTARITA, G. & MARRUCCI, G. (1974). *Principles of Non-Newtonian Fluid Mechanics*, McGraw-Hill, London.

ASTARITA, G. & SARTI, G. C. (1976). An approach to thermodynamics of polymer flow based on internal state variables. *Polym. Engng Sci.*, **16,** 490.

BALLENGER, T. F. & WHITE, J. L. (1971). The development of the velocity field in polymer melts approaching a capillary die. *J. appl. Polym. Sci.*, **15,** 1949.

BATCHELOR, G. K. (1959). Small-scale variation of convected quantities like temperature in turbulent fluid. Part 1. General discussion and the case of small conductivity. *J. Fluid Mech.*, **5,** 113–33. (See also *ibid.*, **5,** 134–9.)

BHATNAGAR, R. K. & GIESEKUS, H. (1970). On the stability of viscoelastic fluid flow. Part II Plane channel flow; Part III Flow in a cylindrical tube and an annulus. *Rheol. Acta*, **9,** 53; 412.

BENBOW, J. J. & LAMB, P. (1963). New aspects of melt fracture. *SPE Trans.*, **3,** 1.

BENKREIRA, H., EDWARDS, M. F. & WILKINSON, W. L. (1981). A semi-empirical model of the forward roll-coating flow of Newtonian fluids. *Chem. Engng Sci.*, **36,** 423.

BEN-SABAR, E. & CASWELL, B. (1981). Heat transfer effects in die swell. *J. Rheol.*, **25**, 537.

BERLIS, A., BROYER, E., MUND, C. & TADMOR, Z. (1973). Flow patterns in a partially filled screw. *Plast. & Poly.*, **41**, 145.

BERNHARDT, E. C. (Ed.) (1959). *Processing of Thermoplastic Materials*, Reinhold/SPE, New York.

BERNSTEIN, B. & ZAPAS, L. J. (1981). Stability and cold drawing of viscoelastic bars. *J. Rheol.*, **25**, 83.

BIALAS, G. A. & WHITE, J. L. (1969). Extrusion of polymer melts and melt flow instabilities. I. Experimental study of capillary flow and extrudate distortion. *Rubber Chem. Technol.*, **42**, 675.

BIRD, R. B. (1982). Kinetic theory and constitutive equations for polymeric liquids. *J. Rheol.*, **26**, 245.

BIRD, R. B., ARMSTRONG, R. C. & HASSAGER, O. (1977). (contracted in the text to BA & H) *Dynamics of Polymeric Liquids*, Vol. I, Wiley, New York.

BIRD, R. B., HASSAGER, O., ARMSTRONG, R. C. & CURTISS, C. F. (1977). *Dynamics of Polymeric Liquids*, Vol. II, Wiley, New York.

BIRD, R. B., STEWART, W. E. & LIGHTFOOT, E. N. (1960). *Transport Phenomena*, Wiley, New York.

BLACK, J. R., DENN, M. M. & HSIAO, G. C. (1975). Creeping flow of a viscoelastic liquid through a contraction. A numerical perturbation solution. *Theoretical Rheology* (Ed. Hutton, Pearson & Walters), Chapter 1, Applied Science Publishers, London.

BLATZ, P. J., SHARDA, S. C. & TSCHOEGL, N. W. (1973). A new elastic potential for rubbery materials. *Proc. Nat. Acad. Sci. USA*, **70**, 3041.

BRETHERTON, F. P. (1961). The motion of long bubbles in tubes. *J. Fluid Mech.*, **10**, 166.

BRISCOE, B. J. & TABOR, D. (1978). Shear properties of thin polymeric films. *J. Adhesion*, **9**, 145.

BYAM, J. D., COLBERT, G. P. & ZIEGEL, K. D. (1979). Practical simulation of mold filling and vulcanization in injection molding of elastomeric materials. Golden Jubilee Meeting, Society of Rheology, Boston, Mass.

CAREY, D. A., WUST, C. J. & BOGUE, D. C. (1980). Studies in non-isothermal rheology: behaviour near the glass transition temperature and in the oriented glassy state. *J. appl. Polym. Sci.*, **25**, 575.

CASWELL, B. & TANNER, R. I. (1978). Wirecoating die design using finite element methods. *Polym. Engng Sci.*, **18**, 416.

CERRO, R. & SCRIVEN, L. E. (1980). Rapid free surface flows. An integral approach. *Ind. Engng. Chem. Fund.*, **19**, 40.

CHANG, H. & LODGE, A. S. (1971). *Rheol. Acta*, **10**, 448.

CHANG, W. V., BLOCK, R. & TSCHOEGL, N. W. (1976*a*). Time dependent response of soft polymers in moderately large deformations. *Proc. Nat. Acad. Sci. USA*, **73**, 981.

CHANG, W. V., BLOCK, R. & TSCHOEGL, N. W. (1976*b*). On the theory of viscoelastic behaviour of soft polymers in moderately large deformations. *Rheol. Acta*, **15**, 367.

CHOI, K. J., WHITE, J. L. & SPRUIELL, J. E. (1980). Orientation development in tubular film extrusion of polystyrene. Polym. Sci. & Engng. Report no. 143, University of Tennessee.

CHONG, J. S. (1968). Calendering thermoplastic materials. *J. appl. Polym. Sci.*, **12,** 191.

CHOO, K. P., HAMI, M. L. & PITTMAN, J. F. T. (1981). Deep channel operating characteristics of a single screw extruder. Finite element predictions and experimental results for isothermal non-Newtonian flow, *Polym. Engng Sci.*, **21,** 100.

CHUNG, C. I. (1970). New ideas about solids conveying in extruders. *Soc. Plast. Engrs J.*, **26,** 32.

CHUNG, T. S. & RYAN, M. E. (1981). Analysis of the packing stage in injection molding. *Polym. Engng Sci.*, **21,** 271.

CLEGG, P. L. (1958). Elastic effects in the extrusion of polythene. *Rheology of Elastomers*, pp. 174–189, Pergamon Press, London.

COGSWELL, F. N. (1973). The influence of pressure on the viscosity of polymer melts. *Plast. & Polym.*, **41,** 39.

COGSWELL, F. N. (1981). *Polymer Melt Rheology*, George Goodwin, London.

COLEMAN, B. D. (1964*a*). Thermodynamics of materials with memory. *Arch. Rat. Mech. Anal.*, **17,** 1.

COLEMAN, B. D. (1964*b*). On thermodynamics, strain impulses and viscoelasticity. *ibid.*, **17,** 230.

COLEMAN, B. D. & GURTIN, M. E. (1968). On the stability against shear waves of steady flows of non-linear viscoelastic fluids. *J. Fluid Mech.*, **33,** 165.

COLEMAN, B. D. & NOLL, W. (1961). *Ann. N.Y. Acad. Sci*, **89,** 672.

COLEMAN, B. D., MARKOVITZ, H. & NOLL, W. (1966). *Viscometric Flows of non-Newtonian Fluids.* Springer, New York.

COX, A. P. D. & FENNER, R. T. (1980). Melting performance in the single screw extrusion of thermoplastics. *Polym. Engng Sci.*, **20,** 562.

COX, A. P. D., WILLIAMS, J. G. & ISHERWOOD, D. P. (1981). The melting behaviour of a low density polyethylene powder in a screw extruder. *Polym. Engng Sci.*, **21,** 86.

COYNE, J. C. & ELROD, H. G. (1970). *J. Lub. Tech.*, **92,** 451. (See also 1971 ASME Paper no. 70-Lub 3.)

CROCHET, M. J. (1975). A non-isothermal theory of viscoelastic material. *Theoretical Rheology* (Ed. Hutton, Pearson & Walters), Chapter 8, pp. 111–22, Applied Science Publishers, London.

CROCHET, M. J. & KEUNINGS, R. (1980). Die swell of a Maxwell fluid: numerical prediction. *J. non-Newt. Fluid Mech.*, **7,** 192.

CROCHET, M. J. & KEUNINGS, R. (1982). Finite element analysis of die swell of a highly elastic fluid. *ibid.*, **10,** 339.

CROCHET, M. J. & NAGHDI, P. M. (1972). A class of non-isothermal viscoelastic fluids. *Inter. J. Engng Sci.*, **10,** 775.

CROCHET, M. J. & NAGHDI, P. M. (1974). On a restricted non-isothermal theory of simple materials. *J. Mecanique*, **13,** 97.

CURTISS, C. F. & BIRD, R. B. (1981). A kinetic theory for polymer melts. *J. Chem. Phys.*, **74,** 2016; 2026. (See also Bird, Saab & Curtiss, 1982, *ibid.*, **86,** 1102.)

DANCKWERTS, P. V. (1952). The definition and measurement of some characteristics in mixtures. *Appl. Sci. Res.*, **3,** 279.

DARNELL, W. H. & MOL, E. A. J. (1956). Solids conveying in extruders. *Soc. Plast. Engrs J.*, **12,** 20.

DASHNER, P. A. & VAN ARSDALE, W. E. (1981). A phenomenological theory for elastic fluids. *J. Non-Newt. Fluid Mech.*, **8,** 59.

DATTA, A. B. & STRAUSS, K. (1976). Die langsame Stromung einer einfacher viskoelastischen Flussigkeit durch einer endlich breiten Spalt. *Rheol. Acta,* **15,** 283.

DE CLEYN, G. & MEWIS, J. (1981). A constitutive equation for polymer liquids: application to shear flow. *J. Non-Newt. Fluid Mech.*, **9,** 91.

DE KEE, D. & CARREAU, P. J. (1979). A constitutive equation derived from Lodge's network theory. *J. Non-Newt. Fluid Mech.*, **6,** 127.

DEKKER, J. (1976). *Kunststoffe*, **66**.

DENN, M. M. (1977). Extensional flows: experiment and theory. *The Mechanics of Viscoelastic Fluids* (Ed. R. S. Rivlin), ASME-AMD **22,** 101.

DENN, M. M. (1980). Continuous drawing of liquids to form fibers. *Ann. Rev. Fluid Mech.*, **12,** 365.

DENN, M. M. (1981). Pressure drop–flow rate equation for adiabatic capillary flow with a pressure- and temperature-dependent viscosity. *Polym. Engng Sci.*, **21,** 65.

DENN, M. M. & PEARSON, J. R. A. (1981). An overview of the status of melt spinning instabilities. *Proc. 2nd World Cong. Chem. Engng*, Montreal, p. 354.

DENN, M. M., PETRIE, C. J. S. & AVENAS, P. (1975). Mechanics of steady spinning of a viscoelastic fluid. *A.I.Ch.E. J.*, **21,** 791.

DENNISON, M. T. (1967). Flow instability in polymer melts: a review. *Plast. & Polym.*, **35,** 803.

DENSON, C. D. & HYLTON, D. C. (1980). A rheometer for measuring the viscoelastic response of polymer melts in arbitrary planar and biaxial extensional flow fields. *Polym. Engng Sci.*, **20,** 535.

DIETZ, W. & BOGUE, D. C. (1978). Isothermal and non-isothermal stress relaxation during steady shear flow. *Rheol. Acta,* **17,** 595.

DIETZ, W. & WHITE, J. L. (1978). A simple model for the calculation of pressure losses during mold filling and the frozen orientation in injection molding of amorphous polymers (in German). *Rheol. Acta,* **17,** 676.

DIETZ, W., WHITE, J. L. & CLARK, E. S. (1978). Orientation development and relaxation in injection molding of amorphous polymers. *Polym. Engng Sci.*, **18,** 273.

DOBBELS, F. & MEWIS, J. (1977). Nonisothermal nip flow in calendering operations. *A.I.Ch.E. J.*, **23,** 224.

DOI, M. (1980). Molecular rheology of concentrated polymer systems. *J. Polym. Sci.* (Poly. Phys.) **18,** 1005.

DOI, M. & EDWARDS, S. F. (1978). Dynamics of concentrated polymer systems. *J. Chem. Soc., Faraday Trans. II,* **74,** 1789; 1802; 1818.

DOI, M. & EDWARDS, S. F. (1979). *ibid.*, **75,** 38.

DUTTA, A. & RYAN, M. E. (1982). A study of parison development in extrusion blow molding. *J. Non-Newt. Fluid Mech.*, **10,** 235.

EDMONDSON, I. R. & FENNER, R. T. (1975). Melting of thermoplastics in single screw extruders. *Polymer,* **16,** 49.

EIRICH, F. R. (Ed.) (1958). *Rheology.* Academic Press, New York and London.

FARBER, R. (1973). Measurement of deformation rates in the film blowing of polyethylene. M.Eng. Thesis, McGill University.

FARBER, R. & DEALY, J. M. (1974). Strain history of melt in film blowing. *Polym. Engng Sci.*, **14,** 435.

FENNER, R. T. (1970). *Extruder Screw Design*, Iliffe, London.

FENNER, R. T. (1975). The design of large hot melt extruders. *Polymer*, **16,** 298.

FENNER, R. T. (1977). Developments in the analysis of steady screw extrusion of plastics. *ibid.*, **18,** 617.

FENNER, R. T., COX, A. P. D. & ISHERWOOD, D. P. (1979). Surging in single screw extruders. *ibid.*, **20,** 733.

FENNER, R. T. & WILLIAMS, J. G. (1967). Analytical methods of wire-coating die design. *Trans. & J. Plast. Inst.*, **35,** 701.

FERRY, J. D. (1980). *Viscoelastic Properties of Polymers*, 3rd edn, Wiley, New York.

FISHER, E. G. (1958). *Extrusion of Plastics*, Iliffe, London.

FISHER, R. J. & DENN, M. M. (1975). Finite-amplitude stability and draw resonance in isothermal melt spinning. *Chem. Engng Sci.*, **30,** 1129. (See also 1975 Appl. Poly. Symp. no. 27, 103.)

FISHER, R. J. & DENN, M. M. (1976). A theory of isothermal melt spinning and draw resonance. *A.I.Ch.E. J.*, **22,** 236.

FISHER, R. J., DENN, M. M. & TANNER, R. I. (1980). Initial profile development in melt spinning. *Ind. Engng Chem. Fund.*, **19,** 195.

FOX, T. G., GRATCH, S. & LOSHACK, S. (1956). Viscosity relationships for polymers, *Rheology*, Vol. 1 (Ed. F. R. Eirich), Chapter 12, p. 431, Academic Press, New York.

FRADOS, J. (Ed.) (1976). *Plastics Engineering handbook*, 4th edn, Van Nostrand-Rheinhold, New York.

FULLER, G. G. & LEAL, L. G. (1981). *J. Non-Newt. Fluid Mech.*, **8,** 271; see also *Rheol. Acta*, **19,** 580 (1980).

FUNT, J. M. (1975). *Polym. Engng Sci.*, **15,** 817.

FUNT, J. M. (1976). *Mixing of Rubber*, R.A.P.R.A., Shawbury.

GALE, G. M. (1970). Dry blend extrusion of rigid PVC. *Plast. & Polym.*, **38,** 183.

GAVIS, J. & LAURENCE, R. L. (1968*a*). Viscous heating in plane and circular flow between moving surfaces. *Ind. Engng Chem. Fund.*, **7,** 232.

GAVIS, J. & LAURENCE, R. L. (1968*b*). Viscous heating of a power law liquid in plane flow. *ibid.*, **7,** 525.

GELDER, D. (1971). The stability of drawing processes. *Ind. Engng Chem. Fund.*, **10,** 534.

GIESEKUS, H. (1961). *Rheol. Acta*, **1,** 395, 404.

GIESEKUS, H. (1972). On instabilities in Poiseuille and Couette flows of viscoelastic fluids. *Prog. Heat & Mass Trans.*, **5,** 187.

GODDARD, J. D. (1979). Polymer fluid mechanics. *Adv. appl. Mech.*, **19,** 143.

GREENER, J. & MIDDLEMAN, S. (1981). Reverse roll coating of viscous and viscoelastic liquids. *Ind. Engng Chem. Fund.*, **20,** 63.

GRIFFIN, O. M. (1977). An integral energy balance model for the melting of solids on a fast moving surface, with application to the transport processes during extrusion. *Int. J. Heat & Mass Trans.*, **20,** 675.

GRIFFITH, R. M. (1962). Fully developed flow in screw extruders. *Ind. Engng Chem. Fund.*, **1,** 180.

GUPTA, R. K. & METZNER, A. B. (1982). Modeling of non-isothermal polymer processes. *J. Rheol.*, **26,** 181.

HALMOS, A. L., PEARSON, J. R. A. & TROTTNOW, R. (1978). Melting in single-screw extruders Part III. Solutions for a power law temperature-dependent viscous melt. *Polymer*, **19,** 1199.

HAPPEL, J. & BRENNER H. (1973). *Low Reynolds Number Hydrodynamics*, Noordhoff, Leiden.

HASSAN, G. A. & PARNABY, J. (1981). Model reference optimal steady-state adaptive computer control of plastics extrusion processes. *Polym. Engng Sci.*, **21,** 276.

HIEBER, C. A. & SHEN, S. F. (1980). A finite element/finite difference simulation of the injection moulding filling process. *J. Non-Newt. Fluid Mech.*, **7,** 1.

HINCH, E. J. & LEAL, L. G. (1975). Constitutive equations in suspension mechanics. *J. Fluid Mech.*, **71,** 481.

HO, T. C., DENN, M. M. & ANSCHUS, B. E. (1977). Low Reynolds number instability from viscous heating. *Rheol. Acta.*, **16,** 61.

HOLSTEIN, H. (1981). A singular finite difference treatment of the re-entrant corner Part 1: Newtonian fluids. *J. Non-Newt. Fluid Mech.*, **8,** 81.

HOOLEY, C. J. & COHEN, R. E. (1978). On the determination of apparent activation energies of viscoelastic deformation processes using the instantaneous temperature-change experiment. *Rheol Acta*, **17,** 538.

HOPKINS, M. R. (1957), *Brit. J. appl. Phys.*, **8,** 442.

HOWELLS, E. R. & BENBOW, J. J. (1962). Flow defects in polymer melts. *Plast. Inst. Trans.*, **30,** 240.

HUANG, D. & WHITE, J. L. (1980). Experimental and theoretical investigation of extrudate swell of polymer melts from small (length) (cross-section) ratio slit and capillary dies. *Polym. Engng Sci.*, **20,** 182.

HUILGOL, R. R. (1971). A class of motions with constant stretch history. *Quart. appl. Math.*, **29,** 1.

HUILGOL, R. R. (1975). *Continuum Mechanics of Viscoelastic Liquids*, Hindustan Publishing Corporation, Delhi.

HUILGOL, R. R. (1981). Remarks on nearly extensional flows. *J. Non-Newt. Fluid Mech.*, **8,** 169.

HULL, A. M. (1981*a*). An exact solution for the slow flow of a general linear viscoelastic fluid through a slit. *J. Non-Newt. Fluid Mech.*, **8,** 327.

HULL, A. M. (1981*b*). Die entry flow of polymers. Ph.D. thesis, University of London.

HULL, A. M., RICHARDSON, S. M. & PEARSON, J. R. A. (1981). A rheological model for rubber extrusion through a converging nozzle (see also Pearson, J. R. A., Richardson, S. M. & Hull, A. M. Models for flow of highly elastic fluids through slits and nozzles). Joint Brit. Soc. of Rheol./Belgian Gp. of Rheol. meeting, Liege, April 1981.

HUXTABLE, J., COGSWELL, F. N. & WRIGGLES, J. D. (1981). Polymer-metal friction as relevant to processing. *Plast. & Rubber Proc. Appl.*, **1,** 87.
INGEN-HOUSZ, J. F. & MEIJER, M. E. H. (1981). The melting performance of single screw extruders. *Polym. Engng Sci.*, **21,** 352.
ISHIHARA, H. & KASE, S. (1975). Studies on melt spinning V. Draw resonance as a limit cycle. *J. appl. Polym. Sci.*, **19,** 557. See also Ishihara, H. 1977. Theoretical analysis of draw resonance in melt spinning and its application to characterization of the tensile rheological properties of molten polymers. Dissertation, Faculty of Engineering, Kyoto University.
ISHIZUKA, O. & KOYAMA, K. (1980). Elongational viscosity at a constant elongational strain rate of polypropylene melt. *Polymer*, **21,** 164.
JANESCHITZ-KREIGL, H. (1979). Injection moulding of plastics II. Analytical solution of heat transfer problem. *Rheol. Acta*, **18,** 693.
JANESCHITZ-KRIEGEL, H. & SCHIJF, J. (1969). A study of radial heat transfer in single-screw extruders. *Plast. & Poly.*, **37,** 523.
JANSSEN, L. P. B. M. (1978). *Twin Screw Extrusion.* Elsevier, Amsterdam.
JEAN, M. & PRITCHARD, W. G. (1980). The flow of fluids from nozzles at small Reynolds number. *Proc. Roy. Soc. A.*, **370,** 61.
JENKINS, A. D. (Ed.) (1972). *Polymer Science*, Vol. I, North Holland, Amsterdam.
JONGSCHAAP, R. J. J. (1981). Derivation of the Marrucci model from transient network theory. *J. non-Newt. Fluid Mech.*, **8,** 183.
JOHNSON, M. W. & SEGALMAN, D. J. (1977). A model for viscoelastic fluid behaviour which allows non-affine deformation. *J. non-Newt. Fluid Mech.*, **2,** 255.
JOSEPH, D. D. (1974). Slow motion and viscometric motion; stability and bifurcation of the rest state of a simple fluid. *Arch. Rat. Mech. & Anal.*, **56,** 99.
JOSEPH, D. D. (1976). *Stability of Fluid Motion*, Springer-Verlag, Berlin.
JOSEPH, D. D. (1979). Perturbations of states of rest and rigid motions of simple fluids and solids. *J. non-Newt. Fluid Mech.*, **5,** 13.
JOSEPH, D. D. & BEAVERS, G. S. (1977). Free surfaces induced by the motion of viscoelastic fluids. *The Mechanics of Viscoelastic Fluids Ch. XII AMD* **22,** ASME, New York.
JOSEPH, D. D. & FOSDICK, R. L. (1973). The free surface in a liquid between cylinders rotating at different speeds. *Arch. Rat. Mech. & Anal.*, **49,** 321, 381.
KAMAL, M. R. & NYUN, H. (1980). Capillary viscometry: a complete analysis including pressure and viscous heating effects. *Polym. Engng Sci.*, **20,** 109.
KASE, S. (1974). *J. appl. Polym. Sci.*, **18,** 3279.
KASE, S. (1981). Transfer function approach to the dynamics of melt spinning. *Proc. 2nd World Congr. Chem. Engng*, Montreal, Canada, 356.
KASE, S. & DENN, M. M. (1978). Dynamics of the melt spinning process. *Proc.* 1978 *Joint Automatic Control Conf.*, **2,** 71.
KASE, S. & IKKO, K. (1981). An analytical transient solution of the non-linear equations governing the isothermal spinning of power-law fluids. *Proc 2nd World Congr. Chem. Engng*, Montreal, Canada, 329.

Kase, S. & Matsuo, T. (1965). *J. Polym. Sci.*, **A3,** 2541.
Kase, S. & Nakajima, T. (1980). Growth of a dent on an isothermal fluid filament in uniaxial extension. *Rheol. Acta,* **19,** 698.
Klenk, P. (1968). Plastfiziermodelle fur die Verarbeitung benetzender und nichtbenetzender Thermoplaste auf Einschnecken Extruden. *Rheol. Acta,* **7,** 74.
Kosel, U. M. (1971). A novel concept of single screw extrusion. *Plast. & Polym.,* **39,** 319.
Kreuger, W. L. & Tadmor, Z. (1980). Injection molding into a rectangular cavity with inserts. *Polym. Engng Sci.,* **20,** 326.
Lamantia, F. P. & Titomanlio, G. (1979). Testing of a constitutive equation with free volume dependent relaxation spectrum. *Rheol. Acta,* **18,** 469.
Lan, H. C. & Schowalter, W. R. (1981). On the use of mixed corotational and codeformational properties in constitutive equations. *J. Rheol.,* **24,** 507.
Langlois, W. E. & Rivlin, R. S. (1959). Steady flow of slightly viscoelastic fluids. Tech. Report no. DA-4725/3 Division of Applied Maths, Brown University, Providence, Rhode Island. (See also 1963 *Rendi. Mat.,* **22,** 109.)
Laun, H. M. (1978). *Rheol. Acta,* **17,** 1.
Lee, D. G. & Zerkle, R. D. (1969). The effect of liquid solidification in a parallel plate channel upon laminar flow heat transfer and pressure drop. *Trans. ASME J. Heat Transf.,* **91,** 583.
Leonov, A. I. (1976). Non-equilibrium thermodyanimcs and rheology of viscoelastic polymer media. *Rheol. Acta,* **15,** 85.
Leonov, A. I., Lipkina, E. H., Pashkin, E. D. & Prokunin, A. N. (1976). Theoretical and experimental investigation of shearing in elastic polymer liquids. *Rheol. Acta,* **15,** 411.
Lin, C. C. (1955). *The Theory of Hydrodynamic Stability.* Cambridge University Press, Cambridge.
Lindt, J. T. (1976). A dynamic melting model for a single screw extruder. *Polym. Engng Sci.,* **16,** 284.
Liu, T. Y., Soong, D. S. & Williams, M. C. (1981). Time dependent rheological properties and transient structural states of entangled polymeric liquids—A kinetic network model. *Polym. Engng Sci.,* **21,** 675.
Locati, C. (1976). A model for interpreting die swell of polymers. *Rheol. Acta,* **15,** 525.
Lodge, A. S. (1964). *Elastic Liquids,* Academic Press, London and New York.
Lodge, A. S. (1974). *Body Tensor Fields in Continuum Mechanics,* Academic Press, London and New York.
Lodge, A. S., Armstrong, R. C., Wagner, M. H. & Winter, M. H. (1981). Constitutive equations from Gaussian molecular network theories in polymer rheology. *Pure appl. Chem.* IUPAC Division Working Party 1979.
Lovegrove, J. G. A. & Williams, J. G. (1973). Solids conveying in a single screw extruder; the role of gravity forces. *J. Mech. Engng Sci.,* **15,** 114.
Lovegrove, J. G. A. & Williams, J. G. (1974). *Poly. Engng Sci.,* **14,** 589.
Lupton, J. M. (1963). A.I.Ch.E./S.P.E. Symp. on Fundamentals of Polymer Processing, Philadelphia.
Lupton, J. M. & Register, R. W. (1965). Melt flow of polyethylene at high rates. *Polym. Engng Sci.,* **5,** 235.

MADDOCK, B. H. (1959). A visual analysis of flow and mixing. *Soc. Plast. Engrs J.*, **15,** 383.

MAILLEFER, C. (1963). A two channel extruder screw. *Modern Plast.*, **40,** 132.

MARRUCCI, G., TITOMANLIO, G. & SARTI, G. C. (1973). Testing of a constitutive equation for entangled networks by elongational and shear data of polymer melts. *Rheol. Acta*, **12,** 269.

MARTIN, B. (1967). Some analytical solutions for viscometric flows of power-law fluids with heat generation and temperature dependent viscosity. *Int. J. Non-Linear Mech.*, **2,** 285.

MARTIN, B. (1969). Numerical studies of steady state extrusion processes. Ph.D. dissertation, University of Cambridge.

MARTIN, B., PEARSON, J. R. A. & YATES, B. (1969). On Screw Extrusion. Part I: Steady Flow Calculations. University of Cambridge Dept. of Chem. Engng Polymer Processing Research Centre Report no. 5.

MARTIN, G. (1972). Einfarben von Kunststoffen. *Kunststoffe Technik*, **12,** 329.

MATOVICH, M. A. (1966). Mechanics of a spinning threadline. Ph.D Thesis, University of Cambridge.

MATOVICH, M. A. & PEARSON, J. R. A. (1969). Spinning a molten threadline: steady state isothermal viscous flow. *Ind. Engng Chem. Fund.*, **8,** 512.

MATSUI, M. & BOGUE, D. C. (1976). Non-isothermal rheological response in melt spinning and idealized elongational flow. *Polym. Engng Sci.*, **16,** 735.

MATSUI, M. & BOGUE, D. C. (1977). Studies in non-isothermal rheology. *Trans. Soc. Rheol.*, **21,** 133.

MATSUMOTO, T. & BOGUE, D. C. (1977). Non-isothermal rheological response during elongational flow. *Trans. Soc. Rheol.*, **21,** 453.

MCKELVEY, J. M. (1962). *Polymer Processing*, Wiley, New York.

MEISSNER, J. (1975). *Rheol. Acta*, **14,** 201.

MEISSNER, J. (1975). *Basic Parameters, Melt Rheology, Processing and End-use Properties of Three Similar Low-density Polyethylene Samples*, Butterworths, London.

MEISSNER, J., DEMARMELS, A. & PORTMAN, P. R. (1981). Elongational flow behavior including planar extension of polyisobutylene. Joint Meeting of British and Italian Societies of Rheology on Stretching Flows and General Rheology, Newcastle upon Tyne, September, 1981.

MENDELSON, M. A., YEH, P-W., BROWN, R. A. & ARMSTRONG, R. C. (1982). Approximation error in finite element calculation of viscoelastic fluid flows. *J. Non-Newt. Fluid Mech.*, **10,** 31.

MENGES, G. & KLENK, P. (1967). *Kunststoffe*, **57,** 598.

METZNER, A. B. (1971). Extensional primary field approximations for viscoelastic media. *Rheol. Acta*, **10,** 434.

MHASKAR, R. D., SHAH, Y. T. & PEARSON, J. R. A. (1977). On the stability of flow in channels with a moving boundary: the mold filling problem. *Trans. Soc. Rheol.*, **21,** 291.

MICHAEL, D. H. (1958). The separation of viscous liquid at a straight edge. *Mathematika*, **5,** 82.

MIDDLEMAN, S. (1977). *Fundamentals of Polymer Processing*, McGraw-Hill, New York.

MOFFATT, H. K. (1964). Viscous and resistive eddies near a sharp corner. *J. Fluid Mech.*, **18,** 1.

MOORE, C. A. & PEARSON, J. R. A. (1975). Experimental investigation into an isothermal spinning threadline: extensional rheology of a Separan AP30 solution in glycerol and water. *Rheol. Acta*, **14,** 436.

MUNSTEDT, H. & LAUN, H. M. (1979). Elongational behavior of a low density polyethylene melt. *Rheol. Acta*, **18,** 492.

MURAYAMA, N. (1981). Network theory of the non-linear behavior of polymer melts. *Colloid & Polym. Sci.*, **259,** 724.

NOLL, W. (1962). Motions with constant stretch history. *Arch. Rat. Mech. Anal.*, **11,** 97.

NOVOZHILOV, V. V. (1964). *Thin Shell Theory*, Noordhoff, The Hague.

OCKENDON, H. (1979). Channel flow with temperature-dependent viscosity and internal viscous dissipation. *J. Fluid Mech.*, **93,** 737.

OCKENDON, H. & OCKENDON, J. R. (1977). Variable-viscosity flows in heated and cooled channels. *J. Fluid Mech.*, **83,** 177.

OKOBO, S. & HORI, Y. (1980). Model analysis of oscillating flow of high density polyethylene melt. *J. Rheol.*, **24,** 253.

OLBRICHT, W. L., RALLISON, J. M. & LEAL, L. G. (1982). Strong flow criteria based on microstructure deformation. *J. non-Newt. Fluid Mech.*, **10,** 291.

OLDROYD, J. G. (1950). On the formulation of rheological equations of state. *Proc. R. Soc., Lond.*, **A200,** 523.

OLDROYD, J. G. (1965). Some steady flows of the general elastico-viscous liquid. *Proc. R. Soc., Lond.*, **A283,** 115.

OTTINO, J. M., RANZ, W. E. & MACOSKO, C. W. (1979). A lamellar model for analysis of liquid-liquid mixing. *Chem. Engng Sci.*, **34,** 877.

PARNABY, J., HASSAN, G. A., HELMY, H. A. A. & ALI, A. (1981). Design of plastics processing machinery using lumped parameter methods. *Plast. & Rubber Proc. Appl.*, **1,** 303. (See also Parnaby *et al.* 1978 *Plast. & Rubber: Processing*, **3,** 89.)

PASLAY, P. R. (1955). The calendering of elastico viscous materials. M.I.T. Mechanical Engineering Doctoral Thesis. See also ASME preprints, Applied Mechanics Division Summer Conference, 1957.

PAWLOWSKI, J. (1967). Zur Theorie der Ahnlichkeitsubertragung bei Transportvorgangen in nicht-Newtonschen Stoffen. *Rheol. Acta*, **6,** 54.

PEARSON, J. R. A. (1960). Instability of uniform viscous flow under rollers and spreaders. *J. Fluid Mech.*, **7,** 481.

PEARSON, J. R. A. (1962). Non-Newtonian flow and die design, Parts I & II. *Trans. & J. Plast. Inst.*, **30,** 230.

PEARSON, J. R. A. (1963). Non-Newtonian flow and die design, Part III. A cross-head die design. *Trans. & J. Plast. Inst.*, **31,** 125.

PEARSON, J. R. A. (1964). Non-Newtonian flow and die design, Part IV. Flat film design. *Trans. & J. Plast. Inst.*, **32,** 239.

PEARSON, J. R. A. (1966). *Mechanical Principles of Polymer Melt Processing*, Pergamon Press, Oxford.

PEARSON, J. R. A. (1967). The lubrication approximation applied to non-Newtonian flow problems: a perturbation approach. In: *Non linear partial differential systems.* (Ed. Ames), Academic Press, New York, p. 73.

PEARSON, J. R. A. (1969). Mechanisms for melt flow instability. *Plast. & Polym.*, **37**, 285.

PEARSON, J. R. A. (1972). Heat transfer effects in molten polymers. *Progr. Heat & Mass Transf.*, **5**, 73.

PEARSON, J. R. A. (1976*a*). On the melting of solids near a hot moving interface, with particular reference to beds of granular polymers. *Int. J. Heat Mass Transf.*, **19**, 405.

PEARSON, J. R. A. (1976*b*). On the scale-up of single-screw extruders for polymer processing. *Plast. & Rubber*: *Processing*, **1**, 113.

PEARSON, J. R. A. (1976*c*). Instability in non-Newtonian fluid flow. *Ann. Rev. Fluid Mech.*, **8**, 163.

PEARSON, J. R. A. (1977). Variable-viscosity flows in channels with high heat generation. *J. Fluid Mech.*, **83**, 191.

PEARSON, J. R. A. & DEVINE, F. (1963). Rheological data and its application to the design of rubber extrusion dies. *Rubber World*, **149**, 49.

PEARSON, J. R. A. & GUTTERIDGE, P. A. (1978). Stretching flows for thin film production. Part I. Bubble blowing in the solid phase. *J. non-Newt. Fluid Mech.*, **4**, 57.

PEARSON, J. R. A. & MATOVICH, M. A. (1969). Spinning a molten threadline: stability. *Ind. Engng Chem. Fund.*, **8**, 605.

PEARSON, J. R. A. & PETRIE, C. J. S. (1970*a*). The flow of a tubular film. Part 1 Formal mathematical representation. *J. Fluid Mech.*, **40**, 1.

PEARSON, J. R. A. & PETRIE, C. J. S. (1970*b*). The flow of a tubular film. Part 2 Interpretation of the model and discussion of solutions. *J. Fluid Mech.*, **42**, 609.

PEARSON, J. R. A. & PETRIE, C. J. S. (1970*c*). A fluid mechanical analysis of the film-blowing process. *Plast. & Polym.*, **38**, 85.

PEARSON, J. R. A. & PETRIE, C. J. S. (1985). Inlet boundary conditions for fibre spinning of viscoelastic fluids. *J. non-Newt. Fluid Mech.* (to be submitted).

PEARSON, J. R. A. & PICKUP, T. J. (1973). Stability of wedge and channel flow of highly viscous and elastic liquids. *Polymer*, **14**, 209.

PEARSON, J. R. A. & RICHARDSON, S. M. (Eds) (1983). *Computational Analysis of Polymer Processing*. Applied Science Publishers, London.

PEARSON, J. R. A. & SHAH, Y. T. (1972). Stability analysis of the fibre spinning process. *Trans. Soc. Rheol.*, **16**, 519.

PEARSON, J. R. A. & SHAH, Y. T. (1974). On the stability of isothermal and non-isothermal fibre spinning of power-low fluids. *Ind. Engng Chem. Fund.*, **13**, 134.

PEARSON, J. R. A., SHAH, Y. T. & MHASKAR, R. D. (1976). On the stability of fibre spinning of freezing fluids. *Ind. Engng.Chem. Fund.*, **15**, 31.

PEARSON, J. R. A., SHAH, Y. T. & VIEIRA, E. S. A. (1973). Stability of non-isothermal flow in channels I. Temperature-dependent Newtonian fluid without heat generation. *Chem. Engng Sci.*, **28**, 2079.

PEARSON, J. R. A. & TROTTNOW, R. (1978). On die swell. *J. non-Newt. Fluid Mech.*, **4**, 195.

PETRIE, C. J. S. (1973). Memory effects in a non-uniform flow: a study of the behavior of a tubular film of viscoelastic fluid. *Rheol. Acta*, **12**, 92.

PETRIE, C. J. S. (1974). Mathematical modeling of heat transfer in film blowing—a case study. *Plast. & Polym.*, **42,** 259.

PETRIE, C. J. S. (1975*a*). A re-interpretation of the spinnability predictions of Chang and Lodge. *Rheol. Acta*, **14,** 955.

PETRIE, C. J. S. (1975*b*). A comparison of theoretical predictions with published experimental measurements on the blow film process. *A.I.Ch.E.J.*, **21,** 275.

PETRIE, C. J. S. (1977). On stretching Maxwell models. *J. non-Newt. Fluid Mech.*, **2,** 221.

PETRIE, C. J. S. (1979). *Elongational Flows*, Pitman, London.

PETRIE, C. J. S. & DENN, M. M. (1976). Instabilities in polymer processing. *A.I.Ch.E. J.*, **22,** 209.

PHAN-THIEN, N. & TANNER, R. I. (1977). A new constitutive equation derived from network theory. *J. non-Newt. Fluid Mech.*, **2,** 353.

PHUOC, H. B. & TANNER, R. I. (1980). Thermally-induced extrudate swell. *J. Fluid Mech.*, **91,** 253.

PICKUP, T. J. F. (1970). Converging flow of viscoelastic fluids. Ph.D dissertation, University of Cambridge.

PIPKIN, A. C. & OWEN, D. H. (1967). *Phys. Fluids*, **10,** 836.

PITTS, E. & GREILLER, J. (1961). The flow of thin liquid films between rollers. *J. Fluid Mech.*, **11,** 33.

POTENTE, H. & FISCHER, P. (1977). Model flows for the design of single-screw plasticizing extruders (in German). *Kunststoffe*, **67,** 242.

POUTNEY, D. C. & WALTERS, K. (1978). *Phys. Fluids*, **21,** 1482.

POUTNEY, D. C. & WALTERS, K. (1979). *ibid.*, **22,** 1007.

POWELL, R. L. & SCHWARTZ, W. H. (1981). Infinitesimal perturbation of extensional motions. *J. non-Newt. Fluid Mech.*, **8,** 139.

PRICHATT, R. J., PARNABY, J. & WORTH, R. A. (1975). Design considerations in the development of extrudate wall-thickness control in blow moulding. *Plast. & Polym.*, **43,** 55.

RICHARDSON, S. (1967). Slow viscous flows with free surfaces. Ph.D dissertation, University of Cambridge.

RICHARDSON, S. (1970). The die swell phenomenon. *Rheol. Acta*, **9,** 193 (see also *Proc. Camb. Phil. Soc.*, **67,** 477).

RICHARDSON, S. M. (1979). Extended Leveque solutions for flows of power law fluids in pipes and channels. *Int. J. Heat Mass Transf.*, **22,** 1417.

RICHARDSON, S. M. (1983). Injection moulding of thermoplastics: Freezing during mould filling. *Rheol. Acta*, **22,** 223.

RICHARDSON, S. M., PEARSON, H. J. & PEARSON, J. R. A. (1980). Simulation of injection moulding. *Plast. & Rubber: Processing*, **5,** 55.

RONCA, G. (1976*a*). A network theory of isothermal spinning. *Rheol. Acta*, **15,** 628.

RONCA, G. (1976*b*). Statistical invariance of an impermanent network I. A renormalization of the theory of Lodge, II. The energetic assumption. *Rheol. Acta*, **15,** 149; 156.

RYAN, M. E. & CHUNG, T-S. (1980). Conformal mapping analysis of injection mold filling. *Polym. Engng Sci.*, **20,** 642.

Sampson, P. & Gibson, R. D. (1981). A mathematical theory of nozzle blockage by freezing. *Int. J. Heat & Mass Trans.*, **24,** 231.

Sarti, G. C. (1977). Thermodynamics of polymeric liquids: simple fluids with entropic elasticity obeying the time-temperature superposition principle. *Rheol. Acta*, **16,** 516.

Savage, M. D. (1977). Cavitation in lubrication. *J. Fluid Mech.*, **80,** 743.

Schenkel, G. (1966). *Plastics Extrusion Technology and Theory* (English translation), Iliffe, London; American Elsevier, New York.

Schneider, K. (1969). *Chemie Ing. Tech.*, **41,** 364; see also Technical Report on Plastics Processing—Processing in the feeding zone of an extruder. I.K.V. Aachen.

Schlultz, W. W. & Davis, S. H. (1982). One dimensional liquid fibers. *J. Rheol.*, **26,** 331.

Schummer, P. (1967). *Rheol. Acta*, **6,** 192.

Shah, Y. T. & Pearson, J. R. A. (1972*a*). On the stability of non-isothermal spinning. *Ind. Engng Chem. Fund.*, **11,** 145.

Shah, Y. T. & Pearson, J. R. A. (1972*b*). On the stability of non-isothermal spinning—general case. *Ind. Engng Chem. Fund.*, **11,** 150.

Shah, Y. T. & Pearson, J. R. A. (1972*c*). Stability of fibre spinning of power-law fluids. *Polym. Engng & Sci.*, **12,** 219.

Shah, Y. T. & Pearson, J. R. A. (1974). Stability of non-isothermal flow in channels. II Temperature dependent power-law fluids without heat generation. III Temperature dependent power-law fluids with heat generation. *Chem. Engng Sci.*, **29,** 737; 1485. (See also Pearson *et al.*, 1973.)

Shapiro, J. (1971). Melting in plasticating extruders. Ph.D Thesis, University of Cambridge.

Shapiro, J., Halmos, A. L. & Pearson, J. R. A. (1976). Melting in single screw extruders. Part I: The mathematical model. Part II: Solution for a Newtonian fluid. *Polymer*, **17,** 905.

Skelland, A. H. P. (1967). *Non-Newtonian Flow and Heat Transfer.* Wiley, New York.

Sokolnikov, I. S. (1964). *Tensor Analysis*, Wiley, New York.

Spearot, J. A. & Metzner, A. B. (1972). Isothermal spinning of molten polyethylenes. *Trans. Soc. Rheol.*, **16,** 495.

Steiber, W. (1933). Das Schwemmlayer, V.D.I. (Berlin).

Stevenson, J. F. (1976). *Chem. Engng. Sci.*, **31,** 1225.

Strauss, K. (1975). Stability and over-stability of the plane flow of a simple viscoelastic fluid in a converging channel. *Theoretical Rheology* (Ed. Hutton, J. F., Pearson, J. R. A. & Walters, K.), Applied Science Publishers, London, p. 56.

Street, L. F. (1961). Plastifying extrusion. *Int. Plast. J*, **1,** 289.

Stuart, J. T. (1963). Hydrodynamic stability. Chapter IX in *Laminar Boundary Layers* (Ed. Rosenhead, L.), Clarendon Press, Oxford.

Sturges, L. D. (1981). A theoretic study of extrudate swell. *J. non-Newt. Fluid Mech.*, **9,** 357.

Swift, H. W. (1932). *Proc. Inst. Civ. Engrs*, **233,** 267.

Tadmor, Z. (1966). *Polym. Engng Sci.*, **6,** 185.

TADMOR, Z. & BIRD, R. B. (1974). Rheological analysis of stabilizing forces in wire coating dies. *Poly. Engng Sci.*, **14,** 124.

TADMOR, Z. & GOGOS, C. G. (1979). *Principles of Polymer Processing,* SPE/Wiley/Interscience, New York.

TADMOR, Z. & KLEIN, I. (1970). *Engineering Principles of Plasticating Extrusion,* Van Nostrand, New York.

TANNER, R. I. (1970). A theory of die swell. *J. Polym. Sci.*, **A28,** 2067.

TANNER, R. I. (1976). A test particle approach to flow classification for viscoelastic fluids. *A.I.Ch.E.J.*, **22,** 910.

TANNER, R I. (1980*a*). A new inelastic theory of extrudate swell. *J. non-Newt. Fluid Mech.*, **6,** 289.

TANNER, R. I. (1980*b*). The swelling of plane extrudate at low Weissenberg numbers. *ibid.*, **7,** 265.

TANNER, R. I. & HUILGOL, R. R. (1975). On a classification scheme for flow fields. *Rheol. Acta*, **14,** 959.

TAYLER, A. B. (1972). Singularities at flow separation points. *Quart. J. Mech. appl. Math.*, **16,** 153.

TAYLOR, G. I. (1953). Dispersion of soluble matter in solvent flowing steadily through a tube. *Proc. R. Soc., Lond.*, **A219,** 186.

TAYLOR, G. I. (1954). Conditions under which dispersion of a solute in a stream of solvent can be used to measure molecular diffusion. *ibid.*, **225,** 474.

TAYLOR, G. I. (1963). Cavitation of a viscous fluid in narrow passages. *J. Fluid Mech.*, **16,** 595.

THOMAS, R. H. & WALTERS, K. (1963). On the flow of an elastico-viscous liquid in a curved pipe under a pressure gradient. *J. Fluid Mech.*, **16,** 228–42. (See also Walters, K. 1962. *Quart. J. Mech. Appl. Math*, **15,** 63.)

TOKITA, N. & WHITE, J. L. (1966). Milling behavior of gum elastomers: experiment and theory. *J. appl. Polym. Sci.*, **10,** 1011.

TRUESDELL, C. & NOLL, W. (1965). *Non-linear Field Theories of Mechanics. Encyclopedia of Physics*, III/3, Springer, Berlin.

TSANG, W. K-W. & DEALY, J. M. (1981). The use of large transient deformations to evaluate rheological models for molten polymers. *J. non-Newt. Fluid Mech.*, **9,** 203.

VANDERBORCK, G. & PLATTEN, J. K. (1977). *Letters in Heat & Mass Transf.*, **4,** 453.

VANDERBORCK, G., PLATTEN, J. K. & CORNET, P. (1979). Stabilité hydrodynamique de l'ecoulement de Poiseuille cylindrique incluant la dissipation visqueuse II Influence de fluctuations de temperature. *Letters in Heat & Mass Transf.*, **6,** 83.

VAN WAZER, J. R., LYONS, J. W., KIM, K. Y. & COLWELL, R. E. (1963). *Viscosity and Flow Measurement,* Interscience, New York.

VINOGRADOV, G. V. (1971). Flow and rubber elasticity of polymeric systems. *Pure & appl. Chem.*, **26,** 423.

VINOGRADOV, G. V. (1972). Characteristics of polymer systems. *Prog. Heat & Mass Transf.*, **5,** 51.

VINOGRADOV, G. V. (1977). Ultimate regimes of deformation of linear flexible chain fluid polymers. *Polymer*, **18,** 1275.

VINOGRADOV, G. V. (1981). Limiting regimes of deformation of polymers. *Polym. Engng Sci.*, **21,** 339.

VINOGRADOV, G. V. & MALKIN, A. Ya. (1966). Rheological properties of polymer melts. *J. Polym. Sci.*, **A24,** 135.

VINOGRADOV, G. V. & MALKIN, A. Ya. (1980). *Rheology of Polymers*, Mir, Moscow.

WAGNER, M. H. (1976*a*). *Rheol. Acta*, **15,** 136.

WAGNER, M. H. (1976*b*). Das Fohenblasverfahren als rheologish- thermodynamischer Prozess. *Rheol. Acta*, **15,** 40. (See also: A combined rheological thermal analysis of the film-blowing process. Proc. VII Intern. Congr. on Rheol. p. 217.)

WAGNER, M. H. (1978). A constitutive analysis of uniaxial elongational flow data of a low-density polyethylene melt. *J. non-Newt. Fluid Mech.*, **4,** 39.

WAGNER, M. H. .(1979*a*). Zur Netzwerktheorie von Polymer- Schmelzen. *Rheol. Acta*, **18,** 33.

WAGNER, M. H. (1979*b*). Elongational behavior of polymer melts in constant tensile stress and constant tensile force experiments. *Rheol. Acta*, **18,** 681.

WAGNER, M. H. & LAUN, H. M. (1978). Non-linear shear creep and constrained elastic recovery of a LDPE melt. *Rheol. Acta*, **17,** 138.

WAGNER, M. H., RAIBLE, T. & MEISSNER, J. (1979). Tensile stress overshoot in uniaxial extension of an LDPE melt. *Rheol. Acta*, **18,** 427.

WAGNER, M. H. & STEPHENSON, S. E. (1979). *J. Rheol.*, **23,** 489.

WALTERS, K. (1965). *Rheometry*, Chapman & Hall, London.

WANG, C. C. (1975). *Arch. Rat. Mech. & Anal.*, **20,** 329.

WANG, K. K., SHEN, S. F., STEVENSON, J. F., HIEBER, C. A., CHUNG, S., COHEN, C. JAHANMIR, S., ISAYEV, A. I., TAYLER, A. & AKIYAMA, T. (1975–1981). *Computer Aided Injection Molding System*, Reprints 1–8, Injection Molding Project College of Engineering, Cornell University, Ithaca, New York.

WHITE, J. L. & DIETZ, W. (1979). Considerations of the 'freezing in' of flow-induced orientation in polymer melts by vitrification with application to processing. *J. non-Newt. Fluid Mech.*, **4,** 299.

WHITE, J. L. & TOKITA, N. (1968). Instability and failure phenomena in polymer processing with application to elastomer mill behavior. *J. appl. Polym. Sci.*, **12,** 1589.

WILLIAMS, J. G. (1970). A method of calculation for thermoforming plastics sheets. *J. Strain Anal.*, **5,** 49.

WILLIAMS, J. G. (1981). On the prediction of residual stresses in polymers. *Plast. & Rubber Proc. Appl.*, **1,** 369.

WILLIAMS, M. C. & BIRD, R. B. (1962). Steady flow of an Oldroyd viscoelastic fluid in tubes, slits and narrow annuli. *A.I.Ch.E. J.*, **8,** 378.

WINEMAN, A. (1979). On the simultaneous elongation and inflation of a tubular membrane of BKZ fluid. *J. non-Newt. Fluid Mech.*, **6,** 111.

WINTER, H. H. (1977). Viscous dissipation in shear flows of molten polymers. *Adv. Heat Transf.*, **13,** 205.

Winter, H. H. & Fischer, E. (1981). Processing history in extrusion dies and its influence on the state of the polymer extrudate at the die exit. *Polym. Engng Sci.*, **21,** 366.

Yates, B. (1968). Temperature development in single screw extruders. Ph.D. dissertation, University of Cambridge.

Yeow, Y. L. (1972). The stability of the film casting and the film blowing processes. Ph.D. dissertation, University of Cambridge.

Yeow, Y. L. (1974). On the stability of extending films: a model for the film casting process. *J. Fluid Mech.*, **66,** 613.

Yeow, Y. L. (1976). Stability of tubular film flow: a model of the film blowing process. *J. Fluid Mech.*, **75,** 577.

Yi, B. & Fenner, R. T. (1976). Scaling-up plasticating screw extruders on the basis of similar melting performance. *Plastics & Rubber: Processing*, **1,** 119.

Yoo, J., Joseph, D. D. & Beavers, G. S. (1979). Higher order theory of the Weissenberg effect. *J. Fluid Mech.*, **92,** 529.

Zamodits, H. J. (1964). Extrusion of thermoplastics. Ph.D. dissertation, University of Cambridge.

Zamodits, H. J. & Pearson, J. R. A. (1969). Flow of polymer melts in extruders. Part 1 The effect of transverse flow and of a superposed steady temperature profile. *Trans. Soc. Rheol.*, **13,** 357.

Ziabicki, A. (1976). *Fundamentals of Fiber Formation*, Wiley, London.

Index